Molecular Biology in Clinical Medicine

Molecular Biology in Clinical Medicine

Ralph C. Williams, Jr., MD
Professor of Medicine and Eminent Scholar
Marcia Whitney Schott Chair in Rheumatoid Arthritis
Division of Rheumatology and Clinical Immunology
Department of Medicine
University of Florida College of Medicine
Gainesville, Florida

With Contributions by
Donna H. Duckworth, PhD
Professor
Department of Immunology and Medical Microbiology
University of Florida College of Medicine
Gainesville, Florida

Elsevier
New York • Amsterdam • London • Tokyo

Elsevier Science Publishing Co., Inc.
655 Avenue of the Americas, New York, NY 10010

Sole distributors outside the United States of America and Canada:

Elsevier Science Publishers B.V.
P. O. Box 211, 1000 AE Amsterdam, The Netherlands

This book is printed on acid-free paper.

Library of Congress Cataloging-in-Publication Data

Williams, Ralph C., 1928-
Molecular biology in clinical medicine / Ralph C. Williams, Jr.; with contributions by Donna Duckworth.
p. cm.
Includes index.
ISBN 0-444-01600-7 (hardcover : alk. paper)
1. Medical genetics. 2. Molecular Biology. 3. Pathology, Molecular.
4. Recombinant DNA—Therapeutic use. 5. Biological response modifiers —Therapeutic use. I. Duckworth, Donna.
II. Title.
[DNLM: 1. Clinical Medicine. 2. DNA, Recombinant.
3. Gene Expression Regulation. 4. Molecular Biology. QH 506 W726m]
RB155.W53 1991
616'.042—dc20
DNLM/DLC 91-20248
for Library of Congress CIP

Current printing (last digit):
10 9 8 7 6 5 4 3 2 1

Manufactured in the United States of America

Contents

Preface

Molecular biology currently represents much of the cutting edge of new biomedical discovery, which seems to be occurring now with increasing frequency. It is impossible for a book like this to cover the entire spectrum of molecular approaches to human disease. Each passing week seems to bring more and more new applications or fresh insight into important clinical medical problems using molecular biological approaches. This volume is intended to be an introduction to the subject in general and to open the door for the medical student, graduate student, budding scientist, houseofficer, practitioner, or seasoned clinician who wants to be able to keep up with modern medicine and scientific thought. Like the revolution in data analysis and organization of information that has spawned our current computer age, application of the basic tools and approaches of molecular biology to basic problems of human disease clearly represents a large part of the future in biomedical thought. It is clear that attempts to treat common as well as uncommon diseases will employ recombinant DNA products and technology. Hopefully this book will provide a means for the previously uninitiated to grasp the basic molecular approaches and thus to be able to follow and keep abreast of the exciting new times ahead.

Ralph C. Williams, Jr., MD

Acknowledgment

I am very grateful to Ms. Kay Adam in the Department of Medicine at the University of New Mexico for typing the manuscript for this book and to Ms. Karen Crandall and Ms. Suzanne Smith in the Department of Medicine at the University of Florida for preparing the many revisions and re-revisions. Their patience and help is much appreciated.

Ralph C. Williams, Jr., MD

Messenger Molecules: From Gene to Protein

1

Donna H. Duckworth, PhD and Ralph C. Williams, Jr., MD

Discussions about which medical discovery has had the greatest benefit for humankind are likely to be lively; would it be anesthesia, antibiotics, antisepsis? Those are just the A's. Recombinant DNA, farther down in the alphabet, must come close to the top of the list in importance, for not only does it offer direct applications—recombinant vaccines, with which immunity to a dozen debilitating diseases can be produced with one swoop of an inoculating needle, a superabundance of growth hormones grown in bacteria, cloned human genes for the detection and correction of genetic abnormalities—but also it has provided dramatic new ways of studying the principles of biological processes to uncover the way life "works." Thus, the results of this discovery will increase exponentially for some time to come. In contrast to most other medical discoveries, no one person can be assigned credit for developing the method for making recombinant DNA. Rather, many scientists, perhaps thousands but certainly hundreds, have contributed to implementing these simple procedures that can produce new organisms in the laboratory in hours or days rather than in eons of evolutionary time. During the last several decades a steady stream of new knowledge has provided increasing insight into the area we now call molecular biology. This new information has already begun to make an immense impact on both the theory and the practice of clinical medicine. Gene cloning, DNA sequencing, and molecular hybridization techniques now provide the tools for examining a vast array of fundamental medical problems. Before we focus on how recombinant DNA technologies may affect clinical medicine or even our understanding of a single human disease, it is important to review some historical aspects along with the basic chemistry and structure of DNA and RNA.

Path to Recombinant DNA: Gregor Mendel's Peas

Many people would say the path to recombinant DNA technology began in Gregor Mendel's garden in the monastery at Brno. However, the process leading to the production of recombinant DNA molecules began centuries earlier when humans began to wonder at the capacity of living organisms to pass on their own qualities to their offspring. Aristotle actually proposed that this took place because of the transmission of a *plan* or of *information* for the development of the individual. This farsighted proposal was forgotten for centuries while people struggled with theories of pangenesis and preformation to explain the ability of humans to reproduce themselves. One such theory, that of the homunculus, envisaged a tiny midget in the father's semen that was planted in the mother's body to grow.[1]

Mendel's work with the ordinary garden pea was the first scientific study of inheritance. He used various strains of plants that differed from each other in only a few, well-defined characteristics such as seed color or morphology. By cross-breeding the plants, he was able to deduce that these characteristics are passed on to the progeny as discrete units and that each pea must possess a homologous pair of such units, which can produce alternative characteristics. He also concluded that of the two homologous units, one is dominant and the other recessive. In the first generation of hybrids, only the character of the dominant unit is observed, but the recessive traits reappear in later generations at a predictable frequency. In addition, he noted that determinants for different characteristics were randomly *segregated* in the offspring; that is, the genes for different traits could be recombined in various ways in the offspring. Later geneticists concluded that the determinants of hereditary characteristics are carried on chromosomes found in the nucleus.[2] However, geneticists seemed little interested in what genes were composed of. James Watson, the then brassy postdoctoral fellow who was one of the codiscoverers of the structure of DNA has said, "You would have thought that with all their talk about genes, they should worry about what they were . . . all that most of them wanted out of life was to set their students onto uninterpretable details of chromosome behavior or to give elegantly phrased, fuzzy-minded speculations over the wireless on topics like the role of the geneticist in this transitional age of changing values." Those who did think about the chemical structure of the gene were, in fact, in error for most of the century that elapsed between Mendel's work and the first recombinant DNA molecule synthesized in the laboratory.

Identification of the Genetic Substance

Since the early 1800s it had been known that cells are composed of three major substances: lipids, carbohydrates, and proteins. The fourth major constituent of living matter, a related group of compounds called nucleic acids, was discovered later, in 1868, by Friedrich Meischer, a chemist studying the composition of the nucleus. Any role for nucleic acids in inheritance was, however, dismissed for almost a century! Ever since the time when it was realized that there *is* a substance conferring heredity, it had been assumed that such a substance would be shown to be protein. Proteins, being made of 20 possible amino acid subunits, were the only structures thought capable of generating enough variability to encode genetic information. Nucleic acids, made up of only four possible subunits, even though known to be found everywhere genes were, were thought more likely to play some structural, scaffolding role. No data existed to indicate that genes were proteins, and in fact very good data were published that strongly implicated nucleic acids as the genetic substance. These data were from the experiments of Avery, MacLeod, and McCarty, who showed that bacteria could be changed or "transformed" with a substance that was almost pure DNA, one major kind of nucleic acid, from a rough colony type that does not produce a polysaccharide capsule to a smooth, polysaccharide-producing colony type.[3]

These experiments were prompted by previous work on encapsulated and non-encapsulated pneumococci by Frederick Griffith in the 1920s.[4] The encapsulated pneumococci were highly virulent because they could resist phagocytosis, whereas organisms unable to make a capsule were readily phagocytosed and hence not virulent. Griffith showed that avirulent nonencapsulated pneumococci injected into living animals could be transformed into virulent encapsulated organisms. He inoculated one group of mice with avirulent pneumococci, a second group with heat-killed virulent cells, and a third group with a mixture of avirulent and *heat-killed* virulent cells. The first two groups of mice were not affected by the inoculations. Mice in the last group, however, became extremely ill and many died. Moreover, even though only dead virulent cells had been injected, living virulent organisms could be isolated from the sick and dead mice. Two alternative explanations for these results were possible: (1) The presence of viable avirulent pneumococci restored the heat-killed cells to viability, or (2) the heat-killed cells somehow conferred the property of virulence on the living avirulent cells.

The latter explanation was subsequently found to be correct, as it was shown that DNA from the heat-killed cells could confer on the avirulent cells the ability to make a capsule and thus become virulent. This process, in which characteristics of bacteria are changed by the introduction of isolated DNA, is still known as transformation and represents one cornerstone of recombinant DNA technology. The term should not be confused with the change of normal eukaryotic cells to cells with oncogenic potential, a process also known as transformation.

A major advance in the study of bacterial transformation came with the demonstration in the early 1930s that transformation could be obtained in vitro, or in test tubes. When nonencapsulated bacterial cells were exposed to the contents of disrupted encapsulated bacteria, a small fraction of the cells were transformed to the encapsulated virulent type.

In 1944, Oswald T. Avery, Colin M. MacLeod, and Maclyn McCarty announced the results of 10 years of work in the analysis of the mechanisms of transformation.[3] They fractionated the debris from disrupted encapsulated pneumococcal cells into various chemical components and determined the transforming activity of each component. Only the DNA fraction caused transformation of unencapsulated cells to encapsulated ones. Later it was shown that DNA could cause the transformation of not just the ability to form a capsule but also other hereditary characteristics. Furthermore, the DNA was shown to be essentially pure and to contain virtually no protein. Even this very strong evidence did not immediately lead to the conclusion that DNA is the genetic substance, however. Rather, it was presumed that DNA could *act on* the genetic material, whatever it was, to mutate it.

In the early 1950s, A. D. Hershey and Martha C. Chase[5] demonstrated that when a bacteriophage or a bacterial virus infects a bacterium, its DNA actually enters the host cell but over 90% of its protein remains attached to the outside surface of the cell. Most of the protein on the outside of the cells could be removed by shear forces vigorously applied throughout a suspension of phage-infected bacteria. Infected bacteria from which the protein moiety of the phage particles had been completely removed produced a normal crop of new phage; thus, the DNA that entered the cell constituted the sole physical thread of hereditary continuity between the infecting and the emerging (progeny) virus particles. This experiment was published in 1952 and, by itself, would probably not have been any more effective than Avery, MacLeod, and McCarty had been in convincing people that DNA is the substance of heredity. In fact, its data were not nearly as good. However, in 1953, James Watson and Francis Crick[6] published their proposed structure of the DNA molecule, which allowed an explanation of how DNA could replicate and also carry genetic information by virtue of an almost infinite number of different base sequences. This, finally, allowed acceptance of the fact that DNA is actually the genetic substance and made the data of the Hershey-Chase experiment seem conclusive.

DNA represents the genetic substance in all cells, but it has been found that another nucleic acid, RNA, can also carry genetic information. This was originally shown for tobacco mosaic virus but is now known to be true for many plant viruses, some bacterial viruses, and some extremely important mammalian viruses, including the viruses for measles, leukemia, and acquired immunodeficiency syndrome (AIDS), among many others.

Structural Components of Nucleic Acids

Although RNA commonly exists as single chains, DNA almost always exists as a double-stranded structure with two antiparallel chains intertwined. The elucidation of the three-dimensional structure of DNA was dependent, first of all, on an understanding of the chemical constituents of these molecules.

Both DNA and RNA are constructed from four nucleotide building blocks (Figure 1.1). Each nucleotide contains a phosphate group linked to a five-carbon sugar (Figure 1.2). The sugar is joined to a base, a planar aromatic compound containing carbon and nitrogen. The sugar-base combination is called a nucleoside. The bases can be either a double-ring purine or a single-ring pyrimidine. Cytosine, uracil, and thymine (5-methyluracil) are the pyrimidines found in nucleic acids. Cytosine and thymine are components of DNA, but in RNA uracil takes the place of thymine. The purines found in both DNA and RNA are adenine and guanine. The two sugars found in nucleic acids are ribose and deoxyribose. Ribose is found in RNA (ribonucleic acid), whereas the sugar in DNA (deoxyribonucleic acid) is always deoxyribose. The structures of the bases, nucleosides, and nucleotides and their respective names are shown in Figure 1.1. Usually, however, just calling them A, T, G, C, and U will suffice.

The Architecture of Nucleic Acids

In both DNA and RNA, the nucleotides are joined by covalent bonds linking the phosphate group of one nucleotide to a hydroxyl group on the sugar of the adjacent nucleotide, forming chains of nucleotides or polynucleotides.

For many years it was thought that these chains were quite short, having only four components, and that the apparent size of nucleic acids was due to aggregation of these tetranucleotides. All were supposed to have the same structure, and all were thought to contain equimolar amounts of A, T (or U), G, and C. Evidence began to accumulate, however, that DNA molecules could be quite large. The first electron micrographs, taken in the late 1940s, showed that DNA is a long, thin molecule with a diameter of only 20 Å.[7] Individual molecules had lengths of many thousands of angstroms, indicating that they contained thousands of nucleotides in very long chains. However, just how these long chains might carry genetic information was not known. Erwin Chargaff provided the first clue.[8] In 1950 he

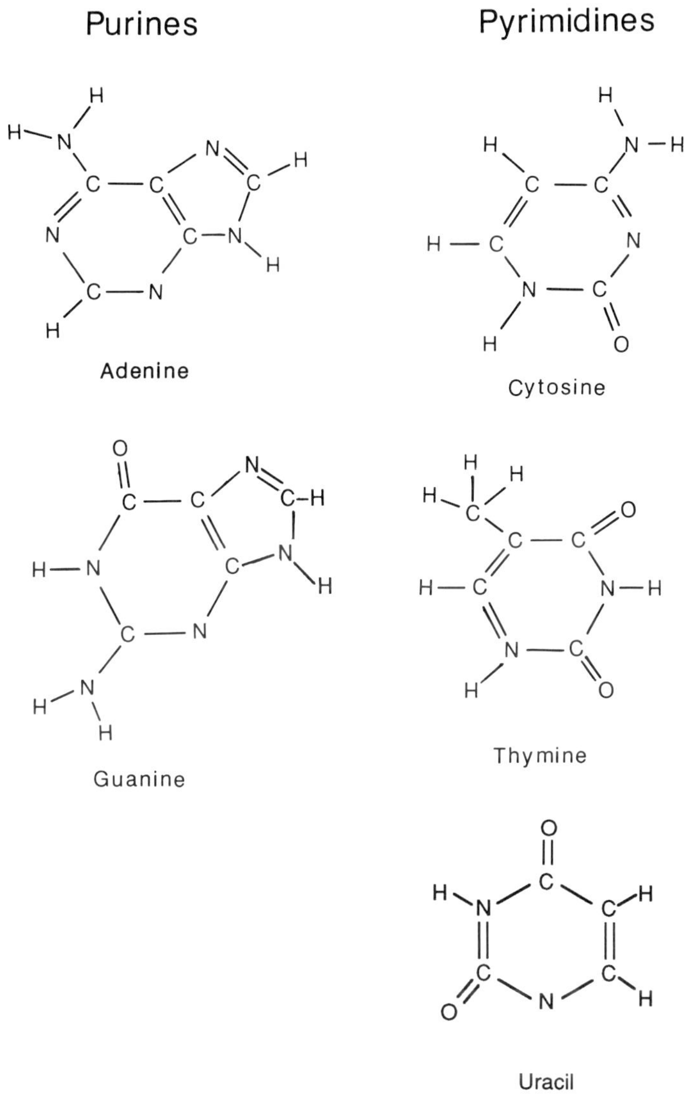

Figure 1.1 Purines and pyrimidines making up DNA and RNA.

noticed that the molar proportions of adenine and thymine and of guanine and cytosine are equivalent in DNA. The relative proportions of A + T and G + C could vary greatly, depending on the source of the DNA, but the amount of A always equaled the amount of T and the amount of G always equaled the amount of C. This was a remarkable observation, as it had previously been assumed that all DNA was alike and that it contained equimolar amounts of A, T, G, and C. Finding differences in composition of DNA from different sources suggested that it is the precise arrangement of nucleotides within a DNA molecule that confers its genetic specificity. The great significance of Chargaff's observation that A = T and G = C was not understood until Watson and Crick deduced the double-helical structure of DNA.

Inherent in any structure of the gene must be a mode of self-replication; that

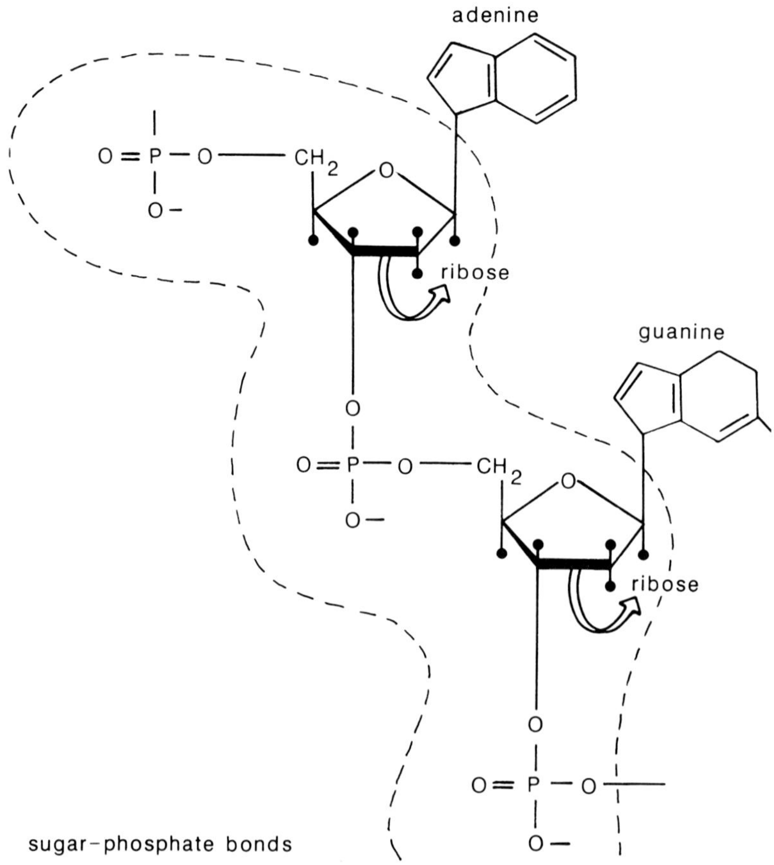

Figure 1.2 Five-carbon sugar (ribose)-phosphate bonds linking polynucleotides.

is, the gene must be copied exactly when the chromosome number doubles during cell division. One possible scheme for gene copying was based on the alternative formation of complementary surfaces. According to this model, gene duplication required the formation of a complementary (negative) image whose shape was related to the original (positive) surface like a lock to a key. The complementary negative image would then function as the mold or template for the synthesis of a new positive image. The alternative hypothesis would be that the gene directly produced an exact copy of itself.

One hypothesis considered by Watson when he was trying to deduce the structure of DNA was that DNA is a two-chain structure in which two chains of *identical* base sequences are held together by hydrogen bonds between pairs of identical bases. Such a structure would immediately suggest how it could replicate itself. A new DNA chain would be an exact copy of the old one. However, a structure with like paired to like would not have a regular backbone, because the purines, adenine and guanine, and the pyrimidines, thymine and cytosine, have different shapes, with the purines being substantially larger. Thus, where there was a pyrimidine pair, the structure would buckle in, and where there was a purine pair, it would bulge out. But x-ray diffraction studies of DNA crystals had shown that DNA molecules are highly regular. The resolution of the structure of DNA came when Watson realized the possibilities of a two-chain structure in which the chains are not alike but *complementary*. "Suddenly I became aware that an adenine-thymine pair held together by two hydrogen bonds was identical in shape to a guanine-cytosine pair held together by at least two hydrogen bonds. All the hydrogen bonds seemed to form naturally; no fudging was required to make the two types of base pairs identical in shape. . . . I suspected that we now had the answer to the riddle of why the number of purine residues exactly equaled the number of pyrimidine residues. Two irregular sequences of bases could be regularly packed in the center of a helix if a purine always hydrogen-bonded to a pyrimidine. Even more exciting, this type of double helix suggested a replication scheme. . . . Given the base sequence of one chain, that of its partner was automatically determined."[9] The new chain would not be identical to but rather a complement of the old chain. Copying the complementary sequence would then give an exact copy of the original. This insight, combined with the experimental results of Rosalind Franklin and Maurice Wilkins, led to the now famous double-helical structure of DNA.[6,9]

X-ray crystallographic studies by Wilkins and Franklin[10,11] had indicated that the underlying DNA structure is helical (spiral). The width of the molecule indicated that it contained more than one chain, perhaps two or three intertwined chains. Watson and his colleague, Francis Crick, deduced the correct structure, a complementary double helix, by building a model of the stereochemically most favorable configuration of the chains compatible with the x-ray diffraction data. In the double helix, the two DNA chains are on the outside and are held together by hydrogen bonds between pairs of bases on opposite strands that face inward (Figure 1.3). Because the outside of the molecule is the sugar-phosphate backbone, with the central base pairs stacked regularly next to each other perpendicular to the helical axis, the structure appears, in crystallographic studies, to be very regular even though the number of sequences is almost infinite. There seem to be no constraints on which bases can be next to each other, although the pairing between the bases on the two chains is very specific with A always paired to T and G to C. The helix is right-handed, and the two chains are headed in opposite directions. One is said to go from the 5′ to the 3′ direction, while the other goes in the 3′ to 5′ direction. The numbers refer to the carbon atoms of the sugar. If a chain ends with an unesterified hydroxyl or phosphate group on the 5′ carbon, this is termed the 5′ end. If the free group is on the 3′ carbon, this is termed the 3′ end. The form of DNA studied by Watson and Crick is the so-called B form, which is by far the most common form. Later, other structures for DNA were discovered. DNA containing alternating purine and pyrimidine residues can fold up into left-handed as well as right-handed helices. The left-handed conformation has a zigzag shape and is known as Z-DNA. Z-DNA is probably only one of several forms of DNA causing structural aberrations along a DNA chain. These are presumed to play a role in key cellular processes. Z-DNA, for instance, is believed to specifically bind certain proteins. It has been postulated that methylation of key cytosine residues acts to control gene function. Methylation is known to potentiate the conversion of B-DNA to Z-DNA, and in this way it could influence which proteins bind to DNA and, thus, what functions the DNA can undergo.[12]

Melting and Reannealing DNA

The double-helical structure of DNA results in one of its most important characteristics for many experiments involving recombinant DNA. This is its ability to "melt" or undergo a hyperchromic shift. By virtue of their conjugated double-bond

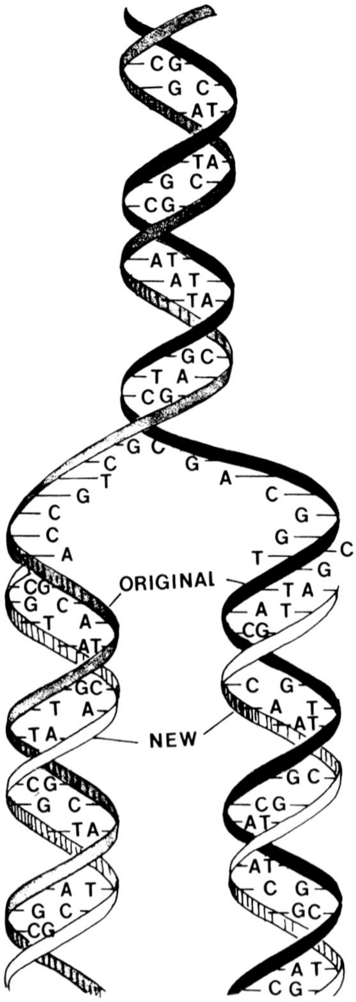

Figure 1.3 Replication of original double-stranded DNA.

systems, the purine and pyrimidine components of DNA have an intense ultraviolet absorption in the wavelength range 260 to 290 nm. This property can be used to detect the presence of nucleic acids and to roughly quantitate purified solutions of nucleic acids. It also, of course, is why ultraviolet light is such a potent mutagen. When a dilute aqueous solution of DNA is heated, the DNA undergoes what is known as a *helix-coil* transition. The heat causes the hydrogen bonds forming the double helix to break so that the two chains separate. This is shown by a sharp increase in absorbance at 260 nm, which is called *hyperchromicity* or the *hyperchromic shift*. The melting point, or T_m, is defined as the midpoint of the increase in absorbance. Because hydrogen bonding between G and C is stronger than that between A and T, T_m is higher in DNAs with a high proportion of G plus C. At temperatures about 5° above T_m, the two strands of the DNA come apart because of Brownian movement. If the solution is cooled rapidly, the two single strands remain separate, but if the solution is cooled slowly specific reannealing or renaturation of the two strands may occur and the double helix configuration is restored.[13] This reannealing is very specific and occurs only if two strands have exactly complementary sequences. This reaction can therefore be used to detect sequences that are complementary to a given sequence.

"Probes" are made of radioactively labeled DNA sequences and are added to segments of an unknown DNA mixture. Only if the unknown mixture contains complementary base-pair sequences will the probe anneal or fix to the unknown DNA. This is an extremely powerful tool and is the second cornerstone of recombinant DNA technology, the first being transformation or the ability to change cells by adding DNA molecules to them. Adjusting the conditions of the annealing reaction can sometimes cause sequences that are very similar but not exactly complementary to anneal, and this provides much useful information as well.

Gene "probing" (hybridization) experiments such as these are used to study the evolution of genes, to determine the relatedness of different species, and to detect foreign genes in clinical diagnostic tests. One test for the AIDS virus, for instance, utilizes such a reaction. Hybrid formation can occur not only between single strands of DNA and complementary strands but also between strands of DNA and RNA. This, in fact, was instrumental in the discovery of messenger RNA and in showing how a sequence of DNA could control all the many and varied processes that go on in a cell.

What Determines Cell Structure and Function?

Most of the key structural features or physiological processes in a cell are directly related to proteins. Protein catalysts regulate and direct the chemical reactions of the cell that result in the synthesis of other proteins and required cell materials, the breakdown of these materials, and the production and utilization of energy. Proteins themselves are composed of 20 different amino acids joined together in polypeptide chains. The actual shape or three-dimensional folding of polypeptide chains within proteins may determine critical biological properties such as the site of enzyme-substrate binding, the physical properties of various structures, or the availability of reactive sites to interact with other structural or regulatory molecules. A good example of this is the role of the size and three-dimensional shape of the antibody combining site that reacts with determinants on lysozyme, as illustrated in Figure 1.4. The way protein polypeptide chains fold or wind up is determined solely by the primary amino acid sequence of their constituent chains. It has been shown that the primary sequence in both enzymatic and structural proteins determines folding and shape, which in turn dictates function. A genetic program that could determine the primary sequence of all proteins in the cell would, thus, completely determine the properties of that cell and, in turn, of multicellular organisms.

One Gene, One Enzyme

The elucidation of how this genetic program worked began in 1902 with Archibald Garrod.[14] Garrod was the first to recognize that certain hereditary human diseases are disorders of metabolism that might be explained by the absence or disordered functioning of a single enzyme. In his studies of alkaptonuria, Garrod noticed that the healthy, and apparently normal, parents of patients with the disease were frequently blood relatives. This could occur if the disease were due to a rare, re-

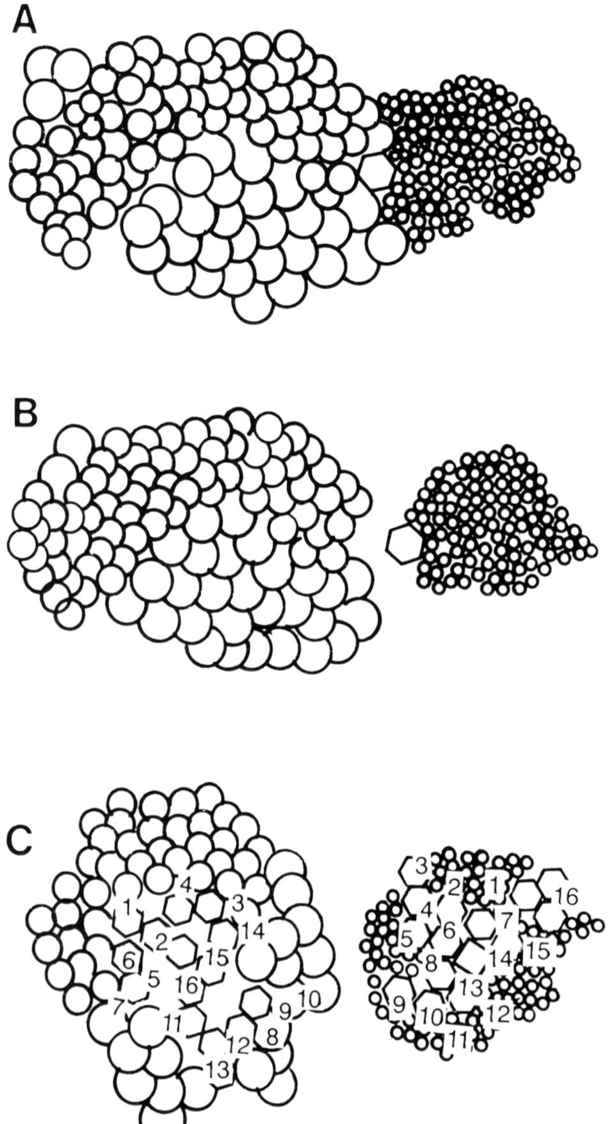

Figure 1.4 Three-dimensional model of lysozyme antigen (right) and the antibody combining with it (left) the lower panel shows an end-on view of the antibody combining site with the amino acids making contact shown as numbered areas. L chain residues that contact the antigen are labeled 1 (His 30) 2 (Tyr 32) 3 (Tyr 49), 4 (Tyr 50), 5 (Phe 91), 6 (Trp 92), and 7 (Ser 93). H chain residues that contact the antigen are labeled 8 (Thr 30), 9 (Gly 31), 10 (Tyr 32), 11 (Trp 52), 12 (Gly 53), 13 (Asp 54), 14 (Arg 99), 15 (Asp 100), 16 (Tyr 101), and 17 (Arg 102). Lysozyme residues that contact the antibody are labeled 1 (Asp 18), 2 (Asn 19), 3 (Arg 21), 4 (Gly 22), 5 (Tyr 23), 6 (Ser 24), 7 (Leu 25), 8 (Asn 27), 9 (Lys 116), 10 (Gly 117), 11 (Thr 118), 12 (Asp 119), 13 (Val 120), 14 (Gln 121), 15 (Ile 124), and 16 (Leu 129). Gln 121 fits into the antibody surface pocket surrounded by V_L and V_H residues 2, 5, 6, 7, and 16. *Reprinted from Amit A, Mariuzza R, Phillips SEV, Poljak RJ. Three-dimensional structure of an antigen–antibody complex at 2.8 Å resolution. Science 1986;233:747–753, with permission.*

cessive mendelian factor. Mendel had shown that recessive traits appeared in subsequent generations of crosses involving related pea plants. Thus Garrod proposed that alkaptonuria is an inherited disease based on a biochemical difference in the tissues of affected patients and that the normal metabolism is blocked because of an inherited lack of a specific enzyme. Other diseases were thought to be due to a lack of other enzymes that were also controlled by Mendelian factors or genes. It was many years before elegant direct chemical proof of Garrod's hypothesis was provided by Beadle and Tatum in their studies of the x-ray– and ultraviolet light–induced *Neurospora* mold mutants.[15] These classical studies showed that certain mutant strains of *Neurospora* required specific growth factors, such as a single amino acid, a vitamin, or a certain pyrimidine—each produced by a metabolic process dependent on a specific enzyme. When the gene coding for a critical enzyme was altered by ultraviolet light or other manipulation, the *Neurospora* strain dependent on its metabolic product could not survive. The single gene, single enzyme concept was later amply extended in work related to single amino acid substitutions in abnormal hemoglobins.[16] This work on the human hemoglobins emphasized the profound physiological importance of changes of primary amino acid sequence in inheritable diseases. It was demonstrated that sickle cell hemoglobin contained the same α chains as normal hemoglobin but that the β chains had a single amino acid substitution (glutamic acid for valine) at one end of the chain. Just how the change in the gene was linked to the change in the amino acid sequence was not immediately obvious, however.

How Does DNA Determine Protein Structure?

Assuming that the gene determines the order of the 20 different amino acids in the protein chain and that a change in the gene causes a change in the amino acid sequence, the question is how? Enzymes had been shown to control all cellular processes, but hypotheses in which enzymes determined the sequence of amino acids led to a paradox. According to Watson[17]

> Such schemes would require, for the synthesis of a single type of protein, as many ordering enzymes as there are amino acids in the respective protein. But since all known enzymes are themselves proteins, still additional ordering enzymes would be necessary to synthesize the ordering enzymes. And so on. This is clearly a paradox, unless we assume a fantastically interrelated series of syntheses in which a given protein has many different enzymatic specificities. With such an assumption, it might be possible (and then only with great difficulty) to visualize a workable cell. It did not seem likely, however, that most proteins would be found to carry out multiple tasks. In fact, all our knowledge pointed to the opposite conclusion of one protein, one function.

Thus a specific *template* that attracts amino acids and lines them up in the correct order was thought to exist. Although DNA carried the information for ordering amino acids, it was doubtful that the double helix itself could be the template. Ruling out a direct role for DNA were experiments showing that protein synthesis occurs at sites where DNA is absent. Although in bacteria the DNA molecule is in the cytoplasm, where protein synthesis takes place, in all eukaryotic cells the cytoplasm is separated by the nuclear membrane from the chromosomally located DNA. Thus, a second information-containing molecule was hypothesized. This molecule would obtain its genetic specificity from DNA in the nucleus, after which it would move to the cytoplasm to function as the template for protein synthesis.

The identity of this "messenger" was, as with many of the major discoveries of molecular biology, elucidated through work with bacteriophage (bacterial viruses). For some years geneticists had suspected that these viruses are (as all viruses are now known to be) a form of naked, traveling genes. In the late 1940s a group of investigators known as the "phage group" coalesced around Salvador Luria and Max Delbruck, whose one abiding notion was to use bacteriophage to understand the basic coding properties of the gene. Hershey was one of the early members of the phage group, and he and Martha Chase showed, as previously indicated,[5] that only the DNA of the T2 phage entered the cell and could cause the synthesis of all the necessary components, DNA and proteins, for approximately 100 new phage. Seymour Cohen, a student of Chargaff (A = T, G = C), also made great contributions to our understanding of virus replication. He showed that the phage DNA coded for not only the structural phage proteins but also a large number of enzymes involved in phage DNA synthesis.[18] Thus, after infection, the bacterial cell becomes a factory for making proteins under the dictates of the phage DNA. This indicated that phage were indeed a logical choice for determining how the DNA molecule gave its orders. Besides many new proteins, phage were known to make a fraction of RNA that differed from the RNA of the host bacteria. It was later shown that this RNA could *hybridize* to T2 DNA but not to bacterial DNA. Thus, the RNA made in the infected cell had a complementary sequence to phage T2 DNA and must have been made by copying a strand of T2 DNA. This then was messenger RNA, the informational interlocutor between DNA and proteins.[19]

Phage have an alternate means of multiplication in which the DNA first integrates into the host genome from where it can be excised to start a cycle of replication. This is shown in Figure 1.5.

The Central Dogma

The findings described above formed the basis of what is now known as the central dogma of molecular biology:

$$\circlearrowleft \text{DNA} \rightarrow \text{RNA} \rightarrow \text{proteins}$$

Here the arrows indicate the directions proposed for the transfer of genetic information, rather than biochemical transformations. The arrow encircling DNA signifies that DNA is the template for its self-replication. The arrow between DNA and RNA indicates that all cellular RNA molecules are made on DNA templates, a process known as transcription. Correspondingly, all proteins are synthesized or translated on RNA templates. Note that the last two arrows are shown as unidirectional, indicating that RNA sequences are never determined by protein templates and that DNA sequences are not determined by RNA. This was the scheme as it was originally proposed, because at the time it was thought that a protein could never determine an RNA sequence and that DNA could never be made on RNA templates. That proteins never serve as templates for RNA has stood the test of time. However, RNA chains sometimes *do* act as templates for DNA chains of complementary sequence. This process is known as *reverse transcription*, and it oc-

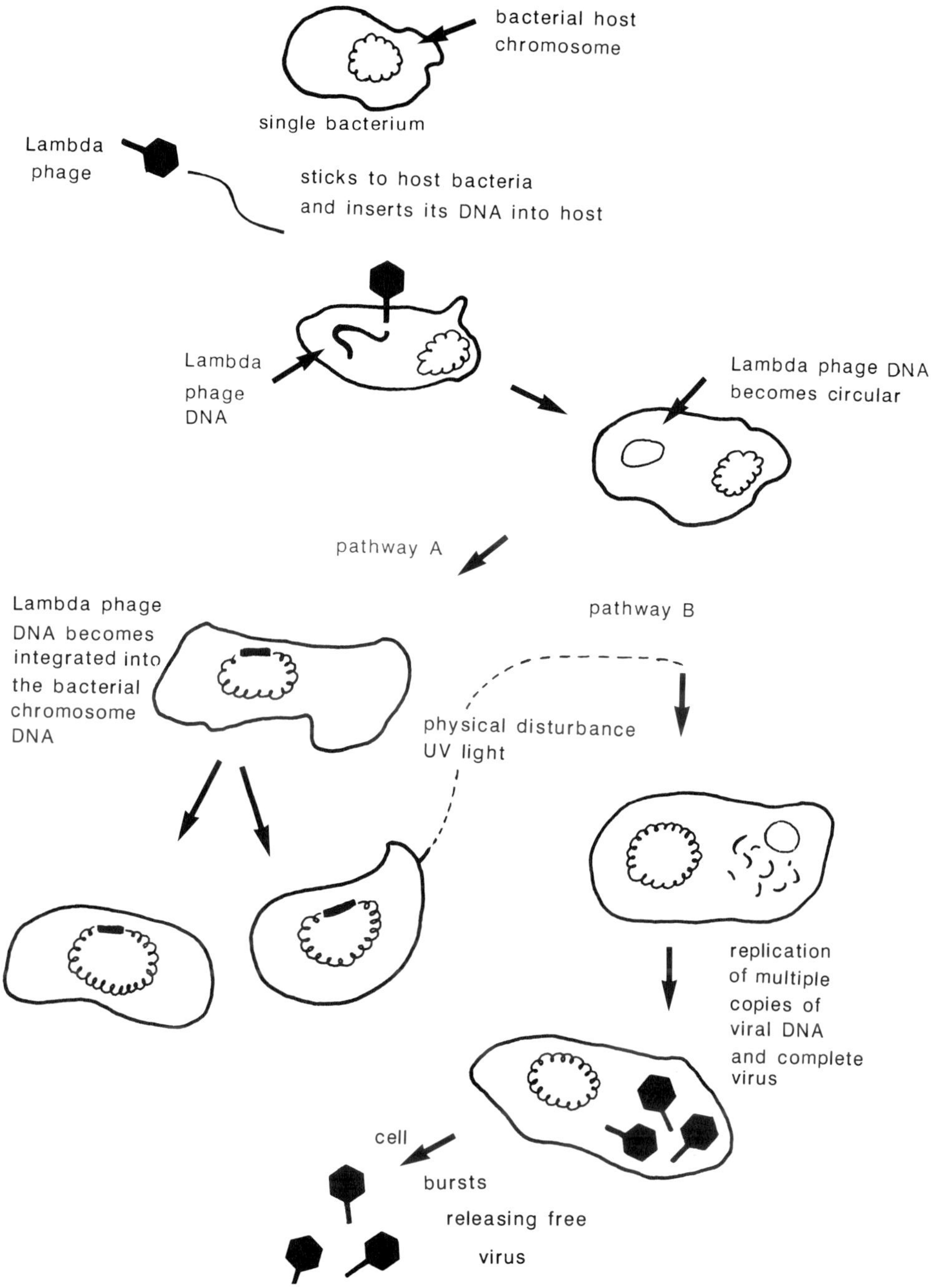

Figure 1.5 Lambda phage replication after entering its bacterial host.

curs when the oncogenic retroviruses, which carry their genetic information in RNA, cause the synthesis of a DNA copy that can become part of the cell's genetic complement.[20,21] However, such reversals of the normal flow of information are very rare events compared with the enormous number of RNA molecules made on DNA templates.

The Adaptor Hypothesis and Its Proof by the Test Tube Synthesis of Proteins

An early hypothesis to explain how RNA actually directed protein synthesis was that the RNA templates were folded up to create cavities on their outer surfaces into which the 20 different amino acids would specifically fit. The cavities would be shaped so that only a particular amino acid would fit, and in this way RNA would provide the information for ordering amino acids during protein synthesis. The basic biochemical properties of amino acids and nucleotide bases argued against this, however. Thus Crick proposed that before incorporation into proteins, amino acids are first attached to specific adaptor molecules, which in turn have unique surfaces that can bind specifically to bases on the RNA templates. Biochemical experiments on the mechanisms of protein synthesis showed that adaptors did, indeed, exist.[22]

The discovery of how proteins are synthesized required the development of cell-free extracts capable of carrying out the essential synthetic steps. These methods became available in the early 1950s and produced some significant results.[23,24] The cellular site of protein synthesis was found to be the ribosomes, small RNA-containing particles in the cytoplasm of all cells actively engaged in protein synthesis. It was also found that before incorporation into proteins, amino acids are first attached to what we now call transfer RNA molecules by a class of enzymes called aminoacyl-tRNA synthetases. This discovery was totally unexpected, except by Crick, who had previously speculated that his proposed "adaptors" might be short RNA chains, since their bases would be able to base-pair with appropriate groups on the RNA molecules that served as the templates for protein synthesis.[22,25] The transfer RNA molecules involved in cell-free protein synthesis were, in fact, Crick's adaptors. They each contain several adjacent bases (the anticodon) that bind specifically during protein synthesis to successive groups of several bases (codons) along the RNA templates, following the rules of complementary base pairing.

Thus, three types of RNA participate in protein synthesis: (1) messenger RNA (mRNA), which carries the information for the order of the amino acids; (2) ribosomal RNA (rRNA), which acts, in conjunction with ribosomal proteins, as the factory where proteins are made; and (3) transfer or adaptor RNAs (tRNAs), which bind specific amino acids and, by virtue of specific codons, line the amino acids up in their proper order. These processes are diagrammed in Figures 1.6 to 1.9. The process of tRNA triplet arrangement and synthesis of individual amino acids making up the polypeptide chain involves a different catalytic enzyme for each of the 20 amino acids. In this process the amino acid is joined to the 3′ end of the tRNA, and the tRNA interacts specifically with mRNA through its complementary sequences (Figures 1.7 to 1.9).

Establishing the Genetic Code

Because protein structure is determined by a sequence of 20 amino acids but there are only four bases in DNA to indicate this sequence of 20 amino acids, groups of several nucleotides must somehow specify a given amino acid. Groups of two, however, would specify only 16 (4×4) amino acids. Beginning in about 1954, with the start of serious thinking about what the genetic code might be like, most attention was given to how triplets (groups of three coding bases) might work, even though they obviously would provide more permutations or code words (4

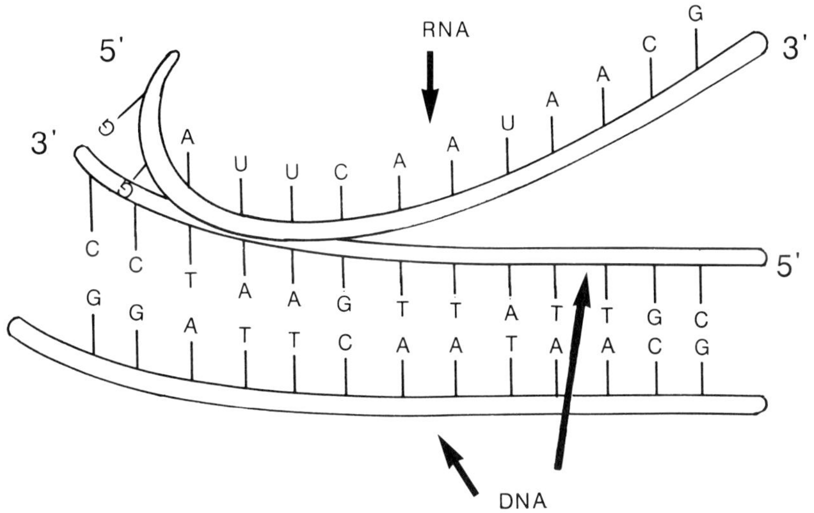

Figure 1.6 A strand of RNA synthesized under the direction of one of the complementary strands of DNA.

Figure 1.7 Messenger RNA directing polypeptide chain synthesis on the ribosome.

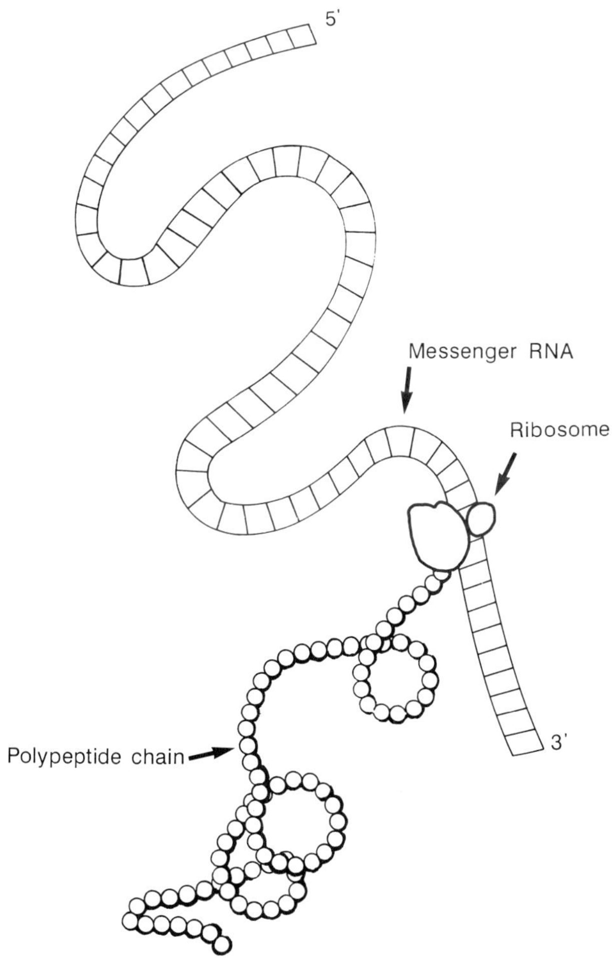

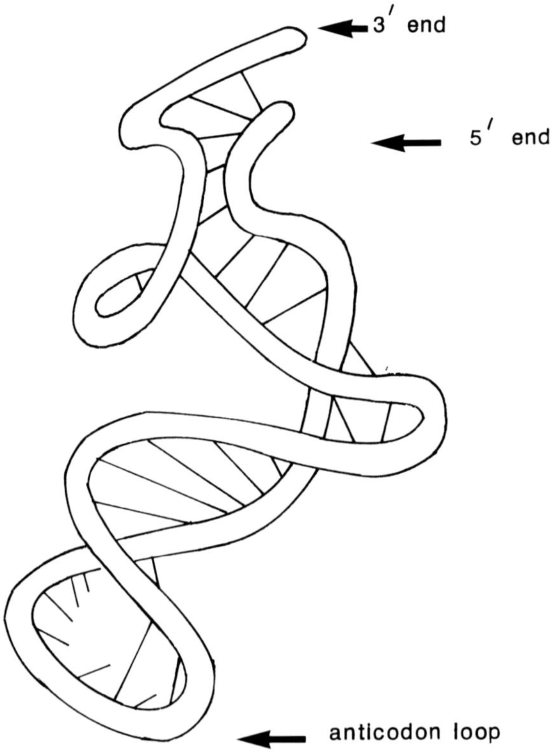

Figure 1.8 Diagrammatic representation of transfer RNA. Complementary base pairing within the structure of the molecule provides three-dimensional shape. A single amino acid attaches at the 3′ end of the polynucleotide chain. The anticodon loop interacts directly with the messenger RNA.

Figure 1.9 Ribosomes provide the stage upon which tRNA anticodons interact with mRNA codons and single amino acids are directed to be linked together to make a polypeptide chain.

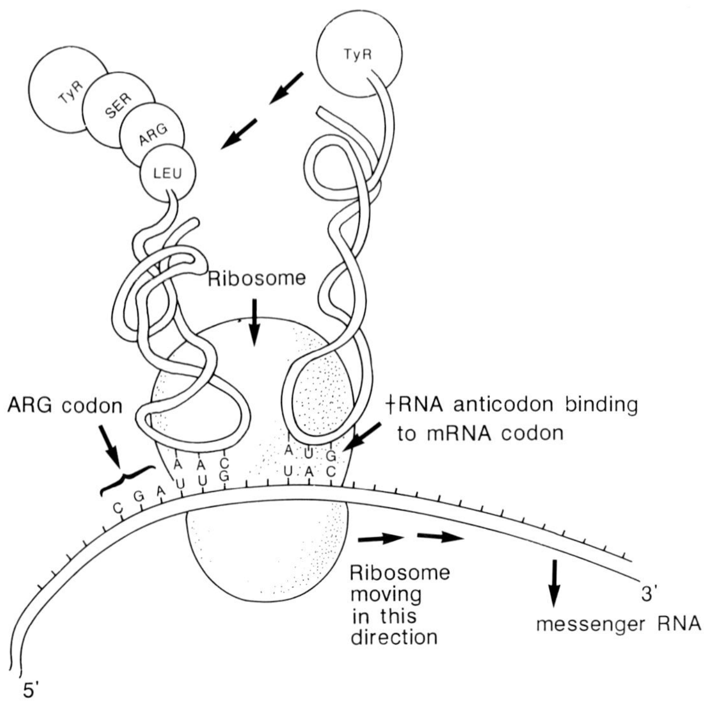

× 4 × 4) than needed, if each amino acid was specified by only a single triplet. The assumption of colinearity in all hypotheses of a code was very important. Colinearity meant that successive groups of nucleotides along a DNA chain must code for successive amino acids along a given polypeptide chain. That colinearity does in fact exist was shown by elegant mutational analysis of the bacterial protein tryptophan synthetase.[26] Experiments with bacteriophages also confirmed the colinearity of DNA and protein, as well as helping to elucidate many of the basic properties of genes and the genetic code. Crick, again, through a set of ingenious experiments using mutants of T2 phage, showed that the genetic code is read from a fixed starting point in groups of three bases and that it is degenerate; that is, one amino acid can be coded for by more than one group of bases. This showed that the theoretical assumption of a triplet code was correct.[27]

Which specific groups of three bases (codons) determine which specific amino acids was learned by biochemical analysis using cell-free protein synthesizing systems. Here the major breakthrough came when Marshall Nirenberg, working with J. H. Matthaei at the National Institute of Health in Bethesda, Maryland, observed that the addition of poly U (UUUUU . . .), a synthetic polynucleotide containing only uridine, to a cell-free system capable of making proteins led to the synthesis of polypeptide chains containing only the amino acid phenylalanine.[28] The nucleotide group UUU thus must specify phenylalanine. Using increasingly complex, defined polynucleotides as synthetic mRNAs and seeing which amino acids were incorporated into a peptide chain dictated by the synthetic message rapidly led to the identification of more and more codons. Final completion of the code in 1966 revealed that 61 of the 64 possible triplets corresponded to amino acids. Furthermore, the code was degenerate, with most amino acids being coded by more than one nucleotide triplet (Table 1.1).

Table 1.1 The Genetic Code[a]

First position 5′ end	Second position U	C	A	G	Third position 3′ end
U(T)	Phe	Ser	Tyr	Cys	U
	Phe	Ser	Tyr	Cys	C
	Leu	Ser	Stop	Stop	A
	Leu	Ser	Stop	Stop	G
C	Leu	Pro	His	Arg	U
	Leu	Pro	His	Arg	C
	Leu	Pro	Gln	Arg	A
	Leu	Pro	Gln	Arg	G
A	Ile	Thr	Asn	Ser	U
	Ile	Thr	Asn	Ser	C
	Ile	Thr	Lys	Arg	A
	Met	Thr	Lys	Arg	G
G	Val	Ala	Asp	Gly	U
	Val	Ala	Asp	Gly	C
	Val	Ala	Glu	Gly	A
	Val	Ala	Glu	Gly	G

[a] Using the table, AUG would specify methionine reading from the 5′ end; similarly, CAU would specify histidine.

A number of observations suggested that the genetic code is *universal*; that is, the same triplet code determines the same amino acid in all organisms. Imagining how anything else could have evolved is very difficult. Experimentally, this universality has held up with only two known exceptions so far. In organelles (mitochondria and chloroplasts) there are slight differences in nucleotide code translation; in addition, mycoplasmas use several different code words. The differences between coding in mitochondrial DNA and in nuclear DNA have been attributed to the much smaller repertoire of tRNAs available in mitochondria than are available for interaction with nuclear-coded messages. Reasons for the unusual codon usages in mycoplasmas are still not known.

Initiation and Termination Signals

At first it was presumed that translation of an mRNA molecule would begin at one end and finish when the entire mRNA message had been read into amino acid sequences. In fact, translation both starts and stops at internal positions; that is, translation does not start or stop at the end of an mRNA molecule but rather both starts and stops in the middle of the molecule. Thus, signals must be present within DNA (and its mRNA products) to indicate where to start and stop translation. The first signals to be worked out were the termination (stop) signals. Three separate codons (UAA, UAG, and UGA) were shown *not* to code for any amino acids but instead to serve as chain-terminating codons. For instance, if an mRNA contains a UAG, the ribosomes will stop incorporating amino acids at that point because there is no amino acid attached to a tRNA with the anticodon for UAG. AUG is a start codon, but other information in the mRNA is necessary to signal initiation. This information is contained in purine-rich blocks of nucleotides that serve to attach mRNA to the ribosomes.

Gene Expression: A Summary

In summary, in protein biosynthesis, DNA (most likely only one strand of the duplex) serves as a template for an mRNA complement made in accordance with the base-pairing rules. The mRNA can then serve as a template on which adaptor RNA (tRNA) molecules are oriented, in a sequence that also is determined by the Watson-Crick pairing rules. As a result of this step, the individual amino acids can be ordered in a specific sequence (or primary structure), for each tRNA molecule carries a specific amino acid. Finally, the amino acids are linked by peptide bonds to form a polypeptide chain. The amino acid sequence of this chain, then, has been dictated by the original DNA code, transcribed through mRNA, and translated by adaptor RNA or tRNA through the mechanism provided by the base-pairing rules.

Antisense Messages

The strand of DNA that is transcribed into mRNA is called the sense strand; the opposite strand is the antisense strand. From time to time, people have wondered what would happen if both strands of DNA synthesized an mRNA molecule. If this happened, presumably the two mRNAs would hybridize to each other and thus be unable to direct protein synthesis. Experimentally, in fact, specific gene functioning can be obliterated by transformation of cells with DNA that will make antisense mRNA for the gene in question.

A few years ago, Bost, Smith, and Blalock[29] developed the interesting hypothesis that complementary mRNAs cause the synthesis of interacting proteins. They used this hypothesis to find the cell receptor for corticotropin (adrenocorticotropic hormone or ACTH). They developed antibodies against the previously undiscovered ACTH receptor by immunizing rabbits with a synthetic peptide whose sequence was specified by the DNA strand opposite to the one that produces the mRNA for bovine ACTH, that is, the antisense strand of the ACTH gene. This "antisense" peptide, called HTCA, was thought to resemble the binding region of the ACTH receptor and was reported to associate specifically with ACTH. The HTCA antibodies, which, interestingly , were found to bind to ACTH antibodies, were used in the purification by affinity chromatography of a protein complex considered to represent the ACTH receptor.[30] This antisense or complementary peptide method was based on the observation that there is a tendency in the genetic code for codons of hydrophilic amino acids to be complemented by codons for hydrophobic amino acids,[6] resulting in peptide structures that might interact specifically through amphiphilic conformations. The concept that specifically interacting proteins may be encoded by nucleic acids containing complementary DNA sequences was initially proposed in 1981 by Biro in his theoretical analysis of specificity in protein-protein interactions.[31] The concept has been very controversial, however, and even its laboratory usefulness has been doubted.[32] But the idea that proteins made by antisense messages may interact with proteins made by the correct message, and that protein receptors may have evolved in this way, remains an interesting hypothesis.

Regulation of Gene Expression

For an organism to maintain a balance or proper proportions in the production of various proteins, some control system of gene regulation is necessary. In higher organisms, most of the protein in one type of cell may constitute a single protein species that is undetectable in other types of cells from the same organism. Also, a particular protein may be made in abundance at one stage during the development of an organism but appear only in very low quantity, or be undetectable, in other stages. This differential expression of gene products in higher organisms is currently only beginning to be understood. Microorganisms also have the ability to express genetic information differentially. In microorganisms, a single protein species may be absent under some conditions of cultivation and yet constitute several percent of the total protein of the cell when the organisms are cultured under different conditions. For example, the enzyme β-galactosidase, which breaks down lactose to glucose and galactose, is synthesized only if lactose is present in the growth medium. There are other proteins that are formed in relatively constant amounts regardless of different conditions of culture.

Although the two major points at which synthesis of a particular protein could be controlled are transcription and translation, most of the control mechanisms in bacteria operate to modify the amount of mRNA that is present in the cell. This is called transcriptional control or control at the level of transcription. For a gene to be transcribed, the enzyme RNA polymerase must be able to bind at or near the beginning of the gene. This binding site is called the promoter, and the RNA polymerase must be able to recognize it. In bacteria, a special subunit of the polymerase that recognizes the promoter is called the sigma factor. By changing this subunit, the cell can change which promotors will be recognized and hence which genes will be transcribed. During sporulation, for instance, a new sigma factor is synthesized, which then "reads" all the genes necessary for the cell to sporulate. The promotors for these genes are not recognized by the normal sigma factor, so only after a change in the sigma factor will sporulation products be made.

Other forms of regulation use a specific protein that is encoded by a regulatory gene. Such proteins can bind to the DNA near the promoter for that gene and influence the attachment or passage of the RNA polymerase. One of the first and most extensively studied examples of this type of gene modulation was that related to the production of β-galactosidase and described by Jacob and Monod.[33–35]

The utilization of lactose by *Escherichia coli* actually involves the production of three new proteins: a transport system to mediate the entry and accumulation of lactose within the cell; the enzyme β-galactosidase, which splits lactose into glucose and galactose; and thiogalactoside transacetylase, whose function is unknown. The genes for these three proteins (called Y, Z, and A) form what is called an operon; that is, they are adjacent to one another and are read from a single promoter into a single mRNA encoding all three proteins. The genes are expressed at a very low level when lactose is not present in the medium, but their expression can be increased by a factor of 1000 following the addition of lactose.

Expression of these genes is subject to negative control, so called because the regulatory gene product, a repressor, coded for by the i gene, acts negatively to turn off production of the mRNA. The repressor protein binds to a specific region of DNA, called the operator, which overlaps the promoter of the lactose operon. The presence of the repressor on the operator prevents the binding of RNA polymerase to the promoter and thereby reduces transcription of the lactose (*lac*) operon to 0.1% of the uninduced level (Figure 1.10).

When lactose enters the cell, it binds to the repressor on the operator site, causing the repressor to change its shape. This change in the repressor reduces its affinity for the operator by a factor of about 1000. The RNA polymerase can then attach to the promoter and transcribe the three genes composing the operon.

Positive control refers to a situation in which a regulatory gene turns on the expression of the controlled genes. In this case, the regulatory gene product is called an activator. When the activator is in an inactive state, which occurs unless an inducer is present, it cannot bind to the regulatory region next to the controlled gene. The promoter of such genes cannot be recognized by RNA polymerase unless the activator protein is bound nearby. However, when the inducer is present, it binds to the activator, altering its structure so that it can now bind near the promoter. The binding of the activator permits the binding of RNA polymerase, possibly through a favorable protein-protein interaction. The best-known example of this type of control is the arabinose operon of *E. coli*, where three catabolic genes and two transport systems are all controlled by the same activator. In the presence of arabinose, transcription of all the genes is turned on. In its absence, none of the protein products that metabolize the sugar are produced.[36]

Another mechanism of control is called attenuation. In this process, which has been extensively studied in the tryptophan operon, RNA polymerase binds and transcription starts, but transcription can be terminated before the RNA polymerase can reach the structural genes.[37] This happens as a result of folding within the mRNA segment, which occurs under certain conditions.

Other control mechanisms may involve the stability of the mRNA once it is made, as well as control of the initiation or rate of translation of the mRNA into protein.

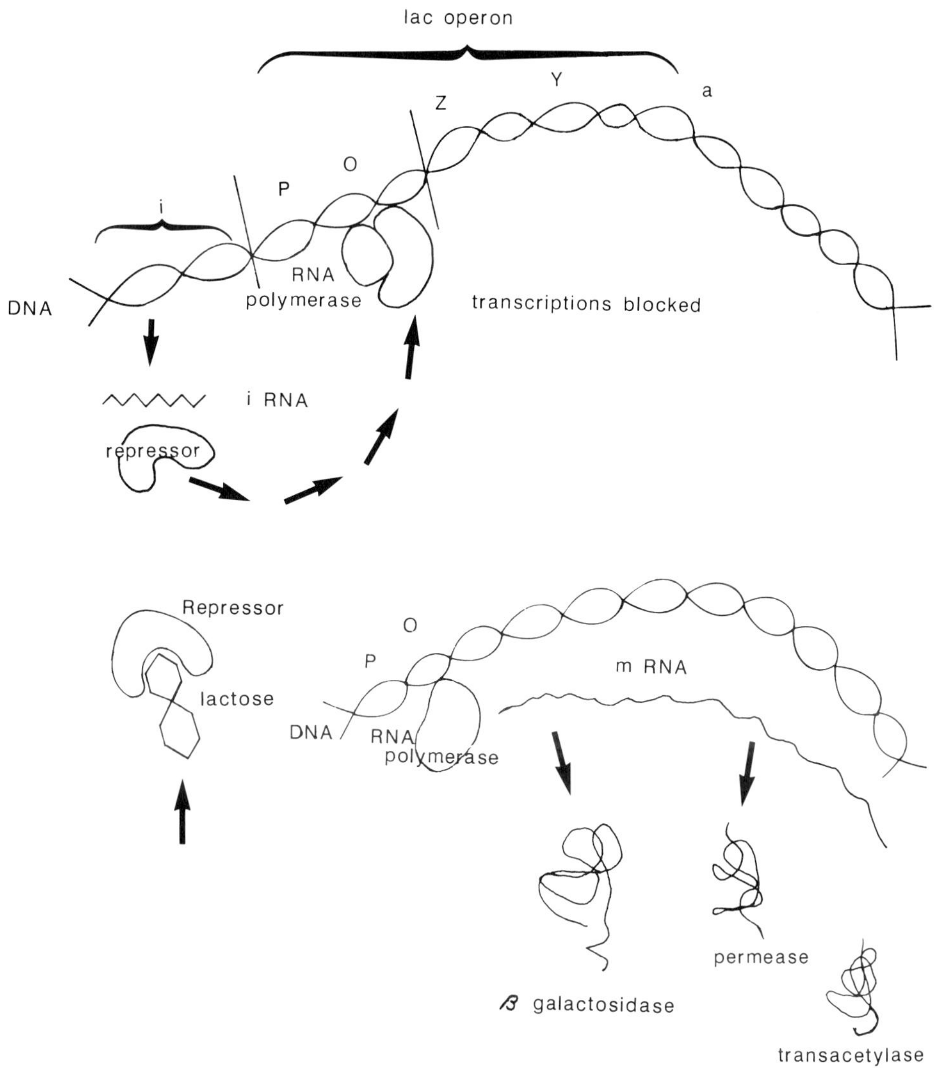

Figure 1.10 Control of the lactose operon. Repressor protein binds to control sites on DNA and prevents mRNA for β-galactosidase from transcribing. When lactose inducer interacts with the repressor (below), the blockage of transcription no longer takes place and the entire *lac* operon mRNA is transcribed with sequential production of β-galactosidase, permease, and transacetylase.

Differences Between Prokaryotes and Eukaryotes

Most of the early studies elucidating the mechanisms of gene duplication and expression were done on prokaryotes or bacteria, usually, in fact, on *E. coli*. This was due to the speed and simplicity with which bacteria could be grown. Cells of higher organisms can now be conveniently grown in cultures, and the study of eukaryotic genetic mechanisms has now taken the forefront. These studies have revealed some surprising differences between prokaryotic and eukaryotic organisms.

One difference, which had been obvious for some time, is that prokaryotes such as bacteria contain as their chromosomal material only a single DNA molecule, which resides in the cytoplasm of the cell. There is no nuclear membrane separating it from the rest of the cell. This allows transcription and translation to be carried out almost simultaneously. The DNA molecule of a typical bacterium such as *E. coli* is a closed circular molecule with a molecular mass of about 1.5×10^9 daltons and contains about 4×10^6 base pairs. This is enough to code for 3000 to 4000 proteins.

Although the bacterial chromosome is a single DNA molecule, bacteria can also contain extrachromosomal DNA in the form of plasmids. Plasmids are almost always relatively small double-stranded closed circular pieces of DNA. Their size can vary tremendously, from only 3000 or 4000 base pairs to several hundred thousand base pairs. The genes that they carry are not necessary for bacterial survival under ordinary conditions but are often useful under stressful situations. Thus, plasmid genes often code for and produce enzymes that can degrade antibiotics (antibiotic resistance genes). Other plasmid genes are involved in the transfer of genes between bacteria and contribute to virulence in pathogenic microorganisms. Plasmids also represent an important cornerstone of genetic engineering and recombinant DNA technology, as they are often used as the vehicles for replicating, storing, and transporting DNA.

In eukaryotic organisms the DNA exists in a special compartment, the nucleus, which is separated from the rest of the cell by the nuclear membrane, which is quite similar in its lipid bilayer construction to the cytoplasmic membrane. In addition, the DNA in eukaryotic cells is contained in a distinct structure—the chromosome—in which the DNA is bound to proteins called histones. Histone proteins can be divided into several main types, H1, H2A, H2B, H3, and H4, which are linked together with DNA in an associative way to constitute nuclear units called nucleosomes. A core of histones composed of two molecules each of H2A, H2B, H3, and H4 is surrounded by 160 base pairs of DNA. Between individual nucleosome units, "linking DNA" is present in association with histone H1. Other DNA-binding proteins besides the histones are also present in chromatin material. It seems likely that many of these DNA-binding proteins have specific functions related to the regulation of gene expression.

These two primary differences between prokaryotes and eukaryotes—the nuclear membrane and the chromosomal structure—have caused a number of other differences in the gene-expressing systems to evolve. Because the structures of the chromosomes are different, the enzymes that transcribe the DNA into RNA are also different. Bacteria have only one RNA polymerase (although it can have several different sigma subunits), whereas eukaryotic cells contain a number of different RNA polymerases.

The ribosomes in prokaryotes and eukaryotes are also slightly different. In both types of organisms the general shape is the same, but the sizes are somewhat different. Eukaryotic mRNA is very different in that it is modified at both ends. Long repetitive sequences of A called poly-A tails are located at the 3′ ends of the molecules. These poly-A tails are tacked on after transcription from DNA has ceased. The 5′ ends of eukaryotic mRNAs are filled out by 7-methylguanosine residues linked to mRNAs by triphosphate bonds, as shown diagrammatically in Figure 1.11. No clear reason for these additional features at the 3′ and 5′ ends of mRNAs has yet been identified, but they may aid in transporting the mRNA through the nuclear membrane or they may afford protection against enzymatic destruction of the messages. In addition, genes in nucleated eukaryotic cells are not in a con-

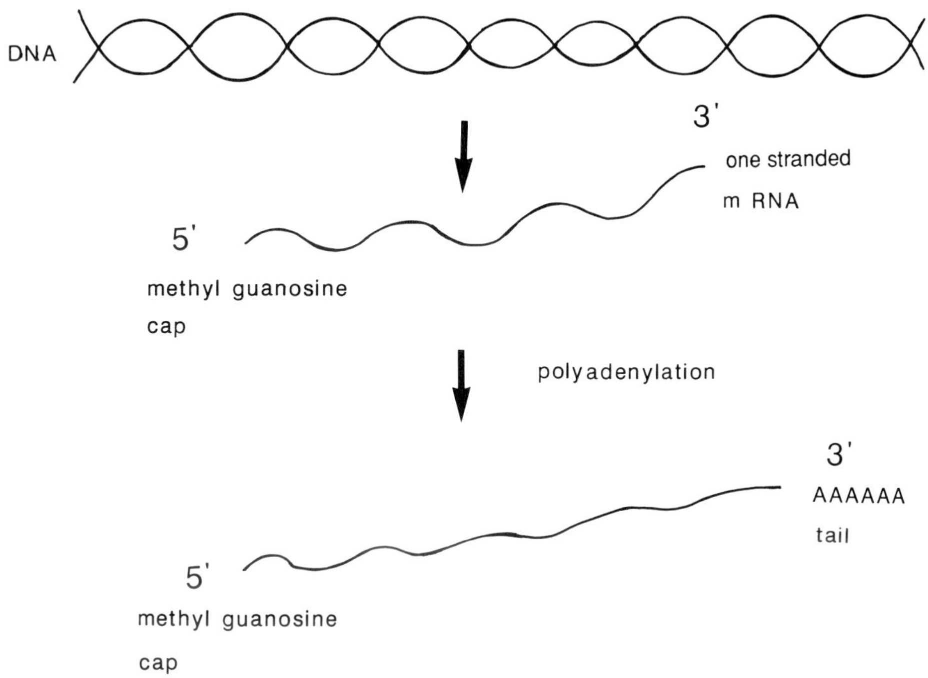

Figure 1.11 Eukaryotic mRNAs code for a single protein. During initial transcription methylguanosine caps are added at the 5′ end; after transcription, a poly-A tail is added at the 3′ terminus.

tinuous sequence, but are divided into pieces by noncoding sequences of DNA. The sequences that are used in coding for proteins are called *exons* (they are expressed) and must be spliced together to form a single continuous sequence, which is then translated. The intervening sequences, or *introns*, interrupt the coding sequences and must be removed or spliced out. This gene splicing occurs not at the DNA level but at the RNA level, after gene transcription has taken place. This *very* surprising finding—interrupted genes—was followed by one that is equally if not more surprising. The enzymes that carry out some of the splicing reactions are different from any other enzyme ever isolated—they are not protein catalysts, as all enzymes were thought to be, but RNA! These interrupted genes and the mechanisms of splicing are discussed further in subsequent chapters.

Differences between prokaryotes and eukaryotes have presented some complications in the development of genetic engineering technology, but they have, in another sense, been a great boon. Almost all the cited differences in the gene expression systems are sites at which antibiotics can act to inhibit bacteria without harming the mammalian host.

References

1. Stent GS, Calendar R: Heredity, in *Molecular Genetics*. G. S. Stent and R. Calendar, San Francisco, W. H. Freeman, 1978, pp 1–3.
2. Weismann A: *Essays on Heredity*, translated by Shipley AE, Schönland S, et al. Oxford, Oxford Univ. Press, 1891–1892, Vols 1 and 2.
3. Avery OT, MacLeod CM, McCarty M: Studies on the chemical nature of the substance

inducing transformation of pneumococcal types: Induction of transformation by a desoxyribonucleic acid fraction isolated from *Pneumococcus* type III. *J Exp Med* 1944; 79:137–158.
4. Griffith F: The significance of pneumococcal types. *J Hyg Cambridge*, England 1928; 27:113–159.
5. Hershey AD, Chase MC: Independent functions of viral protein and nucleic acid in growth of bacteriophage. *J Gen Physiol* 1952;36:39–56.
6. Watson JD, Crick FHC: Molecular structure of nucleic acids: A structure for deoxyribose nucleic acid. *Nature* 1953;171:737–738.
7. Williams RC: Electronmicroscopy of sodium desoxyribonucleate by use of a new freeze-drying method. *Biochim Biophys Acta* 1952;9:237–239.
8. Chargaff E: Chemical specificity of nucleic acids and mechanism of their enzymatic degradation. *Experientia* 1950;6:201–209.
9. Watson JD: *The Double Helix*, New York, Athenum, 1968, pp 194–196.
10. Wilkins MHF, Stokes AR, Wilson HR: Molecular structure of deoxypentose nucleic acids. *Nature* 1953;171:738–740.
11. Franklin RE, Gosling RG: Molecular configuration in sodium thymonucleate. *Nature* 1953;171:740–741.
12. Jaworski A, Hsieh W-T, Blaho JA, et al: Left-handed DNA in vivo. *Science* 1987; 238:773–777.
13. Marmur J, Rownd R, Schildkraut CL: Denaturation and renaturation of deoxyribonucleic acid. *Prog Nucleic Acid Res* 1963;1:231–300.
14. Garrod AE: Inborn errors of metabolism. *Lancet* 1908;2:1–7, 73–79, 142–148, 214–220.
15. Beadle GW, Tatum EL: Genetic control of biochemical reactions in *Neurospora*. *Proc Natl Acad Sci USA* 1941;27:499–506.
16. Ingram VM: A specific chemical difference between the globins of normal human and sickle-cell anaemia hemoglobin. *Nature* 1956;178:792–794.
17. Watson JD: *Molecular Biology of the Gene*. Menlo Park, CA, W. A. Benjamin, 1977, p 142.
18. Cohen SS: *Virus-Induced Enzymes*. New York, Columbia Univ. Press, 1968.
19. Duckworth DH: History and basic properties of bacterial viruses, in Goyal SM, Gerbe CP, Bitton G (eds): *Phage Ecology*. New York, Wiley, 1989, pp 1–43.
20. Temin HM, Mizutani S: Viral RNA-dependent DNA polymerase. *Nature* 1970; 226:1211–1213.
21. Baltimore D: Viral RNA-dependent DNA polymerase. *Nature* 1970;226:1209–1211.
22. Stent GS, Calendar R: RNA translation, in *Molecular Genetics, an Introducing Narrative*. San Francisco, W. H. Freeman, 1978, pp 510–512.
23. Hoagland MB, Keller EB, Zamecnik PC: Enzymatic carboxyl activation of amino acids. *J Biol Chem* 1956;218:345–358.
24. Hoagland MB, Stephenson ML, Scott JF, et al: A soluble ribonucleic acid intermediate in protein synthesis. *J Biol Chem* 1958;231:241–257.
25. Crick FHC: On protein synthesis. Biological replication of macromolecules. *Symp Soc Exp Biol* 1958;12:138–163.
26. Yanofsky C, Carlton BC, Guest JR, et al: On the colinearity of gene structure. *Proc Natl Acad Sci USA* 1964;51:266–272.
27. Crick FHC: The genetic code. *Sci Am* 1962;207(October):66–74.
28. Nirenberg MW, Matthaei JH: The dependence of cell-free protein synthesis in *E. coli* upon naturally occurring or synthetic polyribonucleotides. *Proc Natl Acad Sci USA* 1961; 47:1580–1602.
29. Bost KL, Smith EM, Blalock JE: Similarity between the corticotropin (ACTH) receptor and a peptide encoded by an RNA that is complementary to ACTH mRNA. *Proc Natl Acad Sci USA* 1985;82:1372–1395.
30. Bost KL, Blalock JE: Molecular localization of a corticotropin (ACTH) receptor. *Mol Cell Endocrinol* 1986;44:1–9.
31. Biro, J. Comparative analysis of specificity in protein-protein interactions. *Med Hypoth* 1981;7:969–1007.
32. Eberle AN, Drozdz R, Baumann JB, Girard J: Receptor-specific antibodies by immunization with antisense peptides. *Peptide Res* 1989;2:213–219.

33. Jacob F, Monod J. On the regulation of gene activity. *Cold Spring Harbor Symp Quant Biol* 1961;26:193–211.
34. Jacob F, Monod J. Genetic regulatory mechanisms in the synthesis of proteins. *J Mol Biol* 1961;3:318–356.
35. Jacob F. Genetics of the bacterial cell. *Science* 1966;152:1470–1478.
36. Pastan I, Adhya S. Cyclic adenosine monophosphate in *E. coli. Bacteriol Rev* 1976; 40:527–551.
37. Yanofsky C. Attenuation in the control of expression of bacterial operons. *Nature* 1981; 289:751–758.

Development of Recombinant Methods

2

Donna H. Duckworth, PhD and Ralph C. Williams, Jr., MD

Making new organisms, or at least modifying existing ones, appears to be something humans have always wanted to do. But before the middle of the twentieth century, there were severe constraints on our ability to do so. Animals and plants could be selectively bred—new flower colors could slowly be developed, turkeys could be made to produce a little more white meat, racehorses could become swifter. But the fantasy some children concoct of breeding a lion to a tiger to produce a liger remained just that—a childhood fantasy. When transformation was discovered, it became possible to hasten the process of making new organisms slightly, but only for bacteria; and even this worked only with very closely related organisms.

After the determination of the structure of DNA and the elucidation of its coding mechanism, the creation of ligers came a little closer. It was now *theoretically* possible to join genes from different species—even different genera, or for that matter, different kingdoms! An unknown driving force in humans would have found a way. But the discovery of restriction endonucleases made the matter almost trivial.

Discovery of Restriction Enzymes

Again, it was work with bacteriophage that led directly to the discovery of restriction enzymes, the "genetic scissors." It started in 1952 when Seymour Cohen reported that the DNA of phage T2 was different from any DNA ever before isolated in that it contained no cytosine.[1] In T2 and some of its close relatives hydroxymethylcytosine (H) replaced cytosine, so rather than having a DNA composed of A, T, G, and C, like all other previously characterized DNA, T-even phage had a DNA composed of A, T, G, and H. The base-pairing rules were not overridden; A still paired with T but H now paired with G. The identification of hydroxymethylcytosine led shortly to the discovery that the T-even phage DNA was highly glucosylated and, in addition, led Cohen to the discovery of virus-induced enzymes.

In looking to see whether hydroxymethylcytosine would replace the requirement of a thymine auxotroph for thymine (thymine being a methylated pyrimidine and hence closely related in structure to hydroxymethylcytosine), Cohen made the chance discovery that after phage infection, bacterial cells previously unable to synthesize thymine could now do so. This result suggested to him the idea of virus-induced acquisition of function, a totally novel idea in 1953.

Cohen then searched for and discovered the enzyme responsible for the synthesis of hydroxymethylcytosine and showed, in an extremely rigorous fashion, that the enzyme was synthesized de novo after infection; it was not injected into cells by the phage, nor was it present in an inactive state before infection. He also showed that the synthesis of thymine after phage infection was accomplished by a completely new protein not present in the cell before infection.[1]

This work was the catalyst for a virtual explosion of knowledge about the enzymology of infected cells and led to the discovery of dozens of induced enzymes and to a search for the genes controlling their synthesis. dCTPase, glucosylase, dCMP deaminase, deoxynucleotide kinase, dihydrofolate reductase, DNA polymerase, and DNA ligase were only a few of the enzymes discovered to be induced by the T-even phage. Of these, DNA ligase currently figures very prominently in techniques for making recombinant DNA. Efforts to prove that the phage actually carried the genetic information for the synthesis of these activities led to the discovery of the amber and temperature-sensitive mutants, whose conditional lethality allows them to be grown under certain conditions and then studied under other conditions for their adverse effects on the metabolism of the infected cell. It also allowed for the concept of early versus late protein synthesis and the elucidation of steps in the very orderly process of virus development.

But all these discoveries pale before the discovery of restriction enzymes. Their discovery was a *totally* unforeseen consequence of basic research into a fairly arcane phenomenon involving the behavior of bacteriophage grown on two different strains of bacteria. Sometimes phage grown on one strain grew very poorly on a second strain, and vice versa. It thus appeared that some hosts could "restrict" the process of virus (phage) reproduction.

The phenomenon was first observed by Luria and Mary Human working with phage T2. Because DNA of phage T2 is glucosylated, its restriction is a little different from other cases discovered later. But the observations made by Luria and Human served to establish the existence of the process of host-controlled restriction and modification. Luria had noticed that certain mutants of *Escherichia coli* would lyse when infected by T2 but that the lysate seemed to contain no new virus—at least none that would grow on *E. coli*. Later, Luria and Human wanted to add the lysate to streptomycin-resistant *E. coli*, but because their tube of *E. coli* had been dropped and broken (this was the "broken test tube" in the title of Luria's autobiography), they used streptomycin-resistant *Shigella* instead. T2 had been observed to replicate in this genus of bacteria as well as it did in *E. coli*, so using *Shigella* was quite logical. They found that even though it would not reinfect *E. coli*, the phage produced in their mutant *E. coli* did grow in *Shigella*. Moreover, after one cycle of growth in *Shigella*, the phage could again grow in *E. coli*.[2] It is now known that the mutant *E. coli* was unable to glucosylate (modify) the T2 DNA and that the unglucosylated DNA was broken down (restricted) when injected into other *E. coli*. *Shigella* is apparently missing the enzymes that hydrolyze the unglucosylated DNA, but it has the glucosylating enzymes that can modify the DNA for growth in *E. coli*. Thus, by virtue of having no streptomycin-resistant *E. coli*, Luria and Human serendipitously discovered restriction and modification. Many other phage with nonglucosylated DNA were later found to be subject to a slightly different kind of restriction and modification. This type of restriction and modification involves hydrolysis of DNA at specific sequences and protection or modification by methylation of these sequences, and it was first observed with lambda (λ) phage.

Bertani and Weigle found that if λ phage was grown on *E. coli* K, it could not grow on *E. coli* B strains or on *E. coli* K lysogenized by phage P1.[3] If λ was grown on *E. coli* K lysogenized by P1, it could be grown on *E. coli* K and *E. coli* K(P1). If λ was grown on *E. coli* B, it could not be grown on either lysogenized or unlysogenized K strains. The relationship of these phenomena to what Luria and Human had observed was not immediately obvious, and it was in trying to understand this curious series of observations that the discovery of the site-specific DNA hydrolyzing enzymes (restriction endonucleases) occurred.

The restricted phage were found to adsorb to the restrictive host and inject their DNA normally. In addition, adsorption of restricted λ neither killed nor even delayed the growth of the host cells. However, when the phage DNA was labeled with ^{32}P, the reason for restriction of the phage became apparent. The DNA was degraded soon after injection[4] into a restrictive host. The enzymes found to be primarily responsible for this degradation are the restriction endonucleases or restriction enzymes,[5] and they operate only on foreign DNA. The restrictive host protects its own DNA from the potentially lethal effects of the restriction endonuclease by "modifying" its DNA. Modification involves methylation of certain bases at sequences within DNA that constitute the recognition signals for the restriction endonuclease. This explains why the very small number of phage that survive one cycle of growth in the restrictive host (about 1 in 10^8 do manage to grow) can subsequently reinfect that host. Their DNA has been replicated in the presence of the modifying methylase, so it, like the host DNA, becomes methylated and protected from the endogenous restriction system.

Characterization of Restriction Enzymes

The restriction endonuclease of *E. coli* K was the first of these enzymes to be isolated and studied in detail. It and a similar enzyme from *E. coli* B were found to have some unusual properties. Although they appeared to be able to recognize certain unmodified sequences in DNA, they cut the DNA randomly. It was later found that after interacting with the unmodified "recognition" sequence, they travel along the duplex molecule for a distance corresponding to 1000 to 5000 nucleotides and then make a cut. Enzymes with these properties are now known as type I restriction endonucleases. Like all restriction endonucleases, they recognize specific nucleotide sequences. However, they are not particularly useful for gene manipulation because their cleavage sites are nonspecific.[6]

While these disappointing properties of type I restriction enzymes were being unraveled, a restriction endonuclease from *Haemophilus influenzae* d was discovered[7,8] that was to become the prototype of a large number of restriction endonucleases—now known as type II enzymes—that are fundamentally important in the manipulation of DNA. The type II enzymes recognize a particular target sequence in a duplex DNA molecule and break the polynucleotide chains within that sequence to produce discrete DNA fragments of defined length and sequence.

The sequences recognized by the more commonly used restriction enzymes are usually four to seven base pairs in length. The sequences are frequently said to be palindromic, meaning that they read the same in both directions (Madam I'm Adam). A more correct description of the recognition sequences is that they have an axis of rotational symmetry; that is, the sequence of one strand read forward is the same as the sequence of its *complement* read backward.

G A A T T C
C T T A A G

The terminology is confusing because sequences that read the same in both directions in *one* strand have also been called palindromic.

A G C C G A
T C G G C T

More often, in molecular biology at least, palindromic refers to an axis of rotational symmetry.

To date, restriction enzymes have been found in almost all bacterial genera that have been examined. Although discovered through their ability to prevent viral infection, they would, of course, operate to prevent the entry of any foreign DNA into a cell. Thus, conjugation (plasmid-mediated direct transfer of DNA between bacteria), transduction (DNA transfer by bacterial viruses), and transformation would all be subject to the constraints of host-controlled restriction.

The genes specifying the restriction and modification enzymes may reside in the bacterial chromosome or may be on extrachromosomal elements in bacteria, such as plasmids or prophage. The names of the various restriction enzymes are derived from the bacteria from which they are isolated, using the first letter of the genus name and the first two letters of the species name to form a three-letter abbreviation for the enzyme. This is usually (correctly) italicized; for example, *Escherichia coli* = *Eco* and *Haemophilus influenzae* = *Hin*. The three-letter abbreviation may be followed by a strain identifier, as in *Hind*III. In cases in which the restriction and modification system is coded for by a bacteriophage or a plasmid, the abbreviated name of the bacteria is followed by the name of the extrachromosomal element. For example, the enzyme coded for by the RI plasmid of *E. coli* is called *Eco*RI. A short list of such restriction enzymes, their origins, and selective sites of internucleotide cleavage is shown in Table 2.1. Since their initial description, restriction enzymes have been obtained from over 400 strains of bacteria; these nucleases define more than 100 different nucleotide cleavage sites.[6]

It should be noted that enzymes that recognize four-base sequence groups would have many more sites in a given DNA molecule than enzymes that recognize longer sequences. Thus, fragments resulting from the former (four-base recogni-

Table 2.1 Representative Restriction Nuclease Enzymes That Are Capable of Cleaving DNA

Designated abbreviation	Source of enzyme	Actual nucleotide/ base sequence cleaved[a]
*Bam*HI	*Bacillus amyloliquefaciens* H	G↓GATCC CCTAG↑G
*Eco*RI	*Escherichia coli* RY13	G↓AATTC CTTAA↑G
*Hind*III	*Haemophilus influenzae* Rd	A↓AGCTT TTCGA↑A
*Pst*I	*Providencia stuartii* 164	CTGCA↓G G↑ACGTC

[a] Arrows indicate actual sites of enzymatic cleavage.

tion) enzymes would be shorter and there would be more of them per digested DNA molecule.

Restriction Enzymes Allow Construction of Restriction Maps

The discovery of site-specific endonucleases made possible the physical mapping of DNA molecules. When specific DNA molecules are cut by a selected restriction enzyme, the fragments produced by hydrolysis of the DNA can be separated, primarily on the basis of relative size, using electrophoresis of restriction digests in agarose gels. How far various fragments of DNA move is primarily a function of their respective sizes. Very small DNA fragments move longer distances into the gel than do much larger DNA fragments or undigested DNA. The relative spacing or electrophoretic separation of the restriction fragments of DNA can be adjusted or controlled by varying the time of electrophoresis or the concentration of agarose in the gels. An example of human genomic DNA digested with *Eco*RI restriction enzyme plus markers of known molecular weight made by *Hin*dIII digestion of lambda phage DNA is shown in Figure 2.1. Different restriction enzymes usually

Figure 2.1 Southern blot patterns of the three most common Cλ genotypes noted with the 0.8-kb Cλ probe hybridized to *Eco*RI-digested genomic DNA from normal subjects in parallel with Graves' patients. Relative position of bacteriophage λ *Hin*dIII fragments of 23, 9.4, and 6.6 kb are shown to the left. All Graves' and normal *Eco*RI digests showed a 5-kb pseudogene. *Reproduced from Williams RC, Marshall NJ, Kilpatrick K, Montano J, et al. Kappa/ lambda Ig distribution in Graves' thyroid stimulating antibodies. Simultaneous analysis of Cλ gene polymorphisms. J Clin Invest 1989;82:1306–1312, with permission.*

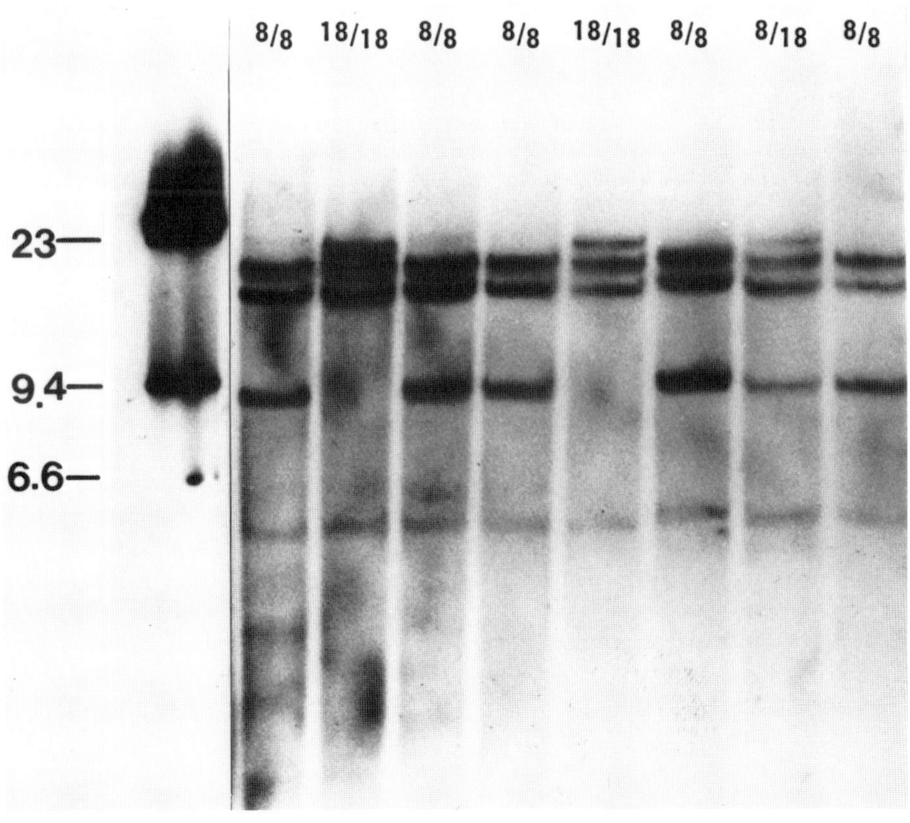

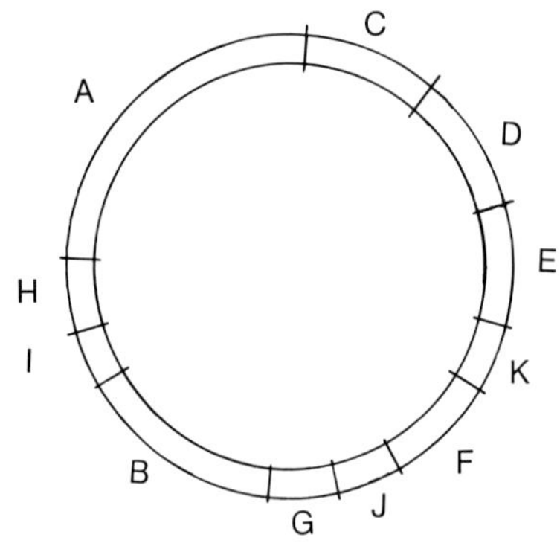

Figure 2.2 Physical order of SV40 DNA fragments produced by *Hin*dIII.

produce fragments of different sizes and agarose gel electrophoresis patterns. Enzymes that have rare recognition sequences (usually the longer recognition sequences) are generally the most useful because they produce smaller numbers of fragments, which are then most easily distinguished by electrophoretic agarose gel separations. The method of restriction mapping of linear segments of DNA that code for certain proteins or functions was introduced in 1971 by Danna and Nathans,[9] using the restriction endonuclease obtained from *H. influenzae* and double-stranded, closed circular DNA from the oncogenic virus SV40. In this instance, the DNA was relatively small, having a molecular mass of 3×10^6 daltons and containing about 5000 base pairs. The investigators used the 11 linear fragments produced by digestion with *Hin*dIII to determine where the small circular SV40 viral DNA (Figure 2.2) was cut. This established reference points for such functions as initiation of DNA replication and transcription and for the localization of specific genes. By using pulse-labeled SV40 DNA, they traced the distribution of label within newly synthesized molecules by examining sequential specific DNA fragments produced by the *Hin*dIII. Through sequential digestion, a temporal order of cleavage and production of the SV40 DNA fragments[9,10] and the initial site (origin) of replication could be determined. Later, maps of sites of other restriction enzymes were used to identify sequences coding for mRNAs for various viral proteins as well as the locations of the promoter sequences.

Recombinant DNA Made Possible by Restriction Enzymes

"Mapping" DNA, however useful, is only one application of restriction enzymes. An even more dramatic use is that of joining unrelated DNA molecules. In cleaving DNA some restriction enzymes produce what are called blunt ends, fragments that retain base-pairing characteristics all the way out to their ends where the double strand has actually been cut. An example of such blunt-ended cleavage by the enzyme *Hpa*I is shown in Figure 2.3. Such blunt ends on each DNA fragment resulting from the cleavage have no propensity at all to adhere or to stick together. Other enzymes such as *Eco*RI produce symmetrical but irregular DNA strand cleavages that leave double-stranded DNA with irregular lengths of each strand at the ends as though it were partially unraveled. In the case of *Eco*RI, four nucleotide or base ends are present on the end of each fragment. Such irregularly cut DNA ends tend to be sticky and adhere to complementary irregular *Eco*RI tails by base pairing (Figure 2.4). This adherence can be made permanent if the enzyme DNA ligase is added, whereupon a permanent chain is put together as usual by phos-

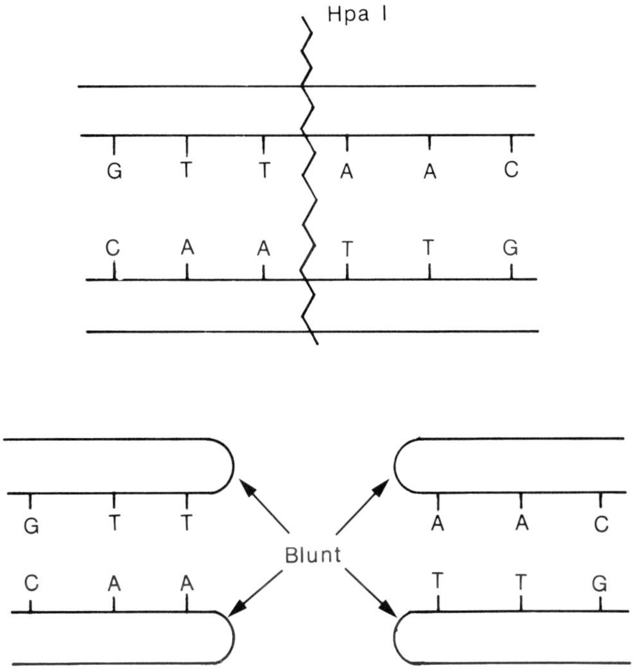

Figure 2.3 Restriction fragment cleavage induced by the enzyme *Hpa*I, which recognizes the six-nucleotide sequences GTTAAC. Cleavage results in a fragmentation directly across the two DNA strands and what are termed blunt ends.

phodiester bonding. These enzymes that produce sticky ends are the ones that have made the production of recombinant DNA molecules almost child's play.

Suppose, now, that some tiger DNA was cut with a certain enzyme, say *Eco*RI, and mixed with some lion DNA cut with the same enzyme. The lion and tiger DNAs would have the same sticky ends and could hybridize to each other. Ligase again could permanently seal this joint, and "liger" DNA would have been produced. This remarkably simple procedure—that of cutting two dissimilar DNAs with the same enzyme and allowing the fragments to hybridize—is the basis for the production of almost all chimeric (recombinant) DNA molecules.[6]

Joining Blunt Ends

The enzyme terminal transferase has been used as a reagent to produce sticky ends at the blunt ends of DNA segments. This enzyme is capable of adding nucleotides directly to 3′ ends of DNA.[11] Thus polydeoxy A can be added to both 3′ ends of a double-stranded DNA chain and polydeoxy T to the 5′ ends. When these are then mixed together, the base-paired poly-A and poly-T ends can be covalently bonded using DNA ligase.[12,13] When DNA chains are joined using the terminal transferase procedure, they do contain regions of $^{\mathrm{TTT}}_{\mathrm{AAA}}$ bases, which may affect the translation of the joined segments. Joining naturally base-pairing cohesive ends, as in the situation described for *Eco*RI, is a much more direct approach.

Sequencing Procedures

Methods for direct sequencing of DNA, and particularly restriction DNA fragments, represent a major breakthrough in DNA technology. The first methods utilized lengthening of DNA chains or their fragments with DNA polymerase[14] or stepwise

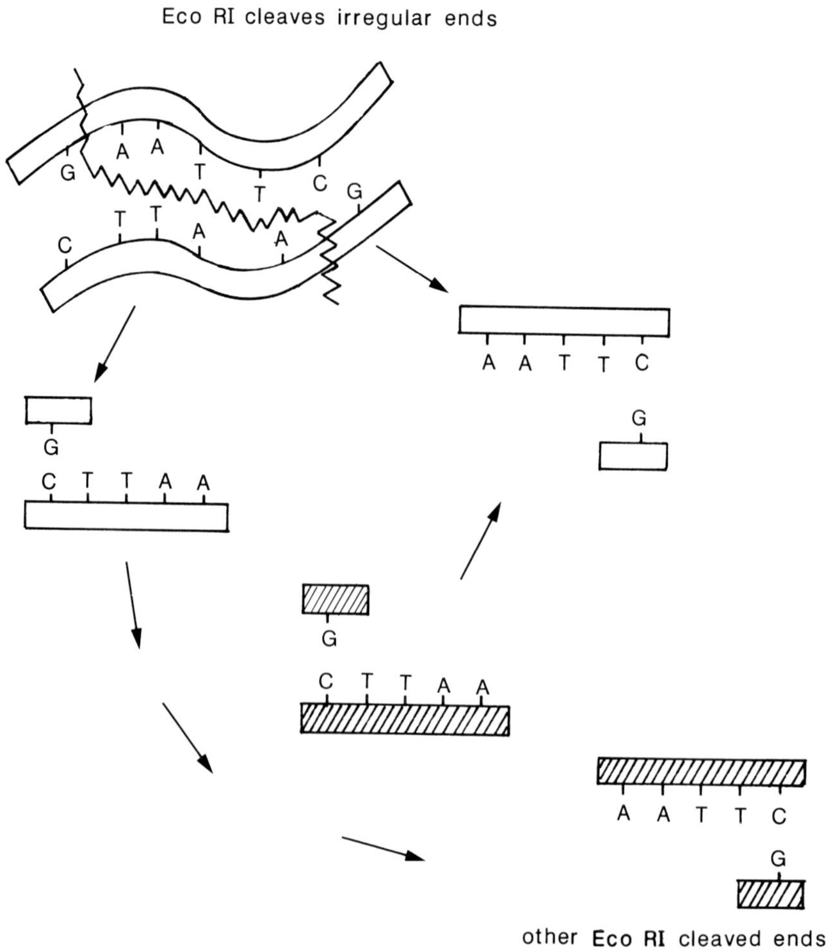

Figure 2.4 *Eco*RI restriction enzyme cleaves DNA leaving irregular ends at the site of chain separation. These staggered cuts are sticky and can adhere to similar complementary ends produced by other *Eco*RI cuts. They can be rejoined chemically using DNA ligase.

chemical degradation of the individual nucleotide members of DNA chains.[15] The chemical degradation technique originally introduced by Maxam and Gilbert is illustrated in Figure 2.5. In essence, this procedure sequences DNA by using chemical treatments that break DNA previously labeled with ^{32}P at one end. The breaks occur at base-specific points, and thus the lengths of the labeled fragments separated by agarose electrophoresis determine the positions of specific bases. Chemical reactions used include those that preferentially cleave DNA at guanines, at adenines and guanines, at cytosines and thymines equally, and at cytosines alone. When products of these four different cleaving reactions are separated by size on agarose or polyacrylamide electrophoresis, the DNA sequence can be read directly from the pattern of radioactive bands of ^{32}P-labeled DNA fragments.

A second method for determining DNA sequence was developed by Sanger and associates.[16,17] This approach is based on interruption of newly synthesized DNA chains rather than chemical reagents that cleave the DNA. It was found that DNA chain production could be interrupted using 2′,3′-dideoxy nucleoside tri-

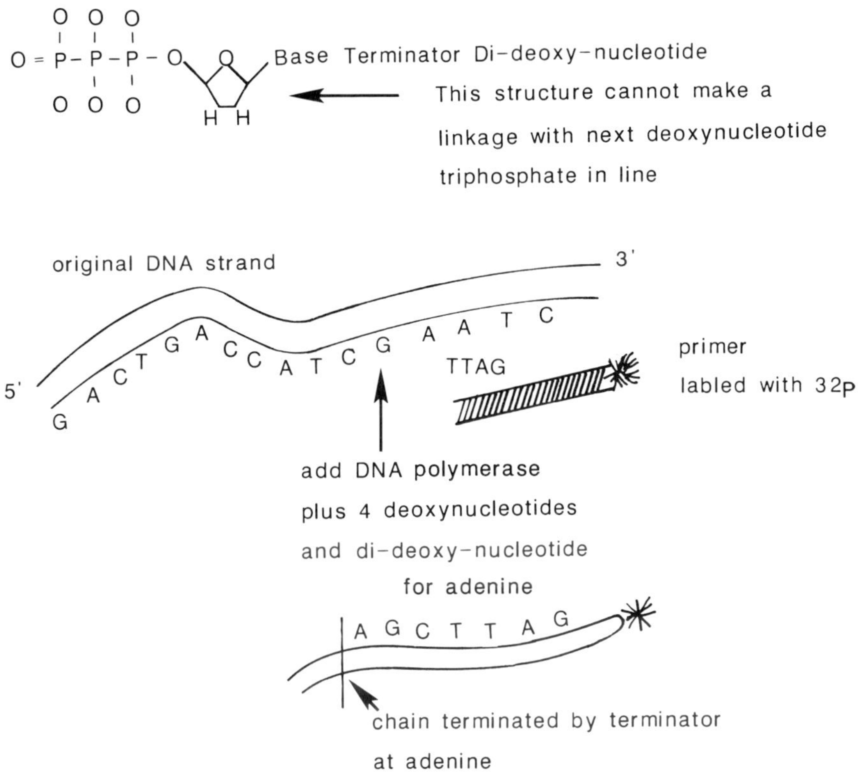

Figure 2.5 Chains terminated at specific nucleotides by the respective dideoxynucleotides have different chain lengths, and these products can be visualized by polyacrylamide gel electrophoresis followed by radioautography to deduce the exact nucleotide sequence of the original DNA strand.

phosphates. These could be incorporated through their 5′ triphosphates into a chain of nucleotides being arranged into linear DNA; however, they could not form phosphodiester linkages with the next deoxynucleotide triphosphates in line, and chain elongation would be halted. This sequencing method therefore employs addition of small amounts of specific nucleotide synthesis terminators (dideoxy nucleoside triphosphates) along with the four deoxy nucleoside triphosphates normally incorporated into the DNA chain by DNA polymerase. The resulting reactions produce a mixture of chains ending with the respective dideoxy group. With the use of labeled DNA primers and four separate DNA polymerase reactions (including the appropriate ddNTPs), a series of DNA fragments is synthesized. These can then be analyzed by gel electrophoresis and radioautography and the original sequence of each DNA strand read directly (Figure 2.6). This technique can produce reproducible sequence data for chains of DNA as long as 5000 to 6000 base pairs.

Using these sequencing methods, it is now possible to sequence rapidly DNA segments of moderate size and for smaller genes to obtain the entire nucleotide sequence of DNA. With the complete sequence of a gene, one can deduce the amino acid sequence of the protein for which the gene codes. This method of sequencing nucleotides and then reading the amino acid sequence of the corresponding polypeptides or proteins is now automated and is often faster and more direct than procedures currently available for protein sequencing.

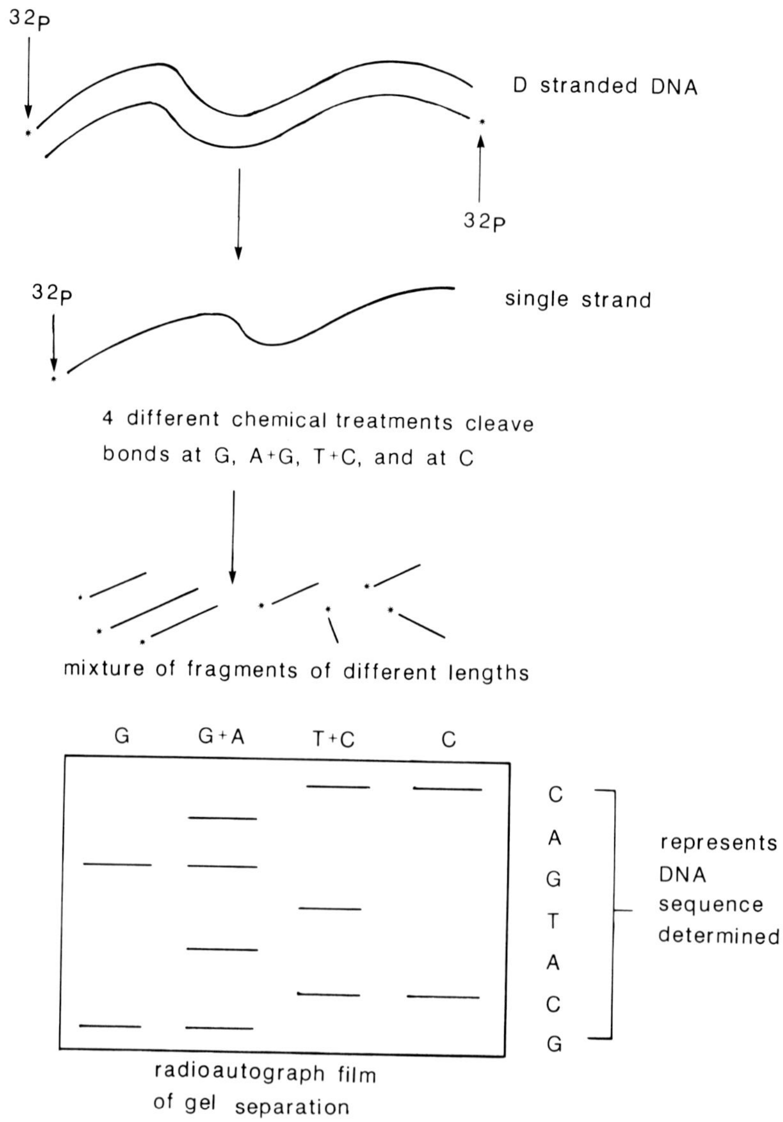

Figure 2.6 Chains terminated at specific nucleotides have different lengths and can be visualized by radioautography on polyacrylamide gel and x-ray film.

In Vitro Synthesis of DNA and Oligonucleotides

After rapid, convenient, and reproducible methods for determining sequence became available, it also became possible to synthesize genetic material or fragments of DNA in vitro. The methodology employed depends on being able to direct the reaction so that both the 5′ and 3′ ends of the polynucleotide chain being constructed remain undisturbed. This is accomplished by shielding the 5′ and 3′ ends of the growing oligonucleotide chain with protective chemical groups, which can be removed later as desired and additional nucleotides added until the final oligonucleotide length and sequence are achieved. The chemical synthesis of long polynucleotide chains has recently been automated and can now be performed rapidly and reproducibly in many laboratories.

Replication of Recombinant Molecules

Producing recombinant molecules is, of course, only one step in the genetic engineering process. To be useful, the recombinant molecule would have to be inserted into an appropriate cell where it could replicate and be expressed. Even before it was possible to produce recombinant DNA molecules, there were a number of attempts to transform both prokaryotic and eukaryotic cells with foreign DNA. However, with rare exceptions such as the pneumococcus, the foreign DNA had no effect on the cells. This could be explained in several ways: the DNA might be unable to enter the cells, or it might enter and immediately be broken down. Or the exogenous DNA might be unable to replicate or be expressed in the transformed cells.

If the exogenous DNA is integrated into (i.e., made an indistinguishable part of the host genome), there is no problem with regard to its replication, but even integrated foreign DNA is not always expressed. The exact mechanism of this integration is not clear, and its occurrence is usually a rare event. If the exogenous DNA fails to be integrated, it will probably be lost during subsequent multiplication of the host cells. In order to be replicated, DNA molecules must contain an *origin of replication,* and in bacteria and viruses there is usually only one per genome. Molecules containing replication origins are called *replicons.* Fragments of DNA are not usually replicons and, in the absence of a replication origin, will be diluted out of their host cells. And even if a DNA molecule contains an origin of replication, it may not function in a foreign host cell.

There were additional problems in making the process of genetic engineering useful. If experiments were to proceed efficiently, a method was required for assessing the fate of the donor DNA. In cases in which the foreign DNA was maintained because it became integrated into the host DNA, a method for mapping the foreign DNA and the surrounding host sequences was required. The solution to the first problem, that of getting foreign DNA into cells, came with the development of a number of cloning "vehicles" or "vectors." The second problem was solved with a molecular hybridization technique known as Southern blotting, which will be discussed later in this chapter.

Cloning Vectors

If fragments of DNA cannot replicate, the obvious solution is to attach them to a suitable replicon. Such replicons are known as *vectors* or *cloning vehicles.* For a vector to be useful, there must be a way to introduce it into the target cell, and it must be able to replicate in the cell. There should also be some easy way to detect the presence of the vector in the cell, and, in order to introduce foreign DNA into it, it should be easily isolable and have single sites for a number of restriction enzymes. Ideally, these "cloning sites" should be in genes whose phenotype will be altered, in a readily detectable way, by insertion or recombination of extra DNA into the sites. Small plasmids and viruses are the most suitable vectors, for they are replicons in their own right, their maintenance does not necessarily require integration into the host genome, and their DNA can be isolated readily in an intact form.

Plasmids: General Properties

Plasmids are widely distributed throughout the prokaryotic kingdom and are found in some lower eukaryotes as well. Usually they will replicate only in closely related genera. They vary in size from less than 1×10^6 daltons to greater than 200 ×

10^6 daltons of DNA. The plasmid DNA is almost always a covalently closed circular molecule and is usually dispensable under ordinary conditions.

Plasmids can be categorized into one of the two major types—conjugative or nonconjugative—depending on whether or not they carry a set of transfer genes, called the *tra* genes, that promote bacterial conjugation and transfer of DNA from one bacterium to another. Plasmids can also be categorized according to whether they are maintained as multiple copies per cell (*relaxed* plasmid) or as a limited number of copies per cell (*stringent* plasmid). Generally, conjugative plasmids are of a relatively high molecular weight and are present as one to three copies per chromosome, whereas nonconjugative plasmids are of low molecular weight and exist in multiple copies per cell.[18]

Plasmids are sometimes classified in *incompatibility* groups, where incompatibility is the inability of some plasmids to coexist in the same host cell. Plasmids that are very different are often compatible, whereas plasmids that are related are incompatible. Related plasmids that are mutually incompatible are said to belong to the same incompatibility group. Currently, over 25 incompatibility groups have been defined among plasmids of *E. coli*, which are the best-studied plasmids. Some plasmids are said to be *promiscuous* because they can transfer to a wide range of gram-negative bacteria and be stably maintained in these diverse hosts. Such promiscuous plasmids have the potential of readily transferring cloned DNA molecules into bacteria with a wide range of genetic backgrounds.[18]

Traits coded for by plasmids often give the host bacteria a selective advantage under threatening circumstances. These traits include antibiotic production, antibiotic resistance, degradation of unusual substrates, heavy metal resistance, bacteriocin production, induction of plant tumors, and a variety of virulence properties, such as hemolysis production, complement resistance, and toxin production. Plasmids to which phenotypic traits have not yet been ascribed are termed *cryptic* plasmids.

Antibiotic Resistance Plasmids

Plasmids were originally discovered because of their ability to cause conjugation and genetic exchange in certain *E. coli* strains. These were the so-called F or fertility factors found in experiments by Tatum and Lederberg demonstrating transfer of genetic information from one bacterium to another.[19] The discovery of antibiotic resistance plasmids very shortly thereafter indicated that plasmids were not just involved in sex and mating in bacteria but were actually important medical entities. In addition, antibiotic resistance plasmids have turned out to be extremely useful as vectors in recombinant DNA experiments.

Antibiotic resistance plasmids were discovered in Japan shortly after the end of World War II. In Japan, dysentery caused by the enteric bacterium *Shigella dysenteriae* was a lingering problem in the years after the war in spite of improved sanitary conditions brought about by modernization. The advent of the commercial production of antibacterials offered some hope that the problem would be eradicated. Starting in about 1945, a large amount of sulfanilamide was used to combat the dysentery bacillus. It was effective for only about 5 years, however, as sulfa-resistant *Shigella* were isolated with increasing frequency beginning in 1949. By 1952, 80% of *Shigella* isolates were resistant to sulfa.

Fortunately, the production of antibiotics such as streptomycin, tetracycline, and chloramphenicol started at that time, and these agents were quite effective against bacillary dysentery. Streptomycin production began in 1950, and by 1953 its production was increasing rapidly to meet the demand. Tetracycline and chloramphenicol were first imported from the United States in 1950 and their production started in Japan soon thereafter. With the increase in antibiotic production, antibiotic-resistant *Shigella* strains appeared and gradually increased.[20]

Isolation of Multiply Resistant Strains

The first isolation of a multiply resistant *Shigella* strain (a strain resistant to two or more antibiotics) from a dysenteric patient occurred in 1952. The multiple resistance involved three drugs, tetracycline, streptomycin, and sulfa. In 1955 a quadruply resistant *Shigella* strain appeared with resistances to tetracycline, streptomycin, sulfa, and chloramphenicol.

From a theoretical standpoint, if 10^{-a} and 10^{-b} are the mutation frequencies of resistance to drugs A and B, respectively, then $10^{-(a+b)}$ is the mutation rate of simultaneous resistance to both drugs, if the development of resistance to one is independent of resistance to the other. Because development of resistance to one antibiotic occurs at a rate of about 10^{-6}, it appears that there is only an infinitesimal chance of simultaneously developing both mutations in a single cell. Hence it was thought that combined chemotherapy might be useful in combating bacteria that were antibiotic resistant. Therefore, the report of a quadruply resistant *Shigella* strain was quite a shock, both from a clinical and a genetic point of view.

Multiple drug–resistant strains were isolated with increasing frequency after 1955. Their occurrence could not be accounted for by an epidemic spread of a single bacterium that by some chance (however small) had become resistant to the four drugs, because the resistant strains were of many different serotypes.

In 1957 it was reported that *Shigella flexneri* and *E. coli*, both resistant to four drugs, were isolated from a single patient. This was a shocking finding, as it would be mathematically impossible for the two different organisms to acquire this rare quadruple resistance independently of each other. It appeared that the drug resistance was being spread from bacteria to bacteria. A *Shigella sensitive* to the four drugs and an *E. coli resistant* to the four drugs were isolated from the patient's wife. After she was treated with chloramphenicol for 5 days, she excreted quadruply resistant *Shigella*. As she was treated with only chloramphenicol, it appeared that the *E. coli* might have transferred its four resistance genes to the *Shigella* all in one "block" and that this had been selected for by the chloramphenicol.

Subsequently, *E. coli, Proteus vulgaris*, and *S. flexneri* resistant to four antibiotics were recovered from another patient. This made it highly likely that the drug resistance was infectious, that is, could be transmitted from one bacterium to another. Later, it was found that the antibiotic resistance is coded for by genes carried on large conjugative plasmids. These plasmids could pick up genes to inactivate antibiotics, by mechanisms that are only beginning to be understood, and transfer them from bacteria to bacteria, species to species, even genus to genus. Antibiotic resistance in such instances is usually due to production of proteins that inactivate or degrade the antibiotic rather than alteration of the bacterial target of the antibiotic.

Gene Engineering with Antibiotic Resistance Plasmids

The utility for genetic engineering of plasmids carrying the resistance genes has been enormous. Cohen, Chang, Boyer, and Helling[21] were the first to use an antibiotic resistance plasmid in what could be considered to be a forerunner of genetic

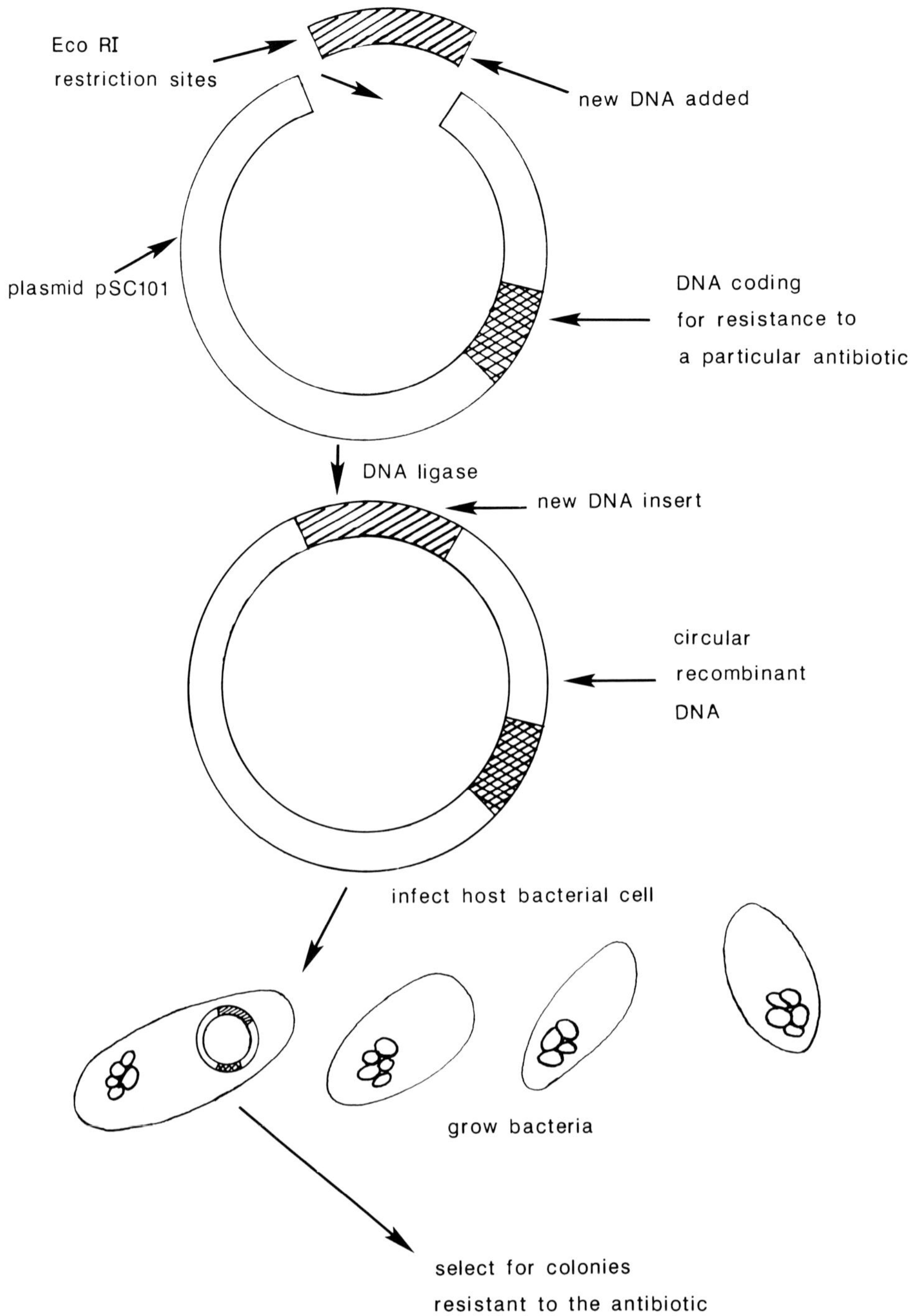

Figure 2.7 The small circular plasmid pSC101 is cut with restriction enzyme *Eco*RI, and a new DNA fragment also cut out at both ends by *Eco*RI is inserted by ligating the sticky ends of both DNA insert and cut plasmid. This results in a circular recombinant DNA plasmid. The plasmid originally contained a gene segment coding for resistance to a particular antibiotic, so when the recombinant plasmid is introduced into its bacterial *E. coli* host and multiplies, cells possessing the newly inserted DNA can be screened on plates by their specific characteristic of antibiotic resistance.

engineering experiments. They utilized a plasmid called pSC101, which was formed by shearing the high-copy-number conjugative plasmid R6K, which codes for resistance to tetracycline, chloramphenicol, streptomycin, neomycin, and sulfonamide. Fragments of R6K were transformed into *E. coli* (*E. coli* is not naturally transformable but can be made so by treatment with $CaCl_2$) and the transformed cells plated on tetracycline agar. Plasmid DNA was isolated from the cells that grew on the tetracycline, and a plasmid about 10% of the size of the original R6K was found. This was a piece of the original plasmid that fortuitously contained the origin of replication of the R6K plasmid plus the gene for detoxifying tetracycline.

Another small plasmid was then made from R6K, this time by cutting the DNA with *Eco*RI and transforming *E. coli* with the fragments produced. This time cells were plated on kanamycin agar to *select* for DNA that coded for kanamycin resistance and had an origin of replication. When plasmid DNA was isolated from the cells that grew on kanamycin, a plasmid with a molecular mass of about 17×10^6 daltons was isolated. Digestion of this new plasmid with *Eco*RI resulted in three fragments on the gels, rather than the original 12 fragments found in pR6K. This meant that the three fragments had hybridized and were ligated in vivo to form the new plasmid. This plasmid was called pSC102.

In the third experiment of Cohen et al., pSC101 and pSC102 were both digested with *Eco*RI and ligated in vitro prior to transformation into *E. coli*. pSC101, when digested with *Eco*RI, contained only one fragment. After transformation with the mixture of four fragments (one from pSC101 and three from pSC102), the bacteria were grown on media containing kanamycin and tetracycline. Plasmid DNA was isolated from the bacteria thus selected. In this case, the isolated plasmid DNA contained pSC101 plus one of the *Eco*RI fragments of pSC102. As pSC101 codes for only tetracycline resistance, the kanamycin resistance must have come from pSC102. This simple experiment showed that full biological activity could be regained by ligating (in vitro) DNA fragments produced in vitro by restriction endonuclease digestion.

In this experiment, pSC101 acted as the vector (the DNA molecule with the origin of replication) for the cloning of the kan^r gene from pSC102. Cells containing the vector could be selected by growth on media containing tetracycline, and the cloned gene could be selected by growth on kanamycin. Growing bacteria on the two antibiotics would select for the vector containing the cloned kan^r gene. Applications of antibiotic resistance genes and their usefulness in cloning DNA fragments are illustrated in Figures 2.7, 2.8, and 2.9.

It was shown that DNA fragments of entirely different origins could be inserted or "cloned" into pSC101 with the formation of new hybrid plasmids (Figure 2.7). This is one of the fundamental strategies in much current recombinant DNA work. In their original experiments, Cohen and co-workers utilized recombination or insertion of DNA from one plasmid into another; however, their discovery paved the way for much more diverse experimental approaches, including restriction fragment cutting out of human or animal genes of interest and their subsequent cloning in various other plasmid and bacteriophage vectors.

Making Plasmid Vectors More Useful

A large number of plasmid vectors have now been synthesized, but one of the earlier vectors developed is still one of the most useful. This is the plasmid pBR322. The usefulness of vectors such as pSC101 was limited because there was no way

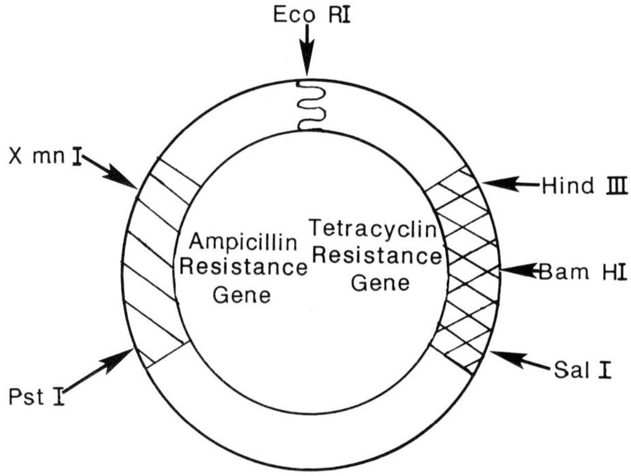

Plasmid pBR322

Figure 2.8 Diagrammatic representation of the circular DNA of plasmid pBR322 showing several restriction enzyme sites. The separate stretches of DNA conferring ampicillin or tetracycline resistance are shown, together with specific enzyme restriction sites present within the nucleotide sequences conferring ampicillin or tetracycline resistance. Insertion of new DNA at any of the sites in the resistance genes will result in loss of that function. This insertional inactivation indicates that DNA has been cloned into the gene.

to tell if the vector contained inserted DNA *unless* that DNA coded for a gene that could be directly selected. Vectors could be made more versatile if insertion caused some change that was easy to detect. A plasmid with two antibiotic resistance genes offered the possibility of doing this. Colonies harboring recombinant plasmids could be selected by *insertional inactivation* of one of the two resistance genes, and the other gene could be used for selection and maintenance of the plasmid in a growing population (see Figure 2.8).

An ampicillin resistance gene was, therefore, inserted into a small plasmid coding for tetracycline resistance, pMB9. The resulting plasmid, pBR312, was further modified by eliminating unnecessary DNA sequences to reduce its size and to increase the probability of finding unique restriction sites. The end result was pBR322, whose structure is shown in Figure 2.8.[22] Utilization of the *Pst*I site in the ampicillin resistance gene for introduction of foreign DNA would inactivate the gene, so that plasmids containing inserts could be distinguished by their altered phenotype from those not containing inserted DNA.

A Simple Cloning Experiment

A simple example of how pBR322 can be used is illustrated by the work of Pinkerton et al. in cloning the colicin gene and the phage-inhibiting gene from the large single copy ColIb plasmid.[23] They wished to determine whether the ColIb plasmid gene that inhibited phage replication was the same gene that produced the lethal colicin molecule. This was a question made to be answered by gene cloning. The large plasmid was digested with *Pst*I and mixed with pBR322 also digested with *Pst*I. The mixture was ligated, and the ligated molecules were transformed into *E. coli* by treatment of the bacteria with $CaCl_2$. Ligation of the mixture would produce three types of molecules:

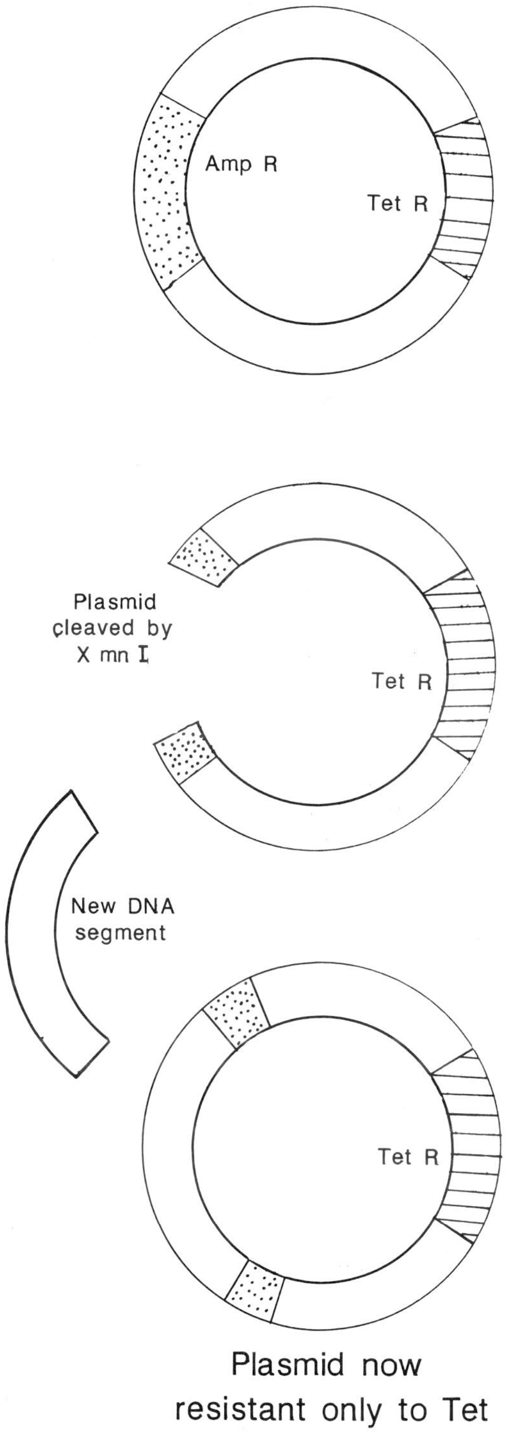

Figure 2.9 Circular plasmid pBR322 with separate antibiotic resistance genes for ampicillin and tetracycline is cleaved within the ampicillin resistance nucleotide sequence by restriction enzyme XmnI and a new DNA fragment is inserted and ligated into the XmnI restriction sites. Plasmids subsequently growing in their bacterial hosts are then resistant only to tetracycline and can be identified by this characteristic.

1. Native pBR322 produced by rejoining the ends of the linear molecule that was formed when pBR322 was digested with *Pst*I.
2. pBR322 with pieces of the ColIb plasmid recombined into it.
3. Circular fragments of the ColIb plasmid.

The first group, native pBR322, would be the most abundant species because the ends of the linear pBR322, being relatively close together, would easily find each other. Cells containing these plasmids would be resistant to both ampicillin and tetracycline. Cells containing the second group of plasmids would be resistant to tetracycline only, as the ampicillin resistance gene would have been inactivated by the insertion of foreign DNA into its middle. The third group of plasmids would be unable to replicate, unless they received the ColIb origin of replication. In any event, cells transformed with these plasmids would be sensitive to both ampicillin and tetracycline and would thus be indistinguishable from untransformed cells.

To find the desired cloned genes, Pinkerton et al. initially plated the transformed cells on tetracycline agar. This would *select for* cells containing plasmids in groups 1 and 2. They then used a replica plating technique in which individual colonies were plated on both ampicillin plates and tetracycline plates and isolated cells resistant to tetracycline but sensitive to ampicillin by comparing the two plates. These "clones" were known to contain recombinant plasmids, as new DNA had been inserted in the amp^r gene. The clones containing the colicin gene were found by overlaying plates with several hundred colonies with a strain of bacteria sensitive to the colicin. Clones containing the colicin gene could be visualized by the zone of killing around them. Clones containing the gene that inhibited phage replication had to be found by testing individual colonies for their ability to produce phage. Hundreds of colonies had to be tested, but because it was necessary to test only the bacteria that were *known* to contain added DNA (because of the insertional inactivation of their ampicillin resistance) the task was manageable. It could be shown that the two genes were different by isolating them into separate clones, because cells containing the colicin gene did not inhibit phage production and cells that inhibited phage production did not produce colicin.

Colorful Clones

Another method of identifying clones containing recombinant DNA molecules involves the enzyme β-galactosidase. This enzyme converts a colorless chromogenic substrate called x-gal into a blue product. Many vectors now include β-galactosidase genes flanking a multiple cloning site. If the gene coding for β-galactosidase is uninterrupted, the x-gal is hydrolyzed to produce a blue color. If foreign DNA is inserted into the gene, the x-gal is not hydrolyzed. Hence a clone of cells containing a recombinant molecule will be white.

Bacteriophage Vectors

Bacteriophage have also been made into very useful vectors. Bacteriophage lambda is one of the most studied and best understood extrachromosomal elements in bacterial genetics. Over the past 20 years, λ has served as a model system for the study of bacteriophage morphogenesis, DNA replication, and gene regulation. Therefore, it is not surprising that it was also called upon to serve as one of the first molecular cloning vectors.

Bacteriophage λ is a temperate phage whose DNA molecule is linear and double stranded (except for a short piece at the ends). Its life cycle can be divided into two distinct phases: lytic and lysogenic. The lytic phase of the cycle begins with the injection of λ DNA into the bacterial host cell. Because of the presence of a 12-base single-stranded cohesive-end site (*cos*) at each end of the linear molecule, the phage DNA assumes a circular conformation and proceeds to replicate. During the infectious process, the λ DNA molecules monopolize the biosynthetic machinery of the host cell to produce progeny phage DNA and various enzymatic and structural (coat) proteins. For the DNA to mature into viable phage, the coat proteins must encapsulate at least 75% (36.8 kb) but no more than 105% (51.5 kb) of the normal phage genome. This is accomplished by the action of site-specific nucleases located at the base of the phage head that cleave at the *cos* sites. As will be discussed later, the requirement for a "headful of DNA" is an essential feature of the λ vector system. Upon completion of the lytic process, the host cell wall disintegrates, releasing about 100 progeny phage particles.

In the lysogenic mode, the infecting bacteriophage DNA inserts itself in a harmless fashion into the host chromosome. Here, the λ genome is maintained in a passive state by chromosomal DNA replication.

There are now over 100 lambda cloning vectors, including the so-called cosmids. When it was found that lambda DNA could be packaged in phage heads in vitro, it became practical to devise plasmid vectors that could be packaged in phage coats. The packaging allowed very efficient delivery of the vector to a host cell. These cosmid vectors were made by introducing the lambda *cos* sites into pBR322. This allows up to 45 kb of DNA to be cloned and the cloned DNA directly injected into a cell by the λ phage coat.

Cloning Eukaryotic DNA into Bacteria

All of the vectors described so far are derived from prokaryotes. The cloned genes have also been of prokaryotic origin. Soon after the development of methods for gene cloning, it was shown that eukaryotic DNA could also be replicated and transcribed in bacteria. One of the first of these experiments involved the construction in vitro of plasmid chimeras composed of both prokaryotic and eukaryotic DNA, and the recovery of recombinant DNA molecules from transformed *E. coli* in the absence of selection for genetic determinants carried by the eukaryotic DNA. The amplified ribosomal DNA (rDNA) from *Xenopus laevis* oocytes was used as the source of eukaryotic DNA for these experiments because it could be readily purified and had been well characterized.

A mixture of *Eco*RI-cleaved pSC101 DNA and *X. laevis* rDNA was ligated and was used to transform *E. coli*. Tetracycline-resistant colonies were isolated and 55 separate transformants were selected. Their plasmid DNA was extracted and analyzed. All 55 plasmids gave a fragment whose molecular weight was shown to be 5.8×10^6 on *Eco*RI digestion. This was the starting pSC101 DNA. Thirteen of the plasmids yielded additional fragments corresponding in size to *Eco*RI-produced fragments of *X. laevis* rDNA. This showed that nothing inherent in the structures of eukaryotic DNA prevented it from being replicated in bacteria—provided it was attached to a suitable vector.[24]

Several years later pBR322 was used to clone an artificial gene for somatostatin, and by "genetically" attaching it to the control genes for β-galactosidase (the *lac* promoter, operator, ribosome binding site, first seven triplets of the structural gene) it was shown that the human hormone somatostatin cold be produced in bacteria.[25] In this experiment, a very sensitive radioimmune assay could be used to detect the somatostatin. In other experiments there might be no assay for the product, the product might be unknown, or the product might not be produced. In these cases, other methods such as Southern blotting[26] are used.

How Human Genes Are Cloned

Much is appearing in the current medical and scientific literature about human gene libraries. These are generally prepared by taking all of the DNA from a human cell source, cutting it into segments with a chosen restriction enzyme, and then putting the cut DNA segments into suitable phage or other vectors and thence into host bacteria.[27] Theoretically, after such a library has been prepared, all of the DNA sequences from the human cell will be incorporated into the vector. In general, this strategy can be applied to most problems of gene cloning, although in some instances incised fragments intended to be ligated within receptor vectors such as phage cannot readily be re-attached. Gene segments of relatively large size may be difficult to insert within the restricted cut ends of smaller recipient vectors. In general, however, libraries of component pieces of restriction enzyme–cut original human genomic DNA now represent one of the standard methodological approaches in molecular biology.

After a library containing various pieces of DNA from the human cell of interest is derived, the next problem is how to identify the particular segments of DNA that carry the message for the protein or polypeptide of interest. The usual approach is to use complementary bases and hybridization as a selection method. Generally, the probes that are most frequently used are radiolabeled mRNAs that are exactly complementary to the genes to be isolated and studied. Thus, if specific mRNA is available, it can be used to pick out the corresponding DNA nucleotide sequences that code for the polypeptide chain of interest. From a technical standpoint, it is usually extremely difficult to isolate single-protein mRNA that is completely uniform and uncontaminated by traces of irrelevant mRNAs. If mRNA that is not completely uniform is employed as a hybridizing probe, a number of false-positive signals will be obtained, particularly with DNA libraries representing entire fractionated segments of complete genomic DNA. When mRNAs are used for screening or sorting through libraries, additional purification steps are often required to ensure complete uniformity and homogeneity of the original mRNA probe. In some instances terminal sequences of mRNA can be cloned and used in further preparative steps.[28]

Use of cDNAs

Cloning and screening techniques involving a variety of human, animal, or even plant genes have repeatedly utilized what are called cDNA clones. It is well established that mRNA molecules derived from eukaryotic cells contain repetitive sequences of adenine nucleotides attached to the 3′ end, which are known as poly-A tails. The poly-A tails can be used to initiate production of DNA that will be exactly complementary to the mRNA. In the presence of the key enzyme reverse transcriptase, short sequences of linked thymidines (oligo-dT) are added to mRNA, inducing AT hybridizations with the mRNA poly-A tail, and act to prime the reverse transcriptase to make a DNA strand exactly complementary to the original mRNA (Figure 2.10). This now provides a cDNA copy of the RNA with the exact message of interest.

The next step is to insert the cDNA segment into a convenient vector like the plasmid pBR322, using coupling of prefabricated restriction enzyme sites to either end of the cDNA. In this regard, restriction enzyme attachment sites have been chemically synthesized in vitro.[29] These manufactured restriction enzyme sites are often referred to as linkers; they can be made in vitro and then stuck to the ends of cDNA strands using DNA ligases. An example of this strategy is shown in Figure 2.11.

Sorting Out cDNA Clones

In practice, the steps described above must be arranged so that the conditions allow only one cDNA-containing plasmid per bacterial cell. Uniform cDNAs of specified types are directly identified using the bacterial vectors as an amplification system. If single bacterial cells containing only one cDNA molecule within a plasmid are picked and this single isolate is expanded by culturing, a clonal expansion of this particular cDNA is available. DNA from this clonal isolate is extracted, denatured to separate the double strands, and linked to nitrocellulose filters. Mixtures of mRNAs can then be exposed to these filters to isolate only mRNAs that hybridize to the uniform population of cDNAs on the filters. In another series of steps, it is possible to detach adherent mRNA molecules from such complementary hybridizations and introduce them into cell-free systems, where they can be translated into the specific polypeptide or protein being studied. This sequence of steps has allowed isolation of a uniform cDNA probe representing the genetic program for a single mRNA coding for a single pure protein. In practice, similar specific hybridization techniques can be used to identify cloned DNA fragments by applying colony blots from agar plates to nitrocellulose membranes, followed by screening with specific RNA probes.[30] This method makes it possible to isolate any gene or DNA sequence for which the corresponding RNA is available.

In practice, ease of application of these techniques is somewhat dependent on the relative amounts of respective mRNAs contained within cells or tissues to be examined. Proteins that are major products or constituents of various cells will be represented by larger amounts of mRNAs than proteins that represent only small fractions of the cell repertoire. When a cDNA probe for a specific protein or polypeptide structure is available, it can be used to screen large numbers of bacteria grown as single colonies on plates containing growth media. This technique frequently employs a number of agar plates covered with bacteria into which a plasmid or bacteriophage containing a cDNA insert has been introduced. After overnight incubations, the plates are dotted with individual bacterial colonies, some containing the cDNA insert and some not. Nitrocellulose filters are then overlaid on the plates and samples of the bacterial colonies are transferred to the filters. Treatment of the filters with NaOH both lyses the bacteria and denatures the DNA, making it stick firmly to the nitrocellulose. These filters can then be treated with specific radiolabeled mRNA and hybridized to the specific complementary cDNA[30] sequences from the filters. Using this technique, individual bacterial colonies con-

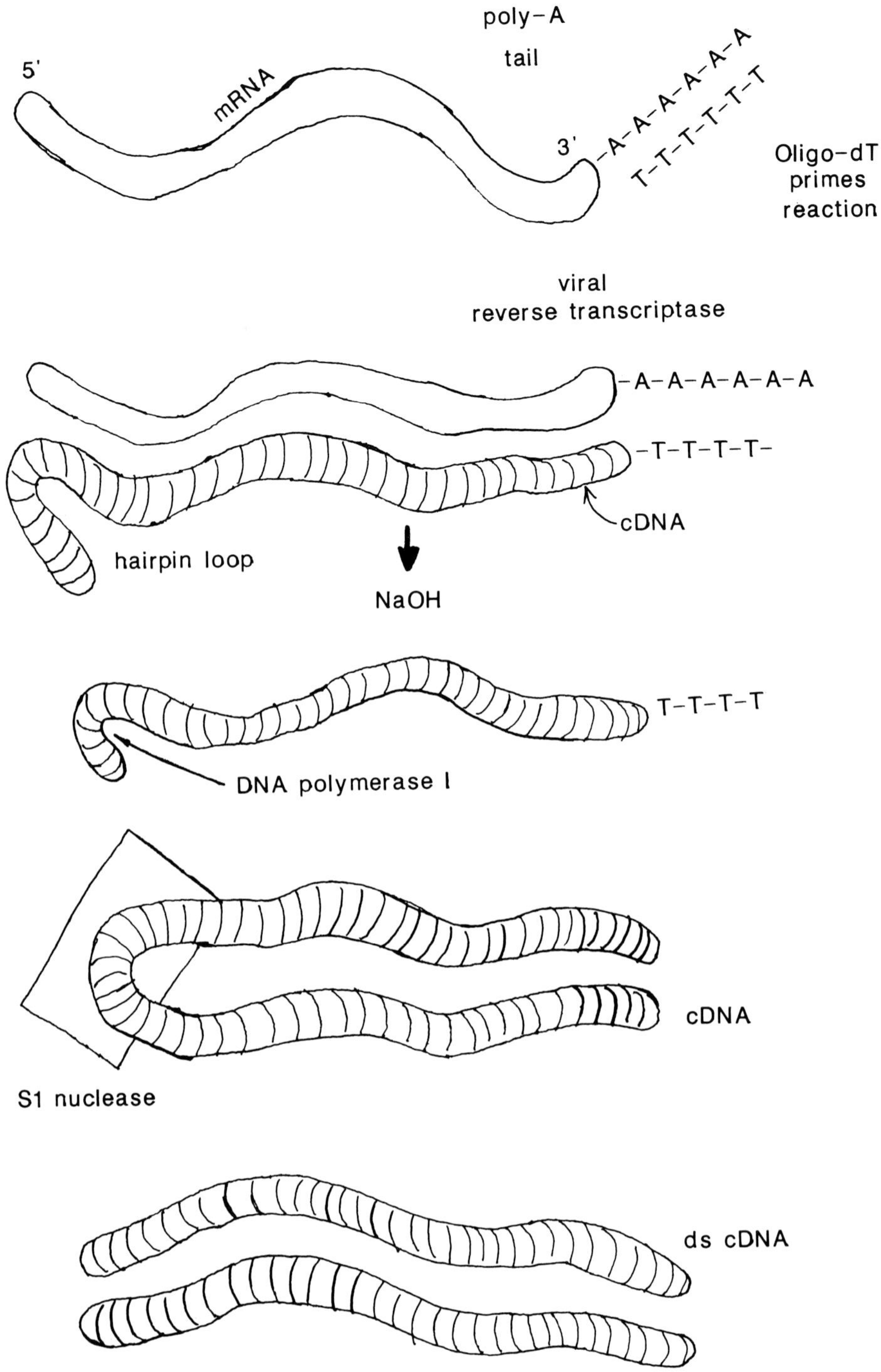

Figure 2.10 Producing cDNA. mRNA contains poly-A tails at the 3′ end. A string of linked thymidines (T-T-T-T oligo-dT) is added in the presence of reverse transcriptase, which then reads backward from the 3′ to the 5′ end and beyond to make what is termed a hairpin loop at the 5′ end, inducing the DNA to copy itself in complementary fashion for a short stretch. mRNA is inactivated by NaOH, and DNA polymerase I uses the initial hairpin loop as a start for completing the second DNA strand. The enzyme S_1 nuclease then cleaves the loop area to provide a double-stranded cDNA molecule.

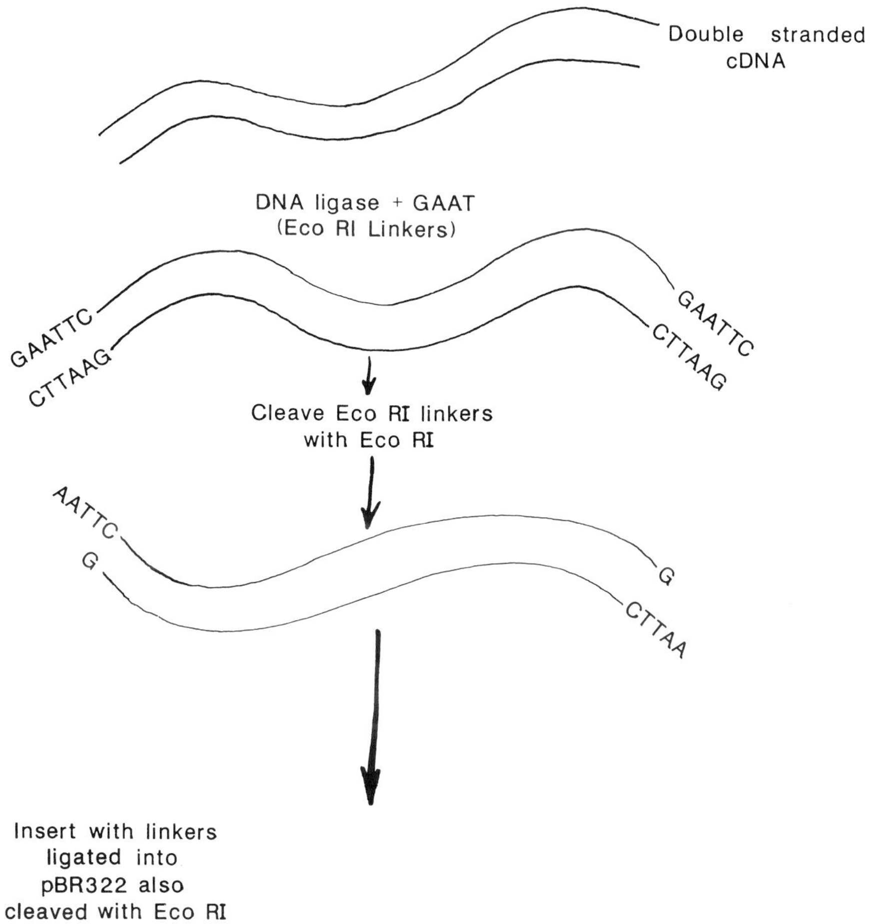

Figure 2.11 cDNA is linked to *Eco*RI linkers which constitute the restriction enzyme site for *Eco*RI. The linkers at each end of the cDNA are then cleaved with *Eco*RI to produce sticky ends and the plasmid vector *Eco*RI site is cleaved as well with the same enzyme. This allows the cDNA to be inserted conveniently into the pBR322 plasmid, which can then be grown in large amounts using propagation in its normal *E. coli* host.

taining plasmids or bacteriophage that have cDNA inserts specific for the hybridizing mRNAs can be identified and amplified further to produce more recombinant cDNA material. This sequence of events is shown diagrammatically in Figure 2.12. By physically examining the complexes formed between the cDNA and the ^{32}P-labeled mRNAs, using size measurements from gradient centrifugation or agarose electrophoresis, it is sometimes possible to ascertain whether the cDNA from the plasmid represents the gene for the entire mRNA message encoding the complete protein sequence or only part of it. It has been found that many mRNA molecules have extra nucleotide sequences at their 5′ ends; these extra sequences code for so-called leader sequences that are used in secretion or further processing of the protein or polypeptide chain product before it assumes its specific function in the organism.

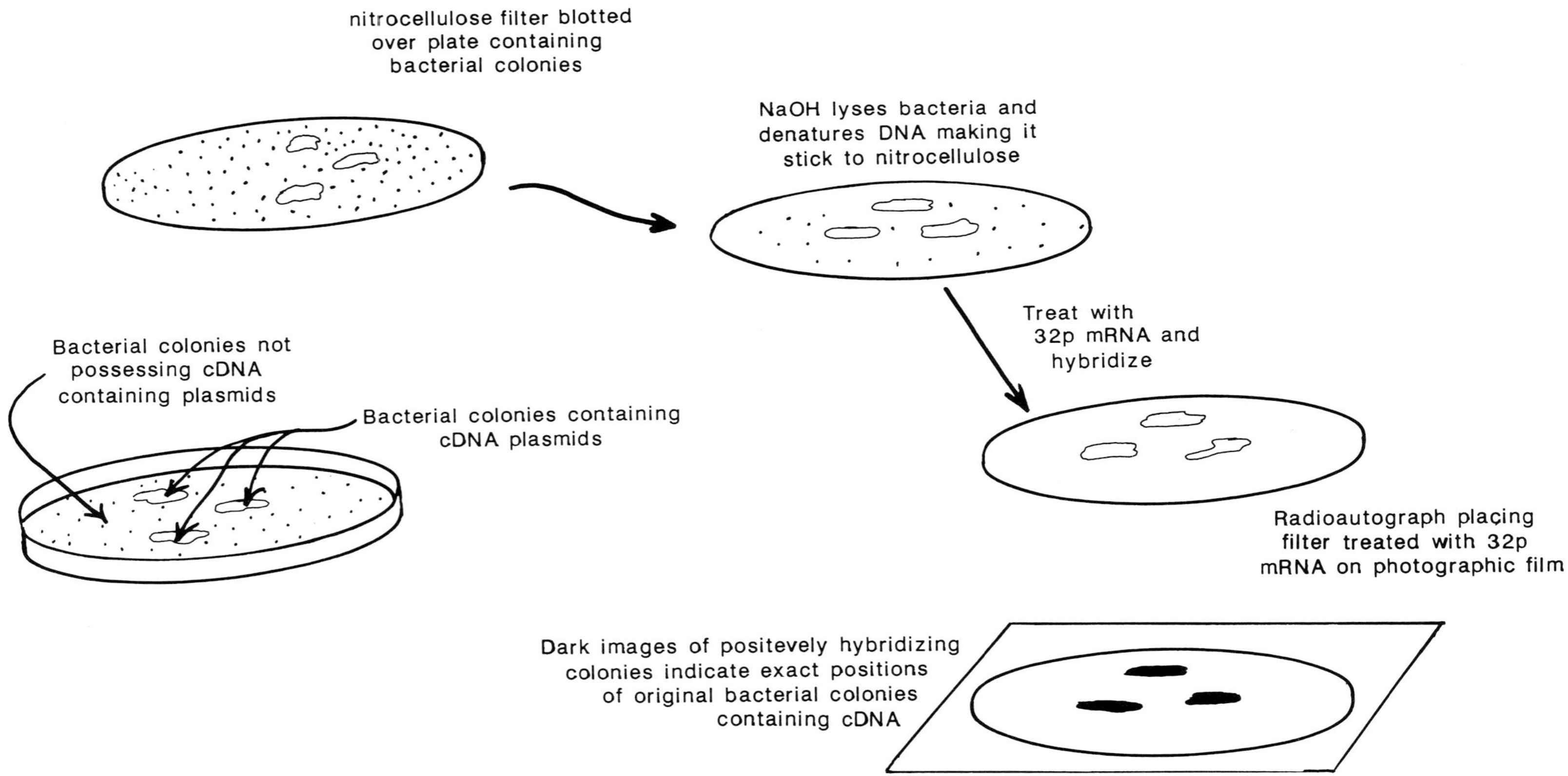

Figure 2.12 Plasmid vector containing cDNA is introduced into *E. coli* and a number of bacterial colony plates is grown. Each colony is then transferred by blotting onto nitrocellulose filters and bacterial DNA denatured by using NaOH. This filter is then hybridized with ^{32}P-labeled specific mRNA, and the exact location of the original bacterial colonies containing cDNA is identified by radioautography.

Different Vectors Used to Clone Other Genetic Material

With the methodology for preparing and cloning DNA sequences of interest into bacterial plasmids established, it seemed possible to study not only the precise nucleotide sequences of specific genes themselves but also adjacent nucleotide sequences at the 3′ or 5′ end that might somehow regulate or control adjacent gene expression. As the methodology was improved and expanded, it became obvious that the size of DNA inserts into various vectors might be an important practical limiting factor, because very large DNA inserts (over 10 to 15 kb) were shown to be unstable. As plasmids or other vectors containing large inserts reproduced, in some instances a progressive deletion of the foreign DNA inserts occurred. This can happen because the foreign DNA inserts are not essential for plasmid reproduction, and it occurs most frequently when cDNA probes are more than 3 to 4 kb in length.

On the other hand, certain vectors such as λ phage are more receptive and stable to large DNA inserts.[31,32] For this reason, phage have often been used to clone human DNA segments of interest. A 15,000-bp DNA fragment can be inserted into λ phage and retain its stability through a number of subsequent passages. After a library or representative collection of λ phage containing specific human or animal genes has been prepared, it can be examined using specific cDNA probes as described above.[33–36]

Analysis of Long DNA Segments

Often the DNA nucleotide messages related to a specific problem or human disease may involve stretches of continuous DNA that are several hundred thousand bases long. With currently available technology, it is not feasible to restrict and store this much DNA in a single phage or cosmid vector. Nevertheless, within the size limits mentioned above, a recombinant λ phage or cosmid can be used to pick out another recombinant sequence that overlaps at the same region of the original genomic DNA. This use of overlapping genomic DNA stretches and complementary hybridizations is referred to as chromosomal walking.[37–41] It resembles a stepwise leapfrogging process whereby a fragment of DNA from one end of a recombinant phage or probe is used to rescreen a library of phage or cosmids to locate a second recombinant complementary to the first DNA, along with the next adjoining portion of the native genomic DNA being analyzed. The second recombinant probe is then used to make a third and again obtain the additional dividend of an adjacent nucleotide sequence intrinsic or obligatory in the complementary hybridization event. This chromosomal walking strategy is illustrated in Figure 2.13. The technique depends on the nucleotide sequences within the gene being unique and not repetitious. If certain gene sequences are repeated periodically, the leapfrogging method cannot be applied. In fact, repeating nucleotide sequences are quite common in the DNA of many animals, and it can be difficult to find unique stretches of sequence that can be used as probes and that do not constitute repeats.

Another method that has been used to study eukaryotic DNA and specific linear arrangements of genes in chromosomes is related to the fact that many mammalian proteins are designated by a group of genes that are physically close to one another on the same chromosome. If one repeatedly screens a λ phage library with a single cDNA probe, most of these physically linked genes will be isolated. If in the original genomic DNA the genes are within a region of 8 to 10 kb, the recombinant DNA

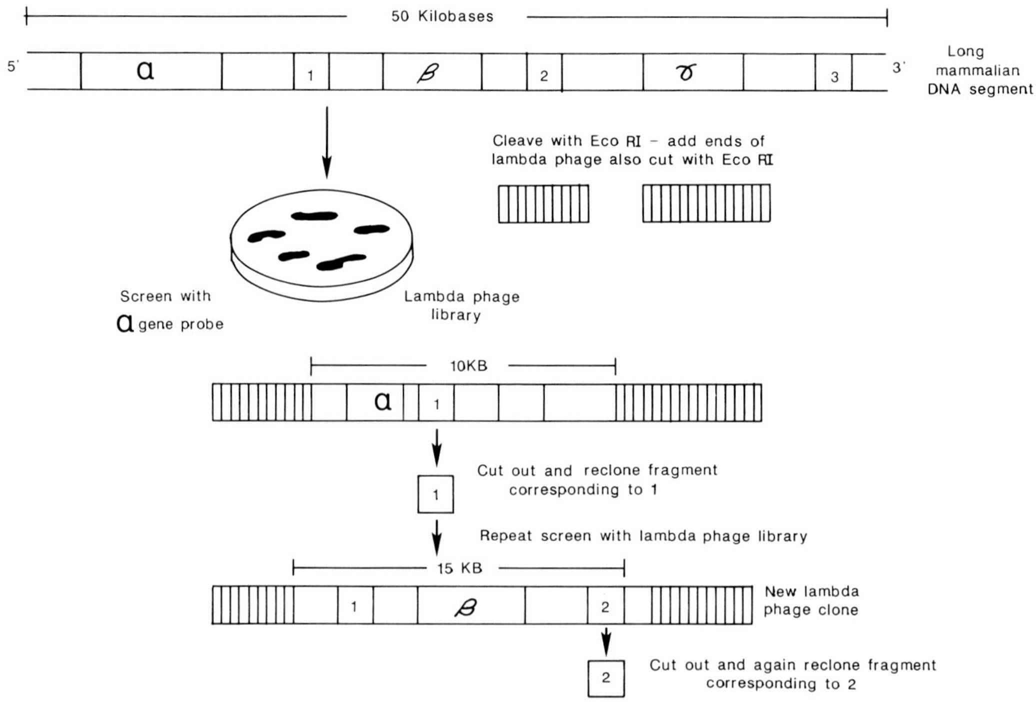

Figure 2.13 Technique of chromosomal walking using ends of lambda phage cut with *Eco*RI containing insert from large 50-kb mammalian DNA segment to be sequenced. Method involved stepwise overlapping of genomic DNA segments using probes chosen at progressively greater distance 3′ to the original probe. At each step a new continuous adjacent DNA sequence is uncovered due to the specific complementary hybridization of each new probe utilized.

obtained in the screen may also contain adjacent overlapping DNA coding for another of the linked family of genes. In a number of instances, particularly involving the immunoglobin or histocompatibility genes, gene sequences or spatial maps have been deduced by analysis of such linked overlaps.[41–44]

M13 Phage and Rapid DNA Analysis

The introduction of phage M13 has provided a unique additional advantage in both cloning and sequencing some genes.[45] M13 phage contains single-stranded DNA. Upon insertion into *E. coli*, the infecting phage DNA makes a double-stranded intermediate form. This is used as a template for the synthesis of new virion strands which are packaged back into progeny phage virus particles. The double-stranded intermediate replicative phase can be utilized to clone genes. This M13 form contains a very useful *Bsu*I restriction enzyme cleavage site where new or foreign DNA segments can be inserted using blunt end ligation. The DNA replicative sequence in M13 is unique in that, when foreign DNA is inserted, only one of the strands is processed into the packaged phage. Thus, when the foreign DNA insert is ligated into M13 in each of the two possible strands, large amounts of one strand or the other can be harvested. The precise sequence at which new DNA is inserted into the M13 vector has been clearly defined. An 8- to 10-base sequence of nucleotides that is complementary to DNA immediately next to the cloning sequence can now be synthesized and used as a primer in vitro. Thus, any DNA insert that has been cloned into the M13 phage can be sequenced using the Sanger method.[46,47] Individual primers for each separate piece of contained foreign DNA are not necessary. Because of this, inserted DNA can be sequenced very rapidly, and the M13 phage has proved extremely useful in this regard.

Southern Blotting Techniques

Another method for analysis of gene or DNA segment structure also makes use of the complementary hybridization behavior of specific cDNA or mRNA. Southern blotting was developed by E. M. Southern in Scotland in 1975.[26] The method is simple and straightforward and employs cutting of genomic or high molecular weight DNA using one or several successive restriction enzymes, followed by electrophoretic separation of DNA fragments on agarose gels. The separated DNA fragments are then blotted onto membranes of nitrocellulose. When the blotted, separated DNA fragments have been fixed to the nitrocellulose membrane, they are probed with radiolabeled ^{32}P DNA or RNA complementary probes hybridizing with the genes being examined, and after specific complementary hybridization the relative positions and amounts of the hybridizing gene fragments are determined by radioautography. The radiolabeled probes used in Southern blotting analyses can be purified RNA, cDNA translated from purified mRNA, or a specific cDNA previously isolated and cloned in appropriate vectors. An example of the application of Southern blotting techniques is illustrated in Figure 2.14. Southern blotting is one of the most powerful and frequently used methods in molecular biology. It is often used to study the number and relative location or positioning of restriction enzyme fragments in isolated whole DNA or fragments of DNA. Moreover, it is the fundamental technique now used to study the allelic and genomic patterns of DNA fragments that may be associated with a wide variety of disease states.[48,49] Other major applications have been in studying the number of copies of certain genes in individual human DNA samples and in the determination of variations within certain genes thought to be related to certain human diseases or metabolic defects.

Northern blotting is a similar analytic technique in which various molecular sizes of RNA are separated by electrophoresis on the basis of size in agarose gels.[50] Again, well-defined labeled probes are reacted with the various sizes of RNA that have been blotted onto nitrocellulose and relative amounts and distribution are subsequently determined by radioautography.

Methods of Selection of Specific Genes

Mixtures of mRNA obtained from living cells can be subjected to physical separation techniques that serve to enrich the concentration of particular mRNAs to be studied. One of the most commonly used techniques employs differential centrifugation through sucrose gradients. This procedure produces banding of different populations of mRNAs on the basis of their size and actual molecular weights. Specific molecular species of RNA initially separated by size on such gradients can then be electrophoretically separated using agarose gels under reducing conditions, which allow the most effective separations of polymeric compounds.

Other approaches to isolating relatively rare mRNAs use some known natural

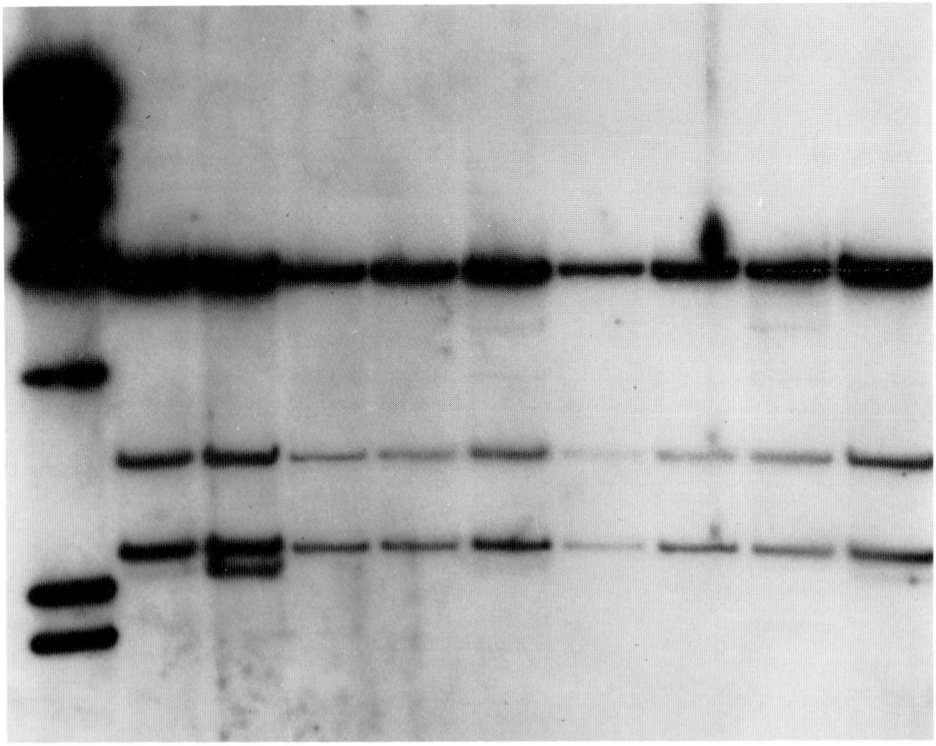

Figure 2.14 PvuII restriction enzyme digestion of several samples of genomic individual human DNA separated by electrophoresis on agarose gel. The separated DNA restriction fragments were then transferred to nitrocellulose membranes overlaid on the agarose gel and the DNA adherent to the membrane was hybridized with ^{32}P-labeled cDNA probe reacting with a known sequence for protein C. The areas of cut DNA reacting with the probe for protein C were determined by radioautography and development of the photographic film. Markers of *Hind*III digested λ phage DNA are shown to the left.

biological stimulus, such as a hormone or polypeptide messenger molecule, that actively raises the levels of the mRNA to be studied before cells are lysed and the specific procedures for preparing mRNAs are initiated.[51–53]

One of the most effective methods for producing specific cDNA probes relies directly on the known amino acid sequence of a protein. When the primary amino acid sequence is known, possible nucleotide sequences of both its RNA and cDNA can be predicted. Such predictions, of course, are dependent on some element of degeneracy in the nucleic acid code. When these nucleotide sequences are established, it is possible to deduce precisely where various restriction enzymes will cut the cDNA. Based on this sort of information, it becomes feasible to produce various fragments of cDNA using the appropriate restriction enzymes and, after insertion and cloning in the appropriate vectors, to develop cDNA probes for well-defined genes. This strategy has been employed for cDNA probes that have been useful in the study of immunoglobulin λ light chain genes,[54,55] human chorionic gonadotropin,[56] and other pituitary tropic releasing factors.[57]

In addition, if the amino acid sequence of a protein or polypeptide is known, synthetic oligonucleotides can be used to probe for specific mRNA or for DNA coding for the particular protein.[58] The strategy of using a sequence of nucleotides artificially synthesized in the laboratory is simple. One first identifies a short sequence of five or six amino acids known to occur in the protein. This sequence is then utilized to predict possible mRNA nucleotide sequences that would translate into the selected amino acid sequence. Using the predicted mRNA sequences, a group of short complementary oligonucleotide sequences are put together in vitro and are used to probe cDNA or genomic libraries for fragments of DNA directly involved in synthesis of the entire protein. This approach is shown in Figure 2.15.

Organization of Sequence Information

With the rapid progress that has been achieved using protein sequence information coupled with parallel approaches in molecular biology, practical applications of artificial intelligence and computerized banks of sequence data have become available. Computer programs and software are now available for a vast number of established protein sequences. As noted earlier, if the protein sequence is known the mRNA sequence can be deduced, and from such mRNA sequence data specific families or types of cloned cDNAs can be identified. When a genomic library has been constructed from cDNAs derived from total mRNA from a tissue of interest, such as liver or human synovium or muscle, cDNA inserts taken out of such vector inserts using restriction enzymes can be sequenced using the Sanger technique[46] and the amino acid sequence coded for by the cDNA insert can be deduced. These sequences can then be directly compared with the known sequences for the liver-specific, synovium-specific, or muscle-specific proteins already characterized and the types of cDNA probes contained within a spectrum of plasmids or a given library can be specifically identified. In such instances, computer analyses of comparative nucleotide or amino acid sequences prove to be a rapid way to analyze and classify data that may already be available in a cDNA plasmid library.

Methods for Amplifying Specific Mammalian cDNAs

Techniques have now been developed for amplification of certain cDNAs initially prepared from highly purified mRNA. In this way, pure proteins of biological interest or even therapeutic usefulness have been synthesized.[59–61] In such instances the mammalian cDNA is often intentionally inserted in close physical proximity downstream from a potent bacterial promoter gene sequence.[61,62] An example of such a strategy is the insertion of cDNA for the insulin polypeptide within the ampicillin resistance gene of the plasmid pBR322. This results in translation of a polypeptide chain containing the ampicillin resistance protein linked to the sequence of insulin. When the ampicillin resistance chain is cleaved off enzymatically, active insulin alone remains. In some instances extremely active promoting factors, such as those that function to maintain the ampicillin resistance gene, may interfere with the process by inhibiting the normal cycle of the bacterial host or even the replication of the containing plasmid itself. The ideal promoters for genes in which high expression but reproducibility and reliability are important are those that can be easily modified and regulated.

An example of the type of exquisite control over promoters can be found in the λ phage pL promoter. When the λ repressor is present, the promoter cannot function. When the λ repressor is not present, the pL promoter is extremely active. The DNA sequence denoting the λ repressor can be placed in the plasmid vector

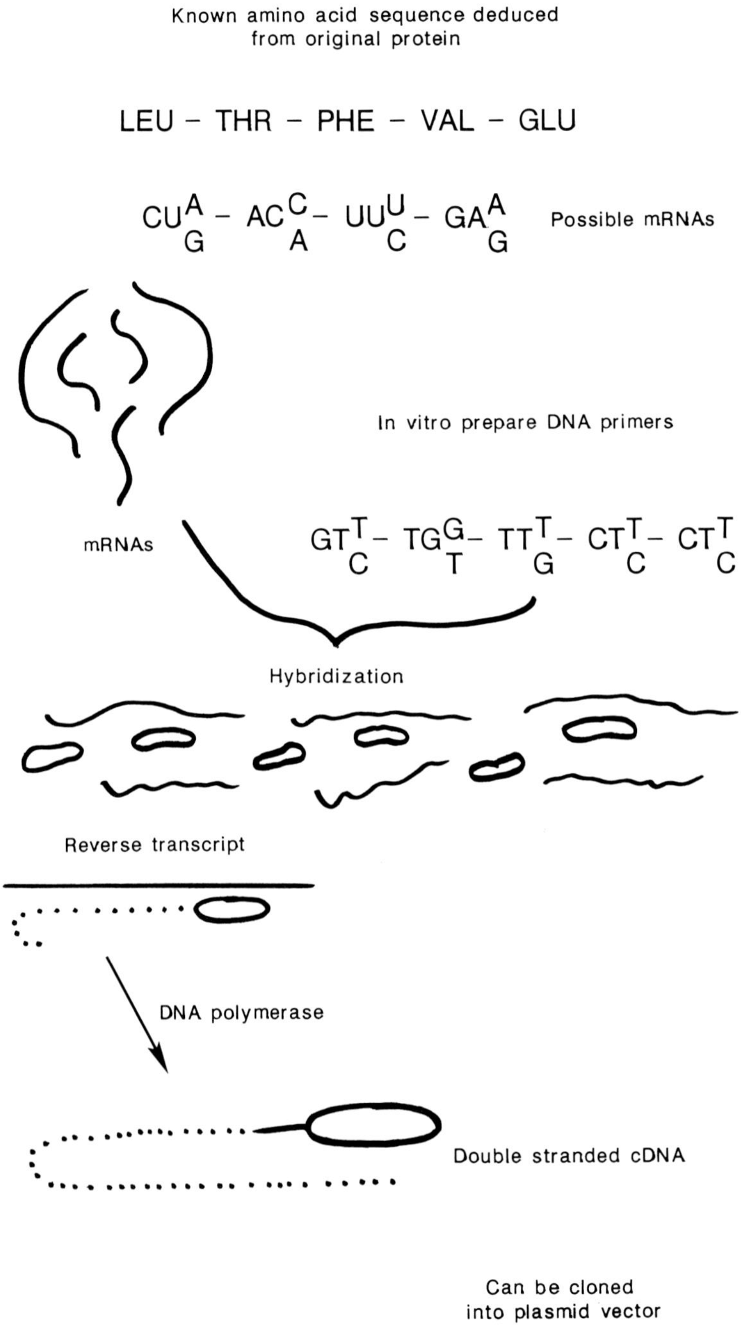

Figure 2.15 When the amino acid sequence of the original protein is known, a short stretch of five or more amino acids in the protein sequence is chosen. This can be used to deduce possible mRNA sequences, and oligonucleotides are synthesized which are complementary to all the possible mRNA sequences. These artificially produced oligonucleotides are annealed with mRNA populations containing mRNA for the original protein. Reverse transcriptase can then be used with its usual oligo dT primer to reverse transcribe the sequences of interest, with the resulting DNA polymerase making a full double-stranded cDNA that can be stored in a plasmid vector.

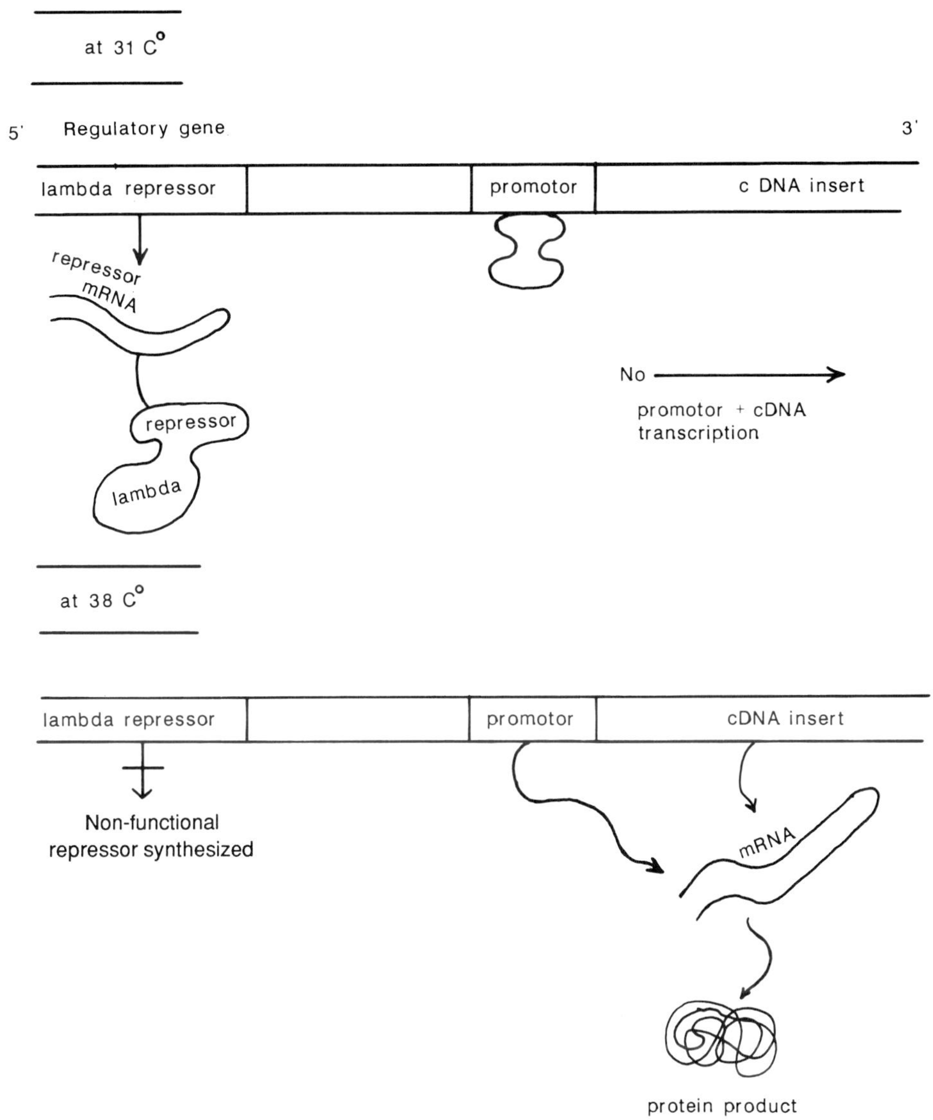

Figure 2.16 When bacteria containing plasmid and cDNA insert with both promoter and repressor are grown at 31°C, the gene transcribing the message for the repressor is active; however, if the organisms are grown at a higher temperature (38°C), the repressor is not activated and the promoter downstream allows a marked increase in cDNA message translation.

or in the bacterial host DNA. In the case of the λ repressor (which is temperature sensitive), no repressor activity is present at a high temperature such as 38°C but repression is complete at 31°C. Strong levels of expression can therefore be conveniently activated by growing the bacteria containing the plasmid at the higher temperature (38°C) and then just as conveniently reversed by bringing the organisms to the λ repressor–activating lower temperature (31°C) (Figure 2.16).

Several specific features influence how effectively a given mRNA insert can interact appropriately with the innate plasmid machinery to achieve reproducible levels of translation. Effective binding between mRNA and ribosomes is related to essential ribosomal binding to the well-defined Shine-Dalgarno sequence. A set distance between this sequence and the AUG initiation codon is also necessary. A number of promoters, including the *lac* promoter or a hybrid tryptophan promoter, can be utilized to ensure proper spacing between the ribosomal binding site and the initiation codon.

Use of Antibody Specificity in Detecting Products of Cloned Genes

The protein or polypeptide products of mRNA coded for by cDNA inserts within various expression vectors can be detected using immunologic techniques that depend on the specificity of the antibodies employed. Thus, if one has a panel of antibodies of known specificity for certain determinants on a protein for which a cDNA library has been constructed, bacterial colonies containing plasmid inserts cultured on plates can be lysed and those making the protein product identified using labeled antibody.[63–65] The general format for such a screening process is shown in Figure 2.17. Variations on these methods have been employed in attempts to characterize some of the nuclear autoantigens such as Sm or La reacting with antibodies in serum from patients with systemic lupus.[66,67] In general, polyclonal antibodies detect a broader range of antigenic epitopes than monoclonal antibodies, and the screening process will depend on the type of specificity required to identify certain products. This general approach has found wide application in studies of proteins or polypeptides that may be useful for protective vaccines or various features of the immune response—particularly with respect to parasitic and bacterial diseases such as malaria, trypanosomiasis, and gonorrhea.

Isolation of the Interferon Gene

Perhaps the best way to integrate all the above, sometimes surely confusing, information would be to describe the identification and cloning of the human interferon gene.[68]

The first step in cloning a gene is to retrieve it from the chromosome of the cell in the form of DNA, or from the cytoplasm in the form of mRNA. There can, however, be a difference between these two forms of nucleic acid sequences. Genes in nucleated eukaryotic cells are not in a continuous sequence but are divided into pieces by noncoding sequences of DNA. The sequences that are used to code for proteins are called exons (they are expressed), and they must be spliced together to form a single continuous sequence, which will be translated. The intervening sequences, or introns, which interrupt the coding sequences, must be removed or spliced out. This gene splicing occurs not at the DNA level, but at the RNA level after it has been transcribed. This difference between the sequences of the initial DNA gene and the final mRNA is very important. Most eukaryotic genes contain introns that must be spliced out, but most gene cloning is done in bacteria, whose genes are continuous and intronless. Bacteria cannot, therefore, process eukaryotic mRNA because they have no splicing machinery or enzymes that can do the job. Because of this, for most genes it is desirable to get the gene sequence from mRNA. With the isolated RNA and a special DNA polymerase—reverse transcriptase—the single-stranded mRNA can be converted into the equivalent double-stranded DNA form that will be needed for cloning.

But first, how do you get the mRNA? Luckily, different cells in multicellular

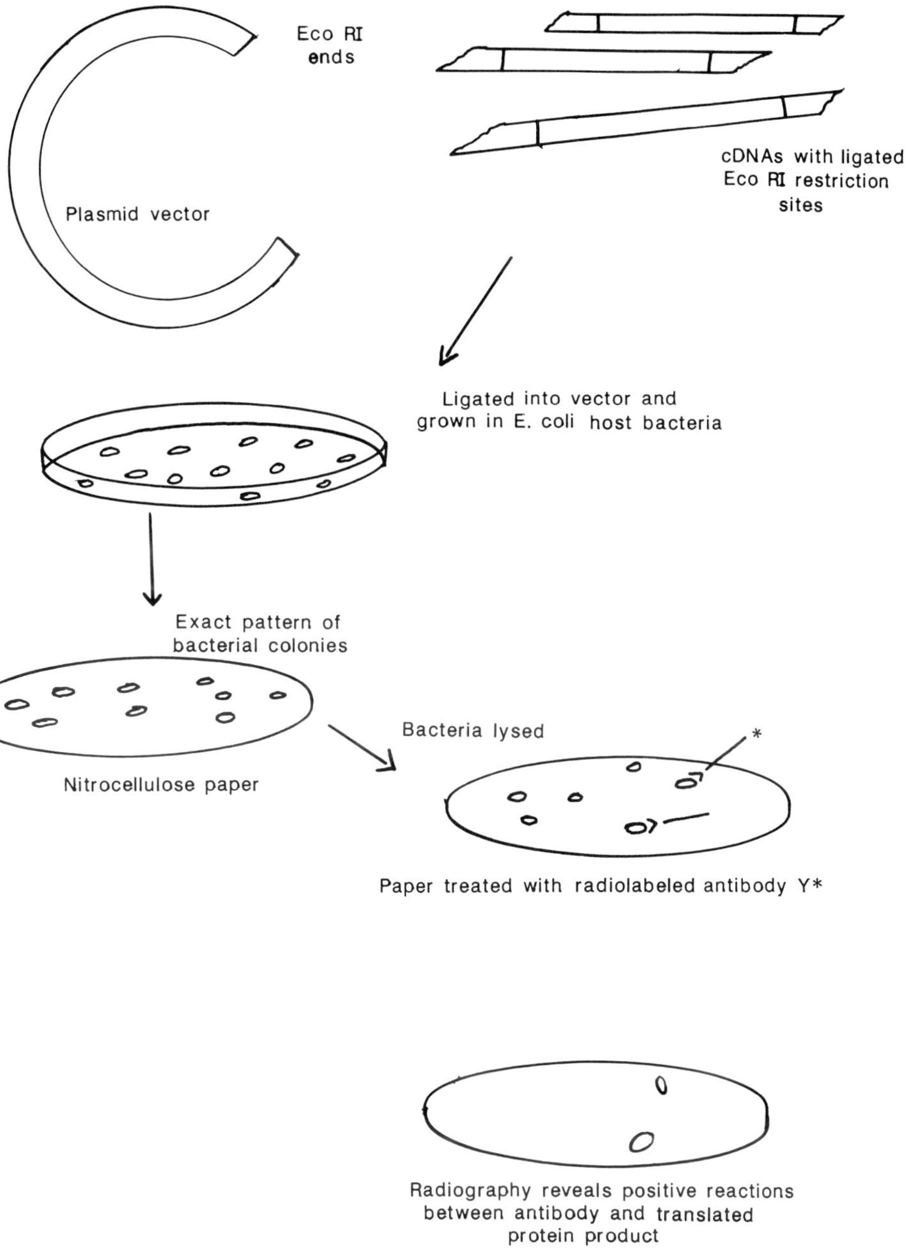

Figure 2.17 From a library of cDNAs inserted into an appropriate plasmid vector, bacterial host colonies are spread on a growth plate. The bacteria are transferred or blotted onto nitrocellulose paper and lysed and the eukaryotic proteins coded for by specific RNA fragments are detected by using labeled antibody and radioautography.

organisms specialize in the synthesis of different proteins. Cells from appropriate tissues will have mRNA populations enriched in the desired mRNA. Of course, messengers of common proteins will also be present. An example of such an isolating technique is the cloning of the insulin gene. This was done by isolating mRNA from beta cells of the pancreas, the only cells in the body that produce insulin.

Interferon is an antiviral protein produced by most vertebrate cells when they are exposed to certain viruses or specific inducers. It is released by virally infected cells and causes an antiviral state in the surrounding target cells. Almost all cells are capable of producing interferon, but there are no specialized cell types that produce interferon continually under normal conditions (as opposed to the beta cells that produce insulin). The strategy used to get interferon mRNA was to induce interferon production in leukocytes by infecting them with a virus. The RNA prepared from these cells would then contain the message for interferon.

Most eukaryotic mRNAs have a long string of adenosines (poly A) on their 3′ ends to help protect them from degradation by exonucleases. This can be exploited to isolate mRNAs from ribosomal RNA and transfer RNA which will also be present in abundance in the crude RNA preparation from the induced leukocytes. A column made of beads—to which polythymidine (poly T) single-stranded DNA is covalently bound—will hold the mRNA by hydrogen bonding between A and T bases, while allowing all other RNAs to pass through. The mRNA can then be eluted off the column separately with a low-salt solution. This results in a large collection of mRNAs, only a small portion of which codes for interferon. How can this portion be isolated from the rest?

The method used was to centrifuge the mRNAs in sucrose gradients, which separated them by size. Fractions of various sizes were then analyzed for interferon-specific mRNA. This was done by injecting the mRNAs into frog oocytes, which translated them, producing interferon whose antiviral activity could be measured.

It was found that mRNA of a certain size, called 12S mRNA (the S stands for Svedberg units), contained the interferon message. But at this point the interferon message was still only a fraction of the 12S mRNA. However, the 12S mRNA was used to make a large number of clones, which could be screened for sequences in the interferon gene. The mRNA was first converted to DNA using the enzyme reverse transcriptase. The poly-A tail of the mRNA was useful here, as a primer (which all DNA polymerases need) of poly T could be used. Because the DNA is made artificially in this way and is an accurate copy of the RNA, it is called copy or *cDNA*. It does not contain the intervening sequences, or introns, that were present in the original DNA from which the 12S RNA was transcribed. The cDNA can be separated from the mRNA template by treating the duplex hybrid molecule with an alkali ($pH > 8.0$), which will degrade the RNA strand. Now the normal replication enzyme, DNA polymerase, can be used to synthesize the complementary strand to give a double-stranded cDNA.

To summarize so far, the 12S RNA (a heterogeneous mixture of a certain size class) was added to reverse transcriptase to make a DNA copy of that RNA. At best, only a few of these DNA fragments code for the interferon gene. To find which cDNA molecule was the one wanted, the DNA was cloned and amplified.

In the case of the interferon gene, the researchers decided to put it into a plasmid, rather than a phage vector, and they chose pBR322. The method used to insert the interferon gene into pBR322 is called *G-C tailing* (or *homopolymer tailing*).[69] It involves use of a commercially available enzyme called *terminal transferase* (TdT for short) that randomly adds bases to the 3′ end, elongating the DNA with any bases available in the reaction mixture. If there are lots of C's in the reaction mixture, the cDNA will have poly-C tails. If G is used, the DNA will have a tail of G's at the 3′ end of each strand. The plasmid was cut with a restriction enzyme to obtain a linear molecule. The enzyme used was *Pst*I, which cuts in the amp^r gene. Then a TdT reaction was performed with the base deoxycytosine (dC). The cDNA was reacted with TdT and deoxyguanosine. Now these two populations of molecules, the poly-G cDNA and the poly-C–tailed plasmid pBR, can be mixed and will anneal (see Figure 2.8). The G's from the cDNA will bind to the C's from the plasmid and form a nicked circular recombinant plasmid. The nicks can be sealed with *ligase*, resulting in a recombinant molecule. This is what was done with the interferon cDNA.

An alternative method takes advantage of restriction enzymes that make staggered cuts. If a small sequence (called a linker) is added to the ends of cDNA molecules (prior to addition to the plasmid) containing the same restriction enzyme used to cut the plasmid, they will automatically base-pair. The end result of either method is a large population of plasmids. Of course, they do not all contain a molecule of cDNA; some are empty (no insert), but these can be screened out because they retain both antibiotic resistances.

The plasmids were transformed into *E. coli* and the bacteria plated on tetracycline (tet) agar. Bacteria without a plasmid are sensitive and die. Only the *E. coli* containing pBR are able to grow. The bacteria that grew on the tet plates were then tested for ampicillin resistance by relica plating, and a set of 5000 tet^r amp^s clones were selected. These were then screened for sequences that might be related to the interferon gene. The research group originally isolated 10,000 clones, but they cut this number down later and decided to screen only 5000 clones. Each colony or clone must be grown on a large scale in order to get enough DNA to examine. This was done by "picking" each colony with a toothpick and inoculating a test tube (a separate one for each colony) with tetracycline-containing medium. In this way each tube represents a pure colony or clone—descendants of *one* bacterial cell per tube.

Which one of the 5000 cDNA clones contains an actual interferon gene? As previously indicated, one could take the amino acid sequence of interferon and make an artificial gene that could be used as a probe. Another method that can be used to find out is called *hybridization selection–oocyte translation*. This method involves first isolating the plasmid cDNA from a single clone and binding it to a nitrocellulose filter. The DNA attaches to the filter via an unknown mechanism, but the attachment is permanent. The cDNA and filter can then be soaked in a solution containing RNA from interferon-induced leukocytes, therefore containing interferon mRNA. Any mRNAs that have sequences homologous with cDNA sequences will hybridize to them and be retained on the filter. Nonhomologous RNA can be washed away, leaving only mRNA that is homologous or at least very similar to the cDNA sequence on the filter. The mRNA can then be eluted off the filter with low-salt solution, resulting in mRNA that presumably was the template from which the cDNA was originally constructed. The mRNA can then be injected into frog oocytes, which will translate the message into proteins. Hopefully, interferon will be the translation product. Of the 5000 interferon cDNA clones, 512 were analyzed in this manner and the one with the best interferon activity was chosen for further study.[68] This clone was analyzed further by restriction enzyme digestion. If it is cut with *Pst*I it should release the cDNA fragment (plus part of the G-C tails), since this is the site of pBR322 into which the cDNA was originally ligated.

The fragment size of the cDNA, or interferon gene, was expected to be about 1000 base pairs. This was determined by the size of the interferon protein molecule and the size of the 12S mRNA from which it was constructed. But when the hybrid molecule (cDNA plus pBR322) was cut and run out on an agarose gel, it was found that the cDNA in this clone was only 320 base pairs long. This was both good and bad news. They had some of the interferon gene but were still missing about two-thirds of it. However, this cDNA fragment could be very useful for analyzing the rest of the 5000 clones for a complete full-length copy.

The analysis was done by the Southern blot technique, using the 320-bp cDNA fragment as a probe. This is a much easier technique than the hybrid selection–oocyte translation method and allowed more rapid analysis of the 5000 clones. When Southern blot analysis was done, it was found that about 4% of the 5000 clones were positive for sequences homologous to the 320-bp interferon cDNA.

The best positives were analyzed by cutting their plasmid DNA with *Pst*I, as was done in the original 320-bp cDNA clone, and running on an agarose gel. The largest cDNA insert was found to be 900 bp long. This was near the size they expected for the complete interferon cDNA, and they hoped that this was the cloned gene they were searching for.

The next step was to express this gene in bacteria. For the cDNA to be expressed, it must be transcribed into mRNA and the mRNA must be translated into protein. This requires that all the proper signals be present for the bacteria to recognize and process the message. Such signals include a promoter, ribosome recognition sequence, translation start codon, translation termination codon, and transcription termination sequence. In addition to having all of these, the cDNA must be in the correct triplicate codon reading frame and correct orientation. This is very important; otherwise the correct amino acid sequence will not be produced, even though the structural gene nucleotide sequence is correct.

This may sound like an almost impossible job to accomplish, but happily it was not! All of the proper signal components were already present in the pBR322 plasmid into which the cDNA was cloned. Luckily, the site in pBR322 into which the cDNA was ligated was inside the β-lactamase gene (which confers ampicillin resistance). And this gene already had all of the correct transcription and translation signals. The bacteria, by expressing the β-lactamase gene, would also express the cDNA interferon gene, producing a hybrid β-lactamase–interferon protein (sandwich-like).

All the proper signals are present, but in order to get the correct reading frame and orientation, the cDNA fragment was cut out and religated (recloned) into three different pBR322 mutants. These mutants had deletions in the β-lactamase (amp^r) gene such that the interferon cDNA ligated back into the *Pst*I site of these three plasmids would give all three possible reading frames by read-through from the β-lactamase gene. By isolating a number of these new clones, it would be possible to get one β-lactamase–interferon cDNA hybrid clone with the correct orientation and the correct reading frame.

The clones were then analyzed for production of interferon activity by growing a culture of bacteria of a particular clone, lysing the cells, isolating the protein, and testing it for antiviral activity. Two of the new clones had very high antiviral activity. The immunological properties of these two were then compared with those of authentic human leukocytic interferon. It was found that sheep anti–human leukocyte interferon antibodies reacted against both the authentic interferon and the two bacterial cloned interferons to a similar degree. The same antibodies reacted only very slightly with human fibroblast interferon (a different type of interferon). Thus, the bacterial cloned interferon was similar in structure to authentic human leukocyte interferon but was distinct from human fibroblast interferon.

The cloned interferon was then compared by molecular weight with authentic human leukocyte interferon by use of a Sephadex G-100 column. This gel column separates proteins based on size. The larger the molecule, the faster it can pass through the column; thus, protein sizes can be mapped. It was found that the cloned interferons were similar in molecular weight to the authentic interferon. This was an unusual, but satisfying, finding because the cloned protein was expected to be much larger. As you recall from earlier in this section, the cloned protein was to have been a hybrid, sandwich-like, protein with β-lactamase polypeptide ends flanking the actual interferon protein. Apparently, though, processing of the hybrid protein was resulting in all or part of the β-lactamase ends being cleaved off. Even if it was not understood why this was happening, it was a highly fortuitous event. Thus the researchers had genetically engineered bacteria (*E. coli*) to produce a close facsimile of human leukocyte interferon which had biological activity.

References

1. Cohen SS: *Virus-Induced Enzymes.* New York, Columbia Univ. Press, 1968.
2. Luria SE: The T2 mystery. *Sci Am* 1955;192:92–98.
3. Bertani G, Weigle JJ: Host controlled variation in bacterial viruses. *J. Bacteriol* 1953; 65:113–121.
4. Dussoix D, Arber W: Host specificity of DNA produced by *Escherichia coli.* II. Control over acceptance of DNA from infecting phage λ. *J Mol Biol* 1962;5:37–49.
5. Lederberg S, Meselson M: Degradation of non-replicating bacteriophage DNA in non-accepting cells. *J Mol Biol* 1964;8:623–625.
6. Old RW, Primrose SB: *Principles of Gene Manipulation.* Blackwell Scientific Publications, Berkeley, CA, 1985.
7. Kelly TJ, Smith HO: A restriction enzyme from *Haemophilus influenzae.* II. Base sequence of the recognition site. *J Mol Biol* 1970;51:393–409.
8. Smith HO, Wilcox KW: A restriction enzyme from *Haemophilus influenzae.* I. Purification and general properties. *J Mol Biol* 1970;51:379–391.
9. Danna K, Nathans D: Specific cleavage of simian virus 40 DNA by restriction endonuclease of *Haemophilus influenzae. Proc Natl Acad Sci USA* 1971;68:2913–2917.
10. Danna K, Nathans D: Bidirectional replication of simian virus 40 DNA. *Proc Natl Acad Sci USA* 1972;69:3097–3100.
11. Chang LMS, Bollum FJ: Enzymatic synthesis of oligodeoxynucleotides. *Biochemistry* 1971;10:536–542.
12. Olivera BM, Hall ZW, Lehman IR: Enzymatic joining of polynucleotides. V. A DNA-adenylate intermediate in the polynucleotide-joining reaction. *Proc Natl Acad Sci USA* 1968;61:237–244.
13. Gumport RI, Lehman IR: Structure of the DNA ligase-adenylate intermediate: Lysine (E-amino)-linked adenosine monophosphoramidate. *Proc Natl Acad Sci USA* 1971; 68:2559–2563.
14. Sanger F, Coulson AR: A rapid method for determining sequences in DNA by primed synthesis with DNA polymerase. *J Mol Biol* 1975;94:444–448.
15. Maxam AM, Gilbert W: A new method of sequencing DNA. *Proc Natl Acad Sci USA* 1977;74:560–564.
16. Sanger F, Nicklen S, Coulson AR: DNA sequencing with chain-terminating inhibitors. *Proc Natl Acad Sci USA* 1977;74:5463–5467.
17. Sanger F, Air GM, Barrel BG, et al: Nucleotide sequence of bacteriophage (X174). *Nature* 1977;265:687–695.
18. Hardy KW: *Bacterial Plasmids.* Washington, D.C., American Society for Microbiology, 1986.
19. Lederberg J, Tatum EL: Novel genotypes in mixed cultures of biochemical mutants of bacteria. *Cold Spring Harbor Symp Quant Biol* 1946;11:113–114.
20. Mitsuhashi S: Epidemiology of bacterial drug resistance. In Mitsuhashi S (ed): *Transferable Drug Resistance Factor R.* Baltimore, University Park Press, 1971, pp. 1–16.
21. Cohen S, Chang A, Boyer H, Helling R: Construction of biologically functional bacterial plasmids in vitro. *Proc Natl Acad Sci USA* 1973;70:3240–3244.

22. Rodrigues RL, Tait RC: *Recombinant DNA Techniques, an Introduction*. London, Addison-Wesley, 1983.
23. Pinkerton TC, Muzyczka N, Walia S, et al: Bacteriophage T5 growth in *Escherichia coli* containing cloned *Pst*I fragments of the colicin Ib plasmid. *Gene*. 1981;16:89–86.
24. Morrow JF, Cohen SM, Chang ACY, et al: Replication and transcription of eucaryotic DNA in *E. coli*. *Proc Natl Acad Sci USA* 1974;71:1743–1747.
25. Itakura K, Hirose T, Crea R, et al: Expression in *E. coli* of a chemically synthesized gene for the hormone somatostatin. *Science* 1977;198:1056–1063.
26. Southern EM: Detection of specific sequences among DNA fragments separated by gel electrophoresis. *J Mol Biol* 1975;98:503–517.
27. Maniatis T, Kee SG, Efstradtiadis A, et al: Amplification and characterization of a β-globin gene synthesized in vitro. *Cell* 1976;8:163–182.
28. Land H, Grez M, Hauser H, et al: 5′-Terminal sequences of eucaryotic mRNA can be cloned with high efficiency. *Nucleic Acids Res* 1981;9:2251–2266.
29. Scheller RH, Dickerson RE, Boyer HW, et al: Chemical synthesis of restriction enzyme recognition sites useful for cloning. *Science* 1977;196:177–180.
30. Grunstein M, Hogness DS: Colony hybridization: A method for the isolation of cloned DNAs that contain a specific gene. *Proc Natl Acad Sci USA* 1975;72:3961–3965.
31. Blattner FR, Williams BG, Blechl AE, et al: Charon phages: Safer derivatives of bacteriophage lambda for DNA cloning. *Science* 1977;196:161–169.
32. Leder P, Tiemeier D, Enquist L: EK2 derivatives of bacteriophage lambda useful in the cloning of DNA from higher organisms: The λgtWES system. *Science* 1977;196:175–177.
33. Benton WD, Davis RW: Screening λgt recombinant clones by hybridization to single plaques in situ. *Science* 1977;196:180–182.
34. Olivera BM, Bonhoeffer F: Replication of *Escherichia coli* requires DNA polymerase I. *Nature* 1974;250:513–514.
35. Maniatis T, Hardison RC, Lacy E, et al: The isolation of structural genes from libraries of eucaryotic DNA. *Cell* 1978;15:687–701.
36. Bahl CP, Marians KJ, Wu R, et al: General method for inserting specific DNA sequences into cloning vehicles. *Gene* 1976;1:81–92.
37. Bender W, Spierer P, Hogness D: Gene isolation by chromosomal walking. *J Supranol Struct*, 1979; Abstract 74, Suppl 3.
38. Steinmetz M, Winoto A, Minard K, Hood L: Clusters of genes encoding mouse transplantation antigens. *Cell* 1982;28:489–498.
39. Shimizu A, Takahashi N, Yaoita Y, Hongo T: Organization of the constant region gene family of the mouse immunoglobulin heavy chain. *Cell* 1982;28:499–506.
40. Hongo T, Kataoka T: Organization of immunoglobulin heavy chain genes and allelic deletion model. *Proc Natl Acad Sci USA* 1978;75:2140–2144.
41. Shimizu A, Takahashi N, Yamawaki-Kataoka Y, et al: Ordering of mouse immunoglobulin heavy chain genes by molecular cloning. *Nature* 1981;289:149–153.
42. Steinmetz M, Hochtl J, Schnell H, et al: Cloning of V region fragments from mouse liver DNA and localization of repetitive DNA sequences in the vicinity of immunoglobulin gene segments. *Nucleic Acids Res* 1980;8:1721–1729.
43. Ploegh HL, Orr HT, Strominger JL: Major histocompatibility antigens: The human (HLA-A, -B, -C) and murine (H-2K, H-2D) class I molecules. *Cell* 1981;24:287–299.
44. Steinmetz M, Frelinger JG, Fisher D, et al: Three cDNA clones endocing mouse transplantation antigens: Homology to imunoglobulin genes. *Cell* 1981;24:125–134.
45. Messing J, Gronenborn B, Muller-Hill B, Hofschneider PH: Filamentous coliphage M13 as a cloning vehicle: Insertion of a *Hind*II fragment of the *lac* regulatory region in M13 replicative form in vitro. *Proc Natl Acad Sci USA* 1977;74:3642–3636.
46. Sanger F, Coulson AR, Barrell BG, et al: Cloning in single-stranded bacteriophage as an aid to rapid DNA sequencing. *J Mol Biol* 1980;143:161–178.
47. Messing J, Crea R, Seeburg PH: A system for shotgun DNA sequencing. *Nucleic Acids Res* 1981;9:309–321.
48. Housman D, Gusella JF: Molecular genetic approaches to neural degenerative disorders. In Schmitt FO, Bird S, Bloom FE (eds): *Molecular Genetic Neuroscience*. New York, Raven, 1982, pp 415–424.
49. Orkin SH, Little PFR, Kazazian HH, Boehm CO: Improved detection of the sickle mutation by DNA analysis. *N Engl J Med* 1982;307:32–36.

50. Alwine JC, Kemp OJ, Stark GR: Method for detection of specific RNAs in agarose gels by transfer to diazobenzyloxymethyl-paper and hybridization with DNA probes. *Proc Natl Acad Sci USA* 1977;74:5350–5354.
51. Kurtz DT, Micodemus CF: Cloning of α_{2u} globulin cDNA using a high efficiency technique for the cloning of trace messenger RNAs. *Gene* 1981;13:145–152.
52. Sippel AE, Jurtz DT, Morris HP, Feigelson P: Comparison of in vivo translation rates and messenger RNA levels of α_{2u} globulin in rat liver and Morris hepatoma 5123 D. *Cancer Res* 1976;36:3588–3593.
53. Kurtz DT, Feigelson P: The multihormonal induction of hepatic α_{2u}-globulin mRNA as measured by hybridization to complementary DNA. *Proc Natl Acad Sci USA* 1977; 74:4791–4795.
54. Hieter PA, Hollis GF, Korsmeyer SJ, et al: Clustered arrangement of immunoglobulin λ constant region genes in man. *Nature* 1981;294:536–540.
55. Taub RA, Hollis GF, Hieter PA, et al: Variable amplification of immunoglobulin λ light-chain genes in human populations. *Nature* 1983;304:172–174.
56. Fiddles JC, Goodman HM: The cDNA for the β-subunit of human chorionic gonadotropin suggests evolution of a gene by readthrough into the 3′-untranslated region. *Nature* 1980;286:684–687.
57. Nakanishi S, Inoue A, Kita T, et al: Nucleotide seqence of cloned cDNA for bovine corticotropin-β-lipotropin precursor. *Nature* 1979;278:423–427.
58. Suggs SV, Wallace RB, Hirose T, et al: Use of synthetic oligonucleotides as hybridization probes: Isolation of cloned cDNA sequenes for human β_2-microglobulin. *Proc Natl Acad Sci USA* 1981;78:6613–6617.
59. Villa-Komaroff L, Efstratiadis A, Broome S, et al: A bacterial clone synthesizing proinsulin. *Proc Natl Acad Sci USA* 1978;75:3727–3731.
60. Gilbert W, Villa-Komaroff L: Useful proteins from recombinant bacteria. *Sci Am* 1980; 242(4):74–94.
61. Guarente L, Lauer G, Roberts TM, Ptashne M: Improved methods for miximizing expression of a cloned gene: A bacterium that synthesizes rabbit β-globin. *Cell* 1980;20:543–553.
62. De Boer HA, Comstock LJ, Vasser M: The *Tac* promoter: A functional hybrid derived from the *Trp* and *Lac* promoters. *Proc Natl Acad Sci USA* 1983;80:21–25.
63. Broome S, Gilbert W: Immunological screening method to detect specific translation products. *Proc Natl Acad Sci USA* 1978;75:2746–2749.
64. Erlich HA, Cohen SN, McDevitt HO: Immunological detection and characterization of products translated from cloned DNA fragments. *Methods Enzymol* 1979;68:443–453.
65. Kemp DJ, Cowman AF: Direct immunoassay for detecting *Escherichia coli* colonies that contain polypeptides encoded by cloned DNA segments. *Proc Natl Acad Sci USA* 1981; 78:4520–4524.
66. Chambers JC, Keene JD: Isolation and analysis of cDNA clones expressing human lupus La antigen. *Proc Natl Acad Sci USA* 1985;82:2115–2119.
67. Wieben ED, Rohleder AM, Nenninger JM, Pederson T: cDNA cloning of a human autoimmune nuclear ribonucleoprotein antigen. *Proc Natl Acad Sci USA* 1985;82:7914–7918.
68. Nagata S, Taira H, Hall A, et al: Synthesis in *E. coli* of a polypeptide with human leukocyte interferon activity. *Nature* 1980;284:316–320.
69. Jackson DA, Symons RH, Berg P: Biochemical method for inserting new genetic information into DNA of SV40. *Proc Natl Acad Sci USA* 1972;69:2904–2909.

Structure, Functional Organization, and Mutation of Mammalian Genes

3

Ralph C. Williams, Jr., MD

About 10 years ago, as restriction fragment analysis and studies of the protein and polypeptide products of both rare and abundant forms of mRNA gained momentum, it was found that identical matchups between actual chromosomal DNA restriction fragments and related cDNA probes were often not present. In view of the strongly established relationship between what mRNA codes for and cDNA, these discrepancies at first were difficult to fit in with the DNA-RNA messages that were supposed to be occurring. At this point, studies of message transcription, principally in the adenoviruses,[1–3] appeared to indicate that a phenomenon of mRNA splicing occurred during viral replication. These studies showed that viral RNA nucleotide sequences within infected cell nuclei had been considerably abbreviated by having one or even multiple sections of prior nucleotide sequence discarded or somehow lost, ultimately resulting in much smaller cytoplasmic active mRNAs that participated directly in linking to ribosomes and prescribing viral protein synthesis. This soon led to the question of whether this viral mRNA splicing mechanism might also be involved in all chromosomal DNAs—viral, bacterial, plant, and eukaryotic.

Careful comparative studies of respective restriction fragment patterns of chromosomal DNA and mRNA nucleotide transcripts in a number of other viral and eukaryotic systems were undertaken to look for stretches of noncoding or nontranslated DNA sequences in a number of genes. This work marshaled a number of lines of evidence showing that gene splicing did indeed occur.[4–6] Direct physical indication that such splicing was going on also came from electron microscopic demonstrations of DNA-RNA complexes in which single-stranded DNA loops without contiguous complementary mRNA could be seen.[7] These and many other similar observations led to the general concept of introns and exons,[8–12] illustrated in Figures 3.1 and 3.2.[13] In virtually all eukaryotic genes that have been sequenced, the nucleotide sequence contained between the specific initiation and terminating codons is composed of coding regions that are often interrupted by noncoding regions. In this arrangement, the actual coding sequences, called exons, are inter-

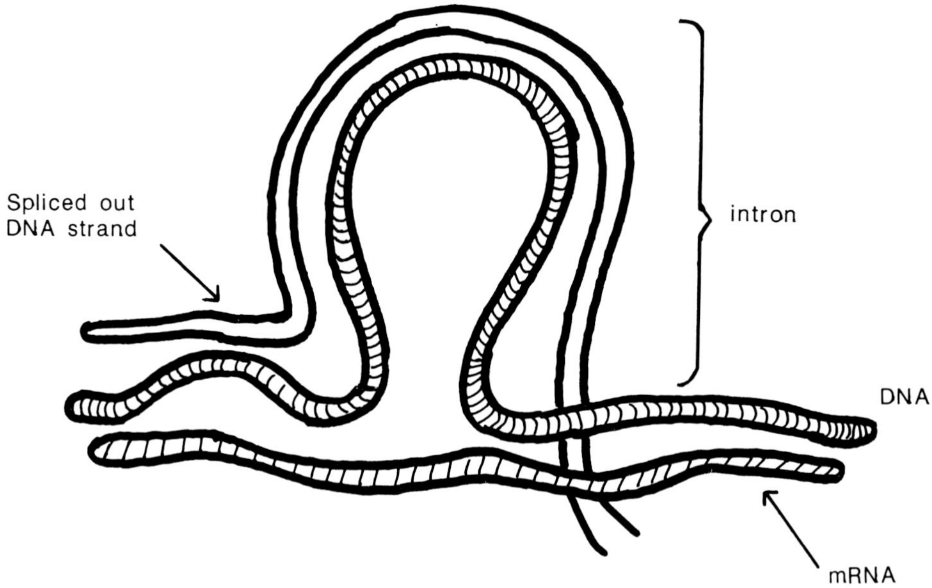

Figure 3.1 DNA-RNA complexes diagramatically showing the noncoding loops or intervening sequences of DNA strands which are spliced out from the final mRNA reading and subsequent translation to protein sequence.

spersed with noncoding intervening sequences, called introns. What then transpires is that mRNA transcribed from the original complete DNA gene nucleotide sequence is processed, presumably within the nucleus before actual translation. Intervening sequences or introns are spliced out and the actual coding exon sequences are spliced together to form the nucleotide sequence that produces the ultimate colinear polypeptide chain or final protein product. In addition, noncoding DNA segments may flank the particular gene at both 5′ and 3′ ends. It is thought that such noncoding flanking sequences may somehow be involved in gene expression.

The number of intervening sequences or introns within the nucleotide sequences of different mammalian genes is quite variable. Thus, there are only two introns between coding exons in the genes for the globin sequences of hemoglobin, but there are more than 50 introns in the gene coding for one of the collagen polypeptide chains. Examples of genes containing multiple intervening sequences

Figure 3.2 Diagrammatic representation of the chicken ovalbumin gene showing a total length of 7464 linked nucleotides of double-stranded DNA. Roman numerals I–VIII indicate positions of coding exons and letters A–G indicate noncoding intervening sequences or introns. *Reproduced from Woo et al.*[13] *with permission.*

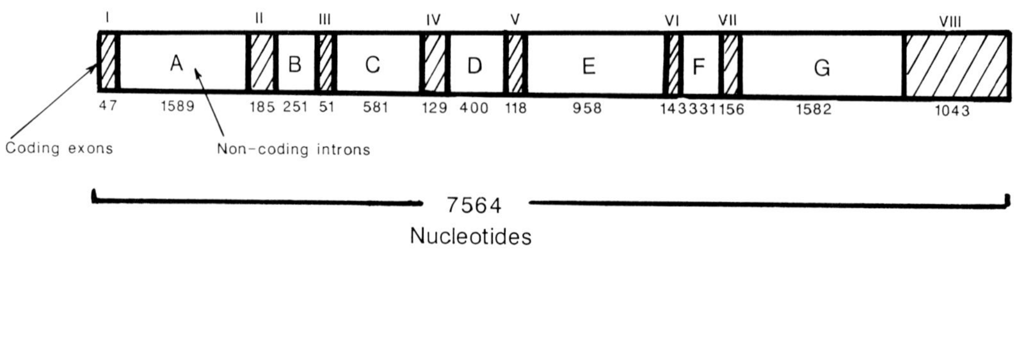

of different lengths are found in the histone proteins that make up an essential element of nuclear chromatin and in the three types of interferon—alpha, beta, and gamma—that are extremely important elements in body defense reactions and messenger activator molecules for many cells of the immune system.

Mammalian genes devoid of introns can be produced using the current in vitro methodology of molecular biology, and the artifically synthesized genes lacking the usual introns can be tested to determine how they function. In certain instances, prior tailoring out of introns appears to have no detectable effect on gene function; in other instances, removal of introns impedes access of active mRNA to cytoplasmic sites and ribosomes. Why this happens is not clear, but it may be related to the physical shape or configuration of the transcribed mRNAs or to production of an unstable form of the mRNA.

In some situations duplicate messages of some genes that are slightly altered in sequence are present within DNA but are never transcribed. These altered or almost duplicate nucleotide sequences are called pseudogenes.[14–21] When active functional coding gene sequences are directly compared with their pseudogenes, it is found that some of the pseudogenes may have lost all of their introns and therefore merely contain continuous runs of coding sequence. Despite this, they are not transcribed. Such observations suggest that the introns as well as pseudogenes may have some importance in transcription of DNA into functional mRNA and hence protein products. The precise role of pseudogenes in the regulation or modulation of gene transcription is still not completely elucidated. The interrupted segments of coding and noncoding sequences described above for eukaryotic genes are also found in viruses that infect mammalian hosts. In many ways, the similarity in organization of genes in eukaryotic hosts and their infecting viruses may indicate why viral infections are successful. Both the mammalian or eukaryotic host and the infecting virus use the same mechanisms to transcribe and translate gene nucleotide sequences. Thus, living cells continuously divide and during this process transcribe and translate their gene exon sequences into mRNA and protein products; the same cells are also capable of transcribing and translating viral genes introduced directly into their cellular machinery as a consequence of viral penetration and infection.

Specific Mechanisms of Gene Splicing

The precise mechanisms involved in splicing out introns and joining active coding sequences of mRNAs have not yet been completely deciphered, but many general guidelines have been established that give some insight into the process.[22–24] When the contiguous or actual sequences at splice points were carefully examined, in a number of instances exact complementary base pairs at each end of the splice sites were not found. Thus, the splicing of the so-called upstream (5′) and downstream (3′) ends of the nucleotide chains did not apparently involve simple base-pair complementarity. However, careful analyses of the splice sites showed an interpretable pattern.[25–27] All sequences at the 5′ or the 3′ ends of various introns that were studied appeared to be related to what were called consensus sequences, including the first two bases at the 5′ end of each intron having the sequence GT and the last two bases at the 3′ end having the sequence AG. Such repetitive short sequences at the opposite ends of each intron suggested that some sort of intervening modulator system might operate—for instance, the attachment of modulator splicing enzymes, possibly to linking transiently acting RNAs. Currently it is

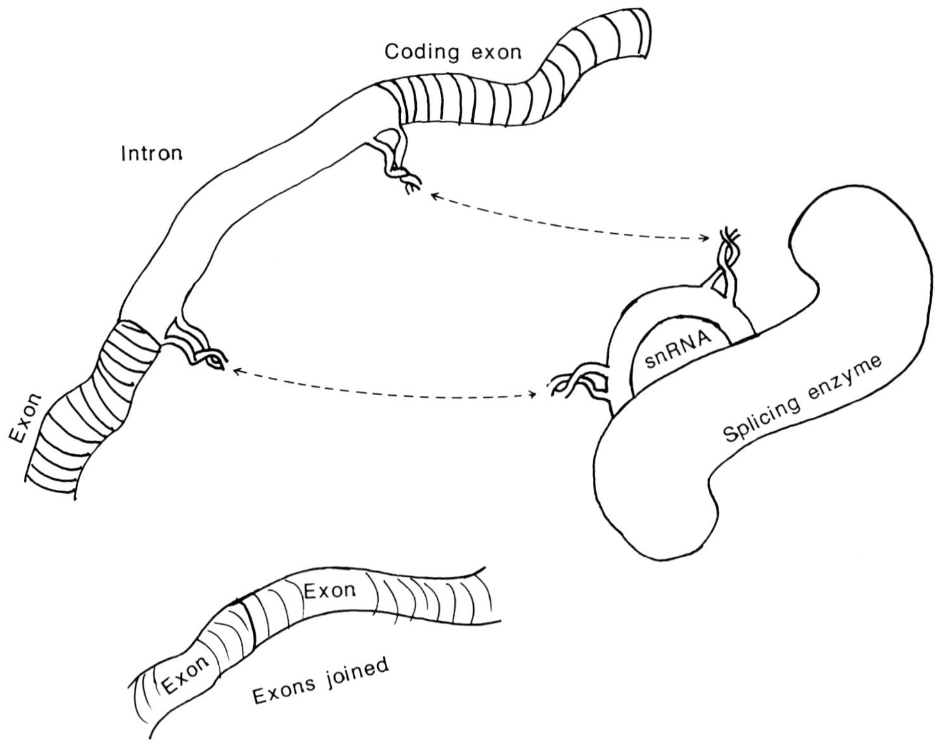

Figure 3.3 One possible model for how snRNAs may modulate splicing reactions. Some snRNAs have sequences that are compatible with the GT or AG consensus sequences at the ends of introns. These mechanisms may be involved in splicing out introns in a reaction where the splicing enzyme is capable of positioning the snRNP in a way that other splicing enzymes can complete the splicing process.

thought that linker small nuclear RNAs (snRNAs) are somehow involved in the splicing process[28,29] (Figure 3.3). For some time it had been known that a host of snRNAs were present within the nucleus. Such molecules—particularly those termed U1, or even possibly Sm, Ro, or La—may be directly involved in splicing out intron loops and subsequently joining consecutive exon coding segments in the actual translational process.

During the past decade there has been great interest in the snRNAs present in normal cellular machinery, and this interest has been intensified by the finding that sera from patients with various presumed autoimmune disorders, such as systemic lupus erythematosus, mixed connective tissue disease, or scleroderma, often have strikingly high levels of autoantibody to antigens related to the small nuclear ribonucleoproteins (snRNPs).[30–35] In fact, individual lupus sera and monoclonal antibodies with apparent anti-Sm or anti-snRNP specificity have in some instances been utilized to examine or dissect out individual steps in the splicing process.[29,34,36–38] Monoclonal antibodies have now been developed which react with snRNP.[39,40,41]

A diagrammatic representation of the possible physical relationships of some of the snRNP antigens, or snurps as they are called, is shown in Figure 3.4.[42] These antigens have been the subject of intense study, because of their apparent importance as antigenic epitopes somehow involved in diseases such as systemic lupus

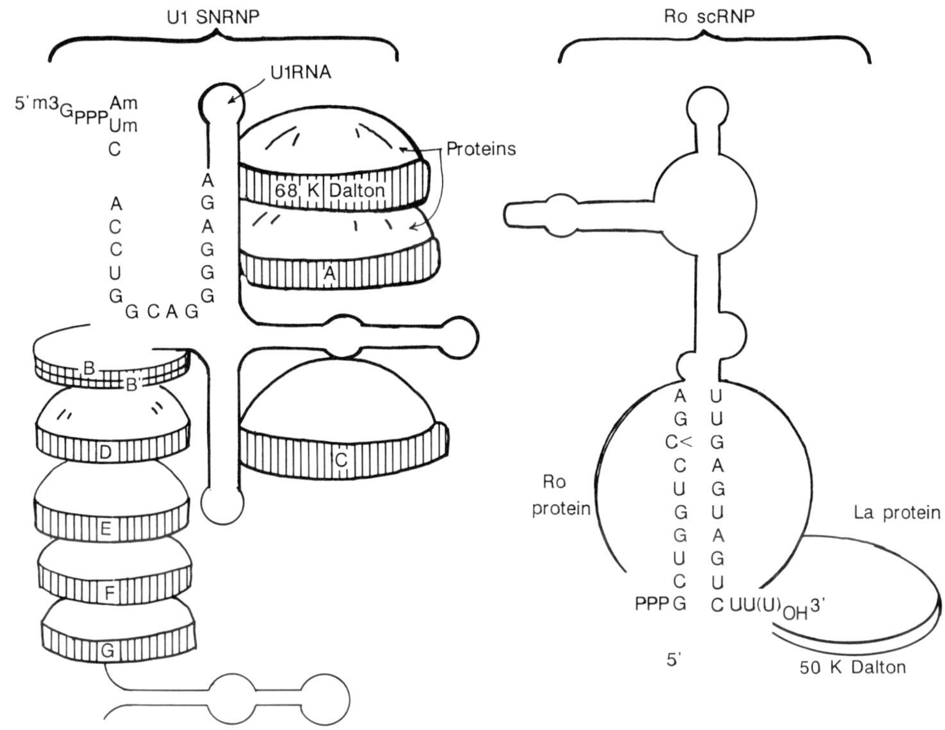

Figure 3.4 The U1 snRNP contains a complex of nine proteins and the U1 RNA. The far 5′ end of this RNA has a base sequence that is complementary to splice junctions within newly transcribed mRNA molecules. Watson-Crick base pairing allows this structure to modulate precise excision of introns during mRNA processing. The RNA secondary structure diagrammed is adapted from that proposed by Mount and Steitz.[36] The associated polypeptides are shown in positions B, B′, D, E, F, and G because the 3′ end of the RNA molecule appears to be protected from RNase digestion by polypeptides D, E, F, and G. In immunoblots, human sera containing anti-U1 RNP antibody usually react with the 68-kd, A, C polypeptides. Sera with anti-Sm antibodies usually react with the B, B′, and D polypeptides. Thus the physical associations diagrammed here may explain the fact that anti-Sm antibodies are usually seen in association with anti-U1 RNP antibodies. In the diagram on the right, the Ro scRNP antigen contains one of five small RNAs that are known as human cytoplasmic 1–5 (hY1–5) RNA and is complexed at the 5′ end and the 3′ end with the Ro polypeptide. The La polypeptide may be physically complexed to Ro. La polypeptide antigen recognizes a binding site consisting of multiple U sequences containing a terminal hydroxyl. *Adapted from Hardin[42] with permission.*

and their relationship to fundamental biological processes involved in RNA splicing.[43]

Other Apparently Self-Driven Splicing Mechanisms

All of the mechanisms governing RNA splicing and attendant elimination of various intron sequences from RNA are by no means clearly understood. A system that appears to be unique has been studied in the primitive ciliate organism *Tetrahymena,* in which introns are removed without any direct protein enzymatic action. In this case, rRNA genes cloned into bacterial plasmid vectors in such a way as to eliminate any contribution of protein enzymatic reactions also showed endogenous RNA

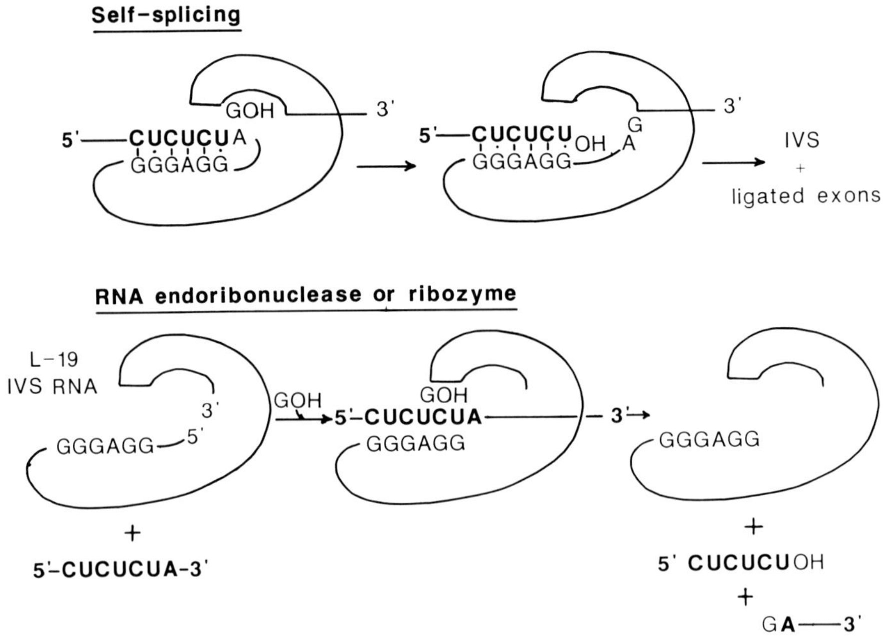

Figure 3.5 A possible model for the endonuclease action of L-19 intervening sequence (IVS) RNA. The mechanism diagrammed is an intermolecular version of the first step of pre-rRNA self-splicing. Lightface letters and lines represent IVS sequences; boldface letters and lines indicate exon sequences (top) or substrate RNA sequences. The G denotes a free guanosine nucleotide or nucleoside. *Reproduced from Zaug et al,*[48] *with permission.*

splicing involving interactions of GMP molecules with phosphodiester bonds between selected nucleotide bases and subsequent correct excision of tRNA introns.[44–46] More recent observations of the *Tetrahymena* system indicate that in this organism a shortened form of the self-splicing rRNA intervening sequence functions autonomously as an endoribonuclease.[47,48] The cleavage of large RNA molecules is thus accomplished by a mechanism that involves guanosine transfer, and the sequence specificity appears to be analogous to that of DNA restriction enzymes. Thus, in this case the ribosomal intervening sequence itself appears to function as a catalytic molecule or ribozyme. This sort of reaction is diagrammed in Figure 3.5.

Establishing Sequences for Human Genes of Interest

With the application of automated methods of sequencing and nucleotide pattern analyses, a number of genes coding for human proteins have now been sequenced, including the immunoglobulins, globins, and ovalbumins. In addition, considerable emphasis has been placed on establishing sequences for many of the polypeptide hormones because of their general biological interest as well as their potential usefulness. Polypeptide hormone genes or genes for so-called releasing factors, such as human growth hormone, proinsulin and insulin, chorionic somatomammotropin, prolactin, and proopiomelanocortin, have now been sequenced. Genes coding for other human messenger proteins, such as the interferons, interleukin 2, clotting factor VIII, tumor necrosis factor, and interleukin 1, have also been sequenced. During a symposium at Cold Spring Harbor there were discussions about the technical feasibility of sequencing the entire human genome and how long it would take for the biotechnology community to do so.[49] The new technical advances, together with interest and financial impetus from the corporate sector, have greatly accelerated the general pace of work in this area. In some human disorders the basic site of the disease has already been well established—for instance, insulin lack in type I adult-onset diabetes mellitus, and a single amino acid substitution in patients with sickle-cell anemia. In other diseases, what might be called reverse genetics may be the most direct application of molecular biology. In these instances identification of stretches of DNA that segregate with the disease would facilitate identification of the gene protein product. Ultimately, the specific malfunction in body economy would then have to be identified and the exact mechanism of disease pathogenesis defined.

Other Methods Used to Predict Important DNA Sequences

With the current precise and practical applications of computer expertise to examination of DNA nucleotide sequence, direct insight into important stretches of DNA that code for proteins can be obtained by looking for what are termed open reading sequences, or open reading frames. These are long stretches of triplet nucleotide codons that do not contain a stop codon. Such an approach is particularly useful when a series of DNA fragments are available as cloned sequences in a convenient plasmid or phage vector and it appears that at least some of the DNA fragments contain nucleotide sequences coding for the actual functional gene. If computer analysis of collections of such DNA sequences clearly reveals an open reading sequence, and if the same stretch of nucleotide sequence shows the important initiation or start-up ATG sequence, it can be predicted that a particular stretch of DNA does in fact contain the gene of interest. Such analyses do not in themselves provide proof or final identification of exact gene localizations, but they are extremely useful in narrowing down the sorting-out process. Conversely, the absence of open reading frames in a particular stretch of nucleotides being examined may indicate that such segments are pseudogenes or not directly related to functional coded protein products.

Signal Peptides and Connected Polypeptides

A great many proteins of biological interest appear to evolve as polypeptide chains that are longer than the eventual active molecule. This has now been established by nucleotide sequence analyses and by comparative direct amino acid analyses of precursors and final product polypeptides. Proteins that are eventually destined for secretion are made on the cytoplasmic polyribosomes in association with membranes within the endoplasmic reticulum of the cytosol. Polypeptides that have just been manufactured pass through the membrane of the endoplasmic reticulum and traverse various cellular organelles making up the secretion route. Finally, as secretory granules fuse with the plasma membrane, the polypeptides contained within the granules are discharged from the cell.[50] Many proteins, such as insulin, immunoglobulin M(IgM), serum albumin, and chymotrypsin, that are transported through various biological membranes or are secreted from cells have precursor sequences that add 15 to 25 leader or initiator amino acids to their amino-terminal ends. Most of these signal sequences are relatively hydrophobic and appear to be biologically active in attaching the polypeptide chains of hormones or secreted

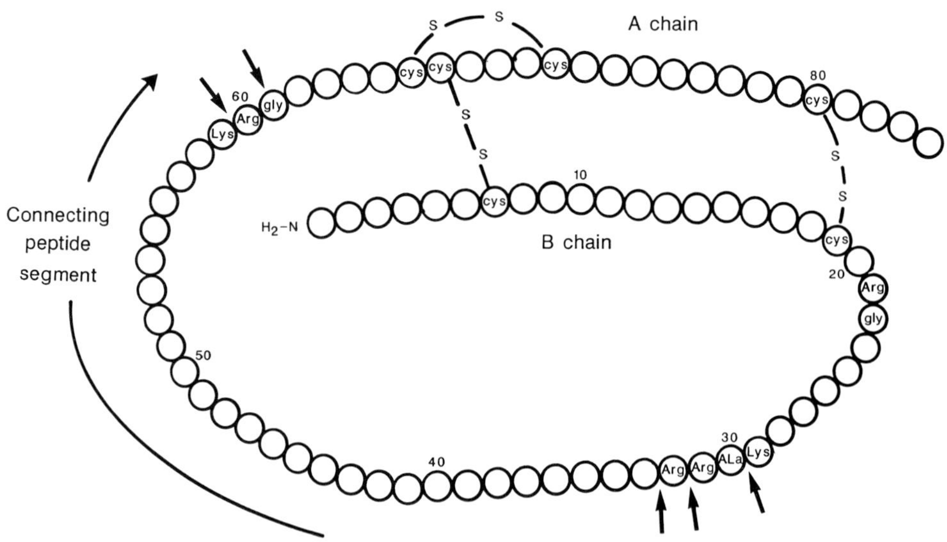

Figure 3.6 Schematic representation of bovine insulin. Arrows indicate sites of tryptic cleavage of connecting peptide leaving the active two-chained insulin molecule.

proteins to cell membranes. Subsequently, they move through the membrane as the protein products are secreted. As the protein is secreted, the signal sequence is generally rapidly cleaved off by proteolytic enzymes called signal peptidases. This may occur within the cell before secretion or after secretion has been completed.[51]

In some secreted proteins, such as various hormone precursors, the role of the precursor sequences or attached peptides was reasonably clear. Thus, when the insulin precursor proinsulin was sequenced, it was found that the alpha and beta insulin chains were initially synthesized as a single amino acid chain that was joined to an additional connecting or C peptide. Apparently, the role of the terminal C peptide is to provide exact spatial orientation of the disulfide bridges in the complete insulin molecule. After the disulfide bridges are formed in the insulin molecule, the C peptide is cleaved off by proteolytic enzymes within secretory granules.[52] A diagrammatic representation of this is shown in Figure 3.6. Similar prohormone forms have now been described for a large number of polypeptide hormones, including parathyroid hormone,[53] glucagon, somatostatin, gastrin, arginine vasopressin, the enkephalins, corticotropin (ACTH, adrenocorticotropic hormone), and β-lipotropin. In some instances, the DNA nucleotide sequences coding for large stretches of amino acid sequence encompass linked, separate polypeptides of different potent biological activities. Thus, the precursor form of ACTH, called proopiocortin, has an estimated molecular weight of 30,000 and is processed by way of glycosylated intermediate forms to ACTH, β-lipotropin, and β-endorphin.[54] A more extended discussion of various families or groups of polypeptide hormones is provided later in the chapter on hormones.

Other Reference Arrangements for Gene Structure

The location of introns within the genes coding for particular proteins such as insulin or hemoglobin has been worked out; however, the actual strategic placement of various introns is still not completely understood in teleological terms. In general, introns are interspersed between general functional domains of proteins, suggesting that the genetic elements coding for certain proteins or polypeptides evolved through repeated recombinations within the interrupted linear arrangement of nucleotide sequences that had previously been on separate chromosomal elements. The long intervening intron sequences may have buffered such recombination events and kept important coding sequences from being disrupted during the evolutionary process.

A single DNA sequence can function to produce several different end product polypeptides or functional proteins by invoking different patterns of splicing in the primary mRNA message. Such alternative splicing patterns result in mRNAs coding for different immunoglobulin classes. They are also involved in the production of two alternative mRNAs and therefore different final gene products from the gene associated with calcitonin.[55] A single gene that codes for calcitonin mRNA produced in the C cells of the thyroid gland also makes a different polypeptide product called calcitonin gene–related product (CGRP) in nerve cells of the hypothalamus. The same basic linked sequence of primary nucleotides can transcribe mRNA for either calcitonin or CGRP because of the existence of two alternative mRNA splicing tracks. The two different mRNAs apparently start out being read as the same at their 5′ ends but splice differently and contain two different messages at their 3′ ends. The precise splicing or alternative selective mechanisms that result in one or the other mRNA product are still not clearly understood. Such alternative mRNA products arising from the same gene sequence are shown in Figure 3.7. In the case of calcitonin, the biological properties and function of the unique 32-amino-acid polypeptide have been extensively studied and some of the effects of this potent hormone on body calcium homeostasis have been defined. However, the biological functions of the alternate mRNA product CGRP are still virtually unknown. When a single gene can be transcribed to make alternate forms of mRNA resulting in different end product polypeptide hormones or messenger molecules, a great deal can sometimes be learned about anatomic and possible functional relationships by using cloned peptide hormone cDNA sequences to localize other tissue sites that contain various specific mRNA populations by in situ tissue section complementary hybridization techniques.[56] The relative abundance of various mRNA populations and the potential effects of hormones or other stimuli on mRNA synthesis can now be measured using convenient dot-blot methodologies.[57,58] These approaches are now being applied with increasing frequency and precision to study the possible relationship of a number of primordial genes and gene clusters to the distribution and interrelationships of many of human hormones, peptides, and lymphokines.

Specific Control Regions on Genes

There is still much to be learned about what actually controls gene expression. It seems likely that certain sequences near the 5′ and 3′ ends that appear to be similar and recur in a number of different genes may somehow be involved in modulation of transcription or translation. Almost all eukaryotic gene sequences and some viral genes have a similar region that occurs approximately 25 to 30 bases upstream (toward the 5′ end) from the site where transcription of DNA to mRNA begins.[59–61] This site is often referred to as the TATA box and is shown in Figure 3.8; currently, the TATA box is thought to guide RNA polymerase II toward the transcription initiation site.[62–64] As yet, no clearly defined function for the CAT box sequence has been discovered. This latter nucleotide sequence is present in most eukaryotic

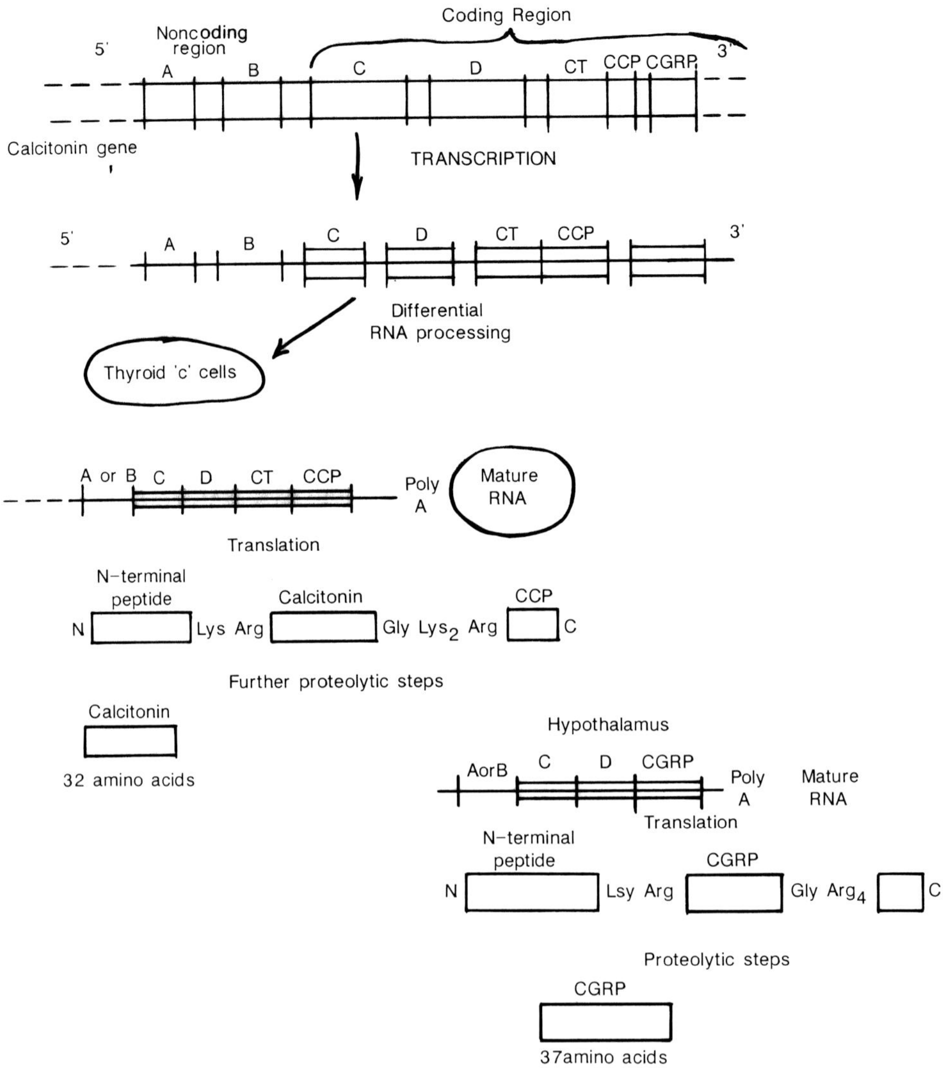

Figure 3.7 Tissue-specific RNA processing of the rat calcitonin gene; in the thyroid C cells mature RNA is translated to a sequential linkage of an n′ terminal peptide followed by calcitonin and a further small 16-amino-acid peptide called CCP. However, the same linear gene sequence in the hypothalamus is differentially spliced so that mRNA codes for N-terminal peptide linked to CGRP. Presumably, differential splicing reactions result in these two different tissue-specific final polypeptide products.

genes; however, when changes are specifically introduced in the CAT sequence, no demonstrable effects on subsequent transcription to mRNA are observed.

The sequence AATAAA is located at the 3′ end of most eukaryotic genes. This six-nucleotide sequence is thought to represent a code for subsequent addition of the poly-A tail to the 3′ end of transcribed mRNA.[65–67] The AATAAA sequence itself does not turn off transcription, but it probably provides a signal for an endogenous nuclease to cleave the mRNA at a position 10 to 15 nucleotides farther down toward the 3′ end. At this point, another important enzyme, poly-A poly-

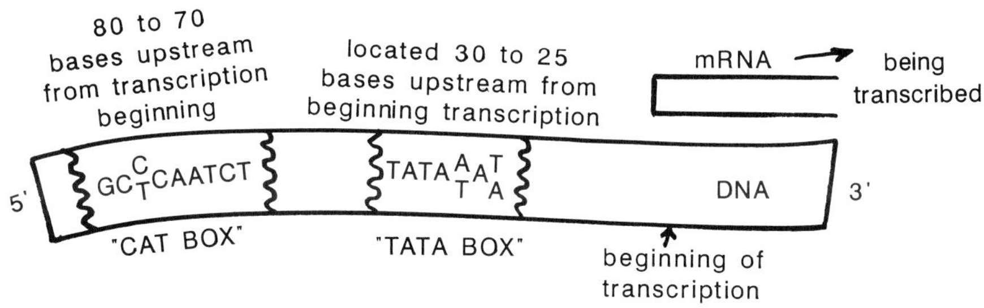

Figure 3.8 Two nucleotide sequences occur toward the 5′ end from the beginning of mRNA transcription of DNA message. The TATA box is thought to help direct RNA polymerase and hence transcription. The function of the second sequence or CAT box has not yet been completely defined.

merase, sticks on the long poly-A tail. This sequence of events in transcription at the 3′ end of mRNA is shown in Figure 3.9.

Conversely, there appear to be no recognizable repetitive sequences at the 5′ ends of mRNAs from eukaryotic cells that can be identified as sites of adherence to ribosomes similar to the specific binding sequences present in prokaryotic mRNAs.

Organization of Groups of Related Genes

The use of cDNA probes for a large number of mammalian genes has provided abundant evidence for clusters or groups of similar genes that are present as multiple, almost identical nucleotide sequences, often spaced close together or in rela-

Figure 3.9 The signal sequence AATAAA occurs toward the 3′ end of most eukaryotic genes. The subsequently transcribed mRNA AAUAAA provides a switch-off signal that first activates a cleaving off endonuclease 10 to 15 bases downstream and then allows poly-A polymerase to add a long poly-A tail at the final 3′ end of the mRNA.

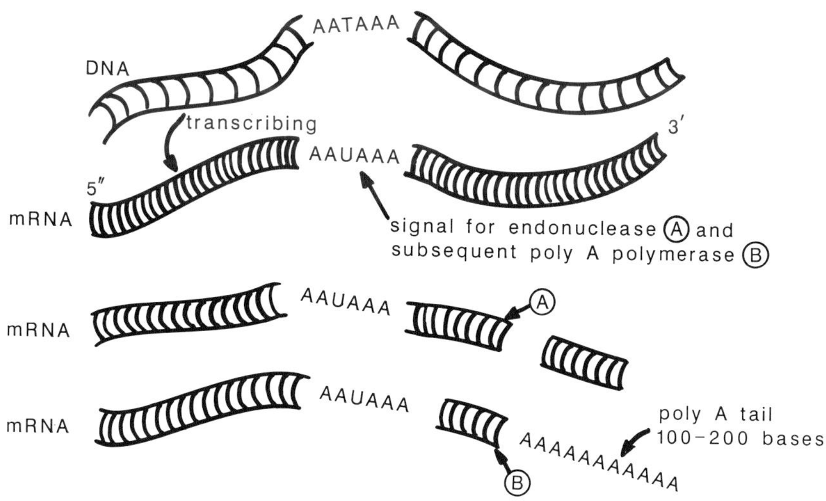

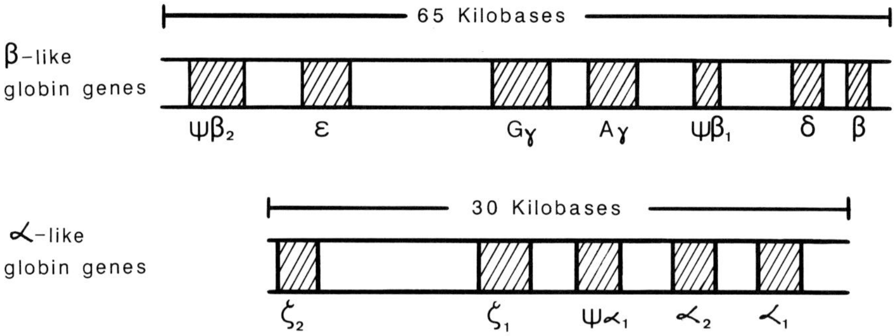

Figure 3.10 Schematic linear diagram of β and α human globin gene groups. Genes designated as $\Psi\beta_1$, $\Psi\beta_2$, and $\Psi\alpha_1$ are evolutionary vestigial pseudogenes that no longer function to make biologically active protein product.

tive proximity. In some instances, related genes that are closely spaced on a particular chromosome are not exact copies of each other but appear to have various degrees of evolutionary relationship. A prominent example of such gene groupings or families is the human globin genes for the various forms of hemoglobin.[68–71] The globin genes contain intervening nucleotide sequences that are not transcribed to functional message due to evolutionary deletions of sequences essential for biological function. A schematic representation of the human globin gene physical groupings indicating functionally active genes along with apparently vestigial or obsolete gene segments is shown in Figure 3.10.

Still not completely understood is the occurrence of "spacer" DNA sequences interspersed between DNA sequences representing true genes. Spacer DNA sequences are not introns, because introns are DNA sequences occurring within genes themselves. Spacer segments frequently encompass longer stretches of DNA nucleotides than genes themselves. It has been estimated that in some chromosomes as little as 10% of the DNA nucleotide sequence actually specifies mRNA coding for specific amino acid sequences; the remaining 90% represents spacer DNA, introns, or pseudogenes. Currently, as is the case with introns and pseudogenes, the physiological function of spacer DNA is unknown.

Alu Repetitive Sequences

When sequence data for a large number of eukaryotic genes were analyzed, it was found that certain sequences are often present as repeats.[72–75] The repeats range from 130 to 300 base pairs and may be present in thousands of copies per genome. In mammalian or human DNA the most common repetitive sequences are referred to as *Alu* repeats because they contain the base pairs AGCT, which are the specific restriction TCGA enzyme cleavage sites for the *Alu* DNA-splitting enzyme. In human genomic DNA the *Alu* repetitive sequences may occur about 300,000 times. The actual *Alu* repetitive nucleotides consist of 300 base pairs. Within the *Alu* sequence, exact repeats of approximately 40 base pairs are present. Flanking the 300 base pairs making up the whole *Alu* segment are other direct repeat sequences of 7 to 10 base pairs. This interesting repetitive arrangement is diagrammed in Figure 3.11.

The *Alu* repeat sequences are dispersed throughout the entire human genome and occur within regular genes as well as within spacer or occasionally intron

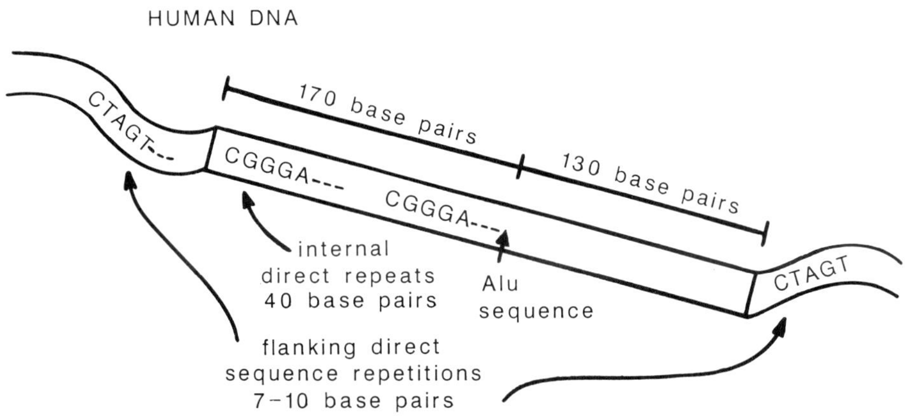

Figure 3.11 *Alu* repeat sequences, which occur up to 300,000 times in human genomic DNA. The overall length of the *Alu* sequence is 300 base pairs. Repetitive 40-bp sequences occur within a single *Alu* sequence unit and the segment is flanked by CTAGT direct repeats up to 7 to 10 base pairs.

stretches of nucleotide sequence. At 3′ ends of the *Alu* repeats are what appear to be remnants of poly-A tails similar to the poly-A tails that regularly occur at 3′ ends of eukaryotic mRNAs.[76,77] This might be interpreted as persistence of *Alu* sequences, possibly through reverse transcription of RNA transcribed from previous *Alu* segments, and subsequent interdigitation of the product cDNA into the human genome at irregular intervals. Ullu and Tschudi have presented evidence that *Alu* sequences are actually processed cytoplasmic RNA genes.[78] The origin and function of *Alu* sequences, as well as other eukaryotic DNA sequence repeats, are still not entirely clear. It has been suggested that some of these DNA segments may code for cytoplasmic snRNAs thought to be involved in splicing or other selective processes.

Mutations

The process of mutation within a single gene results in a change in nucleotide sequence that is inherited by subsequent generations. Studies of mutations have always been of great interest with respect to the precise genetic mechanisms involved in transmission of various heritable characteristics. Mutations have also been of fundamental importance in clinical and basic investigations of human disease states.[79–83] With the advent of molecular biology and the rapid deployment of in vitro methods for mapping genes, cutting parts of them out with restriction enzymes, and identifying normal or abnormal protein sequence products, a great awakening has occurred with respect to gene mutations and their importance in the expression of various human disease states.

A mutation can be described as an abrupt alteration in the genetic material or DNA that produces changes that are inherited by progeny. When the inherited alteration is a change in a single gene, the event is termed a *point mutation*. More generalized mutations involve more widespread changes in chromosomal structure and arrangement within the genome. From the available data concerning the broad spectrum of human diseases in which genes appear to be centrally involved, single point mutations appear to be the most common.[84–89] Mutations are also one of the most important forces producing hereditary alterations during the evolutionary process.

Spontaneous Mutations

Mutations can occur spontaneously within single genes and appear to be random events. The causes of spontaneous mutations are for the most part not well understood; however, it is possible that some mutations represent intermittent and infrequent mistakes in base pairing during the DNA replication process. Various forms of radiation or chemical exposures can cause mutations,[90,91] but these environmental forces are not enough to account for the rate of mutations recorded in most species that have been carefully examined.

Because mutations occur repeatedly and their results can often be observed and recorded, the frequency of occurrence can be calculated and expressed as a mutation rate. In the case of each gene, mutation appears to occur at a particular rate that is characteristic for that gene or specific allele. In many human disease states, the definition of particular genes may have been greatly advanced through recognition of several variants or mutant products. Because mutations appear to be random and unusual but recurrent, they also provide a way to look at various human populations, racial groups, or even individual families for certain types of genetic linkages. Classical genetic analyses can be employed to identify a mutation that produces a dominant trait because the mutation product or final recognizable condition will be expressed even in heterozygous individuals. Such a pattern of inheritance is shown in Figure 3.12. If sex-linked mutations occur, as in the case of hemophilia, the inheritance pattern is characteristic. For instance, careful analyses of the descendants of the British royal family indicate that Queen Victoria probably originated the X-linked hemophilia in that kindred, because none of her male ancestors showed the trait but many of her male descendants had what appeared to be classical hemophilia. This is illustrated in Figure 3.13. In autosomal recessive mutations the particular trait involved can be detected only under special circumstances because full-blown expression is seen only in descendants homozygous for the recessive allele. The trait itself may not be apparent in heterozygous individuals over a number of preceding generations.

Figure 3.12 Typical pattern of autosomal dominant inheritance. Circles designate females and squares males. The third daughter in generation II received a mutant allele in this case and transmitted it to a majority of her offspring.

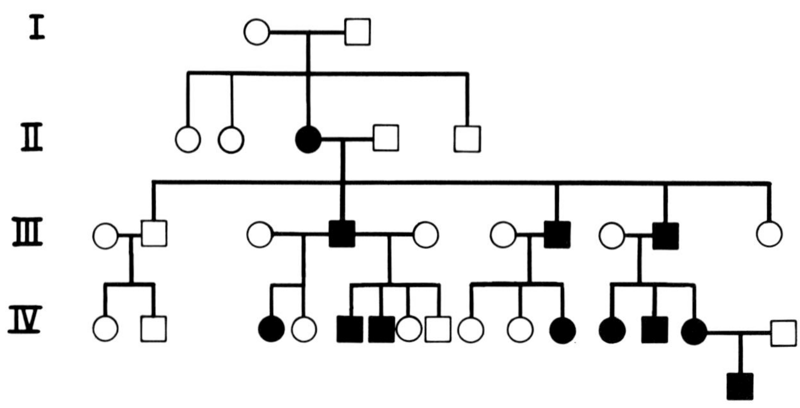

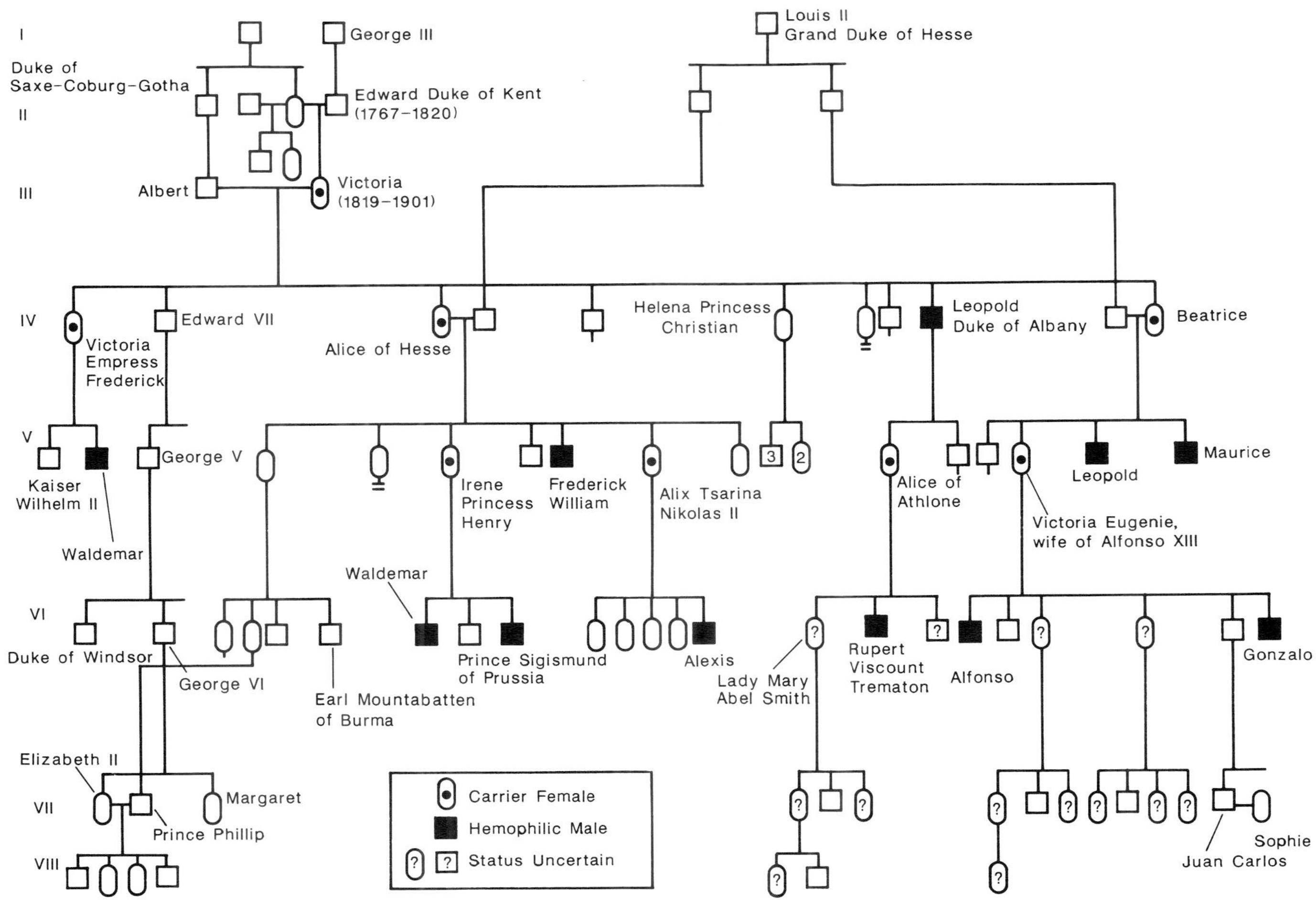

Figure 3.13 Classical X-linked recessive inheritance of hemophilia as illustrated by the descendants of several royal families linked to Queen Victoria. The X-linked recessive pattern shows that all afflicted males were apparently born of maternal carriers and have other afflicted male relatives. Daughters who carry the trait and normal nonaffected sons are descended from afflicted fathers.

Mechanisms Involved in Mutations

With modern molecular biological techniques, the precise chemical mechanisms leading to various mutations can now be examined by two contrasting approaches. One approach is to create mutations of bacteria or other experimental forms in vitro and then sort out progeny displaying a particular recognizable phenotype. These altered progeny can then be studied directly to determine which DNA sequences have been changed. The converse approach is to alter a segment of DNA in vitro and then reintroduce this altered genomic segment back into the organism to study the specific functional, structural, or biochemical changes it induces. There is great interest in the eventual application of this approach to study various human disease states, but a number of practical methodological hurdles must be overcome before this is feasible. Certainly, with the new technology at hand, it is possible to produce intentional mutations by deleting nucleotides, inserting new oligonucleotide sequences, or merely substituting single bases for another one at specific locations in DNA molecules.[92–94] However, such direct applications to humans are neither practical nor ethical.

It has been suggested that some mutational events involve hydrogen shifts, with substitution of hydrogens to different energy positions in certain base pairs. Such a scenario was first suggested by Watson and Crick[95] and is shown diagrammatically in Figure 3.14. In other situations it may be that one or several of the enzymes involved in DNA replication guide the wrong base into line as a new strand of DNA is made, and all subsequent replication copies and progeny retain this change in base sequence. Changes in the nucleotide sequence within DNA produce two general kinds of mutations: base substitutions, such as cytosine for thymine, or more general entire frameshift types of DNA alteration.

Single base substitutions can alter a particular triplet codon so that it calls for an entirely different amino acid. This type of mutational change is referred to as a missense mutation. Such a mutation appears to alter the basic amino acid se-

Figure 3.14 Hydrogen switching to a different position within pairing bases induces a different energy state (tautomer of adenine). In this more stable state, adenine pairs with thymine; however, in its other tautomeric state it pairs with cytosine. This could represent mutation by single base substitution.

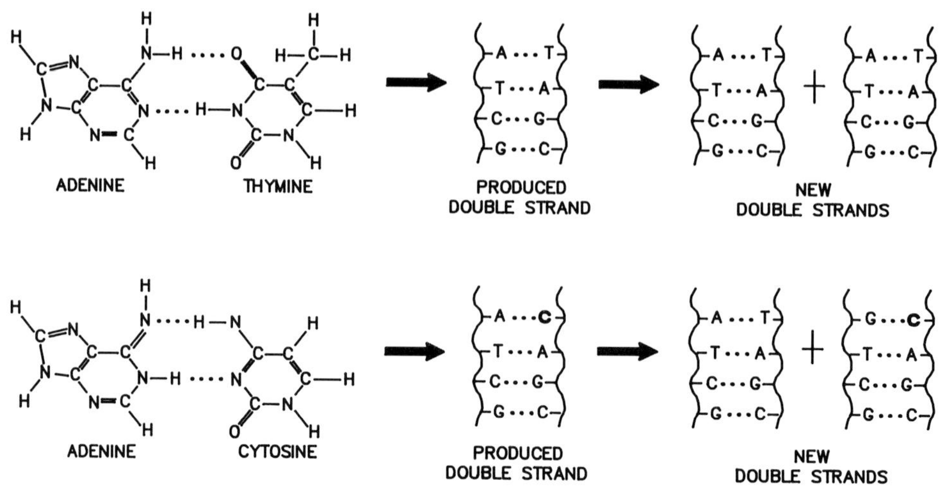

Type of Human Hemoglobin	Triplet Codon Nucleotide Residue Number: 143	144	145	146	Terminator Codon
Normal B-Chain					
Nucleotide sequence	– CAC	– AAG –	UAU	– CAC –	UAA
actual amino acid	– HIS	– LYS –	TYR	– HIS –	COOH
Hemoglobin Rainier (missense mutation)	– CAC	– AAG –	UGU *	– CAC –	UAA
	– HIS	– LYS –	CYS	– HIS –	COOH
Hemoglobin McKees Rocks (nonsense mutation)	– CAC	– AAG * ↓ –	UA A/G		
	– HIS	– LYS – ↑ *	COOH		

Figure 3.15 Different types of mutational changes. In hemoglobin Rainier, a missense mutation specifies a different amino acid at residue 145, but in hemoglobin McKees Rocks, the B chain is actually shortened by two amino acid residues due to a mutation. *Reproduced from C. J. Avers, (ed). Genetics, 2nd ed. Willard Grant Press, Boston, 1984, with permission.*

quence of proteins; however, the entire polypeptide chain specified by the open reading frame of that particular gene is still translated. This is illustrated in Figure 13.15. Another type of mutation illustrated in Figure 13.15 is a substitution of a terminator codon UAGA for the normally occurring codon UGU specifying tyrosine. This nonsense mutation results in premature cutoff of polypeptide chain synthesis, ultimately producing smaller nonfunctional polypeptide segments that are irrelevant and biologically useless.[96,97] A frameshift mutation is one in which the entire reading frame for mRNA synthesis is shifted by either adding on or deleting a base already present within the gene nucleotide sequence. Such frameshift mutations produce scrambled transcription products.[98]

Producing Mutations in Vitro

Several simple techniques can be employed to produce mutations in vitro. Nucleotide sequences can be deleted from large DNA segments by restriction enzyme cleavage and the residual cut ends religated. In the case of circular viruses such as simian virus 40 (SV40), this can be accomplished sequentially.[99,100] As segments of the circular viral genome are cut out with subsequent religation of the cut ends, various functions of the excised elements can be assayed, such as transforming ability or repackaging capacity, and eventually a relative functional map of the various viral DNA elements can be constructed.

Another method that has been used to produce deletions starting with a circular DNA substrate is first to cut the molecule with a restriction enzyme that breaks it only once and leaves it as a single linear DNA segment. Next, an exonuclease is added that begins to degrade the DNA strands from opposite 3′ ends, leaving a single-stranded tail at each end. Subsequently, the single-stranded tails are removed with an S_1 nuclease enzyme, leaving segments of linear double-stranded DNA that have gaps or deletions. Still another specific method for engineering deletions em-

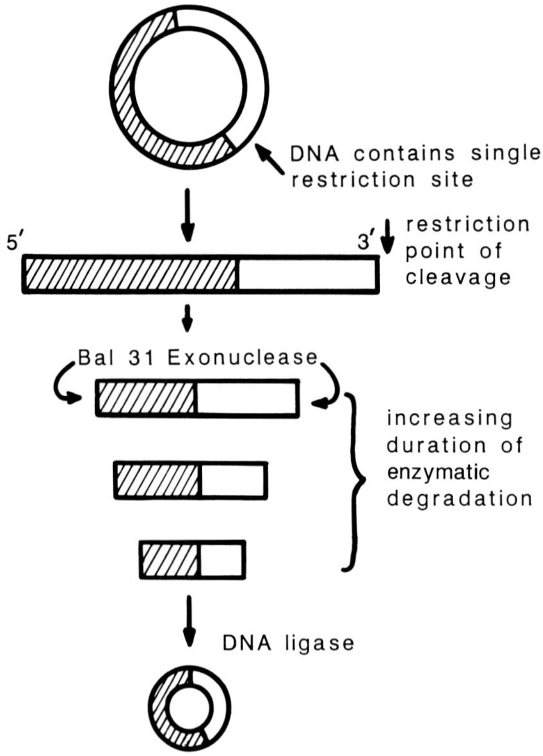

Figure 3.16 A circular DNA molecule is cleaved by a restriction enzyme chosen so that it cleaves at only one site and produces a linear DNA product. The product is treated with *Bal*31 exonuclease, which degrades at both 5′ and 3′ ends. With increasing length of reaction time or increasing concentration of the reactants, more and more nucleotides are removed at each end. Finally, the residual DNA can be religated to a circular form using DNA ligase.

ploys the enzyme *Bal*31 exonuclease, which degrades both strands of the double-stranded DNA from both ends. The conditions of the enzymatic degradation can be controlled in vitro in such a way that a limited number of nucleotide bases (up to several thousand) are cut away before the residual double-stranded DNA is put back together as circular DNA using the enzyme DNA ligase. This sort of deletion strategy is shown in Figure 3.16.

With variations of these procedures it is now possible to construct DNA molecules with precisely defined mutations located exactly where intended. When a library of cDNA mutations resulting from different times of incubation with exonuclease has been prepared, the exact sequences of the residual DNA can then be determined by Maxam-Gilbert DNA sequencing and pairs of fragment molecules can be matched where sequences reading from 3′ or 5′ ends indicate that only a very small segment (8 to 10 bases) is missing. The religated DNA reassembled from such tailored and selected fragments can then be tested to assess its function after a precisely placed deletion has been introduced.

Another recombinant method that can be employed to produce mutants in vitro involves insertion of an oligonucleotide into a normal DNA sequence, either to change the original native DNA sequence or to introduce a selected restriction site that can be used for further intentional changes in DNA nucleotide sequence.[101] Several methods can be utilized. In one, low concentrations of pancreatic DNase plus manganese ions as a cofactor usually induce a series of complete double-stranded interruptions of both DNA strands. This sort of DNA digestion, when performed for brief periods of time at low enzyme concentration, yields linear DNA fragments that can be linked to previously synthesized oligonucleotides and recir-

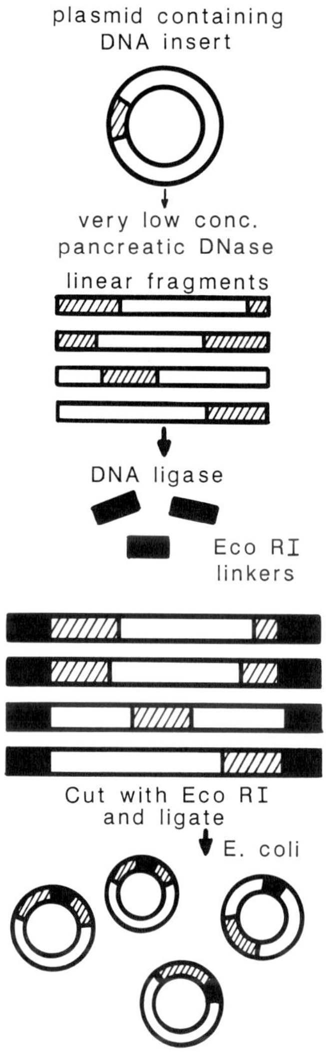

Figure 3.17 Plasmids containing DNA inserts are cut at only one site by using a low concentration of pancreatic DNase. *Eco*RI polynucleotide linkers are then added and, after *Eco*RI digestion, religated together as circles that contain the new inserts at random sites.

cularized with DNA ligase. Another approach is to use a restriction enzyme that ordinarily produces several cuts in the DNA but at a very low concentration so that only one cleavage occurs. Oligonucleotide linker segments are then intentionally inserted at the restriction enzyme sites. These approaches are shown in Figure 3.17.

Chemical Treatments That Can Induce Base Substitutions

In general, there are four groups of chemical mutagens that may be used to alter DNA structure: (1) base analogues, (2) chemical reagents that produce changes in DNA bases, (3) so-called alkylating agents that remove purines from DNA, and (4) acridine dye substances that can produce deletion or addition of bases to DNA.

A base analogue is a molecule in which the primary structure is very similar to that of the naturally occurring base. For this reason, the base analogue can be inserted instead of the regular natural base when the DNA replicates. However,

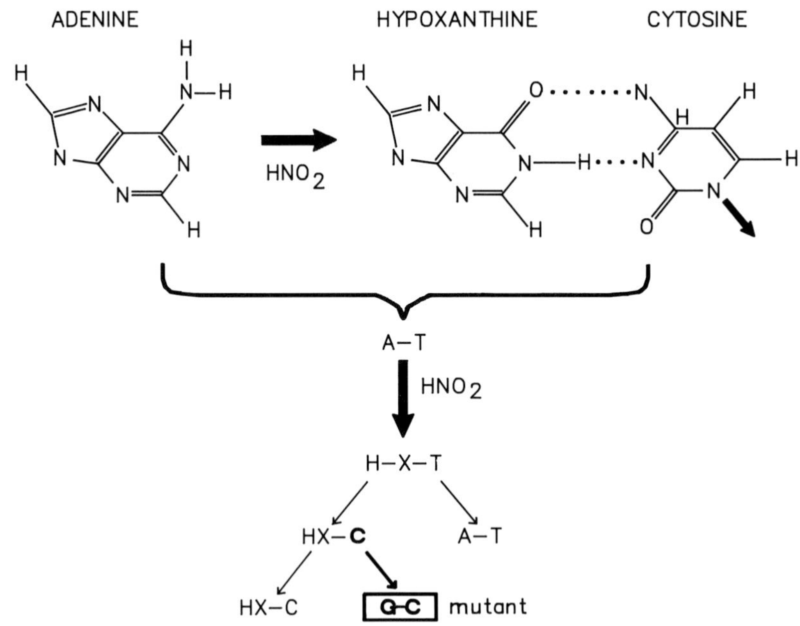

Figure 3.18 Mutation induced through direct chemical reaction of nitrous acid, HNO_2, which oxidatively deaminates adenine to hypoxanthine; this product base-pairs with cytosine, *not* with thymine. The resulting alterations in DNA structure are shown below as an A-T → G-C transition.

once the base analogue is made a part of the DNA, it may produce different subsequent base pairings so that at the next cycle of replication, a base substitution may occur. Chemicals that act directly on bases in DNA include materials such as nitrous acid, which can produce oxidative deamination of adenine, cytosine, and guanine as shown in Figure 3.18. Other chemicals, such as hydroxylamine or nitrosoguanidine, are capable of reacting directly with DNA bases and producing base-pair transitions.[102] These reagents can be regarded as extremely potent mutagens because they are effective when used at low concentration and can react with both replicating and nonreplicating DNA.[91]

In similar fashion, alkylating agents are extremely reactive compounds that put an alkyl group—an ethyl or methyl appendage—in various positions on DNA bases and, like the other agents discussed above, can thereby change base-pairing characteristics. Some of the first alkylating agents used (nitrogen and sulfur mustards) were noted to have a relatively delayed mutagenic effect, with mutations in phage, for instance, arising several replications after the treatment with the alkylating compound.

Acridine dyes are also capable of binding to DNA and interdigitating between adjoining bases. Because of their interposition between normally contiguous bases, they produce a certain amount of distortion at the point of insertion and can therefore produce either addition of an extra base or deletion of a base as replications occur. Acridine dyes, including proflavine and ethidium bromide, are also sometimes involved in frameshift mutations, including gross deletions or actual loss of DNA, because the dye molecules may interfere with effective new strand synthesis during DNA strand replication.

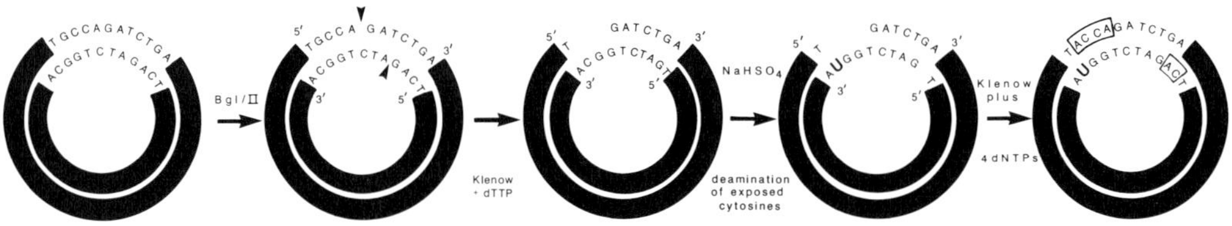

Figure 3.19 Production of a mutant by deamination of exposed cytosine residues after *Bgl*/II restriction enzyme cleavage. After the restriction enzyme has cut the double-stranded DNA, the material is treated with Klenow polymerase with dTTP. The polymerase enzyme degrades single strands in a 3′-to-5′ sequence until it comes upon a T base residue on each strand. Addition of the $NaHSO_4$ and the bases (dNTPs) results in repaired DNA that now contains a single base-pair mutation.

Substitutions

Single base changes, or point mutations, can also be produced by other methods. Cytosines in single-stranded but not ordinarily double-stranded DNA can be treated with bisulfite reagents and deaminated to uracil.[103,104] This method involves production of limited single-stranded regions within cDNA or the DNA segment of interest followed by conversion of all cytosines to uracils (U's). These U's are then read as T's during subsequent replication. One method of producing single-stranded regions near a restriction site uses the enzyme exonuclease III to digest single DNA strands in the 3′-to-5′ sequence.[105,106] This leaves single-stranded DNA tails, depending on how long the exonuclease III reaction is allowed to proceed. Another approach is dependent on the capacity of DNA polymerase, which in the absence of substrate dNTP bases can degrade single DNA strands in a 3′-to-5′ direction. If only one dNTP is added to this sort of reaction, the DNA polymerase will digest single-stranded DNA only until it reaches the juncture where the base of the opposite strand is complementary to the single available nucleotide, at which point the single-strand DNA degradation will stop. Among the DNA polymerases as a general group, phage T4 DNA polymerase and the *Klenow* or large fragment of *E. coli* DNA polymerase have this capacity. An example of application of this method is illustrated in Figure 3.19.

Substitution of Nucleotide Analogues

A method that utilizes an approach similar to that shown in Figure 3.19 also involves a planned set of alterations close to a well-defined restriction site in previously cloned DNA. The DNA is first partially cut with restriction enzyme in the presence of ethidium bromide; this compromises the cleavage of DNA, resulting in the cutting of only one DNA strand rather than both.[107] This single-stranded segment or gap can then be further enlarged using low concentrations of DNA polymerase obtained from the *Mycobacterium luteus* bacterium. If the reaction is carried out in the absence of triphosphates, the *M. luteus* DNA polymerase will digest only five or six bases in a 5′-to-3′ sequence, starting from the region of the single-stranded gap or nick. This treatment produces a DNA containing a five- or six-base single-stranded gap, which can then be repaired using Klenow polymerase and dATP, dCTP, dGTP, and *N*-4-hydroxycytosine instead of dTTP. This chemical nucleotide substitute is replaced within the repaired DNA and can pair almost equally efficiently with either A or G; hence it is capable of inducing small mutations within previously defined gaps or nicked single-stranded segments of DNA.

Misincorporation

DNA molecules that have been nicked or otherwise treated as mentioned above so that they contain small gaps of single-stranded DNA can also be utilized to induce mutations, particularly near known restriction sites, using the process of enzymatic DNA repair and DNA polymerase.[108] DNA molecules in which gaps have been induced near restriction sites are treated with *E. coli* or T4 DNA polymerase in the presence of only three of the four usual dNTPs. Under ordinary circumstances, the polymerase stops when it gets to a point on the DNA strand where it has no proper complementary nucleotide; however, in a small minority of such instances the DNA polymerase makes a mistake and incorporates the wrong nucleotide from the three that it has available instead of the correct one that is missing. The enzyme then carries on with its gap repair assignment until the small single-stranded gap is completely replaced. This mismatch produces circular plasmid DNA with single mismatches of bases. When such plasmids are cycled in *E. coli,* the mismatching base pairs are corrected in the next round of replication, leaving half of the reduplicating DNA strands still with the mutant single base changes. In most such cases, the single base change has occurred at or immediately next to the restriction site where the DNA was originally unraveled to a five- to six-base stretch of single-stranded material. Because the base substitution occurs in the restriction enzyme site, the molecules containing mutant sequences will not be cut by the original enzyme. This sort of strategy is illustrated in Figure 3.20. A problem that can arise in applying this technique for inducing mutations is that most of the DNA polymerases have a fail-safe characteristic; they contain endogenous nuclease activities that are frequently capable of eliminating mismatched nucleotide bases as soon as they are placed in the sequence. This self-monitoring intrinsic corrector mechanism can be overcome by addition of α-thiophosphate nucleotides during the chemical process aimed at repair of the gap. These modified nucleotides cannot be eliminated by the endogenous censor nuclease of the polymerase.

Making Mutations with Well-Defined Synthetic Oligonucleotides

Presynthesized oligonucleotides of known sequence can also be employed to produce mutants; this approach first requires establishment of the exact nucleotide sequence within the cloned DNA to be mutated.[109,110] With the primary DNA sequence established, a site is selected along this DNA segment for introduction of the tentative mutation. A 12- to 15-base oligonucleotide is then synthesized that contains bases complementary to the primary DNA but also includes one or two mismatches.[111,112] The complementary but single site mismatch oligonucleotide is then mixed together with single-stranded DNA from the original primary DNA incorporated in an M13 phage vector. Although the base pairs are not a perfect complementary fit, hybridization can be made to occur provided the physical and chemical conditions are such as not to induce high stringency—that is, the hybridization is performed at relatively low temperature and in high salt concentration. In addition, the mismatched base pairs must be placed in the middle of the strands and not at either end. The result is that the original slightly mismatched oligonucleotide acts as a template for DNA polymerase to go on and make the

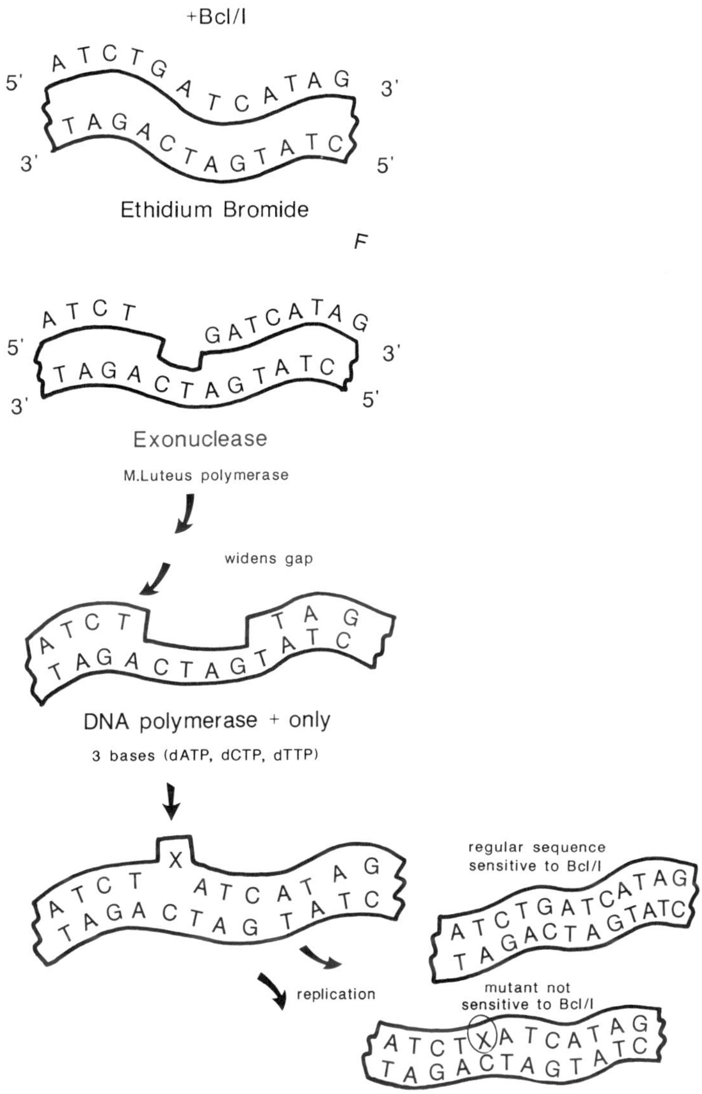

Figure 3.20 Production of mutants by chance misincorporation of a wrong nucleotide. The restriction enzyme *Bcl/*I is employed to cleave the DNA in the presence of ethidium bromide, which is a restriction enzyme modifier that allows the enzyme to cleave or nick only one DNA strand. This nick is then extended using *M. luteus* polymerase to widen the gap. DNA polymerase is added to repair the defect but in the presence of only three nucleotides—adrenine, cytosine, and thymidine. During gap repair, part of the key restriction site TG receives a mismatched nucleotide (X). During subsequent DNA replication one strand replicates having repaired the mismatch. However, the mutation persists in the other replicated strand, and the mutant product is resistant to the original *Bcl/*I restriction enzyme because it has lost the key signal restriction sequence.

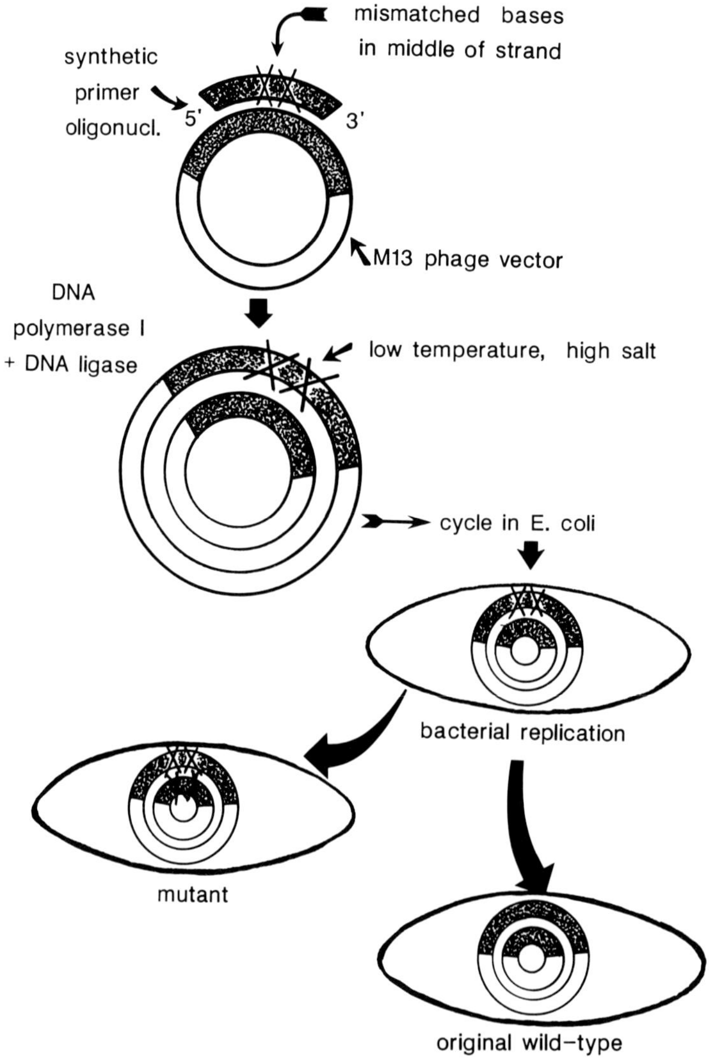

Figure 3.21 Synthetic mutation introduced in putative gene using a preformed synthetic oligonucleotide containing two sets of mismatched bases in the middle of the strand (XX). When this oligonucleotide is hybridized under *nonstringent* conditions (low temperature and high salt concentration), the process is completed despite the two mismatched base pairs. When the resulting double-stranded DNA product is cycled in *E. coli*, the DNA replicates either as the original wild type (without the mutations) or as the mutant containing two base-pair changes.

entire base-pairing DNA strand. Next the double-stranded DNA containing a single or possibly two mismatched base pairs is cycled into *E. coli*, and during this process the single or double base-pair mismatches are repaired, resulting in either the original wild-type DNA or the mutant variety.[112–114] This approach is shown in Figure 3.21. From a theoretical standpoint, it is clear that progeny of such a process should be half original wild-type DNA molecules and half mutants; however, in practice the proportions of DNA molecules containing mutant DNA are much lower

(10–15%) because the DNA polymerase sometimes goes on to an extra cycle of synthesis and, after completing the complementary strand from the oligo primer, makes another complete loop of DNA. This displaces the primer segment and reproduces the nonmutated original wild-type DNA sequence. In addition, it often happens that the native original DNA is individualized through distinctive methyl groups so that at the point where the gene segment containing the intentional mismatched base pair is introduced for cycling in *E. coli,* it is immediately corrected to the original or wild-type sequence rather than the intentional mutant.

After processes such as those discussed above have been completed, mutant progeny can be detected in several ways. If the mutant sequence is intentionally introduced into a recognition sequence for a restriction enzyme, the DNA molecules containing the introduced mutation will no longer cleave with that restriction enzyme and mutants can be separated from wild-type DNA molecules on this basis. Another detection method uses the oligonucleotide sequence that was originally employed to produce the mutation. This oligonucleotide is labeled with ^{32}P, and hybridization conditions of annealing associated with high degrees of stringency are employed (high temperatures and low salt concentration). This can be done using a ^{32}P-labeled oligonucleotide probe applied to nitrocellulose filters that have been pressed over bacterial growth plates containing thousands of bacterial colonies. In the screening process the hybridization temperature is raised in 5°C steps so that a set of conditions is finally reached at which the labeled oligonucleotide hybridizes only to the mutant DNA molecules to which its base pair sequences are exactly complementary. Using this criterion of progressively increasing the conditions of stringency for DNA base-pair annealing, a selective process is thus defined by which colonies containing only mutant DNA molecules are eventually identified. An example of screening bacterial colonies for mutants using a synthetic ^{32}P-labeled probe applied to nitrocellulose filters overlaid on bacterial growth plates to transfer the bacterial growth patterns to the filters is shown in Figure 3.22.

Applications of many of the techniques of intentional mutation in genes using well-defined oligonucleotides have now emerged. When the segments of DNA that code for well-defined proteins of specific biological interest have been identified in a particular gene, it is now possible to introduce oligonucleotide-directed mutations at specific points and study the properties and function of the mutant proteins where a single amino acid change in sequence has been introduced. This approach has wide biological applications and holds great promise.

Mutation Repair

A number of years ago it was found that when bacteria that were irradiated with ultraviolet (UV) light and subsequently incubated in the light to screen for mutants were compared with those that were incubated in the dark, many fewer mutants were noted in bacteria that had been left in the light. This reaction has been termed photoreactivation and represents the phenomenon whereby DNA repair is initiated by light-activated enzymatic activity. It was later found that the most active reactivating light wavelengths were between 310 and 400 nm and that during the light-activated repair process the enzymes activated were capable of splitting dimers to restore single pyrimidines to the site at which mutation had occurred. This is shown in Figure 3.23.

Another form of endogenous DNA repair is referred to as excision repair. In this process (also activated by UV light), the DNA undergoes incision followed by

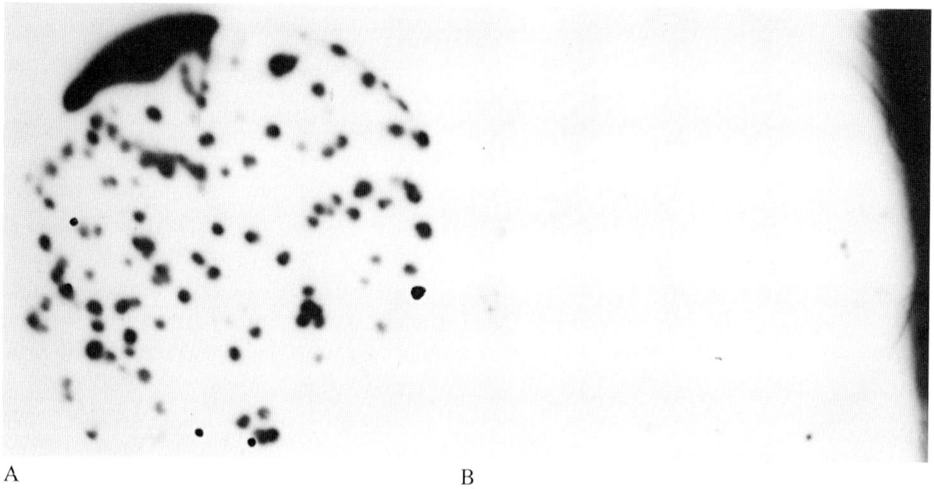

Figure 3.22 Use of ^{32}P-labeled oligonucleotide probe to screen for bacterial colonies containing mutants that can hybridize only with the probe that also contains the mutant base-pair sequences. The conditions of hybridization include high enough stringency (high temperature and low salt concentration) that only nucleotide sequences which are exactly complementary can be annealed. The bacterial colonies containing mutants (A) have been screened by overlaying lawns of bacterial growth with nitrocellulose paper to which the bacteria adhere in the same pattern as on the plates. The bacterial adherence pattern on the nitrocellulose paper is screened using the ^{32}P-labeled oligonucleotide mutant dNA containing colonies thus identified. (A) shows positively reacting mutant colonies and (B) wild-type colonies.

digestion and DNA polymerase repair with final sealing of the duplex by DNA ligase. This endogenous DNA repair process seems to be virtually error free. The rare human disorder xeroderma pigmentosum appears to be caused by inefficient repair of UV-induced damage.[115] Individuals with this inborn error show marked sunlight sensitivity and an abnormally high incidence of skin cancer on sunlight-exposed skin areas. Cell lines derived from xeroderma pigmentosum patients show a broad spectrum of abnormalities in mechanisms for repair of UV-induced mutation. Studies of this rare inborn defect and many others have shown that various mammalian systems have specific DNA repair mechanisms for correcting DNA damage and hence potential mutation by external physical or chemical interactions. Such DNA repair systems have arisen as a natural consequence of evolution because cells must be able to tolerate light and particularly UV wavelengths. The devel-

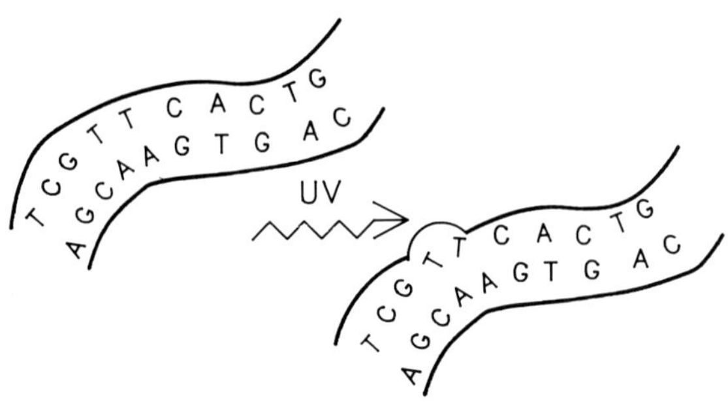

Figure 3.23 After UV irradiation, pyrimidine dimers occur between thymines, causing an outward distortion of the double-stranded DNA. Photoreactivation induces enzymatic repair or splitting apart of the pyrimidine dimers.

opment alternative and therefore foolproof DNA repair systems by mammalian cells was probably essential early in evolution to provide intrinsic mechanisms for protecting against potentially harmful influences of mutagens in the natural environment.[116.]

Mutations and Cancer

At present, there appears to be little direct evidence that changes in DNA can produce neoplasia characterized by uncontrolled growth, aggressive invasion of other tissues, or metastasis. Nevertheless, a number of findings indirectly support the idea that alterations in DNA cause cancer. For instance, most known chemical carcinogens, such as aromatic amines, alkyl halides, and nitrosoamines, can be shown to have mutagenic activity. In most instances, the nature of the alterations in DNA leading to neoplasia has not been defined. The problem is difficult because some of the cells exposed to mutagenic or so-called carcinogenic chemical compounds show DNA changes while others in the same populations do not. A number of carcinogens appear to induce neoplastic transformation only after they have been metabolically converted to other products. An example of this phenomenon is provided by nitrates and nitrites that are used as food preservatives. The original nitrate or nitrite compounds are harmless and nontoxic, whereas their nitrosamine metabolic products are powerful mutagens and carcinogens (Figure 3.24). It seems possible that mutagens in concert with the endogenous influence of oncogenes or antioncogenes may be involved in some cases of neoplasia.

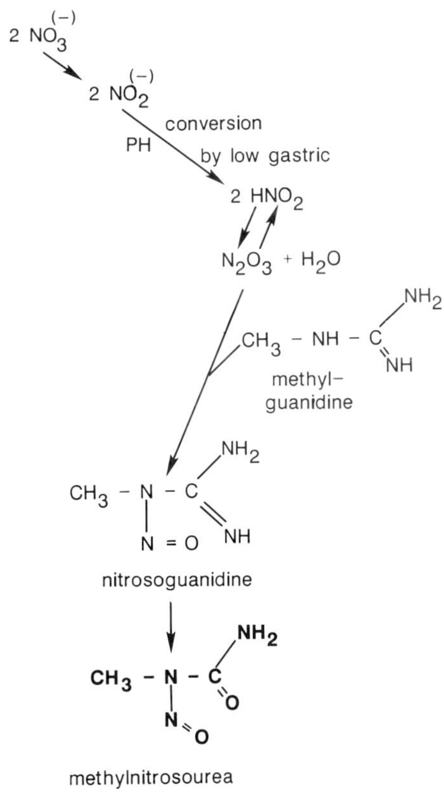

Figure 3.24 Food preservatives such as nitrates or nitrites are metabolized in the body to nitrosamines which can have powerful mutagenic or carcinogenic effects in the body.

There are three human diseases in which apparent inborn errors of metabolism are associated with deficiencies in DNA repair mechanisms and in which there is an apparent increased susceptibility to cancer. They are xeroderma pigmentosum, ataxia telangiectasia, and Fanconi's anemia. In each disorder the specific DNA repair defects appear to be different. It will be fascinating to see how these disorders may lead to a better understanding of the neoplastic process.[117]

References

1. Berget SM, Berk AJ, Harrison T, Sharp PA: Spliced segments at the 5′ termini of adenovirus-2 late mRNA: A role for heterogeneous nuclear RNA in mammalian cells. *Cold Spring Harbor Symp Quant Biol* 1978;42:523–529.
2. Broker TR, Chow LT, Dunn AR, et al: Adenovirus-2 messengers—an example of baroque molecular architecture. *Cold Spring Harbor Symp Quant Biol* 1978;42:531–553.
3. Westphal H, Lai SP: Displacement loops in adenovirus DNA-RNA hybrids. *Cold Spring Harbor Symp Quant Biol* 1978;42:555–558.
4. Doolittle WF: Genes in pieces: Were they ever together? *Nature* 1978;272:581–582.
5. Gilbert W: In Axel R, Maniatis T, Fox CF (eds): *Eucaryotic Gene Regulation* New York, Academic Press, 1979.
6. Crick F: Split genes and RNA splicing. *Science* 1979;204:264–271.
7. Chambron P: Split genes. *Sci Am* 1981;244:48–59.
8. Breathnach R, Mandell JL, Chambon P: Ovalbumin gene is split in chicken DNA. *Nature* 1977;270:314–319.
9. Jeffreys AJ, Flavell RA: The rabbit β-globin gene contains a large insert in the coding sequence. *Cell* 1977;12:1097–1108.
10. Tilghman SM, Tiermier DC, Seidman JG, et al: Intervening sequence of DNA identified in the structural portion of a mouse β-globin gene. *Proc Natl Acad Sci USA* 1978;75:725–729.
11. Gilbert W: Why genes in pieces? *Nature* 1978;271:501.
12. Lewin R: On the origin of introns. *Science* 1982;217:921–922.
13. Woo SLC, Beattie W, Catterall JF, et al: Complete nucleotide sequence of the chicken chromosomal ovalbumin gene and its biological significance. *Biochemistry* 1981;20:6437–6446.
14. Proudfoot NJ, Maniatis T: The structure of a human β-globin pseudogene and its relationship to β-globin gene duplication. *Cell* 1980;21:537–544.
15. Leder P, Hansen N, Konkel D, et al: Mouse globin system: A functional and evolutionary analysis. *Science* 1980;209:1336–1342.
16. Nishioka Y, Leder A, Leder P: Unusual α-globin like gene that has cleanly lost both globin intervening sequences. *Proc Natl Acad Sci USA* 1980;77:2806–2809.
17. Vanin EF, Goldberg GI, Tucker PW, Smithies O: A mouse α-globin–related pseudogene lacking intervening sequences. *Nature* 1980;286:222–226.
18. Little PFR: Globin pseudogenes. *Cell* 1982;28:683–684.
19. Hollis GF, Hieter PA, McBride OW, et al: Processed genes: A dispersed human immunoglobulin gene bearing evidence of RNA-type processing. *Nature* 1982;296:321–325.
20. Reilly JG, Ogden R, Rossi JJ: Isolation of a mouse pseudo +RNA gene encoding CCA—a possible example of reverse flow of genetic information. *Nature* 1982;300:287–289.
21. Lemischka I, Sharp PA: The sequences of an expressed rat α-tubulin gene and a pseudogene with an inserted repetitive element. *Nature* 1982;300:330–335.
22. Breathnach R, Benoist C, O'Hare K, et al: Ovalbumin gene: Evidence for a leader sequence in mRNA and DNA sequences at the exon-intron boundaries. *Proc Natl Acad Sci USA* 1978;75:4853–4857.
23. Breathnach R, Chambon P: Organization and expression of eucaryotic split genes coding for proteins. *Annu Rev Biochem* 1981;50:349–383.
24. Abelson J: RNA processing and the intervening sequence problem. *Annu Rev Biochem* 1979;48:1035–1069.
25. Konarska M, Filipowicz W, Gross HJ: RNA ligation via 2′-phosphomonoester, 3′,5′-

phosphodiester linkage: Requirement of 2′,3′-cyclic phosphate termini and involvement of a 5′-hydroxyl polynucleotide kinase. *Proc Natl Acad Sci USA* 1982;79:1474–1478.
26. Murray V, Holliday R: Mechanism for RNA splicing of gene transcripts. *FEBS Lett* 1979;106:5–7.
27. Rogers J, Wall R: A mechanism for RNA-splicing. *Proc Natl Acad Sci USA* 1980;77:1877–1878.
28. Lerner MR, Boyle JA, Mount SM, et al: Are snRNPs involved in splicing? *Nature* 1980;283:220–224.
29. Yang VW, Lerner MR, Steitz JA, Flint SJ: A small nuclear ribonucleoprotein is required for splicing of adenoviral early RNA sequences. *Proc Natl Acad Sci USA* 1981;78:1371–1375.
30. Lerner MR, Steitz JA: Antibodies to small nuclear RNAs complexed with proteins are produced by patients with systemic lupus erythematosus. *Proc Natl Acad Sci USA* 1979;76:5405–5499.
31. Lerner MR, Andrews NC, Miller G, Steitz JA: Two small RNAs encoded by Epstein-Barr virus and complexed with protein are precipitated by antibodies from patients with systemic lupus erythematosus. *Proc Natl Acad Sci USA* 1981;78:805–809.
32. Lerner MR, Boyle JA, Hardin JA, Steitz JA: Two novel classes of small ribonucleoproteins detected by antibodies associated with lupus erythematosus. *Science* 1981;211:400–402.
33. Rinke J, Steitz JA: Precursor molecules of both human 5S ribosomal RNA and transfer RNAs are bound by a cellular protein reactive with anti-La lupus antibodies. *Cell* 1982;29:149–159.
34. Padgett RA, Mount SM, Steitz JA, Sharp PA: Splicing of messenger RNA precursors is inhibited by antisera to small nuclear ribonucleoprotein. *Cell* 1983;35:101–107.
35. Habets WJ, den Brok JH, Boerbooms AMT, et al: Characterization of the SS-B (La) antigen in adenovirus-infected and uninfected HeLa cells. *EMBO J* 1981;2:1625–1631.
36. Mount SM, Steitz JA: RNA splicing and the involvement of small ribonucleoproteins. In Satir BH (ed): *Modern Cell Biology*. New York, Liss, 1983, pp 249–297.
37. Pettersson I, Hinterberger M, Mimori T, et al: The structure of mammalian small nuclear ribonucleoproteins. Identification of multiple protein components reactive with anti-(U-1) ribonucleoprotein and anti-SM autoantibodies. *J Biol Chem* 1984;259:5907–5914.
38. Chabot B, Black DL, LeMaster DM, Steitz JA: The 3′ splice site of pre-messenger RNA is recognized by a small nuclear ribonucleoprotein. *Science* 1985;230:1344–1349.
39. Pisetsky DS, Semper KF, Eisenberg RA: Idiotypic analysis of a monoclonal anti-Sm antibody. II. Strain distribution of a common idiotypic determinant and its relationship to anti-Sm expression. *J Immunol* 1984;133:2085–2089.
40. Smith PR, Williams DG, Venables PJW, Maini RN: Monoclonal antibodies to the Sjögren's syndrome associated antigen SS-B (La) *J Immunol Methods* 1985;77:63–76.
41. Stefano JE: Purified lupus antigen La recognizes an oligouridylate stretch common to the 3′ termini of RNA polymerase III transcripts. *Cell* 1984;36:145–154.
42. Hardin JA: The lupus autoantigens and the pathogenesis of systemic lupus erythematosus. *Arthritis Rheum* 1986;29:457–460.
43. Turner P: Controlling roles for snurps. *Nature* 1985;316:105–106.
44. Cech TR, Zaug AJ, Grabowski PJ: In vitro splicing of the ribosomal RNA precursor of *Tetrahymena:* Involvement of a guanosine nucleotide in the excision of the intervening sequence. *Cell* 1981;27:487–496.
45. Grabowski PJ, Zang AJ, Cech TR: The intervening sequence of the ribosomal RNA precursor is converted to a circular RNA in isolated nuclei of *Tetrahymena. Cell* 1981;23:467–476.
46. Kruger K, Grabowski PJ, Zaug AJ, et al: Self-splicing RNA: Autoexcision and autocyclization of the ribosomal RNA intervening sequence of *Tetrahymena. Cell* 1982;31:147–157.
47. Cech TR, Bass BL: Biological catalysis by RNA. *Annu Rev Biochem* 1986;55:599–629.
48. Zaug AJ, Been MD, Cech TR: The *Tetrahymena* ribozyme acts like an RNA restriction endonuclease. *Nature* 1986;324:429–433.
49. Lewin R: Molecular biology of *Homo sapiens. Science* 1986;233:157–158.

50. Palade G: Intracellular aspects of the process of protein synthesis. *Science* 1975;189:347–358.
51. Steiner DF, Quinn PS, Chan SJ, et al: Processing mechanisms in the biosynthesis of proteins. *Ann NY Acad Sci* 1980;343:1–16.
52. Steiner DF, Kemmler W, Tager HS, Peterson JD: Proteolytic processing in the biosynthesis of insulin and other proteins. *Fed Proc* 1974;33:2105–2115.
53. Habener JF, Potts JT: Biosynthesis of parathyroid hormone. (First of two parts.) (Second of two parts.) *N Engl J Med* 1978;299:580–585, 635–644.
54. Mains RE, Eipper BA: Biosynthetic studies on ACTH, β-endorphin, and α-melanotropin in the rat. *Ann NY Acad Sci* 1980;343:94–110.
55. Anava SG, Jonas V, Rosenfeld MG, et al: Alternative RNA processing calcitonin gene expression. *Nature* 1982;298:240–244.
56. Hudson P, Penschow J, Shine J, et al: Hybridization histochemistry: Use of recombinant DNA as a "homing probe" for tissue localization of specific mRNA populations. *Endocrinology* 1981;108:353–356.
57. Robins DM, Paek I, Seeburg PH, Axel R: Regulated expression of human growth hormone genes in mouse cells. *Cell* 1982;29:623–631.
58. White BA, Bancroft FC: Cytoplasmic dot hybridization. Simple analysis of relative mRNA levels in multiple small cell or tissue samples. *J Biol Chem* 1982;257:8569–8572.
59. Benoist C, O'Hare K, Breathnach R, Chambon P: The ovalbumin gene sequence of putative control regions. *Nucleic Acids Res* 1979;8:127–143.
60. Benoist C, Chambron P: In vivo sequence requirements of the SV-40 early promoter region. *Nature* 1981;290:304–310.
61. Ghosh PK, Lebowitz P, Frisque RJ, Gluzman Y: Identification of a promoter component involved in positioning the 5′ termini of simian virus 40 early mRNAs. *Proc Natl Acad Sci USA* 1981;78:10–104.
62. Hawley DK, McClure WR: Compilation and analysis of *Escherichia coli* promoter DNA sequences. *Nucleic Acids Res* 1983;11:2237–2255.
63. McKnight SL, Kingsbury R: Transcriptional control signals of a eukaryotic protein-coding gene. *Science* 1982;217:316–324.
64. Lewin B: Promoters for RNA polymerase II are upstream of the start point. In *Genes*. New York, Wiley, 1985, pp 196–201.
65. Lai CJ, Dhar R, Khoury G: Mapping the spliced and unspliced late lytic SV40 RNAs. *Cell* 1978;14:971–982.
66. Nevins JR, Darnell JE: Steps in the processing of AD2 mRNA: Poly (A)+ nuclear sequences are conserved and poly (A) addition precedes splicing. *Cell* 1978;15:1477–1493.
67. Fitzgerald M, Shenk T: The sequence 5′-AAUAAA-3′ forms part of the recognition site for polyadenylation of late SV40 mRNAs. *Cell* 1981;24:251–260.
68. Fritsch E, Lawn R, Maniatis T: Molecular cloning and characterization of the human β-like globin gene cluster. *Cell* 1980;19:959–972.
69. Slauer J, Shen C, Maniatis T: The chromosomal arrangement of human α-like globin genes: Sequence homology and α-globin gene deletions. *Cell* 1980;20:119–130.
70. Maniatis T, Fritsch EF, Lauer J, Lawn RM: The molecular genetics of human hemoglobins. *Annu Rev Genet* 1980;14:145–178.
71. Orkin SH, Michelson A: Partial deletion of the α-globin structural gene in human α-thalassaemia. *Nature* 1980;286:538–540.
72. Britten RJ, and Kohne DE: Repeated sequences in DNA. *Science* 1968;161:529–540.
73. Deininger PL, Jolly DJ, Rubin CM, et al: Base sequence studies of 300 nucleotide renatured repeated human DNA. *J Mol Biol* 1981;151:17–33.
74. Jelinek WR, Toomey TP, Leinward L, et al: Ubiquitous interspersed repeated DNA sequences in mammalian genomes. *Proc Natl Acad Sci USA* 1980;77:1398–1402.
75. Schmid CW, Jelenik WR: The Alu family of dispersed repetitive sequences. *Science* 1982;216:1065–1070.
76. Weiner A: An abundant cytoplasmic 7S RNA is partially complementary to the dominant interspersed middle repetitive DNA sequence family in the human genome. *Cell* 1980;22:209–218.
77. Van Arsdell SW, Denison RA, Bernstein LB, et al: Direct repeats flank three small nuclear RNA pseudogenes in the human genome. *Cell* 1982;26:11–17.

78. Ullu E, Tschudi C: *Alu* sequences are processed 7SL RNA genes. *Nature* 1984;312:171–172.
79. Goldstein JL, Brown MS: The LDL receptor locus and the genetics of familial hypercholesterolemia. *Annu Rev Genet* 1979;13:259–289.
80. Baird M, Driscoll C, Schreiner H, Sciarratta G, et al: A nucleotide change at a splice junction in the human β-globin gene is associated with β°-thalassemia. *Proc Natl Acad Sci USA* 1981;78:4218–4221.
81. Orkin SH, Goff SC, Hechtman RL: Mutation in an intervening sequence splice junction in man. *Proc Natl Acad Sci USA* 1981;78:5041–5045.
82. Myers RH, Madden JJ, Teague JL, Falek A: Factors related to onset age in Huntington disease. *Am J Hum Genet* 1982;34:481–488.
83. Higgs DR, Hill AVS, Nicholls RD, et al: Molecular rearrangements of the human α-globin gene cluster. *Ann NY Acad Sci* 1981;445:45–56.
84. Pauling L, Itano MA, Singer SJ, Wells IC: Sickle cell anemia. A molecular disease. *Science* 1949;110:543–548.
85. Perutz MF, Lehman H: Molecular pathology of human haemoglobin. *Nature* 1968;219:902–909.
86. Perutz MF: Fundamental research in molecular biology: Relevance to medicine. *Nature* 1976;262:449–453.
87. Kan YW: Hemoglobin abnormalities: Molecular and evolutionary studies. *Harvey Lect* 1982;76:75–93.
88. Kurachi K, Chandra T, Degen SJF, et al: Cloning and sequence cDNA coding for α_1 anti-trypsin. *Proc Natl Acad Sci USA* 1981;78:6826–6830.
89. Caskey CT, Kruh GD: The HPRT locus. *Cell* 1979;16:1–9.
90. Stadler LJ, Mutations in barley induced by X-rays and radium. *Science* 1928;68:186–187.
91. Ames BN: Identifying environmental chemicals causing mutations and cancer. *Science* 1979;204:587–593.
92. Lai CJ, Nathans D: Deletion mutants of SV40 generated by enzymatic excision of DNA segments from the viral genome. *J Mol Biol* 1974;89:179–193.
93. McKnight SL, Kingsbury R: Transcriptional control signals of a eukaryotic protein-coding gene. *Science* 1982;217:316–324.
94. Hochstadt J, Ozer HL, Shopsis C: Genetic alteration in animal cells in culture. *Curr Top Microbiol Immunol* 1981;94/95:244–308.
95. Watson JD, Crick FHC: Genetical implications of the structure of deoxyribose nucleic acid. *Nature* 1953;171:964–967.
96. Coulondre C, Miller JH: Genetic studies of the Lac repressor. III. Additional correlation of mutational sites with specific amino acid residues. *J Mol Biol* 1977;117:525–575.
97. Miller JH, Coulondre C, Farabaugh PJ: Correlation of nonsense sites in the lac I gene with specific codons in the nucleotide sequence. *Nature* 1978;274:770–775.
98. Streisinger G, Okada Y, Emrich J, Newton J, et al: Frameshift mutations and the genetic code. *Cold Spring Harbor Symp Quant Biol* 1966;31:77–84.
99. Lai CJ, Nathans D: Deletion mutants of SV40 generated by enzymatic excision of DNA segments from the viral genome. *J Mol Biol* 1974;89:179–193.
100. Mertz JE, Carbon J, Herzberg M, et al: Isolation and characterization of individual clones of simian virus 40 mutants containing deletions, duplications and insertions in their DNA. *Cold Spring Harbor Symp Quant Biol* 1974;39:69–84.
101. Heffron F, So M, McCarthy BJ: In vitro mutagenesis of a circular DNA molecule using synthetic restriction sites. *Proc Natl Acad Sci USA* 1978;75:6012–6016.
102. Chu CT, Parris DS, Dixon RAF, et al: Hydroxylamine mutagenesis of HSV DNA and DNA fragments: Introduction of mutations into selected regions of the viral genome. *Virology* 1979;98:168–181.
103. Shapiro R, Cohen BI, Servis RE: Specific deamination of RNA by sodium bisulfite. *Nature* 1970;227:1047–1048.
104. Hayatsu H: Bisulfite modification of nucleic acids and their constituents. *Prog Nucleic Acid Res Mol Biol* 1976;16:75–124.
105. Shortle D, Botstein D: Single-stranded gaps as localized targets for *in vitro* mutagenesis. In Lemont JF, Generosa WM (eds): *Molecular and Cellular Mechanisms of Mutagenesis.* New York, Plenum 1982, pp 147–156.
106. Ciampi MS, Melton DA, Cortese R: Site-directed mutagenesis of a +RNA gene: box

alteration in the coding region affect transcription. *Proc Natl Acad Sci USA* 1982;79:1388–1392.
107. Greenfield L, Simpson L, Kaplan D: Conversion of closed circular DNA molecules to single-nicked molecules by digestion with DNase I in the presence of ethidium bromide. *Biochim Biophys Acta* 1975;407:365–375.
108. Zakour RA, Loeb LA: Site-specific mutagenesis by error-directed DNA synthesis. *Nature* 1982;295:708–710.
109. Gillam S, Jahnke P, Astell C, et al: Defined transversion mutations of a specific position in DNA using synthetic oligodeoxyribonucleotides as mutagens. *Nucleic Acids Res* 1979;6:2973–2985.
110. Gillam S, Astell CR, Smith M: Site-specific mutagenesis using oligodeoxyribonucleotides: Isolation of a phenotypically silent IX174 mutant, with a specific nucleotide deletion, at a very high frequency. *Gene* 1980;12:129–137.
111. Smith M, Gillam S: Constructed mutants using synthetic oligodeoxyribonucleotides as site-specific mutagens. In Williamson R (ed): *Genetic Engineering, Principles and Methods,* Vol 3. New York, Plenum, 1981, pp 1–32.
112. Wallace RB, Johnson PF, Tanaka S, et al: Directed deletion of a yeast transfer RNA intervening sequence. *Science* 1980;209:1396–1400.
113. Wallace RB, Schold M, Johnson MJ, et al: Oligonucleotide directed mutagenesis of the human β-globin gene: A general method for producing specific point mutations in cloned DNA. *Nucleic Acids Res* 1981;9:3647–3656.
114. Dalbadie-McFarland G, Cohen LW, Riggs AD, et al: Oligonucleotide-directed mutagenesis as a general and powerful method for studies of protein function. *Proc Natl Acad Sci USA* 1982;79:6409–6413.
115. Cleaver JE: Defective repair replication of DNA in xeroderma pigmentosum. *Nature* 1968;218:652–656.
116. Howard-Flanders P: Inducible repair of DNA. *Sci Am* 1981;245:72–80.
117. Setlow RB: Repair deficient human disorders and cancer. *Nature* 1978;271:713–717.

DNA Rearrangements in Both Humoral and Cellular Immunity

4

Ralph C. Williams, Jr., MD

The immune system consists of a complex series of interactive elements that are basically protective and readily adaptable to new external stimuli capable of posing a threat to the organism. If the protective immune response marshaled by the host is to be maximally effective, it must contain both an early warning system and a sustained and modulatable long-term defense mechanism. It must also include the ability to adapt to stimuli that have never been encountered before and that may arise from mutations in potentially harmful external antigens. Implicit in any prescription for an ideal immune response is the requirement that it not readily produce either harmful humoral antibody or T cell–mediated immune reactivity against self or autologous antigens. Given all these important prerequisites for the basic elements of an effective immune response, there is no question that the genes responsible for engineering such a response need great flexibility and capacity for generating a remarkable degree of diversity. How this actually occurs in nature is a story of fascinating yet relatively simple natural adaptation. Several fundamental theoretical questions have troubled those who try to construct theories or models of how a finite number of genes can be organized to produce an appropriate immune response to all the antigenic stimuli an organism or individual will encounter during a lifetime. The concept that a preordained repertoire of all possible B cells is present in an individual at birth was the basis for Burnet's clonal selection theory.[1] One of the recurrent stumbling blocks has always been the question of how the immune system avoids initiating immune responses against itself. A second problem is that of finding appropriate reproducible models for the required degree of diversity that an effective immune response must provide.

A clear perception of genetic mechanisms providing antibody diversity was possible only after the structure and interrelationships of immunoglobulin molecules were firmly established.[2–6] More recently, similar insight has been gained into the molecular basis for T cell–mediated immune responses or cellular immunity. This has become clarified as more basic information has been obtained about the primary structure of the T cell receptor.[7–11]

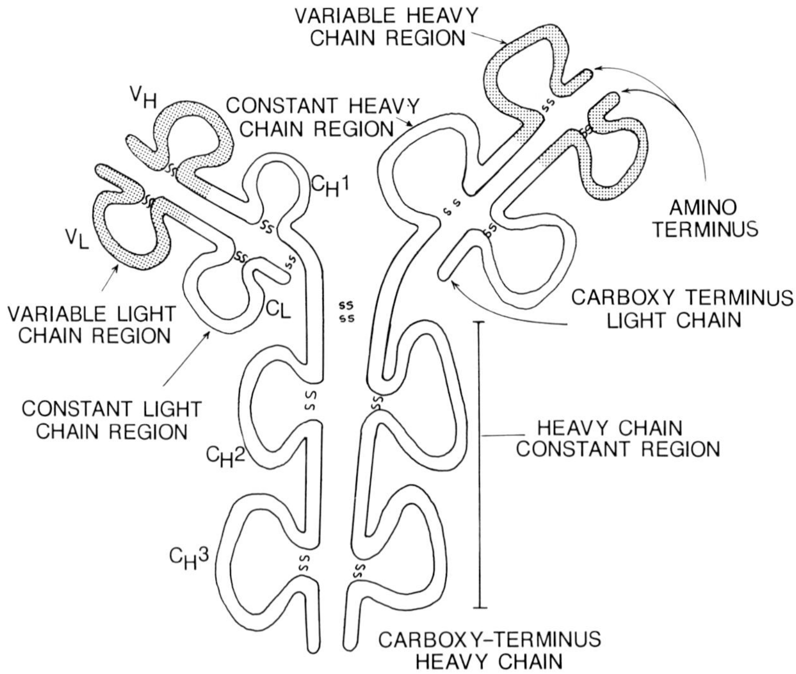

Figure 4.1 Diagram of an IgG molecule indicating constant and variable regions of both heavy and light chains. Both heavy and light chains have domains consisting of intrachain disulfide-linked loops. The variable regions of heavy (V_H) and light chains (V_L) are involved in forming the actual antibody combining sites.

A close look at the basic structure of the immunoglobulin G (IgG) molecule (Figure 4.1) shows that it consists of two identical heavy (H) and two light (L) chains whose molecular masses are about 50,000 and 22,500 daltons, respectively. Although discussion of IgG structure and gene relationships is useful as a model, the same general patterns apply to the other four immunoglobulin classes, IgA, IgM, IgE, and IgD. In immunoglobulin molecules there are two types of light chain (κ or λ), and in IgG molecules there are four major H chain subclasses, IgG1, IgG2, IgG3, and IgG4. Both heavy and light chains of all Ig molecules (IgG, IgA, IgM, IgE, and IgD) have variable (V) sequences at their amino-terminal ends and constant (C) regions at their carboxyl-terminal portions. The constant regions of heavy chains of IgG molecules are involved in functions that are not specific to antibody combining sites, such as complement fixation, interaction with Fc receptors, or skin fixation. The V_H and V_L regions of all classes of antibody molecules are directly involved with the antigen-interlocking function of the antibody and interact to form the antibody combining site. Light chains consist of a constant portion of about 110 amino acids and a variable portion of approximately the same size. Likewise, the variable amino-terminal region of heavy chains consists of approximately 110 amino acids; heavy chain constant regions, however, are much longer (330–440 amino acids in length), varying with H-chain subgroup and major antibody class (IgG, IgA, IgD, IgE, or IgM). Within the amino-terminal variable regions of both heavy and light chains, there are three hypervariable regions that appear to be most closely associated with the actual antigen binding site. This is shown diagrammatically in Figure 4.2. Antigen-binding regions were localized by

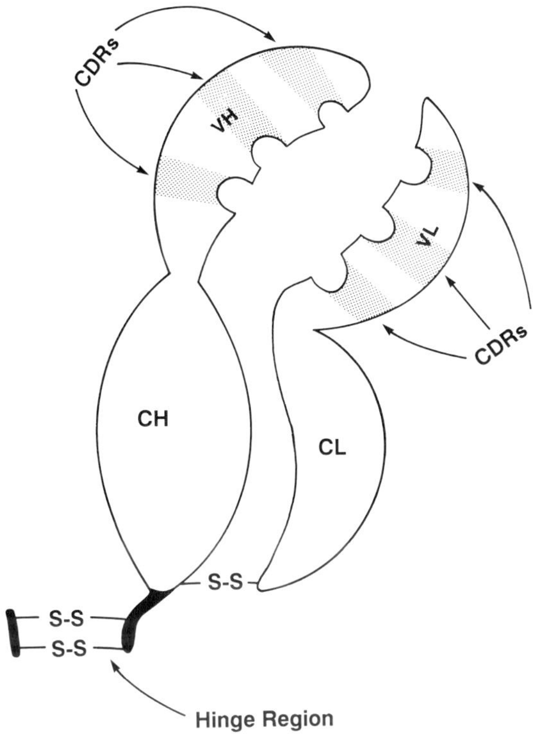

Figure 4.2 Diagram of one half of the IgG molecule showing three hypervariable regions located within the V_H and V_L segments. These regions, designated CDRs for complementary-determining regions, are most directly involved in forming the actual antibody combining site.

affinity labeling of homogeneous antibodies with low molecular weight antigens that bind chemically to adjacent specific amino acid residues.[12–14]

Antibody combining sites vis-à-vis antigen conformational fitting have been analyzed using chemical modification of free and antibody-bound protein antigens.[15] With this technique it was demonstrated that a monoclonal antibody binding to a protein antigen will retard the rate of chemical modification of amino acid residues at the antigenic epitope. Thus, comparison of the degree of acetylation of a limited number of lysine and threonine residues in free and antibody-bound horse cytochrome *c* protein antigen revealed a three-dimensional form of the antigenic epitope on the cytochrome *c* antigen. Since the actual areas on the immunoglobulin molecules that are involved in final antibody specificity and the antigen binding site are determined by the V_H and V_L regions, this quickly reduces the number of genes needed to generate the required degree of diversity for individual antibodies. Thus, any single light chain (V_L) of 2000 genes would include the repertoire for 1×10^6 different combining specificities. It has been variously estimated that the capacity for 1×10^6 to 2.5×10^6 different specificities would probably satisfy the requirements for appropriate antibody diversity based on what is currently estimated to represent the naturally occurring antigenic spectrum.[16,17]

Gene Rearrangements Involved in Antibody Molecule Formation

The division of both heavy and light chains into variable amino-terminal and constant carboxyl-terminal portions initially suggested that genes coding for heavy and light chain polypeptide products were probably divided into two parts.[18] This, in fact, turned out to be the case. It was subsequently found that the variable or V regions were coded for by any one of a set of different V region gene segments and that during differentiation of antibody-forming B cells, a particular V region was connected to an appropriate constant or C region to form the finished antibody molecule product. Actually, it was found that immunoglobulin gene families or clusters interacted to include one or more C gene sequences and sets of V as well as joining (J) gene sequences. This is shown diagrammatically in Figure 4.3. The initial perception of segments of DNA coding for adjacent sequences in the final immunoglobulin molecule was derived from experiments by Tonegawa and coworkers using restriction enzyme fragments of DNA coding for V and C portions of Ig molecules.[19,20] It was found that V and C coding portions of DNA were arranged close together within restriction fragments of DNA, whereas the original DNA segments coding for V and C regions were situated far apart in the embryonic or germ-line DNA. These findings were the first to suggest that during the process of B cell differentiation and organization of the molecular mechanisms controlling final assembly of the various portions of antibody molecules, what is commonly called rearrangement had clearly occurred.[23]

All of the work that eventually eludicated how various discrete segments of DNA became rearranged during the steps leading to assembly of the immunoglobulin molecule was initially based on the study of monoclonal myeloma proteins, for which the entire amino acid sequences of both heavy and light chains had already been established. In this experimental system, a direct comparison

Figure 4.3 Gene groups for κ chains and λ light chains are located on chromosomes 2 and 22, respectively. The gene families making up variable (V), diversity (D), joining (J), constant (C) regions of heavy chains are located on chromosome 14.

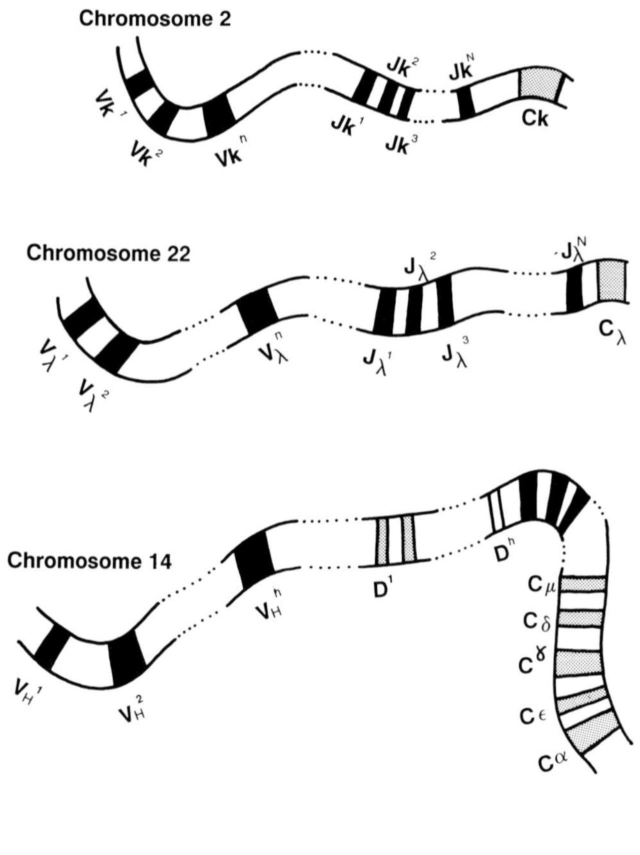

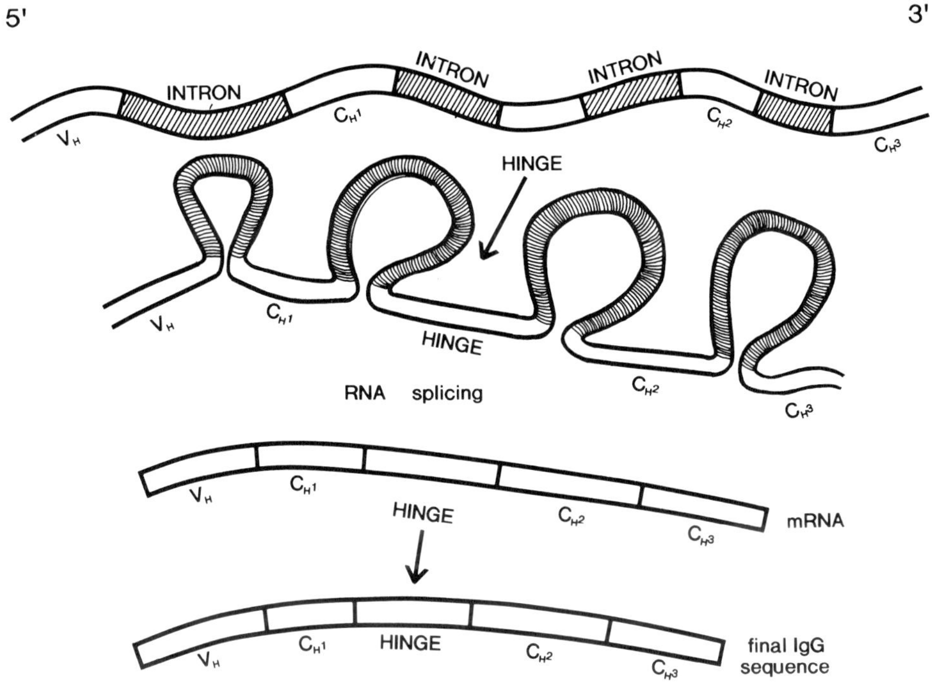

Figure 4.4 Sequential process of splicing out of introns whereby widely separated exons coding for variable and constant region domains, and hinge region of IgG are brought close together and rearranged to make the final mRNA and hence the final IgG heavy chain sequence.

could be made between known amino acid sequences of V or C regions and corresponding nucleotide segments known to code for these specific functional antibody regions. As might be expected from the final rearranged V-D-J-C immunoglobulin product, it was found that a number of introns or intervening sequences were present and that they were localized precisely between various functional immunoglobulin domains. Thus, interspersed between the C_H1, C_H2, and C_H3 domains of the gene for IgG were introns or intervening sequences. In addition, the so-called hinge region of the Ig molecule situated between the C_H1 and C_H2 regions was composed of specific exons. These findings indicated that extensive rearrangement of exons occurred before the final immunoglobulin molecule was spliced together. A scheme diagramming this sort of rearrangement is shown in Figure 4.4.

V, J, and C Segments

Cross-hybridization studies using V region or C region DNA probes have indicated that the actual numbers of C_L (C_λ or C_κ) are limited and, as shown in Figure 4.3, localized to different chromosomes. By contrast, the numbers of the V_L and V_H genes appear to be much higher (200+). With respect to constant region (C_H) genes, a single family specifying constant region sequences for all five major immunoglobulin classes—IgG, IgD, IgM, IgE, and IgA—is grouped together on chromosome 14 as shown in Figure 4.3.

In addition, the relative degree of flexibility of the variable portions of Ig molecules was found to be directly influenced by the existence of multiple joining (J) segments of DNA, which also participate in the basic rearrangement process. The J segments are located upstream (toward the 5′ end) of the C_L or C_H genes. Since any one of a number of J segments may be interposed between C and V sequences, the J segments confer an additional degree of variability or diversity because the different J genes produce slight differences in primary amino acid sequence depending on which J segment is spliced between contiguous V and C region genes.

J regions were initially recognized when it was found that the embryonic V regions coding for both λ and κ light chains did not specify approximately 13 amino acids known to be contained within what was considered the variable (V_L) region. Subsequently, sequence analysis indicated that the DNA segments involved in J region coding were located 1000 to 3000 nucleotides upstream from the C region[21,22] and thus were clearly the products of further gene segment rearrangements. The J regions of immunoglobulin light chains are quite close to the gene segments coding for the third hypervariable light chain region. For this reason, variation in sequences specified by the J regions will influence the structure of the antibody combining site in the final gene product. A schematic model for B cell gene rearrangements that occur during final B cell maturation is shown in Figure 4.5.

Figure 4.5 Progressive steps whereby exons and introns present in germ-line embryonic DNA are deleted or spliced out to leave final mRNA with one variable gene V_1 and one joining segment J_4 to attach to the constant region sequence (C) for production of the final light chain product.

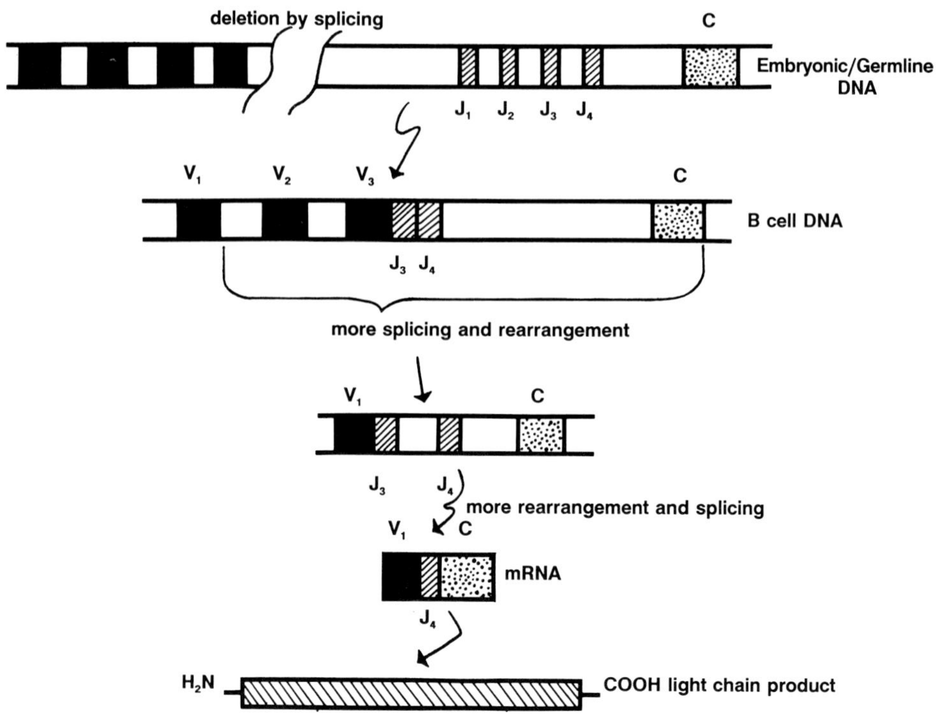

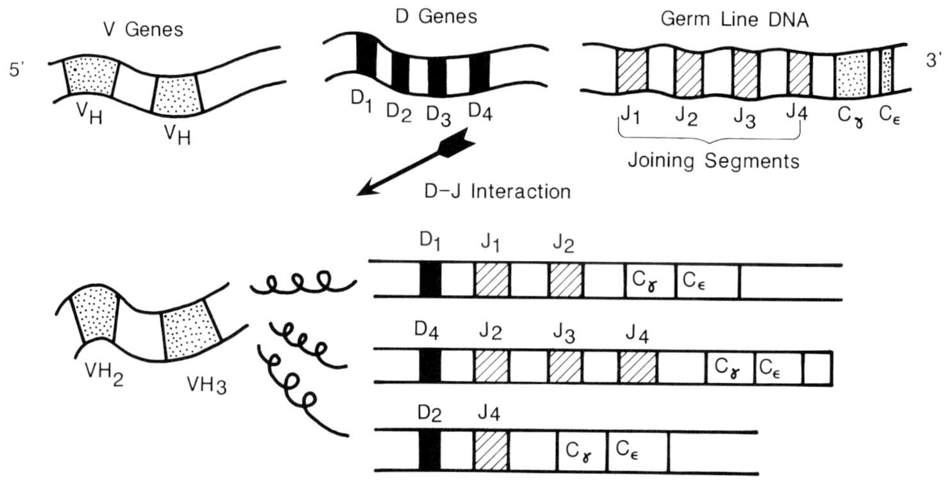

Figure 4.6 The interposition of D segments provides for additional variability within antibody genes. D segments connected to J segments are capable of combining with any variable region gene sequence.

Diversity Segments

In addition to the VJC splicing mechanisms discussed above, a feature that provides still more flexibility in the construction of antibody sequences is related to yet another family of gene segments called diversity (D) sequences. These are tandem multiple gene groups that provide a single D segment between heavy chain V_H and $J_H C_H$ sequences. In a similar fashion to the potential proximity and steric contribution of the J sequences to antibody specificity, the D sequences may also influence the antibody combining site. Diversity segment interposition is shown in Figure 4.6. The D sequences are located near the third hypervariable region of the heavy chain segments.

An interesting aspect of the rearrangement and immunoglobulin gene segment splicing and sorting-out process is that D regions of heavy chains (like the V and J region genes of light chains) have characteristic flanking sequences consisting of seven and nine bases. Similar flanking sequences have also been identified behind heavy chain V regions and just in front of heavy chain J regions. It is thought that such flanking sequences probably serve as built-in signals for various steps that may be essential to the normal process of deletion or splicing. A diagrammatic representation of these flanking sequences is shown in Figure 4.7.

Mutations and Antibody Heterogeneity

B cells can use relatively small numbers of V, J, and D gene segments to produce a vast spectrum of genes programming for antibody molecules. The individual antibody combining site or its specificity is often determined by a conformational three-dimensional structure in which sequences from the amino-terminal variable regions of both heavy and light chains contribute further capacity for diversity. In addition to these intrinsic mechanisms for antibody diversity, an extra dimension of variation is available because of mutations that appear to occur randomly in the

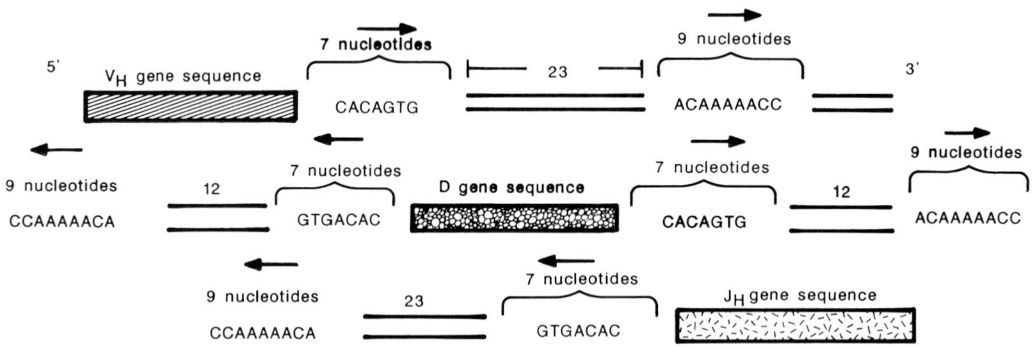

Figure 4.7 Characteristic flanking sequences of seven or nine nucleotides that are found close to V_H, D, and J_H genes. These sequences are thought to play a role in processing and splicing mechanisms.

neighborhood of the V region gene family. Indeed, evidence that such mutations occur was derived from studies of monoclonal myeloma proteins capable of binding to phosphocholine through the V_H and V_L[24,25] or $F(ab')_2$ portions of the IgG molecule. When mRNAs from the myeloma cells producing the monoclonal proteins that bound phosphocholine were studied with respect to RNA coding for light chain variable regions (V_L), it was found that the RNA hybridized only to a single V region DNA. When a number of these V region segments from the mouse myeloma were sequenced, a proportion of them displayed a V region sequence identical to the germ-line V region DNA segment obtained from normal embryonic mouse DNA of the same strain. A smaller proportion of the clonal V region segments showed differences from embryonic or germ-line DNA in one or more than one nucleotide site. These differences were found to be the result of spontaneous mutations. Some mutations quite expectedly produced different amino acid products within the coded polypeptide chains, whereas other mutations appeared to have no effect on the primary amino acid sequence. Interestingly, all such mutations were within 2 kb of the V region gene, affording yet another intrinsic mechanism for antibody diversity.[26–28]

One B Cell Can Make Several Different Ig Classes

Quite early in the studies of B cell production of immunoglobulins, it was recognized that a single antigen-committed B cell or plasma cell could produce immunoglobulin molecules of more than one class—that is, IgM antibody molecules as well as IgG or IgA. In addition, plasma cell myelomas maintained in vitro were capable of switching the class of antibody they produced.[29] One B cell was found to be capable of producing heavy chains of two different classes that still contained the same variable and hence antibody combining site–related sequences.[30] Direct study of the gene sequences in such instances indicated that the V segments, a J sequence, and the two different major immunoglobulin class constant regions (C_μ and C_δ) were situated very close together in the genome. For this reason, the entire gene cluster could be translated directly into an extended RNA segment, followed by selective splicing with elimination of one or the other C_μ or C_δ message. Selective RNA splicing is thus capable of producing two different heavy chain class mRNAs.

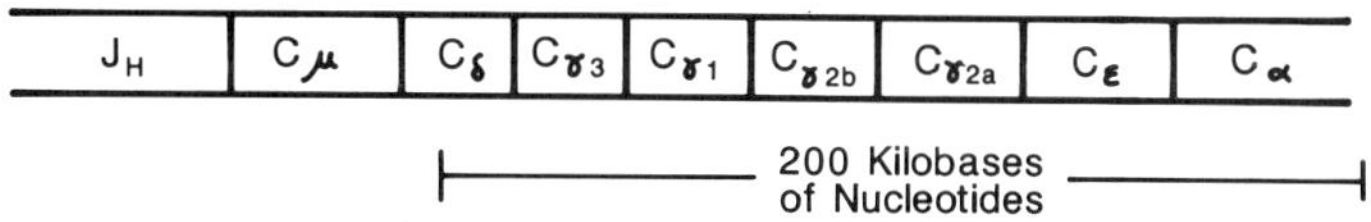

Figure 4.8 Schematic designation of order of constant region (C) genes for Ig heavy chains in mouse. Class switching is frequently accomplished by a mechanism of deletion in which sequences between J_H and C_μ for instance contain sequences that somehow signal the immunoglobulin class switching.

The same or an analogous differential splicing is also involved in selecting genes coding for antibody molecules such as IgM or IgD, which are slightly different in primary structure and length depending on whether they are expressed on B cell membranes or within the B cell cytoplasm.[31,32] In the case of IgM molecules, the difference between the membrane-bound and cytosolic forms of the antibody molecule resides only at the carboxyl-terminal end of the μ chains. Membrane-bound μ chains contain a hydrophobic carboxyl-terminal sequence of amino acids that anchors the molecule within the lipid cell membrane bilamellar structure. On the other hand, μ chains within the B cell cytosol, which will eventually be secreted as humoral antibody, have a strongly hydrophilic carboxyl terminus that facilitates their exitus and ultimate secretion from the cell.[33] Studies of precise gene sequences have indicated that immediately following the J segment nucleotides, the gene contains a group or cluster of eight C region segments beginning with μ for IgM and ending with an α for IgA. The constant region segments encompass a region of more than 200 kb within the DNA and are arranged in an order that is consistent with what is now understood concerning Ig class switching. This immunoglobulin C region cluster or constellation for mouse constant region genes is shown in Figure 4.8.

Participants in Antibody Diversity

The various gene splicing alternatives of V_H, V_K, D, and J nucleotide sequence segments indicate an enormous built-in potential for antibody structure and hence diversity.[33] The protagonists in the final assembled gene include selected sequences from perhaps 100 to 300 V_H and V_K gene segments, selected sequences from 10 or more D segments, and numerous J_H and J_K segments. This large repertoire, the low level of somatic mutations clustered near V region genes, and what is termed junctional diversity (joining of gene segments may occur at a number of places at their ends) all confer characteristics which ensure and preserve antibody diversity.

The cells directly involved in humoral antibody production can be classified into two major groups: B cells or plasma cells, which go through well-characterized controlled activation, maturation, and secretory phases, and memory B cells. The latter cells are an important backup subgroup for the immune system because they are retained and remain capable of reactivation throughout the life of an individual host. Memory B cells retain mRNAs for all previously synthesized antibodies to antigens that have been processed and recognized earlier by the individual host. When the antigen is encountered again, a recall immune reaction or the so-called anamnestic response is quite effective in what might be called an instant replay. Thus, the same but much more antibody product is rapidly produced and secreted.

T Cells and the Cellular Immune Response

The other major arm of the immune system, of course, involves T cells and what is called cell-mediated immunity. This type of immune response is centrally involved in homograft rejection, delayed-type hypersensitivity, as in a positive tuberculin reaction, and many aspects of host defense against potential infectious agents. As an example of the fundamental importance of cell-mediated immune responsiveness and T cells, patients infected with human T lymphotropic virus type III (HTLV-III) or human immunodeficiency virus (HIV), the acquired immunodeficiency syndrome (AIDS) virus, show markedly reduced T cell function, particularly with respect to helper T cell reactivity. When T cell function is severely impaired, many untreated AIDS patients eventually die from superimposed infections against which they cannot mount an effective T cell–mediated immune response.

Like antibody molecules, T cells develop specific antigen receptors that in many ways are analogous to antibody combining sites. Early work comparing the immunochemical and immunological behavior of combining sites on both antibodies and T cells reacting with chemically well-defined homogeneous antigens such as haptens[36–39] indicated that the idiotypes or individual structural features of both types of combining sites might be quite similar.

As more and more precise structural and sequence information has emerged concerning how T cell receptors function and interact, it appears that the genomic nucleotide sequences for many parts of the T cell receptor have striking degrees of homology with their antibody counterparts. An important difference lies in the fact that antibodies exist in the body in two forms, as membrane-bound molecules and free secreted products that circulate. However, T cell receptors appear to exist only on the surface of activated or sensitized T cells. Thus, by their very nature T cell immune responses require interaction with T cell receptors or individual recognition units on T cell membranes.

Much is still to be learned about all of the sequential events and physical interactions that take place as T cells are activated by antigens. In most instances it appears that antigens must be presented or formally served up to T cells by intermediaries called antigen-presenting cells (APCs). In many cases the antigen-presenting cells turn out to be phagocytic monocyte- or macrophage-type cells. A large body of evidence indicates that when antigen-presenting cells serve up a particular antigen to T cells they must do so in juxtaposition to major histocompatability complex (MHC) determinants on the APCs in what might be thought of as a dual recognition event. This is shown diagrammatically in Figure 4.9. During the last several years a great deal of new information has emerged concerning the structure and organization of the T cell receptor, and new insights and refinements are constantly being reported. The physical nature and organization of MHC molecules that play an important role in presenting antigens to T cells have now been extensively defined.[40–42] It appears that the MHC molecules have a distinct cleft 10 Å wide and 20 to 30 Å long into which antigenic peptides can fit, initiating the activation of cell-mediated immunity. How the MHC structure and its antigen- or peptide-binding cleft fit into the basic mechanisms of T cell activation is currently being elucidated at a rapid pace.

T cells themselves have now been extensively divided and subclassified into a number of functional subsets, largely through the application of discrete monoclonal antibodies. Human T helper/inducer (T_H) cells are now recognized using

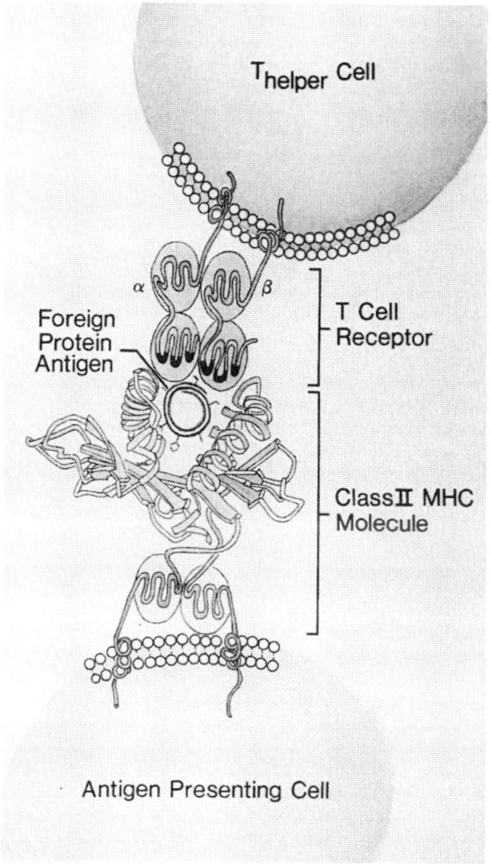

Figure 4.9 Schematic representation of the interaction between the MHC Class II molecule on the antigen presenting cell presenting a foreign protein antigen to the T cell receptor. *Reproduced from Sinha et al Science 1980;248:1381, with permission.*

mouse monoclonals such as OKT4 or Leu-3; human T cytotoxic/suppressor cells can be recognized with OKT8 or Leu-2 on the basis of distinct cell membrane antigens. Helper/inducer OKT4 cells can interact with B cells making humoral antibody to increase antibody production, or they can aid other T cells already committed to specific missions of T cell–mediated immunity, such as homograft rejection or generation of a specific delayed hypersensitivity reaction. In many of these diverse sorts of cell-directed reactions, T cell receptors seem to be involved. Their exact roles and how various parts of the T cell receptor interact have not yet been worked out to the last detail, but much new information about this general type of receptor has been accumulated.

T Cell Receptor Structure

T cell receptors are molecules composed of two chains (α and β) of approximately the same size—40,000 daltons. Carbohydrate residues in these chains are linked through asparagines, and the receptors as a unit form two domains of roughly equal magnitude; one domain is constant and, as in antibody molecules, the other is variable. A schematic model of a generic T cell receptor is shown in Figure 4.10.

Detailed studies of the human T cell receptor[43–45] have now provided extensive biological and immunochemical knowledge about its subunit antigens and their

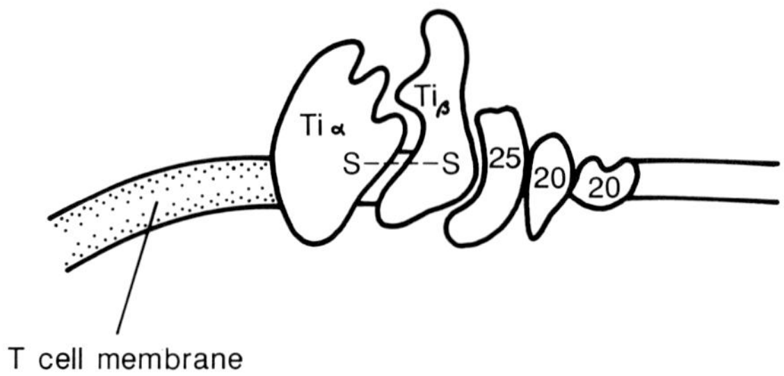

Figure 4.10 Theoretical model of T cell receptor and its component subunits. The Ti_α and Ti_β subunits are stabilized and held together by disulfide bonds; these subunits are closely associated with the 25-kd chain of the T3 molecule. The α and β T cell receptor components are well anchored in the cell membrane via their transmembrane segments. The T3 complex has two additional subunits with molecular masses of 20 kd. *Reproduced from Acuto and Reinherz,*[44] *with permission.*

interactions during various phases of T cell activation. For instance, part of the process of clonal proliferation of T cells in response to a well-defined antigen seems to involve a so-called autocrine facilitating pathway whereby an important autologous lymphokine called interleukin 2 aids in T lymphocyte proliferation.[46,47] As might be expected, the genes involved in expressing immunologic specificity in the T cell receptor in many ways strongly resemble the gene family organization responsible for V genes in B cells.[7] In the primordial or immature fetal thymus, T cell receptor genes exist in embryonic germ-line configuration. As thymic cells develop, the variable T cell receptor sequences rearrange to form parts of the gene segments coding for the α and β chains of the T cell receptor.[48] Like the immunoglobulin genes, the rearranged T cell receptor genes contain variable, diversity, joining, and constant regions. Recognition nucleotide sequences that signal for joining reactions act in a similar fashion to those specifying analogous junction messages in the immunoglobulin gene family. Gene families coding for T cell receptor α and β chains are located on different chromosomes from the immunoglobulin genes. A schematic representation of the structural organization of the T cell receptor β gene locus and its translated product is shown in Figure 4.11.

An entirely different or third gene cluster coding for a T cell receptor γ chain has now been described.[10,49,50] This 35-kd protein is disulfide linked to a 45-kd partner and is also associated with the T3 cell–surface membrane complex. T cell receptor gamma (TCR_γ) genes were identified as early as 1984[51] and were recognized to be composed of V, J, and C region gene segments. The TCR_γ locus was found to be on murine chromosome 13, distinct from the TCR_β locus on chromosome 6 and the TCR_α locus on chromosome 14. In addition, the organization of the TCR_γ locus was completely different from that of the TCR_α or TCR_β locus. Four distinct C region gene segments have been identified along with associated V and J segments,[52] but there is no evidence for diversity (D) gene elements. Human TCR_γ sequences have been shown to constitute a T cell rearranging gene family located on chromosome 7.[53] This locus consists of up to 14 V segments, 5 J segments, and 2 C segments. A number of observations have suggested an important role for TCR_γ early in fetal life, because it is expressed at high levels in early murine fetal thymus development and actually before expression of TCR_α. More recently, a distinct new class of TCR ($TCR_{\gamma\delta}$) has been identified using a series of monoclonal antibodies specific for elements previously shown to react with framework antigens on all lymphocytes bearing αβ T cell receptors.[54] A number of structural features

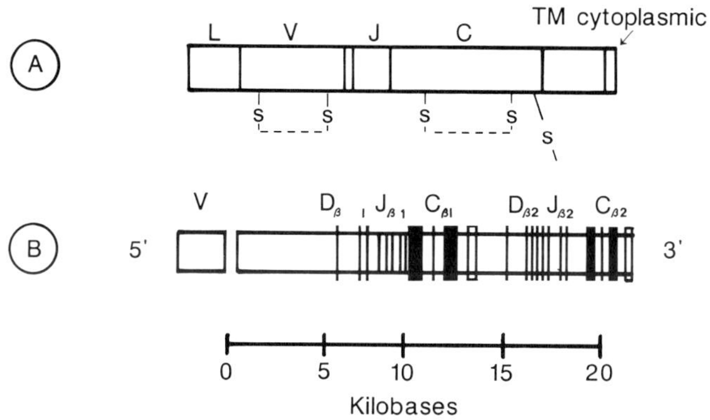

Figure 4.11 *A*, Primary structure of the T cell receptor β chain subunit after translation from complementary DNA. The variable region leader sequence (L) is followed by the variable (V), diversity (D), and joining (J) segments. TM shows a portion of the constant region (C) gene coding for a hydrophobic transmembrane segment that is adjacent to the remaining cytoplasmic (Cy) portion. *B*, Proposed scheme for the order of human T cell receptor β chain genes.

indicate that $TCR_{\gamma\delta}$ receptors are a second major class of T cell receptors quite distinct from the TCRs containing α and β subunits. A striking feature identified quite early was that many lymphocytes expressing $TCR_{\gamma\delta}$ were of double-negative (CD4−, CD8−) phenotype.[54] This population of γδ T cells is distributed in distinct anatomical sites, such as the epidermis[55] and intestine.[56] The full significance of the $T_{\gamma\delta}$ cell population and specific roles of $TCR_{\gamma\delta}$ remain to be defined.[57]

Differences Between T Cell Receptor and Immunoglobulin Genes

A number of intrinsic differences in organization between immunoglobulin and T cell receptor genes have become apparent. The interposition of signal or spacing elements consisting of conserved 7- and 9-nucleotide sequences (see Figure 4.7) generally separated by 12 and 23 nucleotides is different in TCR genes than in Ig heavy chain genes. This allows for potential D-D rearrangements presumably to make a VDDJ exon and V-J rearrangements forming directly.[9,58,59] Apparently, this frequently occurs in T cell receptors but rarely in Ig genes. In addition, there appear to be many more J regions in the α and β chain loci of T cell receptors than there are in immunoglobulin genes. Thus, 11 or 12 functional J_β[60] and 40 or more J_α gene segments have been described.[61,62] This means that the number of possible J_α-J_β combinations ($11 \times 40 = 440$) in T cell receptors would far exceed the possible $J_H - J_K$ combinations ($4 \times 4 = 16$) in immunoglobulin genes.

On the other hand, the T cell receptors have a smaller repertoire of V_β and V_α gene segments than their counterparts in immunoglobulin genes. It appears that as few as 20 V_β sequences are repeatedly used,[63] and though more V_α sequences have been identified[62,64] it seems unlikely that more than 50 V_α genes will eventually be found in any particular genome. This can be contrasted with rough estimates of 100 to 300 for the V_K and V_H immunoglobulin gene sequences identified in the mouse.[65,66] Thus, there would be roughly $15 \times 50 = 750$ possible V_α- V_β combinations for T cell receptor repertoires compared to approximately $100 \times 300 = 30{,}000$ V_H-V_K combinations, providing a 40-fold margin in the opposite direction from what was estimated for J region repertoires. For these reasons, the overall capacity for diversity in terms of combinational pairings for T cell receptor genomic nucleotide sequences is probably about equal to that of the immunoglobulin gene repertoire. The differences lie in more diversity being generated in J region pairings for one and in V region heterogeneity for the other.

One difference between immunoglobulin and T cell receptor genes is that somatic mutation may not be as common a feature leading to diversity in gene expression for T cell receptors as it is in the case of gene segments near V regions in immunoglobulin genes. Several observations indicate a relative lack of frequent somatic mutation in T cell receptor V_β genes[59,61,67] compared with immunoglobulin genes.[35,68] An exception to this general rule is the observation reported by Augustin and Sim,[69] who documented T cell receptor gene mutations in a T cell hybridoma cultured in vitro.

There is yet another difference between the immunoglobulin and T cell receptors in generation of diversity. Careful analysis of nucleotide and deduced amino acid sequences for variable regions of six human Ti_β cDNAs indicates that although the level of homology with V genes of immunoglobulins is high, the Ti_β V regions show a more scattered pattern of hypervariability than the three discrete hypervariable regions of V region immunoglobulin genes.[66,70] The precise degree of β chain variability was reported by Schiffer and co-workers,[71] and two subgroups were identified on the basis of hypervariable V gene subregion analyses.

In addition, there may be differences in T cell receptor hypervariable and framework regions compared with analogous structures in conventional immunoglobulin models. Exact sequence comparisons have indicated that additional hypervariable portions in T cell receptor structures may be formed in a second domain on the outer surface of the T cell receptor.[62,63] Other analyses of V_β sequences and the behavior of their polypeptide chain product[59] have not emphasized additional hypervariability apart from what is considered the probable T cell receptor binding site.

Homology between Immunoglobulin and T Cell Receptor Genes

Despite the differences between immunoglobulin gene behavior and repertoire and the genes coding for the various elements of the T cell receptor, a remarkable general homology exists in the overall schema and three-dimensional projections of the $V_H V_L$ arrangement in immunoglobulin and the $T_\gamma T_\delta$ chain interactions in the T cell receptor. This is seen when the primary gene nucleotide sequences for the VDJC regions of both structures and their actual amino acid sequences are compared. These striking similarities are indicated in Figures 4.12 and 4.13.

Much insight into the biological functioning of important T cell membrane antigens and complex structures such as the T cell receptor has been derived from studies of cultured cell lines that possess complete or partially expressed components of the various membrane complexes. For example, several years ago it was shown that a functionally complete and active T cell receptor could be reconstituted in a mutant human T cell line that lacked a complete β chain mRNA using the technique of transfection, in which a segment of cDNA coding for the complete functional β chain was introduced directly into the mutant T-cell-receptor deficient cell line.[72] In experimental approaches such as this, a facilitator consisting of long terminal repeat (LTR) sequences of the Freund spleen focus–forming virus was used as a transcriptional promoter for the 1.3-kb β chain transcript that was trans-

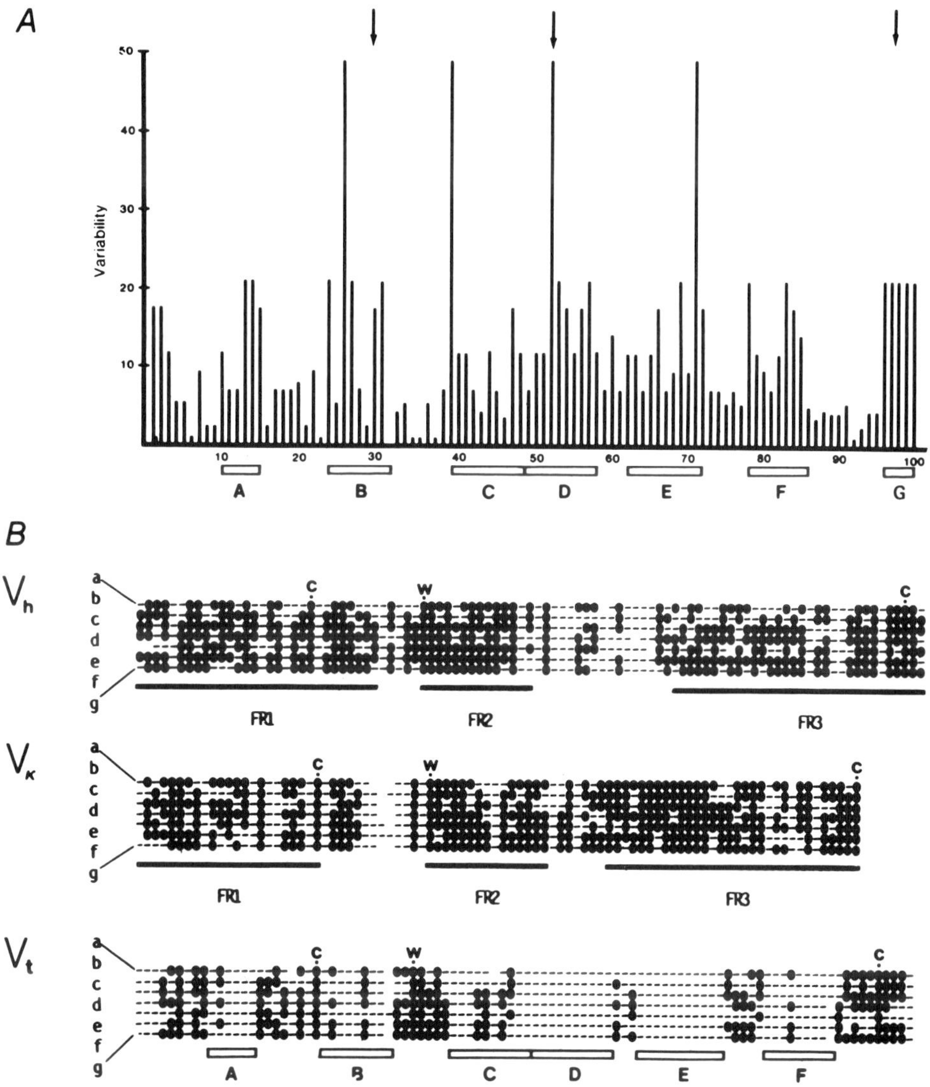

Figure 4.12 Hypervariable and framework regions. *A*, Variability plot of the T cell receptor β chain V region sequences shown in Figure 4.2, using the convention of Wu and Kabat. The positions of the three hypervariable peaks of immunoglobulin light chains are indicated by arrows. Regions of hypervariability are marked by open bars. B, D, and G correspond to the classical immunoglobulin hypervariable regions. *B*, Homologies between a variety of V_H, K_K, and the seven V_β sequences. Homologies of four or better are indicated by a filled circle, homologies of less than four by a dash, and spaces denote gaps inserted to maximize homology. Framework regions of the immunoglobulin sequences are indicated with filled bars, A, C, E, and F are regions of V_β sequence that correspond to conserved V_κ framework residues, but are significantly more variable in V_β. Immunoglobulin heavy chain (V_H): a, mouse group I MOPC 141CL: b, mouse group II MOPC 104E; c, mouse group III TEPC601; d, dog MOO; e, rabbit K25; f, human VH26; g, caiman pC3. Immunoglobulin κ chain (V_K); a, mouse 10K26-12A1; b, mouse S107B CL; c, dog GOM; d, human group 1 ROY; e, human group 111 ITI; f, rabbit K9 335; g, rabbit K25. V_t: a, 86TI; b, 2B4.71; c, YT35; d, HDS11; e, E1; f, LB2; g, Cf. *Reproduced from Patten et al,*[63] *with permission.*

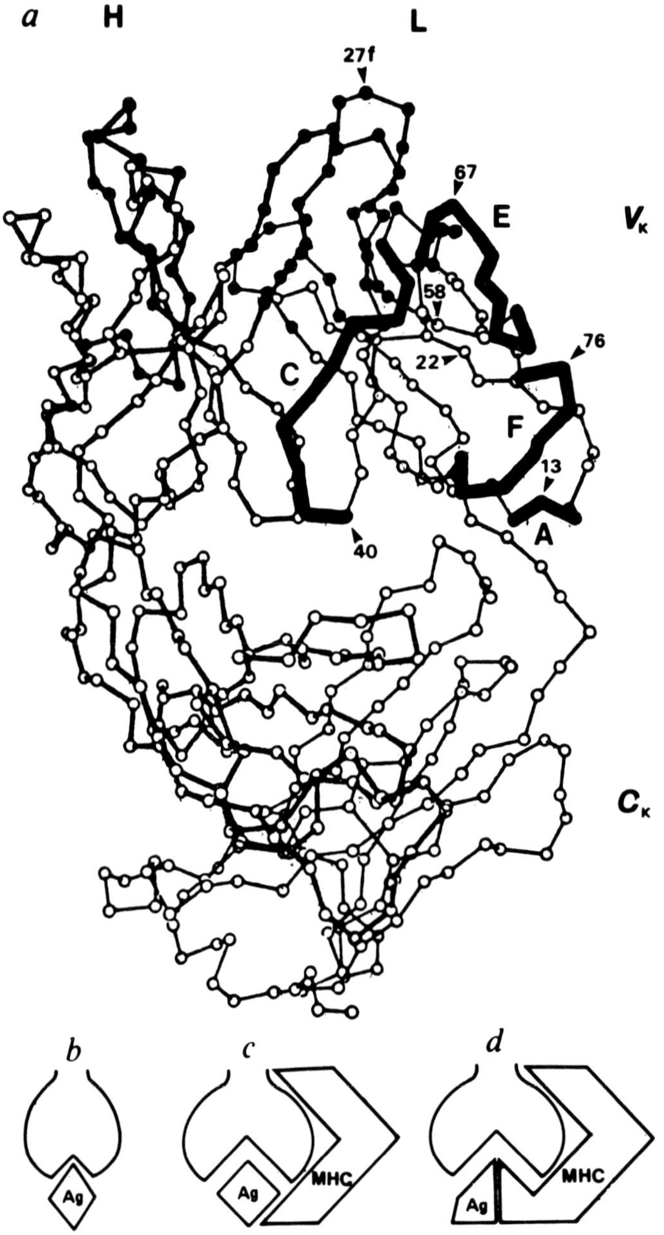

Figure 4.13 Possible T cell receptor structure and implications for recognition. *a,* An α-carbon skeleton of the x-ray crystallographic structure of McPC 603. Thick lines denote heavy chain residues. Filled circles correspond to classical hypervariable residues. The regions of sequence identified in A, C, E, and F that correspond to V_κ framework regions, but are significantly more variable in V_βS, are shown as solid bars. *b,* Schematic diagram of antibody-antigen (Ag) interaction. *c* and *d,* Models of T cell receptor binding. *Reproduced from Patten et al.*[63] *with permission.*

fected. In this instance, transfection reconstitution of the defective mutant T cell line completely restored a functionally active T3-Ti T cell receptor complex.

Genes Related to the Major Histocompatability Loci

For some time it has been recognized that the major histocompatability loci have a profound influence on homograft rejection and on many other aspects of the immune response. Genes coding for histocompability antigens are fundamental to the body because they regulate the mechanisms whereby foreign tissues are rejected or simply serve as markers for distinguishing not self from self. As knowledge of the complexity and relative organization of various types of transplantation antigens has evolved, it has become clear that the MHC includes genes that code for three major groups of histocompatability protein products.[73–75] Class I antigens include what are referred to as H-2 and human leukocyte antigen (HLA) transplantation antigens in the mouse and human systems, respectively. These class I MHC proteins are expressed on almost all cell membranes. Class II MHC antigens are products of a similar but distinct group of genes that code for cell surface molecules somehow linked to the level and selectivity of the immune response.[76] Initially, these gene products were referred to as Ia or Ia-like in the mouse and human. Class II MHC gene products show a much more selective distribution and occur on B lymphocytes, antigen-processing cells such as monocytes or macrophages, Langerhans cells in the skin, and various categories of immune cells following activation. Thus, activated T cells of helper or suppressor type may show cell membrane class II antigen expression. In addition, class II MHC antigens have now been described in a variety of human and animal tumors, including melanomas and other primary cancers. Finally, class III MHC gene products are represented by plasma protein components of the complement system.

The genes coding for MHC protein products, like the immunoglobulin and T cell receptor genes, were more clearly understood after their translated protein products were carefully characterized and identified. Class I and class II MHC proteins in the mouse are anchored inside the cell at their carboxyl terminus, extend through the cell membrane lipid bilayer, and project outside the cell expressing three major general domains of approximately 90 amino acids. The transmembrane protein sequence includes about 40 amino acids, and the anchor end within the cell cytoplasm encompasses approximately 30 amino acids. Attached to each class I polypeptide chain is a 12,000-dalton component (β_2-microglobulin) that is produced by an entirely separate gene on chromosome 2 in the mouse. This general format is shown in Figure 4.14. The actual structure shown is produced by separate exons coding a leader peptide, each of the three outermost domains, and the constant regions. In the mouse the H-2 genes are present on chromosome 17, whereas in humans the HLA genes occupy the short arm of chromosome 6. In humans, class II MHC proteins are composed of two glycosylated polypeptide chains of molecular weights 33,000 (α) and 28,000 (β) (Figure 4.15). Immunochemical, serologic, DNA sequencing, and monoclonal antibody techniques have led to the subdivision of human class II proteins into three separate antigenic groups called DP, DQ, and DR, and it appears that each chromosome 6 has six α genes, including two each for DP and DQ antigens, one for DR, and one additional as yet undesignated nucleotide sequence. The relative arrangements of the major human histocompatability genes on chromosome 6 are shown schematically in Figure 4.16).

A number of genes from the human MHC have been cloned, including large

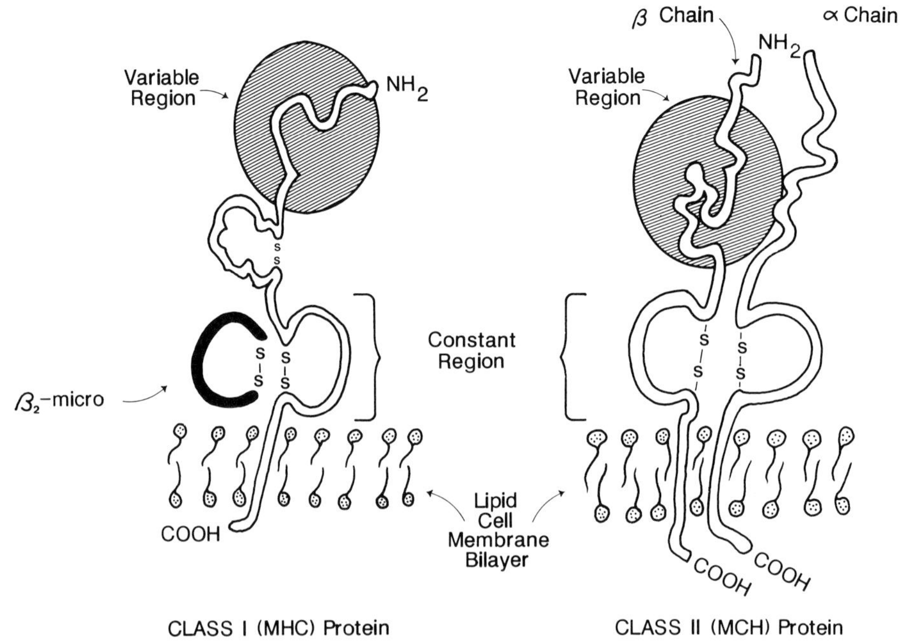

Figure 4.14 Major histocompatibility protein gene products. Both class I (H-2 or HLA) and class II (Ia or DR) molecules show constant and variable regions somewhat analogous to the organization of immunoglobulin gene product structure. Class I MHC proteins are associated with β_2-microglobulin.

Figure 4.15 Class II MHC molecules in the human consist of two chains, α (34,000 daltons) and β (29,000 daltons). The carboxyl ends of each chain are anchored within the cell lipid bilayer. Both proximal portions of the α and β chains correspond to constant or C regions, whereas the terminal amino ends of the chains show structural variability.

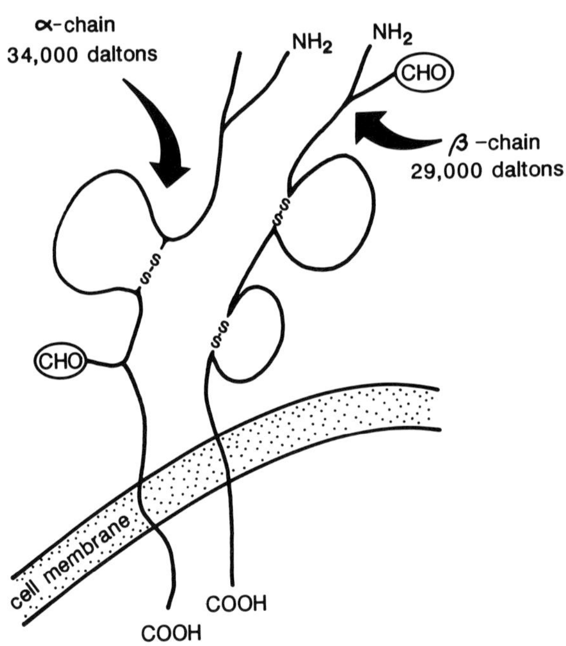

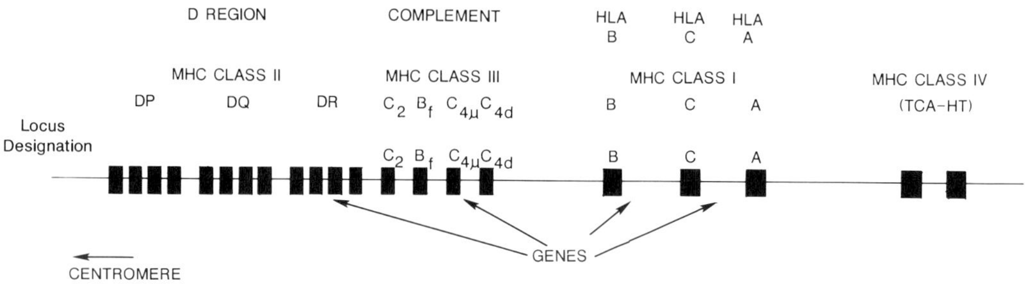

Figure 4.16 Relative locations of MHC class I, II, and III molecules on human chromosome 6. The class IV regions shown at the far right are less well defined; their protein products are present on thymus cells, a subpopulation of T cells, and some leukemic phenotypes.

sequences representing many of the major HLA-A and HLA-B antigenic polypeptides.[77–79] In addition, it has become possible to study the cellular interactions and possible functions of some of the MHC polypeptide chain products by transfecting mouse or human cells with gene sequences capable of coding for the synthesis and appearance of new cell surface MHC antigens.

The Superfamily of Immunoglobulin-Related Proteins

From the discussions already presented, it is clear that a number of interacting systems and their basic genes within the immunoglobulin–T cell receptor and MHC categories have a large degree of structural homology to one another. Immunoglobulin genes are programmed to produce molecules with built-in first, second, and third orders of diversity of way of their V, J, and D regions. Similarly, T cell receptors employ an analogous organizational protocol to rearrange gene segments providing for variable, joining, diversity, and backbone constant region gene segments. The structural organization of MHC products such as class I or class II HLA antigens is based on a roughly analogous series of exons with intervening sequences capable of producing variable as well as constant regions in line with individual HLA specificities. Since the present human immune system is undoubtedly the product of billions of years of evolutionary pressure, Hood and co-workers[80] have proposed the concept of superfamilies of genes—particularly as applied to what has been called the superfamily of immunoglobulin-related proteins. Such a possible evolutionary format is shown in Figure 4.17. It is interesting that the same evolutionary hypothesis may also be applied to many other extensively studied human or mammalian biological systems, including the entire complement component cascade, many of the complex series of interacting proteins making up the blood coagulation system, and many of the endocrine hormones. With respect to the mechanisms by which certain amino acid sequences are conserved and persist in living organisms, a number of remarkable findings have emerged. Polypeptide hormone or hormone precursors have now been identified in unicellular organisms. For example, LeRoith and co-workers[81] have provided evidence that unicellular parasites or microorganisms contain genes coding for ACTH- or β-endorphin–like molecules. Likewise, insulin or insulin-like proteins can be produced by various bacteria, fungi, or protozoans.[81,82] Such findings suggest that polypeptide or protein products containing certain unique sequences might be adaptable to a number of key biological roles and that such sequences may therefore be conserved by ne-

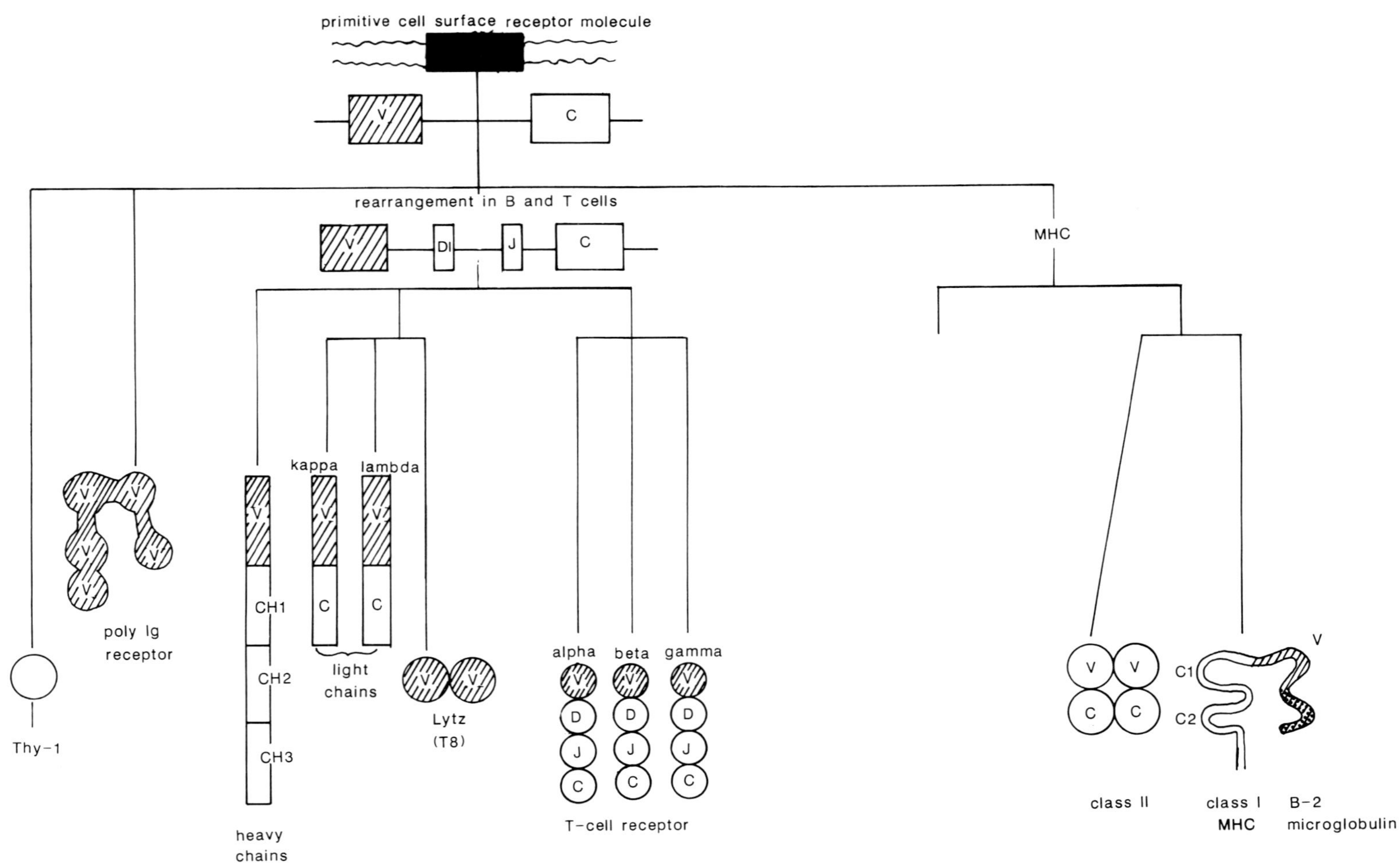

Figure 4.17 Schematic of the superfamily of immunoglobulin-related genes and their protein products. The particular genes coding for the final protein products shown below may have evolved from a single common ancestral gene. *Adapted from Hood et al,*[80] *with permission.*

cessity during evolution—they may contain the polypeptide equivalent of a "jack of all trades." The remarkable conservation of such sequence homologies between primitive forms such as protozoans and bacteria and essential mammalian and human proteins also suggests that such homologies could be involved in various autoimmune diseases. For instance, the cross-reactions between various group A streptococcal components and human tissues including the brain, heart, and heart valves[81–86] implicated in the molecular mimicry of a disorder such as acute rheumatic fever[87] could be related to evolutionary pressures producing common determinants in bacteria and humans. Much is still to be learned about the implications of supergene families and the heterogeneity of their final polypeptide chain products.

References

1. Burnet FM: A modification of Jerne's theory of antibody production using the concept of clonal selection. *Aust J Sci* 1957; 20:67–69.
2. Edelman GM: The structure and function of antibodies. *Sci Am* 1970;223:34–42.
3. Davies DR, Metzger H: Structural basis of antibody function. *Annu Rev Immunol* 1983;1:87–117.
4. Loh DR, Bothwell LM, White-Scharf M, et al: Molecular basis of a mouse strain-specific anti-hapten response. *Cell* 1983;33:153–159.
5. Davies D, Padlan E, Segal D: Three dimensional structure of immunoglobulins. *Annu Rev Biochem* 1975;44:639–667.
6. Poljak R: Three-dimensional structure, function and genetic control of immunoglobulins. *Nature* 1975;256:373–376.
7. Yanagi Y, Yoshikai Y, Leggett K, et al: A human T cell–specific cDNA clone encodes a protein having extensive homology to immunoglobulin chains. *Nature* 1984;308:145–149.
8. Hedrick SM, Cohen DI, Neilsen EA, Davis MM: Isolation of cDNA clones encoding T cell–specific membrane-associated proteins. *Nature* 1984;308:149–153.
9. Chien Y-H, Becker DM, Lindsten T, et al: A third type of T cell receptor gene. *Nature* 1984;312:31–35.
10. Saito H, Kranz DM, Takagaki Y, et al: A third rearranged and expressed gene in a clone of cytotoxic T lymphocytes. *Nature* 1984;312:36–40.
11. Sim GK, Yague J, Nelson J, et al: Primary structure of a human T cell receptor α chain. *Nature* 1984;312:771–775.
12. Metzger H, Wofsy L, Singer S: The participation of A and B polypeptide chains in the active sites of antibody molecules. *Proc Natl Acad Sci USA* 1964;51:612–618.
13. Benjamin DC, Berzofsky JA, East IJ, et al: The antigenic structure of proteins: A reappraisal. *Annu Rev Immunol* 1984;2:67–101.
14. Jemmerson R, Paterson Y: Mapping epitopes on a protein antigen by the proteolysis of antigen-antibody complexes. *Science* 1986;232:1001–1004.
15. Cunningham AJ: *The Generation of Antibody Diversity. A New Look.* London, Academic Press, 1976.
16. Kohler G: Frequency of precursor cells against the enzyme β-galactosidase: An estimate of the BALB/c strain antibody repertoire. *Eur J Immunol* 1976;6:340–347.
    Kabat E: Origins of antibody complementarity and specificity—hypervariable regions and the minigene hypothesis. *J Immunol* 1980;125:961–969.
17. Burnens A, Demotz S, Corradin G, et al: Epitope mapping by chemical modification of free and antibody-bound protein antigen. *Science* 1987;235:780–782.
18. Dreyer WJ and Bennett JC: The molecular basis of antibody formation: A paradox. *Proc Natl Acad Sci USA* 1965;54:864–869.
19. Hozumi N, Tonegawa S: Evidence for somatic rearrangement of immunoglobulin genes coding for variable and constant regions. *Proc Natl Acad Sci USA* 1976;73:3628–3632.
20. Tonegawa S, Brack C, Hozumi N, Pirrotta V: Organization of immunoglobulin genes. *Cold Spring Harbor Symp Quant Biol* 1978;42:921–931.

21. Brack C, Hirama M, Lenhard-Schuller R, Tonegawa S: A complete immunoglobulin gene is created by somatic recombination. *Cell* 1978;15:1–14.
22. Seidman J, Leder P: The arrangement and rearrangement of antibody genes. *Nature* 1978;276:790–795.
23. Sakano H, Huppi K, Heinrich G, Tonegawa S: Sequences at the somatic recombination sites of immunoglobulin light-chain genes. *Nature* 1979;280:288–294.
24. Gearhart PJ, Johnson ND, Douglas R, Hood LE: IgG antibodies to phosphorylcholine exhibit more diversity than their IgM counterparts. *Nature* 1981;291:29–34.
25. Gearhart P, Bogenhagen D: Clusters of point mutations are found exclusively around rearranged antibody variable genes. *Proc Natl Acad Sci USA* 1983;80:3439–3443.
26. Baltimore D: Somatic mutation gains its place among the generators of diversity. *Cell* 1981;26:295–296.
27. Selsing E, Storb U: Somatic mutation of immunoglobulin light-chain variable region genes. *Cell* 1981;25:47–58.
28. Gorski J, Rollini P, Mach B: Somatic mutations of immunoglobulin genes are restricted to the rearranged V gene. *Science* 1983;220:1179–1181.
29. Davis M, Kim S, Hood L: DNA sequences mediating class switching in α-immunoglobulins. *Science* 1980;209:1360–1365.
30. Sledge C, Fair DS, Black B, et al: Antibody differentiation: Apparent sequence identity between variable regions shared by IgA and IgG immunoglobulins. *Proc Natl Acad Sci USA* 1976;73:923–927.
31. Ravetch J, Siebenlist U, Korsmeyer S, et al. Structure of the human immunoglobulin mu locus: Characterization of embryonic and rearranged J and D genes. *Cell* 1981;27:583–591.
32. Cheng H, Blattner F, Fitzmaurice L, et al. Structure of genes for membrane and secreted murine IgD heavy chains. *Nature* 1982;296:410–415.
33. Early P, Rogers J, Davis M, et al: Two mRNAs can be produced from a single immunoglobulin α gene by alternative RNA processing pathways. *Cell* 1980;20:313–319.
34. Huang H, Hood L: Origins of antibody diversity, *in* Messer R, Porter IH (eds): *Recombinant DNA and Medical Genetics*. New York, Academic Press, 1983, pp 97–104.
35. Tonegawa S: Somatic generation of antibody diversity. *Nature* 1983;302:575–581.
36. Jerne NK: Toward a network theory of the immune system. *Ann Immunol (Paris)* 1974;125C:373–389.
37. Harvey MA, Adorini L, Miller A, Sercarz E: Lysozyme-induced T suppressor cells and antibodies have a predominant idiotype. *Nature* 1979;281:594–596.
38. Rowley DA, Kohler H, Cowan JD: An immunologic network. *Contemp Top Immunobiol* 1980;9:205–230.
39. Jerne NK, Roland J, Cazenave PA: Recurrent idiotypes and internal images. *EMBO J* 1982;1:243–247.
40. Bjorkman DJ: Structure of the human class I histocompatibility antigen HLA-A2. *Nature* 1987;329:506–512.
41. Bjorkman DJ: The foreign antigen binding site and T cell recognition regions of class I histocompatibility. *Nature* 1987;329:512–518.
42. Buus S, Sette A, Colon SM, et al: The relation between major histocompatibility complex (MHC) restriction and the capacity of Ia to bind immunogenic peptides. *Science* 1987;235:1353–1358.
43. Marx JL: Structure of MHC protein solved. *Science* 1987;238:613–614.
44. Acuto O, Reinherz EL: The human T cell receptor structure and function. *N Engl J Med* 1985;312:1100–1111.
45. Acuto O, Meuer SC, Hodgdon JC, et al: Peptide variability exists within the α and β subunits of the T cell receptor for antigen. *J Exp Med* 1983;158:1368–1373.
46. Meuer SC, Hussey RE, Hodgdon JC, et al: Surface structures involved in target recognition by human cytotoxic T lymphocytes. *Science* 1982;218:471–473.
47. Smith KA, Baker PE, Gillis S, Ruscetti FW: Functional and molecular characteristics of T cell growth factor. *Mol Immunol* 1980;17:579–589.
48. Royer HD, Acuto O, Fabbi M, et al: Genes encoding the Ti β subunit of the antigen/MHC receptor undergo rearrangement during intrathymic ontogeny prior to surface T3–Ti expression. *Cell* 1984;39:261–266.
49. Garman RD, Doherty PJ, Raulet DH: Diversity, rearrangement, and expression of murine T cell gamma genes. *Cell* 1986;45:733–742.

50. Lew AM, Pardoll DM, Maloy WL, et al: Characterization of T cell receptor gamma chain expression in a subset of murine thymocytes. *Science* 1986;234:1401–1405.
51. Saito H, Kranz DM, Takagaki Y, et al: Complete primary structure of a heterodimeric T cell receptor deduced from cDNA sequences. *Nature* 1984;309:759–762.
52. Iwamoto A, Rupp F, Ohashi PS, et al: T cell-specific γ genes in C57 BL/10 mice. *J Exp Med* 1986;163:1203–1212.
53. Rabbitts TH, Le Franc M-P, Stinson MA, et al: The chromosomal location of T cell receptor genes and a T cell rearranging gene: Possible correlation with specific translocations in human T cell leukemia. *EMBO J* 1985;4:1461–1465.
54. Brenner MB, McLean J, Dialynas DP, et al: Identification of a putative second T cell receptor. *Nature* 1986;322:145–149.
55. Konning F, Stingl S, Yokoyama WM, et al: Identification of a T3-associated γδ T cell receptor on Thy-1+ dendritic epidermal cell lines. *Science* 1987;236:834–836.
56. Goodman T, Le Francois L: Expression of the γ-δ T cell receptor on intestinal CD8+ intraepithelial lymphocytes. *Nature* 1988;333:855–858.
57. Brenner MB, Strominger JL, Krangel MS: The γδ T cell receptor. *Adv Immunol* 1988;43:133–192.
58. Davis MM: Molecular genetics of the T cell-receptor beta chain. *Annu Rev Immunol* 1985;3:537–560.
59. Barth RK, Kim BS, Lan NC, et al: The murine T-cell receptor uses a limited repertoire of expressed V_β gene segments. *Nature* 1985;316:517–523.
60. Gascoigne NRJ, Chien J, Becker DM, et al. Genomic organization and sequence of T cell receptor β-chain constant- and joining-region genes. *Nature* 1984;310:387–391.
61. Hayday AC, Diamond DJ, Tanigawa G, et al: Unusual organization and diversity of T cell receptor α-chain genes. *Nature* 1985;316:828–832.
62. Becker DM, Patten P, Chien J, et al: Variability and repertoire size of T cell receptor V_α gene segments. *Nature* 1985;317:430–434.
63. Patten P, Yokota T, Rothbard J, et al. Structure, expression and divergence of T cell receptor β-chain variable regions. *Nature* 1984;312:40–46.
64. Arden B, Klotz JL, Siu G, Hood L: Diversity and structure of genes of the α family of mouse T cell antigen receptor. *Nature* 1985;316:783–787.
65. Seidman JG, Leder A, Nau M, et al: Antibody diversity. The structure of cloned immunoglobulin genes suggests a mechanism for generating new sequences. *Science* 1978;202:11–17.
66. Broduer PH, Riblet R: The immunoglobulin heavy chains variable region (Igh-V) locus in the mouse. I. One hundred Igh-V genes comprise seven families of homologous genes. *Eur J Immunol* 1984;14:922–930.
67. Chien J, Gascoigne NRJ, Kavaler J, et al: Somatic recombination in a murine T cell receptor gene. *Nature* 1984;309:322–326.
68. Griffiths GM, Berek C, Kaartinen M, Milstein C: Somatic mutation and the maturation of immune response to 2-phenyl oxazonone. *Nature* 1984;312:271–275.
69. Augustin AA, Sim GK: T cell receptors generated via mutations are specific for various major histocompatability antigens. *Cell* 1984;39:5–12.
70. Leiden JM, Strominger JL: Generation of diversity of the β chain of the human T-lymphocyte receptor for antigen. *Proc Natl Acad Sci USA* 1986;83:4456–4460.
71. Schiffer M, Wu TT, Kabat EA: Subgroups of variable region genes of β chains of T cell receptors for antigen. *Proc Natl Acad Sci USA* 1986;83:4461–4463.
72. Ohashi PS, Mak TW, Van den Elsen P, et al: Reconstitution of an active surface T3/T cell antigen receptor by DNA transfer. *Nature* 1985;316:606–609.
73. Hood L, Steinmetz M, Malissne B: Genes of the major histocompatability complex of the mouse. *Annu Rev Immunol* 1983;1:529–568.
74. Klein J: The major histocompatability complex of the mouse. *Science* 1979;203:516–521.
75. Malissen M, Malissen B, Jordan BR: Exon/intron organization and complete nucleotide sequence of an HLa gene. *Proc Natl Acad Sci USA* 1982;79:893–897.
76. Shackelford DA, Kaufman JF, Korman AJ, Strominger JL: HLA-DR antigens: Structure, separation of subpopulations, gene cloning and function. *Immunol Rev* 1982;66:133–187.
77. Das HK, Duceman B, Sood AK, et al: Molecular studies of the genes of the major histocompatability complex, *in* Caskey CT, White RL (eds): *Banbury Report 14: Recom-*

binant DNA Applications to Human Disease. Cold Spring Harbor, NY, Cold Spring Harbor Press, 1983, pp 41–51.

78. Marx JL: A close look at the genes of the MHC. *Science* 1983;220:937–939.
79. Seidman JG, Evans GA, Ozato K, et al: The genes encoding the major histocompatability antigens, *in* Messer A, Porter IH (eds): *Recombination DNA and the Medical Genetics.* New York, Academic Press, 1983, pp 105–115.
80. Hood L, Kronenberg M, Hunkapllar T: T cell antigen receptors and the immunoglobulin supergene family. *Cell* 1985;40:225–229.
81. LeRoith D, Liotta AS, Roth J, et al: Corticotropin and β-endorphin-like materials are native to unicellular organisms. *Proc Natl Acad Sci USA* 1982;79:2086–2090.
82. LeRoith D, Shiloach J, Roth J, Lesniak MA: Insulin or a closely related molecule is native to *Escherichia coli. J Biol Chem* 1981;245:6533–6536.
83. Kaplan MH: Immunologic relation of streptococcal and tissue antigens. I. Properties of an antigen in certain strains of group A streptococci exhibiting an immunologic cross-reaction with human heart tissue. *J Immunol* 1963;90:595–606.
84. Zabriskie JB, Freimer EH: An immunological relationship between the group A streptococcus and mammalian muscle. *J Exp Med* 1966;124:661–678.
85. Husby G, Van de Rijn I, Zabriskie JB, et al: Antibodies reacting with cytoplasm of subthalamic and caudate nuclei neurons in chorea and acute rheumatic fever. *J Exp Med* 1976;144:1094–1110.
86. Goldstein I, Rebeyrotte P, Parlebas J, Halpern B: Isolation from heart valves of glycopeptides which share immunological properties with *Streptococcus haemolyticus* group A polysaccharides. *Nature* 1968;219:866.
87. Williams RC Jr: Molecular mimicry and rheumatic fever, *in* Williams RC (ed): *Clinics in Rheumatic Diseases.* London, W. B. Saunders, 1985;573–590.

Tumor Viruses and Oncogenes

5

Ralph C. Williams, Jr., MD

Studies related to several tumor viruses such as simian virus 40 (SV40), polyomavirus, and certain adenoviruses provided some of the earliest insights into the molecular biology of the neoplastic process. The comparatively small size of SV40 (5243 base pairs) and polyomavirus (5292 base pairs) provided relatively small length segments of DNA for study, fortunately at a time in the evolution of basic DNA analysis techniques when the number and variety of restriction enzymes were somewhat limited. Digestion with restriction enzymes such as *Eco*RI produced relatively limited numbers of DNA fragments, which could be separated by electrophoresis on agarose gels, and separate gene segments that were involved in production of early- and late-transforming proteins as well as other distinct features of the viral replication cycle could be readily identified.

Perhaps one of the most fundamental insights into how viruses might be involved in the process of transforming cells or inducing neoplasia came from detailed studies of retroviruses (RNA viruses containing reverse transcriptase). Using restriction enzymes and cloning approaches, it rapidly became possible to prepare cDNA replicas of retrovirus RNA and to begin a careful analysis of how retroviral nucleotide segments—either proviral DNA or viral RNA itself—could initiate a malignant self-replicating process. This phenomenon could then be examined in vitro.

Pioneering experiments by Gross, Huebner, and others[1–3] indicated that what appeared to be infectious leukemia and sarcoma retroviruses could be demonstrated in murine tumors induced by radiation or chemical carcinogens. These observations suggested that latent tumor viruses might be present in normal cells and that under certain circumstances, when the cellular milieu was perturbed, such latent viruses might be activated in a manner similar to bacteriophage. In parallel, Huebner and Todaro[3] suggested that unexpressed or relatively silent oncogenes might be present in all normal cells and that activation of these relatively silent genes by retroviruses or carcinogenic substances might produce cancer. This proposal subsequently became known as the oncogene theory or hypothesis. After the discovery of reverse transcriptase, Temin[4] proposed that normal cells might contain potential (proto) oncogenes, which could become viral or cellular oncogenes after either somatic mutation or retroviral cellular reverse transcription. The protovirus hypothesis suggested by Temin held that a qualitative rather than a quantitative alteration was necessary to change normal cellular genes to oncogenes. In the early 1970s, the first viral oncogene, *src*, was definitively characterized using *src* deletion mutants.[5–7] At present, the characteristic features of *src* and all other retroviral oncogenes are their specific nucleotide sequences, which seem to be unrelated to essential viral genes. Because they do not appear to be essential for viruses, they could in a sense be considered as vestigial genetic elements. It is thought that viral oncogenes may have originated from rare transductions of normal cellular nucleotide sequences by the action of retroviruses without oncogenes. A number of cellular genes in normal mammalian cells contain oncogene-related sequences and have repeatedly been shown to be involved in a wide spectrum of normal cell functions. Before we examine how oncogenes may be involved in a range of both normal as well as abnormal cellular activation events, it is important to focus briefly on tumor viruses and how their DNA integrates within DNA of the host cells.

Tumor Virus Integrates into Normal Host DNA

Cells transformed by tumor viruses such as SV40 show viral DNA that has been integrated into autologous host DNA.[8,9] When the tumor viral DNA has been intercalated, its expression produces the process of cell transformation and neoplastic cell properties. It now appears that the integration process itself may be random, since viral DNA insertion does not appear to be limited to any particular viral or host sites or sequences. When such integration events occur, nucleotide sequence analysis may show unusual rearrangements of original viral sequences. Also, integration of viral sequences may encompass replication of DNA or DNA segments being transcribed into RNA with the result that the final products may in some instances contain small segments of single-stranded DNA.

Tumor Proteins

Early in the studies of tumor viruses it was thought that nucleotide sequences in viruses such as SV40 or polyoma contained codes for an important tumor-related protein that appeared to be essential for promoting cancerous cell growth.[10,11] Studies primarily with viral mutants indicated that these tumor proteins had an important influence on initiation of viral DNA replication, as well as being essential for continued malignant transformation of affected cells.[12–14] During these early tumor virus studies it was discovered that RNA splicing occurred in the course of adenovirus replication. Thus primary RNA being transcribed could be spliced in such a way that a high molecular weight (100,000 kd) large tumor (T) antigen was produced.[15] This T antigen was localized to the nucleus and bound to the start site of SV40 DNA replication. An alternative type of RNA splicing produced a small t protein of 18,000 kd that was present in the cell cytoplasm. These two mechanisms and ultimate RNA splicing results are shown in Figure 5.1. We still do not fully understand the precise roles of the large T and small t proteins. This has been addressed by introducing mutations at various sites in the SV40, polyoma, or adenovirus DNA that would in turn induce formation of slightly altered T proteins.[16] Currently it is thought that one class of T proteins somehow provides the

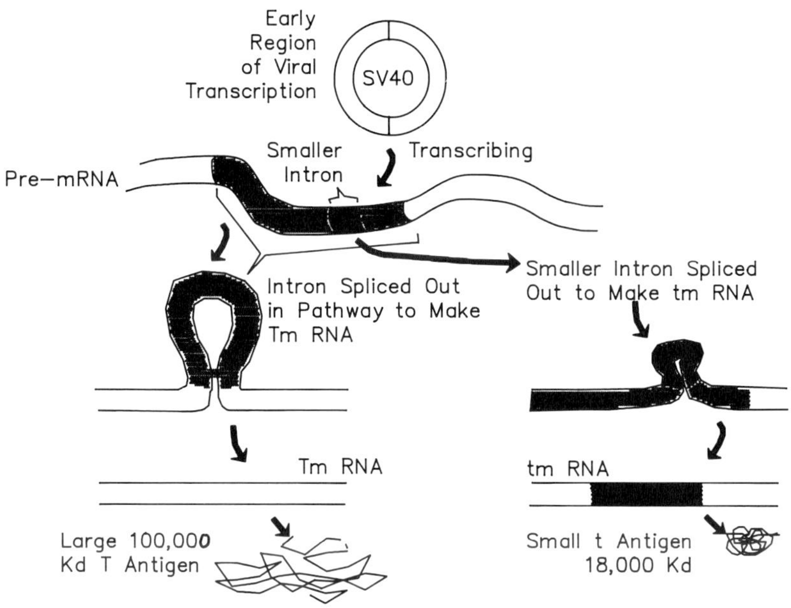

Figure 5.1 Diagram indicating two different splicing pathways involved in making large T or small t antigen. The large intron spliced out to make mRNA for the large T antigen is shown to the left; the smaller intron on the right is spliced out to make small t antigen.

cell with resources so that it can multiply without internal controls, whereas other types of cytoplasmic t proteins on the inside of the cell membrane may function as oncogene products.

Modulation of Different Phases of Polyoma mRNA Synthesis

RNA polymerase II is directly involved in the production of RNA polynucleotide in both the early and late stages of synthesis. However, the precise signals which initiate the actions are different. Early viral SV40 or polyoma mRNA start-up signals are generally immediately preceded by TATA and CAT sequences.[17,18] These latter signals are thought to be related not to the frequency of initiation of synthesis but to the exact nucleotide sites where synthesis of early viral mRNA will commence. By contrast, late viral mRNA synthesis starts after replication of the basic viral SV40 or polyoma templates has begun.[19] There is considerably more variability in nucleotide start-up sites in late mRNA chains. Actually, instead of being limited to a fixed set of 10 joined nucleotides, as is the case with early viral mRNA initiation, the late mRNA can start within a stretch of several hundred nucleotides. This broader range in initiation sites for late viral mRNAs may parallel the relative lack of TATA- or CAT-like sequences in this part of the nucleotide chain. The initial phase in late mRNA synthesis may be related to binding of RNA polymerase II to well-defined points on single-stranded DNA immediately next to DNA replication forks or branching regions. The interrelationship of nucleotide sequences involved in control of SV40 or polyomavirus replication is illustrated in Figure 5.2. The region of about 100 base pairs essential for origin of replication contains a stretch of immediately sequential AT base pairs and a number of inverted repeat sequences that can form hairpin loops if denatured into single strands. The three nucleotide

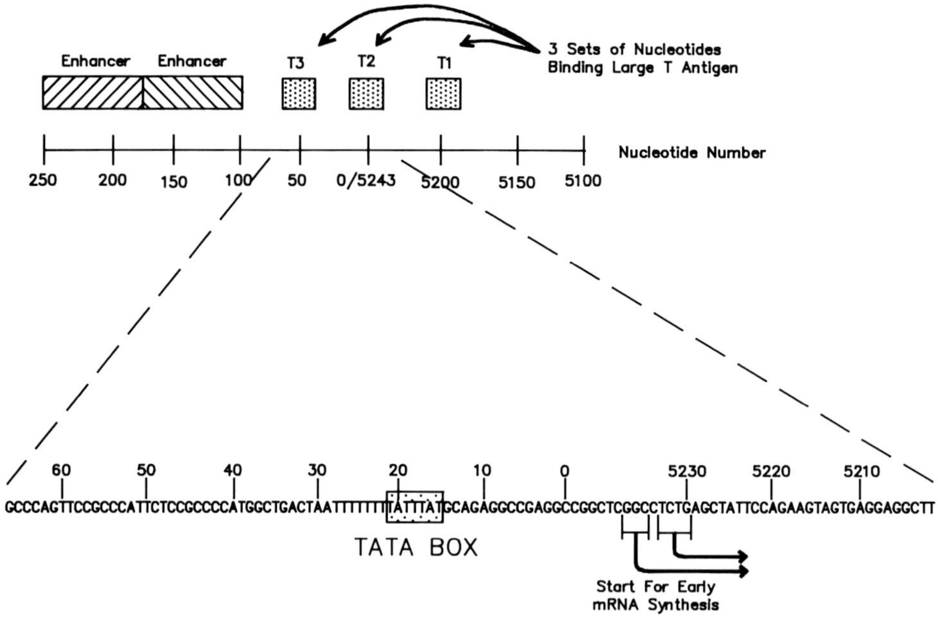

Figure 5.2 Diagrammatic representation of nucleotide sequence of SV40 at its origin or replication. The TATA box controlling the initiation points for early mRNA synthesis is outlined. T1, T2, and T3 represent the three sets of nucleotides that bind the large T antigen. The starting points for early mRNA are shown as brackets.

sequences (T3, T2, and T1) binding large T antigen[20] are also shown. When this entire 100-bp segment or key parts of it are removed by restriction enzymes and put into another site within the SV40 DNA, they can initiate replication at the new site as well.

Anatomy of Retroviruses

The discovery that RNA tumor viruses replicated by making an intermediary DNA copy through the action of reverse transcriptase was followed by the discovery that RNA tumor viruses themselves actually consist of doublets of RNA chains of 8 to 10 kilobases (kb). It is still not clear why the RNA viral genomes exist as dimers. In all cases the viral RNA is organized to code sequentially for three major viral proteins: (1) GAG, which is a structural protein bound to the RNA within the central viral core; (2) POL, which is the reverse transcriptase synthesizing the complementary DNA intermediate; and (3) ENV, which encompasses the lipoprotein viral envelope.[21] All retroviral particles, including human immunodeficiency virus (HIV), the agent of acquired immunodeficiency syndrome (AIDS), contain these three basic viral proteins (Figure 5.3). The dimeric genetic RNA is situated within a central core in close association with the GAG protein and small quantities of reverse transcriptase. The entire viral particle is surrounded by a membrane that contains the ENV protein gene product.

As members of the retrovirus group, the tumor viruses are capable of transforming normal cells into neoplastic cells; in addition, they can produce tumors in experimental animals.[5,22] Besides the basic *gag*, *pol*, and *env* genes many retro-

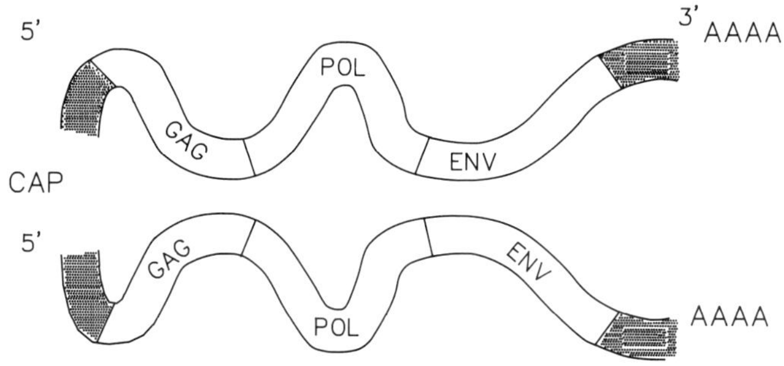

Figure 5.3 Dimeric form of RNA viruses, which exist in pairs instead of a single nucleotide chain.

viruses contain a fourth gene—an oncogene—that codes for a protein associated with transformation of the host cell to a tumor cell.[23–25] In general, retroviruses or tumor viruses that show only weak cell-transforming activity do not possess an oncogene. When the early studies of retroviruses were begun, it was necessary to sequence their RNAs. However, with the newer rapid methods of DNA sequencing, RNA virus is used to make intermediary cDNA, which can then be conveniently sequenced. In general, RNA viruses that replicate with proviruses or DNA intermediate copies are termed retroviruses. As mentioned, this group of viruses includes strongly tumor-promoting or oncogenic RNA viruses such as the Rous sarcoma virus, as well as a large number of viruses that cause a wide spectrum of diseases, including AIDS and the rare form of T cell leukemia caused by human T cell lymphotropic virus type I (HTLV-I).[26,27]

RNA Tumor Virus Infection

When RNA tumor virus infects a susceptible cell, it utilizes its *pol* gene to manufacture reverse transcriptase, which in turn synthesizes double-stranded DNA copies from the viral RNA template. Then the viral DNA copies progress to the infected cell nucleus and exist as either linear or circular DNA. Located within the nucleus, the viral DNA segments become integrated within host chromosomal DNA by a recombination event.[28] This is shown diagrammatically in Figure 5.4. During the integration of the proviral DNA into host chromosomal DNA, the viral DNA maintains its original sequence. It is still not clear how the host DNA is cut to allow insertion of the viral DNA segment; however, mechanisms exist for distinguishing the 3′ from the 5′ end of the viral RNA. This is associated with concurrent insertion of two long terminal repeats (LTRs) containing 100 to 200 base pairs at each end of the inserted proviral DNA segment. Of interest is the finding that within the newly arranged nucleotide sequence, four to six base pairs of host DNA are immediately adjacent to the two viral LTRs containing long tandem duplicates of nucleotides. These tandem repeats are characteristic of genetic elements called transposons, which are segments of DNA capable of moving about the gene.

How Viral RNA Can Subsequently Emerge from the Host Cell

The proviral DNA that has been integrated into host chromosomal DNA provides a built-in message for production of RNA polynucleotide chains duplicating those present in the original virus before it entered the cell. The initiating mechanism or

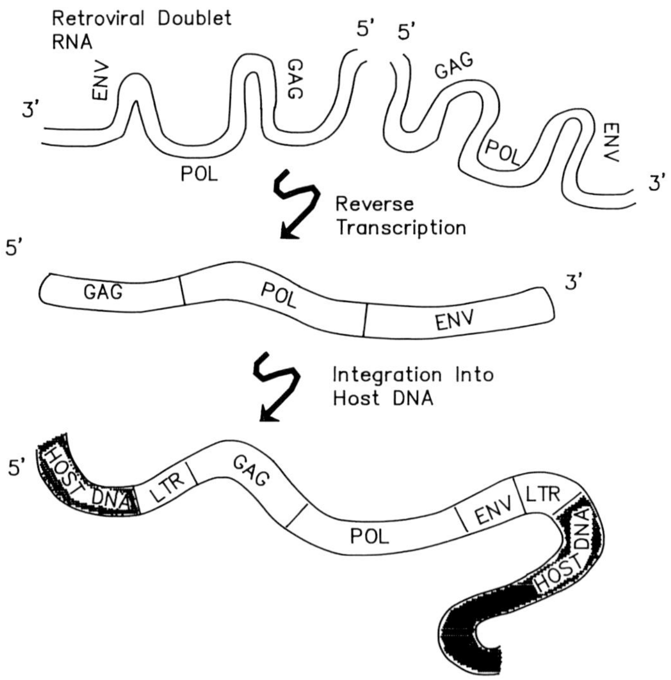

Figure 5.4 Retroviral genes are copied by reverse transcriptase to cDNA, which then integrates into host chromosomal DNA flanked by long terminal repeats (LTRs). The integrated retroviral DNA in this stage is called a provirus.

promoter for this DNA to viral RNA transcription is situated in the 5′ LTR sequence flanking the proviral DNA insert.[29] In this location the promoter can initiate transcription exactly at the first 5′ nucleotide of the original viral RNA. As the various mRNAs are transcribed, splicing occurs so that ENV is translated from its mRNA and other viral mRNAs are spliced and translated to GAG protein (Figure 5.5). The original unspliced mRNA is translated into a combined GAG–reverse transcriptase product, which can then be cleaved off enzymatically to produce free reverse transcriptase. Alternatively, some of the primary transcripts from the integrated proviral DNA may serve as genomic RNA polypeptide chains that will be condensed within new viral particles. These can then be extruded from infected cells to infect new cells. A cell originally infected with retroviral RNA then transfers the genomic message to proviral DNA. The RNA transcripts destined to serve as genomic viral RNA for export to infect other cells are maintained separate from the RNAs that are to function as messengers for ENV, POL, or GAG protein products.

Retroviral Oncogenes

The original Rous sarcoma virus described by Peyton Rous more than seven decades ago represents a vintage type of RNA tumor virus in that it contains an oncogene that can quickly alter normally growing cells to cancerous growth. The protein product of the Rous sarcoma virus oncogene (v-*src*) is called a *sarc* protein; it has a molecular mass of approximately 60,000 daltons and has been demonstrated to have protein kinase activity.[30,31] This biological activity is involved in the phosphorylation of tyrosine residues in protein polypeptide chains. Protein kinase activity has also been associated with a surprisingly large number of other oncogene

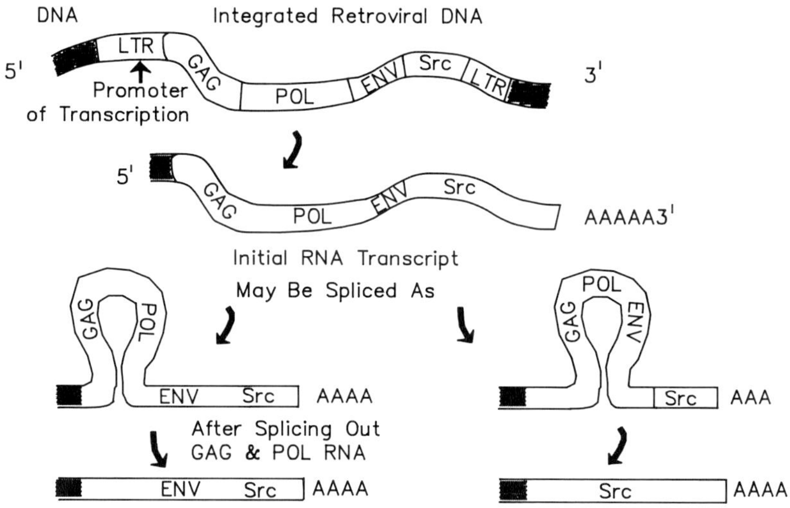

Figure 5.5 When the proviral retrovirus DNA is called upon to make new viral RNA, the promoter sequences within the 5′ long terminal repeat activate the process. Different mRNAs for the GAG, POL, and other viral proteins are produced through selective splicing from the same transcripted message. In some instances a fusion protein GAG-POL can be translated from the primary transcript.

protein products but is by no means the only biochemical or physiological activity now recognized as oncogene related.

After the v-*src* oncogene of the Rous sarcoma virus had been extensively studied, other RNA tumor viruses were examined for the presence of oncogenes distinct from the commonly shared *gag, pol,* and *env* nucleotide sequences. It was found that a considerable proportion of other RNA tumor viruses did indeed possess oncogenes; however, some had the oncogene interdigitated into the basic nucleotide sequence of their *gag, pol,* or *env* genes. Such viruses containing oncogene sequences directly within their basic genetic building blocks are defective and cannot replicate without input from a coinfecting helper virus. This is illustrated diagrammatically in Figure 5.6. The positioning of the oncogene directly within the *gag, pol,* or *env* nucleotide region ablates proper functioning of the respective structural gene and ability to initiate normal replication. In cells infected with one of the highly oncogenic retroviruses, the virus produces a fusion protein as a product of all or part of *gag, pol,* or *env* mRNA spliced together. Again, it is of great interest that many of these fusion proteins have protein kinase enzymatic activity. This observation has led to the idea that the effect of protein kinases on cell growth is related to the phosphorylation of fundamental regulatory cell proteins, which markedly alters their function in terms of orderly control of cell growth. We still have a great deal to learn about which cellular proteins are most important in controlling cell growth. The *sarc* gene product, one of the first oncogene products extensively studied, is located close to the cell plasma membrane,[32] which should be an ideal location for influencing various events in cell activation.

Where Did Retroviral Oncogenes Come From?

An important basic question that arose soon after oncogenes were recognized was where did they originate—from retroviruses themselves or from normal cellular genes? Initially it seemed possible that oncogenes such as v-*src* of Rous sarcoma

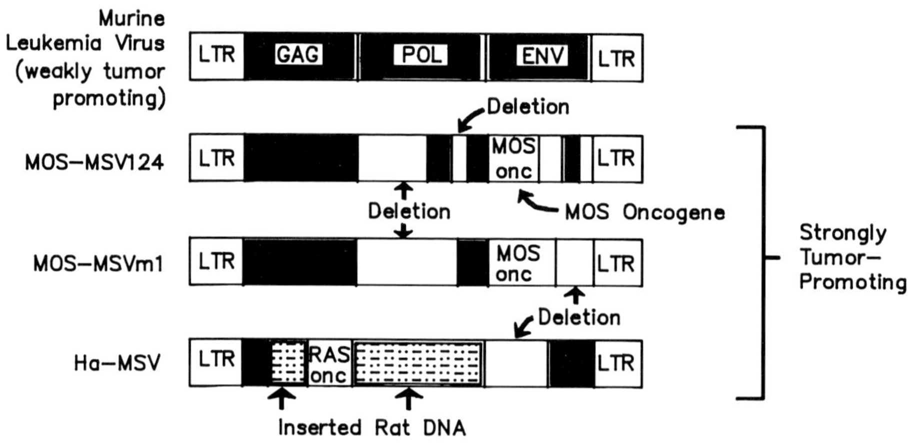

Figure 5.6 Comparative genomic maps of several mammalian tumor retroviruses. MOS and RAS are different cellular oncogenes that have been incorporated by various retroviruses. Insertion of the oncogene sequence into a gene essential for viral replication often induces deletions within adjacent genomic segments. Such defective viruses may require additional helper viruses to initiate effective replication.

virus were originally cellular genes that somehow had been appropriated and integrated into the viral nucleotide sequence. This hypothesis could be rapidly tested by screening DNA from diverse normal mammalian cells for hybridization using cDNA obtained from the native v-*src* RNA or cloned cDNA originally obtained from *src* RNA templates. When this sort of screening was done, it was found that a homologous gene, c-*src*, was present in DNA from normal cells of many species, including birds, mammals, fish, and even the intensively studied fruit fly *Drosophila*. Careful analysis of cloned c-*src* gene sequences indicated that normal mammalian c-*src* genes were very similar to but not exactly the same as the v-*src* genes of Rous sarcoma virus or other retroviruses.[33,34] As a corollary of the idea that retroviral oncogenes could be derived by partial integration of normal host genes, experiments were performed using a strain of Rous sarcoma virus that had a partial v-*src* gene deletion. The virus was introduced into cultures of normal avian cells. Viral progeny included viral representatives that had reconstituted the entire v-*src* genomic sequence by incorporating the appropriate nucleotides from the normal host cell DNA. This reconstitution of complete oncogene genetic potential is shown diagrammatically in Figure 5.7.

After the work with v-*src* of Rous sarcoma virus had been completed, it was shown that many other RNA retroviruses from a wide variety of avian and mammalian primary hosts contained DNA-integrated oncogenes that appeared to be closely related in both organizational structure and primary sequence to normal cellular genes. Again, molecular hybridization studies indicated that a number of these viral oncogenes showed a low but definite level of expression in normal mammalian cells. This finding and the fact that the nucleotide sequence expression has been maintained over much of the evolutionary scale, from fishes through amphibians to mammals, probably indicate that viral oncogenes expressed in normal cells function somehow in regulation of normal cell growth.

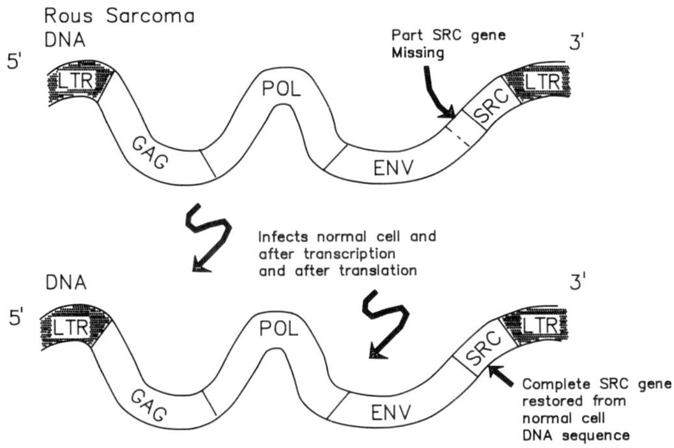

Figure 5.7 The Rous sarcoma gene (top) may have part of its *src* oncogene nucleotide sequence missing, but after infecting a normal cell it can restore its entire *src* nucleotide sequence by complementation with homologous *src*-like sequences provided by the normal cell genome.

Oncogenes and Normal Cell Activation

A number of lines of evidence indicate that oncogene-related events may contribute to normal cell activation. One of the earliest observations in that regard was the report by Kelly et al.[35] that activation of normal human lymphocytes by mitogens such as concanavalin A or lipopolysaccharide induced a remarkable increase in c-*myc* mRNA and eventual c-*myc* protein expression. They also showed that platelet-derived growth factor (PDGF), a substance derived from a different oncogene, c-*sis*, could participate in lymphocyte activation and c-*myc* expression. A subsequent report by Lacy and co-workers[36] indicated that c-*myc* expression could be induced in normal human B lymphocytes by B cell growth factors or anti-immunoglobulin. Studies of the level of c-*myc* expression indicated that c-*myc* was elevated in all phases of the cell cycle. In addition, the protein product of the human c-*myc* oncogene was localized to the nucleus and had a molecular mass of 65 kd.[37,38] The c-*myc* protein could bind to both double- and single-stranded DNA.

C-*myc* Translocation and Burkitt's Lymphoma

During the early studies of the c-*myc* oncogene it was found that Burkitt's lymphoma cells had a remarkable chromosomal translocation that brought the c-*myc* gene into very close proximity with the immunoglobulin heavy chain locus t(8;14) or the immunoglobulin light chain loci t(2;8) and t(8;22).[39–42] The translocation process involved a shift of the c-*myc* gene from chromosome 8 to a new site on human chromosome 14. Because this shift placed the c-*myc* gene close to several important immunoglobulin gene loci, there has been a great deal of speculation about how this might be associated with activation of the malignant process in Burkitt's lymphoma. It seems unlikely that malignant growth potential is related to a parallel mutation in the transposed c-*myc* coding exons (Figure 5.8), since in no case studied to date has either exon 2 or exon 3 been found to be broken. Also, in Burkitt's patients who have been examined in detail, no frequent mutations have

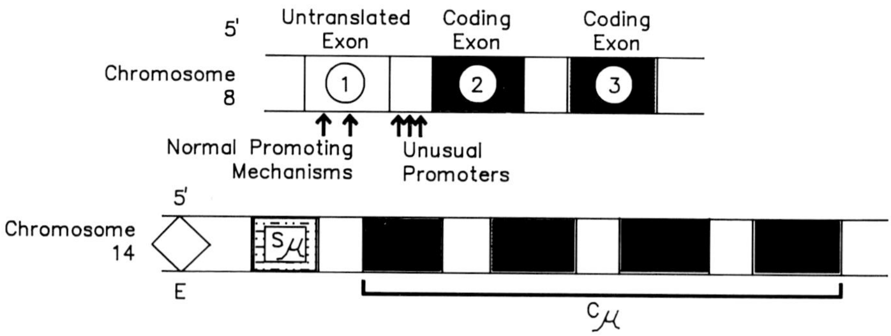

Figure 5.8 (Top) Normal c-*myc* gene is located on human chromosome 8 and has three exons, but the most 5′ exon is untranslated. Normally, the c-*myc* gene is transcribed from two promoters at the 5′ end of exon 1. When exon 1 is removed from the c-*myc* gene during chromosomal translocation, subsequent transcription is started at an area of unusual promoters within the first intron. (Bottom) The immunoglobulin heavy-chain locus is shown on human chromosome 14. This is involved in the Burkitt's lymphoma reciprocal switch or translocation with c-*myc* on chromosome 8. Normally, with no translocation, rearrangements of gene segments on chromosome 14 bring one of several V region segments into close juxtaposition with the C_μ constant segments and therefore under the influence of the enhancer (E) sequence just 5′ to its location. The S_μ switch sequence shown 5′ to the constant region genes participates in subsequent gene rearrangement during final coding for the immunoglobulin product.

been noted in the translocated exons 2 and 3. Thus if appears possible that the nontransposed c-*myc* exon 1 plays a regulatory role in preventing malignant transformation. However, the exact relationship between the extensively studied translocation of c-*myc* to chromosome 14 and the malignant stimulus in Burkitt's lymphoma remains unsolved. One of the difficulties is that the chromosomal break points that allow these interesting interchanges to take place appear to be quite variable. The chromosomal translocation products are shown in Figure 5.9.

A number of studies have attempted to elucidate a relation between the fascinating 14 translocation and Burkitt's lymphoma. Some possible mechanisms include deregulation of the c-*myc* gene allowing constitutively high levels of expression[43]; release of transcriptional repressors[44]; and differential usage of promoters akin to normal and malignant cells or somatic mutations occurring because of the proximity to the immunoglobulin locus,[45,46] where higher spontaneous levels of mutation are known to occur. Studies using antisera to synthetic peptides made from the c-*myc* nucleotide sequence indicate that c-*myc* transcripts are expressed at much higher levels in malignant B cells than in normal B cells.[47] These findings suggest that an increase in the level of the c-*myc* protein product is an important feature of the malignant transformed state.

C-*myb*

In similar fashion to c-*myc*, it has been found that levels of mRNA for the nuclear proto-oncogene c-*myb* also vary considerably in normal immature thymocytes as a function of markedly increased normal cellular transcription,[48] as well as being present in association with proliferation of other cell types. In parallel, c-*myb* has been reported to be markedly amplified in some instances of human acute myelogenous leukemia.[49] Since both c-*myb* and c-*myc* code for nuclear protein products

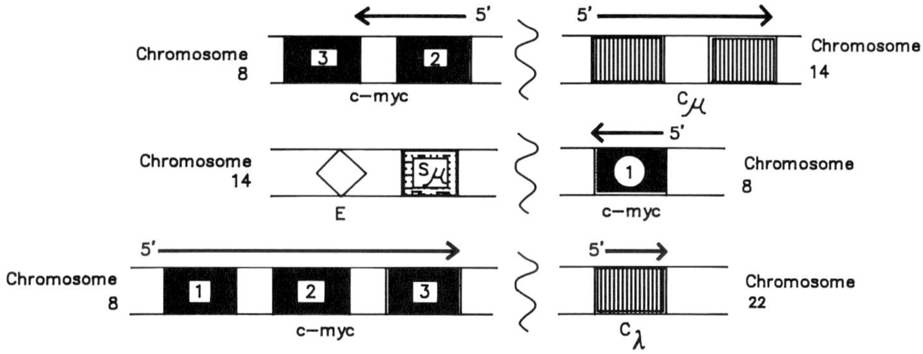

Figure 5.9 (Top) Typical translocation with c-*myc* separated from its exon 1. The most typical translocation seen in Burkitt's lymphoma involves a reciprocal exchange between chromosomes 8 and 14 resulting in exons 2 and 3 of c-*myc* occurring next to the immunoglobulin heavy-chain constant locus C_μ. However, the actual distance of the chromosome 14 breakpoint from the C_μ region is extremely variable. (Middle) The reciprocal of the translocation shown above. (Bottom) Rare variant translocations between chromosome 8 and the immunoglobulin light-chain loci on chromosome 22. In these instances c-*myc* remains intact and does not lose an exon. It is joined to the constant region gene for lambda light chains.

that bind to DNA and have some degree of structural homology[50] and their respective amplification is regulated similarly during the normal myeloid differentiation process, the finding of unequal gene amplification in similar tumors such as acute leukemias suggests that these two oncogenes may be alternatively activated in the same pathogenetic step. Amplification of c-*myb* or c-*myc* may prevent the physiological suppression of these genes that normally occurs when they switch from a proliferative to a differentiating program.

N-*myc*

Another *myc* oncogene that has received considerable attention is N-*myc*, originally identified after it was found to be frequently amplified and overexpressed in several human tumors—particuarly neuroblastomas and retinoblastoma.[51,52] Studies with several assays indicate that N-*myc* has a transforming potential similar to that of c-*myc*. Despite some degree of structural homology, the two genes often appear to be different with respect to amplification in tumors as well as expression in specific tissues.[53] Precise studies of the gene structure of c-*myc* and N-*myc* have indicated considerable homology. This can be appreciated by the comparisons shown in Figure 5.10. It is not clear whether the degree of structural homology between c-*myc* and N-*myc* is related to the gene products sharing similar general regulatory functions. It appears that they represent kindred genes that produce similar protein products. Moreover, both c-*myc* and N-*myc* are amplified in small cell carcinoma[54,55]; however, the proteins produced by these two *myc* genes differ by 68% in their specific amino acids. These comparisons indicate that they are both members of the *myc* family and have similar properties but are uniquely expressed and regulated. N-*myc* appears to contribute to neoplastic transformation of cultured mammalian cells, as monitored in cotransfection studies by Schwab et al.[56]

Many clinical tumors are being studied to relate levels of oncogene expression to clinical progression or staging of disease. In the case of N-*myc* it has been shown

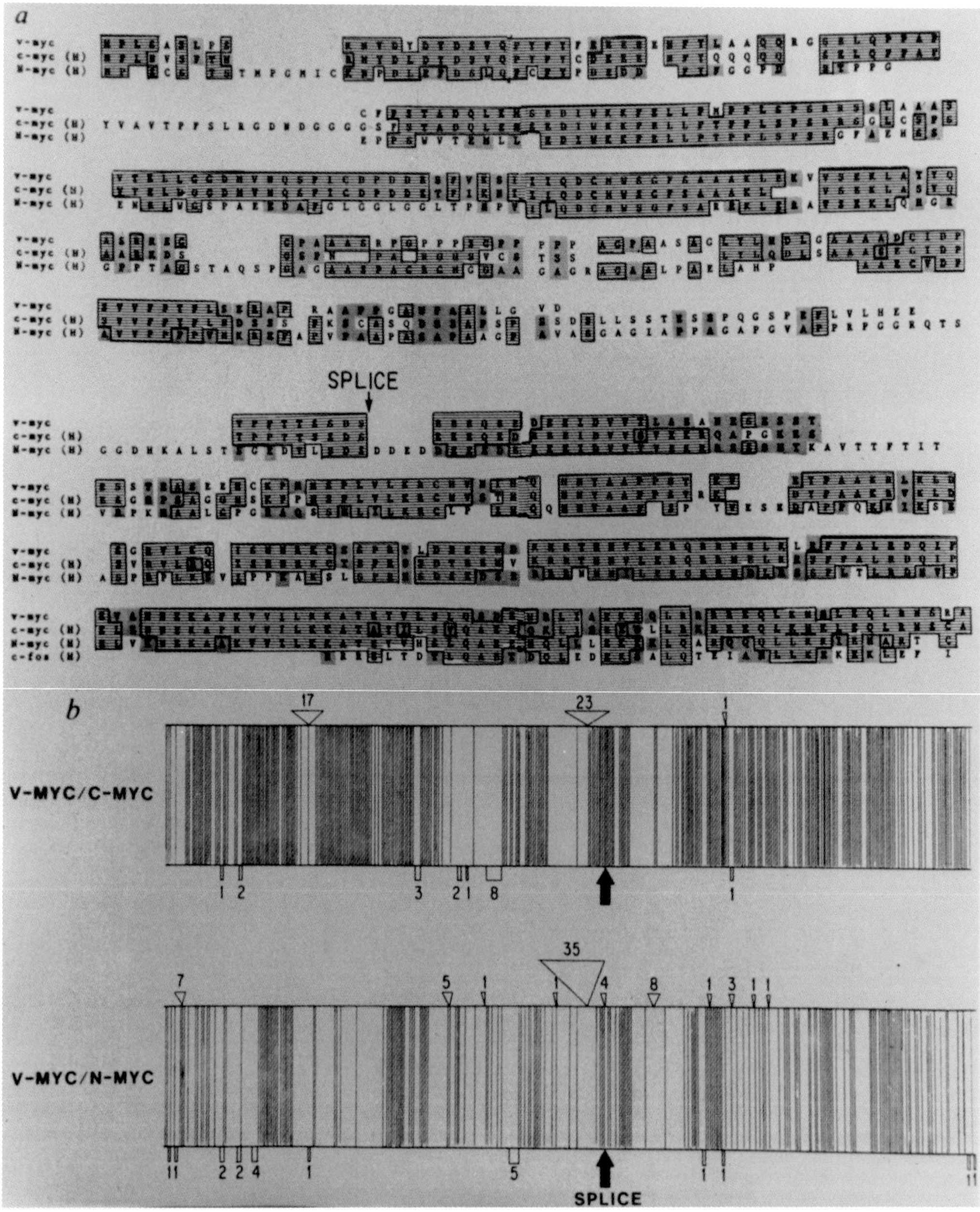

Figure 5.10 (a) Comparison of amino acid sequences of N-*myc*, c-*myc*, and v-*myc* proteins. Homologous amino acids are shown in boxes and amino acid substitutions judged to occur at higher than average frequencies are shown by shading alone. (b) The degree of homology between the v-*myc* protein and the c-*myc* and N-*myc* proteins. Homologous amino acids are indicated by a line. Amino acid insertions in the N-*myc* and c-*myc* proteins relative to the v-*myc* protein are shown by triangles at the top and deletions are shown by rectangles at the bottom. Numbers indicate the size in terms of number of amino acids of the insertion or deletion. *Reproduced from Kohl NE, Legouy E, DePinho R, et al: Human N-myc is closely related in organization and nucleotide sequence to c-myc. Nature 1986;319:73–77, with permission.*

that human neuroblastoma cell lines and tissues obtained directly from neuroblastomas show marked amplification of N-*myc*, 100- to 300-fold in some cases and 3- to 10-fold in other clinical groups.[57] Patients in this study with advanced (stage III or stage IV) disease showed highest proportions of N-*myc* amplification. This, in turn, correlated with poor prognosis.

Parallel amplification of several oncogenes may be important in determining the growth potential of some tumors. Schwab et al.[58] noted amplification of a DNA domain or nucleotide sequence partially homologous to c-*myc* in both cell lines and tissue from a human neuroblastoma tumor. These findings were among the first to suggest that clusters spaced together on domains of DNA might influence the growth and intrinsic malignancy of some tumors.

C-*myc* and Neoplasia

As noted above, a great deal of investigation has been focused on the relationship between c-*myc* and various cancers. Originally, expression or regulation of c-*myc* was recognized as a competence factor[59] that, like PDGF, could promote cells in their growth cycles. As the more sinister ramifications of what oncogenes could do became apparent, direct links with a spectrum of human neoplasms were suspected. Using the technique of transgenic microintroduction of foreign genes, it was shown that transgenic mice bearing the cellular *myc* oncogene coupled to the immunoglobulin μ or kappa enhancer often developed rapidly progressive lymphomas.[60] This remarkable experimental result was reminiscent of the finding in Burkitt lymphoma that c-*myc* translocation near the constant μ immunoglobulin gene locus was directly associated with B cell malignancy. It appeared that in these transgenic mice the c-*myc* gene, aided and abetted by the immunoglobulin gene enhancers, also could provide a powerful stimulus for lymphoid malignancy. An example of the transgenic constructs used in these experiments is shown in Figure 5.11.

Additional studies by several groups have shown the profound influence of transgenic *myc* constructs on the development of spontaneous lymphoma or leukemia in mice,[61] particularly when the c-*myc* gene is driven by an immunoglobulin enhancer such as E-*myc* for the IgG heavy chain. Tumors in such transgenic animals present as rapidly progressive lymphoblastic lymphomas that appear by cell surface marker analysis to be mixtures of pre-B lymphomas, pre-B and B cell lymphomas, or B cell lymphomas. Moreover, transgenic rabbits have been reported to develop B cell lymphocytic leukemia after introduction of the c-*myc* oncogene fused with the E-*myc* IgG heavy chain enhancer.[62] These leukemias in the rabbit model develop extremely rapidly at less than 3 weeks after birth and appear to be oligoclonal in origin.

Subsequent studies have implicated c-*myc* regulation and expression in a wide spectrum of malignant disorders, including human acute myelogenous leukemia,[63] mammary carcinomas,[64,65] and several other tumors.[66] One of the most convincing examples of the influence of c-*myc* in malignant transformation was the finding that a significant proportion of feline leukemias might be triggered or somehow activated by association with c-*myc*. Close association of feline leukemia virus (FeLV) DNA with c-*myc* sequences has now been documented, suggesting that FeLV may actually insert within the c-*myc* gene.[67,68]

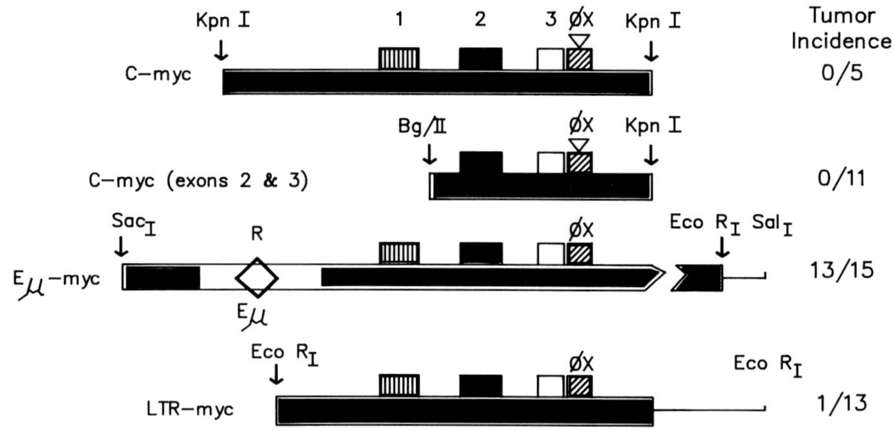

Figure 5.11 The c-*myc* constructs were injected into fertilized mouse ova and tumor incidence was evaluated in the resulting transgenic mice. Incidence of malignant lymphomas is shown to the right. The solid bar represents c-*myc* sequences with exons raised, untranslated regions striped, and ϕX174 marker as a triangle. Vector sequences are represented as a thin line. The linear fragments shown were isolated from plasmid constructs before injection. The E_μ-*myc* construct, which induced lymphomas in 13 of 15 mice, contains 2.3 kb of immunoglobulin DNA encompassing the H chain enhancer (diamond). The LTR-*myc* construct contains 210 base pairs of the Soule murine leukemia virus enhancer. *Adapted from Adams et al., with permission.*[60]

C-*myc* Bypasses IL-2 or IL-3 Dependence

A fascinating aspect of the *myc* oncogene story is related to a series of experiments by Rapp and co-workers.[69] It had been postulated that several oncogenes induce malignant transformation by affecting signal transmission of growth factors, as in the induction of c-*myc* by PDGF.[35,70] Moreover, various recombinant murine retroviruses containing v-*myc* nucleotide sequence inserts could abrogate the usual requirements for growth factors such as interleukin 2 (IL-2) or interleukin 3 (IL-3). Interleukin 2 is generally required for proliferation of antigen-activated mature T cells, whereas IL-3 induces both proliferation and differentiation of early hematopoietic/lymphoid stem cells. In the work of Rapp and co-workers, when the IL-2–dependent cell line was exposed to a murine retrovirus (J-2) containing v-*myc* inserts, IL-2–independent cell lines emerged, indicating that the v-*myc* gene was able to substitute for a basic stimulatory lymphokine (IL-2) in the maintenance of cell viability and continued replication. These findings are illustrated in Figure 5.12.

V-*sis* and Platelet-Derived Growth Factor

One of the most remarkable stories in the oncogene scenario is related to the discovery by two groups at virtually the same time that the oncogene v-*sis*, derived from the simian sarcoma virus, coded for a protein that was closely related to a major component of human PDGF.[71,72] These findings, along with those of Kelly et al. on c-*myc*[35] activation during normal lymphocyte stimulation, were among the first to indicate a definite link between oncogene products and growth factor–mediated or normal cellular differentiation pathways. Subsequent detailed studies showed that the product of the v-*sis* gene was actually a PDGF-related precursor

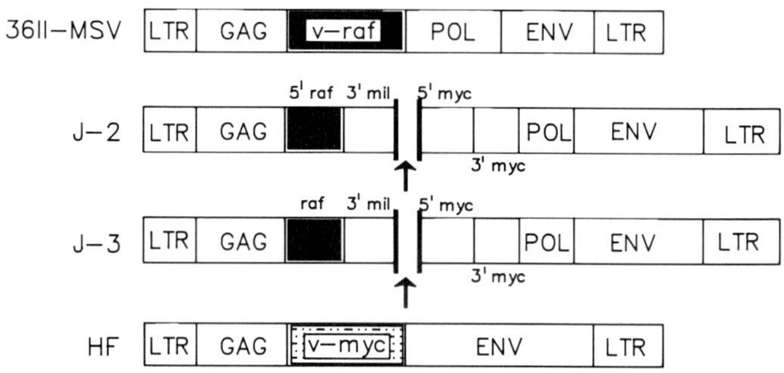

Figure 5.12 A general map of the defective 3611-MSV retrovirus is shown above. This virus has a v-*ras* oncogene insert between the GAG and POL sequences. Specific additional constructs J-2 and J-3 show additional inserts containing v-*mil* and the 5′ and 3′ nucleotide sequences of v-*myc*. Virus HF contains only a v-*myc* insert. Double bars and arrows indicate splice acceptor sites for v-*myc* gene inserts. Using these various constructs, it was shown that J-2 and J-3 or HF viruses containing v-*myc* sequences could make previously IL-3 or IL-2 dependent cell lines independent of these interleukins. *Adapted from Rapp et al.,*[69] *with permission.*

polypeptide that assumed a homodimer configuration shortly after synthesis. By contrast, human PDGF preparations contained two distinct but related polypeptide chains, one of which is encoded by the human *sis* proto-oncogene.[73] However, physical isolation of the v-*sis* gene product from cells transformed with simian sarcoma virus indicated that this polypeptide was capable of specifically binding PDGF receptors and subsequently stimulating tyrosine phosphorylation of the receptors.[74] Furthermore, these studies provided direct support for the idea that the simian sarcoma virus transforming activity was directly turned on by interaction of a virus-coded mitogenic peptide with PDGF receptors. The close relationship between the v-*sis* oncogene product and human PDGF was further emphasized by the finding that antisera produced against a synthetic 17-mer peptide representing residues 139 to 155 of the predicted simian sarcoma virus v-*sis* protein reacted directly with human PDGF.[75] Additional proof that PDGF is an essential oncogene-related substance in normal cell activation was provided by the demonstration that antibodies against PDGF inhibited both proliferation and simian sarcoma virus transformation in human diploid fibroblasts.[76] These results were extremely important because they supported the idea that simian sarcoma virus transformation was due to the autocrine action of a PDGF agonist instead of other spontaneous properties of various target cell lines.

As a parallel example of the potential importance of oncogenes in normal cell differentiation, PDGF has been linked to several other oncogenes and their mRNA expression and final translation, including c-*myc*[35] and subsequently c-*fos*.[77] Because several of the PDGF-inducible oncogenes are members of the competence gene family (c-*myc*, c-*fos*) and most are localized in the nucleus, it is possible that this group of DNA-binding proteins functions together in the nuclear matrix to coordinate transcription and replicative DNA synthesis.[78]

PDGF has long been recognized to be one of the major mitogens for cells of mesenchymal origin[79–81] so its relationship to neoplastic cell transformation is a subject of intense interest. PDGF itself is a normally occurring substance that is stored in α granules of platelets. The original term platelet-derived growth factor is probably a misnomer because PDGF can be secreted by a number of different cells. Its identification and implication in human osteosarcoma cells reemphasized the potential link to an ongoing transformation process.[82] Studies by Bronzert et al.[83] have indicated that human breast cancer cell lines synthesize and secrete PDGF. Hybridization studies using cDNA probes for the A chain of PDGF and the B chain of PDGF/SIS identified complementary transcripts in several such cell lines. Production of PDGF-like materials by breast carcinoma cell lines may be important in paracrine stimulation of tumor growth. It has been suggested that production of PDGF-like materials stimulates a fibrotic response in various surrounding tissues that is important for amplification of breast cancer cell growth through induction of other growth factors derived from fibroblasts.[84] Final resolution of the role of PDGF and similar c-*sis* transcripts in many other neoplasms awaits further work.

Characteristics of PDGF

PDGF isolated from human platelets is a molecule of 30 kd (range 28 to 35 kd). The variability in molecular mass is related to differences in extent of glycosylation. Each active molecule of PDGF consists of two disulfide-linked polypeptide chains (an A chain of 14 kd and a B chain of 17 kd). Reduction of disulfide bonds results in complete loss of mitogenic activity. As noted earlier, PDGF is a powerful mitogenic stimulus for most cells derived from mesenchymal tissues. Exposure of fibroblasts or smooth muscle cells to PDGF can lead to cell number doubling within 30 to 36 hours. After PDGF reacts with its cellular receptors, there is increased turnover of membrane phospholipids[85] and tyrosine phosphorylation. PDGF itself has a number of other physiological effects, including increased cholesterol synthesis, increased binding of low-density lipoprotein to its receptors, increased prostaglandin formation, and potent vasoconstriction.[86] Cells capable of synthesizing and secreting PDGF include platelets, monocytes and macrophages, vascular endothelial cells, arterial smooth muscle cells, embryonic cells, and megakaryocytes.

Role of PDGF in Normal Physiology

Since PDGF provides a natural mitogenic stimulus for mesenchymal cells, it probably plays an important role in wound healing. In addition to stimulating proliferation of connective tissue precursors, PDGF is chemotactic for monocytes, neutrophils, and fibroblasts.[87] The latter cells may well produce additional important growth factors, including epidermal growth factor and fibroblast growth factor, which work together to stimulate connective tissue repair and scar formation.

The fact that PDGF actively stimulates wound healing and connective tissue repair may mean that it is also involved in pathological disease processes. This has been explored by Martinet et al.[88] in patients with idiopathic pulmonary fibrosis. It was found that normal alveolar macrophages constitutively express low levels of c-*sis*. Studies of alveolar macrophages recovered from the lungs of patients with pulmonary fibrosis indicated that cells from these individuals released four times as much PDGF as did alveolar macrophages from normal controls. These findings, illustrated in Figure 5.13, suggest that accumulation of mesenchymal cells such as alveolar macrophages, monocytes, and polymorphonuclear leukocytes chemotactic for PDGF in alveolar tissues of patients with idiopathic pulmonary fibrosis may be secondary to exaggerated PDGF release by mononuclear phagocytic cells in the pulmonary parenchyma. It is possible that other disease conditions associated with localized fibrosis or accumulation of connective tissue, such as cirrhosis of the liver, scleroderma, or chronic deforming rheumatoid arthritis, are activated in part by similar PDGF-related mechanisms.

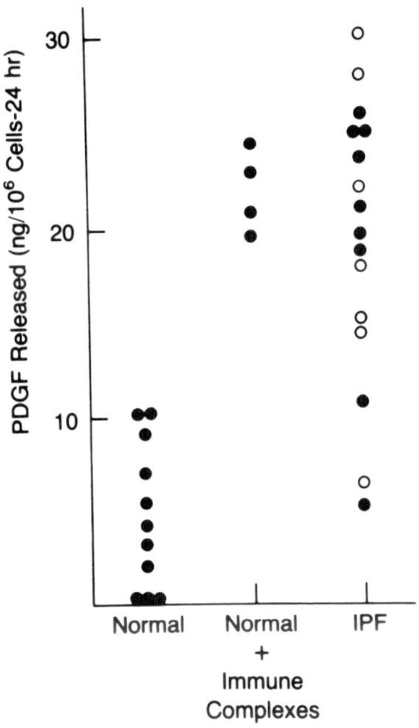

Figure 5.13 Exaggerated spontaneous release of PDGF by alveolar macrophages from patients with idiopathic pulmonary fibrosis. Alveolar macrophages recovered by bronchoalveolar lavage were cultured for 24 hours, and supernatants were evaluated for the presence of PDGF using an enzyme-linked immunoassay. Normal controls show low PDGF release, but normals stimulated by the presence of immune complexes show elevated PDGF in supernatant. Alveolar macrophages from patients with idiopathic pulmonary fibrosis (IPF) show marked spontaneous release of PDGF. *Reproduced from Martinet et al.,*[88] *with permission.*

PDGF and Atherosclerosis

Atherosclerosis, the most common cause of death in the United States and Western Europe, has been linked to PDGF through the basic lesion of the fibrous plaque as the initiation of atherosclerotic lipid accumulation. Proliferations of smooth muscle cells in such lesions are frequently juxtaposed to large numbers of lipid-filled macrophages. The macrophages, activated endothelial cells, or even platelets themselves could be the source of PDGF and thus the stimulus for smooth muscle proliferation during atherogenesis. A number of experimental results seem to support such a hypothesis. Increased PDGF mitogenic activity has been reported in plasma from young patients with coronary atherosclerosis.[89] Furthermore, in this study the PDGF activity was blocked by specific anti-PDGF antibody. These results suggested a connection between increased levels of free PDGF and the intimal or vascular smooth muscle proliferation characteristic of early atherosclerotic lesions. It has also been shown that arterial smooth muscle cells[90] and cultured endothelial cells[91] produce a PDGF-like protein. Human atherosclerotic lesions show elevated *sis*–B chain RNA transcripts in comparison with normal arteries,[92] and human iliac artery endothelial cells express the genes coding for the A and B chains of PDGF.[93] Production of a PDGF-like protein by human arterial endothelial cells could function as a paracrine stimulus to arterial smooth muscle proliferation in fibrous plaque or more advanced atherosclerotic vascular lesions. A more direct approach to this general problem has been described by Penn and co-workers,[94] who transfected DNA from human coronary artery plaques into NIH 3T3 cells. The transfected cells produced transformed foci, whereas DNA from noncancerous normal human tissues including coronary artery were negative. DNA from cloned transformed foci was used successfully in a second round of transfection. These experiments suggested that human coronary atherosclerotic plaques contained transforming genes that could behave like oncogenes in cancer cells. These findings, if confirmed and extended, may fit with the concept of additional paracrine stimuli generated by

PDGF. Of interest in this respect is a report by Kobilka et al.,[95] who found that cDNA coding for the human β_2-adrenergic receptor was very closely juxtaposed on chromosome 5 to the genes encoding the PDGF receptor as well as the FMS proto-oncogene, which codes for macrophage colony-stimulating factor receptor (Figure 5.14). In view of the fundamental importance of β-adrenergic receptors to cardiac function, the proximity of this gene to that coding for the PDGF receptor is of considerable interest.

Studies of vascular smooth muscle cells cultured from human atherosclerotic plaques indicated that these cells released materials into their culture media that were capable of stimulating proliferation of aortic smooth muscle cells and that this activity appeared in part to be due to PDGF.[96] The cells cultured from human atheromas contained mRNAs for the PDGF A chain but not for the B chain. Such localized tissue production of an endogenous potentially self-stimulating autocrine growth factor may eventually help to explain replication of smooth muscle cells early in atherogenesis.

A definitive study of the process of incisional wound healing has indicated that it is hastened by PDGF and other recombinant c-*sis* gene homodimeric proteins.[97] Thus, such materials administered as recombinant molecules in slow-release ve-

Figure 5.14 Diagram of human chromosome 5 indicating by dots maximal sites for hybridization of labeled probe for β_2-adrenergic receptor. Proximity of platelet-derived growth factor receptor gene, macrophage colony-stimulating factor, and FMS proto-oncogene is shown. *Reproduced from Kobilka et al.,[95] with permission.*

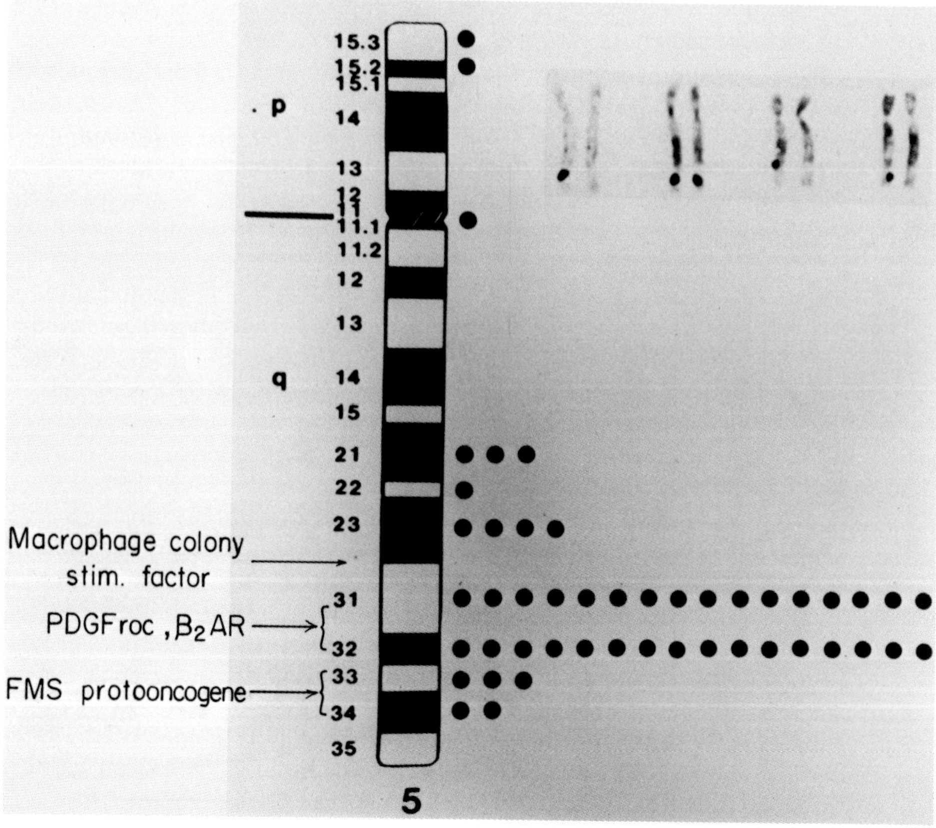

hicles may be capable of accelerating wound healing. As the maps of genes unfold in a virtual avalanche of new discoveries and new data, a much better multidimensional view of both normal function and pathological states may emerge.

Werner's Syndrome

A report by Bauer et al.[98] indicates that patients with Werner's syndrome, a rare autosomal recessive disorder associated with accelerated aging and so-called progeria, are hyporesponsive to PDGF. Fibroblasts from patients with Werner's syndrome manifest markedly decreased mitogenic responsiveness to PDGF, although PDGF receptors on cells from these patients appear to be unaltered. In this case, shifts in normal PDGF-mediated pathways of response may be associated with progeria and accelerated premature aging.

Ras Genes

The term *ras* gene was derived from the words *rat sarcoma* because these genes were first identified as the transforming principle of the Harvey and Kirsten strains of rat sarcoma virus.[99,100] These agents are two acute transforming retroviruses generated by transduction of the rat H-*ras*-1 and K-*ras*-2 cellular genes. Recently, *ras* oncogenes have been identified in tumors induced by retroviruses that appear to lack oncogene sequences. These particular viruses are capable of integrating in close proximity to cellular proto-oncogenes, thereby disrupting their normal regulatory mechanisms.[101]

Deciphering how the *ras* genes are related to neoplasia will have a major impact on our understanding of cancer and malignant transformation. Moreover, the remarkable conservation of *ras* genes throughout evolution attests to their apparent fundamental role in basic cell function. Evolutionary studies, for instance, clearly identify *ras* homology in *Drosophila melanogaster*[102] and other eukaryotic forms. The basic cell functions in which they participate indicate that *ras* genes code for proteins that bind guanine nucleotides,[103] have GTPase activity,[104] and are closely associated with the plasma membrane.[105] Moreover, their sequence homology with G proteins[106] suggests that *ras* protein products may be involved in signal transmission across the cell membrane. A number of basic functions of *ras* protein products have been suggested, including a regulatory role in the activation of adenyl cyclase,[107] a function as the elusive G proteins of the inositol-lipid signaling pathway,[108] and involvement in definitive cellular differentiation pathways. Although the basic function of *ras* proteins still eludes us, they all appear to contain consensus regions for binding to GTP and they are localized to the inner surface of the cell plasma membrane.

Mammalian *ras* genes include H-*ras*-1, K-*ras*-2, and N-*ras*. These three functional *ras* genes code for highly related proteins referred to as p21. Mammalian *ras* genes acquire transformation-inducing potential by single point mutations in their coding sequences. Particular sites on the *ras* gene, when involved in single point mutations, result in oncogenic activation.[109] Thus, a change in the glycine at position 12 and substitution of valine or aspragine at position 13 yield *ras* oncogene products. Various biological properties of *ras* proteins are shown in Figure 5.15. One of the most frequently utilized reagents in the study of *ras* proteins is rat monoclonal antibody Y-13-259, which binds to a highly conserved domain of the *ras* protein. Like antibodies to other oncogene-related proteins, this antibody has

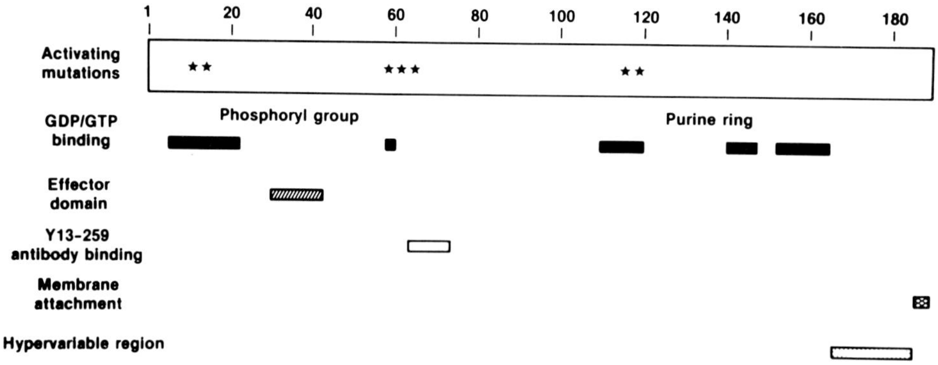

Figure 5.15 Schematic representation of the structural and functional domains in mammalian p21 *ras* proteins. *Reproduced from Barbacid,*[122] with permission.

been shown to inhibit morphological transformation induced by mutated *ras* proteins.[110]

Although, as noted above, it is not yet completely known how *ras* proteins and *ras* oncogenes act, a model can be suggested for the possible function of the *ras* proteins. This is shown in Figure 5.16. In addition, a more detailed model of the three-dimensional structure of the human c-H-*ras* protein p21 has been defined by crystal structure analysis.[111] The actual biological function of *ras* genes in mammalian cells is nevertheless poorly understood. *Ras* oncogenes can induce terminal differentiation in certain cells and may also play an important role in normal cell proliferation. Immunohistochemical analyses of normal human fetal and adult tissues have indicated p21 *ras* expression in virtually all tissues. *Ras* genes can become transforming oncogenes by point mutations in certain critical domains. In addition, increased levels of expression of normal *ras* proto-oncogenes have been associated with the neoplastic process.[112]

Ras Gene Activation and Possible Function

Because p21, the oncogene protein product of the *ras* gene, has been implicated in mediating the effects of a number of growth factors and hormones, several studies have focused on its possible specific cell-regulatory functions. Monoclonal antibody to p21 has been shown to inhibit *Xenopus* oocyte maturation induced by insulin, and this may involve phosphorylation of p21, which was found to be a substrate of the insulin receptor kinase.[113] The importance of transmembrane activity of *ras* was further supported by the observation that elevated levels of glucose transport and transporter messenger RNA were induced by *ras* oncogenes.[114] These effects appeared to involve increased expression of the gene coding for the glucose transport protein.

A number of studies have examined change in *ras* oncogene expression determined by gene promotion mechanisms. For instance, activated Ha-*ras* oncogene could be expressed by a milk protein gene promoter, leading, after a long latency period, to tumors in mammary or salivary glands.[115] Nucleotide sequence studies of the promoter region of the human Ha-*ras* proto-oncogene[116] showed that the promoter region contained neither a TATA nor a CAAT box in the usual characteristic upstream positions and had an extremely high G + C content (80%). The promoter region contained multiple GC boxes and seven CCGCCC repeats and

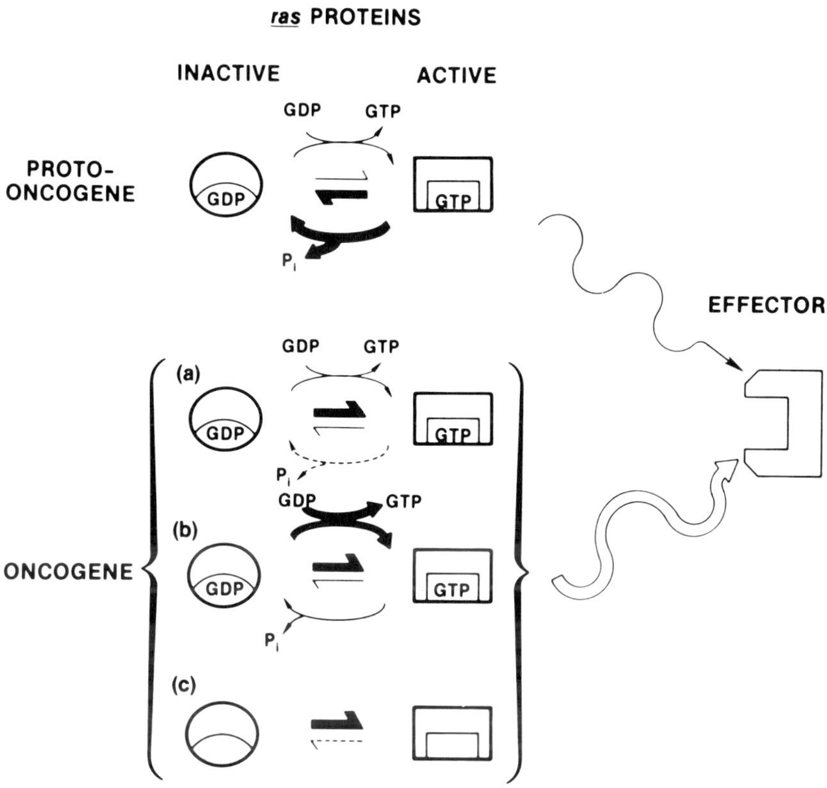

Figure 5.16 Model for the possible mechanism of action of both normal and transforming *ras* proteins. This model proposes that *ras* proteins are present in an equilibrium between an active and an inactive state. The inactive state is accompanied by a shape that permits binding of GDP. When stimulated, GDP is exchanged for GTP, accompanied by activation of the *ras* protein. After activation and transduction of messages across the cell membrane, the active *ras* protein would be deactivated by hydrolysis of GTP. *Adapted from Barbacid.*[122]

repeats of the inverted GGGCGG complement. These findings indicated a strong similarity to the previously described gene promoter for epidermal growth factor (EGF) receptor, another oncogene-related product. It is possible that promoters of many of the genes involved in growth control may have similar structural features that coordinate their expression and possibly facilitate the binding of similar factors.

In contrast to the usually recognized tumor-promoting capacity of some oncogenes, it has been shown that introduction of the viral Harvey *ras* oncogene (v-Ha-*ras*) into a cultured medullary thyroid carcinoma cell line could induce endocrine differentiation of the tumor cells associated with markedly increased expression of the calcitonin gene,[117] a situation most unusual in either the cell line or the primary tumor.

Ras Oncogenes and Malignant Tumors

Activated *ras* oncogenes have been demonstrated in the majority of carcinogen-induced cancers studied.[118,119] Moreover, the activating mutations in these chemically induced tumors are similar to those observed in activated *ras* genes of human tumors. Doniger et al.[120] showed that chemical carcinogens with widely varying

Table 5.1 Specific Mutagenesis of H-*ras*-1 Oncogenes in Chemically Induced Tumors

Animal studied	Type of tumor	Chemical carcinogen	Activating mutation	Incidence of tumor induction
Rat	Mammary carcinoma	NMU	$G^{35} \rightarrow A$	61/61
		DMBA	$A^{182}, A^{183} \rightarrow N$	5/5
Mouse	Skin papilloma or carcinoma	DMBA	$A^{182} \rightarrow T$	33/37
Mouse	Hepatocarcinoma	None	$C^{181} \rightarrow A$	6/11
			$A^{182} \rightarrow T$	3/11
			$A^{182} \rightarrow G$	2/11
		Ho-AFF	$C^{181} \rightarrow A$	7/7
		VC	$A^{182} \rightarrow T$	6/7

NMU = nitroso-methylurea, DMBA = dimethyl-benz(a) anthracine, HOAAF = N-hydroxy-2-acetyl-aminofluoride, VC = vinyl carbamate.
(Adapted from Barbacid,[122] with permission.)

mutagenic activities could produce neoplastic transformations in several different cell lines, which acquired the same N-*ras* point mutation. Because many chemical carcinogens are mutagens that form adducts with specific DNA bases, direct insight into this process has been gained by comparing mutations induced by specific carcinogens with *ras* oncogenes and several experimental tumors. This type of analysis is illustrated in Table 5.1.

In parallel with these observations on chemical carcinogens, products of the human c-*ras* oncogene were studied in both normal tissues and bladder cancer cells from the same individual.[121] Both DNA from the patient's malignant bladder tumor and DNA from normal bladder cells representing germ-line DNA were heterozygous for the c-$ras_1{}^H$ oncogene, in contrast to homozygosity for the normal allele in a number of normal subjects. This observation was one of the first to link possible point mutations involving well-characterized proto-oncogene sequences with subsequent development of tumors.

Ras oncogenes have now been identified in a number of human cancers, including carcinomas of the bladder, breast, colon, kidney, liver, lung, ovary, pancreas, and stomach. They have also been found in association with acute lymphocyte leukemia, B cell lymphoma, Burkitt's lymphoma, and acute and chronic myelogenous leukemia. Other tumors, including melanoma, teratocarcinomas, neuroblastomas, and gliomas, have also been shown to possess *ras* oncogenes. Thus, far, transforming *ras* genes are the oncogenes that have been most frequently identified in human cancers, with an overall estimated incidence of 10 to 15%.[122] The exact incidence is difficult to determine because the sensitivity of detection of *ras* gene expression varies from study to study, depending on whether transfection or hybridization techniques are utilized.

In studies using quantitation of Harvey *ras* p21, 66% of human breast tumors and 100% of human colon cancers showed enhanced expression, based on using quantitative radioimmunoassay and cDNA probe hybridization.[123,124] An example of findings based on monoclonal antibody identification of *ras* p21 gene product is shown in Figure 5.17.

Parallel studies of *ras* oncogene p21 product were performed by Nonomura et al.[125] in cirrhosis of the liver and hepatocellular carcinoma. A significant increase in expression of p21 was noted in macronodular cirrhosis as well as in adenomatous hyperplasia. Of patients with hepatic cell carcinoma, 62.4% showed p21 tumor

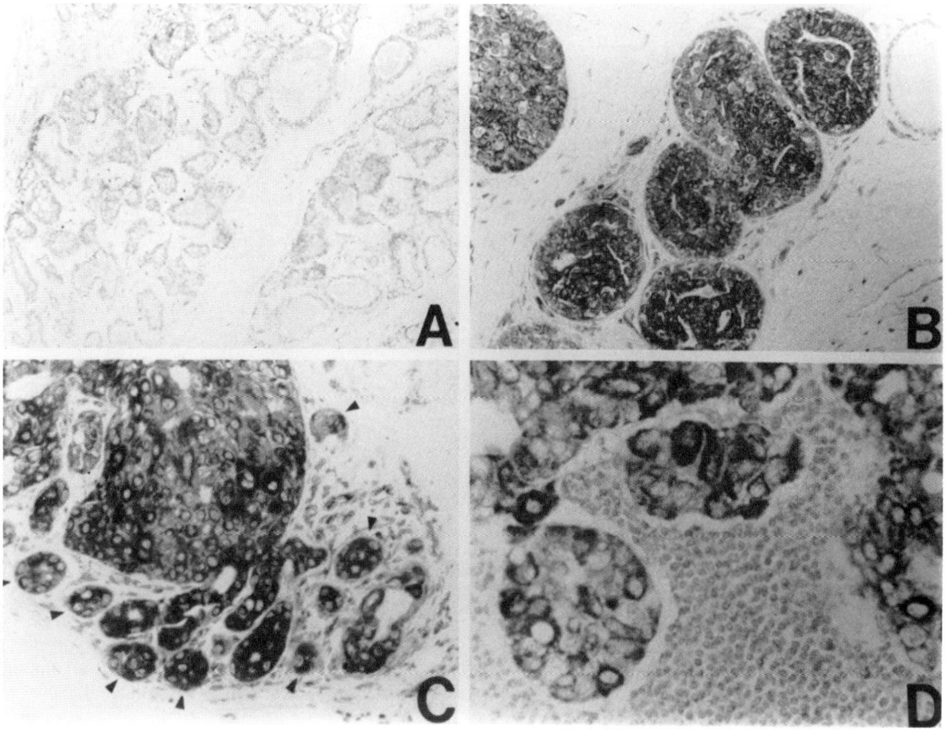

Figure 5.17 Immunohistochemical reactivity of benign and malignant breast tissues with monoclonal antibody RAP-5, counterstained with hematoxylin. (A) Lactating breast, non-reactive; (B) hyperplasia with atypia showing cytoplasmic reactivity; (C) intraductal carcinoma with microscopic invasion (arrows); (D) metastatic breast carcinoma (reactive) and lymph node (nonreactive). A, ×130; B and C, ×220; D, ×330. *Reproduced from Thor et al.,*[124] *with permission.*

cell expression, and this was inversely correlated with extent of tumor cell differentiation.

As in the studies of bladder carcinoma mentioned earlier, comparative analysis of malignant and normal tissues from a patient with lung carcinoma indicated that a single point mutation was to be responsible for the malignant activation of K-*ras* oncogene.[126] It appeared that arginine instead of glycine had been substituted at amino acid position 12 in a mutational event thought to be related to the development of squamous cell lung cancer in the patient studied.

Studies of human colorectal cancers using a combination of DNA hybridization and tissue-sectioning techniques have demonstrated frequent *ras* gene mutations of codon 12 of the c-Ki-*ras* gene[127]; furthermore, parallel studies of adenomatous premalignant tissues from the same patients indicated that the *ras* mutations usually preceded the development of malignancy. Similar findings emphasizing the frequency of mutations at position 12 in K-*ras* genes were reported by others.[128] The remarkably discrete focus of these codon 12 mutations is illustrated in Table 5.2 reproduced from Forrester et al.[128] Expression of the Ha-*ras* oncogene product p21 was studied by Czerniak and co-workers[129] during phases of the cell cycle in a human colonic cancer cell line. These studies indicated that the Ha p21 was cell cycle dependent but was not likely to be a primary regulator of cell cycle progression in this model.

Table 5.2 RNA·RNA Mismatches Generated by Mutations at Codon 12 of the Human c-K-*ras* Gene[a]

	Glycine GGU	Cysteine UGU	Serine AGU	Arginine CGU	Valine GUU	Aspartic GAU	Alanine GCU
Glycine	5' G G U 3' C C A	C G U U G A	A G U C C A	C G U C C A	G U U C G A	G A U C C A	G C U 3' C C A 5'
Cysteine	5' G G U 3' A C A	U G U A C A	A G U A C A	C G U A C A	G U U A C A	G A U A G A	G C U 3' A C A 5'
Serine	5' G G U 3' U C A	U G U U C A	A G U U C A	C G U U C A	G U U U C A	G A U U C A	G C U 3' U C A 5'
Valine	5' G G U 3' C A A	U G U C A A	A G U C A A	C G U C A A	G U U C A A	G A U C A A	G C U 3' C A A 5'

[a] Amino acids encoded by the normal (Gly) and mutant codon 12 of the human c-K-*ras* gene and the corresponding nucleotide sequence of the gene transcripts (reading 5′ to 3′) are indicated at the top. Nucleotide sequences of the 12th codon in antisense RNA (reading 3′ to 5′) corrsponding to the normal or mutant c-K-*ras* transcripts are shown paired to the codon 12 mRNA sequences. Perfectly paired RNA·RNA duplexes are indicated by circles. All single or double mismatches are boxed. Hatched boxes represent mismatches recognized and cleaved by RNase A; empty boxes, mismatches that have not be analysed; double box, the G-U mismatch that is not recognized by the enzyme.

(Reprinted from Forrester et al,[128] with permission.)

Leukemia

In some patients, presence of a transforming N-*ras* gene has been identified in acute myeloblastic leukemia,[130] whereas the gene was not present in normal fibroblasts from the same subjects. Later, using synthetic ologonucleotide probes, mutations of N-*ras* genes were found to occur randomly in various myeloid leukemia subtypes, whereas the N-*ras* mutations at codon 12 were more likely than other mutations to occur in lymphoid leukemias.[131]

Melanoma

Although initial restriction fragment length polymorphism studies suggested that some Ha-*ras* allele selection might occur in affected members of familial melanoma kindreds,[132,133] further studies of the distribution of rare Ha-*ras*-1 alleles has not confirmed this association.[134]

Metastases and *ras* Expression

Some degree of correlation may be present between H-*ras* expression and metastatic potential. Transfection with *ras* gene sequences, particularly in conjunction with *ras* activation, significantly increased metastatic potential in NIH 3T3 cells.[135] However, the precise characteristics necessary for induction of metastatic capability in a particular tumor are complex and probably involve more than one factor or activated oncogene.

*Erb*B and Epidermal Growth Factor

The concept that oncogenes are derived from normal cellular genes that regulate cell differentiation and normal cell growth received more momentum when it was discovered that one of the lesser known viral oncogenes—*erb* B—appeared to be derived from the gene coding for the receptor for EGF.[136] Like the finding that the product of the *sis* gene resembled the sequence and structure of PDGF, the *erb* B viral oncogene and its similarity to EGF provided evidence linking oncogenes to previously described normal growth factors.

The *erb* B oncogene was initially identified in avian erythroblastosis virus. The initial studies of *erb* B oncogene–EGF receptor peptide homologies suggested that the EGF receptor gene gave rise to *erb* B, since *erb* B and the EGF receptor gene map to similar positions on the long arm of human chromosome 7. The *erb* B gene sequence codes for a protein polypeptide chain product of 604 amino acids, which is about half the size of the 1250-amino-acid chain constituting the backbone of the entire EGF receptor.[137]

Previous extensive studies of the EGF receptor had indicated that it consists of three major regions. One region projects outside the cell and contains the binding site for EGF. The second region appears to be embedded in the cell membrane, and the third region extends inward into the cytosol. Initial studies of *erb* B–EGF receptor amino acid sequence homologies suggested that the *erb* B product corresponded to the transmembrane and cytoplasmic domains of the EGF receptor. It was originally proposed that transformation of cells by the avian erythroblastosis virus might result in part from inappropriate acquisition of a truncated EGF receptor as a product of the viral c-*erb* B gene.

After the similarities between the v-*erb* B and EGF receptor products were recognized, additional parallel studies identified an apparent close structural relationship between the domain organization of the human glucocorticoid receptor and the v-*erb* A oncogene product.[138] It was found that both the α and β forms of the human glucocorticoid receptor were closely related in domain structure and organization to the v-*erb* A proto-oncogene, indicating that they were probably derived from a common ancestral gene. These findings also suggested that the transforming activity of avian erythroblastosis virus could be related to inappropriate stimulation of a truncated or short-circuited steroid receptor or a related regulatory transducing molecule encoded by the homologous v-*erb* A. The amino acid similarities of the carboxyl-terminal half of the human glucocorticoid receptor and the v-*erb* A oncogene product are shown in Figure 5.18. As noted recently, the similarities in structure between oncogene products and important cell membrane hormone receptors may eventually help to elucidate the precise functioning of steroid receptors.[139] As the v-*erb* A and v-*erb* B products may differ with subsequent important functional differences, so too might steroid receptors on tumors or normal tissues show important differences in primary structure and function.[140]

The product of the human c-*erb* B2 gene has been shown to be a glycoprotein with tyrosine kinase activity[141]; however, EGF itself did not stimulate similar tyrosine kinase activity either in vivo or in vitro. The relationship of c-*erb* B2 and tyrosine kinase activity confirms the frequent association noted between oncogene products and tyrosine kinase enzymatic function.

C-*erb* B2 and Tumors

EGF receptors are expressed at high levels on the surfaces of some human squamous cell carcinomas[142] and brain tumors.[143] In some instances the c-*erb* B1/EGF receptor gene is translocated, amplified, or rearranged, suggesting that abnormalities of this gene family may be related to tumor development. When 101 fresh samples from a variety of human tumors were examined, c-*erb* B2 was noted to be amplified in 5 of 63 adenocarcinomas and the c-*erb* B1/EGF receptor gene was amplified in

```
hGR     S N G Y S S P S M R P D V S S P P S S S S T A T T G P P P K L -   420
v-erbA  E D T R W L D G K H K R K S S Q C L V K S S M S G Y I P S C L D    32

hGR     - - - - C L V C S D E A S G C H Y G V L T C G S C K V F F K R -   447
v-erbA  K D E Q C V V C G D K A T G Y H Y R C I T C E G C K S F F R R T    64

hGR     - A V E G Q H N Y L C A G R N D C I I D K I R R K N C P A C R Y   478
v-erbA  I Q K N L H P T Y S C T Y D G C C V I D K I T R N Q C Q L C R F    96

hGR     R K C L Q A G M - - - - - - - - - - - - - - - - - - - - - N L E A   490
v-erbA  K K C I S V G M A M D L V L D D S K R V A K R K L I E E N R E R   128

hGR     R K T K K K I K G I Q Q A T T G V S Q E T S E N P G N K T I V P   522
v-erbA  R R K E E M I K S L Q H R P S P S A E E W E L I H V V T E A H R   160

hGR     A T L P Q - - - L T P T L V S L L E V I E P E V L Y A G Y D S S   551
v-erbA  S T N A Q G S H W K Q R R K F L L E D I G Q S P M A S M L D G D   192

hGR     V P D S T W R I M T T L N M L G G R Q V I A A V K W A K A I P G   583
v-erbA  K V D L E A F S E F T - - K I I T P A I T R V V D F A K N L P M   222

hGR     F R N L H L D D Q M T L L Q Y S W M F L M A F A L G W R S Y R Q   615
v-erbA  F S E L P C E D Q I I L L K G C C M E I M S L R A A V R Y D P E   254

hGR     S S A N L L C F A P D L I I N E Q R M T L P C M Y D Q C K H M L   647
v-erbA  S E T L T L S G E M A V K R E Q L K N G G L G V V S D A I F D L   286

hGR     Y V S S E L H R L Q V S Y E E Y L C M K T L L L L S S V P K D G   679
v-erbA  G K S L S A F N L D D T E V A L L - - - Q A V L L M S S D R T G   315

hGR     L K S Q E L F D E I R M T Y I K E L G K A I V K R E G N S S Q N   711
v-erbA  L I C V D K I E K C Q E S Y L L A F E H Y I N Y R K H N I P H F   347

hGR     W Q R F Y Q L T K L L D S M H E V V E N L L N Y C F Q T F L D K   743
v-erbA  W - - - - - - - - S K L L M K V A D L R M I G A Y H A S R F L H M   372

hGR     T M S I E F P E M L A E I I T N Q I P K Y S N G N I K K L L F H   775
v-erbA  K V E C P T E L P P R R C R A L Q I L G S I L P F V   398

hGR     Q K   777
```

Figure 5.18 Amino acid sequence alignment of the carboxyl-terminal half of the human glucocorticoid receptor (hGR) and v-*erb* A oncogene product. Amino acid residues 370–777 from hGR were aligned with residues 1–398 from $075^{gag\text{-}erb\text{-}A}$; identical residues are bossed and gaps are indicated by dashes. Matching cysteine residues are indicated by dots above the sequence. *Reproduced from Weinberger et al.,*[138] *with permission.*

only 1 of 8 squamous cell carcinomas.[144] Other studies have indicated that c-*erb* B2 gene amplification is absent or very infrequent in sarcomas and malignant hematopoietic disorders.

HER-2/*neu*

Another oncogene, HER-2/*neu*, which is a member of the *erb* B–like oncogene family,[145] is amplified in various human breast carcinoma cell lines.[146] The HER-2/*neu* gene encodes yet another member of the tyrosine kinase family and differs from the EGF receptor (EGFR) gene in being located on band q21 of chromosome 17, compared to band p11–13 of chromosome 7 for the EGFR gene. HER-2/*neu*

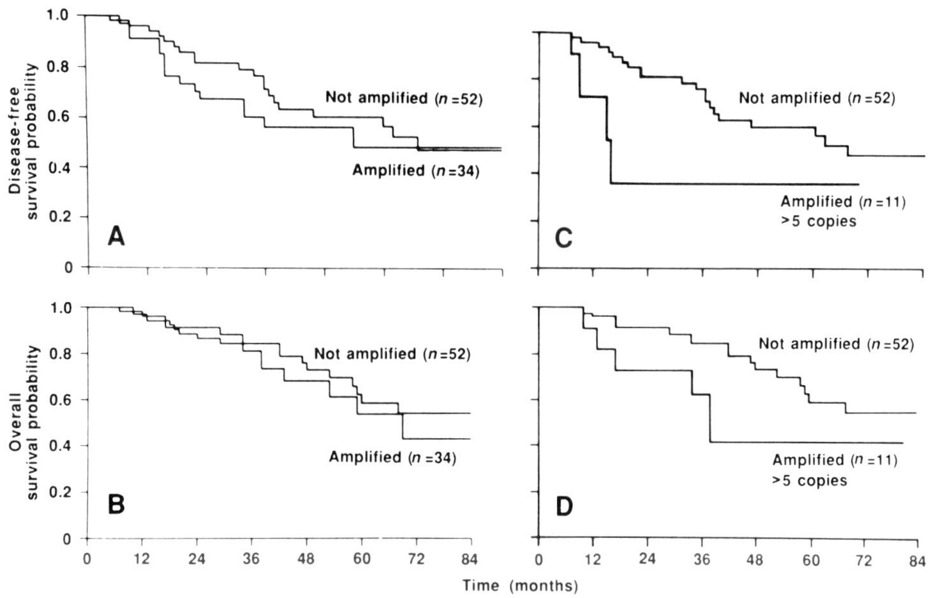

Figure 5.19 Actuarial curve for relapse in (A) node-positive patients with no amplification versus node-positive patients with any amplification (>two copies) of HER-2/*neu* and (C) node-positive patients with no amplification versus node-positive patients with greater than five copies of HER-2/*neu*. Actuarial curve for overall survival in (B) node-positive patients with no amplification versus node-positive patient with any amplification (>two copies) of HER-2/*neu* and (D) node-positive patients with no amplification versus node-positive patients with greater than five copies of HER-2/*neu*. *Reproduced from Slamon et al.,*[147] *with permission.*

produces a protein product that, like the EGFR protein, has an extracellular domain, a transmembrane segment containing two cysteine-rich clustered repeats, and an intracellular kinase domain,[145] indicating that this oncogene product probably functions as a cellular receptor for an as yet unidentified binding ligand.

Studies of HER-2/*neu* gene expression in a large number of tissues from human primary breast cancers indicate that gene amplification is frequently associated with disease relapse and overall survival.[147] Amplification of the HER-2/*neu* gene was a significant predictor of both overall survival and time to relapse in patients with breast cancer. Of interest in this study was the finding that HER-2/*neu* gene amplification appeared to have more prognostic value than other currently used features in breast cancer, such as hormonal receptor status or degree of regional lymph node involvement. Results of these studies are shown in relative actuarial curves for relapse in Figure 5.19. The authors also examined the tissues to determine whether HER-2/*neu* gene amplification was accompanied by similar increases in the EGFR genes; however, no parallel increment in the latter was detected. Similar studies of two other tyrosine kinase–related oncogenes, *abl* and *fes*, did not show significant amplification. These initial studies of HER-2/*neu* gene expression in human breast cancer may eventually provide the basis for an excellent new method for predicting outcome in one of the most common of human cancers. The results of this study are similar to those regarding neuroblastoma, where N-*myc* gene amplification also appeared to correlate with disease prognosis.[57,148] There are several possible mechanisms for oncogene involvement in cancer. One possibility is that cellular oncogenes in tumors are either overexpressed or expressed at an inappropriate time, thus greatly enhancing uncontrolled growth. A second possibility involves the mutational alteration of oncogenes (proto-oncogenes/oncogenes), producing gene products that cause cells to lose control of their own growth. As we have seen in the many examples already discussed, there is considerable evidence to support both hypotheses.

C-*abl* and Chronic Myelocytic Leukemia

Localization of cellular oncogenes near break points of tumor-specific chromosomal abnormalities suggests direct involvement of these genes in the production of neoplasia. Certainly this has been the case in the widely investigated translocations of gene sequences previously recognized in Burkitt's lymphoma. In 1982 it was demonstrated that chronic myelogenous leukemia (CML) involved a translocation from chromosome 9 to the Philadelphia chromosome and c-*abl*, related to the transforming sequence of the Abelson murine leukemia virus.[149] The Philadelphia chromosome (Ph) results from the specific translocation t(9;22) (q34;q11) and occurs in approximately 92% of adult patients with CML. The presence of Ph in CML is of prognostic importance because it indicates a better overall outlook and response to chemotherapy than in Ph-negative CML patients. The fact that the translocation also involves movement of the c-*abl* nucleotide sequence is of considerable interest. Additional studies of this c-*abl* translocation have confirmed its correlation with the presence of the Ph chromosome.[150] The mechanism whereby translocations of oncogene sequences could be directly related to malignant transformation is still not understood. In the case of the c-*abl* translocation to the Ph chromosome, the c-*abl* sequence is transferred from chromosome 9 to a site on chromosome 22 adjacent to the band containing the λ immunoglobulin light chain gene cluster. One mechanism that could also apply to the c-*myc* translocation in Burkitt's lymphoma would be that the new position of the oncogene nucleotide sequence allows it to function in an environment of active transcription[147,151] where it can initiate neoplastic transformation. Subsequent studies of the relationship of c-*abl* to CML have indicated that increased gene expression is present in cell lines representing CML blast crisis,[152] suggesting that the c-*abl* gene product may be directly involved in the transition from CML to blast crisis.

Although the exact relationship between c-*abl* and leukemia is not clear, HeisterKamp and co-workers[153] have reported that acute myeloid or lymphoid leukemia was induced in mice made transgenic for a *bcr*/*abl* p190 DNA construct. The *bcr*/*abl* fusion protein translated by the chimeric messenger RNA showed abnormal tyrosine kinase activity and strong in vitro autophosphorylation. Again, these experimental results reemphasize the causal relationship between the Ph chromosome and human leukemia.

DNA constructs capable of detecting residual evidence of cells positive for the Ph chromosome have been used to screen patients several years after bone marrow transplant treatment for chronic myelogenous leukemia.[154] In this practical application of recombinant techniques, patients considered cured 5 to 7 years after bone marrow transplant were studied using the polymerase chain reaction aimed at amplifying any remaining traces of the *bcr*/*abl* hybrid gene, and no evidence of leukemic mRNA was found.

Although relatively more is known about *abl* than some of the other oncogene products, its precise function in normal cells is unknown. The v-*abl* oncogene can

Table 5.3 Retroviral Oncogenes and their Cellular Homologs

Onco-gene	Viral origin[a]	Viral gene product	Cellular homolog	Activity	Subcellular localization
sis	Simian sarcoma virus	$p28^{sis}$	PDGF B chain	PDGF agonist	Cytoplasm
src	Rous sarcoma virus	$pp60^{v\text{-}src}$	$pp60^{c\text{-}src}$	Tyrosine kinase	Plasma membrane
yes	Y73 avian sarcoma virus	$p90^{gag\text{-}yes}$		Tyrosine kinase	
fps[b]	Fujinami sarcoma virus	$p140^{gag\text{-}fps}$	$p98^{c\text{-}fps}$	Tyrosine kinase	Cytoplasm, plasma membrane
fes[b]	Snyder-Theilen feline sarcoma virus	$p85^{gag\text{-}fes}$	$p92^{c\text{-}fes}$	Tyrosine kinase	
ros	UR2 avian sarcoma virus	$p68^{gag\text{-}ros}$		Tyrosine kinase	
abl	Abelson murine leukemia virus	$p120^{gag\text{-}abl}$	$p150^{c\text{-}abl}$	Tyrosine kinase	Plasma membrane
fgr	Gardner-Rasheed feline sarcoma virus	$p70^{gag\text{-}fgr}$		Tyrosine kinase	
erb B	Avian erythroblastosis virus	$gp65^{erb\,B}$	Truncated EGF receptor		Plasma membrane
mos	Moloney murine sarcoma virus	$p37^{mos}$			Cytoplasm
myc	Avian myelocytomatosis virus MC29	$p110^{gag\text{-}myc}$	$p58^{c\text{-}myc}$	Binds DNA	Nucleus
H-*ras*	Harvey murine sarcoma virus	$p21^{v\text{-}H\text{-}ras}$	$p21^{c\text{-}K\text{-}ras}$	Threonine kinase, binds GDP or GTP	Plasma membrane
K-*ras*	Kirsten murine sarcoma virus	$p21^{v\text{-}K\text{-}ras}$	$p21^{v\text{-}K\text{-}ras}$		
N-*ras*	[c]		$p21^{c\text{-}N\text{-}ras}$		

[a] Only one example of a virus strain is given for each oncogene.
[b] *fps* and *fes* are homologous genes from chicken and cat, respectively.
[c] The transforming gene product $p21^{N\text{-}ras}$ identified by transfection experiments.
(Reprinted from Heldin and Westermark,[155] with permission.)

be divided into two distinct regions, of which the amino-terminal 1.2 kb codes for a tyrosine-specific protein kinase. A comparative sequence analysis of v-*abl* indicates that, like v-*src*, v-*yes*, v-*fes*/*fps*, v-*fgr*, v-*erb* B, v-*ras*, and v-*fns*, it is a member of the general family of protein kinases.

Oncogenes, Growth Factors, and Their Interrelationships

The preceding discussion has covered several but by no means all of the known oncogenes or proto-oncogenes. A listing of several of the most widely studied retroviral oncogenes is shown in Table 15.3. How oncogenes participate in normal or abnormal body economy will be the subject of a great deal of exciting work in the future. The extensively demonstrated relationship between oncogene products and related growth factors, such as PDGF, osteosarcoma growth factor, fibroblast-derived growth factor, EGF, transforming growth factor, and insulin, as well as the several classes of insulin-like growth factors,[155] will eventually open many new pathways to our understanding of human disease processes.

Several main lines of evidence appear to link proto-oncogenes with carcinogenesis. Proto-oncogenes such as c-*myc* and c-*erb* appear to be activated in animal tumors in some instances by adjacent insertion of a tumor-promoting retrovirus, as in avian leukemia virus–induced lymphoid leukosis.[156] When coupled with the proper enhancer sequences, some viral oncogenes show remarkably selective tumor-inducing capacity. Thus the gene for SV40 large T antigen placed under control of a major histocompatibility complex class I enhancer produced choroid plexus papillomas and also lymphoid hyperplasia and multiple endocrine neoplasias involving the pancreas, pituitary, thyroid, adrenals, and testes in transgenic mice.[157] Some human and animal tumors contain proto-oncogenes, for instance, of the *ras* family, that can be activated by single point mutations and assayed by their ability to transform indicator cells such as the NIH 3T3 cell line. In other instances, oncogenes like c-*myc* and c-*abl* are located at chromosomal break points of specific translocations and gain access perhaps to fortuitous transforming opportunities. Finally, proto-oncogene amplification as noted in N-*myc* in neuroblastoma and retinoblastoma or Hc-2/*neu* in carcinoma of the breast may be directly related to inducing the neoplastic process.

Anti-Oncogenes or Tumor Suppressor Influences

Because as normal cellular genes proto-oncogenes appear in many instances to play an important role in normal body economy and because they appear to be related to an alternative pathway leading in some instances to neoplasia, it follows that normal cellular homeostasis must involve a control mechanism for regulating proto-oncogenes.[158] By implication, if such normal regulator-control genes are present, a mutation in such controlling regions could itself lead to carcinogenesis. Several lines of evidence do indeed suggest that genetic mechanisms exist to suppress neoplastic transformation. For instance, in various hereditary tumors associated with chromosomal deletions, there is loss of mutation of both copies of a specific allele.[159] Moreover, fusion of transformed malignant cells with normal cells frequently results in hybrids with a normal nonmalignant phenotype indicating that controlling genetic elements of the normal fusion partner can suppress the malignant phenotype. Genes that can produce negative regulation of neoplasia have been called anti-oncogenes or tumor suppressor genes. Conversely, dysfunction of anti-oncogenes may be important in the development of cancer. Thus, amplification of an oncogene may be able to overwhelm the anti-oncogene control. Alternatively, if the oncogene is integrated as part of a retrovirus, powerful viral promoter sequences could overpower intrinsic anti-oncogene control. Work in this area is proceeding very rapidly.

Intensively studied examples of what might be considered anti-oncogenes or oncogene control mechanisms are related to several childhood tumors including retinoblastomas and Wilms' tumor, and there is growing evidence that inactivation of both alleles of certain genes actually triggers cancerous growth.[160–163] In retinoblastoma, for instance, the genomic location determining susceptibility has been mapped to chromosome 13q14, and cDNA clones have been derived from this region which identify a putative retinoblastoma (*R*b) gene.[163] Interactions between oncogenes and anti-oncogenes have been discussed extensively, and protein-protein complexes between the oncogene adenovirus E1A protein and the retinoblastoma gene product have been demonstrated.[164] These findings indicate that two classes of genes—one stimulating various aspects of cell growth and the other inducing negative growth signals—may actually confront each other through interaction of their coded protein products. Of interest in this regard is the finding of apparent inactivation of the retinoblastoma susceptibility gene in human breast cancers.[165] Such observations indicate that loss of cancer suppressor genes or anti-oncogenes may be important in many other tumors besides retinoblastomas or Wilms' tumor.

A Series of Gene Changes May Result in Cancer

Rather than a single event, such as carcinogen oncogene activation, or even a double-hit scenario involving single oncogene activation and a single suppressor gene mutation, which may be relevant to retinoblastoma, evidence from a number of laboratories now appears to favor a stepwise accumulation of mutations that affect both oncogenes and suppressor genes.[166] Observations of many types of tumors support this concept. Much work in this area was stimulated by a series of observations of mutations in what is called the p53 gene. Since its discovery,[167,168] the p53 gene has received increasing attention. Although it was originally considered to be an oncogene, subsequent observations have shown that the wild-type gene product functions as a tumor suppressor gene.[169–171] Thus, cancer and the neoplastic cell phenotype in experimental models are inhibited rather than accelerated when rat cells are transfected with the murine wild-type p53 gene together with other oncogenes.[169] Moreover, in human tumors the short arm of chromosome 17, where p53 resides, is often deleted.[172] Mutations have now been reported in the p53 gene associated with diverse human tumor types.[173] It was found that most human tumors showing allelic deletions at chromosome 17p13.1 contained p53 point mutations resulting in amino acid substitutions. Moreover, such mutations were not confined to tumors with allelic deletions but were also found in at least some tumors that had retained both parental 17p alleles. The p53 gene mutations were clustered in four "hot spots" that coincided with the four most highly conserved regions of the gene. It had previously been noted that allelic deletions of chromosome 17p occurred in at least 60% of tumors of the colon, breast, lung, ovaries, cervix, adrenal cortex, bone, and bladder and in at least 30% of human brain tumors. The findings of mutations in p53 led to the hypothesis that during tumorigenesis mutations occurred in the p53 gene and, through a dominant cumulative effect, caused progression of the respective tumors.

Increased expression of mutant forms of p53 oncogene has been documented in many instances of primary lung cancer.[174] Direct evidence for homozygous expression of mutant p53 mRNA in various lung cancers was obtained using an asymmetric polymerase chain reaction mRNA sequencing approach that permitted sequencing without an intervening cloning step. Changes in p53 expression were detected using both monoclonal and polyclonal antibodies and immunochemical studies in sections of tumor tissues. Since p53 mutations were not restricted to a single site, it was thought that an immunochemical rather than a nucleic acid–based approach was most suitable for identification of p53 mutations.

Tumor Suppressor Genes in Human Colorectal Cancer

The concept that some cancers may be secondary to a loss of genetic function has been extended, particularly with respect to colorectal cancer.[175–177] Colorectal cancer has been the most extensively studied tumor with respect to molecular genetic changes linked to its progression. This has occurred because tissue changes that

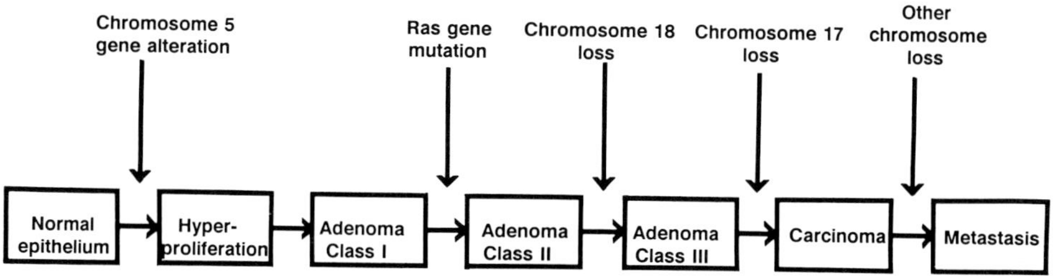

Figure 5.20 Model proposed by Vogelstein and co-workers for series of genetic alterations resulting in development of carcinoma of the colon. These changes include *ras* gene activation and subsequent loss of presumed tumor suppressor genes on chromosomes 5, 17, and 18.

represent a sequence of differentiation and proliferative events spanning hyperplasia of cells to development of benign adenomas to carcinoma in situ and finally metastasis are readily defined. This sort of progression is illustrated in Figure 5.20. The molecular events that parallel transitions from hyperplasia to adenomas and finally to colonic cancer have been studied using DNA from representative samples and assays for activated oncogenes and loss of restriction fragment length polymorphism (RFLP) heterozygosity, which might indicate loss of tumor suppressor gene activity. These studies showed activation of Ki-*ras* oncogene in 40% to 50% of patients and loss of genetic information from chromosomes 5q21–q22 in 35% of cases. Moreover, 17p12–13 (>70%) and 18q21–qter (>70% of cases) also showed loss of genetic material.[178–181] Identification of 5q21–q22 in sporadic cases of colorectal cancer is of interest because this general region had been identified by linkage studies as the locus for autosomal dominant familial adenomatous polyposis, which appears to predispose to colorectal cancer.

Identification of the spectrum of tumor suppressor genes related to colon cancer first highlighted the p53 gene on chromosome 17p. Each tumor assayed showed partial or complete deletion of one allele of the p53 gene.[171] Sequence analysis of the other allele (genomic or cDNA) showed that every case examined had at least one point mutation. Thus, the p53 gene met the criteria for a tumor suppressor gene; inactivation by a combination of allele loss and point mutation resulted in development of tumor. The p53 protein is a DNA-binding protein that functions as a homodimer. It is postulated that interaction of one mutant subunit with a wild-type subunit may be enough to subvert normal cell function. Recently, the field has been extended by cloning of the candidate tumor suppressor gene on chromosome 18q21–qter.[182] This gene is expressed in many normal tissues, including normal colonic mucosa; however, absent or reduced expression is uniformly noted in colorectal carcinomas. The gene itself is quite large, with an RNA transcript size of 10 to 12 kb. Somatic mutations have been observed within the gene in association with many colorectal cancers, but their significance remains to be determined. Interestingly, this last tumor suppressor gene, designated DCC (deleted in colorectal carcinomas), has a significant homology to the gene for neural adhesion molecules. Whether homology to neural adhesions or fibronectin-like structures has any relevance to the process of metastasis has not yet been fully explored. Work now going on has been directed to extending these initial observations.

Caveats Concerning Oncogenes

An opposite point of view about oncogenes has been advanced by Duesberg,[183] who suggested that regarding the 20 known transforming *onc* genes of retroviruses as latent cancer genes has been seriously overdone. The proto-oncogene hypothesis predicts that (1) viral *onc* genes and proto-oncogenes are isogenic, (2) activated proto-oncogenes transform diploid cells upon transfection, and (3) expression of proto-oncogenes produces tumors. Duesberg points out that, as yet, none of these predictions has been confirmed. Thus far, structural comparisons between retroviral oncogenes, essential retroviral genes, and proto-oncogenes have shown that all viral oncogenes are indeed new genes. Moreover, proto-oncogenes are often expressed in normal cells, and not a single activated proto-oncogene has been isolated that can directly transform diploid cells. Finally, no diploid tumors with activated proto-oncogenes have been identified. Duesberg suggests that the hypothesis that proto-oncogenes are latent cellular oncogenes may be an overinterpretation of sequence homologies between normal cellular genes and viral oncogenes. Alternatively, only very rare genetic events, such as truncations or illegitimate recombinations that permanently alter the germ-line sequence of cellular genes, are capable of producing viral or intrinsic cancer genes. Moreover, chromosome clonal abnormalities consistently recorded in tumor cells may be direct evidence for production of intrinsic cancer genes. This alternative point of view certainly must be considered.

Whatever the eventual outcome of work in this area, it is clear that the application of modern molecular biological techniques to the study of growth factors, oncogenes, tumor suppressor genes, and cell differentiation will vastly increase our understanding of both normal and abnormal body function.

References

1. Tooze J: *The Molecular Biology of Tumor Viruses*. New York, Cold Spring Harbor Laboratory, 1973.
2. Gross L: *Oncogenic Viruses*. New York, Pergamon, 1970.
3. Huebner RJ, Todaro GL: Oncogenes of RNA tumor viruses as determinants of cancer. *Proc Natl Acad Sci USA* 1969;64:1087–1094.
4. Temin HM: The protovirus hypothesis: Speculations on the significance of RNA directed DNA synthesis for normal development and for carcinogenesis. *J Natl Cancer Inst* 1971;46:215–226.
5. Duesberg PH, Vogt PK: Differences between the ribonucleic acids of transforming and nontransforming avian tumor viruses. *Proc Natl Acad Sci USA* 1970;67:1673–1680.
6. Duesberg PH, Vogt PK: RNA species obtained from cloned lines of avian sarcoma and from avian leukosis virus. *Virology* 1973;54:207–219.
7. Lai MM, Duesberg PH, Horst J, Vogt PK: Avian tumor virus RNA: A comparison of three sarcoma viruses and their transformation-defective derivatives by oligonucleotide fingerprinting and DNA-RNA hybridization. *Proc Natl Acad Sci USA* 1973;70:2266–2270.
8. Sambrook J, Green R, Stringer J, et al: Analysis of the sites of integration of viral DNA sequences in rat cells transformed by adenovirus 2 or SV40. *Cold Spring Harbor Symp Quant Biol* 1980;44:569–584.
9. Stringer JR: DNA sequence homology and chromosomal deletion at a site of SV40 DNA integration. *Nature* 1982;296:363–366.
10. Hassell JA, Topp WC, Rifkin DB, Moreau P: Transformation of rat embryo fibroblasts by cloned polyoma virus DNA fragments containing only part of the early region. *Proc Natl Acad Sci USA* 1980;77:3978–3982.
11. Ito Y, Spurr N: Polyoma virus T antigens expressed in transformed cells: Significance

of middle T antigen in transformation. *Cold Spring Harbor Symp Quant Biol* 1980; 44:149–159.
12. Lania L, Gandini-Attardi G, Griffiths M, et al: The polyoma virus 100K large T antigens is not required for the maintenance of transformation. *Virology* 1980;101:217–232.
13. Novak Y, Dilworth SM, Griffin BE: Coding capacity of a 35% fragment of the polyoma virus genome is sufficient to initiate and maintain cellular transformation. *Proc Natl Acad Sci USA* 1980;77:3278–3282.
14. Rassoulzadegan M, Cowie A, Carr A, et al: The roles of individual polyoma virus early proteins in oncogenic transformation. *Nature* 1982;300:713–718.
15. Crawford LV, Cole CN, Smith AE, et al: The organization and expression of simian virus 40's early genes. *Proc Natl Acad Sci USA* 1978;75:117–121.
16. Cole CN, Landers T, Goff SP, et al: Physical and genetic characterization of deletion mutants of simian virus 40 constructed in vitro. *J Virol* 1977;24:277–294.
17. Ghush PK, Lebowitz P, Frisque RJ, Gluzman Y: Identification of a promoter component involved in positioning the 5′ termini of simian virus 40 early mRNA. *Proc Natl Acad Sci USA* 1981;78:100–104.
18. Mathis DJ, Chambon P: The SV40 early region TATA box is required for accurate in vitro initiation of transcription. *Nature* 1981;290:310–315.
19. Contrerase R, Gheysen D, Knowland J, et al: Evidence for the direct involvement of DNA replication origin in synthesis of late SV40 RNA. *Nature* 1982;300:500–505.
20. Soeda E, Arrand JR, Smolar N, Griffin BE: Sequence from early region of polyoma virus DNA containing the viral replication origin and encoding small, middle and (part of) large T antigens. *Cell* 1979;17:357–370.
21. Varmus HE: Form and function of retroviral proviruses. *Science* 1982;216:812–820.
22. Martin GS: Rous sarcoma virus: A function required for the maintenance of the transformed state. *Nature* 1970;227:1021–1023.
23. Roussel M, Saule S, Lagrou C, et al: Three new types of viral oncogene of cellular origin specific for haematopoietic cell transformation. *Nature* 1979;281:452–455.
24. Hanafusa T, Wang L-H, Anderson SM, et al: Characterization of the transforming gene of Fujinami sarcoma virus. *Proc Natl Acad Sci USA* 1980;77:3009–3013.
25. Bishop JM, Courtneidge SA, Levinson AD, et al: Origin and function of avian retrovirus transforming genes. *Cold Spring Harbor Symp Quant Biol* 1980;44:919–930.
26. Ueshima Y, Fukuhara S, Hattori T, et al: Chromosome studies in adult T-cell leukemia in Japan: significance of trisomy 7. *Blood* 1981;58:420–425.
27. Miyamoto K, Sato J, Kitajima K-I, et al: Adult T-cell leukemia chromosome analysis of 15 cases. *Cancer* 1981;52:471–478.
28. Bishop JM: Oncogenes. *Sci Am* 1982;246(3):80–92.
29. Temin HM: Function of the retrovirus long terminal repeat. *Cell* 1982;28:3–5.
30. Purchio AF, Erikson E, Brugge JS, Erikson RL: Identification of a polypeptide encoded by the avian sarcoma virus *src* gene. *Proc Natl Acad Sci USA* 1978;75:1567–1571.
31. Collet MS, Erikson RL: Protein kinase activity associated with the avian sarcoma virus *src* gene product. *Proc Natl Acad Sci USA* 1978;75:2021–2024.
32. Willingham MC, Jay G, Pastan I: Localization of the ASV *src* gene product to the plasma membrane of the transformed cells by electron microscopic immunocytochemistry. *Cell* 1979;18:125–134.
33. Langbeheim H, Shih TY, Scolnick EM: Identification of a normal vertebrate cell protein related to the P^{21} *src* of Harvey murine sarcoma virus. *Virology* 1980;106:292–300.
34. Oskarsson M, McClements WL, Blair DG, et al: Properties of a normal mouse cell DNA sequence (*sarc*) homologous to the *src* sequence of Moloney sarcoma virus. *Science* 1980;207:1222–1224.
35. Kelly K, Cochran BH, Stiles CD, Leder P: Cell-specific regulation of the c-*myc* gene by lymphocyte mitogens and platelet-derived growth factor. *Cell* 1983;35:603–610.
36. Lacy J, Sarkar SN, Summers WC: Induction of c-*myc* expression in human B lymphocytes by B-cell growth factor and anti-immunoglobulin. *Proc Natl Acad Sci USA* 1986;83:1458–1462.
37. Persson H, Leder P: Nuclear localization and DNA binding properties of a protein expressed by human c-*myc* oncogene. *Science* 1984;225:718–721.
38. Persson H, Hennighausen L, Taub R, et al: Antibodies to human c-*myc* oncogene product evidence of an evolutionarily conserved protein induced during cell proliferation. *Science* 1984;225:687–693.

39. Dalla-Favera R, Bregni M, Erickson J, et al: Human c-*myc onc* gene is located on the region of chromosome 8 that is translocated in Burkitt lymphoma cells. *Proc Natl Acad Sci USA* 1982;79:7824–7827.
40. Neel BG, Jhanwar SC, Chaganti RSK, Hayward WS: Two human c-*onc* genes are located on the long arm of chromosome 8. *Proc Natl Acad Sci USA* 1982;79:7842–7846.
41. Taub R, Kirsch I, Morton C, et al: Translocation of the c-*myc* gene into the immunoglobulin heavy chain locus in human Burkitt lymphoma and murine plasmacytoma cells. *Proc Natl Acad Sci USA* 1982;79:7837–7841.
42. Hamlyn PH, Rabbitts TH: Translocation joins c-*myc* and immunoglobulin q1 genes in a Burkitt lymphoma revealing a third exon in the c-*myc* oncogene. *Nature* 1983; 304:135–139.
43. Nishikura K, Ar-Rushdi A, Erikson J, et al: Differential expression of the normal and of the translocated human c-*myc* oncogenes in B cells. *Proc Natl Acad Sci USA* 1983; 80:4822–4826.
44. Leder P, Battey J, Lenoir G, et al: Translocations among antibody genes in human cancer. *Science* 1983;222:765–771.
45. Taub R, Moulding C, Battey J, et al: Activation and somatic mutation of the translocated c-*myc* gene in Burkitt lymphoma cells. *Cell* 1984;36:339–348.
46. Rabbitts TH, Hamlyn PH, Baer R: Altered nucleotide sequences of a translocated c-*myc* gene in Burkitt lymphoma. *Nature* 1983;306:760–765.
47. Giallongo A, Apella E, Ricciardi R, et al: Identification of the c-*myc* oncogene product in normal and malignant B cells. *Science* 1983;222:430–432.
48. Thompson CB, Challoner PB, Nieman PE, Groudine M: Expression of the c-*myb* protooncogene during cellular proliferation. *Nature* 1986;319:374–380.
49. Pelicci P-G, Lanfrancone L, Brathwaite MD, et al: Amplification of the c-*myb* oncogene in a case of human acute myelogenous leukemia. *Science* 1984;224:1117–1121.
50. Ralston R, Bishop JM: The protein products of the *myc* and *myb* oncogenes and adenovirus E1a are structurally related. *Nature* 1983;306:803–806.
51. Kohl NE, Kanda N, Schreck RR, et al: Transposition and amplification of oncogene-related sequences in human neuroblastomas. *Cell* 1983;35:359–367.
52. Schwab M, Alitalo K, Klempnauer K-H, et al: Amplified DNA with limited homology to *myc* cellular oncogene is shared by human neuroblastoma cell lines and a neuroblastoma tumor. *Nature* 1983;305:245–248.
53. Kohl NE, Gee CE, Alt FW: Activated expression of the N-*myc* gene in human neuroblastomas and related tumors. *Science* 1984;226:1335–1337.
54. Nau MM, Carney DN, Battey J, et al: *Curr Top Microbiol Immunol* 1984;113:172–177.
55. Little CD, Nau MM, Carney DN, et al: Amplification and expression of the c-*myc* oncogene in human lung cancer cell lines. *Nature* 1983;306:194–196.
56. Schwab M, Varmus HE, Bishop JM: Human N-*myc* gene contributes to neoplastic transformation of mammalian cells in culture. *Nature* 1985;316:160–162.
57. Brodeur GM, Seeger RC, Schwab M, et al: Amplification of N-*myc* in untreated human neuroblastomas correlates with advanced disease stage. *Science* 1984;224:1121–1124.
58. Schwab M, Alitalo K, Klempnauer KH, et al: Amplified DNA with limited homology to *myc* cellular oncogene is shared by human neuroblastoma cell lines and a neuroblastoma tumor. *Nature* 1983;305:245–248.
59. Kaczmarek L, Hyland JK, Watt R, et al: Microinjected c-*myc* as a competence factor. *Science* 1985;228:1313–1315.
60. Adams JM, Harris AW, Pinkert CA, et al: The c-*myc* oncogene driven by immunoglobulin enhancers induces lymphoid malignancy in transgenic mice. *Nature* 1985; 318;533–538.
61. Harris AW, Pinkert CA, Crawford M, et al: The E-*myc* transgenic mouse: A model for high-incidence spontaneous lymphoma and leukemia for early B cells. *J Exp Med* 1988; 167:353–371.
62. Knight KL, Spieker-Polet H, Kazdin DS, Oi VT: Transgenic rabbits with lymphocytic leukemia induced by the c-*myc* oncogene enhancer. *Proc Natl Acad Sci USA* 1988; 85:3130–3134.
63. Alitalo K, Winqvist R, Keski-Oja J, et al: Acute myelogenous leukemia with c-*myc* amplification and double minute chromosomes. *Lancet* 1985;2:1035–1038.

64. Stewart TA, Pettengale PK, Leder P: Spontaneous mammary adenocarcinomas in transgenic mice that carry and express MTV/*myc* fusion genes. *Cell* 1984;38:627–637.
65. Croce CM, Kozbor D: Amplification of the c-*myc* oncogene in one of five human breast carcinoma cell lines. *Cancer Res* 1984;44:438–441.
66. Slamon DJ, de Kermon JB, Verma IM, Cline MJ: Expression of cellular oncogenes in human malignancies. *Science* 1984;224:256–264.
67. Levy LS, Gardner MB, Casey JW: Isolation of a feline leukemia provirus containing the oncogene *myc* from a feline lymphosarcoma. *Nature* 1984;308:853–856.
68. Mullins JI, Brody DS, Binari RC Jr, Cotter SM: Viral transduction of c-*myc* gene in naturally occurring feline leukemias. *Nature* 1984;308:856–858.
69. Rapp UR, Cleveland JL, Brightman K, et al: Abrogation of IL-3 and IL-2 dependence by recombinant murine retroviruses expressing v-*myc* oncogenes. *Nature* 1985; 317:434–438.
70. Cochran BH, Reffel AC, Stiles CD: Molecular cloning of gene sequences regulated by platelet-derived growth factor. *Cell* 1983;33:939–947.
71. Doolittle RF, Hunkapiller MW, Hood LE, et al: Simian sarcoma virus *onc* gene, v-*sis* is derived from the gene or (genes) encoding a platelet-derived growth factor. *Science* 1983;221:275–277.
72. Waterfield MD, Scrace GT, Whittle N, et al: Platelet-derived growth factor is structurally related to the putative transforming protein p38 *sis* of simian sarcoma virus. *Nature* 1983;304:35–39.
73. Josephs SF, Guo C, Raner L, Wong-Staal F: Human proto-oncogene nucleotide sequences corresponding to the transforming region of simian sarcoma virus. *Science* 1984;223:487–491.
74. Leal F, Williams LT, Robbins KC, Aaronson SA: Evidence that the v-*sis* gene product transforms by interaction with the receptor for platelet-derived growth factor. *Science* 1985;230:327–330.
75. Niman HL: Antisera to a synthetic peptide of the *sis* viral oncogene product recognize human platelet-derived growth factor. *Nature* 1984;307:180–183.
76. Johnsson A, Betsholtz C, Heldin C-H, Westermark B: Antibodies against platelet-derived growth factor inhibit acute transformation by simian sarcoma virus. *Nature* 1985; 317:438–440.
77. Cochran BH, Zullo J, Verma IM, Stiles CD: Expression of the c-*fos* and of a *fos*-related gene is stimulated. *Science* 1984;226:1080–1082.
78. Curran T, Miller AD, Zokas L, Verma IM: Viral and cellular *fos* proteins. *Cell* 1984; 36:259–268.
79. Balk SD, Whitfield JF, Youdale T, Braun AC: Roles of calcium, serum, plasma, and folic acid in the control of proliferation of normal and Rous sarcoma virus–infected chicken fibroblasts. *Proc Natl Acad Sci USA* 1973;70:675–679.
80. Kohler N, Lipton A: Platelets as a source of fibroblast growth-promoting activity. *Exp Cell Res* 1974;87:297–301.
81. Ross R, Glomset J, Kariya B, Harker L: A platelet-dependent serum factor that stimulates the proliferation of arterial smooth muscle cell in vitro. *Proc Natl Acad Sci USA* 1974;71:1207–1210.
82. Graves DT, Owen AJ, Barth RK, et al: Detection of c-*sis* transcripts and synthesis of PDGF-like proteins by human osteosarcoma cells. *Science* 1984;226:972–974.
83. Bronzert DA, Pantazis P, Antoniades HN, et al: Synthesis and secretion of platelet-derived growth factor by human breast cancer cell lines. *Proc Natl Acad Sci USA* 1987; 84:5763–5767.
84. Clemmons DR, Underwood LE, Van Wyck JJ: Hormonal control of immunoreactive somatomedin production by cultural human fibroblasts. *J Clin Invest* 1981;67:10–16.
85. Habenicht AJR, Glomset JA, King WC, et al: Early changes in phosphatidylinositol and arachidonic acid metabolism in quiescent Swiss 3T3 cell stimulated to divide by platelet-derived growth factor. *J Biol Chem* 1981;256:12329–12335.
86. Ross R: Platelet-derived growth factor. *Annu Rev Med* 1987;38:71–79.
87. Deuel TF, Senior RM, Huang JS, Griffin GL: Chemotaxis of monocytes and neutrophils to platelet-derived growth factor. *J Clin Invest* 1982;69:1046–1049.
88. Martinet Y, Rom WN, Grotendorst GR, et al: Exaggerated spontaneous release of platelet-derived growth factor by alveolar macrophages from patients with idiopathic pulmonary fibrosis. *N Engl J Med* 1987;317:202–209.

89. Nilsson J, Svensson J, Hamsten A, de Faire U: Increased platelet-derived mitogenic activity in plasma of young patients with coronary atherosclerosis. *Atherosclerosis* 1986; 61:237–243.
90. Nilsson J, Sjolund M, Palmberg T, et al: Arterial smooth muscle cells produce a platelet-derived growth factor protein. *Proc Natl Acad Sci USA* 1985;82:4418–4422.
91. DiCorleto PE, Bowen-Pope DF: Cultured endothelial cells produce a platelet-derived growth factor–like protein. *Proc Natl Acad Sci USA* 1983;80:1919–1923.
92. Barrett TB, Benditt EP: *Sis* (platelet-derived growth factor B chain) gene transcript levels are elevated in human atherosclerotic lesions compared to normal artery. *Proc Natl Acad Sci USA* 1987;84:1099–1103.
93. Sitaras NM, Sariban E, Pantazis P, et al: Human iliac artery endothelial cells express both genes encoding the chains of platelet-derived growth factor (PDGF) and synthesize PDGF-like mitogen. *J Cell Physiol* 1987;132:376–380.
94. Penn A, Garte SJ, Warren L, et al: Transforming gene in human atherosclerotic plaque DNA. *Proc Natl Acad Sci USA* 1986;83:7951–7955.
95. Kobilka BK, Dixon RAF, Frielle T, et al: cDNA for the human β_2-adrenergic receptor: A protein with multiple membrane-spanning domains and encoded by a gene whose chromosomal location is shared with that of the receptor for platelet-derived growth factor. *Proc Natl Acad Sci USA* 1987;84:46–50.
96. Libby P, Warner SJC, Salomon RN, Birinyi LK: Production of platelet-derived growth factor–like mitogen by smooth-muscle cells from human atheroma. *N Engl J Med* 1988; 318:1493–1498.
97. Pierce GF, Mustoe TA, Senior RM, et al: In vivo incisional wound healing augmented by platelet-derived growth factor and recombinant c-*sis* gene homodimeric proteins. *J Exp Med* 1988;167:974–987.
98. Bauer EA, Silverman N, Busiek DF, et al: Diminished response of Werner's syndrome fibroblasts to growth factors PDGF and FGF. *Science* 1986;234:1240–1243.
99. Harvey JJ: An unidentified virus which causes the rapid production of tumours in mice. *Nature* 1964;204:1104–1105.
100. Kirsten WH, Mayer LA: Morphologic responses to a murine erythroblastosis virus. *J Natl Cancer Inst* 1967;39:311–335.
101. Hayward WS, Neel BG, Astrin SM: Activation of a cellular *onc* gene by promoter insertion in ALV-induced lymphoid leukosis. *Nature* 1981;290:475–480.
102. Brock H: Sequence and genomic structure of *ras* homologues Dm *ras* 85D and Dm *ras* 64B of *Drosophila melanogaster*. *Gene* 1987;51:129–137.
103. Scolnick EM, Papageorge AG, Shih TY: Guanine nucleotide–binding activity as an assay for *src* protein of rat-derived murine sarcoma virus. *Proc Natl Acad Sci USA* 1979; 76:5355–5359.
104. Temeles GL, Gibbs JB, D'Alonzo JS, et al: Yeast and mammalian *ras* proteins have conserved biochemical properties. *Nature* 1985;313:700–703.
105. Willumsen BM, Christensen A, Hubbert NL, et al: The p21 *ras* C-terminus is required for transformation and membrane association. *Nature* 1984;310:583–586.
106. Hurley JB, Simon MI, Teplow D, et al: Homologies between signal transducing G proteins and *ras* gene products. *Science* 1984;226:860–862.
107. Nurse P: Oncogenes yeast aids cancer research. *Nature* 1985;313:631–632.
108. Michell B: Oncogenes and inositol lipids. *Nature* 1984;308:770.
109. Seeburg PH, Colby WW, Capon DJ, et al: Biological properties of human c-Ha-*ras* 1 genes mutated at codon 12. *Nature* 1984;312:71–75.
110. Kung HF, Smith MR, Bekesi E, et al: Reversal of transformed phenotype by monoclonal antibodies against Ha-*ras* p21 proteins. *Exp Cell Res* 1986;162:363–371.
111. deVos AM, Tong L, Milburn MV, et al: Three dimensional structure of an oncogene protein: Catalytic domain of human c-H-*ras* p21. *Science* 1988;239:888–893.
112. Change EH, Furth ME, Scolnick EM, Lowry DR: Tumorigenic transformation of mammalian cells induced by a normal human gene homologous to the oncogene of Harvey murine sarcoma virus. *Nature* 1982;297:479–483.
113. Korn LJ, Siebel CW, McCormick F, Roth RA: *Ras* p21 as a potential mediator of insulin action in *Xenopus* oocytes. *Science* 1987;236:840–843.
114. Flier JS, Mueckler MM, Usher P, Lodish HF: Elevated levels of glucose transport and transporter messenger RNA are induced by *ras* or *src* oncogenes. *Science* 1987; 235:1492–1495.

115. Andres A-C, Schonenberger C-A, Groner B, et al: Ha-*ras* oncogene expression directed by a milk protein gene promoter: Tissue specificity, hormonal regulation, and tumor induction in transgenic mice. *Proc Natl Acad Sci USA* 1987;84:1299–1303.
116. Ishii S, Merlio GT, Pastan I: Promoter region of the human Harvey *ras* proto-oncogene: Similarity to the EGF receptor proto-oncogene promoter. *Science* 1985;230:1378–1381.
117. Nakagawa T, Mabry M, deBustros A, et al: Introduction of v-Ha-*ras* oncogene induces differentiation of cultured human medullary thyroid carcinoma cells. *Proc Natl Acad Sci USA* 1987;84:5923–5927.
118. Sukumar S, Notario V, Martin-Zanca D, Barbacid M: Induction of mammary carcinomas in rats by Nitrosomethylurea involves malignant activation of H-*ras*-1 locus by single point mutations. *Nature* 1983;306:658–661.
119. Doniger J: Molecular characterization of oncogenes in guinea pig lines chemically initiated *in vitro* acquisition of tumorgenicity is associated with activated *ras* related oncogenes. In Barrett J, Tennant RW (eds): *Carcinogenesis: A Comprehensive Survey*. Vol 9. *Mammalian Cell Transformation, Mechanisms of Carcinogenesis and Assays for Carcinogens*. New York, Raven, 1985, pp 51–55.
120. Doniger J, Natario V, DiPaolo JA: Carcinogens with diverse mutagenic activities initiate neoplastic guinea pig cells that acquire the same N-*ras* point mutation. *J Biol Chem* 1987;262:3813–3819.
121. Maizels M, Simpson SB: The human c-ras_1^H oncogene: A mutation in normal and neoplastic tissue from the same patient. *Science* 1982;219:509–510.
122. Barbacid M: *Ras* genes. *Annu Rev Biochem* 1987;56:779–827.
123. Hand PH, Vilasi V, Thor A, et al: Quantitation of Harvey *ras* p21 enhanced expression in human breast and colon carcinomas. *J Natl Cancer Inst* 1987;79:59–65.
124. Thor A, Ochuchi N, Hand PH, et al: *Ras* gene alterations and enhanced levels of *ras* p21 expression in a spectrum of benign and malignant human mammary tissues. *Lab Invest* 1986;55:603–615.
125. Nonomura A, Ohta G, Hayashi M, et al: Immunohistochemical detection of *ras* oncogene p21 product in liver cirrhosis and hepatocellular carcinoma. *Am J Gastroenterol* 1987;82:512–518.
126. Santos E, Martin-Zanca D, Reddy EP, et al: Malignant activation of a K-*ras* oncogene in lung carcinoma but not in normal tissue from the same patient. *Science* 1984; 223:661–664.
127. Bos JL, Fearon ER, Hamilton SR, et al: Prevalence of *ras* gene mutations in human colorectal cancers. *Nature* 1987;327:293–297.
128. Forrester K, Almoguera C, Han K, et al: Detection of high incidence of K-*ras* oncogenes during human colon tumor genesis. *Nature* 1987;327:298–303.
129. Czerniak B, Herz F, Wersto RP, Koss LG: Modification of Ha-*ras* oncogene p21 expression and cell cycle progression in the human colonic cancer cell line HT-29. *Cancer Res* 1987;47:2826–2830.
130. Gambke C, Signer E, Moroni C: Activation of N-*ras* in bone marrow cells from a patient with acute myeloblastic leukaemia. *Nature* 1984;307:476–478.
131. Nishida J, Hirai H, Takaku F: Activation mechanism of the N-*ras* oncogene in human leukemias detected by synthetic oligonucleotide probes. *Biochem Biophys Res Commun* 1987;147:870–875.
132. Sutherland C, Shaw HM, Roberts C, et al: Harvey-*ras* oncogene restriction fragment alleles in familial melanoma kindreds. *Br J Cancer* 1986;54:787–790.
133. Krontiris TG, DiMartino NA, Colb M, Parkinson DR: Unique allelic restriction fragments of the human Ha-*ras* locus in leukocyte and tumor DNAs of cancer patients. *Nature* 1985;313:369–374.
134. Gerhard DS, Dracopoli NC, Bales SJ, et al: Evidence against Ha-*ras*-1 involvement in sporadic and familial melanoma. *Nature* 1987;325:73–75.
135. Egan SE, McLarty GA, Jarolim L, et al: Expression of H-*ras* correlates with metastatic potential: Evidence for direct regulation of the metastatic phenotype in 10T1/2 and NIH 3T3 cells. *Mol Cell Biol* 1987;7:830–837.
136. Downward J, Yarden Y, Mayes E, et al: Close similarity of epidermal growth factor receptor and v-*erb*-B oncogene protein sequences. *Nature* 1984;307:521–527.

137. Yamamoto T, Nishida T, Miyajima N, et al: The *erb* B gene of avian erythroblastosis virus is a member of the *src* gene family. *Cell* 1983;35:71–78.
138. Weinberger C, Hollenberg SM, Rosenfield MG, Evans RM: Domain structure of human glucocorticoid receptor and its relationship to the v-*erb*-A oncogene product. *Nature* 1985;318:670–672.
139. Parker MG: Oncogenes from hormone receptors. *J Endocrinol* 1987;112:1–2.
140. Krust A, Green S, Argos P, et al: The chicken oestrogen receptor sequence: homology with V-*erb*A and the human oestrogen and glucocorticoid receptors. *EMBO J* 1986; 5:891–897.
141. Akiyama T, Sudo C, Ogawara H, et al: The product of the human c-*erb* B-2 gene: A 185-kilodalton glycoprotein with tyrosine kinase activity. *Science* 1986;232:1644–1646.
142. Coeley G, Smith JA, Gusterson B, et al: The amount of EGF receptor is elevated on squamous cell carcinomas. *Cancer Cells* 1984;1:5–10.
143. Libermann TA, Razon N, Bartal AD, et al: Expression of epidermoid growth factor receptors in human brain tumors. *Cancer Res* 1984;44:753–760.
144. Yokota J, Toyoshima K, Sugimura T, et al: Amplification of c-*erb*-B-2 oncogene in human adenocarcinomas in vivo. *Lancet* 1986;1:765–766.
145. Schechter AL, Stern DF, Vaidyanathan L, et al: The *neu* oncogene: An *erb*-B–related gene encoding of a 185,000-M_r tumor antigen. *Nature* 1984;312:513–516.
146. King CR, Kraus MH, Aaronson SA: Amplification of a novel v-*erb* B–related gene in a human mammary carcinoma. *Science* 1985;229:974–976.
147. Slamon DJ, Clark GM, Wong SG, et al: Human breast cancer: Correlation of relapse and survival with amplification of the HER-2/*neu* oncogene. *Science* 1987;235:177–181.
148. Seeger RC, Brodeur GM, Sather H, et al: Association of multiple copies of the N-*myc* oncogene with rapid progression of neuroblastoma. *N Engl J Med* 1986;313:1111–1116.
149. deKlein A, van Kessel AG, Grosveld G, et al: A cellular oncogene is translocated to the Philadelphia chromosome in chronic myelocytic leukemia. *Nature* 1982;300:765–767.
150. Bartram CR, deKlein A, Hagemeijer A, et al: Translocation of c-*abl* oncogene correlates with the presence of a Philadelphia chromosome in chronic myelocytic leukaemia. *Nature* 1983;306:277–280.
151. Rowley JD: Identification of the constant chromosome regions involved in human hematologic malignant disease. *Science* 1982;216:749–751.
152. Collins SJ, Kubonishi I, Miyoshi I, Groudine MT: Altered transcription of the c-*abl* oncogene in K-562 and other chronic myelogenous leukemia cells. *Science* 1984; 225:72–74.
153. HeisterKamp N, Jenster G, ten Hoeve J, et al: Acute leukaemia in *bcr/abl* transgenic mice. *Nature* 1990;344:251–253.
154. Morgan GJ, Janssen JWG, Guo AP, et al: Polymerase chain reaction for detection of residual leukaemia. *Lancet* 1989;1:928–929.
155. Heldin C-H, Westermark B: Growth factors: Mechanism of action and relation to oncogenes. *Cell* 1984;37:9–20.
156. Hayward WS, Neel BG, Astrin SM: Activation of a cellular *onc* gene by promoter insertion in ALV-induced lymphoid leukosis. *Nature* 1981;290:475–480.
157. Reynolds RK, Hoekzema GS, Vogel J, et al: Multiple endocrine neoplasia induced by the promiscuous expression of a viral oncogene. *Proc Natl Acad Sci USA* 1988;85:3135–3139.
158. Green AR, Wyke JA: Anti-oncogenes: A subset of regulatory genes involved in carcinogenesis? *Lancet* 1985;2:475–477.
159. Knudson AG: Hereditary cancer, oncogenes and anti-oncogenes. *Cancer Res* 1985; 45:1437–1443.
160. Knudson AG Jr: Mutation and cancer: Statistical study of retinoblastoma. *Proc Natl Acad Sci USA* 1971;68:820–823.
161. Hansen MF, Koufos A, Gallie BL, et al: Osteosarcoma and retinoblastoma: A shared chromosomal mechanism revealing recessive predisposition. *Proc Natl Acad Sci USA* 1985;82:6216–6220.

162. Friend SH, Dryja TP, Weinberg RA: Oncogenes and tumor-suppressing genes. *N Engl J Med* 1988;318:618–622.
163. Lee WH, Bookstein R, Hong F, et al: Human retinoblastoma susceptibility gene: Cloning, identification, and sequence. *Science* 1987;235:1394–1399.
164. Whyte P, Buchkovich KJ, Horowitz JM, et al: Association between an oncogene and an anti-oncogene: The adenovirus E1A proteins bind to the retinoblastoma gene products. *Nature* 1988;334:124–129.
165. Lee EY-HP, To H, Shew J-Y, et al: Inactivation of the retinoblastoma susceptibility gene in human breast cancers. *Science* 1988;241:218–221.
166. Marx J: Many gene changes found in cancer. *Science* 1989;246:1386–1388.
167. Lane DP, Crawford LV: T-antigen is bound to a host protein in SV40 transformed cells. *Nature* 1979;278:261–263.
168. Linzer DIH, Levine AJ: Characterization of a 54k dalton cellular SV40 tumor antigen present in SV40-transformed cells and uninfected embryonal carcinoma cells. *Cell* 1979;17:43–52.
169. Finlay CA, Hinds PW, Levine AJ: The p53 proto-oncogene can act as a suppressor of transformation. *Cell* 1989;57:1083–1093.
170. Munroe DG, Rovinski B, Bernstein A, Benchimol S: Loss of a highly conserved domain on p53 as a result of gene deletion during Friend virus-induced erythroleukemia. *Oncogene* 1988;2:621–624.
171. Baker SJ, Fearon ER, Nigro JM, et al: Chromosome 17 deletions and p53 gene mutations in colorectal carcinomas. *Science* 1989;244:217–221.
172. Cavanee WK, Hastie ND, Stanbridge EJ (eds): *Current Communications in Molecular Biology: Recessive Oncogenes and Tumor Suppression*. New York, Cold Spring Harbor Press, 1989.
173. Nigro JM, Baker SJ, Preisinger AC, et al: Mutations in the p53 gene occur in diverse human tumor types. *Nature* 1989;342:705–708.
174. Iggo R, Gatter K, Bartek J, et al: Increased expression of mutant forms of p53 oncogene in primary lung cancer. *Lancet* 1990;1:675–679.
175. Yokota J, Wada M, Shimosato Y, et al: Loss of heterozygosity on chromosomes 3, 13, and 17 in small-cell carcinoma and on chromosome 3 in adenocarcinoma of the lung. *Proc Natl Acad Sci USA* 1987;84:9252–9256.
176. Derilee P, van den Broek M, Kuipers-Kijkshoorn N, et al: At least four different chromosomal regions are involved in loss of heterozygosity in human breast carcinoma. *Genomics* 1989;5:554–650.
177. Vogelstein B, Fearon ER, Hamilton SR, et al: Genetic alterations during colorectal tumor development. *N Engl J Med* 1988;319:525–532.
178. Bos JL, Fearon ER, Hamilton SR, et al: Prevalence of *ras* gene mutations in human colorectal cancers. *Nature* 1987;327:293–297.
179. Forrester K, Almoguera C, Han K, et al: Detection of high incidence of K-*ras* oncogenes during human colon tumorigenesis. *Nature* 1987;327:298–303.
180. Vogelstein B, Fearon ER, Kern SE, et al: Allelotype of colorectal carcinomas. *Science* 1989;244:207–211.
181. Solomon E, Voss R, Hall V, et al: Chromosome 5 allele loss in human colorectal carcinomas. *Nature* 1987;328:616–619.
182. Fearon ER, Cho KR, Nigro JM, et al: Identification of a chromosome 18q gene that is altered in colorectal cancers. *Science* 1990;247:49–56.
183. Duesberg DH: Cancer genes: rare recombinants instead of activated oncogenes (a review). *Proc Natl Acad Sci USA* 1987;84:2117–2124.

Recombinant DNA in Infectious Disease

6

Ralph C. Williams, Jr., MD

Application of recombinant DNA methodology to problems of human disease has perhaps received no greater direct emphasis than in the work during the past several years on infectious diseases. The direct applicability of gene cloning and selected recombinant cDNA as well as oligonucleotide probes to practical problems related to serious human infections has been great and undoubtedly will continue to grow. Therefore, the material presented in this chapter is only a preliminary basis for a grasp of the applications of recombinant methods in a selected number of microbial, parasitic, and viral infections. Molecular biology appears to offer unique tools for diagnostic, epidemiological, and therapeutic approaches to important human infections, including malaria, typanosomiasis, leprosy, gonorrhea, and human immunodeficiency virus (HIV)-related acquired immunodeficiency syndrome (AIDS).

Parasitic Disease

Malaria

Malaria, which is the most widely distributed disease of humans, remains a serious threat to people in nearly half the population of the earth. It is estimated that malaria presently occurs at a rate of 300 million acute cases per year. The numbers of individuals who remain chronically infected is staggering. Periodic efforts to eradicate the disease, first with insecticides and more recently with newly developed parasiticidal drugs, have been only partially successful. Such control measures are seriously limited because of the increasing drug resistance of malarial parasites and the widespread DDT resistance of their mosquito vector. For these reasons an attractive alternative solution in the way of an effective battery of vaccines has long been championed by public health scientists and biologists worldwide.

General Immune Response in Malaria

Earlier work on malarial vaccines usually concentrated on attempting to decipher the types of parasitic antigens that might be protective, as well as the stages of blood or tissue parasites most vulnerable to immune surveillance and therefore candidates for vaccine component preparation. How the host might acquire some element of natural immunity to malaria has been studied throughout the world for decades.[1–3] The importance of serum antibody in acquired resistance to malaria has repeatedly been demonstrated by passive transfer in monkey, human, and rodent infection models,[4,5] and the parallel role of cell-mediated immunity has been carefully examined.[5] It appears that an effective vaccine against *Plasmodium falciparum*, the agent producing the uniformly most lethal form of human malaria, will probably have to include protective antigens from both the sporozoite and blood stages of the malarial parasite.[2,6] An abbreviated schematic life cycle of the parasite is shown in Figure 6.1.

Initially, by using monoclonal antibodies it was found that Mab directed against the sporozoite forms were protective both in vivo and in vitro.[7,8] Studies of the protective parasitic antigens reacting with such monoclonal antibodies indicated that they reacted with epitopes that were uniformly distributed over the entire surface of the sporozoite as determined by immunofluorescence and that the Mab reacted with sporozoites obtained from different geographic isolates of the malarial strains (*P. falciparum* and *P. vivax*) being studied.[8] Moreover, using the monoclonal antibodies and Western blotting techniques with immunoprecipitation of electrophoretically separated malarial sporozoite antigens, it was possible to identify polypeptides of 67 to 58 kd from *P. falciparum* and 51 to 45 kd from *P. vivax* that contained the important protective antigens.[8,9] Thus, by using monoclonal antibodies directed against protective sporozoite antigens the antigens themselves could be further purified and examined. One of the most important aspects of this work on identifying immunodominant and at the same time protective malarial antigens was based on the discovery by Trager and Jensen 10 years before of methodology

Figure 6.1 Sporozoites injected into the bloodstream by *Anopheles* are quickly taken up by the liver, alter their form, and become schizonts. They then enter the blood again and live in red blood cells, where they multiply as merozoites → trophozoites and → gametocytes.

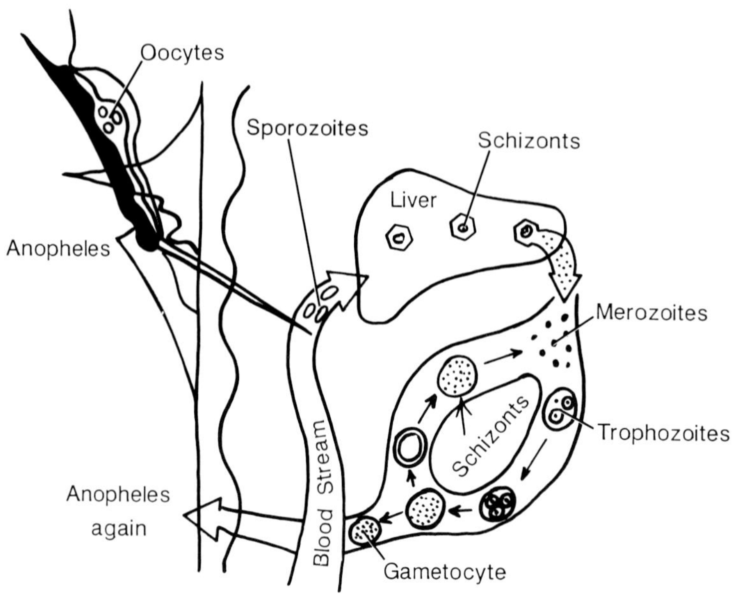

whereby human malarial parasitic forms could be grown in the laboratory in continuous culture.[10] Without this important method, much of the current work on malaria could not proceed.

Recombinant Approaches

Concurrently, Kemp and co-workers in Australia[11] had begun to examine libraries of *P. falciparum* cDNA sequences for expression of antigenic epitopes capable of reacting with antibodies from malaria immune or immunized human subjects. A diagram of the basic strategy for cloning malarial parasite cDNA that expresses potentially important antigenic polypeptide sequences is shown in Figure 6.2. This general procedure was later used by the same group in attempting to define the most common malarial antigens identified by human convalescent malarial sera.[12]

After the initial recognition and further detailed studies of the physical size and antigenic composition of a number of prime candidate malarial sporozoite or other antigens, DNA cloning of a number of promising antigens was rapidly accomplished.[12,13] In the work by Enea et al.[13] the cDNA isolated encoded a circumsporozoite protein that reacted with Mab directed against what appeared to be a

Figure 6.2 Schematic representation of construction of *P. falciparum* cDNA expression library. Malarial mRNA (top left) was used to make double-stranded DNA segments, which were then cloned into λgt10 phage vector to amplify and methylate the cDNA. The amplified cDNA isolated from the λgt10 library using *Eco*RI restriction enzyme was ligated into λgt11 ampicillin-resistant vector, and bacteria containing the cDNA of interest were generated by infecting *E. coli* with the λgt11 phage vector. β-Lactam Amp R refers to the fusion product inducing ampicillin resistance.

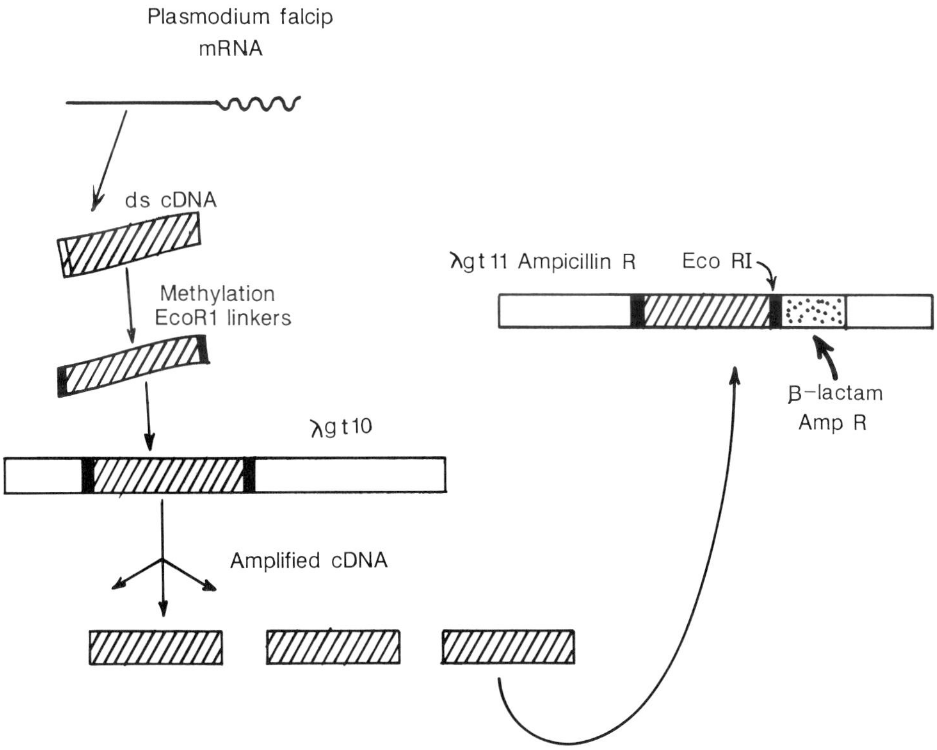

protective sporozoite antigen. There is great interest in the sporozoite as a protective antigenic source because sporozoite-derived material is capable of eliciting a protective immune response in human subjects.[14] Moreover, the immunogenicity of sporozoites is largely related to a single antigen—the circumsporozoite protein—which is evenly distributed over the entire parasite surface.[15,16] The gene product identified as the circumsporozoite antigen by Enea et al.[13] had a 4-amino-acid sequence, proline-asparagine-alanine-asparagine, that was tandemly repeated 23 times. The relationship of this short sequence to the dominant antigens of the circumsporozoite material may be fundamental, because synthetic trimers of the tetra-amino acid (Pro-Asn-Ala-Asn) can inhibit the binding of monoclonal antibodies to a sporozoite extract.[13]

Other Malarial Antigens

Additional work on cloning and defining important malarial antigens has continued to develop. The primary structure of the precursor molecule for three major surface antigens on *P. falciparum* merozoites was identified.[17] The merozoite antigen precursor appeared to be of relatively high molecular mass (195 kd), considerably larger than the circumsporozoite material (59 kd) intensively studied by the group at New York University.[9–15] Also of interest was the fact that the 195-kd protein appeared to be the precursor for three immunodominant merozoite proteins of 83,000, 42,000, and 19,000 kd.[18] The complete gene for the precursor protein contained 5920 nucleotides, and analysis with various Mab indicated that the 83-kd antigen was derived from the NH_2 terminus and the 42-kd antigen was located on the COOH terminus of the polypeptide chain product.

Initial Vaccine Trials

In 1987 the initial report of use of a synthetic peptide malaria vaccine was published by Herrington et al.[19] These workers had prepared a vaccine by conjugating acetylcysteine-$(NANP)_3$-OH[Ac-Cys-$(NANP)_3$-OH] prepared by solid-phase peptide synthesis to tetanus toxoid (TT) It had previously been demonstrated that synthetic NANP peptides (which represented the original repetitive Asn-Ala-Asn-Pro tetrapeptide) conjugated to protein carriers were capable of producing specific antisporozoite antibodies in animals.[19] These preliminary findings had paved the way for eventual testing of similar synthetic vaccine preparations in humans.

In the initial studies of Herrington et al.[19] normal volunteers were given 20 to 160 mg of $(NANP)_3$-TT and booster vaccinations administered 28, 56, and in some instances 133 days later. No significant untoward side effects were noted, and the humoral antibody responses elicited appeared to parallel the doses of vaccine given. Moreover, no evidence of T cell stimulation by $(NANP)_3$ was obtained. Vaccine efficacy was then assessed by inoculating volunteers with infectious sporozoites of *P. falciparum* strain NF54, which was known to be sensitive to chloroquine. The intentional sporozoite inoculations were performed through the bites of infected mosquitoes. All four control unvaccinated volunteer subjects developed parasitemia 7 to 10 days after challenge. Two vaccinees developed malaria infection after challenge but did not have detectable asexual blood stage parasites until 11 days after challenge. The remaining vaccinee remained free of parasitemia. All subjects were given curative courses of chloroquine following this vaccine trial. The complete protection in one of the three vaccinated subjects attests to the potential usefulness of this approach. A possible disadvantage of this particular peptide-conjugated vaccine was that it apparently did not prime parasite-specific T cells. Therefore, using this particular reagent, a boosting of immune responsiveness and associated lymphokine production may not occur during repeated exposure to infectious sporozoites under natural exposure conditions.[20]

An additional report of an early trial of a recombinant DNA *P. falciparum* vaccine containing multiple NANP repeated sequences also recorded induction of antibody production coincident with vaccine administration.[21] In this study, a 32 four-amino-acid repeat of the tetrapeptide immunogen was utilized. Complete protection against challenge was achieved in one of six vaccinated volunteers and occurred in the subject with the highest antisporozoite antibody level.

The results of these two experimental trials of recombinant DNA engineered vaccines are of considerable interest and provide a great deal of new information about prospects for the future as well as problems that must still be overcome. Antisporozoite antibodies in malarial infection probably protect by preventing entry of the parasite into liver cells, where ordinarily it can multiply and develop into the merozoite stage that subsequently infects erythrocytes. To achieve unqualified efficacy, the antibodies must have high titers and must persist sufficiently long to prevent multiple cycles of infection, particularly in the natural endemic malarial habitat. With the initial vaccines 100% protection and universal high antisporozoite antibody levels have not been achieved. Moreover, no booster effects of repeated injections were noted.

Immunological Requirements for an Ideal Vaccine

An important aspect is that antisporozoite antibodies may require T cell help, which is currently being provided in the foreign carrier (tetanus toxoid).[22] Ideally, an effective vaccine should contain T-cell–activating sites also derived from the malarial sporozoite antigenic repertoire. Experiments in mice indicate that the circumsporozoite protein has only a few dominant T helper cell activating sites, which must interact specifically with the major histocompatibility system of the vaccinated individual.[23,24] It may eventually be important to include T cell–reactive antigenic epitopes from the sporozoite other than the circumsporozoite protein to achieve potent T cell activation and thus a successful vaccine.

Furthermore, an important protective effect of the sporozoite vaccine may actually be antibody independent,[25] suggesting that T cells may be able to inhibit parasite development in the liver.[26] This additional T cell–mediated immune effect is another reason for effective T cell stimulation in any eventual vaccine preparation. During work on relative effects of humoral and cellular immunity to malaria in animals, it was found that the circumsporozoite protein antigen in the murine malarial strain of *Plasmodium berghei* had a repetitive sequence of two tandemly repeated octapeptides.[27,28] Thus, repeated sequences in these parasitic forms may afford a useful handle or tool for exploring immune responsiveness in the host at risk. As we shall see later in this chapter, tandem repeats or repetitive sequences may also be useful eventually in devising recombinant vaccines against other infectious agents, including gonorrhea and syphilis.

Antigenic Analysis of Plasmodium Falciparum *Using Molecular Techniques*

A concerted effort has been made to define *P. falciparum* antigens that would stimulate effective T cell immune responses. In one study antigenic epitopes revealed by deletion analysis stimulated T cell clones, and well-defined peptides from these *falciparum*-derived antigens stimulated both T cell clones and peripheral blood mononuclear cells from malaria-infected individuals.[29] Additional studies using *falciparum* antigens derived from the COOH-terminal amino acid repeat region showed T cell responses in human subjects primed to *P. falciparum* by natural exposure.[30] Moreover, 65% of the responders also reacted to short synthetic peptides representing overlapping T cell epitopes.

Several other reports have provided additional possible approaches to malaria vaccine development. Patarroyo et al.[31] used two polymeric synthetic hybrid proteins derived from three synthetic peptides (83.1, 55.1, and 35.1 kd) related to *P. falciparum* merozoite–specific proteins in human volunteers and achieved partial protection in some subjects. This approach was of interest because it utilized synthetic hybrid molecules containing epitopes from different stages of the parasite. These materials were highly immunogenic and did not require foreign carrier proteins. Yet another approach might be directed at inducing antibody that would prevent infection of mosquitoes when they ingest a blood meal containing the sexual stages of the malaria parasite. A gene has been cloned that encodes a 25-kd surface protein expressed on zygotes and ookinetes of *P. falciparum*,[32] and this may represent another reasonable starting material for vaccine development. Of interest in the latter study was that the peptide produced contained four tandem epidermal growth factor–like domains.

Plasmodium vivax

Considerable progress has also been made in the knowledge and application of similar recombinant methods with respect to *P. vivax* malaria. The circumsporozoite protein has been cloned and sequenced.[33,34] The protein product of this gene includes 373 amino acids with a large number of tandem repeats of 9 amino acids. Plasmids have been constructed that include expression of important gene segments encoding *P. vivax* antigens.[35] The strategy involved in cloning this gene is shown in Figure 6.3. Initial studies using the recombinant *P. vivax* protein injected into mice indicated strong immunogenicity. Of interest was the finding that the repeat domain represented by the sequence $(DGQPAGDRA)_2$ appeared to be one of the immunodominant epitopes in the polypeptide gene product. Also of practical importance in this report was the successful cloning of the *P. vivax* gene into engineered yeast cultures. This procedure provided high yields and involved relatively simple steps in the eventual purification process. Experience with other yeast expression systems has already demonstrated ready applicability to the production of other proteins of clinical usefulness.[36–38] The yeast expression system used for *P. vivax* can also be used to express other portions of the circumsporozoite molecule that might be more antigenic in terms of T cell recognition than the repeat units that have proved so potent in stimulating humoral antisporozoite immune responsiveness.

Some progress in understanding the stimulation of T cell–dependent immunity by malaria vaccines was reported when it was found that T cell immunity protecting mice from *P. berghei* sporozoite challenge was genetically restricted, indicating that malaria antigens on infected hepatocytes were recognized by immune effector cells.[39]

Diagnosis of Malaria

Further potential usefulness of recombinant DNA techniques has been reported with oligonucleotide probes using a 21-base-long repetitive sequence in the diagnosis of *falciparum* malaria.[40] This report may signal a major technical advance

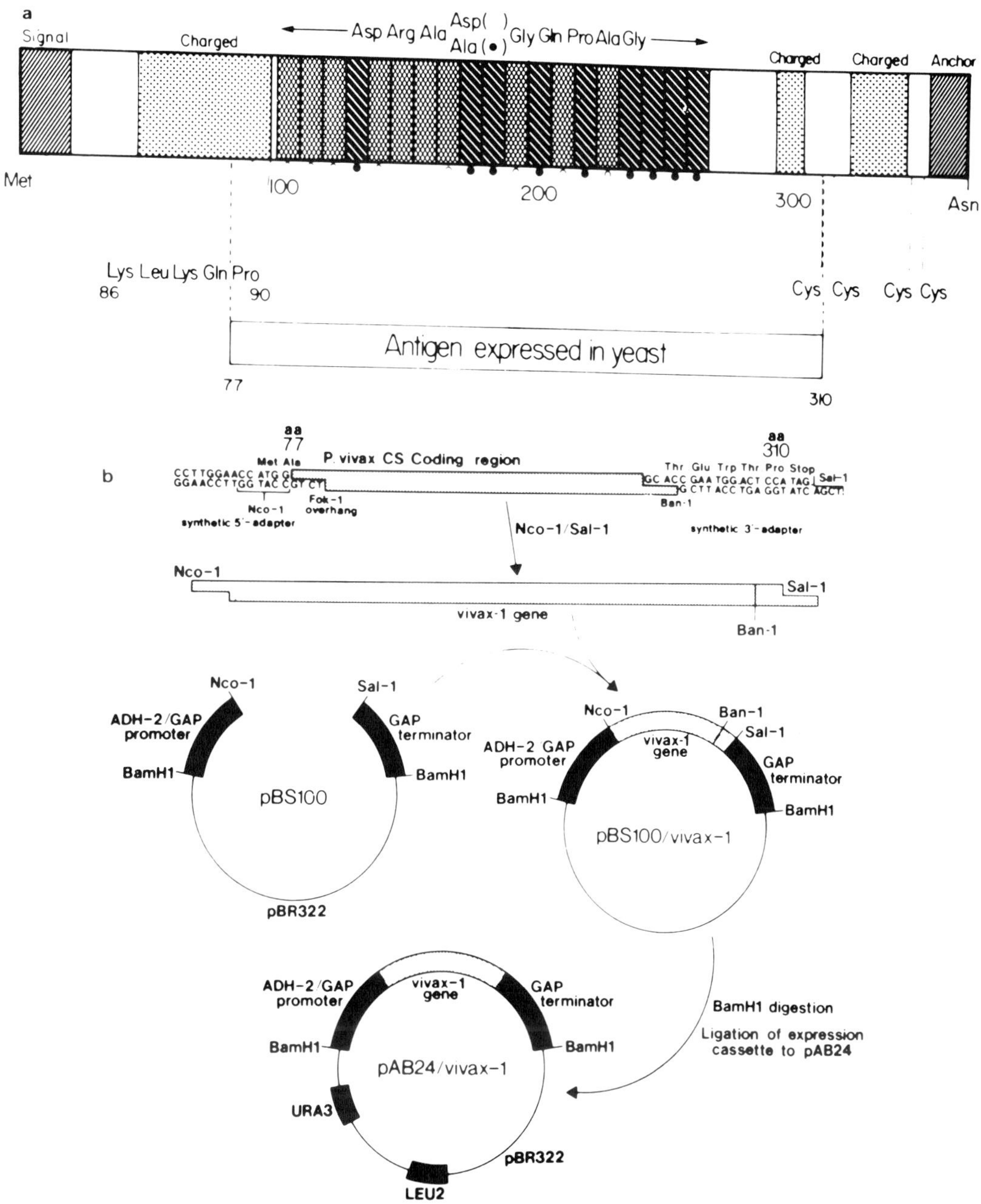

Figure 6.3 (a) Schematic representation of the circumsporozoite (CS) protein gene and of the segment expressed in yeast. The assigned amino acid numbers correspond to those of Arnot et al.[34] In addition to signal and presumed anchor sequences and areas containing a large number of charged amino acids, a large portion of the CS molecule consists of tandem repeats of nine amino acids. Repeats bearing alanine in the fourth position (●) predominate at one end of the repeat domain. (b) Plasmid constructions are diagrammed for the expression of *P. vivax* C5 proteins in yeast. *Reproduced from Barr et al.,*[36] *with permission.*

in the diagnosis of malarial infection, which traditionally required rather laborious microscopic examination of Giemsa-stained blood smears. The use of DNA hybridization for detection of *P. falciparum* infection may result in a new level of sensitivity in screening samples from patients. Instead of using a radioactive label for the DNA probe, the synthetic oligonucleotide containing a 21-base-long tandem repeat characteristic of the *P. falciparum* genome was initially linked to biotin. The biotinylated probes proved not to be sensitive enough for malaria diagnosis, so the same 21-base-long DNA probe was covalently coupled to alkaline phosphatase. A positive reaction was indicated by a blue color resulting from activation of the added alkaline phosphatase substrate. When this test was applied to erythrocytes infected with *P. falciparum, P. malariae, P. ovale, P. vivax, P. inui, P. fragile, P. cynomolgi*, or *P. knowlesi*, it was specific for the *P. falciparum* samples. The alkaline phosphatase–DNA conjugate detected serial dilutions of *P. falciparum* DNA but not the other seven *Plasmodium* DNA species. This method appeared capable of detecting 10–30 pg of *P. falciparum* DNA, which was equivalent to about 10^3–10^4 parasite nuclei recognized on ordinary blood films. The reliable sensitivity of this method was estimated to be 100 parasites/ml or 0.002% infected erythrocytes and appeared to be equivalent to the sensitivity of detection with isotopically labeled genomic, plasmid-derived, or synthetic *P. falciparum* DNA probes with diagnostic samples in the field.[40,41] Clearly, in underdeveloped countries in which radioactive labeling and counting equipment may be in short supply, this method may be a real asset for diagnosis or detection of active malaria infection.

Molecular Analysis of Chloroquine Resistance

Chloroquine is presumed to act against *falciparum* malaria by accumulating in the acid vesicles of the parasite and interfering with their functions. When it was found that verapamil reverses chloroquine resistance in vitro, it was postulated that this effect might involve an ATP-mediated P-glycoprotein pump similar to that observed in multidrug-resistant (mdr) tumor cell lines. It was found that *P. falciparum* contained at least two *mdr*-like genes,[42] one of which might be involved in induction of chloroquine resistance. Genetic crosses between chloroquine-resistant and chloroquine-susceptible clones of *P. falciparum* showed that the rapid chloroquine efflux or resistance gene was not directly linked to the *mdr*-like loci.[43] In this way molecular analysis could be compared with the most important drug resistance problem related to *P. falciparum* infection.

Malarial Antigenic Variation

With increasing interest in synthetic peptide vaccines and their possible usefulness in prevention of malaria, concern about natural variability in the antigenic structure of the parasite has surfaced. Natural genetic variation occurs in the Th2R region of the *P. falciparum* circumsporozoite protein, which may function to some extent as a T cell recognition epitope. Studies of ThR variants isolated from different areas of the Amazon basin have shown very limited variation.[44] Whether antigenic variation in this region is important as a mechanism for escaping cytotoxic $CD8^+$ T cells generated by the host remains to be determined.

Schistosomiasis

Observations of endemic schistosomiasis caused by *Schistosoma mansoni, S. haematobium*, or *S. japonicum* indicate that acquired immunity is very slow to develop and relatively ineffective. This group of disorders currently affects 250 million people worldwide, as well as countless animals including cattle and wild buffalo. Schistosomal infections start when the larval worm in its cercarial form bores directly through the skin of the host and enters blood vessels, where it will eventually mature. During its life in the circulation, the worm actually acquires or borrows a wide variety of host antigens, which render it invisible or undetectable from a immunological standpoint. Thus the infected host generates very little in the way of an effective immune response to the adult worm. Both larval forms and adult worms produce antigens that do elicit an immune response, and this is directed at subsequent waves of invading larvae. The mechanisms of immune destruction are complex and involve several immunoglobulin classes, including IgE, and macrophages, lymphocytes, mast cells, and eosinophils.[45] Therefore, it was presumed that an antischistomsomal vaccine, to be effective, must be similarly directed at larval antigens rather than adult worm forms.

Experimental Infection

Because of the relative ease of maintaining it in a functional life cycle in an experimental situation, *S. mansoni* has been the organism most commonly used in laboratory studies of acquired immunity, and rodents, particularly mice, have often served as the experimental animals. It was found that resistance to infection could be produced by immunization with cercarial forms attenuated by prior irradiation.[46] Effective resistance or immunity in the murine model is dependent on the thymus and can also be abrogated by anti-μ suppression, which aborts B lymphocyte responsiveness and subsequent humoral immunity.[47] Transfer of sera from multiply vaccinated mice can induce partial protection against subsequent challenge, indicating a role for IgG antibody in the immune state. Lymphokine-activated macrophages act as effector cells of immune responsiveness.[48,49] Experimental evidence from animals indicates that effective resistance to schistosomiasis is mediated by mechanisms involving IgG or IgE humoral antibody, which interacts with macrophages or eosinophil effector cells. In addition, T cell–mediated immune mechanisms, including lymphokine production, appear to be important in cell-mediated immunity to the parasitic forms. The direct apparently protective function of IgE antibody has been most convincingly demonstrated in the rat model.[46]

Vaccine Strategy

From a theoretical viewpoint, a vaccine based on T cell–mediated immune responsiveness would offer a number of distinct advantages over one that stimulated primarily humoral immunity.[50] Thus, T cell sensitization against virtually any recognizable parasitic antigen could on reexposure be of potential protective value by inducing lymphokine production. For instance, interferon γ and lymphotoxin or tumor necrosis factor can act synergistically to induce macrophages to kill schistosomula of *S. mansoni*.[51] Moreover, T cell–mediated immune responses induced by a vaccine would be antigen specific during the immunizing phase but later could provide nonspecific protection in the effector phase by activating macrophage effector cells, which can kill schistosome forms without being involved in specific antigen recognition. This advantage of the macrophage cytotoxic arm of cell-mediated immunity might be able to counterbalance the schistosomal evasion strategies of antigen masking and antigen borrowing.

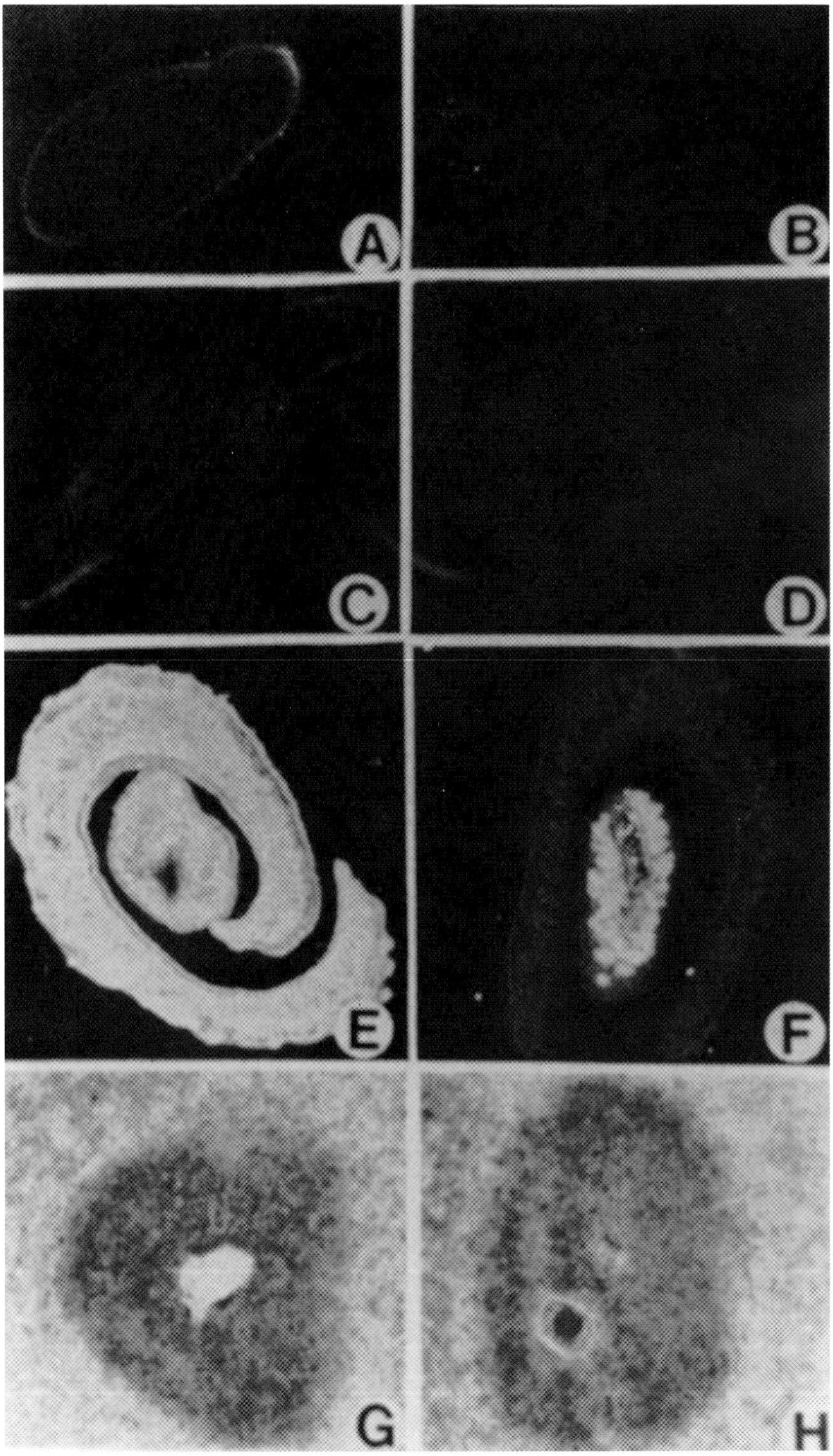
A
B
C
D
E
F
G
H

Vaccine Antigen Cloning

The first application of molecular cloning strategies to developing protective vaccine elements was reported by Ballou et al. in 1987.[52] These workers cloned and expressed in *E. coli* the cDNA encoding a 28-kd antigen present in the extracts of adult worms. The same 28-kd antigen was demonstrated to be present in *S. haematobium, S. bovis,* and *S. japonicum*. The cloning strategy involved preparation of RNA from adult worms and its use to synthesize cDNA. After nuclease fragmentation, the cDNA was fractionated on gradients and fragments between 500 and 4000 base pairs were pooled. The DNA fragments were inserted into λgt10 using a synthetic adapter, and phage containing inserts were amplified in *E. coli*. The cDNA products were screened using both rat and rabbit polyclonal monospecific antibodies against a 28-kd antigen known to be associated with protective immunity. It appeared that this antigen was not glycosylated because no potential glycosylation sites were noted in the predicted amino acid sequence.

When the recombinant antigens were tested in immunization protocols, significant levels of protection were induced in rats subsequently challenged with live cercariae. Additional experiments were undertaken with baboons. Immunization protocols appeared to induce a comparable cell-mediated and humoral immune response, which included both IgG and IgE antibodies. This was demonstrated to be associated with relatively high levels of eosinophil-dependent cytotoxicity. The induction of both IgG and IgE antischistosomal isotypes by immunization with the 28-kd recombinant antigen was encouraging and undoubtedly will be followed by more extended work in primate models. Also of interest in this initial attempt to apply recombinant techniques to schistosomiasis was the fact that analysis of the 211-amino-acid primary sequence of the 28-kd recombinant protein showed no repetitive elements or similarity to sequence data for other parasites. The report was encouraging because it provided positive results in terms of the ability of the recombinant protein to afford some level of protective immunity.

Several other studies have provided additional data on the applicability of this general approach. Spleen cells of mice immunized with radiation-attenuated *S. mansoni* cercariae were used to produce monoclonal antibodies that reacted with schistosomular surface antigens.[53] By using a Mab that reacted with an 18-kd schistosomal polypeptide, as well as a polyclonal antibody to the same material, recombinant clones from a cDNA expression library produced in λgt11 were analyzed. One clone with an insert of 0.53 kilobases produced a fusion protein of 125 kd that reacted with sera of immune mice. This work demonstrated that by cloning techniques it was possible to obtain a potentially important schistosomal antigen that was expressed on the surface of schistosomula. The reactivity of both monoclonal and polyclonal antibodies with this recombinant protein is illustrated in Figure 6.4. The protein recognized by mAb 654.2 has subsequently been further

←

Figure 6.4 Indirect immunofluorescence localization of the target antigen(s) of monoclonal antibody 654B2/1 and polyclonal rabbit anti-654 at various developmental stages of *S. mansoni*. Polyclonal rabbit anti-654 showed strong staining of the surface of newly transformed schistosomula (A) and of the more elongated lung-stage schistosomula (C); no surface staining of these stages was observed with 654B2/1 (not shown) or with serum from a noninfected rabbit (B and D). Monoclonal 654B2/1 and polyclonal anti-654 showed intense staining within the tissues of male and female adult worms (E) and in schistosome eggs but not in the egg granuloma (G); no specific staining was observed with sera from noninfected mice, but autofluorescence of the female eggshells was seen (F and H). ×100. *Reproduced from Dalton et al.,*[53] *with permission.*

characterized by these same workers as a glycosylphosphatidylinositol-linked surface protein which shares strong homology with a novel family of cysteine rich hydrophobic proteins, most of which are expressed on human cells of hematopoietic origin!

Reports have also appeared on cloning of other potentially important schistosomal antigens. An *S. japonicum* genomic DNA fragment expressing a 70-kd protein has been derived again from a λgt11 expression library.[54] The protective effects of this component have not yet been studied, but the material appears to react strongly with antibodies from immune rabbits.

Schistosoma mansoni *Antigens and Heat Shock Protein*

One of the most interesting by-products of comparative molecular biological approaches in modern parasitology is the occasional striking sequence homology that becomes apparent between a protein or gene of interest and another component previously studied or known in a different context. As pointed out by Hedstrom et al.,[55] an antigen of 70 kd that represents an immunodominant epitope in humans and animals infected with *S. mansoni* appears to have major sequence homology with the heat shock protein hsp70, which is present in humans as well as ancient evolutionary forms such as *Drosophila*. The 70-kd *S. mansoni* antigen was identified as a major homologue of the eukaryotic heat shock protein by sequence analysis of DNA from a cDNA insert obtained from a λgt11 clone expressing the schistosomal antigen and located on the COOH-terminal, immunodominant end of the molecule. In many other eukaryotic organisms the heat shock protein gene is present in multiple copies, usually dispersed in the genome. Some members of this multigene family appear to have a spectrum of functions, as indicated by different patterns of expression in normal and heat-shocked cells.[56] The relationship of heat shock protein to schistosome infections is not clearly understood. Some element of heat shock may be the fate of schistosomes as they make their transition from the invertebrate snail to the vertebrate host; studies indicate that production of hsp70 is more likely to be constitutive than heat induced. It is possible that the heat shock protein is involved in multiple biological functions.[57] The striking degree of sequence homology between the *S. mansoni* 70-kd antigen and hsp70 is illustrated in Figure 6.5.

Cellular Immunity and Major Histocompatibility Complex Restriction

The final answer is not yet in on the feasibility of using recombinant molecules as immunogens or components of a vaccine against schistosomal infection. From the preliminary work it appears that several components may eventually be necessary in order to ensure both humoral and cell-mediated immunity against the parasite. A cocktail of recombinant antigens may be needed to ensure efficient humoral T cell–mediated and macrophage- or eosinophil-activated immune protection. Finally, the MHC constituents of the individual host appear to play an important role in immune responsiveness. Studies of the responsiveness of various well-defined human cell lines have indicated that nonresponder as well as high-responder HLA-DR, DG human haplotypes have a marked quantitative influence on the level of immune response in vitro to several well-defined schistosomal antigens.[58] In this study the HLA-DR and DQ molecules, which are controlled by different genes, were shown to be related to both immune response and immune suppression. As in the case of malarial infection and individual components of a

Figure 6.5 Amino acid homologies between the *S. mansoni* 70-kd peptide sequence (Sm) and the corresponding region of the hsp70 protein sequence from human (H) and *Drosophila* (DM). The amino acid sequence for the *S. mansoni* peptide was derived from the cDNA sequence of clone F7; the human and *Drosophila* sequences were from Hunt and Moromoto.[214] Regions of identity are enclosed, and numbers refer to the amino acid residues of the human sequence. The F4 clone encodes the COOH-terminal region beginning with the aspartic acid residue at position 507 (arrow). *Reproduced from Hedstrom et al.,*[55] *with permission.*

potential malaria vaccine, it will eventually have to be determined whether certain phenotypic HLA profiles in the population result in hyporesponders or nonresponders based on their individual MHC constituency.

Trypanosomiasis

Cloning approaches to the problem of trypanosomiasis have provided some hope that again a recombinant component or group of components may be identified that will be useful in developing a protective vaccine. The parasitic protozoan *Trypanosoma cruzi* infects humans and other vertebrate hosts by direct transmission of the trypomastigote forms of the parasite through the bite of the insect vector tsetse fly.[59] Only the trypomastigotes present in the insect vector and the vertebrate host's bloodstream are capable of directly invading vertebrate cells. Trypomastigote-specific surface glycoproteins of 90 and 85 kd that may influence immune responsiveness of the infected host to trypanosome infestation have now been characterized.[59–61] The 85 kd glycoprotein appears to be necessary for effective trypomastigote entry into host cells.

Cloning Antigens

The polypeptide chain expressed on the 85-kd epitopes has been cloned by partially digesting trypomastigote DNA with the restriction enzyme *Hae*III and later *Bal*31, followed by repair of digested DNA strand ends with the Klenow fragment and ligation via *Eco*RI linkers into λgt11. This cloning strategy is illustrated in Figure 6.6.

Immune sera from a *T. cruzi*–infected host showed strong reactivity with the 85-kd recombinant molecule. Sequence analysis of the 500-bp DNA fragment encoding this antigen indicated that a 27-bp repeat unit occurred five times in tandem within the entire sequence. These tandem repeating units appeared to be strikingly similar to tandem repeating units previously described in both the S antigen and

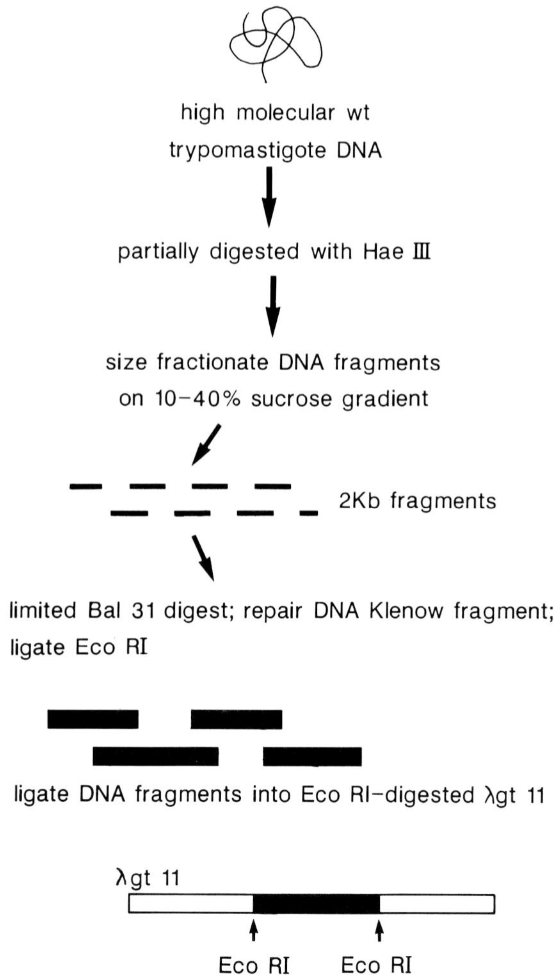

Figure 6.6 High molecular weight DNA obtained from trypomastigote forms of *Trypanosoma* was digested initially with restriction enzyme *Hae*III, followed by size fractionation and separation of ~2-kb DNA fragments by sucrose gradient centrifugation. This was followed by limited *Bal*31 nuclease digestion and repair of DNA strand ends using the Klenow fragment DNA polymerase. Next, *Eco*RI linkers were added to DNA strand ends and the resulting fragments were ligated into λgt11 phage at *Eco*RI sites.

circumsporozoite of *Plasmodium*.[62] Work on the relationship of this cloned 85-kd antigen to acquired immunity is ongoing. However, a polypeptide product also showing 22 tandem repeats of the dipeptide Glu-Pro has now been reported to be present in the insect form of the trypanosome.[63] This finding suggests that the tandem amino acid sequence repeats probably did not evolve in parasitic protozoans to provide protection exclusively for their mammalian hosts.

Bacterial Diseases

A number of bacterial diseases have been newly approached with the idea that recombinant DNA methodology may offer a new dimension in both understanding and treatment or prevention. Bacterial infections that are only moderately or poorly responsive to antibiotic agents have been most intensively studied. In addition, considerable effort has been devoted to organisms that often escape adequate or curative treatment, such as *Neisseria gonorrhoeae* or *Treponema pallidum*.

Leprosy and Gene-Cloned Antigens

Leprosy is a chronic infectious disease that is caused by Hansen's bacillus or *Mycobacterium leprae*. In probably no other bacterial infectious disease are the involved tissues so heavily populated by actively growing bacteria. As with most other intracellular parasites, truly protective immune reactions are probably largely dependent on cell-mediated immune responses.[64] The disease has a broad clinical spectrum. Patients with lepromatous leprosy show a *M. leprae*–specific T cell unresponsiveness, whereas patients with tuberculoid leprosy show both acquired immunity and delayed hypersensitivity reactivities, which are both thought to be secondary to helper T cells.[65] The disease is distributed worldwide and probably afflicts 10 million to 15 million people. Therapeutic approaches to this chronic disorder have been severely limited because *M. leprae* is one of the few human pathogens that has not yet been successfully cultured in vitro.

By using monoclonal antibodies against *M. leprae* antigens, genes and their specific polypeptide products have been isolated from a recombinant DNA expression library by Young and co-workers.[66] They utilized a recombinant DNA library constructed in λgt11, a bacteriophage vector capable of driving inserted foreign DNA with *E. coli* transcription and translation signals. The work on leprosy gene expression followed a pattern that had proved successful in analysis of a similar gene library derived from *M. tuberculosis*.[66,67] DNA fragments derived from a single *M. leprae* bacterial strain cultivated in an armadillo were expressed as a fusion protein with β-glactosidase in λgt 11. A total of 13 monoclonal antibodies against a number of individual antigenic epitopes on various *M. leprae* proteins were used to screen gene products. These antibodies recognized a group of five major proteins of 65, 36, 28, 18, and 12 kd. Various recombinant DNA clones were isolated in this way and restriction maps were prepared.

This initial report was very promising, since it indicated that such an approach could be used to clearly define more important *M. leprae* antigens, which could then be used to study both humoral and cell-mediated immune responsiveness in the underlying disease itself. Subsequently, an 18-kd protein antigen also obtained from a λgt 11 expression library was cloned and its gene nucleotide sequence was determined.[68] It was thought that the 18-kd protein might contain peptide fragments of potential importance as vaccine candidates because this polypeptide contains epitopes specific for *M. leprae*. Since some *M. leprae*–specific helper T cell clones were found to proliferate in response to λgt11 recombinant material containing the carboxyl-terminal 111-amino-acid sequence of the original 18-kd protein, an analysis of the final deduced amino acid sequence was made. This indicated five short peptides (15-mers to 20-mers) whose sequences indicated that they would form amphipathic helices. These might be potentially good helper T cell antigens, based on the analyses of Margalit et al.[69]

Specific antigenic analyses of *M. leprae* epitopes fundamental in stimulating helper T cells as well as suppressor T cells will eventually be crucial in designing an effective vaccine. Work along these lines by Mustafa et al.[70] using lysates of recombinant proteins related to the 18-kd protein of *M. leprae*, indicated that specific clones of helper T cells could be stimulated. Such an approach can now be expanded and specific sequences examined for direct induction of *M. leprae* suppressor T cells. A thorough knowledge of this repertoire will be essential in the final preparation of any trial vaccine. Two different groups have isolated and cloned from leprosy patients suppressor T cells that are capable of directly suppressing the proliferation of helper T cells in an antigen-specific, genetically restricted manner.[71,72] These findings make it imperative to establish ways to examine recombinant *M. leprae* constructs for helper-inducing or suppressor-inducing sequences. Furthermore, they illustrate a much broader point—namely the marked increment in precision and depth which the use of recombinant proteins allows in specific studies of various interacting cell populations within the immune system.

Recombinant Interferon γ in Lepromatous Leprosy

The antimycobacterial drugs currently used in the treatment of lepromatous leprosy are only partially or slowly effective, so a number of possible additional avenues have been explored for effective adjuvant treatment modalities. Because interferon γ was thought to act as an intermediary lymphokine capable of activating macrophages and thereby increasing immune elimination of the billions of infecting mycobacteria present in tissue lesions, it was used in a therapeutic trial of recombinant interferon γ given in doses of 1 or 10 μg daily by jet gun injection to six patients with lepromatous leprosy.[73] Histologic examination of infected skin sites showed a number of features compatible with induction of delayed-type hypersensitivity, including increased tissue induration, T cell and monocyte infiltration, diminution of epidermal Langerhans cells, and both dermal and epidermal cell HLA-DR antigen expression. In parallel, studies of monocyte functions before and after interferon γ administration showed a marked increase in hydrogen peroxide release in response to the secretagogue phorbol myristate acetate and after exposure of monocytes to *M. leprae*. Representative data from this therapeutic trial are shown in Figure 6.7. These apparently positive effects on stimulating macrophage-monocyte activity and functional capacity were encouraging, particularly since it appeared that there was a rapid decrement in numbers of acid-fast bacilli in the areas of interferon γ injection after the short course of treatment. These observations furthermore support the idea that interferon γ released normally from antigen-reactive T cells may induce many of the manifestations of delayed-type hypersensitivity and cell-mediated immunity. Longer periods of therapeutic trial in larger numbers of patients seem appropriate. Also of importance, as pointed out by the group conducting this trial,[73] would be attempts to determine whether administration of recombinant interferon γ is capable of enhancing macrophage-mediated

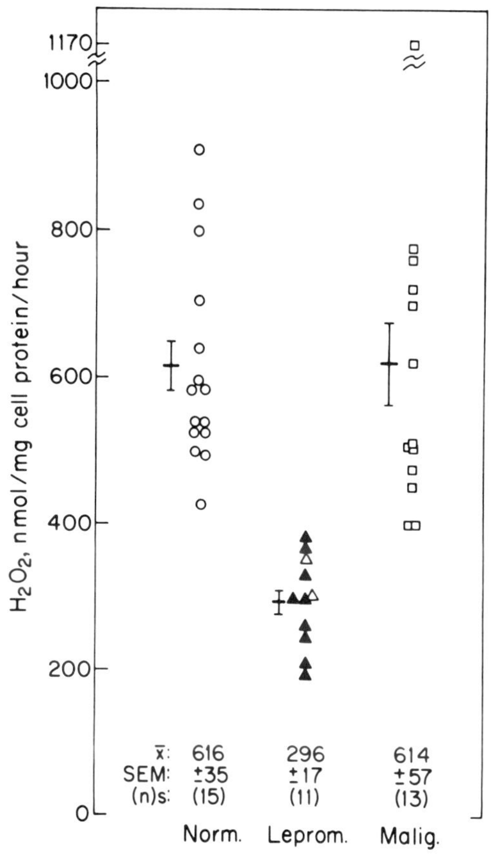

Figure 6.7 Hydrogen peroxide (H_2O_2) released from monocytes of normal controls (Norm.), patients with lepromatous leprosy (Leprom.), and controls with advanced neoplasms (Malig.) after challenge with phorbol myristate acetate. Filled symbols represent patients taking dapsone or rifampin. Values for cancer patients were taken from Nathan et al. for comparison and were obtained by the same method over the same period. Bars denote group means ± SEM. *Reproduced from Nathan et al.,[73] with permission.*

microbicidal activity against *M. lepare* in lepromatous leprosy without reversing the apparent concomitant action of T cells toward the organism itself. The use of gene-engineered product such as recombinant interferon γ in a chronic recalcitrant bacterial infection such as lepromatous leprosy would, if successful, represent a striking and appropriate application of recombinant DNA technology in a terrible human disease (Figure 6.8).

Treponema Pallidum and Syphilis

Despite the availability of penicillin, syphilis continues to cause significant morbidity throughout the world. Failure to culture the organism readily has limited in vitro studies; the organism can be passaged in the rabbit testis, but *T. pallidum* is not readily available in its motile virulent form. Studies of the biochemistry and physiology of *T. pallidum* have had to rely on relatively small numbers of organisms that can be harvested from infected animals. A number of studies over the past several decades have provided some insight into the antigenic makeup and pathogenic features of *T. pallidum.*[74,75] Because of the difficulties of in vitro propagation, it was recognized that expression of *T. pallidum* antigens cloned into phage libraries would provide a logical, precise way to study antigenic composition. The first successful cloning attempts were made in 1982.[76,77] Expression vectors produced gene products that reacted with human antibodies derived from syphilitic sera. As the

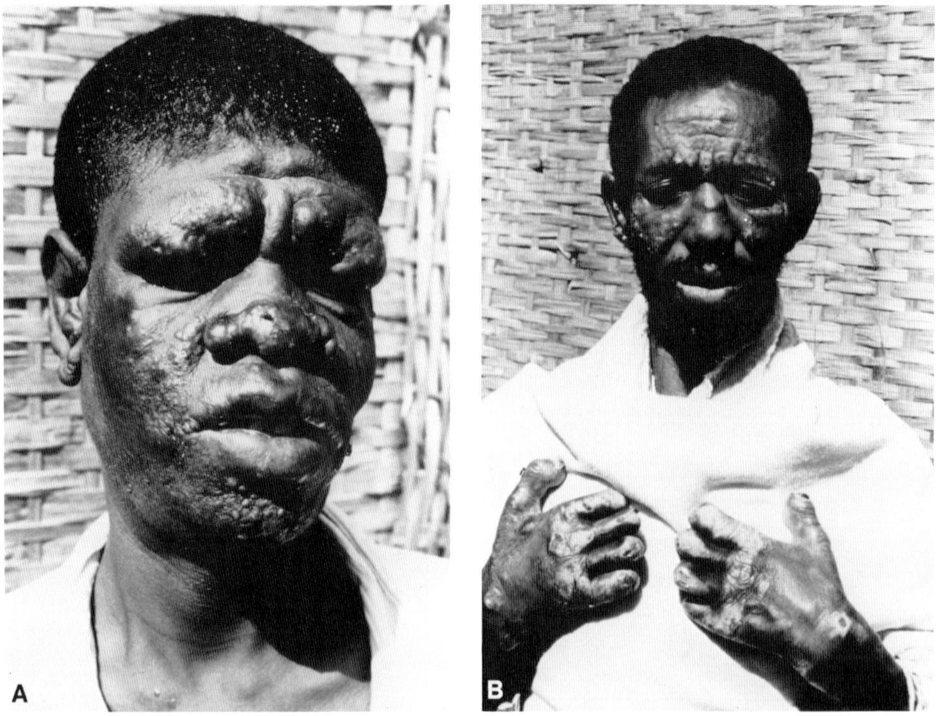

Figure 6.8 (A) Lepromatous leprosy in a patient being treated in Addis Ababa, Ethiopia. Marked induration of skin folds is notable. (B) Patient in Ethiopia with mixed leprosy combining features of both lepromatous and tuberculoid disease. Loss of several terminal digits has occurred.

gene cloning techniques improved and the immunochemical definition of various *T. pallidum* antigens became more refined,[78–80] potential biological significance was tentatively assigned to a major immunodominant surface antigen of *T. pallidum* with a molecular mass of 47 kd.[80] A number of in vivo and in vitro assays subsequently demonstrated that this 47-kd antigen is associated with the surface of the organism and is highly immunogenic in both rabbits and humans.[79–81] In addition, monoclonal anti–47 kd antigen can bind to *T. pallidum* in the *T. pallidum* immobilization assay and can effectively neutralize *T. pallidum* in in vivo neutralization tests.[82] For these reasons it appeared that the 47-kd antigen might be a primary candidate for both vaccine and diagnostic potential.[83]

Further Recombinant Approaches

The 47-kd surface immunogen was successfully cloned in *E. coli* by Norgard et al. in 1986.[84] Moreover, the antigen expressed in *E. coli* reacted with human immune syphilitic sera but not with normal control sera. When the cloning of this important *T. pallidum* antigenic material was initially reported, it was suggested that cDNA from the 47-kd sequence might be useful diagnostically to identify treponemal materials in genital ulcers, skin lesions, and other body fluids. This was based on the previous successful detection of enterotoxigenic *E. coli* by DNA colony hybridization in a search for ways to identify toxigenic strains.[85]

Additional studies have provided more information on similar *T. pallidum* antigens isolated by gene cloning and recombinant techniques[86,87]; however, to date there has been no convincing evidence that these materials could be utilized effectively in a protective vaccine.

Neisseria gonorrhoeae

After a decade of increase, the number of cases of gonorrhea in the United States declined by 12% between 1975 and 1984.[88] However, this infection is still the most frequently reported communicable disease in the United States.[89] Shortly after the clinical effectiveness of antibiotics for the treatment of *N. gonorrhoeae* infections was firmly established, the organism began to show resistance. Sulfonamides were introduced in the 1930s, and their initial use in the therapy of gonorrhea was followed by what appeared to be increasing rates of treatment failure within less than 10 years. Penicillin therapy was initiated in 1943, and for three succeeding decades a gradual increase in chromosomally mediated penicillin resistance was countered by using increasingly large doses of the antibiotic itself.[90] In 1976 *N. gonorrhoeae* acquired a plasmid producing penicillinase,[91] and penicillin became clinically ineffective against the penicillinase-producing strains. The clear need for alternative antibiotic agents provided the stimulus necessary for the development of spectinomycin. During the past 5 years the prevalence of antibiotic resistance has rapidly increased, and organisms with plasmid-mediated resistance to tetracycline and decreased susceptibility to many other antibiotics have been identified. Spectinomycin-resistant strains, originally recognized in the Far East and in the Republic of Korea, have been responsible for a substantial proportion of treatment failures.[92,93] The ability of the gonococcus to develop antibiotic resistance continues to keep pace with the various antimicrobials directed at its eradication. Resistance to spectinomycin has caused a shift to ceftriaxone treatment of uncomplicated gonorrheal infections in geographic regions with increased prevalence of resistant *Neisseria* strains. Periodic recognition in the United States of penicillinase-producing *N. gonorrhoeae*[89] also shows the magnitude of the recurrent treatment and public health problem presented by this organism.

Vaccine Strategy

All of these potential difficulties and the incidence of gonorrhea have repeatedly spurred scientific and public health efforts to engineer a vaccine or more effective protection against this pathogen. Studies of the organism have indicated that gonococci display considerable antigenic heterogeneity between strains as well as within a single strain. Much of the success of gonococci as virulent pathogens may be ascribed to their capacity to switch expression of a number of separate virulence determinants or antigens.[94,95] The ability to vary important antigens remains the major obstacle to the development of an effective vaccine and has repeatedly focused attention on conserved or potentially invariant antigenic structural domains. Monoclonal antibodies directed at important variable antigenic structures such as specific anti-pilus antibodies do indeed protect in various model systems, whereas antibodies to relatively conserved epitopes that may be common to all gonococci do not.

Initial strategies in vaccine development concentrated on preventing the binding of gonococci to epithelial cell surfaces and on selectively stimulating bactericidal antibodies.[96] As more detailed structural studies of gonococcal surface and pili components were completed, the remarkable diversity and complexity of the antigenic structure of this organism have become apparent. For some time it had been known that *N. gonorrhoeae* strains that resist complement-dependent killing by normal human serum are sometimes killed by immune convalescent serum from patients recovering from disseminated gonococcal infection. Studies by Rice and co-workers[97] indicated that human antibodies directed against the gonococcal protein III (an antigenically relatively conserved outer membrane protein) contained most of the blocking IgG antibody preventing effective killing. Antibodies that were reactive with the gonococcal porin (protein I), the major outer gonococcal membrane protein, showed no blocking activity.

Gonococcal Proteins as Potential Vaccine Candidates

The gonococcal outer membranes were thought to be possible prime sites of antigens that could be cloned into appropriate vectors and used for a vaccine. Protein I is present in all strains of gonococci and is the most abundant outer membrane protein.[98] Moreover, only a limited number of proteins I exist, and the type of protein I found on a particular gonococcal strain appears to be a stable characteristic. For these reasons protein I has been used to develop a serotyping profile for gonococci.[99,100] Moreover, antibodies against protein I have been demonstrated to be bactericidal, and limited evidence indicates that these antibodies may afford protection against salpingitis.[101] Protein I can be divided into two structurally distinct molecules, proteins IA and IB. Protein IB molecules appear to have both termini buried within the cell membrane, but the central portion of protein IB is exposed on the cell surface. Based on studies of both proteolytic enzyme and cyanogen bromide cleavage of protein IB, a tentative model of its arrangement has been proposed,[102] which is shown in Figure 6.9. It now appears that the middle CB2 peptide of protein IB may be located in the portion exposed on the bacterial cell surface and provide important baseline data for studies of protective immunity as well as for an effective vaccine.

Cloning Approaches

Recombinant techniques have begun to be applied in earnest toward isolation and amplification of *N. gonorrhoeae* antigens. Bacterial DNA cloned into a λgt11 bank was screened using rabbit antiserum to protein I. Additional studies indicated that several clones obtained by this technique expressed antigens related to the protein III (PIII) antigen as well as the gonococcal H.8 antigen.[103] Strittmatter and Hitchcock[104] found that the H.8 antigen was extractable by phenol-chloroform-petroleum ether and therefore was associated with lipid components. Whether the

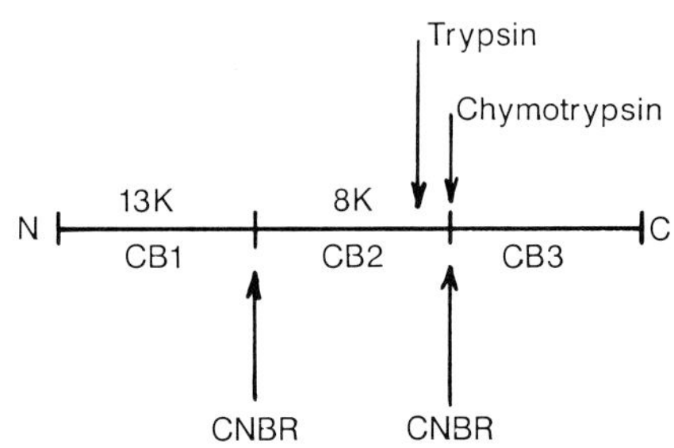

Figure 6.9 Preliminary suggested peptide order of protein 1B. The cleavage sites for trypsin, chymotrypsin, and cyanogen bromide (CNBR) are indicated by arrows. Peptides obtained after cyanogen bromide degradation are numbered CB1 to CB3 beginning at the NH_2 terminus. *Reproduced from Teerlink et al.,[102] with permission.*

H.8 antigen is therefore functionally analogous to lipoprotein anchoring the cell outer membrane to the peptidoglycan layer is not yet clear. It does not appear that current recombinant methods will amplify our biomedical knowledge of this important gonococcal material.

Additional studies of the structural gene for the gonococcal PIII by Gotschlich and co-workers[105] have indicated the presence of a repetitive sequence in the flanking region of the PIII gene. This sequence was repeated at least 20 times in the genome; however, unlike the situation encountered with repetitive sequences in malaria, no clear relationship has been identified between these gene products and fortuitously immunodominant antigens of the gonococcal organism. Moreover, since the PIII protein, whose complete gene sequence is now known, is always expressed by gonococci and appears to exhibit heterogeneity of either molecular weight or proteolytic patterns, it may eventually be useful in studies related to protection and vaccine development. However, as demonstrated by the studies of Rice et al.,[97] the reactivity of immune blocking IgG antibodies with PIII interfering with bacterial killing makes it unlikely that it could be used as the major constituent in any putative protective vaccine.

Recombinant Studies of Antigenic Variation

As noted earlier, the gonococcus has been effective in eluding host immune response because of its ability to alter its antigenic makeup and thus its interface with the host immune system. The genetic mechanisms whereby this occurs have been studied in detail by Swanson et al.[106] particularly with respect to transitions in pilin antigens. The ability of the gonococcus to change its pilin antigens is related directly to the diversity of the pilin gene repertoire. An estimated 12 to 16 partial genes compose this system; these genes contain nucleotide sequences analogous to the 3′ half to two-thirds of the complete pilin gene. When the complete pilin gene of the gonococcus undergoes what is termed *gene conversion*, a portion of the partial gene sequence is shifted into the complete pilin gene with the result that the gene product represents a distinct or unique new epitope. In some ways, gene conversion allows the organism to generate diversity analogous to that of the VDJC genes of mammalian immunoglobulin. The genetic mechanisms involved in the variation of the gonococcal pilin antigenic repertoire are illustrated in Figure 6.10. Also analogous to the diversity expressed by immunoglobulin genes is the hypervariable region of the pilin gene product illustrated here. Further studies of the genetic mechanisms of pilin antigenic variation indicated that there are two major pathways of variation: transformation-mediated recombination, which is influenced by growth conditions and inhibited by DNase I, and intragenomic reciprocal recombination, which is not affected by DNase I.[107]

Applications of recombinant methodology have continued to provide new understanding of both the anatomy and the products of important pathogens such as the gonococci. Analysis of the genes for the IgA protease produced by *N. gonorrhoeae*[108] revealed evidence for three functional domains of this important enzyme: an amino-terminal leader, which initiates the transport of the precursor; the protease itself; and a carboxyl-terminal helper domain, which is apparently required for extracellular export. This study is an excellent illustration of how knowledge of the complete gene structure of a protein can be related to its activity and specific function. It has been thought for some time that the IgA protease

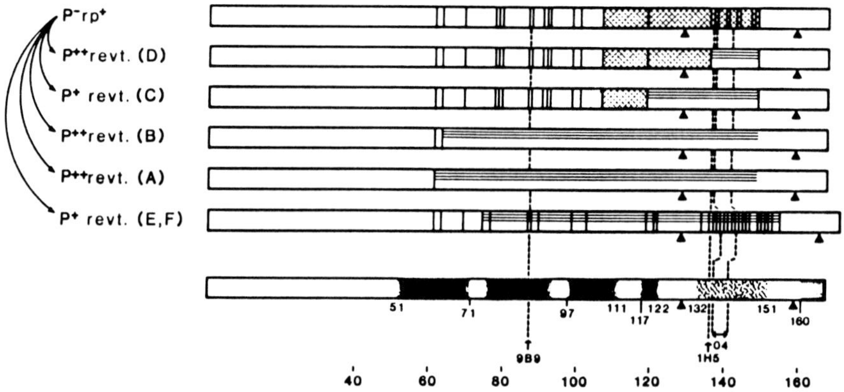

Figure 6.10 Gonococcal pilin amino acid sequences of $P^- rp^+$ parental strain and pilus$^+$ revertant gonococci. Differences in amino acid sequences are indicated by vertical lines at appropriate locations. Portions of the P:151 copy 5 found in pilin mRNA of $P^- re^+$ are indicated with crosshatching. Cysteine residues are noted by ▲. Locations of gonococcal pilin polypeptide constant regions are shown as open bars, semivariable regions as shaded bars and hypervariable regions as crosshatched bars. *Adapted from data presented by Swanson et al.*[106]

produced by the gonococcus is important as a virulence factor in being able to split protective IgA molecules present on secretory surface epithelium.

Viral Diseases

Viral Hepatitis

Perhaps there is no better illustration of the practical importance and ultimate clinical usefulness of recombinant DNA technology than the evolution of our knowledge and understanding of several forms of viral hepatitis, particularly hepatitis B and more recently hepatitis C. Not long after hepatitis B was recognized as a major cause of viral hepatitis, isolation and ultimate purification of the virus and its various constituent antigens resulted in successful production of a vaccine,[109] which has now been utilized as a protective measure throughout the world. After the production of the original vaccine,[110] based on the 22-nm particle form of hepatitis B surface antigen (HBsAg), hepatitis B surface and core antigens were synthesized by cloning the appropriate genes in *E. coli.*[111] This discovery was welcome because theoretically application of recombinant DNA methods to this problem could provide a limitless source of vaccine. After the initial demonstration that the viral genes coding for HBsAg and hepatitis B core antigen (HBcAg) could be cloned and synthesized in bacteria,[112,113] more efficient plasmids were constructed using the tryptophan operon regulatory gene, which considerably enhanced gene product synthesis.[111] The strategy involved in the synthesis of the hepatitis B core antigen is outlined in Figure 6.11. This construct and others like it have been utilized in the preparation of recombinant hepatitis B vaccine.

A great deal of work is being done concerning the practical and theoretical advantages of using the formalinized or the recombinant forms of the vaccine. The information gained during the initial hepatitis B studies provided an additional impetus; it suggested that smaller vaccine subunits readily retaining immunostim-

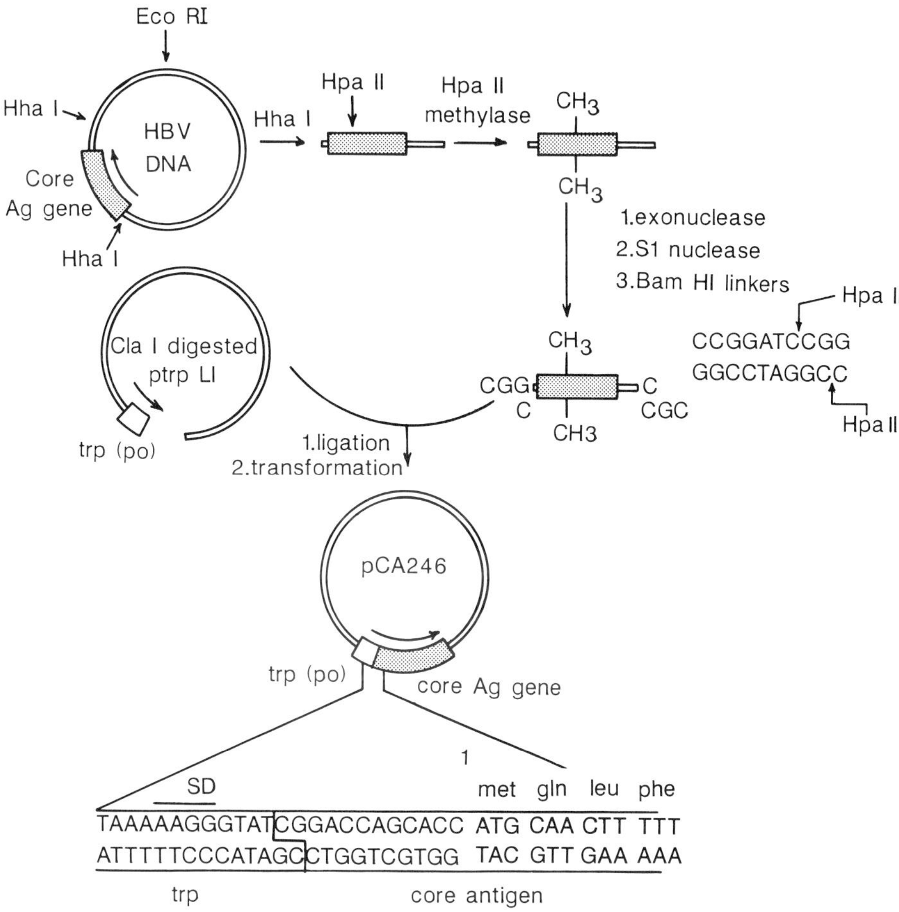

Figure 6.11 Construction of plasmids expressing HBcAg. Cloned HBV DNA was digested with *Hha*I and a 1005-bp fragment containing the HBcAg gene was isolated by polyacrylamide gel electrophoresis (PAGE); 20 μg of this fragment was then treated with *Hpa*II methylase. After phenol-chloroform extraction and ethanol precipitation, the fragment was treated with 28 U of T4 DNA polymerase in the absence of dNTPs (to allow the 3′ exonuclease activity to function) in 200 μg of 30 mM Tris acetate, pH 7.8, 67 mM KOAc, 10 mM MgOAc, 0.5 mM dithiothreitol, and 100 mg/ml bovine serum albumin for 30 s at 37°C. The reaction was stopped with phenol, extracted with chloroform, and precipitated with ethanol. The fragments were then treated with S_1 nuclease, and *Bam*HI linkers were added using T4 DNA ligase. The fragments were digested with *Hpa*II and purified from the digested linkers by PAGE, 150 ng was ligated with 200 μg of *Cla*I-digested and calf intestine alkaline phosphatase–treated ptrpL1 (10 U Sigma type VII phosphatase incubated with 20 μg DNA at 68°C for 30 min in 400 μl 10 mM Tris, pH 7.5, 0.5 mM EDTA) in 20 μl, and the ligation mixture was used to transform HB101. Transformants were selected on L plates containing 20 μg/ml ampicillin and screened for recombinant plasmid DNA. Forty such recombinants were then tested for HBcAg using a double-antibody radioimmunoassay with human anti-HBcAg IgG; 17 were HBcAg positive. Analysis of DNA after digestion with restriction enzymes showed that all 17 contained HBcAg DNA sequences in a *trp*-dependent expression orientation. Four were chosen for DNA sequence analysis. The sequence of the plasmid producing the highest levels of HBcAg, pCA246, is shown.

ulating potency might also be prepared as synthetic peptides derived from the original gene nucleotide or amino acid product sequences. In fact, it was demonstrated that antibody to hepatitis B surface antigen could be produced after a single injection of an uncoupled synthetic HBsAg peptide containing only 14 to 15 amino acids.[114] These observations were some of the first to show the potential usefulness of relatively small synthetic peptides as immunogens against important infectious agents.

Extensions of Medical Scope Using Recombinant Approaches

The recombinant technology that arose from the original studies of the hepatitis B genome was rapidly applied in fundamental investigations of the relationship between hepatocellular carcinoma and chronic active hepatitis. For many years it had been suspected that hepatitis B could be involved in the etiology of primary cancer of the liver,[115] since the geographic distribution of chronic carriers of the HBsAg is similar to that of the prevalence of hepatocellular carcinoma. Also, there is an increase in hepatitis B markers in serum of patients with primary liver cancer and it was often noted that hepatitis B infection preceded onset of the tumor. Therefore, the relationship of hepatitis B to hepatoma was assessed by using in situ hybridization of segments of cDNA obtained from the hepatitis B genome with DNA in tumor tissues from patients with hepatocellular carcinomas.[116–118] These studies showed integration of hepatitis B virus (HBV) DNA sequences in cellular DNA of fresh tumor tissue and in DNA from a hepatoma cell line, using cloned radiophosphorus-labeled HBV DNA as a probe. The demonstration of hepatitis B DNA in the cellular genome of the human hepatocellular carcinoma provided a strong argument for HBV as an agent that somehow activated the cancer. The only previous demonstration of such a direct virus-cancer relationship was the identification of Epstein-Barr viral genomic material in tissues from patients with Burkitt's lymphoma and some nasopharyngeal carcinomas.[119]

An example of the hybridization experiments clearly indicating the presence of HBV DNA in the genome of human hepatic cancer cells is shown in Figure 6.12. In examining the relationship between hepatitis B viral gene expression and the parallel process of neoplasia in hepatocellular carcinoma, it is important to remember that transformation, or the neoplastic process, could have resulted from altered cellular gene expression rather than integration of HBV genes. Work on these relationships is continuing in many laboratories throughout the world. Yee[120] has demonstrated that an 88-bp fragment of the core promoter of hepatitis B virus contains a potent liver-specific enhancer and a functional promoter. This enhancer also functions in human hepatoma cells and may be one part of the link between hepatitis B and hepatocellular carcinoma.

A number of additional recombinant approaches have been successfully applied in the study of hepatitis B infection. The polymerase chain reaction (PCR) has been used to demonstrate small quantities of hepatitis B viral DNA in serum or plasma.[121] Because serum factors inhibit DNA polymerase, it was necessary to extract nucleic acid from sera before performing the PCR. Using these sensitive methods, it was noted that surface antigenemia may precede viremia during acute hepatitis B infection. Again using the PCR, mother-to-infant transmission of hepatitis B has been documented.[122] In this study, HBV-specific DNA from 11 known carrier mothers was detected in 10 maternal serum samples, 8 samples of colostral whey, 8 samples of cells from colostrum, and 1 cord blood sample. The infant whose cord blood

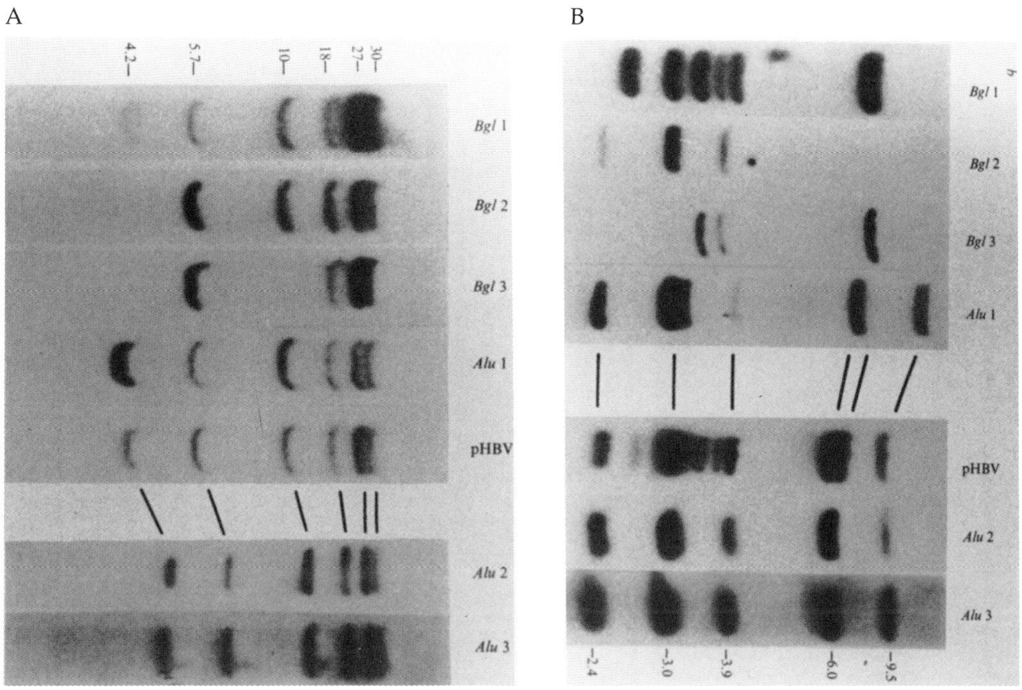

Figure 6.12 Hybridization of *Hin*dIII (A) and *Eco*RI (B) digests of Alexander (hepatoma cell line) DNA with probes representing fragments of HBV DNA. Digestion products were run on 0.8% agarose gels and blotted to nitrocellulose filters. Individual lanes were cut out and hybridized to probes derived from cloned HBV DNA. The various probes represent different regions prepared from restriction fragments of HBV DNA. pHBV represents probe prepared from the entire HBV sequence. Fragment sizes are indicated in kilobase pairs. *Reproduced from Edman et al.,*[117] *with permission.*

tested positive for HBV DNA showed a low level of responsiveness to hepatitis B vaccine and subsequently became an HBV carrier.

Direct insight into the pathogenesis of liver injury in hepatitis B has been gained by using hepatitis B transgenic mice.[123] HBV envelope transgenic mice were used to show that HBV-encoded antigens were expressed at the hepaptocyte surface in a form recognized by MHC-restricted $CD8^+$ cytotoxic T cells. These interactions led to the death of the hepatocyte in vivo and provided direct evidence that liver cell injury in human HBV infection may also be mediated by immune mechanisms.

Alternative Vaccines

After the success with the hepatitis B subunit vaccines prepared initially from plasma of chronically infected patients, several alternative strategies were introduced that illustrate novel potentially useful recombinant techniques. An effective live recombinant vaccinia virus containing a 1350-bp DNA fragment of HBV coding for the HBsAg has now been produced.[124] The hepatitis gene encoding surface antigen was fused to an early vaccinia virus promoter, ensuring transcription and eventual successful translation into HBV product. The presentation of HBsAg in the form of a live virus vaccine product ensured ongoing replication and antigenic presentation within the immunized host. In the study, live vaccinia containing the HBV insert protected challenged chimpanzees against hepatitis B viral infection.[124]

In the study of hepatitis B and other forms of chronic liver disease, ample use has been made of hybridization procedures to detect footprints of the infecting virus. Thus, 134 patients with HBsAg-negative chronic liver disease were examined for hybridization with HBV genomic sequences.[125] HBV hybridizing sequences were detected in 52 of 88 liver tissue samples (59%), including 17 of 20 samples from patients with hepatocellular carcinoma. A number of similar etiologic and epidemiologic investigations have confirmed the impression that HBV is a factor in a substantial proportion of HBsAg-negative patients with various forms of chronic liver disease.

Delta Infection and Hepatitis B

Delta virus infection is a serious epidemiologic and health problem in many widely scattered areas of the world. Delta is a hepatotropic agent whose antigenic system depends for its expression on concurrent hepatitis B infection.[126] The putative delta agent appeared initially to be a small ribonucleic acid with no direct molecular relation to the DNA genome of HBV.[127] Delta appears to be highly pathogenic and has frequently been implicated as the cause of acute and chronic liver disease. In a study of 432 patients with viral hepatitis B from France, England, and Italy,[128] patients with fulminant hepatitis showed a significantly higher prevalence of delta markers than did those with benign hepatitis. The human delta virus (HDV) is apparently a defective virus consisting of a 36-nm particle that contains an envelope with the HBV antigen and a nucleocapsid with the hepatitis delta antigen (HD Ag). The agent contains an RNA genome of 1.75 kb. RNA from the HDV has been shown to be a single-stranded circular molecule with properties similar to those of a viroid.[129,130] Molecular cloning and sequencing of a human hepatitis delta virus RNA were reported by Makino et al.,[131] who obtained samples directly from a patient with acute delta hepatitis. In this study the complete sequence of the HDV DNA was obtained by starting with HDV RNA primers and using reverse transcriptase to produce complementary DNA. When the HDV sequence was examined, it became apparent that this first human isolate differed considerably from the strains isolated and previously sequenced from serial passage in chimpanzees.[129] Because the human-isolated HDV had a circular RNA, it was suggested that it might represent a viroid or virusoid.[132] It was also noted that HDV contained the viroid consensus sequences GAAAC and GAUUUU, thought to be particularly important for the replication of viroids. The GAAC sequence is necessary for self-cleavage of the virusoid RNA. Furthermore, in the human HDV isolate the RNA sequence contained open reading frames in both normal genomic sense and antigenomic sense (orientation). If these open reading frames are used for the translation of proteins, this strange viroid associated with HDV would be the first example of a single-stranded RNA with bidirectional translation capability.

The discovery of the circular RNA of human HDV is probably an important milestone in our understanding of viruses and how they cause infection in human or other animal hosts.[133] The circular shape of the delta virus has been confirmed by Kos et al.[134] by electron microscopy. Hepatitis delta virus appears to be similar to agents called plant viroids, which have an RNA genome formed from a single-stranded, closed circle. Having 1678 bases in its genome, the delta virus is about four times larger than typical plant viroids; also, plant viroids generally do not have protein coats, as does the delta virus. The identification of two consensus sequences of five and six nucleotides that exist in virtually all plant viroids and in the human hepatitis delta virus was a finding of great interest and puts this class of agents somewhere between animals and plants in terms of host population.

Group I Introns

Another fascinating aspect of the discovery of the human delta virus and its apparent circular RNA is related to a second parallel and what have been termed group I introns. These nucleotide sequences include those found in mitochondrial and ribosomal RNA genes, chloroplast transfer RNA genes, and nuclear ribosomal RNA genes. These particular introns, which have long puzzled cell biologists, are characterized by a series of relatively conserved sequences that in turn dictate certain constraints on secondary and tertiary structure. One of the most distinctive aspects of this class of introns is that some of them are able to undergo self-splicing reactions, thereby excising themselves and in the process ligating the adjacent parental strands, apparently in the absence of enzymes. This process in group I introns releases small circular RNA molecules, which suggests that viroids and even the human delta hepatitis agent could be escaped introns. Sequence homologies have now been identified between certain group I introns and both plant viroids and the human hepatitis delta agent. More work in this exciting area is clearly needed.

Diagnostic and Epidemiologic Applications

As the in vitro Southern blot gel and tissue hybridization techniques have become more readily available and their exquisite sensitivity and specificity have been recognized, they have been applied to specific diagnostic or epidemiologic questions. A good example of their usefulness was the study of fulminant hepatitis B in successive female sexual partners of two anti-HBs–positive males.[135] In this study two unrelated female sexual partners of a symptom-free HBsAg-positive carrier male died of fulminant hepatitis B. Using molecular hybridization techniques with cloned HBV DNA labeled by nick translation, it was shown that the asymptomatic HBsAg carrier males both carried HBV DNA in seminal fluid, sputum, saliva, peripheral blood, and liver tissues. These findings were of particular interest in view of the lack of serologic evidence of HBV replication, such as DNA polymerase activity. It seems likely that the approaches available with cDNA probes and similar hybridization techniques will prove to be important diagnostic procedures in a number of other clinical situations.

Non-A, Non-B Hepatitis and Identification of Hepatitis C

For many years, it was recognized that a proportion of patients with viral hepatitis did not fit into the category of either hepatitis B or hepatitis A. The term non-A, non-B (NANB) hepatitis was applied to this group of patients, and it was estimated that the NANB form accounted for 20% to 40% of all cases of viral hepatitis.[136] With the recent identification of hepatitis C and the isolation of genetic probes for detecting hepatitis C infection,[137,138] our knowledge of the prevalence and epidemiology of this form of viral hepatitis should increase rapidly. The hepatitis C virus was discovered by constructing a random-primed cDNA library from plasma known to contain NANB hepatitis and screening the library with serum from a patient with NANB hepatitis. A complementary clone was isolated and then shown to encode an antigen specifically associated with NANB hepatitis infections. It was shown that the clone isolated was *not* derived from host DNA but from an RNA molecule that consisted of at least 10,000 nucleotides and that was positive stranded with respect to the NANB hepatitis–encoded antigen. These findings indicated that the new agent is similar to the togaviruses or flaviviruses and contains a positive-stranded RNA molecule. Using these initial findings, it was possible to synthesize a polypeptide in recombinant yeast clones that could be used to detect anti–hepatitis C viral antibodies.[138] Subsequent assays of blood from 10 transfusions that resulted in NANB hepatitis showed that there was at least one positive blood donor in nine of these cases and that all 10 blood recipients seroconverted, showing anti–hepatitis C virus (HCV) during their subsequent illnesses. About 80% of patients who had chronic posttransfusion NANB hepatitis from Japan and Italy had circulating HCV antibody, and 58% of NANB hepatitis patients from the United States without an identifiable parental source of exposure also were positive for HCV antibody. Whether all cases of NANB hepatitis can be explained by hepatitis C remains problematic. It seems likely that other subgroups of non-A, non-B, and even non-C hepatitis will also be defined—in fact, yet another agent causing enterically transmitted NANB hepatitis has been described, again using a molecular strategy.[139] This fourth hepatitis virus has tentatively been named hepatitis E.

AIDS Viruses and Their Relatives

The public awareness of and scientific interest in the AIDS problem in the United States and throughout the world have focused a great deal of attention on the human immunodeficiency virus (HIV). Since we live in an era of molecular biology and widespread applications of molecular biology, it is appropriate that the major part of the AIDS research and epidemiology has unfolded through extensive applications of gene cloning, hybridization, and comparative nucleotide sequence data for the original immune deficiency isolates as well as related viral species. The key recognition of and research on the AIDS virus has been the application of recombinant DNA techniques.

Human T Lymphotropic Virus Type I

The story of AIDS probably began when human T lymphotropic virus type 1 (HTLV-I) was isolated in association with a rare adult form of human T cell leukemia.[140–143] This malignancy represents a monoclonal proliferation of CD4-positive T cells. Despite epidemiologic evidence indicating a close correlation between signs of HTLV-I infection and disease incidence in parts of Japan and in the Caribbean,[143] it has been difficult in the final analysis to link HTLV-I directly with the genesis of this form of leukemia. Less than 0.1% of individuals infected by the virus ever develop the disease, and malignancy, when it does occur, may arise after a latent period of as long as 20 years after initial virus exposure.[144,145] Even more difficult to fathom is the fact that freshly isolated peripheral blood leukemic cells frequently do not show detectable viral antigen expression.[146] Based on present knowledge, it seems that an independent second activation event after the initial viral infection may be necessary for leukemic transformation to occur. HTLV-I may somehow initiate a process that later progresses to development of the leukemia.

Work directed at understanding the gene structure of the HTLV-I retrovirus indicated that the usual *gag*, *pol*, and *env* genes were expressed. In addition, several other domains expressed new protein products of interest, in particular a 40-kd protein termed $p40^x$, which is encoded by a doubly spliced 2-kb RNA segment.

This material has been referred to as the *tat* protein and apparently is capable of trans-activating the HTLV-I long terminal repeat sequences that regulate viral transcription. Since the *tat* protein appears to be involved in viral regulation of gene expression, it may also function to regulate other cellular genes and to alter growth characteristics. Studies using transgenic mice in which the entire *tat* gene and its flanking long terminal repeats were excised with restriction enzymes and microinjected into fertilized mouse eggs have indicated that successful takes of the injected *tat* genes are accompanied by either thymic inoculation or mesenchymal tumors.[147] Much of the study of HTLV-I is ongoing, but it has laid important groundwork for further understanding of the genomic structure and patterns of behavior of the entire HTLV viral group.

HTLV-I has also been shown to be the etiologic agent of tropical spastic paraparesis (TSP) in diverse geographic regions, including several Caribbean islands, South America, the Seychelles islands, parts of Africa, and southern Japan. Clinically, the disease is characterized by paraparesis and lower limb spasticity, sometimes associated with posterior column and superficial sensory changes. An HTLV-I isolate from a Jamaican female with typical TSP was found to be a prototype HTLV-I type C retrovirus; however, it appeared to differ somewhat from other HTLV-I isolates from patients with adult T cell leukemia when examined by restriction enzyme mapping.[148] These initial findings may indicate that two subgroups of HTLV-I are detectable by restriction enzyme mapping and comparative analysis.

Human T Lymphototropic Virus Type II

Human T lymphototropic virus type II (HTLV-II) is a human retrovirus with a nucleic acid sequence homology to HTLV-I of approximately 60%. Like HTLV-I, HTLV-II has been associated with malignant disease, since the virus has been identified in and cloned from peripheral blood and splenic cells of several patients with a T cell variant of hairy-cell leukemia.[149–151] These findings suggest a role for HTLV-II in the pathogenesis of atypical hairy-cell leukemia. In general, hairy-cell leukemia is thought to be a distinct lymphoid malignancy associated with pancytopenia, splenomegaly, and the presence of morphologically distinct cells in the peripheral blood.[152] Phenotypic analysis of hairy cells indicated that most have a B cell precursor; however, use of various cell surface markers indicated that a minority are derived from T cells or show a hybrid T-B phenotype. In cases in which cells infected with HTLV-II were identified, probes derived from 5′ or 3′ ends of an HTLV-II cDNA showed clear-cut hybridization with DNA from the patients' peripheral blood or splenic tissues. Examples of such positive hybridization results are shown in Figure 6.13. As with HTLV-I, viral RNa expression was absent in fresh leukemic tissues despite the evidence in leukemic cells for integrated HTLV-II provirus. These findings suggest that HTLV-I and HTLV-II may induce leukemic transformation by similar mechanisms and that continued viral replication may not be required for maintenance of the leukemic state.

AIDS and HTLV-III or HIV

AIDS has now spread worldwide and appears to be an acute as well as a chronic, even subclinical, problem in many different areas.[153,154] In 1981–1982 several groups of workers independently began to recognize what appeared to be a new disease, AIDS, that occurred predominantly in young homosexual men.[155–159] Epi-

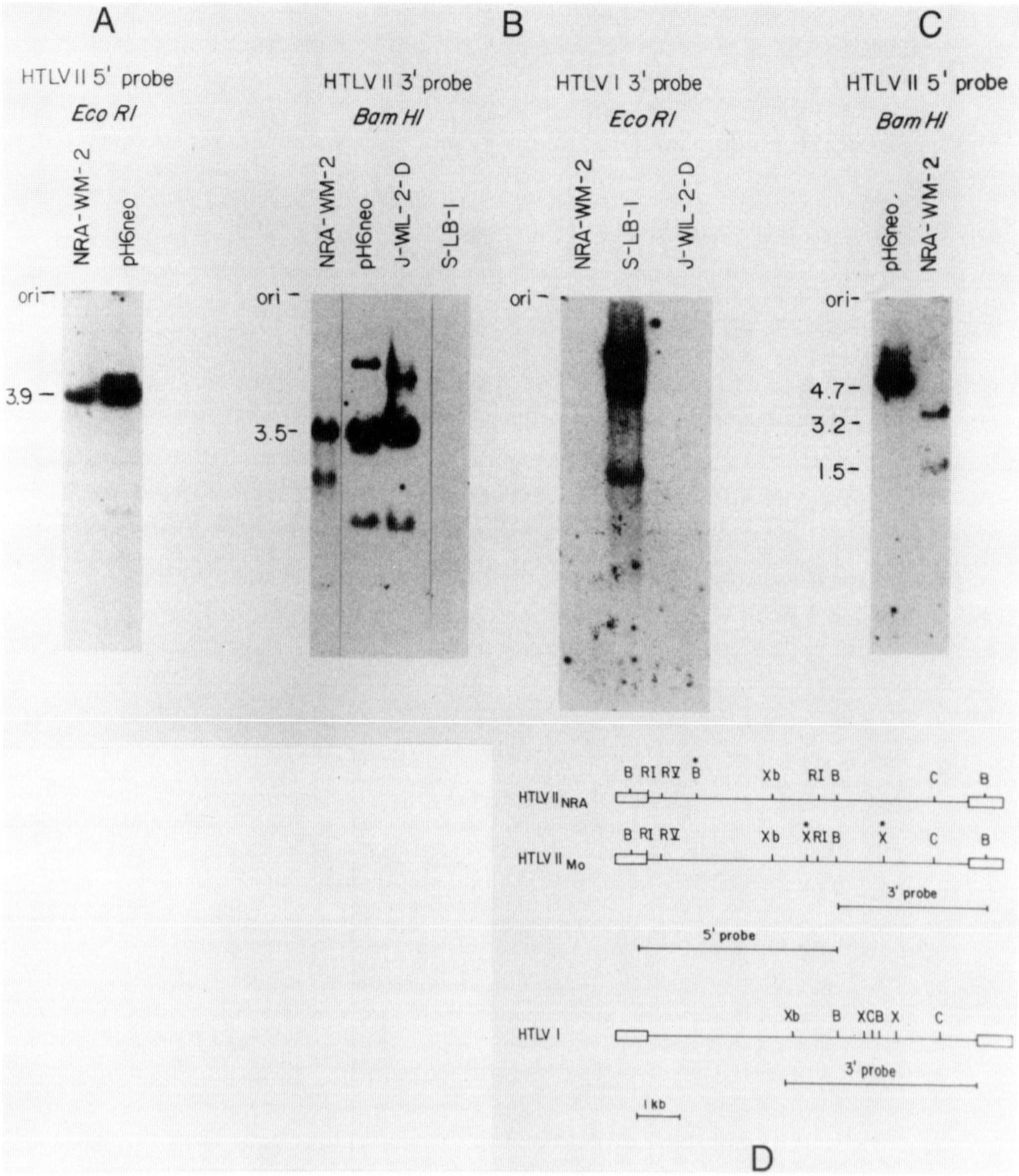

Figure 6.13 Comparison of the HTLV-II$_{nra}$ provirus to the HTLV-II$_{mo}$ and HTLV-I genomes by Southern hybridization analysis. (A) Hybridization to 5′ HTLV-II sequences. DNA from NRA-WM2 cells (lymphocytes from a normal donor infected and transformed by the HTLV-II$_{nra}$ isolate), which was digested with *Eco*RI, hybridized strongly to the HTLV-II 5′ probe, resulting in a conserved 3.9-kd internal fragment. A control digest of the plasmid pH6neo (a clone containing the HTLV-II$_{mo}$ genome) is shown for comparison. The 5′ HTLV-II probe is drawn to scale in part D. (B) Hybridization to 3′ HTLV-II sequences and lack of hybridization to 3′ HTLV-I sequences. DNA from NRA-WM2 cells, which was digested with *Bam*HI, hybridized strongly to the HTLV-II 3′ probe, resulting in a conserved 3.5-kb band. Control digests of the plasmid pH6neo (containing HTLV-II$_{mo}$), the B cell line infected with HTLV-II, J-WIL-2D, and the S-LB-I cell line infected with HTLV-I are shown. DNA from NRA-WM2 and J-WIL-2D cells did not hybridize to an HTLV-I 3′ probe, whereas a control digest of DNA from an S-LB-I cell (infected with HTLV-I) shows strong hybridization to the HTLV-I probe. The HTLV-I and HTLV-II 3′ probes are drawn to scale in part D. (C) Presence of an additional *Bam*HI site in the 5′ half of the HTLV-II$_{nra}$ provirus. A *Bam*HI digest of DNA

demiologic investigations suggested that the new disease was concentrated in male homosexuals and patients with a history of intravenous drug abuse or blood transfusions. In 1983 members of Gallo's and Montagnier's groups identified a T cell retrovirus as being associated with the AIDS syndrome.[160,161] These early studies reported isolation of a new nontransforming retrovirus that was clearly different from HTLV-I or HTLV-II in cultures from patients with lymphadenopathy syndrome; they also noted its continuous passage during its short-term growth in T lymphocytes from normal blood donors. A major protein (p25) not immunologically cross-reactive with those of HTLV-I was detected; this protein (antigen) was reactive with antibodies present in serum from infected patients. The group at the National Institutes of Health (NIH) initially found evidence for the presence of the viral genome of HTLV-I or an HTLV-I variant in 2 of 33 AIDS patients. The basic scientific investigation of the HTLV (AIDS) virus then began to unfold rapidly and still continues. It was found that the virus had a selective affinity for $CD4^+$ helper T cells; moreover, the actual CD4 epitope represented the initial viral attachment site before infecting cell penetration.[162,163] Antibodies reacting with HTLV-I were identified in 10% of AIDS patients, and isolates of it were successful in fewer than 10% of cases. As work progressed, permanent cell lines containing HTLV-III were derived,[164] and the Western blotting technique demonstrated antibodies in 42 of 48 patients with AIDS and 11 of 14 homosexuals with pre-AIDS syndrome.[165] The glycoprotein gp41 was identified as a major viral antigen in this study. The AIDs virus itself is now referred to as HTLV-III or HIV. An example of the positive Western blot identification of anti-gp41 antibody is shown in Figure 6.14. With technical improvement in antibody assays, Brun-Vezinet et al.[166] later reported that 34 of 34 or 100% of AIDS patients were positive for HTLV-III antibodies, as were 84% of patients with lymphadenopathy (pre-AIDS) syndrome and 0% of controls.

Recombinant Definition of AIDS

In 1984, Hahn et al. reported the molecular cloning of the entire HTLV-III virus.[167] Shortly thereafter the genomic heterogeneity of HTLV-III was defined,[168] and later that year (1985) the entire nucleotide sequence of the AIDS virus genome was established independently by three groups.[169–171] This sequence established that the HTLV-III gene had four long open reading frames, the first two corresponding to the retroviral *gag* and *pol* genes and the fourth coding for two functional polypeptides—a large precursor of the major viral envelope glycoprotein and a smaller protein derived from the 3′-terminal long open reading frame and analogous to a similar product in HTLV-I and HTLV-II. This format of gene organization is shown in Figure 6.15. Work on the immunobiology of the AIDS virus continues in many laboratories worldwide.[172,173]

Since progress toward cloning effective antigens or immunogens seemed slow, innovative attempts have been made using molecular biological approaches to

from NRA-WM2 cells results in two fragments rather than the single 4.7-kb fragment seen in the digest of pH6neo (representing HTLV-II_{mo}). (D) Restriction enzyme maps comparing HTLV-II_{nra}, HTLV-II_{mo}, and HTLV-I proviral genomes. Marked similarity between HTLV-II_{nra} and HTLV-II_{mo} is seen, whereas the map of HTLV-I is substantially different. Asterisks indicate restriction cleavage sites that differ between the DNA of HTLV-II_{nra} and that of HTLV-II_{mo}. The size and location of the 5′ and 3′ HTLV-II probes and the 3′ HTLV-I probe are shown beneath their respective provirus maps. B, *Bam*HI; C, *Cla*I; RI, *Eco*RI; RV, *Eco*RV; Xb, *Xba*I; X, *Xho*I. *Reproduced from Rosenblatt et al.,*[151] *with permission.*

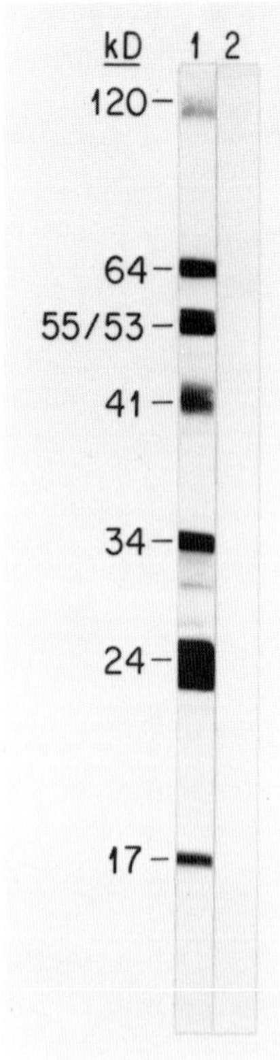

Figure 6.14 Typical Western blot showing multiple antibodies reacting with HIV-1 viral antigens. Lane 2 represents an uninfected control serum and lane 1 the serum from a patient infected with HIV-1. The viral antigens detected are: gp120 (glycoprotein 120), the external membrane of glycoprotein; p64, a *pol* (polymerase) gene product; p55, a *gag* (core) gene precursor product that is broken down to p24 and p17; p53, a *pol* gene product that is usually seen in association with p64; gp41, the transmembrane glycoprotein; p34, a *pol* gene product (endonuclease); p24, the major *gag* protein seen; and p17, *gag* protein product. *Western blot kindly provided by Dr. R. Marlink, Department of Cancer Biology, Harvard School of Public Health.*

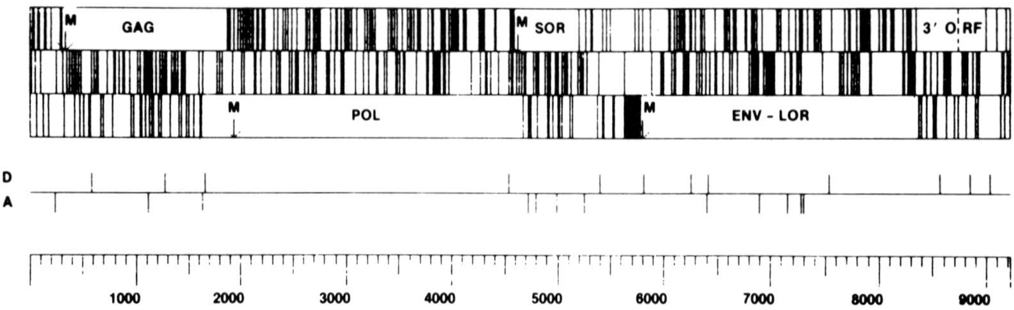

Figure 6.15 Distribution of open reading frames in HTLV-III. The positions of termination codons are indicated for each of the three possible reading frames as a vertical line within each box. Thicker lines represent clusters of closely spaced termination codons. The positions of consensus splice donor (D) and acceptor (A) sites are shown below. The nucleotide sequence positions are also indicated below, starting from the transcription initiation site. The positions of the four largest open reading frames are identified as *gag, pol, sor,* and *env-lor,* and the position of the first ATG codon (M) in each reading frame is indicated. A fifth open reading frame (3′-*orf*) is found in clone BH8 but not in BH10; the dashed vertical line indicates the position of a termination codon in BH10. *Reproduced from Ratner et al.,*[169] *with permission.*

control virus replication through site-specific mutagenesis of the AIDS virus reverse transcriptase.[174] Because a crucial step in the virus life cycle involved host cell invasion and conversion of the viral RNA genome to double-stranded DNA by reverse transcriptase, it was proposed that site-directed mutagenesis of this enzyme might effectively block all subsequent viral replication. Six regions of the reverse transcriptase gene were identified as possible prime targets for mutagenesis; all six of these domains showed striking similarity to corresponding regions in other viral reverse transcriptases. The process of site-directed mutagenesis (described in Chapter 4) was undertaken using selected oligonucleotides. Each mutation either introduced a new restriction site into the gene or deleted a new one. Each mutation was also designed to introduce a single amino acid substitution and thus encode a polypeptide chain identical in length to the original gene product. The mutation strategy and relative positions of induced single point mutations are shown in Figure 6.16. The most dramatic effect was achieved with the Asp → His mutation at residue 185, which destroyed the crucial Asp-Asp doublet and thereby completely abolished reverse transcriptase activity of the protein. A mutation in a second major domain, Asp → Glu at residue 110, also resulted in complete loss of enzymatic activity. Theoretically, it might be possible to insert mutant reverse transcriptase back into the infectious virus and thus completely shut off its continuing infective cycle. Such a general strategy is clearly worth considering in parallel with

Figure 6.16 Positions of mutations and conserved regions in the HIV reverse transcriptase (RT). The HIV RT polypeptide is represented as a linear molecule with amino acid residue numbers based on the sequence of the lymphadenopathy-associated virus (LAV) strain of HIV and conserved regions (A–F) shown as solid bands with (expanded) sequences in single-letter code for amino acids. These were identified because they were previously shown to have amino acid sequences similar to those of other retroviral *pol* genes, and in the case of region E (residues 180–190) there is also homology with a number of nonretroviral polymerases. Regions in which single amino acid substitutions have been introduced (Table 6.1) are shown above the reverse transcriptase polypeptide with the amino acid changes illustrated. Two other identified conserved regions are shown below the reverse transcriptase polypeptide. The putative location of the RNase H activity associated with the reverse transcriptase is depicted at the carboxyl terminal of the polypeptide (shaded area). *Reproduced from Larder et al.,*[174] *with permission.*

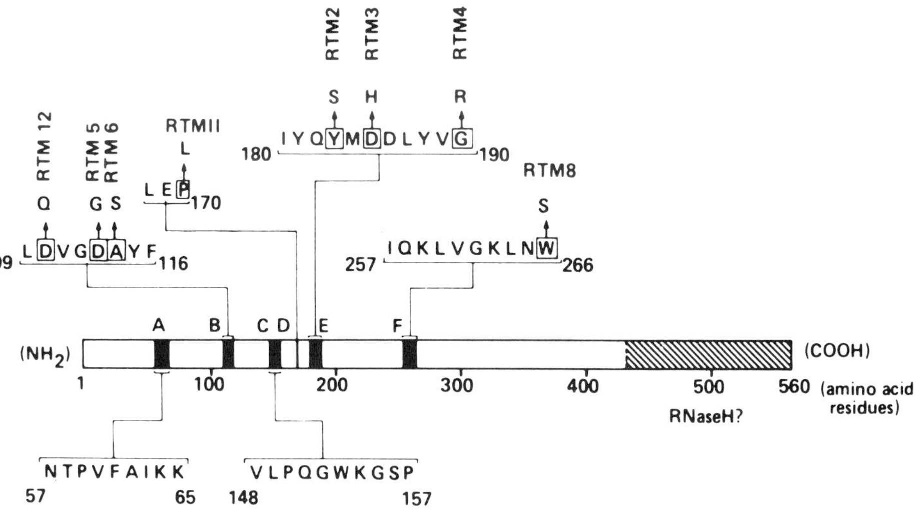

chemotherapeutic agents that might be capable of selectively completely inactivating HTLV reverse transcriptase.

The HIV-1 reverse transcriptase has continued to be of interest as a possible vulnerable point at which therapeutic strategies might inactivate the virus. Cytotoxic T lymphocytes have been described in blood samples from HIV-1–infected individuals. These T cells showed selective killing for the reverse transcriptase as well as MHC restriction.[175] These observations suggest that epitopes representing the HIV-1 reverse transcriptase may eventually prove useful in construction of a vaccine.

Additional AIDS-Like Agents

Since the basic molecular biological investigations and comparative antigenic and genomic comparisons of the human T cell lymphotropic retroviruses HTLV-I, HTLV-II, and finally HTLV-III, several additional viruses have been described that appear to be related to or similar to HTLV-III in that they are also associated with induction of immune deficiency. One of these agents was initially recognized as simian immunodeficiency virus.[176] This T cell lymphotropic virus appears to be immunologically related to several other viruses isolated from macaque or African green monkeys. A human virus isolated from healthy humans and immunologically related to the simian T lymphotropic virus (STLV-III_{mac}) was shown to be virtually identical to STLV-III_{agm} (African green monkey strain). Extensive genomic and sequence analyses of this group of viruses indicated that STLV-III has a gene organization very similar to that of HTLV-I and HTLV-II. STLV-III_{agm} and HTLV-II also appeared to have a close genomic relationship. Parallel studies of the interrelationships of HTLV and simian immunodeficiency virus[177] must be carefully considered in terms of the possible existence of an animal reservoir for human AIDS and its possible role in the AIDS pandemics in many parts of the world. Precise comparisons of nucleotide sequences indicate that HTLV-II and African isolates of HTLV-I, although obviously derived from a common ancestral gene, are not simple evolutionary links between simian immunodeficiency viruses and European or U.S. isolates of HIV-1. Therefore, even if it is presumed that monkeys could be an accessory reservoir for AIDS-like retroviruses, nucleotide sequence data for comparable genes that have now been completely sequenced[176,177] do not support the idea that AIDS came about through recent transmission of simian immunodeficiency virus to humans followed by rapid viral evolution or mutation to HIV-1.

Additional virus isolations and serum antibody as well as genomic hybridization assays continue to add more diversity and thereby fascination to the molecular biology of AIDS. A new human T lymphotropic virus (HTLV-IV) was described in apparently healthy people from Senegal.[178,179] This virus was similar to STLV-III in electron microscopic appearance, CD4 tropism, and major viral protein constituents. When this new virus, termed HTLV-IV, infected CD4 cells, it induced the formation of multinucleated syncytia in vitro. In humans it does *not* appear to be associated with immunodeficiency.

A survey of 4248 individuals in six West African countries for evidence of HTLV-IV infection showed infection in five major regions. Despite their close geographic proximity, patients or subjects in the different countries studied showed marked variations in seropositivity for HTLV-IV regardless of the health category of the individual subjects. The prevalence of seropositivity varied from zero to 6.9–

16%. In all areas, the prevalence of HTLV-IV positivity was much higher (14.9–64%) in a high-risk population (prostitutes) than in control groups. This study was of particular interest since it indicated that together HTLV-IV and HTLV-III (HIV-1) were present in these regions of West Africa but that the HTLV-III incidence was low in most urban centers studied. HTLV-IV infections appeared to be more widespread, with rates of infection elevated in sexually active risk categories but not closely associated with disease. Apparently, in vitro studies have indicated that HTLV-IV may inhibit some of the pathogenic effects of HTLV-III. As epidemiologic and serological data accumulate, we will learn considerably more about the whole human T lymphotropic virus spectrum.

New observations continue to expand the spectrum of the original HTLV-I and HTLV-II associations with leukemia. A new human retrovirus, HTLV-V, has been described after isolation from a patient with $CD4^+$ Tac^- cutaneous T cell lymphoma/leukemia.[180] Using fragments of provirus cloned from one patient with T cell leukemia, closely related sequences were found in cell lines and tumor cells from seven additional patients. None of the patients studied appeared to be at any risk for HIV-1 infection. It will be of interest to determine whether this new member of the related retroviruses turns out to be a major etiological factor in all cutaneous T cell lymphomas and leukemias, including mycosis fungoids.

Chronic Progressive Myelopathy and Tropical Spastic Paraparesis

As mentioned earlier, another disorder that has been linked to HTLV-I infection is known as tropical spastic paraparesis. This slowly progressive myelopathy is associated with increased levels of serum and cerebrospinal fluid antibodies to the human T lymphotropic retrovirus HTLV-I.[181] Isolation of an HTLV-I–like virus from T cell lines derived from peripheral blood and spinal fluid of patients with TSP has been reported.[182] Studies of other patients with chronic progressive myelopathy and other chronic neurological syndromes have provided both serologic and recombinant evidence for a link with HTLV-I.[183,184] A similar disorder observed with clustering of HTLV-I–seropositive patients showing chronic progressive myelopathy has been described from Japan.[185] The exact connection with HTLV-I infection has not been determined because the low rate of myelopathy does not correlate with a much higher seropositivity in certain areas, such as parts of the Caribbean and Japan.

Immune Approaches to AIDS

Several promising approaches have emerged as possible eventual solutions to the AIDS problem. Recombinant techniques have been used to produce an envelope glycoprotein antigen (gp120) in *E. coli*.[186] A precursor, gp160, is proteolytically cleaved to generate the external envelope glycoprotein gp120 and the transmembrane glycoprotein gp41 (Figure 6.17). The possibility that gp120 might be used as a protective component in a vaccine was suggested by the findings of antibodies to gp120 capable of neutralizing the AIDS virus in patients with AIDS and AIDS-related complex (ARC) as well as in asymptomatic individuals infected with the virus.[187,188] Purified gp120 stimulates production of apparently neutralizing antibodies,[189] and direct binding of gp120 to the T4 receptor has been demonstrated.[190,191] Putney et al.[186] undertook a further immunochemical evaluation of gp120 as an isolated antigen. About half of the apparent weight of gp120 consisted of carbohydrate. With this degree of glycosylation, it was problematic whether

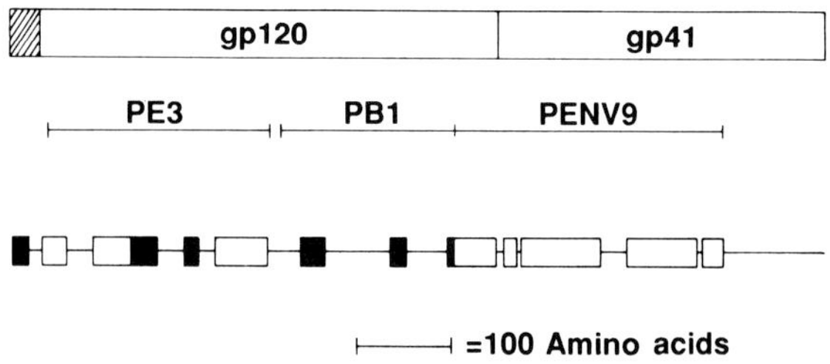

Figure 6.17 Location of proteins, PB1, PE3, and PENV9 in relation to gp120 and gp41. The envelope is synthesized as a precursor, gp160, that is cleaved to gp120 and gp41. The secretion signal sequence is hatched. The locations of the variable and conserved regions are shown as filled or open boxes, respectively. *From Putney et al.,*[186] *with permission.*

recombinant techniques would be effective in producing a vaccine component. It was found that the PB1 polypeptide subunit of the gp120 protein (see Figure 6.17), which contained 180 of the 481 amino acids in the whole gp120 molecule, contained antigenic epitopes capable of inducing completely neutralizing antibody. However, viral isolates from different individuals showed some gp120 antigenic variability. Amino acid sequence variations in the PB1 region could be a serious obstacle to the development of a vaccine using this approach. Nevertheless, positive or promising results have been reported by Hu et al.[192] using a recombinant HTLV envelope vaccine to protect chimpanzees. In these trials the animals received a recombinant vaccinia virus vaccine that had been combined with the entire HTLV envelope (*env*) gene sequence. Immunoblotting analyses of serum from immunized chimpanzees showed antibodies to both gp41 and gp120. After challenge with AIDS virus, there were no signs or symptoms of disease in eight of nine chimpanzees, but within 6 weeks one of two control animals developed lymphoadenopathy and virus was isolated from an enlarged node. It appeared that AIDS envelope virus immunization did not prevent HIV infection, since subsequent studies showed establishment of HIV-1 in the prevaccinated animals as well.

A number of other research groups are testing various AIDS virus components as candidates for vaccine preparation. A summary of some of these materials is shown in Table 6.1. The idea of using an anti-idiotypic reagent that reacts with either the HTLV-III envelope epitope (which in turn reacts with the CD4 receptor) or the CD4 receptor itself is an attractive one and is being pursued in several research laboratories. With many of the materials shown in Table 6.1, both neutralizing antibody and cell-mediated immune response need to be thoroughly evaluated.[193] Toxicity is another important consideration because there is some evidence that binding of gp120 to CD4 on lymphocytes may actually damage the helper cells.

Goals of Vaccine Therapy

Another important question concerns immunoprotection. Should a candidate vaccine protect against initial infection or only prevent the progression of infection to significant disease? In general, because of the high mortality caused by AIDS, prevention of infection seems to be the primary goal. Clinical trials of potential vaccines have already been initiated. The homosexual risk group has been extremely cooperative, but intravenous drug abusers and prostitutes (two other high-risk groups) may pose ethical and legal problems because their behavior patterns interfere with adequate follow-up and evaluation. The most difficult problem is that of devising strategies to overcome the potential extensive genomic variation in HIV-

Table 6.1 Possible Vaccine Candidates for Use in Protection against HTLV/HIV AIDS

Immunizing component	Degree of expression or yield	Carbo-hydrate content	Advantages	Comment
gp120	HIV-producing cell lines, low yield	High	Native material	Poor yield, ? toxic effect on T4 cells
Rec gp130	Mammalian cells, high yield	High	Native-like	? Toxic effect on T4 cells
Rec gp160	Insect cells, high yield	Partial	Whole envelope protein may offer advantages	Different glycosylation
PB1	*E. coli*, high yield	None	May provide hidden epitopes, does not bind CD4	Does not block fusion
Pept 735	Synthetic peptide of gp41 region, high yield	None	May provide hidden epitodes, does not bind CD4	? Limited specificity
HGP-30	Synthetic, high yield	None	Very conserved antigen	Does not block fusion
Vaccinia gp160	Vaccinia, high yield	High	Improved antigen presentation	Live virus ? associated side effects
Anti-idCD4	Immunized rabbit	None	Correct combining site conformation	No experience in patients

1. On the other hand, if only a limited number of extremely virulent subtypes can be associated with most cases of the disease, an effective vaccine can be achieved with greater dispatch.

Inducible Transcription Factor

As more and more scientific talent has focused on the AIDS problem, a number of parallel and potentially important discoveries have emerged. An interesting control protein for the AIDS virus has been identified by Nabel and Baltimore.[194] The elements in the AIDS virus genome that control activation are not completely known, but it was found that activation of T cells increases HIV-1 expression up to 50-fold in conjunction with induction of a DNA-binding protein that appears to act as a transcription factor binding to the viral enhancer sequences. Activation of a T cell line by mitogens such as phytohemagglutinin or phorbol myristate acetate led to expression of an inducible DNA-binding protein, NF-kB. It was thought that this protein might activate virus by binding directly to its enhancer nucleotide sequences and increasing transcription of the whole viral genome. These findings are of particular interest because they suggest that T cell activation, by any mechanism (including other viral, bacterial, or protozoan infections such as occur in those at risk of developing AIDS), could stimulate T cells and activate dormant HIV-1 in the process.

CD4 Binding of HIV-1

A good deal of attention has been directed at precise localization of the attachment site on human helper $CD4^+$ T cells for initial HIV-1 entry. Studies using purified soluble forms of CD4 in conjunction with proteolytic fragmentation and sequencing have indicated that the virus binding sites are in the NH_2-terminal regions of $CD4^+$ cells.[195] Moreover, synthetic peptides derived from known $CD4^+$ sequences inhibited infection of CD^+ T cells by the AIDS virus.[196] Similarly, dextran sulfate (molecular weight 8000) was shown to be capable of blocking attachment of HIV-1 to the target cell receptor on the CD4 molecule.[197] Because sulfate had been used by patients as both an anticoagulant and an antiepileptic for some time, the gay community enthusiastically supported immediate clinical trials of this compound.

A number of experimental protocols involving immunization of humans against HIV-1 have begun, and some initially faintly promising results have been reported.[198] However, the challenges to mastery of effective treatment and prevention are still formidable. As alluded to earlier, many important questions remain to be answered. Among these are: What factors are most important in governing viral tropism? What is the natural degree of biological heterogeneity in HIV, including genetic diversity of important components such as the envelope glycoprotein? What mechanisms actually determine pathogenesis and eventual progression from the carrier state to clinical AIDS? These areas have been addressed in detail by Levy.[199]

A number of different strategies are being considered and discussed. One approach involves the use of the so-called scid mouse, a strain that is susceptible to a genetic disease called severe combined immunodeficiency, which leaves them open to opportunistic infections very much like those affecting AIDS patients. By transplanting human immune cells into scid mice, human immune responses can be studied almost directly.

A more radical approach has also been considered, namely to create a strain of HIV that is even more virulent than natural strains but is incapable of establishing long-term infections.[200] If such a strain were allowed more replication potential, it might produce greater stimulation of the immune system and thus possibly a protective response. A number of other genetic changes would have to be engineered in such strains to make them nonpathogenic. Work along these lines continues.

Molecular Modeling of the HIV-1 Protease

Perhaps one of the most fundamental contributions to AIDS research has been the description of the three-dimensional structure of aspartyl protease from HIV-1.[201,202] This monumental work has defined the crystal structure of the protease of HIV-1 that releases proteins and enzymes from viral polyprotein products. With this new insight into an important functional protein of the virus, it may be possible to design an effective antiviral therapy, possibly in the form of specific protease inhibitors. A great deal has been accomplished in a few years of work on the AIDS problem. Much of the knowledge concerning this serious infection has come from direct application of recombinant techniques.

Recombinant Probes in Other Clinical Viral Conditions

Molecular hybridization techniques have provided important new dimensions for diagnostic and epidemiologic aspects of a host of human diseases of known and unknown etiology. A few of many applications and current examples will be described in this section; however, the applications of these techniques are occurring so rapidly in clinical medicine that no discussion of this subject can hope to be complete.

Cytomegalovirus

This infectious agent can be life-threatening to newborns or immunodeficient patients. The presently available diagnostic assays for cytomegalovirus (CMV) infection are time consuming and relatively slow (24 to 48 days). Using cDNA probes, positive hybridization results indicating CMV infection can now be obtained within 24 hours. Urine samples from patients at risk for CMV infection were screened using a ^{32}P-labeled restriction fragment of CMV, and the results correlated well with eventual tissue culture identification.[203] The sensitivity of the probe hybridization assay has been progressively improved, and it may be applicable in monitoring studies of susceptible populations.[204] There is considerable interest in the possibility that particular strains of CMV identified by gel fragment restriction patterns might be associated with induction of mental retardation and other adverse reactions. Patterns of such specific CMV virus strains identified by labeled cDNA probes are shown in Figure 6.18.[205]

Herpes Simplex

It is now well established that after a primary exposure, herpes simplex virus types 1 and 2 can reside as a lifelong latent infection in trigeminal and dorsal root ganglia, where they can be reactivated to produce recurrent mucocutaneous infections. There are differences of opinion about whether the latent herpes virus is totally quiescent in the ganglia without any viral gene expression between attacks or whether limited transcription of the viral genome occurs in latent infected ganglion cells. This question was addressed using labeled single-stranded RNA probes encompassing 72% of the herpes simplex genome and human trigeminal ganglia recovered at autopsy from 24 subjects.[206] Ganglia from 16 subjects showed 0.2% to 4.3% of neuronal cells with strong nuclear signals for hybridizing viral RNA. Transcription was detected only from a part of the herpes simplex gene encoding a very early protein; however, actual normal transcripts of this gene were not present. By contrast, a novel antisense transcript was detected that read in the opposite direction to the sense 5′-to-3′ message. It was thought that this novel product might play a role in maintaining viral latency between attacks. The types of signals obtained with this in situ hybridization technique are shown in Figure 6.19.

Genital Warts and Papillomaviruses

Another successful application of recombinant DNA techniques to diagnosis and epidemiology is related to the papillomaviruses, genital warts, and cervical cancer. The association between wart virus infection and carcinoma of the cervix has been recognized for some time.[207,208] Since the original association was recorded, a number of advances have been made in studying this relationship, particularly on the basis of DNA sequences associated with specific human papillomavirus (HPV) types. The types thus far identified include 6, 10, 11, 16, 18, and 31. Correlation of virus type with morphology of the cervical cancer lesions indicates that sequences homologous to HPV 16 and 18 can be identified in over 80% of invasive squamous cell cancers of the cervix, vulva, and penis. By contrast, types 6 and 11 are more often associated with benign genital warts or mild cervical cell dysplasia. Human

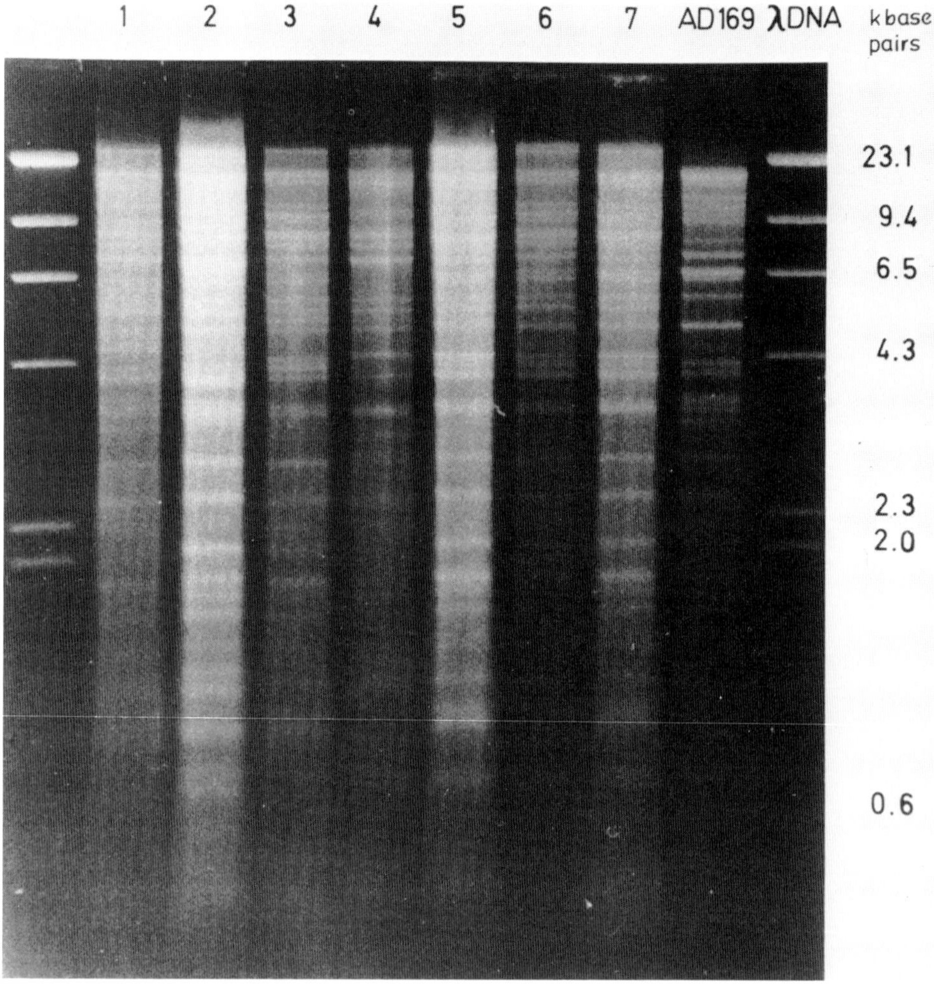

Figure 6.18 *Eco*RI cleavage patterns of CMV DNA from seven strains (columns 1 to 7) isolated in day care center A and the prototype strain AD169. *Hin*dIII-cleaved lambda DNA is used as molecular weight standard. The fragments have been separated on an 0.8% agarose gel at 40 V for 16 to 18 hours. *Reproduced from Grillner and Strangert,*[205] *with permission.*

papillomavirus type 16 has also been demonstrated in cervical adenocarcinomas but not in endometrial malignancies. Rapid and reproducible methods for DNA probe hybridization are needed to cope with the large number of cervical scrapings that should be assayed for human papillomavirus sequences. Newer techniques, such as sandwich hybridization or in situ hybridization using biotinylated probes that can be subsequently developed with specific avidin generating detectable color, could be applied to this important area. High sensitivity of such DNA hybridization techniques with cervical scrapings from women attending a sexually transmitted diseases clinic has been reported.[209] Use of recombinant techniques has also uncovered a fascinating connection with respect to human papillomavirus; it has been found that E6 proteins from papilloma types 16 and 18 bind to p53, the tumor suppressor gene product.[210] This may help to explain pathways of malignant transformation in this instance.

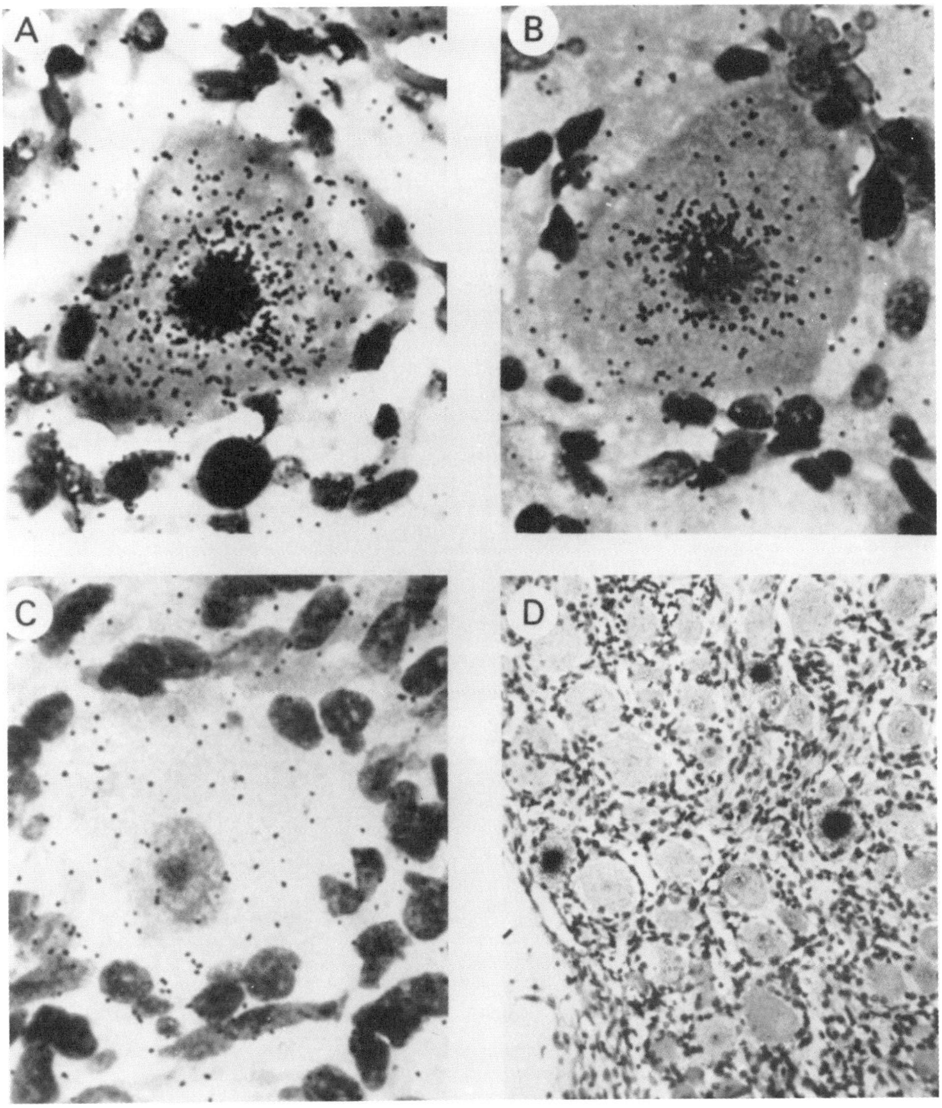

Figure 6.19 In situ hybridizations of human trigeminal ganglia for herpes simplex virus type 1. (A and B) Positive autoradiographic signals overlying ganglion cell nuclei after hybridization with the probe for an antisense transcript (×300). (C) Negative autoradiograph of a ganglion cell and surrounding satellite cells; this is typical of results seen when any probe other than the one for the antisense transcript is used (×300). (D) Another positive autoradiograph at lower magnification (×50), showing three isolated positive ganglion cells and numerous negative cells in a single section. All the autoradiographs were exposed for 4 days at 4°C. *Reproduced from Croen et al.,*[206] *with permission.*

Recombinant Probes in Disorders of Unknown Cause

The success of studies of known pathologic entities such as herpes simplex or human papillomavirus has prompted application of recombinant probes to disorders of unknown etiology. The basic cause of cardiomyopathy or chronic myocarditis remains obscure in many patients. Because coxsackie B viruses are probably the most common viral agents producing myocarditis, cDNA complementary to a 6.3-kb fragment of coxsackie B viral RNA was used as a hybridization probe in myocardial biopsy samples.[211] Positive results were obtained in 9 of 17 samples from patients with histological evidence of active or healing myocarditis or dilated cardiomyopathy with inflammatory changes. The methodology employed in this study was simple and relied on straightforward dot blot techniques using nucleic acid extracted from biopsy tissues and applied in serial dilutions to nitrocellulose membranes. Of interest in this study was the finding of coxsackie B virus–specific sequences in patients with apparent healed myocarditis as well as patients with ensuing congestive cardiomyopathy. It was thought that the persistence of viral RNA late in the disease process implied continuing viral replication. Previous failure to demonstrate infectious virus or viral antigens in such patients might be ascribed to a lack of posttranslational processing of viral capsid proteins in cells harboring the agents.

Molecular Epidemiology

One of the most interesting applications of recombinant methods to a common problem in modern infection control took place during a widely recognized outbreak of diphtheria in Sweden.[212] This infectious disease is of interest because despite mass vaccination, many people in various parts of the world have antibody titers below what is considered the protective level. The authors analyzed 36 strains of *Corynebacterium diphtheriae* isolated in Sweden and Denmark over a 10-year period (1976–1986) and, using a new genetic diphtherial probe, found that all of the clinical and fatal cases were caused by a single strain that possibly had a virulence factor separate from its toxin-producing capacity. They used a probe encompassing the whole part of the bacterial gene coding for diphtheria toxin, as well as a separate probe to examine additional bacterial expression elements. The excellent correlation in this epidemiologic study between contact history and hybridization patterns was striking, in that the strains isolated from alcohol and drug abusers and from patients with clinical diphtheria showed the same hybridization patterns. Strains obtained from carriers who indicated no contact with alcohol or drug abusers were entirely different in their hybridization patterns. This use of recombinant hybridization specificity in an important epidemiologic investigation further underscores the potential usefulness of this approach in a number of other human disease states.[213]

References

1. Cohen S: Mechanisms of malarial immunity. *Trans R Soc Trop Med Hyg* 1977;71:283–286.
2. Cohen S: Progress in malaria vaccine development. *Br Med Bull* 1982;38:161–165.
3. Brown CV, Anders RF, Stace JD, et al: Immunoprecipitation of biosynthetically-labelled proteins from different Papua New Guinea *Plasmodium falciparum* isolates by sera from individuals in the endemic area. *Parasite Immunol* 1981;3:283–298.
4. Brown IN: Immunological aspects of malaria infection. *Adv Immunol* 1969;11:267–349.
5. Cohen S: Immunoprophylaxis of protozoal diseases, in Gell PGH, Coombs RRA, Lachmann PJ (eds): *Clinical Aspects of Immunology*, 3rd ed. Oxford, Blackwell Scientific Publications, 1974, p 1649.
6. Brown KN: Resistance to malaria, in Cohen S, Sadun EH (eds): *Immunology of Parasitic Infections*. Oxford, Blackwell Scientific Publications, 1976, p 268.
7. Protocnjak P, Yoshida N, Nussenzweig RS, Nussenzweig V: Monovalent fragments

(Fab) of monoclonal antibodies to a sporozoite surface antigen (P644) protect mice against malarial infection. *J Exp Med* 1980;151:1504–1513.
8. Yoshida N, Nussenzweig RS, Potocnjak P, et al: Hybridoma produces protective antibodies directed against the sporozoite stage of malaria parasite. *Science* 1980;207:71–73.
9. Nardin EH, Nussenzweig V, Nussenzweig RS, et al: Circumsporozoite proteins of human malaria parasites *Plasmodium falciparum* and *Plasmodium vivax*. *J Exp Med* 1982; 156:20–30.
10. Trager W, Jensen JB: Human malaria parasites in continuous culture. *Science* 1976; 193:673–675.
11. Kemp DJ, Coppel RL, Cowman AF, et al: Expression of *Plasmodium falciparum* blood-stage antigens in *Escherichia coli*: Detection with antibodies from immune humans. *Proc Natl Acad Sci USA* 1983;80:3787–3791.
12. Stahl HD, Coppel RL, Brown GV, et al: Differential antibody screening of cloned *Plasmodium falciparum* sequences expressed in *Escherichia coli*: Procedure for isolation of defined antigens and analysis of human antisera. *Proc Natl Acad Sci USA* 1984;81:2456–2460.
13. Enea V, Ellis J, Zavala F, et al: DNA cloning of *Plasmodium falciparum* circumsporozoite gene: Amino acid sequence of repetitive epitope. *Science* 1984;225:628–630.
14. Hall R, Hyde JE, Goman M, et al: Major surface antigen gene of a human malaria parasite cloned and expressed in bacteria. *Nature* 1984;311:379–382.
15. Cochrane AH, Nussenzweig RS, Nardin EH: In Kreier JP (ed): *Malaria*, Vol. 3. New York, Academic Press, 1980, pp 163–202.
16. Zavala F, Cochrane AH, Nardin EH, et al: Circumsporozoite proteins of malaria parasites contain a single immunodominant region with two or more identical epitopes. *J Exp Med* 1983;157:1947–1957.
17. Aikawa M, Yoshida N, Nussenzweig R, Nussenzweig V: The protective antigen of malarial sporozoites (*Plasmodium berghei*) is a differentiation antigen. *J Immunol* 1981; 126:2494–2495.
18. Holder AA, Lockyer MJ, Odink KG, et al: Primary structure of the precursor to the three major surface antigens of *Plasmodium falciparum* sporozoites. *Nature* 1985; 317:270–273.
19. Herrington DA, Clyde DF, Losonsky G, et al: Safety and immunogenicity in man of a synthetic peptide malaria vaccine against *Plasmodium falciparum* sporozoites. *Nature* 1987;328:257–259.
20. Zavala F, Tam JP, Cochrane AH, et al: Rationale for development of a synthetic vaccine against *Plasmodium falciparum* malaria. *Science* 1985;228:1436–1440.
21. Miller LH, Russell JH, Carter R, et al: Research toward malaria vaccines. *Science* 1986; 234:1349–1356.
22. Ballou WR, Hoffman SL, Sherwood JA, et al: Safety and efficacy of a recombinant DNA *Plasmodium falciparum* sporozoite vaccine. *Lancet* 1987;1:1277–1281.
23. Perlmann P: Malaria vaccines: Are they approaching reality? *Nature* 1987;328:205–206.
24. Good MF, Berzofsky JA, Maly WL, et al: Genetic control of the immune response in mice to a *Plasmodium falciparum* sporozoite vaccine. *J Exp Med* 1986;164:655–660.
25. Del Giudice G, Cooper JA, Merino J, et al: The antibody response in mice to carrier-free synthetic polymers of *Plasmodium falciparum* circumsporozoite repetitive epitope is I-A restricted: Possible implication for malaria vaccine. *J Immunol* 1986;137:2952–2955.
26. Egan JE, Weber JL, Ballou WR, et al: Efficacy of murine malaria sporozoite vaccines: Implications for human vaccine development. *Science* 1986;236:453–456.
27. Ferreira A, Schofield L, Enea V, et al: Inhibition of development of exoerythrocytic forms of malaria parasites by γ-interferon. *Science* 1986;232:881–884.
28. Eichinger DJ, Arnot DE, Tam JP, et al: Circumsporozoite protein of *Plasmodium berghei*: Gene cloning and identification of the immunodominant epitopes. *Mol Cell Biol* 1985; 6:3965–3972.
29. Crisanti A, Muller H-M, Hilbich C, et al: Epitopes recognized by human T cells map within conserved part of the GP190 of *P. falciparum*. *Science* 1988;240:1324–1326.
30. Kabilan L, Troye-Blomberg M, Perlmann H, et al: T cell epitopes in Pf155/RESA, a

major candidate for *Plasmodium falciparum* malaria vaccine. *Proc Natl Acad Sci USA* 1988;85:5659–5663.

31. Patarroyo ME, Amador R, Clavijo P, et al: A synthetic vaccine protects humans against challenge with asexual blood stages of *Plasmodium falciparum* malaria. *Nature* 1988; 332:158–161.
32. Kaslow DC, Quakyi IA, Syin C, et al: A vaccine candidate from the sexual stage of human malaria that contains EGF-like domains. *Nature* 1988;333:74–76.
33. Young JF, Hockmeyer WT, Gross M, et al: Expression of *Plasmodium falciparum* circumsporozoite proteins in *Escherichia coli* for potential use in a human malaria vaccine. *Science* 1985;228:958–962.
34. Arnot DE, Barnwell JW, Tam JP, et al: Circumsporozoite protein of *Plasmodium vivax*: Cloning of the immunodominant epitope. *Science* 1985;230:815–818.
35. McCutchan TF, Altaf AL, de la Cruz VF, et al: Sequence of the immunodominant epitope of the surface protein on sporozoites of *Plasmodium vivax*. *Science* 1985; 230:1381–1383.
36. Barr PJ, Gibson HL, Enea V, et al: Expression in yeast of a *Plasmodium vivax* antigen of potential use in a human malaria vaccine. *J Exp Med* 1987;165:1160–1171.
37. Travis J, Owen M, George P, et al: Isolation and properties of human alpha-1-proteinase inhibitor variants produced in yeast. *J Biol Chem* 1985;260:4384–4389.
38. Valenzuela P, Medina A, Rutter WJ, et al: Synthesis and assembly of hepatitis B virus surface antigen particles in yeast. *Nature* 1982;298:347–350.
39. Hoffman SL, Isenbarger D, Long GW, et al: Sporozoite vaccine induces genetically restricted T cell elimination of malaria from hepatocytes. *Science* 1989;244:1078–1081.
40. McLaughlin GL, Ruth JL, Jablonski E, et al: Use of enzyme-linked synthetic DNA in diagnosis of *Falciparum* malaria. *Lancet* 1987;1:714–715.
41. Franzen L, Shabo R, Perlman H, et al: Analysis of clinical specimens by hybridization with a probe containing repetitive DNA from *Plasmodium falciparum*. A novel approach to malaria diagnosis. *Lancet* 1984;1:525–528.
42. Maberti S: Desarrollo de resistencia a la pirimetamina: Presentacion de 15 casos estudiados en Trujillo, Venezuela. *Arch Venez Med Trop Parasitol Med* 1960;3:239–259.
43. Wellems TE, Panton LJ, Gluzman IY, et al: Chloroquine resistance not linked to *mdr*-like genes in a *Plasmodium falciparum* cross. *Nature* 1990;345:253–255.
44. Nussenzweig V, Nussenzweig RS: Progress toward a malaria vaccine. *Hosp Pract* 1990; 25:45–57.
45. Barker RH, Suebsaeng L, Rooney W, et al: Specific DNA probe for the diagnosis of *Plasmodium falciparum* malaria. *Science* 1986;231:1434–1436.
46. Capron M, Capron A: Rats, mice, and men—models for immune effector mechanisms against schistosomiasis. *Parasitol Today* 1986;2:69–75.
47. Dean DA: A review: *Schistosoma* and related genera: Acquired resistance in mice. *Exp Parasitol* 1983;55:1–104.
48. Sher A, Hieny S, James SL, Asofsky R: Mechanisms of protective immunity against *Schistosoma mansoni* in mice vaccinated with irradiated cercariae. II. Analysis of immunity in hosts deficient in T lymphocytes, B lymphocytes, or complement. *J Immunol* 1982;128:1880–1884.
49. James SL, Sher A: Mechanisms of protective immunity against *Schistosoma mansoni* in mice vaccinated with irradiated cercariae. III. Identification of a mouse strain, P/N, that fails to respond to vaccination. *Parasite Immunol* 1983;5:567–575.
50. James SL, Sher A: Towards a schistosomiasis vaccine: Prospects for a nonliving vaccine against schistosomiasis. *Parasitol Today* 1986;2:134–137.
51. Esparza I, Mannel D, Ruppel A, et al: Interferon γ and lymphotoxin or tumor necrosis factor act synergistically to induce macrophage killing of tumor cells and schistosomula of *Schistosoma mansoni*. *J Exp Med* 1987;166:589–594.
52. Ballou JM, Sondermeyer P, Dreyer D, et al: Molecular cloning of a protective antigen of schistosomes. *Nature* 1987;326:149–153.
53. Dalton JP, Tom TD, Strand M: Cloning of a cDNA encoding a surface antigen of *Schistosoma mansoni* schistosomula recognized by sera of vaccinated mice. *Proc Natl Acad Sci USA* 1987;84:4268–4272.
54. Scallon BJ, Bogitsh BJ, Carter CE: Cloning of a *Schistosoma japonicum* gene encoding a major immunogen recognized by hyperinfected rabbits. *Mol Biochem Parasitol* 1987; 24:237–245.

55. Hedstrom R, Culpepper J, Harrison RA, et al: A major immunogen in *Schistosoma mansoni* infections is homologous to the heat-shock protein hsp 70. *J Exp Med* 1987; 165:1430–1435.
56. Lanar DE, Pearce EJ, Sher A: Expression in *Escherichia coli* of two *Schistosoma mansoni* genes that encode major antigens recognized by immune mice. *Mol Biochem Parasitol* 1985;17:45–60.
57. Lindquist S: The heat shock response. *Annu Rev Biochem* 1986;55:1151.
58. Hirayama K, Matsushita S, Kikuchi I, et al: HLA-DQ is epistatic to HLA-DR in controlling the immune response to schistosomal antigen in humans. *Nature* 1987; 327:426–430.
59. Brener ZA: Biology of *Trypanosoma cruzi*. *Rev Microbiol* 1973;27:347–383.
60. Andrews NW, Katzin AM, Colli W: Mapping of surface glycoproteins of *Trypanosoma cruzi* by two dimensional electrophoresis: A correlation with the cell invasion capacity. *Eur J Biochem* 1984;140:599–604.
61. Beard CA, Wrightsman RA, Manning JE: Identification of monoclonal antibodies against the trypomastigote stage of *Trypanosoma cruzi* by use of immunobiotinylated surface polypeptides. *Mol Biochem Parasitol* 1985;16:199–212.
62. Dame JB, Williams JL, McCutchen TF, et al: Structure of the gene encoding the immunodominant surface antigen on the sporozoite of the human malaria parasite *Plasmodium falciparum*. *Science* 1984;225:593–599.
63. Roditi I, Carrington M, Turner M: Expression of a polypeptide containing a dipeptide repeat is confined to the insect stage of *Trypanosoma brucei*. *Nature* 1987;325:272–274.
64. Bloom BR, Godal T: Selective primary health care: Strategies for control of disease in the developing world. V. Leprosy. *Rev Infect Dis* 1983;5:765–780.
65. Lowe C, Brett SJ, Rees RJW: Adaptive cell transfer of the resistance to *Mycobacterium leprae* in mice. *Clin Exp Immunol* 1985;61:336–342.
66. Young RA, Mehra V, Sweetser D, et al: Genes for the major protein antigen of the leprosy parasite *Mycobacterium leprae*. *Nature* 1985;316:450–452.
67. Young RA, Bloom BR, Grosskinsky CM, et al: Dissecting *Mycobacterium tuberculosis* antigens using recombinant DNA. *Proc Natl Acad Sci USA* 1985;82:2583–2587.
68. Booth RJ, Harris DP, Love JM, Watson JD: Antigenic proteins of *Mycobacterium leprae*. Complete sequence of the gene for the 18-kDa protein. *J Immunol* 1988;140:597–601.
69. Margalit H, Sponge JL, Cornette JL, et al: Prediction of immunodominant helper T-cell antigenic sites from the primary sequence. *J Immunol* 1987;138:2213–2229.
70. Mustafa AS, Gill HK, Nerland A, et al: Human T cell clones recognize a major *M. leprae* protein antigen expressed in *E. coli*. *Nature* 1986;319:63–66.
71. Modlin RL, Kato H, Mehra V, et al: Genetically-restricted suppressor T-cell clones derived from lepromatous leprosy lesions. *Nature* 1986;322:459–461.
72. Ottenhoff THM, Elferink DG, Klatser PR, de Vries RRP: Cloned suppressor T cells from a lepromatous leprosy patient suppress *Mycobacterium leprae* reactive helper T cells. *Nature* 1986;322:462–464.
73. Nathan CF, Kaplan G, Levis WR, et al: Local and systemic effects of intradermal recombinant interferon-γ in patients with lepromatous leprosy. *N Engl J Med* 1986; 315:6–15.
74. Sell S, Norris SJ: The biology, pathology, and immunology of syphilis. *Int Rev Exp Pathol* 1983;24:203–276.
75. Penn CW: Pathogenicity and immunobiology of *Treponema pallidum*. *J Med Microbiol* 1987;24:1–9.
76. Walfield AM, Hanff PA, Lovett M: Expression of *Treponema pallidum* antigens in *Escherichia coli*. *Science* 1982;216:522–523.
77. Stamm LV, Folds JD, Bassford PJ: Expression of *Treponema pallidum* antigens in *Escherichia coli* K-12. *Infect Immun* 1982;36:1238–1241.
78. Moskophidis M, Muller F: Molecular analysis of immunoglobulins M and G immune response to protein antigens of *Treponema pallidum* in human syphilis. *Infect Immun* 1984;43:127–132.
79. Thornburg RW, Baseman JB: Comparison of major protein antigens and protein profiles of *Treponema pallidum* and *Treponema pertenue*. *Infect Immun* 1982;42:623–627.
80. Van Eijk RVW, van Embden JDA: Molecular characterization of *Treponema pallidum*

proteins responsible for the human immune response to syphilis Antonie van Leeuwenhoek. *J Microbiol* 1982;48:486–487.
81. Marchitto KS, Jones SA, Schell RF, et al: Monoclonal antibody analysis of specific similarities among pathogenic *Treponema pallidum* subspecies. *Infect Immun* 1984; 45:660–666.
82. Strugnell RA, Williams WF, Drummond L, et al: Development of increased serum immunoblot reactivity against a 45,000-dalton polypeptide *Treponema pallidum* (Nichols) correlates with establishment of chancre immunity in syphilitic rabbits. *Infect Immun* 1986;51:957–960.
83. Marchitto KS, Selland-Grossling CK, Norgard MV: Molecular specificities of monoclonal antibodies directed against virulent *Treponema pallidum*. *Infect Immun* 1986; 51:168–176.
84. Norgard MV, Chamberlin NR, Swancutt MA, Goldberg MS: Cloning and expression of the major 47-kilodalton surface immunogen of *Treponema pallidum* in *Escherichia coli*. *Infect Immun* 1986;54:500–506.
85. Moseley SL, Huq I, Alim ARMA, et al: Detection of enterotoxigenic *Escherichia coli* by DNA colony hybridization. *J Infect Dis* 1980;142:892–898.
86. Hindersson P, Cockayne A, Schouls LM, van Emden JD: Immunochemical characterization and purification of *Treponema pallidum* antigen TpD expressed by *Escherichia coli* K12. *Sex Transm Dis* 1986;13:237–244.
87. Peterson KM, Baseman JB, Aldrete JF: Isolation of a *Treponema pallidum* gene encoding immunodominant outer envelope protein P6 which reacts with sera from patients at different stages of syphilis. *J Exp Med* 1986;164:1160–1170.
88. Rice RJ, Aral SO, Blount JH, Zaidi A: Gonorrhea in the United States 1975–1984: Is the giant only sleeping? Sex Transm Dis 1987;14:83–87.
89. Centers for Disease Control. Gonorrhea annual summary, 1983. *MMWR* 1984;32:22–25.
90. Altshuler LN: Gonorrhea in World War II. *Am J Syph Gonorrhea Vener Dis* 1948; 32:115–123.
91. Antibiotic resistance in *Neisseria gonorrhoeae*. *Med Clin North Am* 1972;56:1133–1144.
92. Elwell LP, Roberts M, Mayer LW, Falkow S: Plasmid-mediated beta-lactamase production in *Neisseria gonorrhoeae*. *Antimicrob Agents Chemother* 1977;11:528–533.
93. Boslego JW, Tramont EC, Takafuji ET, et al: Effect of spectinomycin use on the prevalence of spectinomycin-resistant and of penicillinase-producing *Neisseria gonorrhoeae*. *N Engl J Med* 1987;317:272–278.
94. Multiple strain outbreak of penicillinase producing *Neisseria gonorrhoeae*—Denver, Colorado, 1986. *JAMA* 1987;258:1713–1714.
95. Heckels JE: Gonococcal antigenic variation and pathogenesis, *in* Birkbeck H, Penn CW (eds): *Antigenic Variation in the Course of Infectious Diseases: A Survival Strategy for Pathogenic Microoganisms*. Cambridge, England, Cambridge Univ. Press, 1988.
96. Gotschlich EC: Development of a gonorrhoea vaccine: Prospects, strategies, and tactics. *Bull WHO* 1984;62:671–680.
97. Rice PA, Vayo HE, Tam MR, Blake MS: Immunoglobulin G antibodies directed against protein III block killing of serum resistant *Neisseria gonorrhoeae* by immune serum. *J Exp Med* 1986;164:1735–1748.
98. Johnson KH, Gotschlich EC: Isolation and characterization of the outer membrane of *Neisseria gonorrhoeae*. *J Bacteriol* 1974;119:250–257.
99. Johnson KH, Holmes KK, Gotschlich EC: The serological classification of *Neisseria gonorrhoeae*. I. Isolation of the outer membrane complex responsible for serotypic specificity. *J Exp Med* 1976;143:741–758.
100. Sandstrom EG, Knapp JS, Buchanan TM: Serology of *Neisseria gonorrhoeae*: W-antigen serogrouping by coagglutination and protein I serotyping by enzyme-linked immunosorbent assay both detect protein I antigens. *Infect Immun* 1982;35:229–239.
101. Buchanan TM, Eschenbach DA, Knapp JS, Holmes KK: Gonococcal salpingitis is less likely to recur with *Neisseria gonorrhoeae* of the same principal outer membrane protein antigenic type. *Am J Obstet Gynecol* 1980;138:975–980.
102. Teerlink T, Versantvoort H, Beuvery EC: Antigenic and immunogenic properties of cyanogen bromide peptides from gonococcal outer membrane protein IB. Evidence for the existence of a surface-exposed conserved epitope. *J Exp Med* 1987;166:63–76.

103. Gotschlich EC, Blake MS, Koomey JM, et al: Cloning of the structural genes of three H8 antigens and of protein III of *Neisseria gonorrhoeae. J Exp Med* 1986;164:868–881.
104. Strittmatter W, Hitchcock PJ: Isolation and preliminary biochemical characterization of the gonococcal H.8 antigen. *J Exp Med* 1986;164:2038–2048.
105. Gotschlich EC, Seiff M, Blake MS: The DNA sequence of the structural gene of gonococcal protein III and the Hanking region containing a repetitive sequence homology of protein III with enterobacterial OmpA proteins. *J Exp Med* 1987;165:471–482.
106. Swanson J, Robbins K, Barrera O, Koomey JM: Gene conversion variations generate structurally distinct pilin polypeptides in *Neisseria gonorrhoeae. J Exp Med* 1987; 165:1016–1025.
107. Gibbs CP, Reimann B-Y, Schultz E, et al: Reassortment of pilin genes in *Neisseria gonorrhoeae* occurs by two distinct mechanisms. *Nature* 1989;338:651–652.
108. Pohlner J, Halter R, Beyreuther K, Meyer TF: Gene structure and extracellular secretion of *Neisseria gonorrhoeae* IgA protease. *Nature* 1987;325:458–462.
109. McAuliffe VJ, Purcell RH, Gern JL: Type B hepatitis: A review of current prospects for a safe and effective vaccine. *Rev Infect Dis* 1980;2:470–492.
110. Szmuness W, Stevens CE, Harley EJ, et al: Hepatitis B vaccine. Demonstration of efficacy in a controlled clinical trial in a high-risk population in the United States. *N Engl J Med* 1980;303:833–841.
111. Edman JC, Hallewell RA, Valenzuela P, et al: Synthesis of hepatitis B surface and core antigens in *E. coli. Nature* 1981;291:503–506.
112. Pasek M, Goto T, Gilbert W, et al: Hepatitis B virus genes and their expression in *E. coli. Nature* 1979;282:575–579.
113. Charnay P, Gervais M, Louise A, et al: Biosynthesis of hepatitis B virus surface antigen in *Escherichia coli. Nature* 1980;286:893–895.
114. Dreesman GR, Sanchez Y, Ionescu-Matiu I, et al: Antibody to hepatitis B surface antigen after a single inoculation of uncoupled synthetic HBsAg peptides. *Nature* 1982; 295:158–160.
115. Kew MC: Hepatoma and the HBV. in Vyas GN, Cohen SN, Schmid R (eds): *Viral Hepatitis*. Philadelphia, Franklin Institute Press, 1978, pp 439–450.
116. Brechot C, Pourcel C, Louise A, et al: Presence of integrated hepatitis B virus DNA sequences in cellular DNA of human hepatocellular carcinoma. *Nature* 1980;286:533–535.
117. Edman JC, Gray P, Valenzuela P, et al: Integration of hepatitis B virus sequences and their expression in a human hepatoma cell. *Nature* 1980;286:535–538.
118. Chakraborty PR, Ruiz-Opazo N, Shouval D, Shafritz DA: Identification of integrated hepatitis B virus DNA and expression of viral RNA in an HBsAg-producing human hepatocellular carcinoma cell line. *Nature* 1980;286:531–533.
119. Kaschka-Dierich C, Falk L, Fjursell G, et al: Human lymphoblastoid cell lines derived from individuals without lymphoproliferative disease contain the same latent forms of Epstein-Barr virus DNA as those found in tumor cells. *Int J Cancer* 1977;20:173–180.
120. Yee J-K: A liver-specific enhancer in the core promoter region of human hepatitis B virus. *Science* 1989;246:658–661.
121. Zeldis JB, Lee JH, Mamish D, et al: Direct method for detecting small quantities of hepatitis B virus DNA in serum and plasma using the polymerase chain reaction. *J Clin Invest* 1989;84:1503–1508.
122. Mitsuda T, Mori T, Ookawa N, et al: Demonstration of mother-to-infant transmission of hepatitis B virus by means of polymerase chain reaction. *Lancet* 1989;2:886–888.
123. Moriyama T, Guilhot S, Klopchin K, et al: Immunobiology and pathogenesis of hepatocellular injury in hepatitis B virus transgenic mice. *Science* 1990;248:361–364.
124. Moss B, Smith GL, Gerin JL, Purcell RH: Live recombinant vaccinia virus protects chimpanzees against hepatitis B. *Nature* 1984;311:67–69.
125. Brechot C, Degos F, Lugassy C, et al: Hepatitis B virus DNA in patients with chronic liver disease and negative tests for hepatitis B surface antigen. *N Engl J Med* 1985; 312:270–276.
126. Rizzetto M, Canese MG, Gerin JL, et al: Transmission of hepatitis B virus–associated delta antigen to chimpanzees. *J Infect Dis* 1980;141:590–602.
127. Rizetto M, Hoyer B, Canese MG, et al: Delta antigen: The association of delta antigen

with hepatitis B surface antigen and ribonucleic acid in the serum of delta infected chimpanzees. *Proc Natl Acad Sci USA* 1980;77:6124–6128.
128. Snedile A, Verme G, Cargnel A, et al: Influence of delta infection on severity of hepatitis B. *Lancet* 1982;2:945–947.
129. Wang KS, Choo Q-L, Weiner AJ, et al: Structure, sequence and expression of the hepatitis delta viral genome. *Nature* 1986;323:508–513.
130. Wang KS, Choo Q-L, Weiner AJ, et al: Structure, sequence and expression of the hepatitis delta (δ) viral genome. *Nature* 1987;328:456.
131. Makino S, Chang M-F, Shieh, C-K, et al: Molecular cloning and sequencing of a human hepatitis delta (δ) virus RNA. *Nature* 1987;329:343–346.
132. Chen DJ, Kalpana G, Goldberg J, et al: Structure and replication of the genome of the hepatitis (δ) virus. *Proc Natl Acad Sci USA* 1986;83:8774–8778.
133. Lewin R: New class of animal virus found in virulent form of human hepatitis. *Science* 1986;234:423–424.
134. Kos A, Dijkema R, Arnberg AC, et al: The hepatitis delta (δ) virus possesses a circular RNA. *Nature* 1986;323:558–560.
135. Fagan EA, Davison F, Smith DM, Williams R: Fulminant hepatitis B in successive female sexual partners of two anti-HBe-positive males. *Lancet* 1986;2:538–540.
136. Francis DP, Hadler SC, Prendergast TJ: Occurrence of hepatitis A, B, and non-A, non-B in the United States. CDC Sentinel County hepatitis study I *Am J Med* 1984;76:69–74.
137. Choo Q-L, Kuo G, Weiner AJ, et al: Isolation of a cDNA clone derived from a blood-borne non-A, non-B viral hepatitis genome. *Science* 1989;244:359–362.
138. Kuo G, Choo Q-L, Alter HJ, et al: An assay for circulating antibodies to a major etiologic virus of human non-A, non-B hepatitis. *Science* 1989;244:362–364.
139. Reyers GR, Purdy MA, Kim JP, et al: Isolation of a cDNA from the virus responsible for enterically transmitted non-A, non-B hepatitis. *Science* 1990;247:1335–1339.
140. Poiesz BJ, Ruscetti FW, Reitz MS, et al: Isolation of a new type C retrovirus (HTLV) in primary uncultured cells of a patient with Sezany T-cell leukemia. *Nature* 1981; 294:268–271.
141. Yoshida M, Miyoshi I, Himuma Y: Isolation and characterization of retrovirus from cell lines of human adult T cell leukemia and its implication in the disease. *Proc Natl Acad Sci USA* 1981;79:2031–2035.
142. Hinuma Y, Nagata K, Hanoaka M, et al: Adult T-cell leukemia: Antigen in an ATL cell line and detection of antibodies to the antigen in human serum. *Proc Natl Acad Sci USA* 1981;78:6476–6480.
143. Ito Y: The epidemiology of human T-cell leukemia/lymphoma virus. *Curr Top Micribiol Immunol* 1985;115:99–112.
144. Gallo RC: The first human retrovirus. *Sci Am* 1986;255(6):88–98.
145. Ichimaru M, Kinoshita K, Kamihira S, et al: Familial disposition of adult T-cell leukemia and lymphoma. *Hematol Oncol* 1986;4:21–29.
146. Franchini G, Wong-Staal F, Gallo R: Human T-cell leukemia virus (HTLV-I) transcripts in fresh and cultured cells of patients with adult T-cell leukemia. *Proc Natl Acad Sci USA* 1984;81:6207–6211.
147. Nerenberg M, Hinrichs SH, Reynold RK, et al: The *tat* gene of human T-lymphotropic virus type 1 induces mesenchymal tumors in transgenic mice. *Science* 1987;237:1324–1329.
148. Sarin PS, Rodgers-Johnson P, Sun DK, et al: Comparison of a human T cell lymphotropic virus type I strain from cerebrospinal fluid of a Jamaican patient with tropical spastic paraparesis with a prototype human T cell lymphotropic virus type I. *Proc Natl Acad Sci USA* 1989;86:2021–2025.
149. Kalyanaraman VS, Sarngadharam MG, Robert-Guroff M, et al: A new subtype of human T-cell leukemia virus (HTLV-II) associated with a T-cell variant of hairy leukemia. *Science* 1982;218:571–573.
150. Gelmann EP, Franchini G, Manzari V, et al: Molecular cloning of a unique human T-cell leukemia virus (HTLV-IImo). *Proc Natl Acad Sci USA* 1984;81:993–997.
151. Rosenblatt JD, Golde DW, Wachsman W, et al: A second isolate of HTLV-II associated with atypical hairy-cell leukemia. *N Engl J Med* 1986;315:372–377.
152. Bouroncle BA, Wiseman BK, Doan CA: Leukemia reticuloendotheliosis. *Blood* 1958; 13:609–630.

153. Marx JL: The AIDS virus—well known but a mystery. *Science* 1987;236:390–392.
154. Marx JL: Probing the AIDS virus and its relatives. *Science* 1987;236:1523–1525.
155. Gottleib MS, Schanker HM, Fran PT, et al: Pneumocystis pneumonia—Los Angeles. *MMWR* 1981;30:250–252.
156. Friedman-Kien AE, Laubenstein L, Marmor M, et al: Kaposi's sarcoma and pneumocystis pneumonia among homosexual men—New York City and California. *MMWR* 1981;30:305–308.
157. Siegel PP, Lopez C, Hammer GS, et al: Severe acquired immunodeficiency in male homosexuals, manifested by chronic perianal ulcerative herpes simplex lesions. *N Engl J Med* 1981;306:1439–1444.
158. Masur H, Michelis MA, Greene JB, et al: An outbreak of community-acquired *Pneumocystis caranii* pneumonia. Initial manifestation of cellular immune dysfunction. *N Engl J Med* 1981;305:1431–1438.
159. Mildvan D, Mathur U, Enlow RW, et al: Opportunistic infections and immune deficiency in homosexual men. *Ann Intern Med* 1982;96:700–704.
160. Barre-Sinoussi F, Chermann JC, Rey F, et al: Isolation of a T-lymphotropic retrovirus from patients at risk for acquired immune deficiency syndrome (AIDS). *Science* 1983; 220:868–871.
161. Gelman EP, Popovic M, Blayney D, et al: Proviral DNA of a retrovirus, human T-cell leukemia virus, in two patients with AIDS. *Science* 1983;220:862–865.
162. Klatzman D, Champagne E, Chamaret S, et al: T-lymphocyte T4 molecule behaves as the receptor for human virus LAV. *Nature* 1984;312:767–768.
163. Dalgleish AG, Beverley PCL, Clapham PR, et al: The CD4 (T4) antigen is an essential component of the receptor for the AIDS retrovirus. *Nature* 1984;312:763–766.
164. Popovic M, Sarngadharan MQ, Read E, Gallo RC: Detection, isolation and continuous production of cytopathic retroviruses (HTLV-III) from patients with AIDS and pre-AIDS. *Science* 1984;224:497–500.
165. Sarnagadharan MG, Popovic M, Bruch L, et al: Antibodies reactive with human T-lymphotropic retroviruses (HTLV-III) in the serum of patients with AIDS. *Science* 1984; 224:506–508.
166. Brun-Vezinet F, Barre-Sinoussi F, Saimot AG, et al: Detection of IgG antibodies to lymphadenopathy-associated virus in patients with AIDS or lymphadenopathy syndrome. *Lancet* 1984;1:1253–1256.
167. Hahn BH, Shaw GM, Arya SK, et al: Molecular cloning and characterization of the HTLV-III virus associated with AIDS. *Nature* 1984;312:166–169.
168. Wong-Staal F, Shaw GM, Hahn BH, et al: Genomic diversity of human T-lymphotropic virus type III (HTLV-III). *Science* 1985;229:759–762.
169. Ratner L, Haseltine W, Patarca R, et al: Nucleotide sequence of the AIDS virus, HTLV-III. *Nature* 1985;313:277–284.
170. Meusing MA, Smith DH, Cabradilla CD, et al: Nucleic acid structure and expression of the human AIDS/lymphoadenopathy retrovirus. *Nature* 1985;313:460–468.
171. Sanchez-Pescador R, Power MD, Barr PJ, et al: Nucleotide sequence and expression of an AIDS-associated retrovirus (ARV-2). *Science* 1985;227:484–492.
172. Redfield RR, Markham PD, Salahuddin SZ, et al: Frequent transmission of HTLV-III among spouses of patients with AIDS-related complex and AIDS. *JAMA* 1985; 253:1571–1573.
173. Commentary—the chronology of AIDS research. *Nature* 1987;326:435–436.
174. Larder B, Purifoy DJM, Powerll KL, Darby G: Site-specific mutagenesis of AIDS virus reverse transcriptase. *Nature* 1987;327:716–717.
175. Walker BD, Flexner C, Paradis TJ, et al: HIV-1 reverse transcriptase is a target for cytotoxic T-lymphocytes in infected individuals. *Science* 1988;240:64–66.
176. Franchini G, Gurgo C, Guo H-G, et al: Sequence of simian immunodeficiency virus and its relation to the human immunodeficiency viruses. *Nature* 1987;328:539–542.
177. Chakrabarti L, Guyader M, Alizon M, et al: Sequence of simian immunodeficiency virus from macaque and its relationship to other human and simian retroviruses. *Nature* 1987;328:543–547.
178. Barin F, Denis F, Allan JS, et al: Serological evidence for virus related to simian T-lymphotropic retrovirus III in residents of West Africa. *Lancet* 1985;2:1387–1389.
179. Kanki PJ, Barin F, M'Boup S, et al: New human T-lymphotropic retrovirus related to simian T-lymphotropic virus type II (STLV-III AGM). *Science* 1986;232:238–243.

180. Manzari V, Gismondi A, Barillari G, et al: HTLV-V: A new human retrovirus isolated in a Tac-negative T cell lymphoma/leukemia. *Science* 1987;238:1581–1583.
181. Johnson RT, McArthur JC: Myelopathies and retroviral infections. *Ann Neurol* 1987; 21:113–116.
182. Jacobson S, Raine CS, Mingioli ES, McFarlin DE: Isolation of an HTLV-I–like retrovirus from patients with tropical spastic paraparesis. *Nature* 1988;331:540–543.
183. Bhagavati S, Ehrlich G, Jula RW, et al: Detection of human T cell lymphoma/leukemia virus type I DNA and antigen in spinal fluid and blood of patients with chronic progressive myelopathy. *N Engl J Med* 1988;318:1141–1147.
184. Reddy EP, Mettus RV, DeFreitas E, et al: Molecular cloning of human T cell lymphotropic peripheral lymphocyte DNA of a patient with chronic neurologic disorders. *Proc Natl Acad Sci USA* 1988;85:3599–3603.
185. Osame M, Matsumoto M, Usuku K, et al: Chronic progressive myelopathy associated with elevated antibodies to human T-lymphotropic virus type I and adult T cell leukemia–like cells. *Ann Neurol* 1987;21:117–121.
186. Putney SD, Matthews TJ, Robey WG, et al: HTLV-neutralizing antibodies to an *E. coli*–produced fragment of the virus envelope. *Science* 1986;234:1392–1395.
187. Robert-Guroff M, Bronn M, Gallo RC: HTLV-III neutralizing antibodies in patients with AIDS and AIDS-related complex. *Nature* 1985;316:72–74.
188. Weiss RA, Clapham PR, Cheingsong-Popov R, et al: Neutralization of human T-lymphotropic virus type III by sera of AIDS and AIDS-risk patients. *Nature* 1985;316:69–74.
189. Robey WG, Arthur LO, Matthews TJ, et al: Prospect for prevention of human immunodeficiency virus infection: Purified 120-kDa envelope glycoprotein induces neutralizing antibody. *Proc Natl Acad Sci USA* 1986;83:7023–7027.
190. Seligmann M, Chess L, Fahey J, et al: AIDS—an immunologic reevaluation. *N Engl J Med* 1984;311:1286–1292.
191. McDougal JS, Kennedy MS, Sligh JM, et al: Binding of HTLV-III/LAV to $T4^+$ T cells by a complex of the 100K protein and the T4 molecule. *Science* 1986;231:382–385.
192. Hu S-L, Fultz PN, McClure HM, et al: Effect of immunization with a vaccinia–HIV *env* recombinant on HIV infection in chimpanzees. *Nature* 1987;328:721–723.
193. Homsy J, Steimer K, Kaslow R: Towards an AIDs vaccine: Challenges and prospects. *Immunol Today* 1987;8:193–196.
194. Nabel G, Baltimore D: An inducible transcription factor activates expression of human immunodeficiency virus in T cells. *Nature* 1987;326:711–713.
195. Richardson NE, Brown NR, Hussey RE, et al: Binding site for human immunodeficiency virus coat protein gp120 is located in the NH_2-terminal region of T4(CD4) and requires the intact variable-region–like domain. *Proc Natl Acad Sci USA* 1988;85:6102–6106.
196. Lifson DJ, Hwang KM, Hara PL, et al: Synthetic CD 4 peptide derivatives that inhibit HIV infection and cytotoxicity. *Science* 1988;241:712–716.
197. Mitsuya H, Looney DJ, Kuno S, et al: Dextran sulfate suppression of viruses in the HIV family: Inhibition of virion binding to $CD4^+$ cells. *Science* 1988;240:646–649.
198. Zagury D, Bernard J, Cheynier R, et al: A group specific anamnestic immune reaction against HIV-1 induced by a candidate vaccine against AIDS. *Nature* 1988;332:728–731.
199. Levy JA: Mysteries of HIV: Challenges for therapy and prevention. *Nature* 1988; 333:519–522.
200. Cherfas J: Hope for AIDS vaccines. *Science* 1989;246:23–24.
201. Navia MA, Fitzgerald PMD, McKeever BM, et al: Three-dimensional structure of aspartyl protease from human immunodeficiency virus HIV-1. *Nature* 1987;337:615–620.
202. Weber IT, Miller M, Jaskolski M, et al: Molecular modeling of the HIV-1 protease and its substrate binding site. *Science* 1989;243:928–931.
203. Chou S, Merigan JC: Rapid detection and quantitation of human cytomegalovirus in urine through DNA hybridization. *N Engl J Med* 1983;308:921–925.
204. Huang ES, Alford CA, Reynolds DW, et al: Molecular epidemiology of cytomegalovirus infections in women and their infants. *N Engl J Med* 1980;303:958–962.
205. Grillner L, Strangert K: Restriction endonuclease analysis of cytomegalovirus DNA from strains isolated in day care centers. *Pediatr Infect Dis* 1986;5:184–187.

206. Croen KD, Ostrove JM, Dragovic LJ, et al: Latent herpes simplex virus in human trigeminal ganglia: Detection of an intermediate early gene "anti-sense" transcript by in situ hybridization. *N Engl J Med* 1987;317:1427–1432.
207. Meisels A, Fortin R: Condylomatous lesions of the cervix and vagina. I. Cytologic patterns. *Acta Cytol* 1976;20:505–509.
208. Purola E, Savia E: Cytology of gynaecologic condyloma acuminata. *Acta Cytol* 1977; 21:26–31.
209. Wickenden C, Malcolm ADB, Steele A, Coleman DV: Screening for wart virus infection in normal and abnormal cervices by DNA hybridisation of cervical scrapes. *Lancet* 1985; 1:65–67.
210. Werness BA, Levine AJ, Howley PM: Association of human papillomavirus types 16 and 18 E6 proteins with p53. *Science* 1990;248:76–79.
211. Bowles NE, Olsen EGJ, Richardson PJ, Archard LC: Detection of coxsackie-B-virus specific RNA sequences in myocardial biopsy samples from patients with myocarditis and dilated cardiomyopathy. *Lancet* 1986;1:1120–1123.
212. Rappuoli R, Perugini M, Falsen E: Molecular epidemiology of the 1984–1986 outbreak of diphtheria in Sweden. *N Engl J Med* 1988;318:12–14.
213. Karzon DT, Edwards KM: Diphtheria outbreaks in immunized populations. *N Engl J Med* 1988;318:41–43.
214. Hunt C, Morimoto R: Conserved features of the eukaryotic hsp 70 genes revealed by comparison with the nucleotide sequence of human hsp 70. *Proc Natl Acad* 1985;82:6455.

Messenger Molecules—Polypeptide Hormones, Peptides, and Neuroendocrine Transmitters

7

Ralph C. Williams, Jr., MD

Molecular Endocrinology Comes of Age

The field of endocrinology and biologically important peptides as well as neurotransmitter molecules has undergone major changes in direction and emphasis during the past few years, in part based on depths of perception afforded by molecular biology. As the primary structure and additional fine structure of many of the major polypeptide hormones became known, common primordial origins in a single gene or gene family were recognized in some and messenger RNA could be isolated in pure enough preparations to ensure success with cell-free protein synthesis systems. Knowledge of primary structure and chain interactions in relatively small polypeptide hormones such as insulin allowed rapid application of gene cloning techniques, resulting in the production of human insulin in *E. coli* or other recombinant systems. The fact that this was accomplished within a relatively short time after recombinant technology began to be applied to the production of important human proteins and polypeptides can be considered a remarkable medical achievement. In parallel with work on human insulin, the progress that has been made on larger, more complex polypeptide hormones such as human growth hormone has been rapid and encouraging. As workers in endocrinology and neurochemistry began to explore the operation of various messenger molecules in the body, it soon became clear that merely correlating the rise and fall of levels of hormones or prohormones with biologically visible end points such as exophthalmos, obesity, or ovulation provided limited information. Additional kinds of information would provide a deeper level of understanding of how hormones and other messenger molecules actually worked. Thus, if a set of genes were present in competent cells, what exact mechanisms were involved in beginning transcription and translation into polypeptide prohormone gene products and then active hormones? Moreover, the close interrelationships of groups of hormones that had different biological functions but apparently arose from similar genes was underscored in a number of studies. In the past, monoclonal antibodies that reacted with specific antigenic or functional sites on hormone polypeptides were manufactured and fluctuations of hormones were measured by radioimmunoassay or enzyme-linked immunosorbent assay (ELISA) and correlated with other measurable biologic phenomena. Now, attention has shifted to what activates prohormone genes and what sites on the nucleotide sequences of such genes are built in as initiators of transcription or off signals. Insights provided by the molecular biological approach have accumulated rapidly and have already been applied to gut hormones and neuroendocrine transmitter systems, as well as to the more classically established polypeptide hormones such as insulin, somatostatin, and growth hormone.

As hormone gene activation was being examined, a new field related to receptor physiology and cell receptor gene activation became important. If complex or even simple polypeptides had such powerful biological effects, what controlled their ability to interact directly with cell surfaces? Clearly, highly developed receptor molecules—many of them glycoproteins with complex genomic precursors—were required. There has been a surge of new information and some beginning insight into how cell surface receptors such as the *c43E*c74E*c109 glucocorticoid receptor, the insulin receptor, and the growth hormone receptor are assembled by direct synthesis involving highly complex gene families. The present chapter is intended to provide a general overview of what is clearly a new science. As more new information is obtained about messenger molecule genes, their activation and deactivation, and their receptors, the emphasis in medicine and biology may shift from the emperor's clothes to the emperor's genes and their initiating nucleotide sequences.

The Peptide Hormone

Basic work in the 1960s and 1970s provided knowledge of the structure and sequence of a number of small peptide hormones, including information about their synthesis and fundamental action. This type of information, particularly concerning amino acid sequence,[1,2] and the increasing availability of antibodies against synthetic peptides proved to be a tremendous boost to the molecular biologist interested in hormone or messenger molecule action. It appeared that polypeptide hormones, cell growth factors, neurotransmitters, and simple peptides could be grouped according to structural, functional, and genomic similarities.[3] Analysis of these molecules required precise knowledge of their size, unit composition, secondary structure, and primary amino acid sequence and knowledge of various additional structural features such as glycosylation or carbohydrate appendages necessary for biological function.

Large Glycoprotein Hormones

The gonadotropins—pituitary luteinizing hormone (LH) and follicle-stimulating hormone (FSH), thyroid-stimulating hormone (TSH), and placental chorionic gonadotropin (CG)—are each composed of two different noncovalently linked subunits designated as α and β chains. Each of these chains is glycosylated and its shape is stabilized by intrachain disulfide bonds.[4] Amino acid sequence examination indicates that the α chain or subunit is common to or shared by all four hormones (LE, FSH, TSH, and CG), whereas the β chains vary in size (from 118 to 147 amino acids) and contain unique sequences peculiar to their individual biological activity. Moreover, there is evidence that the carbohydrate side chains are important in the biologic activities of these large, multichain pituitary hormones.[5]

Other hormones that are composed of single polypeptide chains include the placental hormone chorionic somatotropin or placental lactogen and the pituitary hormones prolactin and growth hormone. These hormones, made up of 191 to 198 amino acids of established sequence, show some degree of sequence homology as well as biological activity.[6] Parathyroid hormone is a single 84-amino-acid polypeptide chain.[7] Smaller multichain hormones whose structures have been extensively studied include insulin and the related hormone relaxin, both of which contain two nonidentical polypeptide chains linked together by key cysteine-cysteine disulfide bonds.[8] When the sequence homologies between relaxin and insulin were recognized, it was thought that these structural similarities might form the basis for categorizing relaxin as a member of an extended family of insulin-like growth factors with functional similarities in modulation of cell growth activity. Other insulin-like hormones included nerve growth factor,[9] somatomedins A and C,[10] and multiplication-stimulating factor.[11] As structure and sequence knowledge of the large, medium, and small hormones increased, a wide spectrum of small peptides and other messenger molecules were characterized; these include the opiate pentapeptides such as Met- and Leu-enkephalin[12] and larger peptides such as calcitonin, which contains 32 amino acids.[13] The wide distribution of many of the small peptide hormones (gut, brain, and other central nervous system tissues) has evoked an extensive variety of physiological studies related to mechanisms of activation and transcription of genes coding for these potent messenger molecules.

Synthesis and Transport

As knowledge of the primary structure and organization of hormones advanced, it appeared that several common synthetic steps were involved in the production and eventual release of the active polypeptide product. Because individual polypeptides were read from their respective gene codons and translated into products on intracellular ribosomes, it became clear that the polypeptide products were transported out through cell membranes via secretory pathways for eventual utilization as active products. Moreover, when the secretory process was examined, it was clear that the initial polypeptide gene products destined for export were often precursor molecules of higher molecular weight than the eventual hormone product. In some instances the smaller active product was cleaved from larger precursors by proteolytic enzymes after secretion, whereas in others the active hormone was cleaved from its larger precursor within the cell before secretion.[14]

In many instances precise knowledge of the primary structure of hormone precursors or prohormones turned out to be very useful in analysis of their physiological action. When proinsulin from the pancreatic β cells was analyzed, it was found that the α and β chains of insulin were synthesized from mRNA as a single polypeptide chain linked together through an additional C peptide. The intervening C peptide served to maintain correct orientation of the eventual disulfide bonds of the product insulin molecule. When the disulfide bridges were formed, the C peptide was cleaved off by enzymes within β cell secretory granules (Figure 7.1). A similar process of initial synthesis of prohormone has been documented for proparathyroid hormone, cleaved to its active residues by proteolytic enzymes in the Golgi vesicles.[15,16] As yet no biological activity has been associated with the con-

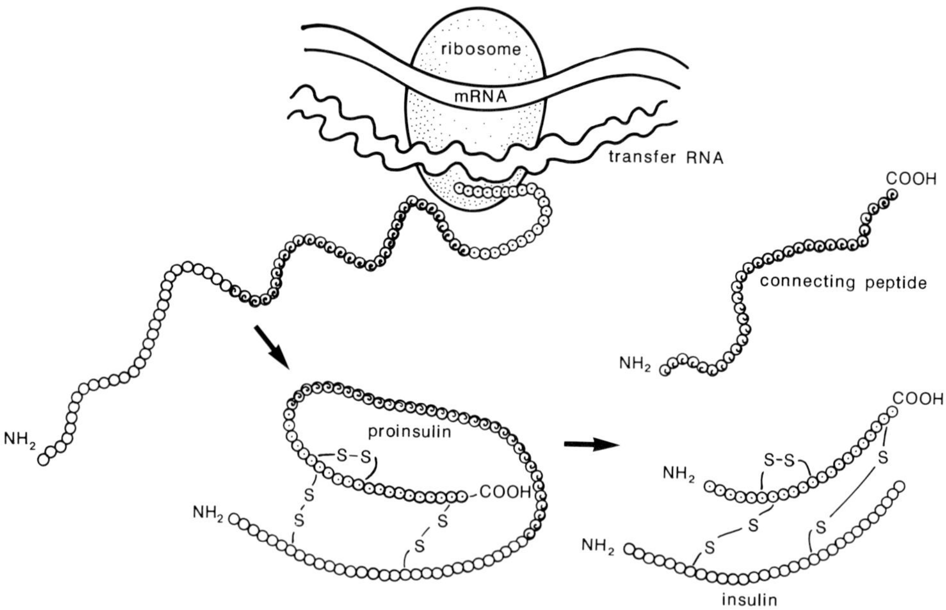

Figure 7.1 Biosynthetic steps in insulin synthesis. Proinsulin includes the C or connecting peptide, which is cleaved off to produce the native insulin molecule.

necting C peptide of insulin or the hexapeptide prohormone sequence cleaved from parathyroid hormone.

Similar analyses subsequently identified higher molecular weight prohormone gene products for a number of other hormones; these include glucagon, somatostatin, gastrin, arginine vasopressin, ocytocin, the enkephalins, corticotropin (adrenocorticotropic hormone, ACTH), and β-lipotropin. In several cases precursor hormone gene products may be cleaved proteolytically to several active products of different biological activities. For example, the precursor 30-kd polypeptide chain of ACTH is processed to produce ACTH, β-lipotropin, β-endorphin, and other peptides not yet completely characterized.[17] Prohormone proteases that are capable of producing specific cleavage products resulting in the active hormones have frequently been identified.[18–20] A number of basic studies have indicated common gene product polypeptide precursors for vasopressin, oxytocin, and their respective neurophysin proteins.[21] Moreover, a common polypeptide precursor for Leu-enkephalin and Met-enkephalin has been localized to the chromaffin cells of the adrenal medulla.[22,23] The various routes of synthesis and processing of hormones, including prohormone processing, cleavage and/or assembly, and addition of key carbohydrate groups, are shown in Figure 7.2.

Cell-Free Protein Synthesis

A great technical boost was provided to the field of hormone biology when systems for cell-free protein synthesis became available. The two earliest approaches to this problem involved the use of wheat germ extracts[24] and lysates of rabbit reticulocytes.[25] This approach was given additional important practical impetus by the development of relatively simple techniques for isolation of active mRNA.[26] The

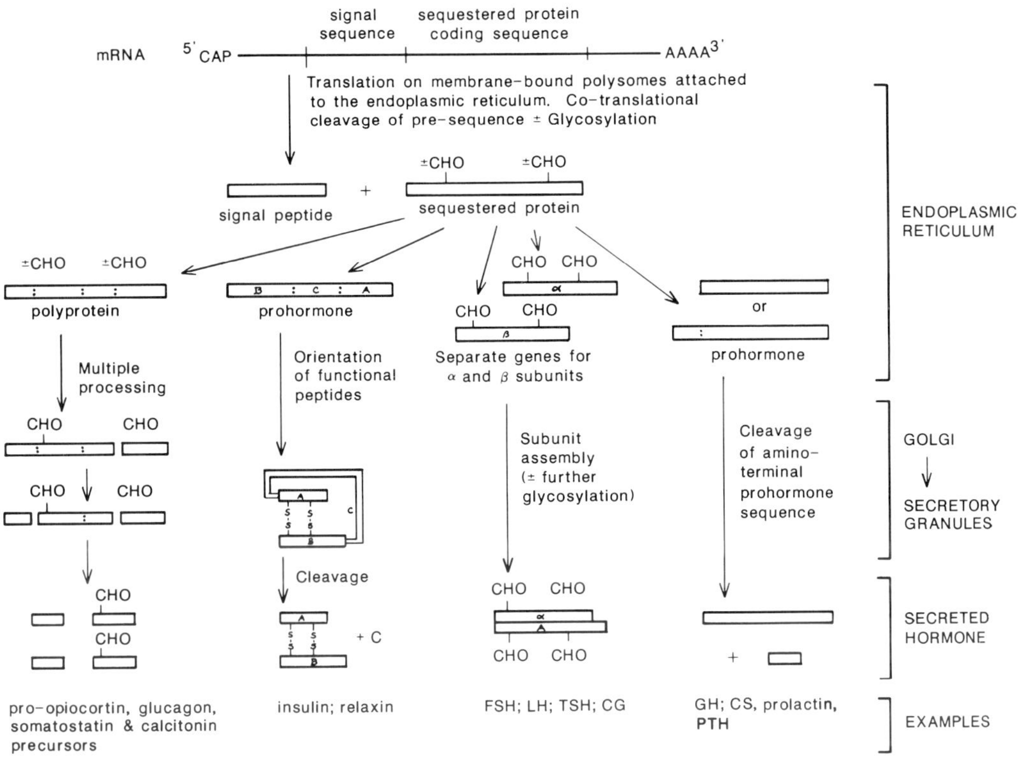

Figure 7.2 Several schemes whereby various groups of hormones are assembled from signal peptides and subsequent prohormone products.

use of in vitro protein synthesis coupled with protein sequencing techniques then rapidly demonstrated that all large polypeptide hormones and their prohormones are made with relatively hydrophobic amino-terminal signal peptide extenders. Following their synthesis, the polypeptide hormones undergo segregation within the endoplasmic reticulum, where signal peptides are removed or important carbohydrate groups are added.

Isolation and Use of Peptide Hormone mRNA Sequences

The polypeptide hormones are produced and eventually secreted by highly specialized cellular tissues in selected endocrine tissues. Peptide hormones may also be produced in cells of the gut or central nervous system. Within a particular endocrine tissue, the amount of the mRNA species needed to produce a particular polypeptide hormone is usually 0.1% to 10% of the total cellular mRNA. The physiological timing of the tissue may be important, of course, in determining the relative abundance of the mRNA in question. Thus, the mRNA for the subunits of human chorionic gonadotropin would be expected to be a major mRNA species in placental tissues obtained during the first trimester. Similarly, mRNA enrichment for TSH might be expected in animals intentionally made hypothyroid by thyroid ablation.

For studies of human mRNA or of mRNA sequences that are present in very low quantities or produced by only a small subset of the normal cell population, tumor tissues or transplantable tumors that produce the particular hormones have been used.[27] Thus, human calcitonin is synthesized by the thyroid C cells, which constitute only 1% to 2% of the entire thyroid cellular tissue. Tissue from patients with medullary carcinoma of the thyroid (in which circulating calcitonin levels may be markedly elevated) may provide an enriched source of human calcitonin prohormone.[28] Similarly, pituitary tumor tissues from patients operated on for acromegaly or hyperprolactinoma may be markedly enriched sources of mRNA for growth hormone or prolactin.[29] Cell lines derived from similar tumor tissues have also been used as a source of peptide hormone mRNA; however, using such cell lines sometimes introduces an additional problem related to continued synthesis of the hormone of interest, that is, instability of the hormone.

Preparation of mRNA

When cloning techniques are directed at in vitro synthesis of polypeptide hormones, the approach to isolation of the particular mRNA usually depends on the relative abundance of the mRNA in the starting tissue or cell line source. One has to choose between isolating total cellular RNA, followed by isolation of the poly(A)-containing RNA populations, or purifying membrane-bound polyribosomes enriched in mRNA of the desired secretory proteins. The latter approach often involves subcellular fractionation steps, which in some cases may increase the opportunity for endogenous ribonucleases to degrade the poly(A)-containing RNA of interest. In general, cell fractionation procedures can avoid endogenous ribonuclease activation by use of sterile glassware and addition of ribonuclease inhibitors such as heparin or diethyl pyrocarbonate. Subcellular fractionation for enrichment of polypeptide hormone mRNA is probably most useful when a relatively low-abundance peptide hormone is to be cloned. Poly(A)-containing RNA can be isolated from polyribosomal RNA preparations by using columns of oligo(dT)-cellulose or poly(U)-Sepharose. The poly(A)-containing RNA is allowed to bind at 100 mM NaCl, and elution is accomplished at 60°C in low-salt buffers.

If the poly(A)-containing RNA preparations are active, they should be capable of directing synthesis of a wide spectrum of protein or polypeptide products detectable by [^{35}S]methionine labeling. The usual approach is to identify the primary translation products produced in the wheat germ or rabbit reticulocyte system by antibody precipitation followed by sodium dodecylsulfate (SDS)–polyacrylamide gel electrophoresis and autoradiography. This approach is illustrated in Figure 7.3. The Western blot or autoradiographic identification of translated and synthesized polypeptide hormone in this situation is highly dependent on the antiserum used. Thus, antibodies produced against the final secreted hormone may not react with epitopes in the precursor hormone. Alternatively, antisera against subunits of the hormone may not be suitable for detecting the final secreted polypeptide product, which may have acquired carbohydrate side chains and thus changed its basic antigenic structure during secretion.

Another in vitro method of mRNA expression and synthesis utilizes injection of mRNA into *Xenopus* oocytes. These oocytes can sequester, glycosylate, and actually secrete a number of animal proteins after mRNA injection.[30,31] Also, the oocyte system can function with great efficiency and can glycosylate secreted polypeptide hormone products, although apparently it cannot activate proteolytic enzymes such as those necessary for cleavage of proinsulin to active insulin. After microinjection of mRNA in the oocyte system, the surrounding medium can be

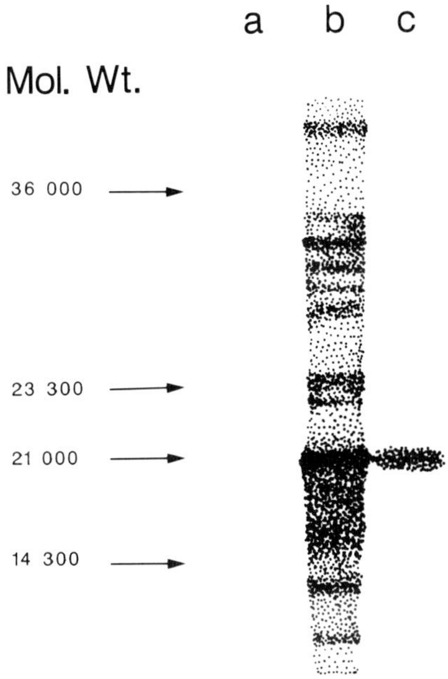

Figure 7.3 Messenger RNA–directed cell-free protein synthesis of human calcitonin precursor polyprotein as judged by immunoprecipitation and SDS–polyacrylamide gel electrophoresis. Total poly(A)-containing RNA was isolated from frozen human thyroid medullary carcinoma tissue translated in a wheat germ cell-free protein-synthesizing system; the resulting [^{35}S]methionine-labeled proteins were separated by SDS-polyacrylamide gel electrophoresis and then visualized by fluorography. (a) No added mRNA; (b) human thyroid medullary carcinoma mRNA; (c) immunoprecipitation of proteins synthesized in (b) using an antiserum raised against synthetic human calcitonin.

assayed for mRNA polypeptide product. This simple technique has been successfully adapted for production of interferon[32] as well as granulocyte-macrophage colony-stimulating factor.[33]

Synthetic Oligonucleotide Hybridization

Where the mRNA species to be studied is scarce or in only moderate abundance, a different strategy has been followed that involves the use of synthetic oligonucleotide probes. This approach requires prior knowledge of the amino acid sequence of at least part of the polypeptide hormone. An oligodeoxynucleotide is synthesized which is complementary by base pairing to the mRNA sequence based on the known amino acid sequence. This strategy is effective provided that the degeneracy of the primary genetic code is only slight.[34]

Synthetic oligodeoxynucleotide primers were first employed for the isolation of peptide hormone mRNA subpopulations in identifying gastrin mRNA from porcine antral musoca total mRNA tissue preparations. Thus, a 12-mer peptide was produced with complementarity to the predicted mRNA sequence which encoded a tetrapeptide known to be present in gastrin; this is illustrated in Table 7.1. It can be seen that the occurrence of G or C was considered in the third position of the

Table 7.1 Use of Synthetic Oligodeoxynucleotide Primers in mRNA Identification

	Amino end			Carboxyl end
Gastrin tetrapeptide	Trp	Met	Glu	Glu
			A	A
Predicted mRNA	5′-UGG	AUG	GAG	GAG-3′
Synthesized cDNA	3′-ACC	TAC	CTC	CTC-5′

codons for glutamic acid. However, other potential effects of genetic code degeneracy were not a problem, since only a single codon is employed in mammalian genes for methionine and tryptophan. The synthetic oligonucleotide was then used to make cDNA by reverse transcription. Finally, the cDNA transcripts were isolated and nucleotide sequences were determined.[35] When code degeneracy is a potential problem, a mixture of possible oligodeoxynucleotides can sometimes be used to select out the mRNA species of interest, provided that stringent hybridization conditions are employed.[36] The synthetic oligodeoxynucleotide hybridization technique is a good example of how limited knowledge of amino acid sequence can be used to identify a specific mRNA in a given RNA population and to define its size and partial nucleotide sequence.

Production of Peptide Hormones Using Recombinant Techniques

Two similar methods have generally been used for production of peptide hormones in prokaryotic systems. One approach requires complete chemical synthesis of the coding DNA based on the previously established amino acid sequence, with careful attention, as noted above, to possible alternative codons in the genetic code. This approach results in a final active hormone product, usually a small peptide normally produced as a higher molecular weight precursor or prohormone. Direct synthesis from the gene without flanking polynucleotide sequences avoids the necessity for the proteolytic cleavage that usually occurs during the secretory process.

The second approach relies on a well-defined cDNA clone. This method is sometimes limited by positions of restriction sites in the genes themselves, and the cDNAs chosen may encode a precursor protein or prohormone that must either be processed in vivo by the recipient or tailored in vitro before final use.

Synthesis of Small Peptide Hormones Initially as Hybrid Fusion Proteins

The first polypeptide hormone to be synthesized in bacteria was somatostatin, a 14-amino-acid hormone that is important in the modulation of other hormones and that may play a role as a neurotransmitter. It is now recognized that two different somatostatins are present and that each is synthesized as a higher molecular weight prohormone.[37] In the initial work, a synthetic gene was put together from a group of overlapping oligodeoxynucleotide segments containing the codons for 15 amino acids (the 14 amino acids of somatostatin plus an additional amino-terminal methionine) followed by a termination or stop codon. In making the synthetic gene the possible degeneracy of the code was taken into account, and since *E. coli* was to be used, *E. coli*–preferred codons were employed. The strategy is illustrated in Figure 7.4. In this instance, the β-galactosidase and somatostatin gene sequences were oriented in phase with each other, resulting in synthesis of a β-galactosidase–somatostatin fusion protein. The somatostatin could then be directly recovered by in vitro treatment with cyanogen bromide, which cleaved the polypeptide chain at methionine.

There were several interesting aspects of this cloning strategy. The synthetic gene was controlled by the powerful *lac* operon, which meant that production of the final fusion protein could be up-regulated by simple chemical inducers so that the fusion product represented 3% of total cellular protein. Also, in this situation,

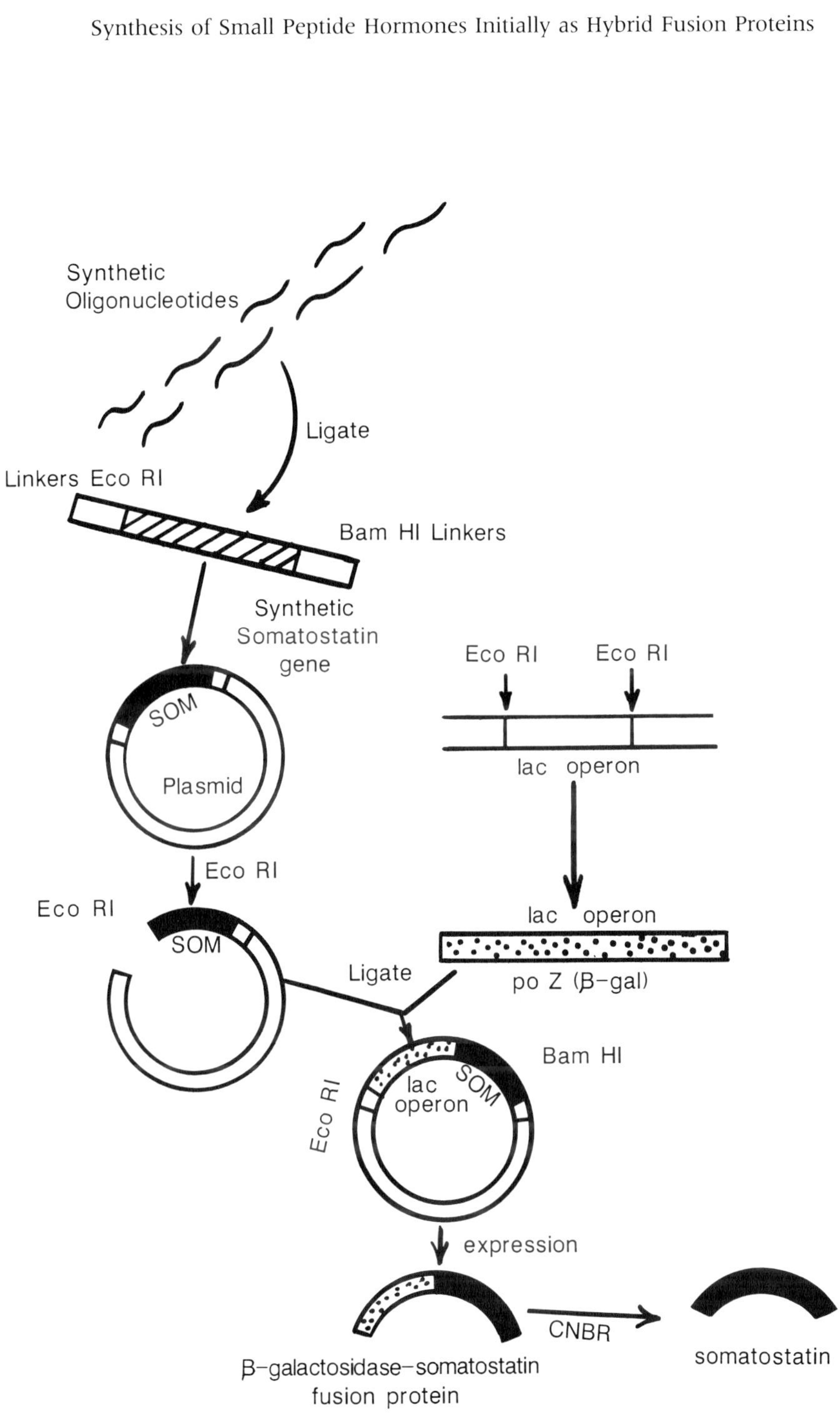

Figure 7.4 Strategy for bacterial expression of a cloned synthetic somatostatin gene. A plasmid was made with the synthetic somatostatin gene placed next to a fragment of the *E. coli lac* operon that contained the promoter and operator and a portion of the β-galactosidase gene. When the fusion protein was expressed, it could be cleaved by cyanogen bromide to free somatostatin.

levels of expression of the somatostatin were considerably higher than when the somatostatin gene was placed more downstream of the *lac* promoter. Similar strategies have now been employed successfully with variations in the amounts of *lac* operon sequence used for production of other small peptide hormones, including thymosin α_1, Leu-enkephalin, and α-*neo*endorphin. Biological activity in these latter instances proved to be well preserved.

A similar approach was used in one method aimed at the production of human insulin. In this case, the genes for the A and B chains were chemically synthesized in such a way that each contained an amino-terminal methionine. This approach involved cloning the B chain in two parts and is illustrated diagrammatically in Figure 7.5. Cleavage of fusion proteins with cyanogen bromide again released the A and B chains of insulin, which could be oxidized to active insulin. In this process a yield of 10% to 15% was achieved,[38] as established by reactivity in radioimmunoassay.

The production of fusion proteins that require active peptide release by cyanogen bromide cleavage is clearly limited to peptide hormones that do not contain methionine residues in their active chains. A number of different strategies have been used to express smaller fusion proteins and free the active peptides by limited

Figure 7.5 Cloned synthetic insulin sequences in separate plasmids containing A chain and left and right halves of B chains. In both cases the insulin nucleotide sequences were placed adjacent to the *E. coli* operon, which resulted in the expression of fused β-galactosidase–insulin hybrid proteins. After cleavage with cyanogen bromide, air oxidation resulted in disulfide bond formation and an active insulin product.

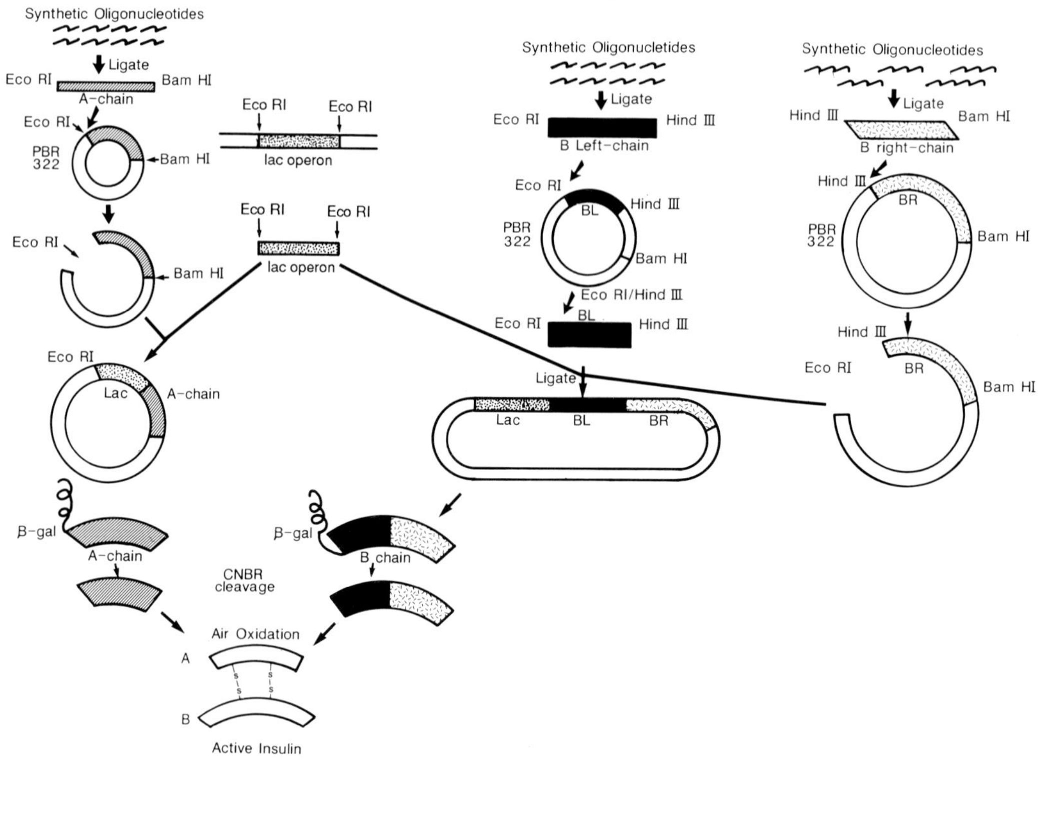

or protected proteolytic digestion. In spite of the successful production of individual insulin A and B chains as shown in Figure 7.5, the resulting fusion protein occurs in low yields. Another strategy involves synthesis of a β-galactosidase–Met-proinsulin–like protein. In this case the insulin A and B chain genes are linked together by an 18-nucleotide sequence representing an abbreviated connecting peptide of six amino acids that include Arg-Arg-Gly-Ser-Lys-Arg in the intermediate plasmid vector. This sequence includes two amino acids necessary for proteolytic cleavage. The final gene product in this case was similar to the insulin prohormone when separated from β-galactosidase by cyanogen bromide; however, it was not clear that correctly placed disulfide bonds had been formed.[39] Most of these technical difficulties have now been overcome, and the recombinant method now used to produce human insulin has been markedly improved with respect to both its adaptability to large batch production and the stability and fidelity of the final product.[40]

The current method utilizes recombinant plasmids containing genes that code for the separate A and B chains of insulin; the plasmids are then inserted into a bacterial strain (*E. coli* K12). These plasmids are constructed in such a way that the A and B chains are produced as chimeric or fusion products in which they are attached through a methionine to a modified *E. coli* protein called LE′. This is done by linking the A and B chain genes directly to the truncated bacterial gene for LE′ protein and including a nucleotide sequence coding for methionine between the LE′ protein and the A and B chains. When these chimeric genes are inserted into plasmids and the plasmids are cloned in bacteria, the presence of the methionine makes it possible to cut the insulin chains off of the LE′ protein. Subsequent combination of the A and B chains produces biologically active human insulin.

Production and Expression of Large Peptide and Prohormone Sequences

The methods used to synthesize large peptide hormones differ considerably from those employed for the small peptides because the larger the polypeptide hormone, the less likely it is to be recoverable from a fusion protein product. Recovery occurs only in favorable situations, where the polypeptide hormone contains no methionine or other indirect blocking techniques may be used. Also, the larger the polypeptide chain, the more difficult it becomes to synthesize the appropriate gene. These intrinsic problems have been overcome by using various combinations of chemically synthesized oligonucleotides and cloned cDNA. This approach has been used, for instance, in establishing a successful expression system for producing human growth hormone. In this case, a gene was put together that was part cDNA and part chemically manufactured DNA and then ligated to an efficient bacterial promoter. This resulted in the expression of a growth hormone polypeptide with an extra initial methionine. The general cloning strategy is illustrated in Figure 7.6. The large *Hae* restriction fragment containing 551 base pairs was ligated in correct approximation to the chemically manufactured adapter fragment containing codons for amino acids 1 to 24 with an amino-terminal methionine. The expressed final product was human growth hormone (Met-HGH), which showed potent biological activity in spite of the extra methionine.[41] This method for producing recombinant human growth hormone provides material that accumulates in the bacterial cell and may require considerable separation and purification from con-

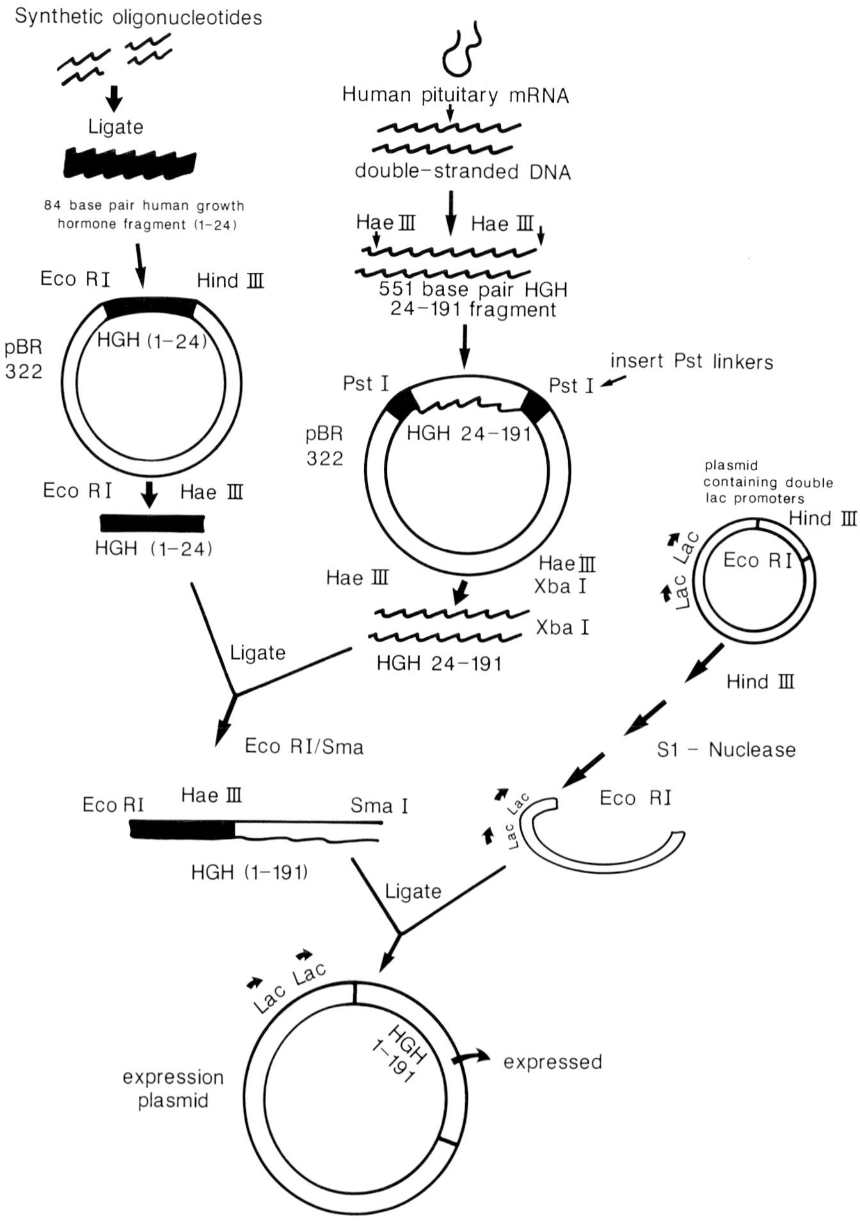

Figure 7.6 Strategy for using a combination of synthetic nucleotide (upper left) coding for residues 1 to 24 in human growth hormone carried into plasmid vector pBR322 and eventually ligated with a 551-bp cDNA for residues 24 to 191 of human growth hormone. The ligated genes are put next to a tandem *lac* promoter and expressed as human growth hormone residues 1 to 191.

stitutive bacterial components. Additional ingenious cloning strategies employed in similar bacterial vector systems have used signal peptides of such materials as β-lactamase to enhance processing of prohormone or precursor fusion products through the bacterial wall to the periplasmic space, thus avoiding an amino-terminal methionine or fusion to another bacterial protein.

Production of Protein Hormones by Host-Vector Activation

Early attempts to adapt bacterial plasmid vector systems revealed the lack of special proteolytic cleavage enzymes so often needed for posttranslational modification or processing of prohormones. Various other methods were subsequently explored aimed at introducing and stimulating the expression of mammalian genes in cultured cells. These have included the use of simian virus 40 (SV40) viral vectors, which provide efficient promoters for inserted polypeptide hormone genes. However, the infected cell lines, like bacteria, lacked specialized proteolytic cleavage enzymes. Nevertheless, successful production of human growth hormone using an SV40 viral inducer indicated that processed growth hormone minus the signal peptide could be synthesized and actually secreted.[42] These studies have also been useful in terms of learning more about gene expression in vivo, because in some cases a group of similar genes capable of producing growth hormone–like polypeptides has been discovered. Much additional work is needed to understand the gene selection and activation process within such complex gene families.

Application of Recombinant Techniques to Understanding Messenger Molecules

It is now possible to manufacture representative cDNA libraries from virtually any mRNA that contains the message for a peptide hormone, growth factor, or other molecule using the basic methods discussed above. During the past decade important insights have been gained into the structure and means of expression of a number of polypeptide hormones, particularly with respect to nucleotide sequences and how they make a final prohormone or hormone product.

Recombinant approaches have also revealed that families of peptides or biologically related hormones sometimes arise from a single precursor polypeptide gene product. Examples of multiple peptides derived from a single precursor polypeptide gene product are illustrated in Figure 7.7. In the case of the pre-proenkephalin gene product, there are a number of copies of the same Met-enkephalin peptide. In some instances precise knowledge of the gene arrangements has been useful in defining the order of synthesis of biologically active but different peptide hormones. In other cases, peptides are produced whose function still remains to be clearly defined. Calcitonin and the carboxyl-terminal peptide are a good example of the latter situation. In many instances carboxyl-flanking or amino-flanking peptides have been shown to have ancillary biological activities. Thus recombinant techniques provide a way to identify and sequence flanking peptides that can then be directly examined for potential biological activity.

In Situ Hybridization

An increasingly frequent application of recombinant methodology has involved the use of cloned peptide hormone cDNA sequences as probes to localize areas with high concentrations of specific mRNAs by in situ hybridization in frozen tissue sections. The relative abundance of a specific mRNA can be determined by using blotting techniques with reproducible accuracy.[43] This sort of approach has proved useful in determining tissue localization of hormone gene precursors and has, for instance, shown a surprising distribution of pregastrin or precalcitonin mRNAs in various cells of the central nervous system.[44] As knowledge of RNA splicing mech-

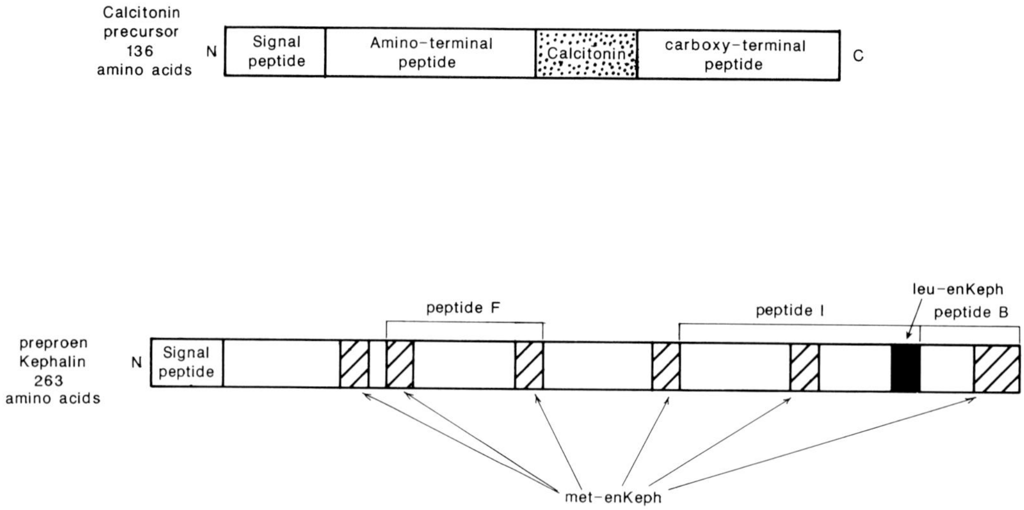

Figure 7.7 Polypeptide hormone precursors of some peptide hormones derived from DNA sequence analysis of cloned mRNA species. The signal peptides are located at the amino terminus of the original gene product.

anisms increases, it should be possible to elucidate why the calcitonin gene produces the calcitonin polypeptide product in the thyroid, whereas in the hypothalamus different splicing mechanisms occur so that the calcitonin gene–related peptide (whose function is still not entirely clear) is produced instead.

As repeatedly noted above, many of the expression vector systems with which peptides or polypeptide hormones are produced lack the tissue-specific processing enzymes which, in the natural state, cleave off active products from their prohormones. As technical facility in this general area advances, it should be possible to include inserted gene segments that will code for the glycosylating or processing enzymes essential for active hormone production. This would greatly simplify some of the current problems in producing a final active hormone product. In the same vein, considerable work has been focused on specific signal nucleotide sequences upstream from prohormone genes in order to try to understand how external stimuli such as other hormones, regulatory proteins, or direct chemical reactions control initial transcription and translation of hormone gene products.

Cloning of Hormone Receptors

After the initial work on understanding the structure and interrelationships of various families of polypeptide hormones, attention was naturally turned to the cellular receptors for many of these hormones. One of the first of these receptors to be studied in great detail was the human insulin receptor. The human insulin receptor is an integral membrane glycoprotein of 350 to 400 kd consisting of two α subunits of 120 to 135 kd and two β subunits of $\approx$95 kd linked by disulfide bonds.[45,46] In intact cells, insulin stimulated the phosphorylation of the β subunit on serine and tyrosine residues. Thus, it appeared that the insulin receptor itself exhibited insulin-activated tyrosine kinase activity.[47,48] As noted in Chapter 5, a number of studies indicated that the specificity of the insulin-dependent protein kinase was very sim-

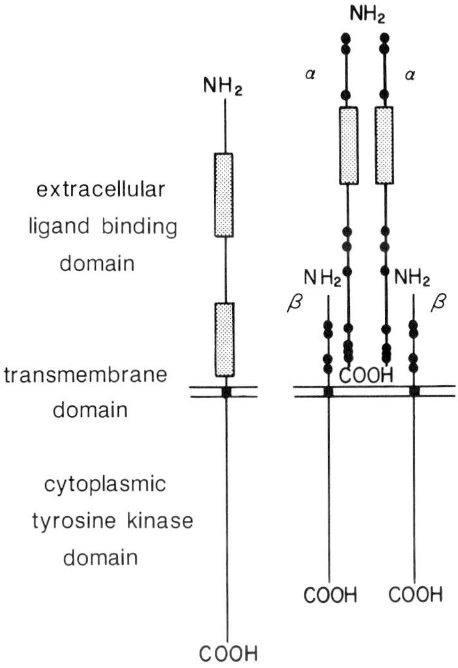

Figure 7.8 Schematic comparison of insulin and EGF receptors. Regions of high Cys residue concentration are shown as stippled boxes, transmembrane domains as black boxes, and single cysteine residues, possibly involved in formation of the $\alpha_2\ \beta_2$ insulin in receptor complex, as black circles.

ilar to that of the epidermal growth factor (EGF) receptor kinase and therefore the *src* family of tyrosine-active protein kinases.[49,50] Thus, the functional cell surface receptor for insulin contained both an insulin-binding function and a specific tyrosine protein kinase function. As physical and chemical knowledge of the insulin receptor increased, it became clear that it contained the ATP-binding site within its β subunit and that portions of both the α and β subunits were exposed on the cell surface. Again as previously noted, structural and sequence studies indicated that the human EGF receptor contained both transmembrane and cytoplasmic domains homologous to the v-*erb* B oncogene product representing the corresponding cellular proto-oncogene. Shortly thereafter, the complete amino acid sequence of the human insulin receptor was deduced from human placental cDNA clones.[51] The strategy employed in this remarkable scientific achievement was to design long single synthetic DNA probes for hybridization screening of a DNA library. A total of more than 5100 nucleotides was finally defined in the complete gene sequence. Moreover, Southern blot and restriction fragment analyses[51] showed that only one insulin receptor gene is present within the human genome.

When nucleotide and deduced amino acid sequences for the insulin and EGF receptors were compared, striking similarities were apparent. Both had a membrane-spanning 230-amino-acid hydrophobic sequence and similar sequences near the cell membrane. The structures of the insulin and EGF receptors are compared in Figure 7.8. When the insulin receptor β subunit was extensively compared with analogous segments of other members of the *src* family of tyrosine-specific kinases, extensive homology was documented. These remarkable similarities included a lysine residue essential for kinase activity located 25 amino acids from the ATP binding site. These extensive homologies clearly placed the insulin receptor in the *src* family of tyrosine-specific protein kinases and proto-oncogene products.

The precise genetic and structural knowledge generated by these studies provided an excellent basis for examination of functional alterations of the insulin receptor in common disease states such as diabetes and rare disorders such as the heritable syndromes of insulin resistance.

Several descriptions of insulin-resistant diabetes due to mutations affecting the insulin receptor were reported.[52,53] In one instance a single point mutation altered the tetrabasic processing site of the proreceptor molecule from Arg-Lys-Arg-Arg to Arg-Lys-Arg-Ser. This resulted in synthesis of an insulin receptor precursor that was inserted into the cell plasma membrane but cleaved to mature α and β subunits.[52] The second example involved an insulin-resistant patient with leprechaunism whose receptors showed multiple abnormalities in insulin binding.[53] Two mutant alleles of the insulin receptor gene were identified—one in the α chain of the receptor and the second deleting both the transmembrane and tyrosine kinase domains of the receptor. Other reports followed documenting mutations or deletions in the tyrosine kinase domain of the insulin receptor associated with diabetes.[54,55] Deletion of the tyrosine kinase domain was associated with acanthosis nigricans and insulin resistance,[54] and the same clinical picture was recorded in a patient with a mutation in the receptor tyrosine kinase domain.[55] Most of the other instances of deletions or molecular defects of the insulin receptor gene have been associated with insulin resistance and various degrees of glucose intolerance.[56]

Now that the exact sequence and structure of the insulin receptor are known, it should be possible to dissect out the exact action of insulin itself using the technique of site-directed mutagenesis and expression of the resulting mutants in appropriate cell lines. The striking homology of the insulin receptor to other tyrosine kinase-activated receptors and oncogene-related products is of great interest, since it suggests that there are important functional similarities between a number of structurally similar cell surface receptors involved in various sorts of signal transduction. This is illustrated schematically in Figure 7.9.

Thus, the presence of the cysteine-rich cross-linking regions in three distinct functional receptors (insulin, EGF, and low-density lipoprotein [LDL]) probably indicates that these regions are not directly associated with ligand binding. The strong homology between the EGF receptor and the insulin receptor in the cyto-

Figure 7.9 Schematic comparison of the structures and possible functional domains of peptide or hormone receptors and related oncogenes with tyrosine kinase activity.

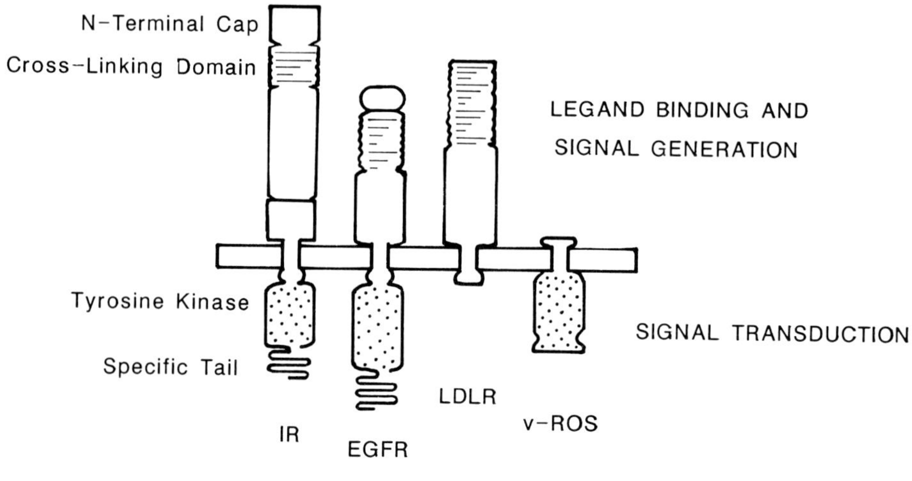

plasmic domains also probably indicates that these cellular regions are most closely associated with their specific biological activities. For some time it has been surmised that the ATP-tyrosine phosphokinase activity is directly related to cellular growth control[57,58]; however, exactly how this is accomplished is still not clear. The similar structural and phosphorylating features of growth factors, oncogene products, and receptors suggests that a common means of producing their intracellular signals may be involved.[59] How this transmembrane signal actually operates remains to be determined.

β-Adrenergic Receptor

β-Adrenergic impulses are important in the control of the coronary artery circulation and the maintenance of cardiac muscular tone and excitation. The basic regulatory pathway for a number of hormones and pharmacologically active substances is now well known to be the adenylate cyclase system, composed of a catalytic functional unit and regulatory quanine nucleotide–binding proteins (G proteins). In the β-adrenergic system two distinct subtypes of receptors (β_1 and β_2) stimulate adenylate cyclase after the receptors bind to specific catecholamines.[60] The avian erythrocyte β_1 receptor and the amphibian erythrocyte β_2 and mammalian lung β_2 receptors have been extensively purified and shown to retain potent specific binding activity.[60–62] The mammalian β-adrenergic receptor has now been cloned using hamster lung material purified to homogeneity as starting material,[63] with eventual production of cyanogen bromide–cleaved peptide fragments. After sequencing portions of the latter, oligonucleotides complementary to the DNA encoding deduced amino acid sequences were used in hybridization experiments. When the β-adrenergic receptor was successfully cloned, the translation product was found to be a polypeptide of 418 amino acids, which was in close agreement with previous immunochemical analyses. When the complete amino acid sequence was examined, considerable sequence homology with the visual pigment rhodopsin became apparent. Since both the β-adrenergic receptor and rhodopsin are directly involved in signal transduction processes that require interactions with guanine nucleotide regulatory proteins, the similarity in their primary sequences is of interest.

Following cloning of the β-adrenergic receptor, subsequent elegant studies subdivided specific structure-function relationships for the α_2 and β_2 adrenergic receptors, which can be differentiated by use of selective drugs with opposite effects on the adenylyl cyclase system. Insight into the functional portions of these receptors was gained after construction and expression of a series of chimeric α_2-, β_2-adrenergic receptor genes by Kobilka and co-workers.[64] It was found that specificity for coupling to the stimulatory guanine nucleotide regulatory protein resides within a region extending from the amino terminus of the fifth hydrophobic domain to the carboxyl terminus of the sixth. The major determinants of α_2- and β_2-adrenergic receptor agonist and antagonist ligand-binding specificity were found to be present within the seventh membrane-spanning domain.

Additional subclassifications of β-adrenergic receptors have now been made using molecular approaches. A human gene encoding a third β-adrenergic receptor, called the β_3-adrenergic receptor, has been isolated.[65] Exposure of eukaryotic cells transfected with this gene to epinephrine or norepinephrine promotes accumulation of adenosine 3′,5′-monophosphate. Only 2 of 11 classical β-adrenergic blocking agents inhibited this effect, and two others functioned as β_3-adrenergic agonists.

Of particular interest was the possible role of these novel β_3-adrenergic receptors in mediating high thermogenic, antiobesity and antidiabetic activities in animal models.

These findings have suggested a possible basis for the way in which a number of other receptors function, including those for prostaglandins, dopamine, and histamine. Clearly, the cloning of the mammalian β-adrenergic receptor has provided considerable insight into the exact structural makeup of an important class of molecules that are a focus of investigation and pharmacological manipulation by way of a large array of β-blockers in coronary artery disease or selective β-stimulators in bronchial asthma.

Human Dopamine Receptor

Insight into other types of G protein–coupled receptors, such as the dopamine D_1 and D_2 receptors, was considerably extended with the isolation and characterization of the gene encoding a human D_1 dopamine receptor.[66–68] Alterations in the number and level of activity of these receptors may contribute to Parkinson's disease, schizophrenia, drug addiction, and alcoholism. Of interest was the finding that the coding region of the human D_1 receptor gene was intronless, unlike the gene that encodes the D_2 receptor. The D_1 receptor gene encoded a protein of 446 amino acids with a transmembrane topology similar to that of other G protein–coupled receptors. Messenger RNA for this receptor was most abundant in the caudate, nucleus accumbens, and olfactory tubercle, and there was little or no activity in the substantia nigra, liver, kidney, or heart. Chromosome mapping indicated that the D_1 receptor locus is on chromosome 5 near the structurally similar genes for β_2-adrenergic receptors and the gene for α_1-adrenergic receptor.[68]

Corticosteroid Receptors

Corticosteroid hormones exert their effects on growth and development as well as homeostasis of cells by interaction with two different corticosteroid receptors—the glucocorticoid receptor (GR) and the mineralocorticoid receptor (MR). Comparative studies of corticosteroid effects showed that glucocorticoids exert their effects by promoting glycogen deposition in the liver, whereas mineralocorticoids promote sodium retention by the kidney. However, some overlap was clearly present, and each steroid class was not limited to interaction with its own receptor. Glucocorticoids may have substantial mineralocorticoid activity,[69,70] and MRs may have considerable affinity for both glucocorticoids and mineralocorticoids.[71]

Molecular cloning of the glucocorticoid,[72] estrogen,[73,74] and progesterone[75] receptors has now been successfully completed. This has permitted definition of their primary amino acid sequences and indicated how their various functional domains interact.

Studies of glucocorticoid[76] and estrogen[77] receptors have shown a centrally located DNA-binding domain with enrichment for cysteine, lysine, and arginine, as well as a carboxyl-terminus region where steroid hormones interact. The functional studies of GRs indicate that glucocorticoid hormone binding to the carboxyl terminus exposes the DNA-binding regions to allow interaction of the receptor with DNA and initiation of transcription.[78,79] Moreover, a certain amount of structural similarity of the steroid-binding regions in the steroid receptor family suggests that they evolved from a common ancestral gene.[80] Using the fact that many steroid receptors may show some degree of relatedness and structural homology, the human MR has also been cloned.[81] These workers used nonstringent hybridization with a human GR probe to isolate a human genomic DNA fragment highly related to the human GR cysteine-rich sequence. Then, using the latter DNA as another probe, complementary DNAs were found that coded for a polypeptide chain product showing strong but not exact homology with the original human GR from the cysteine-rich segment to the carboxyl terminus. This cDNA segment, when expressed in cells, showed high-affinity binding for aldosterone, which identified it as the MR. Complete sequencing and structural analysis have identified this molecule as a 107-kd protein of 984 amino acids. Sequence comparisons also showed 94% amino acid homology in the centrally located DNA region over 68 residues with human GR. In vitro assays established hormone-dependent modulation of gene transcription for this newly cloned MR. Moreover, Northern blots from tissues showed that mRNA homologous to human MR was expressed, as expected, in aldosterone-sensitive cells in the kidney as well as in brain, pituitary, and heart tissues. Of great interest in this study was the localization of the MR gene to human chromosome 4. This was done by using a panel of rodent-human somatic cell hybrids that retained different numbers of human chromosomes. The assignment to chromosome 4 was made by Southern blotting of DNA from such hybrids and probing with an appropriate cDNA from MR.

The extensive structural homology noted during this study again emphasizes the relatedness of MR, GR and progesterone receptors. These similarities are illustrated in Figure 7.10. Now that the mineralocorticoid receptor has been cloned and expressed in several systems, including transfected mammalian cells, it should be possible to examine its tissue expression and function in a number of human disease states, including essential hypertension and malignant hypertension. Rarer disorders, such as pseudohypoaldosteronism, which is apparently related to complete loss of high-affinity aldosterone MR binding sites,[82] may also be more effectively studied using the MR probe.

Figure 7.10 Amino acid sequence comparisons of the human GR, human MR, and rat progesterone receptor (PR) structures. Primary amino acid sequences have been aligned schematically with the percentage amino acid identity indicated for each region of homology in the intervals between dashed lines. The amino acid position of each domain boundary is shown for each receptor. N and C represent the amino and carboxyl termini, respectively. Cys corresponds to the cysteine-rich region encoding the putative DNA-binding domain, and Steroid (cortisol, aldosterone, or progesterone) designates the steroid-binding domain. The immunogenic region (IMM) of the hGR is also indicated. *Reproduced from Arriza et al.,*[81] *with permission.*

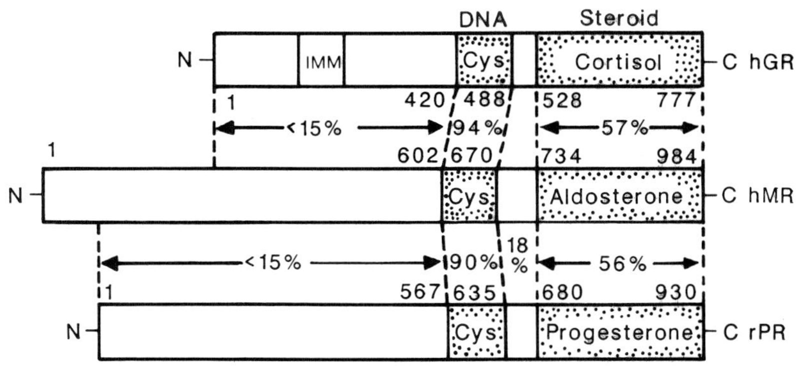

Table 7.2 Amino Acid Sequences for the Rabbit

				5					10					15					
GHR-N:	F	S	G	S	E	A	(T)	P	A	T	L	G	R	A	S	E	[S	V	Q
SBP-N1:	X	X	X	S	E	A	T	P	A	T	L	G	R	A	S	E	S	V	Q
SBP-N2:	-	-	-	-	-	-	-	-	-	-	-	-	-	X	X	E	(S)	V	Q
SBP-N3:	-	-	-	-	-	-	-	-	-	-	-	-	-	-	-	-	-	-	-
GHR-T4:	V	E	P	S	F	N	Q	E	D	I	Y	I	T	T	E	S	L	T	T

[a] GHR-N, GH receptor amino-terminal sequence; SBP-N1–3, three simultaneous amino-terminal sequences found for the serum binding protein; G/HR-T4, sequence of the tryptic fragment of the GH receptor used for cDNA cloning. Sequences were determined by gas-phase sequencer analysis of electroeluted intact proteins or of HPLC-purified tryptic fragment. Brackets indicate residues deter-

Growth Hormone Receptor

After the cloning and expression in bacteria of growth hormone (GH), attention was directed at the cellular receptors for this large polypeptide hormone. For some time the precise mechanisms by which the GH-receptor complex exerted its intracellular effects were poorly understood. In part, this was due to apparent receptor heterogeneity and the presence of soluble GH-binding proteins in both the cytosol and plasma.[83,84] More insight has been afforded by the cloning and expression of the GH receptor by Leung et al.[85] For some time it had been recognized that high-affinity GH binding sites existed in a wide spectrum of tissues,[86,87] indicating that the tropic effects of GH on particular cells might be mediated not only by stimulating the liver to release insulin-like growth factor, resulting in bone and cartilage growth,[88] but also by directly interacting with the tissues themselves. In addition to tissue GH receptors ranging in size from 50 to 110 kd, a GH-binding protein has been described in the serum of humans.[84,89] The cloning strategy employed for the GH receptor started with purification of receptor from rabbit liver membranes monitored by the binding of ^{125}I-labeled human GH to receptor preparations. In parallel, the serum GH-binding protein was isolated and purified 400,000-fold from rabbit serum using human GH affinity columns. The amino acid sequence from the amino terminus of the liver membrane GH receptor was determined, and another peptide (ubiquitin) was found to be covalently linked to the GH receptor polypeptide. When the serum GH-binding protein was also sequenced, it was found that at least the first 37 residues from the amino terminus were identical to those of the liver membrane GH receptor. This is illustrated in Table 7.2. GH receptor clones were finally isolated by screening a liver cDNA library using a single 57-mer oligonucleotide probe based on the 19-residue amino acid sequence of a tryptic peptide of the receptor (Table 7.2). The rest of the sequence of the GH receptor was then established by rescreening the same libraries with fragment probes obtained from the initial set of clones.

The complete cloning and sequencing of the GH receptor gene through this strategy represents a remarkable feat in the application of recombinant techniques. This is particularly noteworthy because the cDNAs used ultimately deduced a gene sequence of over 4300 nucleotides. The complete DNA and translated amino acid sequence of the GH receptor is illustrated in Figure 7.11. After successful establishment of the GH receptor sequence, the human and rabbit GH cDNA was expressed in COS-7 monkey kidney cells[90,91] used as a mammalian expression vector. In this system, transcription of the receptor sequence was directed by a human cytomegalovirus promoter. Using this expression vector, it was shown that ^{125}I-

Liver GH Receptor and Serum Binding Protein[a]

20					25					30					35		
R	V	H	P	G	L	G	T	S		S	G	K	P	K	F	T	K]
X	V	H	P														
R	V	H	P	G	L	(G)	T	X	S	(S)	G	K	P	K	F	T	X
-	X	X	P	X	L	X	(T)	X	S	(S	G)	K	P	K	F	X	K
X	(A)	(E)															

mined from the cDNA sequence (Figure 7.3). X shows residues that could not be identified. Parentheses show uncertain residues. The absence of Asn at position 28 in the serum binding protein sequence may indicate N-linked carbohydrate attached at this position.

labeled human GH bound to the expressed GH receptor with high avidity. In these studies it also became apparent that several GH receptor types could be identified—possibly based on small differences in posttranslational modification or processing. The successful identification and cloning of the GH receptor also established that the serum GH-binding protein was in effect the extracellular, hormone-binding domain of the membrane-derived GH receptor. Human populations that might be expected to manifest defective or abnormal GH receptor gene products include Laron dwarfs[92] and African pygmies,[93] both of which show normal GH levels. Moreover, human serum GH-binding activity is absent in both liver and serum of Laron dwarfs. Besides its application in the understanding of these rare and unusual defects in normal growth, the definition of the sequence and organization of the GH receptor promises to open up a new vista in understanding how this important hormone interacts with cells of many tissues and types. Also of interest in the work on GH receptor structure was the finding that the sequence was not related to those of other known tyrosine kinase growth factor receptors, such as the insulin receptor or the EGF receptor. Thus, if the GH receptor is directly related to important signal transduction, other as yet undefined mechanisms must be involved.

Basic insight into human GH biological effects was provided by Cooke and coworkers,[94] who demonstrated that the major 22-kd and minor 20-kd forms of pituitary human GH were encoded by alternative splicing products of a single human GH-N gene transcript. This alternative splicing was neither species nor tissue specific and appeared to be regulated by at least two separate regions that were rather remote from the splice-acceptor site. Alternate gene splicing to produce two biologically different hormone polypeptide products is a familiar theme in many families and superfamilies of hormones and was alluded to earlier in this chapter with respect to calcitonin and calcitonin gene-related products.

Laron-Type Dwarfism

Laron-type dwarfism is an autosomal recessive disorder that is characterized by high levels of GH and low levels of insulin-like growth factor I in the circulation. Several lines of evidence have suggested that this disease is caused by a defect in the GH receptor. Two groups have now provided direct confirmation of genetic defects—partial gene deletions and mutation—in the GH receptor gene in this disorder.[95,96] One study documented a deletion of a large portion of the extracellular, hormone-binding domain of the receptor gene.[95] This deletion included nonconsecutive exons, suggesting that an unusual rearrangement had occurred. Molecular analysis was thought to implicate either two independent deletion events

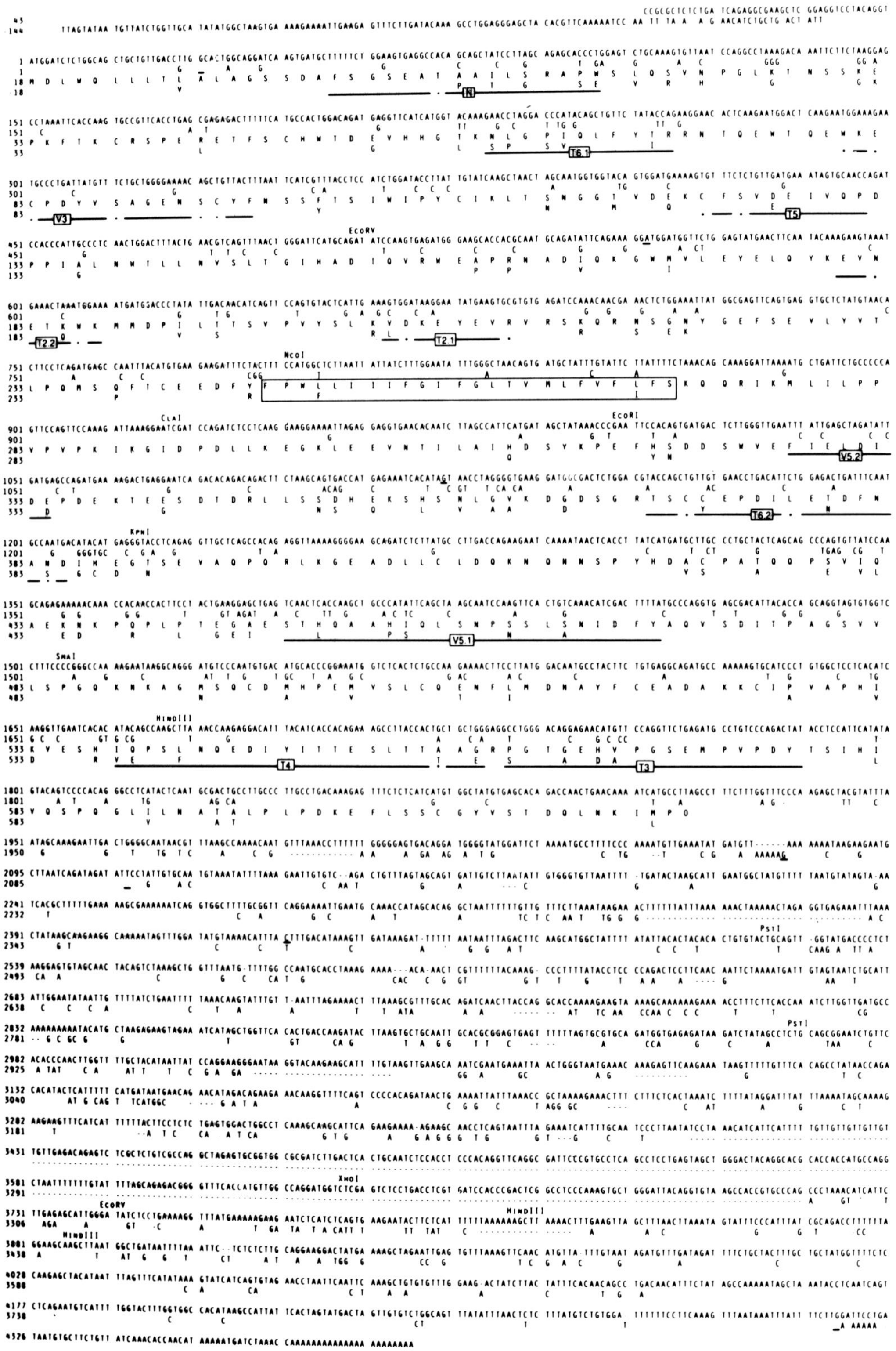

or a complex deletion and rearrangement. Because two prominent mRNAs for GH receptor have now been cloned[97] which encode two forms of receptor (a high molecular weight, membrane-bound type and a lower molecular weight type that diverges before the transmembrane domain), a mutation or deletion in the general region near an alternative splicing site may be likely.

A second molecular genetic analysis of Laron dwarfism took advantage of an experiment of nature involving two Mediterranean families containing consanguineous parents and some children with Laron dwarfism.[96] It was found that the disease trait and DNA polymorphisms were inherited together. Analysis of GH receptor gene RNA transcripts in lymphocytes from one family indicated a thymidine-to-cytosine substitution generating a serine in place of a phenylalanine at position 96 in the extracellular domain of the GH receptor protein. Moreover, the mutation identified in this family was not identified in GH receptor genomic sequences of seven other unrelated patients with Laron dwarfism who belonged to different population groups. These studies identified mutations and/or deletions in GH receptor genes as the fundamental defect in Laron dwarfism and emphasized that the disorder may encompass a spectrum of different individual genetic events within affected families.

Vitamin D Receptor

In addition to common hormones such as insulin, growth hormone, estrogen, and their receptors, a number of other biological compounds have been analyzed using cloning techniques. Vitamin D as D_3 is a remarkably potent substance that is metabolized in both liver and kidney to 1,25-dihydroxyvitamin D_3 [1,25$(OH)_2D_3$], which is the active vitamin D product principally involved in regulation of calcium and phosphorus homeostasis.[98,99] The biological response to 1,25$(OH)_2D_3$ in both birds and mammals involves interaction with an intracellular receptor protein and subsequent initiation of effects via the receptor as in the events stimulated by the classical steroid hormones.[100] Besides the familiar functions of 1,25$(OH)_2D_3$ related to calcium and phosphorus metabolism, this material has been shown to regulate cellular proliferation and differentiation, secretion of polypeptide hormones, and modulation of certain aspects of the immune system.[101–103] All of these diverse biological functions are the result of activation of the vitamin D receptor (VDR).

Monoclonal antibodies that react with purified chicken intestinal VDR were produced and used to screen a library of cDNA clones from a chicken intestinal λgt11 expression library.[104] The cloned VDR structure revealed a cysteine-, lysine-, and arginine-rich region showing great sequence homology with both the steroid receptor family and the avian erythroblastosis *gag-erb* A oncogene product.[72,80] This homologous chicken VDR sequence showed 46% overall homology with the chicken estrogen receptor and the v-*erb* A product, as well as considerable homology with the chicken progesterone and glucocorticoid receptor proteins. These

←

Figure 7.11 DNA and translated amino acid sequence of human and rabbit growth hormone receptor cDNA clones. Line 1, human DNB sequence; line 3, human translated amino acid sequence; line 4, differences for the rabbit amino acid sequence. Dashes show deletions. The DNA sequences are numbered starting with 1 at the initiating ATG, the protein sequences with 1 at the mature amino-terminal phenylalanine. The locations of the sequenced rabbit peptides (Table 7.1) are underlined. The putative transmembrane domain is boxed. *Reproduced from Leung et al.,*[85] *with permission.*

ARG	ILE	CYS	GLY	VAL	CYS	GLY	ASP	ARG	ALA	THR	GLY	PHE	HIS	PHE	ASN	ALA	MET	THR	CYS	GLU	GLY	CYS	LYS
35								43								51							
GLU	GLN	CYS	VAL	VAL	CYS	GLY	ASP	LYS	ALA	THR	GLY	TYR	HIS	TYR	ARG	CYS	ILE	THR	CYS	GLU	GLY	CYS	LYS
177								185								193							
ARG	TYR	CYS	ALA	VAL	CYS	ASN	ASP	TYR	ALA	SER	GLY	TYR	HIS	TYR	GLY	VAL	TRP	SER	CYS	GLU	GLY	CYS	LYS
LYS	ILE	CYS	LEU	ILE	CYS	GLY	ASP	GLU	ALA	SER	GLY	CYS	HIS	TYR	GLY	VAL	LEU	THR	CYS	GLY	SER	CYS	LYS
	I																						

											*												
GLY	PHE	PHE	ARG	ARG	SER	MET	LYS	ARG	—	—	LYS	ALA	MET	PHE	THR	CYS	PRO	PHE	ASN	GLY	ASP	CYS	LYS
59								67								75							
SER	PHE	PHE	ARG	ARG	THR	ILE	GLN	LYS	ASN	LEU	HIS	PRO	THR	TYR	SER	CYS	THR	TYR	ASP	GLY	CYS	CYS	VAL
201								209								217							
ALA	PHE	PHE	LYS	ARG	SER	ILE	GLN	GLY	—	—	HIS	ASN	ASP	TYR	MET	CYS	PRO	ALA	THR	ASN	GLN	CYS	THR
VAL	PHE	PHE	LYS	ARG	ALA	MET	GLU	GLY	—	—	GLN	HIS	ASN	TYR	LEU	CYS	ALA	GLY	ARG	ASN	ASP	CYS	ILE
										II													

ILE	THR	LYS	ASP	ASN	ARG	ARG	HIS	CYS	GLN	ALA	CYS	ARG	LEU	LYS	ARG	CYS	VAL	ASP	ILE	GLY	MET	MET	LYS
		85						93								101							
ILE	ASP	LYS	ILE	THR	ARG	ASN	GLN	CYS	GLN	LEU	CYS	ARG	PHE	LYS	LYS	CYS	ILE	SER	VAL	GLY	MET	ALA	MET
		225						233								241							
ILE	ASP	LYS	ASN	ARG	ARG	LYS	SER	CYS	GLN	ALA	CYS	ARG	LEU	ARG	LYS	CYS	TYR	GLU	VAL	GLY	MET	MET	LYS
VAL	ASP	LYS	ILE	ARG	ARG	LYS	ASN	CYS	PRO	ALA	CYS	ARG	LEU	ARG	LYS	CYS	CYS	GLN	ALA	GLY	MET	VAL	LEU

Figure 7.12 Comparison of the deduced amino acid sequence of chicken VDR with v-*erb* B and other selected members of the steroid receptor family. A 2-amino-acid insertion was allowed in v-*erb* B to permit its exact alignment with each receptor protein; amino acids common to three or more proteins are boxed. Regions corresponding to two putative DNA binding "fingers" are underlined (I and II). An alternative second-finger structure is indicated by the dashed line. An asterisk indicates a nonconserved residue in one of the possible second-finger structures. A numerical designation as indicated for chicken estrogen receptor and v-*erb* B has not been assigned to chicken VDR or progesterone receptor because the orientation of these clones with respect to their encoded products has yet to be confirmed. *Reproduced from McDonnell et al.,*[104] *with permission.*

striking similarities are illustrated in Figure 7.12. The amino acid sequence analysis showed a region including nine cysteine residues in the same general relative location as similar stretches of amino acid sequence in the steroid receptor family. It was presumed that this relatively conserved region represented the DNA-binding domain of the gene family. Further studies with the VDR system have shown that $1{,}25(OH)_2D_3$ appears to be able to up-regulate its own receptor by increasing receptor-specific mRNA.

The high degree of structural similarity between the VDR and other steroid receptors suggests that $1{,}25(OH)_2D_3$ belongs to the general class of true steroid hormones. Moreover, the remarkable homology with the avian erythroblastosis *gag-erb* A fusion peptide could be evidence that this family of genes arose from a single ancestral precursor. The fact that the greatest degree of homology exists between the v-*erb* A product and the VDR suggests that the VDR may have arisen by gene reduplication or other mechanisms at an extremely primitive stage in evolution. The cloning of the avian VDR should lead to new insights into several important human diseases, including type II vitamin D–dependent rickets, which has often been regarded as a heterogeneous disease related to abnormalities in VDR domains.[105]

Thyroid Hormone and TSH Receptors

In 1986 Weinberger and co-workers[106] reported that the c-*erb* A gene encoded a thyroid hormone receptor. The connection between c-*erb* A, the cellular counterpart of the viral oncogene v-*erb* A, and receptors for this and a number of thyroid-related hormones was of considerable interest. Again, it began to appear that the steroid and thyroid hormone receptors had evolved from a primordial receptor gene. Additional heterogeneity within thyroid hormone receptors was subsequently identified using rat brain mRNA with homology to the human thyroid hormone receptor gene.[107] These findings suggested that a family of thyroid hormone receptors might coordinate regulation of overlapping networks of genes to implement developmental and homeostatic functions.

The thyroid hormone receptor expressed in human kidney and other tissues was described by Nakai and co-workers.[108] This form of human thyroid hormone receptor was identified using cDNA encoding the thyroid hormone receptor expressed in human liver, kidney, placenta, and brain from a human kidney library. The expressed protein, designated human TR type α_2, had the same domain structure as other members of the v-*erb* A–related superfamily but also had an extension of the COOH terminal hormone-binding domain that made it 80 amino acids longer than the previously described rat thyroid hormone receptor of the α_1 type.[107]

Subsequently, the pituitary thyrotropin receptor was cloned by Parmentier and co-workers[109] using selective amplification with the polymerase chain reaction of DNA segments having sequence similarity to genes for G protein–coupled receptors. The polypeptide encoded by this thyroid-specific transcript included a 398-amino-acid residue amino-terminal segment (presumably the extracellular domain) connected to a 346-residue carboxyl-terminal domain containing seven transmembrane segments. Expression of the cDNA in *Xenopus* oocytes conferred TSH responsiveness. These findings placed both the TSH receptor and the receptor for luteinizing hormone in a subfamily of G protein–coupled receptors with distinct sequence characteristics.

The Steroid and Thyroid Hormone Receptor Superfamily

It is generally agreed that the steroid and thyroid hormone receptors are members of what appears to be a larger superfamily of receptor for a number of regulatory proteins, including the glucocorticoids, mineralocorticoids, progesterone, estrogen, vitamin D_3, and thyroid hormones. A number of comparative and structural studies have supported this impression.[110] Comparison of the amino acid sequences in DNA-binding domains of many of these hormone receptors shows striking similarity between many divergent molecules. Common features of structure and function remain to be elucidated.

Cloning of Neuroreceptors

The relationship between the chemical and neuroendocrine events involved in the transmission of various nerve signals and gene activation has received increasing attention. This is a natural consequence of our attempts to relate knowledge of central and autonomic nervous system function and what must be engineered through specific gene activation. The most extensively studied function of neurotransmitters is their regulation of transsynaptic communication in the central and peripheral nervous systems. However, neurotransmitters also influence changes in the properties and behavior of neurons that persist considerably longer than the evanescent events involved in synaptic transmission. For example, they influence levels of neuropeptide synthesis[111] and of synthetic enzymes such as tyrosine hydroxylase.[112] In 1985 Greenberg and co-workers[113] reported that activation of cholinergic receptors caused extremely rapid induction of specific gene expression. Two genes—the c-*fos* proto-oncogene and actin—that were previously noted to be readily induced by various growth factors were induced within minutes after nicotinic agonists bound to their receptors. Pheochromocytoma cell line PC12, which expressed both nicotinic and muscarinic acetylcholine receptors, was used. This cell line contained nondividing sympathetic neuron–like cells. When the cells were treated with nerve growth factor (NGF) for several days, exposure of differentiated PC12 cells to 100 μM nicotine (an agonist of nicotinic acetylcholine receptors) produced a 10-fold increase in c-*fos* transcription within 5 minutes. This dramatic result is shown in Figure 7.13. The rapid stimulation of c-*fos* and actin by nicotine appeared to be specific because this cholinergic agonist did not significantly affect the expression of several other genes, including α-tubulin (Figure 7.13) or c-*raf*, another proto-oncogene with sequence homology to tyrosine kinases. Also, the level of transcription of another growth factor–inducible nuclear proto-oncogene, c-*myc*, was unaffected for the first hour after treatment but increased slightly after 2 hours of nicotine addition. Other experiments were directed at understanding how nicotine activated c-*fos* and actin gene expression. When PC12 cells were stimulated in the presence of EGTA (which blocked calcium channels), the c-*fos* and actin transcription was not present. These findings indicated that calcium influx through voltage-sensitive Ca^{2+} channels was necessary for nicotine stimulation of gene expression. The fact that cholinergic agonists could trigger the rapid activation of specific gene expression in nondividing neuronal cells was of considerable interest.

Parallel observation of c-*fos* expression in central nervous system tissue after pentylenetetrazole (Metrazole)-induced seizures has now been reported.[114] One hour after administration of Metrazole, c-*fos* mRNA reached a maximum, and it declined to basal levels of expression after 3 hours. The effect was abolished by prior treatment with the anticonvulsant drug diazepam or pentobarbital. Although Metrazole still elicited seizures during the 3- to 16-hour refractory period, reinduction of c-*fos* was relatively refractory. The c-*fos* gene, representing the homologue of the oncogene v-*fos*, encodes a nuclear protein (*fos* protein) that is associated with chromatin and shows DNA-binding activity in vitro.[115,116] Although initial work on c-*fos* expression suggested that it was associated with mitogenesis, it appears that it may be more directly related to cell differentiation than cell division per se. Because *fos* protein appeared in brain cells after seizures, it cannot be involved in the direct production of convulsions but rather may be involved in longer-term adaptation to seizure activity. Since the expression of c-*fos* can be determined with exquisite specificity by Northern blotting or immunofluorescence techniques in single cells, in the future it may serve as a marker to follow the spatial involvement of certain subsets of neurons during and after seizures.

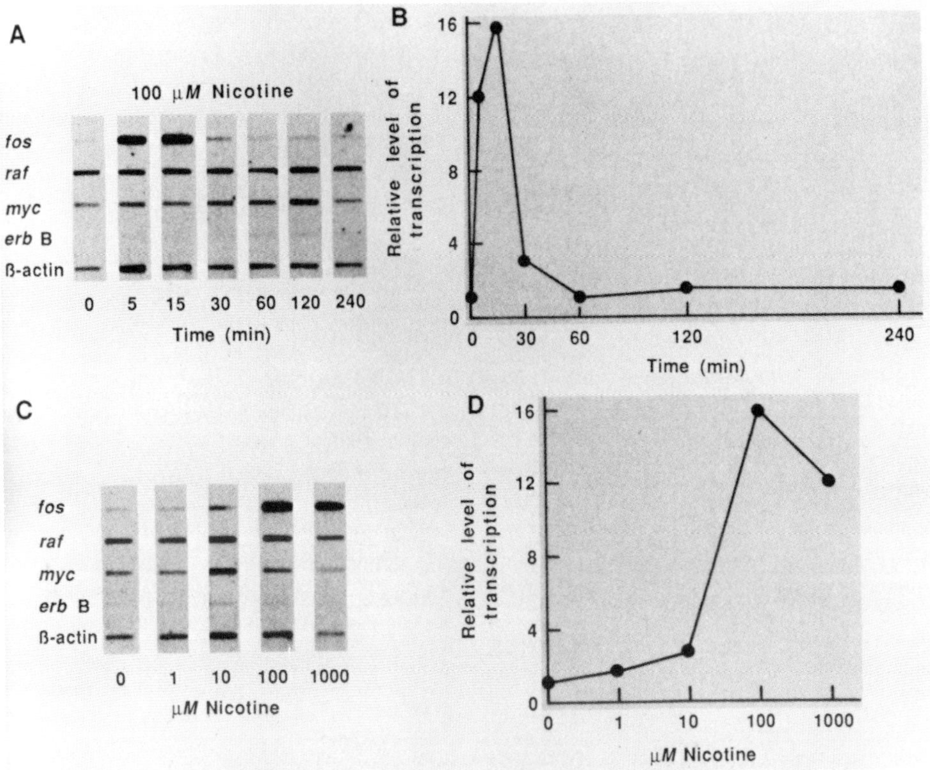

Figure 7.13 Effect of nicotine on transcription in neuronally differentiated PC12 cells. In (A) and (B) nuclei were isolated from neuronally differentiated PC12 cells at various times after treatment with 100 μM nicotine. In (C) and (D) nuclei were isolated from neuronally differentiated PC12 cells that had been incubated for 15 minutes with various concentrations of nicotine. Nicotine was added directly to the complete growth medium from a 1000× stock solution. In this and all subsequent experiments, incubation was at 37°C in an atmosphere containing 7.5% CO_2. Cells were harvested and used for preparation of nuclei. The ^{32}P-labeled RNA runoff transcripts were prepared at approximately 1×10^6 count/min in 1.8 ml of hybridization buffer and incubated for 36 hours at 65°C with 10 μg each of probes pv-*fos*-1, pv-*raf*, pM-c-*myc* 54, pv-*erb*, and Ph-β-actin DNA immobilized on nitrocellulose. The overall level of incorporation of [α-^{32}P]UTP into RNA was unchanged by addition of nicotine to PC12 cell cultures. In (B) and (D), relative signal intensities for c-*fos* were calculated by densitometric scanning of the autoradiographs shown in (A) and (C). PC12 cells were grown on collagen-coated 100-mm dishes in culture medium containing 85% RPMI 1640, 10% horse serum, and 5% fetal bovine serum. To obtain neuronally differentiated cells, cultures were treated for 2 to 4 weeks with complete medium containing nerve growth factor (50 ng/ml). In all experiments, approximately 20×10^6 to 40×10^6 cells were harvested for each determination. *Reproduced from Greenberg et al.,*[113] *with permission.*

Acetylcholine Receptor Genes

Muscarinic acetylcholine receptors that mediate many of the actions of acetylcholine in the central and peripheral nervous systems have also been cloned.[117] Two distinct receptors with pharmacologically different actions (M1 and M2) had previously been defined on the basis of their affinities for the antagonist pirenzepine.[118]

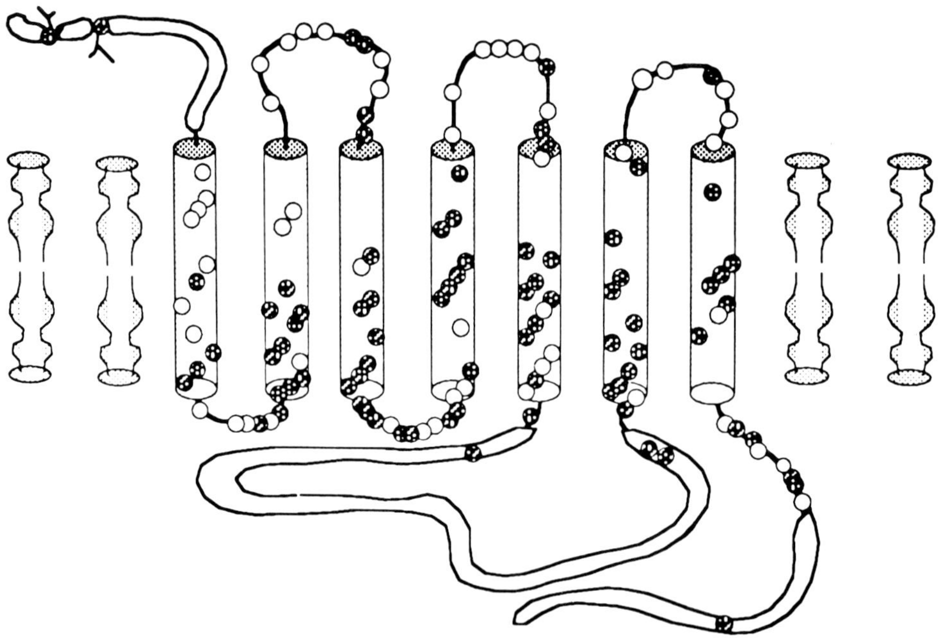

Figure 7.14 Schematic diagram of several receptors with dark circles representing the residues conserved in rhodopsin and β-adrenergic receptor and open circles or boxes representing positions at which there are nonconservative changes in the muscarinic sequences. *Reproduced from Bonner et al.,[117] with permission.*

However, the spectrum of biological functions of the receptors has not always coincided with their classification on the basis of pirenzepine avidity. Muscarinic receptors are members of a large group of neurotransmitter, hormone, and light receptors that function by binding to and activating G proteins. The stimulated G proteins can then either activate or inhibit adenylate cyclase, stimulate phosphatidylinositol turnover, and regulate ion channels—all of which are important steps in a variety of cellular responses. Other general members of this class of receptors are the visual pigment proteins or opsins and the β-adrenergic receptors discussed earlier. Because of the general similarity of functions and structures in this large family of receptors—particularly the fact that they all appear to possess seven transmembrane domains—it seemed reasonable to utilize the remarkable sequence conservation within these domains to probe for other receptor genes. With this approach the muscarinic receptor genes were identified from a rat cerebral cortex cDNA library.[117] The cloning strategy used a sequence that was located near the 5′ end of the coding region, so that cDNAs that hybridized were more likely to contain the entire receptor coding sequences in question. Of interest was the finding that the cloned muscarinic receptor genes contained no introns even with 466-amino-acid gene products. A schematic diagram of the proposed general analogous structures in this gene family, including the rhodopsin, β-adrenergic, and muscarinic receptors, was prepared by Bonner et al.[117] and is shown in Figure 7.14. This illustrates the remarkable structural similarity of these closely related neurotransmitter receptors.

When the cloned muscarinic receptor genes were expressed in mammalian cells using an SV40 viral promoter and the pcD vector, followed by transfection into mouse fibroblasts or other convenient cell recipients, it was found that the transfected cells possessed functional receptors. This was established using labeled muscarinic antagonists including pirenzepine and other competing compounds. When a genomic blot of human DNA was examined with a 516-base probe representing transmembrane regions II to V of the cloned rat muscarinic receptor gene, five to six bands hybridized. Thus a closely related gene family was identified within the muscarinic groups. In situ hybridizations using various tissues such as brain and heart showed different patterns of distribution for these genes and supported a clear microheterogeneity including at least four muscarinic receptors among the group as a whole. This study opened up an exciting new approach to this segment of neuroreceptor physiology. Clearly, using single point mutation techniques and expression of single M1, M2, M3, or M4 receptor genes in mammalian cells will provide precise ways to study neuroreceptor function.

Nicotinic acetylcholine receptor functions have also been studied using transgenic techniques to introduce all four acetylcholine receptor cDNA subunits into the genome of a mouse fibroblast cell line.[119] The nicotinic acetylcholine receptor is a ligand-gated ion channel that binds acetylcholine and mediates transmission of nerve impulses to muscle at the motor end plate. It is composed of four different polypeptide chains with the makeup of $\alpha_2\beta\lambda\delta$. For many years studies of the nicotinic acetylcholine receptor have utilized the electric organ of the marine ray *Torpedo* as an abundant source of acetylcholine receptor.[120,121] Transient expression of acetylcholine receptor in the *Xenopus laevis* oocyte system has allowed the study of several important functional mutations.[122,123]

When *Torpedo* cDNAs for the nicotinic acetylcholine receptor were introduced into a mouse fibroblast cell line by transfection,[119] a clonally stable cell that produced high concentrations of correctly aligned cell surface acetylcholine receptors was formed. This remarkable result was achieved with an SV40 expression vector and an enzymatic gene marker that allowed identification of the successfully transfected cells. This was necessary because DNA-mediated gene transfer by transfection in the calcium phosphate precipitation method usually produces stable integration of the injected genes in only 1 per 10^3 to 10^5 cells treated.

When the transfected cells were studied, it was found that the gene-transferred acetylcholine receptors were functional in terms of ligand binding and other chemical behaviors. This was particularly remarkable in view of the fact that the transfected receptor consisted of multiple subunits that required stabilization in the cell membrane. A schematic diagram of the receptor is shown in Figure 7.15. The ability to study the behavior and structure of cells in which multiple copies of this receptor are available will clearly advance basic understanding of acetylcholine receptor function. During the past 15 years there has been a marked increment in our knowledge of this receptor because it is thought to be the source of antigens and therefore autoimmune response in myasthenia gravis. To be able to take the genes coding for this receptor out of their DNA home and put them into a living cell that will then express them on its own cell membrane in such a way as to stimulate the natural functional receptor is in itself an amazing achievement. The same approach may also be successful in learning more about the subunit topology and interaction of other related receptors such as those for ligand-gated glycine[124] or γ-aminobutyric acid (GABA).

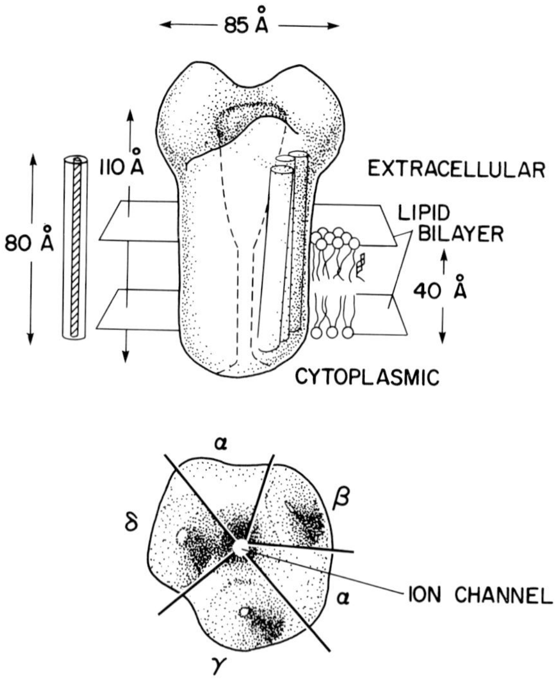

Figure 7.15 Schematic representation of the nicotinic acetylcholine receptor, composed of two α units and separate β, γ, and δ units around a central ion channel.

The GABA Receptor Gene and Its Ligand-Gated Receptor Superfamily

GABA is the major inhibitory neurotransmitter in the vertebrate brain. It mediates neuronal inhibition by binding to the GABA/benzodiazepine receptor and opening up an integral chloride channel. The structure of this receptor has been the subject of extensive study because it contains a spectrum of binding regions for many pharmacologically active drugs, including the benzodiazepines, barbiturates, β-carbolines, and convulsant agents such as picrotoxin. It is known that the GABA receptor is composed of two subunits as $\alpha_2\beta_2$. Purified GABA receptors were obtained from bovine cerebral cortex and isolated by benzodiazepine affinity chromatography. Subsequent cleavage of the receptor protein at the methionine residues with cyanogen bromide produced peptides that could be separated by high-performance liquid chromatography (HPLC) and whose amino acid sequences could be derived.[125] Three of the cyanogen bromide fragments were used to design synthetic oligodeoxyribonucleotide probes. Using these probes, bovine brain and calf cerebral cortex cDNA libraries constructed in λgt10 phage were screened by hybridization. Through this screening one probe was identified that coded for portions of the GABA α subunit and one for part of the β subunit sequence. The polypeptides expressed for the α and β chains were 456 and 474 amino acids, respectively. The complexity of the genes for each subunit deduced is illustrated in Figure 7.16.

Comparative analysis of the sequences and organization of the GABA subunits in parallel with subunits of the nicotinic acetylcholine receptor showed considerable homology (34–62%) in some selected regions. Moreover, a major structural feature shared by both types of receptor was an extracellular β-structural loop formed by cysteines at positions 139 and 153 in the GABA receptor and previously known to be present in the acetylcholine receptor as well. These structural similarities in

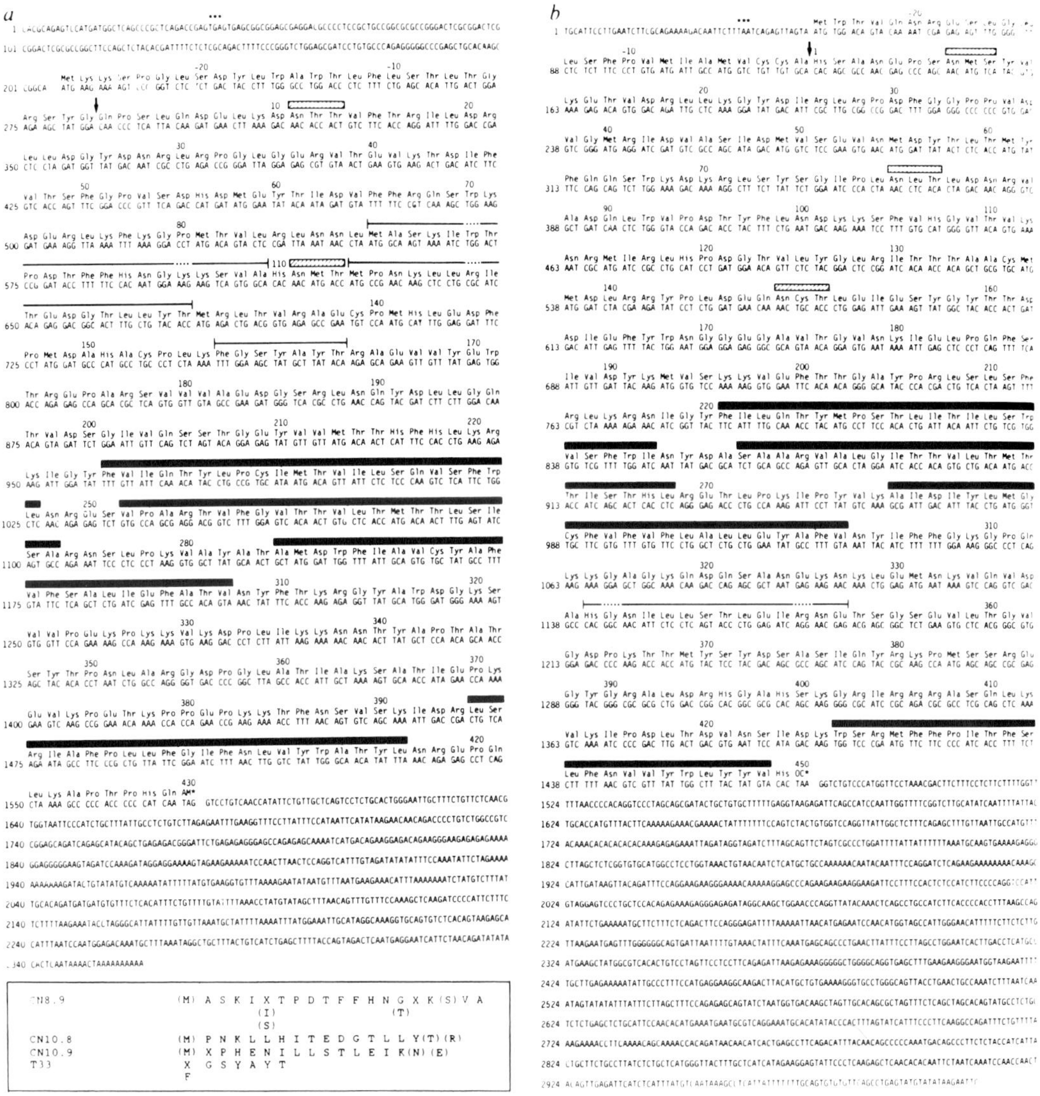

Figure 7.16 DNA sequence and predicted amino acid sequence of the bovine $GABA_A$ receptor subunits: (a) α subunit; (b) β subunit. Chemically determined peptide sequences are outlined; residues that differ from those predicted are indicated by dotted lines. Potential signal sequence cleavage sites are indicated by arrows. Amino acid sequence numbering starts at the proposed mature amino-terminal residue, the presumptive signal sequences being indicated by negative numbering. The proposed membrane-spanning hydrophobic sequences are indicated by solid bars and putative extracellularly located N-glycosylation sites are indicated by hatched bars. An asterisk denotes the first upsteam in-frame stop codon in the 5′ untranslated sequence. *Methods*: Affinity-purified $GABA_A$ receptor protein was subjected to cyanogen bromide cleavage and the gel-eluted α subunit was digested with trypsin; peptides were resolved by HPLC and subjected to gas-phase microsequencing. The sequences used for probe construction and clone identification are shown in the boxed insert. For peptides cleaved by cyanogen bromide, the initial Met is assumed. *Reproduced from Schofield et al.,*[125] *with permission.*

receptors that bind mutually exclusive ions suggest that a superfamily of chemically gated receptors and their genes exists.

As in similar studies, mRNA encoding GABA receptors was efficiently translated in the *Xenopus* oocyte system. Based on the structural and functional similarities to the nicotinic acetylcholine receptor system, a schematic projection for the topology of the $GABA_A$ receptor was proposed. This model is shown in Figure 7.17. A cluster of positively charged side chains is present at both ends of the postulated membrane-spanning domains. These areas are presumed to make up the channel mouth. Such clusters of positively charged residues at the channel mouth would act as an anion-selective filter and increase the driving impetus for

Figure 7.17 Schematic model for the topology of the $GABA_A$ receptor in the cell membrane. Four membrane-spanning helices in each subunit are shown as cylinders. The structure in the extracellular domain is drawn in an arbitrary manner, but the presumed β loop formed by the disulfide bond predicted at cysteines 139 and 153 is shown. Potential extracellular sites for N-glycosylation are indicated by triangles. Charged residues that are located within or close to the ends of the membrane domains are indicated as small circles with charge marked. The site for cAMP-dependent serine phosphorylation, present only in the β subunit, is marked by an encircled P. It is proposed that two copies of each of these subunit structures are complexed in the receptor molecule so as to align the membrane-spanning domains, only some of which will form the inner wall of a central ion channel. *Reproduced from Schofield et al.,*[125] *with permission.*

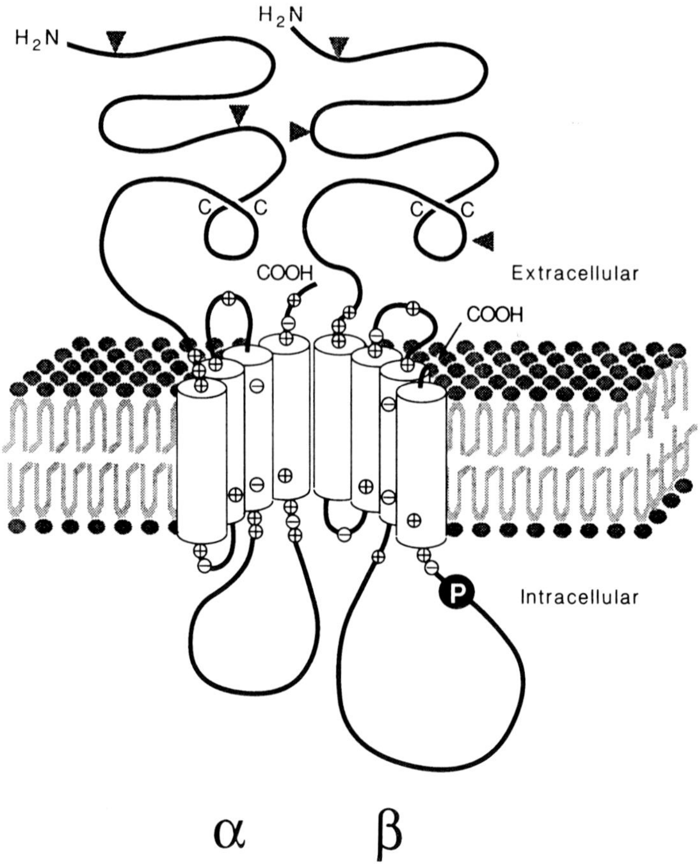

chloride flow when the channel is open. Other allosteric features of the receptor function suggested by this model are open to experimental exploration.

A novel $GABA_A$ receptor subunit that displays high-affinity binding for central benzodiazepine receptor ligands has been described.[126] The primary structures of $GABA_A$ receptor α and β subunits had been deduced from cloned cDNAs,[125,127] and coexpression of these subunits in heterologous systems produced receptors that show much of the pharmacology of their natural neuronal counterparts. These coexpression systems, however, lacked binding sites for benzodiazepines. The isolation of a cloned cDNA encoding a new $GABA_A$ receptor subunit, designated γ_2,[126] has provided a method of extending pharmacological studies in this area.

It is not difficult to perceive what a major leap has been accomplished by cloning the genes for such receptors as those for nicotinic acetylcholine and GABA. When primary amino acid sequence can be correlated with the tremendous amount of knowledge concerning function, much insight will eventually be achieved. Tailor-made, effective drugs or other strategies may be developed by working directly from well-established molecular models such as those discussed here.

Olfactory cDNA

Among the most remarkable events in neurotransmission are the primary events involved in sensory transduction as in the olfactory response—odor recognition and odor signal transmission. This process occurs in the receptor neurons of the olfactory neuroepithelium, a specialized tissue composed of cells enclosing bipolar neurons with apical processes extending to the ciliated luminal surface. A mucous layer overlying the ciliated mucosal surface is thought to be important in solubilizing and concentrating airborne odors. Transduction of odorant-specific messages appears to be performed by odorant-specific protein receptors in the ciliary membranes.[128] Odorant-binding activities have been identified in olfactory tissue extracts.[129] Molecular cloning techniques have also been directed at characterization of olfactory-specific tissues. cDNA was constructed from mRNA obtained from the olfactory mucosa of the frog *Rana pipiens*.[130] Of 2400 clones screened, 10% appeared to be olfactory specific in that they showed strong hybridization to a probe derived from olfactory mRNA and negligible hybridization to mRNA from respiratory epithelium. The clones with the highest binding to olfactory mRNA were further analyzed and showed similar restriction maps. An apparent olfactory-specific protein identified during this study bore great similarities to a family of serum transport proteins, including bovine β-lactoglobulin and retinol-binding protein. These proteins are thought to bind and transport small molecules such as retinol or vitamin B–like molecules in serum or milk. The olfactory protein identified by this technique appears to be present in mucus and presumably plays a general role in presenting olfactory stimuli to the underlying olfactory epithelium. These studies again demonstrate the power of recombinant techniques for the investigation of complex biological problems.

Localization of the Malignant Hyperthermia Locus to Human Chromosome 19q12–13.2

One of the most remarkable achievements in modern molecular pharmacology is the apparent identification of the genetic locus responsible for malignant hyperthermia (MH), an inherited human skeletal muscle disorder and one of the major causes of death due to anesthesia.[131,132] The incidence of MH varies from 1 in 12,000 in children to 1 in 40,000 in adults. It is triggered in susceptible individuals by all commonly used inhalational anesthetics and is accompanied by profound acceleration of muscle metabolism, contractures, extreme hyperthermia, and tachycardia. An almost identical disorder has been recorded in pigs.[133,134] The porcine syndrome has been extensively studied, and it is known that the locus controlling expression of porcine MH is linked to the glucose phosphate isomerase (GPI) locus. In humans, GPI has been mapped to the q12–13.2 region of chromosome 19.[135] McCarthy and co-workers[132] studied genetic linkage in several extended Irish pedigrees in which MH syndrome segregated as an autosomal dominant trait. Results showed a clear-cut linkage between MH and DNA markers from the GPI region of human chromosome 19 with a maximum log likelihood ratio (lod score) of 5.65.

In parallel with this study, MacLennan and co-workers[136] showed that the ryanodine receptor gene may be a candidate for predisposition to MH. Because Ca^{2+} is one of the chief regulators of muscle contraction and metabolism, the principal defect in MH is postulated to lie in Ca^{2+} regulation, and a number of studies have provided evidence for a basic defect in the Ca^{2+} release channel of the sarcoplasmic reticulum[137–139] in the disorder. The ryanodine receptor is thought to represent the Ca^{2+} release channel of sarcoplasmic reticulum, so it was of great interest that the receptor mapped to region q13.1 of chromosome 19, in close proximity to the MH susceptibility locus. These studies show the predictive and correlative power of molecular genomic analysis applied to a clinical pharmacological problem.

DNA Analysis in Endocrine Disorders

Steroid 21-Hydroxylase Deficiency

A number of endocrine disorders have been recognized as inherited or genetically determined diseases. Congenital adrenal hyperplasia is a relatively common inherited disorder, estimated to occur once in every 5000 births. In 95% of these patients the disorder is associated with 21-hydroxylase deficiency, which results in impairment of cortisol synthesis and accumulation of androgens associated clinically with virilization. The genetic defect has been localized to the short arm of chromosome 6, within the human leukocyte antigen (HLA) gene complex. However, the deficiency is in cytochrome P-450_{c21} and not in the gene for 21-hydroxylase itself.[140] Two genes appear to be involved, but lack of one is enough to produce congenital adrenal hyperplasia. White et al.[140] studied six heterozygous affected patients and found that 21-hydroxylase deficiency results from the deletion of a specific cytochrome P-450 gene. Moreover, studies of HLA locus linkage have indicated that the C4 and P-450_{c21} genes are closely linked but not similar in sequence, which does not support ancestral gene reduplication. The significance of their close association is not clear, and there is no obvious reason for the presence of genes coding for the steroidogenic enzyme within the major histocompatibility complex region.

Subsequent studies of steroid 21-hydroxylase deficiency identified two 21-hydroxylase genes, CYP21 A and B, which are adjacent to the complement C4A and C4B gene loci. It was demonstrated that deletions of CYP21 B account for about one-fourth of the classical 21-hydroxylase deficiency alleles.[141,142] Such alleles were presumably produced by unequal crossing over during meiosis due to presence of the adjacent CYP21 A pseudogene. A different mechanism—gene conversion—may also result in steroid 21-hydroxylase deficiency.[143]

Premature Ovarian Failure

Familial premature ovarian failure has been associated with an interstitial deletion of the long arm of the X chromosome.[144] A family was studied in which four women showed menstrual irregularities and partial deletion of the long arm of the X chromosome. Three of the four women had premature ovarian failure. Chromosomal examination suggested a terminal Xq deletion, but DNA hybridization studies with X chromosomal probes showed an interstitial Xq deletion. It was of some interest in this situation to consider how the Xq deletion could impair ovarian function. One possibility is that one or more genes in the deleted segment must be present in two copies to ensure normal function; however, because of the inactivation of the X chromosome, only one copy of these genes functions even in normal (XX) females. Alternatively, Xq deletions might impair ovarian function through a position or alignment effect, that is, a change in gene expression due to an alteration in the position of neighboring or flanking DNA sequences. Also, the effect of Xq deletion could be induced by altered X chromosome activity during mitosis or meiosis. Thus heterozygosity for the Xq deletion could reduce the ability of oocytes to complete meiosis and possibly lead to premature ovarian follicle atresia. Whatever the final association, the studies of the XXq defect emphasize the necessity for both careful chromosomal and DNA hybridization studies using X chromosomal probes for a better understanding of such rare disorders.

Hyperparathyroidism and Parathyroid Adenomas, and Multiple Endocrine Neoplasia

Primary hyperparathyroidism is a common disorder, but its underlying etiology is still not clear. Most patients show overactivity in the form of an adenoma affecting a single parathyroid gland. Whether parathyroid adenomas associated with hyperparathyroidism were clonal or of multicellular origin was examined by Arnold and co-workers,[145,146] who found evidence for clonal proliferation of cells in each case examined. These findings indicated that DNA alterations involving the parathyroid hormone (PTH) locus might determine the production of such tumors. The technique of restriction fragment length polymorphism (RFLP) in an X-linked gene (hypoxanthine phosphoribosyltransferase) was used to examine clonality of eight parathyroid adenomas in women. Six of the eight showed DNA hybridization patterns of monoclonality and two manifested equivocal patterns. The clonal abnormality demonstrated in this study implies that the underlying disturbance resulting in adenoma or tumor formation proceeded from a rare event such as a somatic mutation.

Subsequent studies by the same group revealed that in one parathyroid adenoma studied, a DNA rearrangement occurred at the PTH locus and the rearrangement separated the PTH gene's 5′ flanking region from its coding exons, possibly placing a newly adjacent gene under the influence of PTH regulatory elements.[147] Furthermore, the DNA that recombined with PTH mapped to 11q13, the known chromosomal location of several oncogenes as well as the gene for multiple endocrine neoplasia type 1.[148,149] The identification of the same sort of DNA abnormality—loss of heterozygosity—along with allelic deletions on chromosome 11 in the parathyroid adenomas associated with multiple endocrine neoplasia type 1 was a major breakthrough in the eventual understanding of this disorder. A linkage to the oncogene INT2 was also demonstrated in these studies.[149]

A subsequent study identified the genetic locus for multiple endocrine neoplasia type 2A near the centromere of chromosome 10.[150] This involved RFLP analysis of 130 members of 11 families of European and North African origin. The RFLP analysis was much more useful in predicting carrier status of subjects than either the conventional endocrine challenge or search for medullary carcinoma of the thyroid and/or pheochromocytoma.

Identification of Hormonal mRNAs in Tissues

As mentioned in the section on polypeptide hormones, there has been considerable interest in where hormones or active peptides are produced and what chemical or other activating stimuli are required for expression and translation of their mRNAs. One group of hormones that has not been completely defined in terms of basic function are the somatomedins or insulin-like growth factors. These hormones are produced in many organs and tissues, but the cells that make them in vivo have not been well defined. Using in situ hybridization, somatomedin C and insulin-like growth factor I mRNAs were localized to connective tissue cells or cells of mesenchymal type in organs and tissues from human fetuses.[151] The mRNAs were present in perisinusoidal cells of liver, perichondrium of cartilage, sclera of eye, and other connective tissue locations. A number of findings indicate that the somatomedins or insulin-like growth factors are synthesized in many fetal tissues. Their wide distribution in the fetal tissues examined suggests that they may function as local paracrine growth factors that are capable of acting on multiple cell types and do not merely affect skeletal growth as previously postulated.[152]

The levels of mRNA for several related glycoprotein hormones, including human LH, CG, TSH, and FSH, were examined in both normal and pituitary adenoma tissues.[153] Subunit mRNAs were expressed at relatively low levels in normal pituitary tissues but at higher levels in pituitary gonadotrope and thyrotrope adenomas. In placental tissues there was abundant expression of the CG β gene but no expression of the LH β gene, indicating that there were tissue-specific regulatory sequences in the upstream promoter recognition site of the CG β gene. These studies demonstrated that oligonucleotide probes could distinguish levels of expression of mRNAs for structurally related glycoprotein hormone subunits and could add to eventual understanding of tissue-specific gene expression. A similar study using cDNA probes for growth hormone was performed with five samples of pituitary adenoma cells using in situ hybridization.[154] As expected, growth hormone mRNA showed marked elevations.

Tissue localization of hormonal mRNA in vitro has also been applied to angiogenin, a 14-kd polypeptide that was originally isolated from a human adenocarcinoma cell line in conditioned media.[155] This peptide appears to be an inducer of vascular growth and might be useful in understanding the mechanisms of new blood vessel formation. Using a cDNA probe for angiogenin, the tissue distribution of mRNA was examined in the developing rat.[156] Angiogenin mRNA was found predominantly in adult liver; expression of the angiogenin gene was apparently regulated by development, since levels were low in the fetus, moderate in the neonate, and maximal in the adult. No increment of mRNA was found in human colon or hepatoma cancer cells, and it was thought that the pattern of mRNA for angiogenin was not temporally related to vascular development in the rat. It therefore appeared that the primary function of this peptide in vivo is related to processes other than the regulation of blood vessel growth. This study illustrates how in situ tissue hybridization studies of mRNA distribution can help to elucidate the possible function and evolution of newly discovered hormones or polypeptides.

C-*myc* mRNA in Thyroid Tissue

As originally found by Kelly et al.[157] when normal human lymphocytes were stimulated with mitogens, Hirayu et al.[158] found that c-*myc* mRNA expression was enhanced when normal thyroid follicular cells were established in monolayer culture. In parallel, β-actin mRNA was also enhanced. However, there was no similar increase in expression of another oncogene, c-*ras*, during initiation of primary thyroid cell culture. These findings were assumed to be manifestations of thyroid cell differentiation initiated by the culture technique, and they are another example of proto-oncogene expression in normal cells during activation or differentiation. The c-*myc* proto-oncogene was also studied in thyroid adenomas and thyroid carcinomas for levels of mRNA transcripts.[159] The c-*myc* mRNA level was increased in these tumor tissues compared with that in normal thyroid. Cultured normal thyroid cells did not appear to contain detectable c-*myc* mRNA, but TSH stimulated a marked increase in c-*myc* in time-dependent fashion.

Insulin-Dependent Diabetes Mellitus—Molecular Approaches

Type I or insulin-dependent diabetes mellitus (IDDM) affects primarily children and young adults and is the most common endocrine metabolic disorder of childhood. Pathogenesis of this disorder is still unknown, although virus or other environmental factors have been implicated.[160–162] The disease seems to run in families, but the exact mode of inheritance is not completely known. A number of studies suggest that what appears to be an adverse autoimmune reaction against self β cells may be controlled by HLA genes in the major histocompatibility complex. More than 90% of IDDM patients are HLA-DR3 and/or HLA-DR4, and these groups account for the vast majority of IDDM. Their combined frequencies in the normal population are nearly 60%.[163,164] The HLA-DR3/4 genotype itself is associated with the highest risk of developing the disease. However, the concordance rate for IDDM among monozygotic twins remains below 50%,[165] and the chance of an HLA-DR–identical sibling developing IDDM is only 12% to 24%.[164] There is a great deal of interest in these associations, with the idea that there may be other genes closely linked to HLA-DR that confer an even greater risk for IDDM development. With this in mind, efforts have been focused on obtaining higher resolution than merely HLA-DR3/DR4 typing; HLA-DQ region β-chain DNA fragments have been identified that differ between HLA-DR–identical healthy and IDDM patients.[166–169] These findings suggested that such DNA polymorphism might best define susceptibility to IDDM. Based on differential expression of a 3.7-kb *Bam*HI fragment between HLA-DR–identical controls and IDDM patients, cloning of this fragment from an HLA-DR4–positive chromosome of a healthy subject was undertaken in the hope of defining a better probe to use in RFLP analysis.[170] This approach established that the polymorphism was due to a structural difference in an HLA-DQ β chain gene. Using more discrete probes, it was found that the IDDM susceptibility gene was situated closer to HLA-DQ than HLA-DR. Class II antigens in the DR and DQ designations confer restriction in antigen presentation, perhaps by direct binding of processed antigen.[171] A map of class II MHC genes on the short arm of chromosome 6 is shown in Figure 7.18. Precise definition of the molecular structure of HLA-D class II molecules may eventually explain how some antigens induce an abnormal autoimmune response to pancreatic β cells resulting in IDDM. Similar analyses of *H-2* genetic linkage of diabetes in the nonobese diabetic mouse (a possible model for human IDDM) have indicated that several separate recessive *H-2*–linked loci appear to be required for disease expression.[172] All of these observations support polygenic control of diabetes rather than a single diabetogenic gene in this animal model.

An intriguing aspect of the possible autoimmune mechanisms involved in diabetes is that 60% to 85% of newly diagnosed patients with type I diabetes have islet cell–specific autoantibodies.[173] The appearance of islet cell antibodies implies that β cells are undergoing destruction, and this may occur as much as 8 years before clinical onset of disease.[174] Using transgenic mice, several lineages have been established that carry integrated copies of a hybrid gene representing the upstream regulatory nucleotide sequences of the rat insulin II gene linked to protein coding for the SV40 large T antigen.[175,176] Direct immunohistochemical analysis of pancreatic tissue from these transgenic mice indicated that large T antigen was expressed in a tissue site–specific way, being present exclusively in the β cells of the islets of Langerhans. Expression of large T antigen in these areas in the transgenic animals results in β cell tumors. Interestingly, even though the large T SV gene was integrated within the transfected part of the rat insulin gene, no tolerance to the SV40 large T gene occurred and the mice produced serum autoantibodies against the integrated large T antigen gene product. Moreover, no strict linkage to any mouse *H-2* haplotype was observed. Since these transgenic mice often developed highly vascularized β cell tumors, it seemed possible that the autoantibodies

Figure 7.18 Map of class II MHC genes on the short arm of chromosome 6. The exact orientation of individual genes with respect to each other is not known, although the direction of transcription is indicated by the arrows. The location of the DO subregion is not known, and it may be telomeric to the DR subregion rather than between the DP and DQ subregions as shown in this diagram.

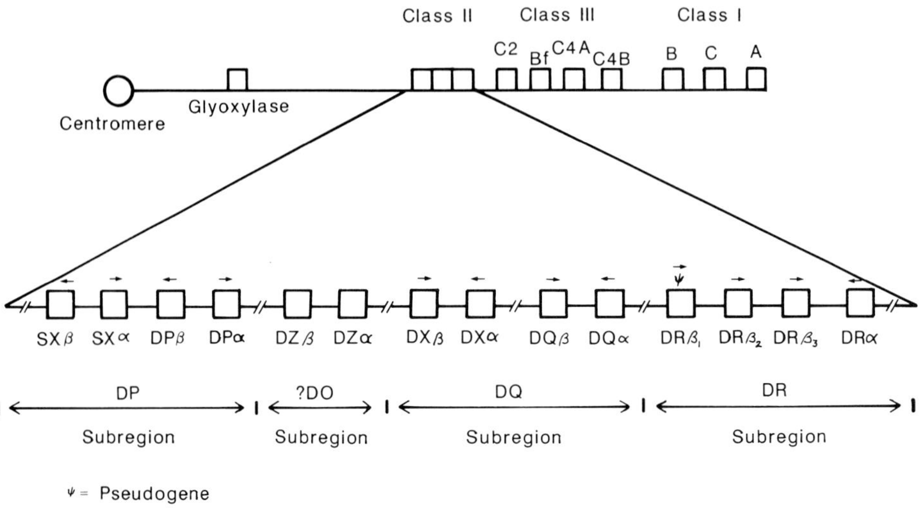

to SV40 large T antigen were related to the expression of the latter in the β-cell tumors; however, no such relationship could be established. Furthermore, the same transgenic mice frequently developed vascularization, fibrosis, and extensive lymphocytic infiltration of the islets of Langerhans. Actually, a wide spectrum of islet morphological changes ranging from normal appearance to extensive destruction within islets was observed in these animals and was not directly correlated with development of β cell tumors. The histological changes were thought to be cell-mediated immune reactions against large T antigen expressed in the β cells of the islets of Langerhans. Examples of the histological findings are shown in Figure 7.19. These studies show that transgenic introduction of new genes into experimental mice can provide evidence for lack of tolerance to introduced gene products and an apparent autoimmune reaction to the original SV40-induced large T antigen. Why some transgenic gene products and not others are treated as self is still not clear; however, the approaches outlined in this study represent a fascinating new direction for the study of autoimmune reactions in both mouse and human.

Studies of transgenic mice to determine how the MHC affects diabetes and the immune response continue to produce exciting new insights. It has been demonstrated that a class I histocompatibility gene, $H\text{-}2k^b$, linked to the rat insulin promoter was overexpressed in pancreatic β cells of transgenic mice. The mice, whether syngenic or allogenic for the transgene, developed insulin-dependent diabetes without detectable T cell infiltrates.[177] These observations suggested a direct and possibly nonimmune role for the transgenic class I molecules in the impairment of β cell function.

As more insight was gained into the possible relevance of structural features of the presumed peptide-binding groove in the major histocompatibility complex, attention was directed to the role that the amino acid at position 57 might play in diabetes.[178] This was tested using transgenes encoding modified 1-A β chains or normal I-E α chains.[179] A similar approach was used by other workers.[180] The effects of the introduced transgenes in preventing diabetes appeared to be related not to a single amino acid in the peptide-binding groove but rather to differences in overall structure produced by the various transgenes.

Atriopeptin: A Cardiac Hormone

Perhaps no other recently discovered hormone owes its scientific and clinical attention more to recombinant DNA techniques than atriopeptin, the cardiac hormone. This peptide, which is stored in the atrial cardiocyte, is capable of potent, selective effects on fluid and electrolyte balance as well as on blood pressure. Hundreds of studies of its physiological effects have appeared since its discovery in 1983.[181] The atrial gene codes for a pre-prohormone of 153 amino acids. Relatively large amounts of peptide are produced; 1% to 5% of the total atrial tissue mRNA codes for the atrial peptides. The hydrophobic 24-amino-acid leader sequence is cleaved off within the perinuclear granules of the atrial myocytes, producing the 126-amino-acid prohormone, atriopeptigen, which is stored in the perinuclear granules of atrial myocytes.[182–184] Detectable levels of atriopeptin are generally present in the circulation, suggesting a steady state of release at relatively low levels. Increments in plasma atriopeptin levels have been observed with atrial stretch induced by volume expansion,[185] constrictor agents that increase atrial pressure,[186] atrial tachycardia,[187] and high-salt diets. Although the peptide is stored as the high molecular weight prohormone, the primary form in plasma is the low

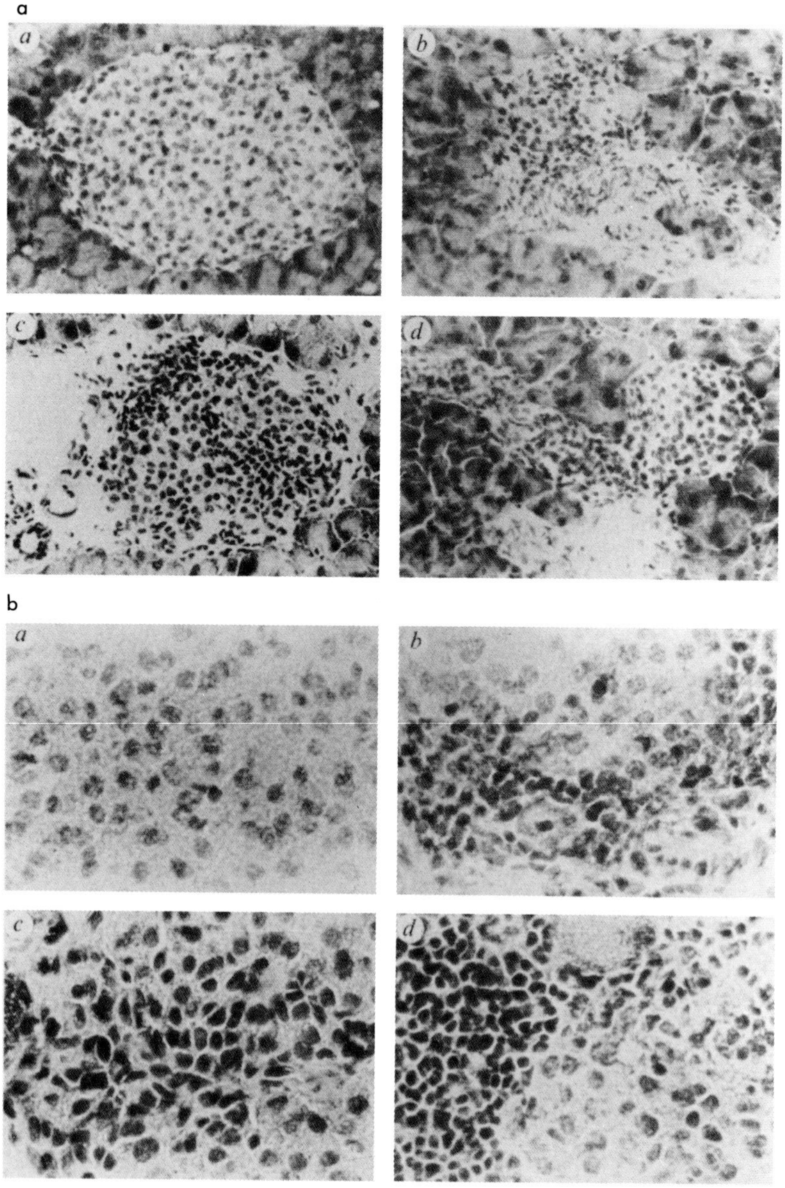

Figure 7.19 Pancreas histopathology in mice producing antibodies against T antigen. Two mice bearing autoantibodies (2365 and 3217) and one with induced antibodies (1812) against SV40 large T antigen were analyzed by cryostat sectioning of the pancreas followed by histochemical staining with hematoxylin and eosin. Mouse 2365 was 8½ months old at the time of this analysis. The pancreas had cell hyperplasia but no distinct tumors (either visibly or histologically). Mouse 3217 was 6 months old and did not have significant β cell hyperplasia or solid tumors, although enlarged lymph nodes were present. Mouse 1812 was immunized with 20 mg of purified SV40 large T protein at 6 weeks of age and analyzed at 11½ months of age. No tumors or hyperplasia was noted in mouse 1812. (A) Characteristic morphology of pancreatic islets in these mice; a, normal, nontransgenic mouse islet; b, islet from mouse 1812, who was immunized with Tag; c, islet from mouse 2365; d, islet from mouse 3217, showing an adjacent aggregation of lymphocytes. All magnifications were 160×. (B) High-magnification (400×) photos showing the interior regions of pancreastic islets; a, normal islet full of β cells; b, islet from mouse 1812, immunized with exogenous large T at a different location (intraperitoneally); c, islet from mouse 2365; d, segment of a lymph node found within a hyperplastic islet in the pancreas of mouse 2365. *Reproduced from Adams et al.,*[176] *with permission.*

PEPTIDE

	Sequence (residues 1–28)
a.	SerLeuArgArgSerSerCysPheGlyGlyArgIleAspArgIleGlyAlaGlnSerGlyLeuGlyCysAsnSerPheArgTyr
b.	(5) Ser ———————— Tyr
c.	(3) ArgArgSer ———————— Tyr
d.	(4) ArgSer ———————— Tyr
e.	(1) SerLeuArgArgSer ———— (12) Met ———— Tyr

Current Terminology:	Proposed Nomenclature:
PEPTIDE	
a. Cardionatrin I	Atriopeptin 28 (AP-28)
b. Atriopeptin III	Atriopeptin 24 (AP-24)
c. $ANF_{101\text{-}126}$	Atriopeptin 26 (AP-26)
d Auriculin B	Atriopeptin 25 (AP-25)
e. α-hANP	h-Atriopeptin 28 (hAP-28)

Figure 7.20 Comparative amino acid sequences and nomenclature of the low molecular weight peptides isolated from atrial extracts. All isolated peptides contain the identical core 17 amino acids within a cysteine disulfide bridge.

molecular weight carboxyl-terminal fragment atriopeptin-28.[188] The structure of atrial peptide was elucidated by peptide sequence analysis and by subsequent isolation of cDNA to express active products in bacterial vectors. The structural and physiological knowledge of atriopeptin was thus greatly accelerated by direct application of recombinant techniques.

A number of peptides with natriuretic, diuretic, and vasodilator properties have now been described and their structures identified. These are shown in Figure 7.20. However, the demonstration of a single form of circulating peptide[188] suggests that this product is the major active form of the hormone, atriopeptin-24. The hormone itself has a number of potent physiological effects, including rapid and transient natriuresis, diuresis, and moderate kaliuresis. It also suppresses elevated plasma levels of renin and relaxes blood vessels directly. It is thought that the major actions of atriopeptin involve loss of fluid and electrolytes and a reduction in vascular tone. Because of the potency of atriopeptin as a diuretic and natriuretic as well as a vasodilator, a number of studies have been initiated in humans. Atriopeptin-induced natriuresis would be expected to spare potassium excretion because it inhibits aldosterone synthesis. Intravenous administration of human atriopeptin to normal or hypertensive subjects caused a moderate but transient decrease in arterial blood pressure with an increase in heart rate.[189,190] The immediate hypotensive effect suggests that the peptide acts directly on arterial tone rather than decreasing vascular volume by its diuretic effect. The potential therapeutic role for this newly discovered hormone might be expanded, particularly in critical care situations. The rapid titration of therapeutic responses and the transient duration of action (3 to 6 minutes) could be advantageous in certain clinical situations. The potential sparing of unwanted potassium excretion might be useful in treating severe volume overload. Also, the usefulness of some antibiotics and chemotherapeutic agents is limited by their nephrotoxicity, and concurrent atriopeptin administration might limit such nephrotoxicity.[191]

Systems for in vitro assay using primary cultures of atrial myocytes have now been established with the aim of examining factors that stimulate specific mRNA synthesis of atriopeptin.[192] Atrial and ventricular tissue levels of atrial natriuretic peptide mRNA are markedly elevated in cardiac tissues of patients with dilated cardiomyopathy. These observations show the physiological basis for the marked elevations of plasma atrial natriuretic peptide in patients with advanced heart failure.[193,194]

Gastrointestinal Hormones

Gastrin

Gastrin, one of the most extensively studied of the gastrointestinal hormones, is another example of the contributions of recombinant DNA techniques during the past two decades to fundamental understanding of biological functions. Gastrin stimulates gastric acid secretion, activates gastric movement, and can serve as a mitogenic factor for gastrointestinal epithelial cells.[195] Initial studies of the hormone identified several biologically active peptide forms, including gastrins G-34, G-17, and G-14 components, which contain 34, 17, and 14 amino acids, respectively, and have the same COOH-terminal polypeptide. A number of recombinant DNA approaches were used by several groups to establish the sequence and eventually clone the human gastrin gene.[196–199] Oligo(dT)-primed cDNA copies of mRNA coding for the human gastrin precursor were used in one study,[197] in which the probe was constructed from RNA derived from a human gastrinoma. A primer for cDNA synthesis was synthesized based on the known gastrin-specific amino acid sequence Try-*met*-Asp-Phe. Using this approach, the entire 303-nucleotide sequence of the coding region of the gene plus the complete 3′ untranslated region of 102 nucleotides was established. These studies showed that G-34 (amino acids 59 to 92) and G-17 (amino acids 76 to 92) are homologous and suggested that the gastrin gene itself could have evolved through gene duplication from an ancestral gene coding for an original stretch of 38 amino acids. Also of great interest was the observation that G-34 is approximately 50% homologous to the transforming protein ("middle T") of polyomavirus. Since both gastrin and the middle T protein can stimulate cellular growth, it is possible that the homology arose by incorporation during evolution of animal DNA sequences into simian virus 40–like genomes. The gastrin gene isolated from a human library derived from fetal liver was found to contain 4100 base pairs with two long intervening sequences, one of which was 3500 base pairs in length[198] and situated in the noncoding region. The second intron was close to the DNA sequence corresponding to the amino-terminal part of the region containing the processing site for gastrin 17, the principal hormone form. A diagram of the arrangement of introns and exons in the gastrin gene is shown in Figure 7.21.

Subsequently, mechanisms of gastrin production in human gastrinomas have been compared with that in normal gastric antrum by analysis of transcription initiation codon points for mRNA.[200] There was no evidence for genomic rearrangements associated with gastrinoma formation, and the same gene product was translated in normal and tumor tissues.

Additional studies with gastrin and other peptides have indicated that the active hormonal peptide is one of many gastrointestinal peptides that require carboxyl-terminal amidation for biological activity.[201,202] Posttranslational forms of gastrin

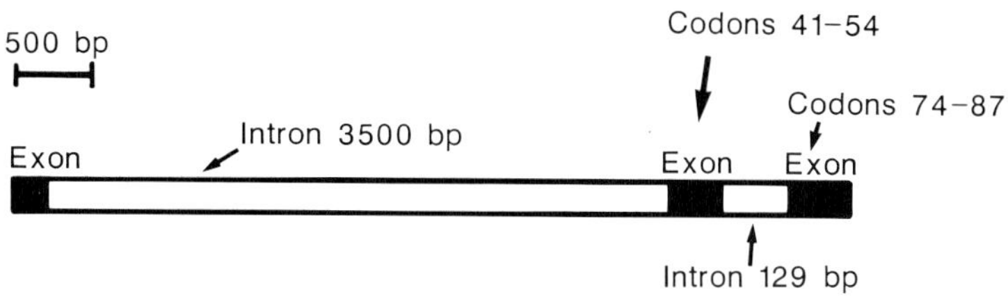

Figure 7.21 Schematic representation of gastrin gene showing large intervening segment (intron) separating transcribed regions of the gene. The exons for codons 41 to 54 and 74 to 87 code for active hormone products.

from patients with Zollinger-Ellison tumors included multiple molecular forms of immunoreactive glycine-extended gastrin that appeared to be altered in this condition.[203] The changes did not represent differences in translated gene products but rather alterations in the ratios and proportions of posttranslational modifications by amidation. The amidation reaction involves addition of glycine residues after the carboxyl-terminal amino acid. How the posttranslational abnormalities are related to the clinical manifestations in Zollinger-Ellison syndrome is not yet clear.

Cholecystokinin and Other Gastrointestinal Peptides

The two polypeptide hormones gastrin and cholecystokinin are structurally and functionally related because they share a common COOH-terminal pentapeptide (GWMDF) that contains the biological active site. Whereas gastrin regulates gastric acid release and possibly growth of the gastrointestinal mucosa, cholecystokinin controls gallbladder contractions and secretion of pancreatic enzymes. The structural genes for human cholecystokinin (like those for gastrin) have now been isolated and characterized.[204] Sequence comparisons have shown extensive homology between pre-progastrin and pre-cholecystokinin,[205] supporting the concept that the genes coding for gastrin and cholecystokinin may have had a common ancestral precursor. To examine linkage between the two genes, Lund et al.[206] examined a panel of human-rodent somatic cell hybrids to determine the chromosomal locations of the genes for cholecystokinin and gastrin. They used small 0.6-kb and 1.6- or 0.9-kb restriction fragments form cDNA sequences specific for the gastrin and cholecystokinin genes. Hybridization with genomic DNAs from the human-rodent hybrid cells, in which the chromosomal origin of the human DNA was known, established that the genes for gastrin were on human chromosome 17 and those for cholecystokinin were on chromosome 3. It is thought that after their derivation from a common ancestral precursor gene, the genes could have been separated by gene duplication followed by translocation of one or both. The region in chromosome 3 where the cholecystokinin gene is located contains an alteration consistently detected in small cell lung carcinoma,[207] a neoplasm often associated with activation of genes coding for biologically active peptide hormones.[208] Because cholecystokinin production is occasionally seen with small cell lung cancer, it would be of interest to explore tumor-associated gene activation mechanisms.

Human Vasoactive Intestinal Polypeptide

Vasoactive intestinal polypeptide (VIP) is a 28-amino acid peptide first isolated from porcine small intestine and characterized as a vasodilatory hormone.[209] Subsequently, this peptide has been identified throughout the central and peripheral nervous systems and has been postulated to act as a neurotransmitter and possibly also as a neuroendocrine releasing substance.[210] Sequence analyses of cDNAs encoding VIP have shown that the peptide is produced as part of a larger precursor molecule, and the primary structure of the human VIP gene has now been established.[211] Although the peptide has been characterized in a wide variety of tissues, including the central nervous system, very little is known about factors that regulate its gene expression. However, it is known that in certain human neuroblastoma cell lines, VIP gene expression is stimulated by cAMP. These baseline observations provided the impetus for a remarkable investigation linking VIP gene elements to responsiveness to cAMP.[212] Various fusion genes were constructed that contained portions of the 5′-flanking region of the human VIP gene ligated to nucleotides coding for the structural sequences of the bacterial chloramphenicol acetyltransferase (CAT) gene. These fusion genes were then transfected into cell line PC12 and the transcriptional activity of the fusion genes was assayed. Forskolin, an adenylate cyclase–activating agent, was used to increase transcription of fusion genes containing the human VIP 5′-flanking sequences. Analysis of specifically prepared deletions showed that the 17-bp region of nucleotides −86 to −70 upstream from the VIP transcription origin control led to cAMP responsiveness in the VIP gene product. This study is an excellent example of the precision and depth of insight into fundamental biological process that can be gained by the use of recombinant approaches.

Peptide-Binding and Complementary RNAs

A novel idea has been suggested by Blalock and Smith[213] in relation to various patterns in the genetic code. When read in the 5′-to-3′ direction, codons for hydrophilic and hydrophobic amino acids are generally complemented inversely by codons for hydrophobic and hydrophilic amino acids, respectively. The average tendency of codons for uncharged or neutral amino acids is generally complemented by other codons for uncharged or neutral amino acids. The same general pattern appears when the complementary codon is read "backward" in the 3′-to-5′ direction.[214] Moreover, it appears that this pattern results in the interaction of peptides specified by complementary RNAs regardless of whether amino acids are assigned in the 5′-to-3′ or 3′-to-5′ direction. Experiments were done in which peptides were synthesized according to either the 5′-to-3′ or 3′-to-5′ direction. The synthetic peptide products were then tested in vitro using a solid-phase binding assay with the results shown in Figure 7.22. Peptides synthesized by reading the codons in either direction showed approximately equivalent binding avidities. Since, regardless of the reading direction, the second nucleotide or base of the codon never changes, it apparently is responsible for the hydropathic anticomplementarity of amino acids that are specified by complementary codons. There appears to be a virtually perfect interchange of hydrophobic and hydrophilic amino acids whether the complementary codon is read 5′ to 3′ or 3′ or 5′. These observations raise some interesting questions concerning the genetic code and evolution. Perhaps before the beginning of a single direction (5′ to 3′) of nucleic acid reading frames, an amino acid from the same hydropathic group would be present and thus the resulting peptides would be roughly similar in conformation whether nucleic acids were read 5′ to 3′ or 3′ to 5′. Although an exact mechanism or enzyme cannot be directly implicated at present, it is conceivable that the same general principle could have been involved in inducing hormone receptors on cell mem-

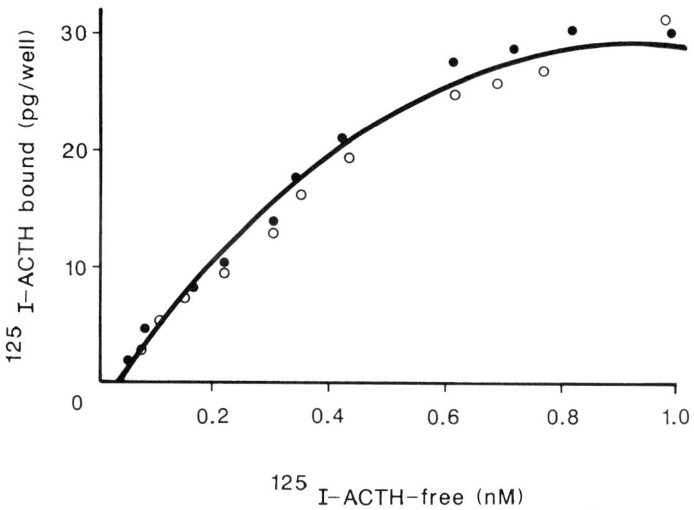

Figure 7.22 Binding of ^{125}I-labeled ACTH to the 5′-to-3′ or 3′-to-5′ complementary-mRNA-strand peptides. Dilutions of ^{125}I-labeled ACTH were incubated in wells coated with 5′-to-3′ complementary-mRNA-strand peptide (●) or 3′-to-5′ complementary-mRNA-strand peptide (○), and the amount of specifically bound radiolabel was determined. Unlabeled ACTH blocked more than 84% of the ^{125}I-labeled ACTH binding.

branes for particular peptides or hormones. Certainly, if such a mechanism were at work during the evolutionary process, it would by the hydrophilicity-hydrophobicity phenomenon ensure a tight fit between the hormone or peptide and its respective receptor.

References

1. Stein S: Application of fluorescent techniques to the study of peptides, in Griner H (ed): *Peptides in Neurobiology*. New York, Plenum Press, 1978, pp 9–37.
2. Tatemoto K, Mutt V: Isolation and characterization of the intestinal peptide porcine PHI (PHI-27), a new member of the glucagon-secretin family. *Proc Natl Acad Sci USA* 1981;78:6603–6607.
3. Blundell TL, Humbel RE: Hormone families: Pancreatic hormones and homologous growth factors. *Nature* 1980;287:781–787.
4. Pierce JG, Parsons TF: Glyco-protein hormones: Structure and function. *Annu Rev Biochem* 1981;50:465–495.
5. Bahl OP: Human chorionic gonadotropin, its receptor and mechanism of action. *Fed Proc* 1977;36:2119–2127.
6. Niall HD, Hogan ML, Sauer R, et al: Sequences of pituitary and placental lactogenic and growth hormones: Evolution from a primordial peptide by gene reduplication. *Proc Natl Acad Sci USA* 1971;68:866–869.
7. Habener JF, Rosenblatt M, Kemper B, et al: Pre-proparathyroid hormone: Amino acid sequence, chemical synthesis, and some biological studies of the precursor region. *Proc Natl Acad Sci USA* 1978;75:2616–2620.
8. James R, Niall H, Kwok S, Bryant-Greenwood G: Primary structure of porcine relaxin: Homology with insulin and related growth factors. *Nature* 1977;267:544–546.
9. Frazier WA, Hogue-Angeletti R, Bradshaw RA: Nerve growth factor and insulin. *Science* 1972;176:482–488.
10. Van Wyk JJ, Underwood LE, Baseman JB, et al: Explorations of the insulinlike and growth-promoting properties of somatomedin by membrane receptor assays. *Adv Metab Dis* 1975;8:127–150.

11. Smith G, Temin HM: Purified multiplication-stimulating activity from rat liver cell conditioned medium: Comparison of biological activities with calf serum, insulin, and somatomedin. *J Cell Physiol* 1974;84:181–192.
12. Hughes J, Smith TW, Kosterlitz HW, et al: Identification of two related pentapeptides from the brain with potent opiate agonist activity. *Nature* 1975;258:577–579.
13. MacIntyre I, Evans IM, Hobitz HHG, et al: Chemistry, physiology and therapeutic applications of calcitonin. *Arthritis Rheum* 1980;23:1139–1147.
14. Steiner DF, Quinn PS, Chan SJ, et al: Processing mechanisms in the biosynthesis of proteins. *Ann NY Acad Sci* 1980;343:1–16.
15. Habener JF, Potts JT: Biosynthesis of parathyroid hormone. *N Engl J Med* 1978; 299:580–585.
16. Habener JF, Potts JT: Biosynthesis of parathyroid hormone. *N Engl J Med* 1978; 299:635–644.
17. Mains FE, Eipper BA: Biosynthetic studies on ACTH, β-endorphin, and α-melanotropin in the rat. *Ann NY Acad Sci* 1980;343:94–110.
18. Loh YP, Gainer H: Characterization of pro-opiocortin–converting activity in purified secretory granules from rat pituitary neurointermediate lobe. *Proc Natl Acad Sci USA* 1982;79:108–112.
19. Fletcher DJ, Quigley JP, Bauer GE, Quigley JP: Characterization of the conversion of a somatostative precursor to somatostatin by islet sercretory granules. *Diabetes* 1980; 29:593–599.
20. Fletcher DJ, Quigley JP, Bauer GE, Noe BD: Characterization of proinsulin- and proglucagon-converting activities in isolated islet secretory granules. *J Cell Biol* 1980; 90:373–378.
21. Brownstein MJ, Russell JT, Gainer H: Synthesis, transport, and release of posterior pituitary hormones. *Science* 1980;207:373–378.
22. Kimura S, Lewis RV, Stern AS, et al: Probable precursors of [Leu]enkephalin and [Met]enkephalin in adrenal medulla: Peptides of 3–5 kilodaltons. *Proc Natl Acad Sci USA* 1980;77:1681–1685.
23. Rossier J, Trifaro JM, Lewis RV, et al: Studies with [^{35}S]methionine indicate that the 22,000 dalton [Met]enkephalin-containing protein in chromaffin cells is a precursor of [Met]enkephalin. *Proc Natl Acad Sci USA* 1980;77:6899–6891.
24. Roberts BE, Paterson BM: Efficient translation of tobacco mosiac virus RNA and rabbit globin qsRNA in a cell-free system from commercial wheat germ. *Proc Natl Acad Sci USA* 1973;70:2330–2334.
25. Pelham HRB, Jackson RJ: An efficient mRNA-dependent translation system from reticulocyte lysates. *Eur J Biochem* 1976;67:247–256.
26. Aviv H, Leder P: Purification of biologically active globin messenger RNA by chromatography on oligothymidylic acid–cellulose. *Proc Natl Acad Sci USA* 1972;69:1408–1412.
27. Vamvakopoulos NC, Kourides IA: Identification of separate mRNAs for the α and β subunits of thyrotropin. *Proc Natl Acad Sci USA* 1979;76:3809–3813.
28. Allison U, Hall L, MacIntyre I, Craig RK: The construction and partial characterization of plasmids containing complementary DNA sequences to human calcitonin precursor polyprotein. *Biochem J* 1981;199:725–731.
29. Martial JA, Hallewell RA, Baxter JD, Goodman HM: Human growth hormone: Complementary DNA cloning and expression in bacteria. *Science* 1979;205:602–607.
30. Lane CD, Colman A, Mohun T, et al: The *Xenopus* oocyte as a surrogate secretory system. The specificity of protein export. *Eur J Biochem* 1980;111:225–235.
31. Colman A, Lane CD, Craig R, et al: The influence of topology and glycosylation on the fate of heterologous secretory proteins made in *Xenopus* oocytes. *Eur J Bioehcm* 1981;113:339–348.
32. Colman A, Morser J: Export of proteins from oocytes of *Xenopus laevis*. *Cell* 1979; 17:517–526.
33. Lusis, AJ, Golde DW, Quon DH, Lasky LA: Translation of mRNA for human granulocyte-macrophage colony stimulating factor. *Nature* 1982;298:75–77.
34. Itakura K, Riggs AD: Chemical DNA synthesis and recombinant DNA studies. *Science* 1980;209:1401–1405.
35. Noyes BE, Mevarech M, Stein R, Agarwal KL: Detection and partial sequence analysis

of gastrin mRNA by using an oligodeoxynucleotide probe mRNA by using an oligodeoxynucleotide probe. *Proc Natl Acad Sci USA* 1979;76:1770–1774.
36. Kakidani H, Funutani Y, Takahashi H, et al: Cloning and sequence analysis of cDNA for porcine β-neo-endorphin/dinorphin precursor. *Nature* 1982;298:245–249.
37. Hobart P, Crawford R, Shen L, et al: Cloning and sequence analysis of cDNAs encoding two distinct somatostatin precursors found in the endocrine pancreas of angelfish. *Nature* 1980;288:137–141.
38. Goeddel DV, Kleid DG, Bolivar F, et al: Expression in *Escherichia coli* of chemically synthesized genes for human insulin. *Proc Natl Acad Sci USA* 1979;76:106–110.
39. Wetzel R, Kleid DG, Crea R, et al: Expression in *Escherichia coli* of a chemically synthesized gene for a "mini-C" analog of human proinsulin. *Gene* 1981;16:63–71.
40. Johnson IS: Human insulin from recombinant DNA technology. *Science* 1983; 219:632–637.
41. Olson KC, Fenno J, Lin N, et al: Purified human growth hormone from *E. coli* is biologically active. *Nature* 1981;293:408–411.
42. Pavlakis GN, Hizuka N, Gorden P, et al: Expression of two human growth hormone genes in monkey cells infected by simian virus 40 recombinant. *Proc Natl Acad Sci USA* 1981;78:7398–7402.
43. White BA, Bancroft FC: Cytoplastic dot hybridization. Simple analysis of relative mRNA levels in multiple small cell or tissue samples. *J Biol Chem* 1982;257:8569–8572.
44. Amara SG, Jonas V, Rosenfeld MG, et al: Alternative RNA processing in calcitonin gene expression generates mRNAs encoding different polypeptide products. *Nature* 1982;298:240–244.
45. Massague J, Pilch P, Czech MP: Electrophoretic resolution of three major insulin receptor structures with unique subunit stoichiometries. *Proc Natl Acad Sci USA* 1980; 77:7137–7141.
46. Hedo J, Kasuga M, Van Obberghen E, et al: Direct demonstration of glycosylation of insulin receptor subunits by biosynthetic and external labeling: Evidence for heterogeneity. *Proc Natl Acad Sci USA* 1981;78:4791–4795.
47. Petruzzelli LM, Ganguly S, Smith CJ, et al: Insulin activates a tyrosine-specific protein kinase in extracts of 3T3-L1 adipocytes and human placenta. *Proc Natl Acad Sci USA* 1982;79:6792–6796.
48. Van Obberghen E, Ross B, Kowalski A, et al: Receptor-mediated phosphorylation of the hepatic insulin receptor: Evidence that the Mr 95,000 receptor subunit is its own kinase. *Proc Natl Acad Sci USA* 1983;80:945–1040.
49. Stadtmauer L, Rosen OM: Phosphorylation of exogenous substrates by the insulin receptor–associated protein kinase. *J Biol Chem* 1983;258:6682–6685.
50. Pike LJ, Kuenzel EA, Caswellie JE, Krebs EG: A comparison of the insulin- and epidermal growth factor–stimulated protein kinases from human placenta. *J Biol Chem* 1984;259:9913–9921.
51. Ullrich A, Bell JR, Chen EY, et al: Human insulin receptor and its relationship to the tyrosine kinase family of oncogenes. *Nature* 1985;313:756–761.
52. Yoshimasa Y, Seino S, Whittaker J, et al: Insulin-resistant diabetes due to a point mutation that prevents insulin proreceptor processing. *Science* 1988;240:784–787.
53. Kadowaki T, Bevins CL, Cama A, et al: Two mutant alleles of the insulin receptor gene in a patient with extreme insulin resistance. *Science* 1988;240:787–790.
54. Taira M, Taira M, Hashimoto N, et al: Human diabetes associated with a deletion of the tyrosine kinase domain of the insulin receptor. *Science* 1989;245:63–66.
55. Odawara M, Kadowaki T, Yamamoto R, et al: Human diabetes associated with a mutation in the tyrosine kinase domain of the insulin receptor. *Science* 1989;245:66–68.
56. Shimada F, Taira M, Suzuki Y, et al: Insulin-resistant diabetes associated with partial deletion of insulin-receptor gene. *Lancet* 1990;1:1179–1181.
57. Varmus HE: Form and function of retroviral proviruses. *Science* 1982;216:812–820.
58. Bishop JM: Cellular oncogenes and retroviruses. *Annu Rev Biochem* 1983;52:301–354.
59. Ebina Y, Ellis L, Jarnagin K, et al: The human insulin receptor cDNA: The structural basis for hormone-activated transmembrane signalling. *Cell* 1985;40:747–758.
60. Lefkowitz RJ, Stadel JM, Caron MG: Adenylate cyclase–coupled beta-adrenergic re-

ceptors: Structure and mechanisms of activation and desensitization. *Annu Rev Biochem* 1983;52:159–186.
61. Shorr RG, Strohsacker MW, Lavin TN, et al: The β_1-adrenergic receptor of the turkey erythrocyte. Molecular heterogeneity revealed by purification and photoaffinity labeling. *J Biol Chem* 1982;257:12341–12350.
62. Homcy CJ, Rockson SG, Countaway J, Egan DA: Purification and characterization of the mammalian β_2-adrenergic receptor. *Biochemistry* 1983;22:660–668.
63. Dixon RAF, Kobilka BK, Strader DJ, et al: Cloning of the gene and cDNA for mammalian β-adrenergic receptor and homology with rhodopsin. *Nature* 1986;321:75–79.
64. Kobilka BK, Kobilka TS, Daniel K, et al: Chimeric α_2-, β_2-effector coupling and ligand specificity. *Science* 1988;240:1310–1316.
65. Emorine LJ, Marullo S, Briend-Sutren M-M, et al: Molecular characterization of the human β_3-adrenergic receptor. *Science* 1989;245:1118–1121.
66. Dearry A, Gingrich JA, Falardeau P, et al: Molecular cloning and expression of the gene for a human D_1 dopamine receptor. *Nature* 1990;347:72–76.
67. Zhou Q-Y, Grandy DK, Thambi L, et al: Cloning and expression of human and rat D_1 dopamine receptors. *Nature* 1990;347:76–80.
68. Sunahara RK, Niznik HB, Weiner DM, et al: Human dopamine D_1 receptor encoded by an intronless gene on chromosome 5. *Nature* 1990;347:80–83.
69. Marver D: The mineralocorticoid receptor, in Litwack G (ed): *Biochemical Actions of Hormones*, Vol. 12. Orlando, Florida, Academic Press, pp 385–431.
70. Fanestil DD, Park CS: Steroid hormones and the kidney. *Annu Rev Physiol* 1981; 43:637–649.
71. Beaumont K, Fanestil DD: Characterization of rat brain aldosterone receptors reveals high affinity for corticosterone. *Endocrinology* 1983;113:2043–2051.
72. Hollenberg SM, Weinberger C, Ong ES, et al: Primary structure and expression of a functional human glucocorticoid receptor cDNA. *Nature* 1985;318:635–641.
73. Green S, Walter P, Kumar V, et al: Human estrogen receptor cDNA: Sequence, expression and homology to v-*erb*-A. *Nature* 1986;320:134–139.
74. Green GL, Gilna P, Waterfield M, et al: Sequence and expression of human estrogen receptor complementary DNA. *Science* 1986;231:1150–1153.
75. Loosfelt H, Atger M, Misrahi M, et al: Cloning and sequence analysis of rabbit progesterone-receptor complementary DNA. *Proc Natl Acad Sci USA* 1986;83:9045–9049.
76. Giguere V, Hollenberg SM, Rosenfeld MG, Evans RM: Functional domains of the human glucocorticoid receptor. *Cell* 1986;46:645–648.
77. Green S, Chambon P: Oestradiol induction of a glucocorticoid-responsive gene by a chimaeric receptor. *Nature* 1987;325:75–78.
78. Godowski PJ, Rusconi S, Miesfeld R, Yamamoto KR: Glucocorticoid receptor mutants that are constitutive activators of transcriptional enhancement. *Nature* 1987;325:365–368.
79. Hollenberg SM, Giguere V, Segui P, Evans RM: Colocalization of DNA-binding and transcriptional activation functions in the human glucocorticoid receptor. *Cell* 1987; 49:39–46.
80. Weinberger C, Hollenberg SM, Rosenfeld MG, Evans RM: Domain structure of human glucocorticoid receptor and its relationship to v-*erb*-A oncogene product. *Nature* 1985; 318:670–672.
81. Arriza JL, Weinberger C, Cerelli G, et al: Cloning of human mineralocorticoid receptor complementary DNA: Structural and functional kinship with the glucocorticoid receptor. *Science* 1987;237:268–275.
82. Armanini D, Kuhnle U, Strasser T, et al: Aldosterone-receptor deficiency in pseudohypoaldosteronism. *N Engl J Med* 1985;313:1178–1181.
83. Ymer SI, Stevenson JL, Herington AC: Identification of a rabbit cytosolic binding protein for human growth hormone. *Biochem J* 1984;221:617–622.
84. Baumann G, Stolar MW, Amburn K, et al: A specific growth hormone–binding protein in human plasma: Initial characterization. *J Clin Endocrinol* 1986;62:134–141.
85. Leung DW, Spencer SA, Cachianes G, et al: Growth hormone receptor and serum binding protein: Purification, cloning and expression. *Nature* 1987;330:537–543.
86. Hughes JP, Friesen HG: The nature and regulation of receptors for pituitary growth hormone. *Annu Rev Physiol* 1985;47:469–482.

87. Isaksson OGP, Eden S, Jansson J-O: Mode of action of pituitary growth hormone on target cells. *Annu Rev Physiol* 1985;47:483–499.
88. Salmon WD, Jr, Daughaday WH: A hormonally controlled serum factor which stimulates sulfate incorporation by cartilage in vitro. *J Lab Clin Med* 1957;49:825–836.
89. Herington AC, Ymer SI, Stevenson JL: Identification and characterization of specific binding proteins for growth hormone in normal human sera. *J Clin Invest* 1986; 77:1817–1823.
90. Gluzman Y: SV40-transformed simian cells support the replication of early SV40 mutants. *Cell* 1981;23:175–182.
91. Gorman C: High efficiency gene transfer into mammalian cells, in Glover D (ed): *DNA Cloning*, Vol. II. Oxford, IRL Press, 1985, pp 143–190.
92. McKusick VA: *Mendelian Inheritance in Man*, 6th ed. Baltimore, Johns Hopkins Press, 1987, pp 895–896.
93. Merimee TJ, Zapf J, Hewlett B, Cavalli-Sforza LL: Insulin-like growth factor in pygmies. *N Engl J Med* 1987;316:906–911.
94. Cooke NE, Ray J, Watson MA, et al: Human growth hormone gene and the highly homologous growth hormone variant gene display different splicing patterns. *J Clin Invest* 1988;82:270–275.
95. Godowski PJ, Leung DW, Meacham LR, et al: Characterization of the human growth hormone receptor gene and demonstration of a partial gene deletion in two patients with Laron-type dwarfism. *Proc Natl Acad Sci USA* 1989;86:8083–8087.
96. Amselem S, Duquesnoy P, Attree O, et al: Laron dwarfism and mutations of the growth hormone–receptor gene. *N Engl J Med* 1989;321:989–995.
97. Smith WC, Kuniyoshi J, Talamantes F: Mouse serum growth hormone (GH) binding protein has GH receptor extracellular and substituted transmembrane domains. *Mol Endocrinol* 1990;3:984–990.
98. Haussler MR, McCain TA: Basic and clinical concepts related to vitamin D metabolism and action. *N Engl J Med* 1977;297:974–983.
99. DeLuca HF, Schnoes HK: Vitamin D: Recent advances. *Annu Rev Biochem* 1983; 52:411–439.
100. Haussler MR, Manolagos SC, Deftos LJ: Evidence for a 1,25-dihydroxyvitamin D_3 receptor–like macromolecule in rat pituitary. *J Biol Chem* 1980;255:5007–5010.
101. Abe E, Miyaura C, Sagagami H, et al: Differentiation of mouse myeloid leukemia cells induced by 1α,25-dihydroxyvitamin D_1. *Proc Natl Acad Sci USA* 1981;78:4990–4994.
102. Bar-Shavit Z, Teitelbaum S, Reitsma P, et al: Induction of monocytic differentiation and bone resorption by 1,25-dihydroxyvitamin D_3. *Proc Natl Acad Sci USA* 1983; 80:5907–5911.
103. Mangelsdorf DJ, Koeffler HP, Donaldson CA, et al: 1,25-Dihydroxyvitamin D_3–induced differentiation in a human promyelocytic leukemia cell line (HL-60): Receptor-mediated maturation to macrophage-like cells. *J Cell Biol* 1984;98:391–398.
104. McDonnell DP, Mangelsdorf DJ, Pike JW, et al: Molecular cloning of complementary DNA encoding the avian receptor for vitamin D. *Science* 1987;235:1214–1217.
105. Marx SJ, Liberman UA, Eil C, et al: Hereditary resistance to 1,25-dihydroxyvitamin D. *Recent Prog Horm Res* 1983;40:689–615.
106. Weinberger C, Thompson CC, Ong ES, et al: A gene encodes a thyroid hormone receptor. *Nature* 1986;324:641–646.
107. Thompson CC, Weinberger C, Lebo R, Evans RM: Identification of a novel thyroid hormone receptor expressed in the mammalian central nervous system. *Science* 1987; 237:1610–1614.
108. Nakai A, Seino S, Sakurai A, et al: Characterization of a thyroid hormone receptor expressed in human kidney and other tissues. *Proc Natl Acad Sci USA* 1988;85:2781–2785.
109. Parmentier M, Libert F, Maenhaut C, et al: Molecular cloning of the thyrotropin receptor. *Science* 1989;246:1620–1622.
110. Evans RM: The steroid and thyroid hormone superfamily. *Science* 1988;240:889–895.
111. Chen CLC, Dionne TT, Roberts JL: Regulation of the pro-opiomelanocortin mRNA levels in rat pituitary by dopaminergic compounds. *Proc Natl Acad Sci USA* 1983; 80:2211–2215.
112. Zigmond R: The long-term regulation of ganglionic tyrosine hydroxylase by preganglionic nerve activity. *Fed Proc* 1980;39:3003–3008.

113. Greenberg ME, Ziff EB, Greene LA: Stimulation of neuronal acteylcholine receptors induces rapid gene transcription. *Science* 1985;234:80–83.
114. Morgan JI, Cohen DR, Hempstead JL, Curren T: Mapping patterns of c-*fos* expression in the central nervous system after seizure. *Science* 1987;237:192–197.
115. Curran T, Miller AD, Zokas L, Verma IM: Viral and cellular *fos* proteins: A comparative analysis. *Cell* 1984;36:259–268.
116. Sambucetti LC, Curran T: The *fos* protein complex is associated with DNA in isolated nuclei and binds to DNA cellulose. *Science* 1986;234:1417–1419.
117. Bonner TI, Buckley NJ, Young AC, Brann MR: Identification of a family of muscarinic acetylcholine receptor genes. *Science* 1987;237:527–532.
118. Hammer R, Berrie CP, Birdsall NJM, et al: Pirenzepine distinguishes between different subclasses of muscarinic receptors. *Nature* 1980;283:90–92.
119. Claudio T, Green WN, Hartman DS, et al: Genetic reconstitution of functional acetylcholine receptor channels in mouse fibroblasts. *Science* 1987;238:1688–1694.
120. McCarthy MP, Earnest JP, Young EF, et al: The molecular neurobiology of the acetylcholine receptor. *Annu Rev Neurosci* 1986;9:383–413.
121. Claudio T: Recombinant DNA technology in the study of ion channels. *Trends Pharmacol Sci* 1986;7:308–312.
122. Mishina M, Tobimatsu T, Imoto K, et al: Location of functional regions of acetylcholine receptor α-subunit by site-directed mutagenesis. *Nature* 1985;313:364–369.
123. White MM, Mixter-Mayne K, Lester HA, Davidson N: Mouse-*Torpedo* hybrid acetylcholine receptors: Functional homology does not equal sequence homology. *Proc Natl Acad Sci USA* 1985;82:4852–4856.
124. Greeningloh G, Rienitz A, Schmitt B, et al: The strychnine-binding subunit of the glycine receptor shows homology with nicotinic acetylcholine receptors. *Nature* 1987; 328:215–220.
125. Schofield PR, Darlison MG, Fujita N, et al: Sequence and functional expression of the $GABA_A$ receptor shows a ligand-gated receptor super-family. *Nature* 1987;328:221–227.
126. Pritchettt DB, Sontheimer H, Shivers BD, et al: Importance of a novel $GABA_A$ receptor subunit for benzodiazepine pharmacology. *Nature* 1989;338:582–585.
127. Levitan ES, Schofield PR, Burt DR, et al: Structural and functional basis for $GABA_A$ receptor heterogeneity. *Nature* 1988;335:76–79.
128. Rhein LD, Cagan RH: Role of cilia in olfactory recognition in Cagan RH, Kare MR (eds): *The Biochemistry of Taste and Olfaction*. New York, Academic Press, 1981, pp 47–68.
129. Pevsner J, Trifiletti RR, Strittmatter SM, Snyder SH: Isolation and characterization of an olfactory receptor protein for odorant pyrazines. *Proc Natl Acad Sci USA* 1985; 82:3050–3054.
130. Lee K-H, Wells RG, Reed RR: Isolation of an olfactory cDNA: Similarity to retinol-binding protein suggests a role in olfaction. *Science* 1987;235:1053–1056.
131. Britt BA: Astiology and pathophysiology of malignant hyperthermia in Britt BA (ed): *Malignant Hyperthermia*. Amsterdam, Martinus Nijhoff, 1987, pp 11–42.
132. McCarthy TV, Healy JMS, Heffron JJA, et al: Localization of the malignant hyperthermia susceptibility locus to human chromosome 19q12–13.2. *Nature* 1990; 343:562–564.
133. Hall LW, Woolf N, Bradley JWP, Jolly DW: Unusual reaction to suxamethonium chloride. *Br Med J* 1966;2:1305.
134. Harrison GG, Biebuyck JF, Terblanche J, et al: Hyperpyrexia during anesthesia. *Br Med J* 1968;3:594–595.
135. Ingram PH, Bruns GAP, Regina VM, et al: Expression of a α-D-mannosidase in man-hamster somatic cell hybrids. *Biochem Genet* 1977;15:455–476.
136. MacLennan DH, Duff C, Zorzato F, et al: Ryanodine receptor gene is a candidate for predisposition to malignant hyperthermia. *Nature* 1990;343:559–561.
137. Ohnishi ST, Taylor S, Gronert GA: Calcium-induced Ca^{2+} release from sarcoplasmic reticulum of pigs susceptible to malignant hyperthermia. The effects of halothane and dantrolene. *FEBS Lett* 1983;161:103–107.
138. Nelson TE: Abnormality in calcium release from skeletal sarcoplasmic reticulum of pigs. *J Clin Invest* 1983;72:862–873.
139. Mickelson JR, Gallant EM, Litterer LA, et al: Abnormal sarcoplasmic reticulum ryanodine receptor in malignant hypotheramia. *J Biol Chem* 1988;263:9310–9315.

140. White PC, New MI, Dupont B: HLA-linked congenital adrenal hyperplasia results from a defective gene encoding a cytochrome P-450 specific for steroid hydroxylation. *Proc Natl Acad Sci USA* 1984;81:7505–7509.
141. Werkmeister JW, New MI, Dupont B, White PC: Frequent deletion and duplication of the steroid 21-hydroxylase genes. *Am J Hum Genet* 1986;39:461–469.
142. Rumsby G, Carroll MC, Porter RR, et al: Deletion of the steroid 21-hydroxylase and complement C4 genes in congenital adrenal hyperplasia. *J Med Genet* 1986;23:204–209.
143. Globerman H, Amor M, Parker KL, et al: Nonsense mutation causing steroid 21-hydroxylase deficiency. *J Clin Invest* 1988;82:139–144.
144. Krauss CM, Turksoy RN, Atkins L, et al: Familial premature ovarian failure due to an interstitial deletion of the long arm of the X chromosome. *N Engl J Med* 1987;317:125–131.
145. Arnold A, Staunton CE, Kim HG, et al: Monoclonality and abnormal parathyroid hormone genes in parathyroid adenomas. *N Engl J Med* 1988;318:658–662.
146. Marx SJ: Genetic defects in primary hyperparathyroidism. *N Engl J Med* 1988; 318;699–701.
147. Arnold A, Kim HG, Gaz RD, et al: Molecular cloning and chromosomal mapping of DNA rearranged with the parathyroid hormone gene in a parthyroid adenoma. *J Clin Invest* 1989;83:2034–2040.
148. Friedman E, Sakaguchi K, Bale AE, et al: Clonality of parathyroid tumors in familial multiple endocrine neoplasia type 1. *N Engl J Med* 1989;321:213–218.
149. Thakker RV, Bouloux P, Wooding C, et al: Assoiciation of parathyroid tumors in multiple endocrine neoplasia type 1 with loss of alleles on chromosome II. *N Engl J. Med* 1989;321:218–224.
150. Sobol H, Narod SA, Nakamura Y, et al: Screening for multiple endocrine neoplasia type 2a with DNA-polymorphism analysis. *N Engl J Med* 1989;321:996–1001.
151. Han VKM, D'Ercole AJ, Lund PK: Cellular localization of somatomedin (insulin-like growth factor) messenger RNA in the human fetus. *Science* 1987;236:193–197.
152. Froesch ER, Schmid C, Schwander J, Zapf J: Actions of insulin-like growth factors. *Annu Rev Physiol* 1985;47:443–467.
153. Lameson JL, Lindell CM, Habener JF: Gonadotropin and thyrotropin α- and β-subunit gene expression in normal and neoplastic tissues characterized using specific messenger ribonucleic acid hybridization probes. *J Clin Endocrinol Metab* 1986;64:319–327.
154. Pixley S, Weiss M, Melmed S: Identification of human growth hormone messenger ribonucleic acid in pituitary adenoma cells by in situ hybridization. *J Clin Endocrinol Metab* 1987;65:575–580.
155. Fett JW, Strydon DJ, Lobb RR, et al: Isolation and characterization of angiogenin, an angiogenic protein from human carcinoma cells. *Biochemistry* 1985;24:5480–5486.
156. Weiner HL, Weiner LH, Swann JL: Tissue distribution and developmental expression of the messenger RNA encoding angiogenin. *Science* 1987;237:280–282.
157. Kelly K, Cochran BH, Stiles CD, Leder P: Cell specific regulation of the c-*myc* gene by lymphocyte mitogens and platelet-derived growth factor. *Cell* 1983;35:603–610.
158. Hirayu H, Dere WH, Rapoport B: Initiation of normal thyroid cells in primary culture is associated with enhanced c-*myc* messenger ribonucleic acid levels. *Endocrinology* 1987;120:924–928.
159. Yamashita S, Ong J, Fagin JA, Melmed S: Expression of the *myc* cellular proto-oncogene in human thyroid tissue. *J Clin Endocrinol* 1986;63:1170–1173.
160. Cahill GF, McDevitt HO: Insulin-dependent diabetes mellitus: The initial lesion. *N Engl J Med* 1981;304:1454–1465.
161. Rossini AA, Mordes JP, Like AA: Immunology of insulin-dependent diabetes mellitus. *Annu Rev Immunol* 1985;3:289–320.
162. Yoon JW, Austin M, Onodera T, Notkins AL: Virus-induced diabetes mellitus. *N Engl J Med* 1979;300:1173–1179.
163. Platz P, Jakobsen BK, Morling M, et al: A genetic analysis of insulin-dependent diabetes mellitus. *Diabetologia* 1982;21:108–115.
164. Wolf E, Spencer KM, Cudworth AG: The genetic susceptibility to type I (insulin-dependent) diabetes: Analysis of the HLA-DR association. *Diabetologia* 1983;24:224–230.
165. Johnston C, Pyke DA, Cudworth AG, Wolf E: HLA-DR typing in identical twins with

insulin-dependent diabetes: Difference between concordant and discordant pairs. *Br Med J* 1983;286:253–255.

166. Owerbach D, Hagglof B, Lernmark A, Holmgren G: Susceptibility to insulin-dependent diabetes defined by restriction enzyme polymorphism of HLA-D region genomic DNA. *Diabetes* 1984;33:958–965.
167. Cohen D, Cohen O, Marcadet A, et al: Class II HLA-DC β-chain DNA restriction fragments differentiate among HLA-DR2 individuals in insulin-dependent diabetes and multiple sclerosis. *Proc Natl Acad Sci USA* 1984;81:1774–1778.
168. Cohen-Haguenauer O, Robbins E, Massart C, et al: A systematic study of HLA class II-α DNA restriction fragments in insulin-dependent diabetes mellitus. *Proc Natl Acad Sci USA* 1984;82:3335–3339.
169. Owerbach D, Lernmark A, Platz P, et al: HLA-D region β-chain DNA endonuclease fragments differ between HLA-DR identical healthy and insulin-dependent diabetic individuals. *Nature* 1983;303:815–817.
170. Michelsen B, Lernmark A: Molecular cloning of a polymorphic DNA endonuclease fragment associates insulin-dependent diabetes mellitus with HLA-DQ. *J Clin Invest* 1987;79:1144–1152.
171. Babbitt BP, Allen PM, Matzueda G, et al: Binding of immunogenic peptides to Ia histocompatibility molecules. *Nature* 1985;317:359–360.
172. Prochazka M, Leiter EH, Serreze DV, Coleman DL: Three recessive loci required for insulin-dependent diabetes in nonobese mice. *Science* 1987;237:286–289.
173. Baekkeskov S, Christie M, Lernmark A: in McGregor A (ed): *Immunology of Endocrine Diseases*. Lancaster, MTP Press, 1986, pp 73–88.
174. Srikanta S, Ganda OP, Eisenbarth GS, Soeldner JS: Islet-cell antibodies and beta cell function in monozygotic triplets and twins initially discordant for type I diabetes mellitus. *N Engl J Med* 1983;308:322–325.
175. Hanahan D: Heritable formation of pancreatic β-cell tumours in transgenic mice expressing recombinant insulin/simian virus 40 oncogenes. *Nature* 1985;315:115–122.
176. Adams TE, Alpert S, Hanahan D: Non-tolerance and autoantibodies to a transgenic calf antigen expressed in pancreatic β cells. *Nature* 1987;325:223–228.
177. Allison J, Campbell IL, Morahan G, et al: Diabetes in transgenic mice resulting from over-expression of class I histocompatibility molecules in pancreatic β cells. *Nature* 1988;333:529–533.
178. Todd JA, Bell JI, McDevitt HO: HLA-DQβ gene contributes to susceptibility and resistance to insulin-dependent diabetes mellitus. *Nature* 1987;329:599–604.
179. Lund T, O'Reilly L, Hutchings P, et al: Prevention of insulin-dependent diabetes mellitus in non-obese diabetic mice by transgenes encoding modified I-A β-chain or normal I-E α-chain. *Nature* 1990;345:727–729.
180. Slattery RM, Kjer-Nielsen L, Allison J, et al: Prevention of diabetes in non-obese diabetic I-A^K transgenic mice. *Nature* 1990;345:724–726.
181. Flynn TG, de Bold ML, de Bold AJ: The amino acid sequence of an atrial peptide with potent diuretic and natriuretic properties. *Biochem Biophys Res Commun* 1983;117:859–865.
182. Currie MG, Geller DM, Cole BR, et al: Purification and sequence analysis of bioactive atrial peptides (atriopeptins). *Science* 1984;223:67–69.
183. Atlas SA, Kleinert HD, Camargo MJ, et al: Purification, sequencing and synthesis of natriuretic and vasoactive rat atrial peptide. *Nature* 1984;309:717–719.
184. Needleman P, Adams SP, Cole BR, et al: Atriopeptins as cardiac hormones. *Hypertension* 1985;7:469–482.
185. Lang RE, Tholken H, Gante D, et al: Atrial natriuretic factor—a circulating hormone stimulated by volume loading. *Nature* 1985;314:264–266.
186. Manning PT, Schwartz D, Katsube NC, et al: Vasopressin-stimulated release of atriopeptin: Endocrine antagonists in fluid homeostasis. *Science* 1985;229:395–397.
187. Schiffrin EL, Gutkowska J, Kuchel O, et al: Plasma concentration of atrial natriuretic factor in a patient with paroxysmal atrial tachycardia. *N Engl J Med* 1985;312:1196–1197.
188. Schwartz D, Geller DM, Manning PT, et al: Ser-Leu-Arg-Arg-atriopeptin III: The major circulating form of atrial peptide. *Science* 1985;229:397–400.
189. Richards AM, Nicholls MG, Ikram H, et al: Renal, haemodynamic, hormonal effects of human alpha atrial natriuretic peptide in healthy volunteers. *Lancet* 1985;1;545–548.

190. Richards AM, Nicholls MG, Espiner EA, et al: Effects of alpha-human atrial natriuretic peptide in essential hypertension. *Hypertension* 1985;7:812–817.
191. Needleman P, Greenwald JE: Atriopeptin: A cardiac hormone intimately involved in fluid, electrolyte, and blood homeostasis. *N Engl J Med* 1986;314:828–834.
192. Glembotski CC, Oronzi ME, Li X, et al: The characterization of atrial natriuretic peptide (ANP) expression by primary cultures of atrial myocytes using an ANP-specific monoclonal antibody as an ANP messenger ribonucleic acid probe. *Endocrinology* 1987; 121:843–852.
193. Burnett JC Jr, Kao PC, Hu DC, et al: Atrial natriuretic peptide elevation in congestive heart failure in the human. *Science* 1986;231:1145–1147.
194. Saito Y, Nakao K, Nishimura K, et al: Clinical application of atrial natriuretic polypeptide in patients with congestive heart failure: Beneficial effects on left ventricular function. *Circulation* 1987;76:115–124.
195. Nilsson G: Gastrin: Isolation, characterization, and functions, in Glass GBJ (ed): *Gastrointestinal Hormones*. New York, Raven, 1980, pp 127–167.
196. Noyes BE, Mevarech M, Stein R, Agarwal KL: Detection and partial sequence analysis of gastrin mRNA by using an oligodeoxynucleotide probe. *Proc Natl Acad Sci USA* 1979; 76:1770–1774.
197. Boel E, Vuust J, Norris F, et al: Molecular cloning of human gastrin cDNA: Evidence for evolution of gastrin by gene duplication. *Proc Natl Acad Sci USA* 1983;80:2866–2869.
198. Wiborg O, Berglund L, Boel E, et al: Structure of a human gastrin gene. *Proc Natl Acad Sci USA* 1984;81:1067–1069.
199. Kato K, Hayashizaki Y, Takahashi Y, et al: Molecular cloning of the human gastrin gene. *Nucleic Acids Res* 1983;11:8197–8203.
200. Kariya Y, Kato K, Hayashizaki Y, et al: Expression of human gastrin gene in normal and gastrinoma tissues. *Gene* 1986;50:345–352.
201. Sugano K, Aponte GW, Yamada T: Identification and characterization of glycine extended post-translational processing intermediates of progastrin in porcine stomach. *J Biol Chem* 1985;260:11724–11729.
202. Bradbury AF, Finnie MDA, Smyth DG: Mechanism of C-terminal amide formation by pituitary enzymes. *Nature* 1982;298:686–688.
203. Del Valle J, Sugano K, Yamada T: Progastrin and its glycine-extended posttranslational processing intermediates in human gastrointestinal tissues. *Gastroenterology* 1987; 92:1908–1912.
204. Takahashi Y, Kato K, Hayashizaki Y, et al: Molecular cloning of the human cholecystokinin gene by use of a synthetic probe containing deoxyinosine. *Proc Natl Acad Sci USA* 1985;82:1931–1935.
205. Deschenes RJ, Narayana SVL, Argos P, Dixon JE: Primary structural comparison of the prehormones cholecystokinin and gastrin. *FEBS Lett* 1985:189:135–138.
206. Lund T, Geurts van Kessel AHM, Haun S, Dixon JE: The genes for human gastrin and cholecystokinin are located on different chromosomes. *Human Genet* 1986;73:77–80.
207. Whang-Peng J, Bunn PA Jr, Kao-Shan CS, et al: A nonrandom chromosomal abnormality, del3p(14–23), in human small cell lung cancer (SCLC). *Cancer Genet Cytogenet* 1982;6:119–139.
208. Sorenson GD, Pettengill OS, Brinck-Johnson T, et al: Hormone production by cultures of small cell carcinoma of the lung. *Cancer* 1981;47:1289–1296.
209. Said SI, Mutt V: Polypeptide with broad biological activity: Isolation from small intestine. *Science* 1970;169:1217–1218.
210. Said SI (ed): Vasoactive intestinal peptide, in *Advances in Peptide Hormone Research*. New York, Raven Press, 1982.
211. Tsukada T, Horovitch SJ, Montminy MR, et al: Structure of the human vasoactive intestinal polypeptide gene. *DNA* 1985;4:293–300.
212. Tsukada T, Fink JS, Mandel G, Goodman RH, Identification of a region in the human vasoactive intestinal polypeptide gene responsible for regulation by cyclic AMP. *J Biol Chem* 1987;262:8743–8747.
213. Blalock JE, Smith EM: Hydropathic anti-complementarity of amino acids based on the genetic code. *Biochem Biophys Res Commun* 1984;121:203–207.
214. Blalock JE, Bost KL: Binding of peptides that are specified by complementary RNAs. *Biochem J* 1986;234:679–683.

Hematopoeitic System, Recombinant Products, and Clinical Applications

8

Ralph C. Williams, Jr., MD

Hemoglobinopathies

In 1956, by identifying sickle hemoglobin through peptide mapping, Ingram made fundamental observations that formed the basis for a true molecular approach to human disease.[1] Subsequently, the sickle cell hemoglobin abnormality was traced to an abnormal peptide and eventually to a single amino acid change in the β chain of adult hemoglobin. Later it became apparent that this single amino acid substitution had resulted from a single base change in the coding DNA.[2] In the decade that followed, increasing medical and scientific attention was focused on a form of anemia known as thalassemia, which was first described in 1925 by Cooley, a pediatrician.[3] Quite early, its association with patients of Mediterranean heritage was apparent. When techniques for the in vitro study of globin and globin chain synthesis became available, it was evident that an inherited defect in hemoglobin synthesis formed the basis for this condition. As systems for cell-free RNA translation and hybridization became available, it was found that α-thalassemia was associated with deficiency in α-globin production related to actual deletion of α-globin structural genes. β-Thalassemia, associated with deficiency of β-globin, was apparently almost never caused by a deletion but was due to other structural β-globin changes. The various thalassemias, particularly β-thalassemia, are a major public health problem throughout the world. The application of new molecular approaches had made their etiology easier to study as knowledge about these disorders has increased.

Human Globin Genes

Human globins are coded for by one or two structural genes.[4] All normal human hemoglobin molecules are composed as tetramers of two α-like and two β-like globin polypeptide chains. Genes for α-like chains are grouped together on the short arm of chromosome 16, whereas the β-like genes are situated together on the short arm of chromosome 11. The β gene complex from the 5′ to the 3′ direction includes the single embryonic ϵ gene; two fetal (γ) genes, Gγ and Aγ; a pseudogene 4β1; the δ gene; and the adult β gene.[5] The order of genes within the DNA reflects embryonic–fetal–adult expression. During very early development, the ϵ-globin gene is expressed, followed by the γ-globin genes during fetal life. At 32 weeks of gestation, expression of the β gene is augmented and that of the γ genes is down-regulated except for a very low level of γ-gene expression that extends even into adult life.[6] The exact mechanisms that control these changes in hemoglobin gene expression are still not clearly understood. The complete β-globin gene complex extends over 60 kb of DNA and has been isolated and examined by cloning techniques.[5] Two intervening sequences, or introns, separate the coding regions of the β-globin gene into three exons.

β-Thalassemias

The general term β-thalassemia indicates a rather diverse group of conditions in which there is a deficiency of β-globin production relative to α-globin within erythrocytes. Subjects who are heterozygous for β-thalassemia—namely those in whom only one β-globin allele is abnormal—are clinically normal. Patients carrying two β-thalassemia genes are designated as homozygous and generally are severely anemic, requiring lifelong medical supervision. A spectrum of severity can be recognized in which patients severely handicapped by the disorder may have β^0-thalassemia, in which no β-globin is detected, and β^+-thalassemia, in which β chains are detectable but in markedly reduced amounts.

The thalassemia syndromes can be considered in terms of the specific mutations involved. At least 30 different mutations have been identified on the basis of molecular cloning, DNA sequence analysis, and studies of mutant gene segments. Most of these mutations are relatively simple and involve single base substitutions or small deletions or insertions either directly within or just upstream of the β-gene sequences. Mutations influence major aspects of gene function, including transcription, RNA processing, and RNA translation to final gene products. Several sorts of mutations that result in β-thalassemia are diagrammed in Figure 8.1. It can be seen that these mutations involve transcription, nonsense, frameshift, deletions, insertions, and substitutions involving splicing regions of the gene.

In contrast to simple thalassemias, the complex β-thalassemias involve abnormalities in expression of other β-like genes, either with a thalassemic component or with persistent expression of fetal hemoglobin into adult life. Complex β-thalassemias can be due to rather extensive DNA deletions. For instance, in typical persistence of fetal hemoglobin into adult life, (termed HPFH) (as recognized in blacks) deletion of both β- and δ-globin genes is found. At the same time both γ-globin genes are expressed at a high level. In δβ-thalassemia the β gene is completely deleted and most, but not all, of the δ gene is absent. Several examples of globin gene deletions with designations of comparable clinical syndromes are shown in Figure 8.2. The precise mechanisms leading to these various deletions are not yet known. Rarely, in some clinical syndromes associated with an increase in fetal hemoglobin production, all genes in the β complex are inactive or deleted in a clinical picture termed γδβ-thalassemia. The entire β-globin gene complex may be deleted from the chromosome,[7] or the μ, γ, and δ genes and the 5′ region of the β gene may be deleted.[8,9] In such instances there is some evidence for altered methylation of DNA in the region upstream from the β-globin sequences.[10]

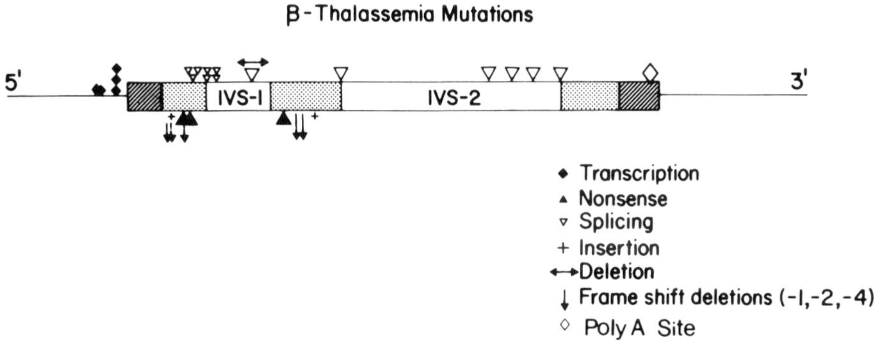

Figure 8.1 Location of β-thalassemia mutations in the β-globin gene. Specific mutations are indicated along the linear map of the β-gene with a symbol corresponding to its effect on β-gene expression. Key: ◆ , transcription mutant; ↓, frameshift deletion; ↔, 25-bp deletion; ◇ , 3′ RNA cleavage mutant; ▲, nonsense mutant; +, frameshift insertion; ▽, RNA splicing mutant. *Reproduced from Orkin and Kazazian*[14] *with permission.*

As noted above, most cases of β-thalassemia result from deletions or point mutations of β-globin genes as shown in Figure 8.1. In some patients the mutations can be identified directly using conventional β-region probes that recognize such deletions or mutations by changes in restriction enzyme recognition sites.[11–13] Direct analyses of chorionic villous tissue for fetal DNA in the first trimester permits an early diagnosis. Because more than 30 different mutations are now known to be associated with β-thalassemia,[14] in order to identify a mutation it is necessary to synthesize a pair of oligonucleotide probes, one of which is completely complementary in sequence to the normal gene and the other complementary to its abnormal counterpart. Such an approach has been applied to several Mediterra-

Figure 8.2 Deletions in the β-globin gene cluster. The respective regions of deletion for the clinical designations listed at the right are shown as dark bars. *Adapted from Orkin and Kazazian,*[14] *with permission.*

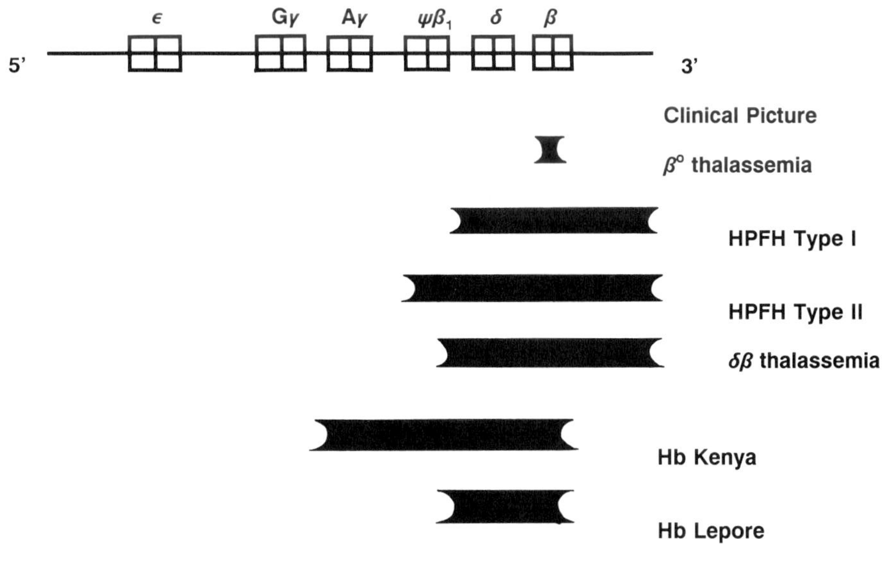

nean populations at risk.[15] In this analysis the two oligonucleotide probes synthesized were for the G → A mutation at position 110 within the first intervening sequence and for the C → T mutation at codon 39 of the β-globin gene. Each oligonucleotide consisted of 19 bases, half of which were completely homologous to the normal β-globin gene sequence and the other half to the thalassemia gene sequence. DNAs prepared from peripheral blood samples obtained from a large number of families of Mediterranean origin known to contain thalassemia cases were cut by restriction enzymes and hybridized under stringent conditions to ^{32}P-labeled oligonucleotide probes. Using these two probes for the most common thalassemic mutations in the populations being studied, it was found that analysis of 75% of the parents by this technique could predict the likelihood of the children being homozygous for the β-thalassemia defects. In such instances, the success of this technique in prenatal diagnosis varies with the region or country in which it is applied. A single oligonucleotide probe for the $\beta^0 39$ mutation would detect 95% of all thalassemia mutations in Sardinia, where the genetic heterogeneity for this gene presumably was more restricted.[16] Extension of this basic screening technique has indicated that specific β-thalassemia mutations are correlated with certain well-defined β-globin haplotypes.[17] Thus, to be complete, a screening process should probably include probes for all possible mutations, which would of necessity include probes for all rare β-thalassemia mutations. These exploratory studies provide important guidelines for prenatal diagnostic screening for thalassemias using fetal DNA analyses. In most populations at risk, there are usually one or two particularly common mutations together with a number of rarer forms of β-thalassemia. At present, unless automation systems become more widely available, covering the latter possibility with a large enough battery of oligonucleotide probes may prove too expensive and time consuming for widespread implementation. Even if the multiple oligonucleotide probe approach eventually becomes feasible, it probably would have to be combined with fetal blood sampling in order to detect the relatively small number of cases in which one or both parents are carriers for rare β-thalassemia mutations.

Amplified Single-Copy DNA

A great technical step forward was made in studies of human genomic mutations with the introduction and application of the polymerase chain reaction (PCR), which allows tremendous amplification of specific DNA regions of interest.[18–20] This method utilizes the *Taq* DNA polymerase enzyme and results in marked expansion of amplified gene products, which can then be sequenced and analyzed in comparative studies (Figure 8.3). The PCR method, which has now been applied to numerous human genetic diseases, is capable of amplifying selected DNA sequences 220,000-fold by 20 cycles of DNA polymerase I reactivity.[18] Unusual β-thalassemia mutations have been studied by this method using direct nucleotide sequencing of amplified single-copy DNAs.[21] These analyses uncovered two previously unrecognized mutations and clearly defined additional mutations present among other patients studied. Because of its speed and the exact information it yields, direct genomic sequencing can in some instances bypass the need for determination of haplotypes in the prenatal diagnosis of disorders such as β-thalassemia.

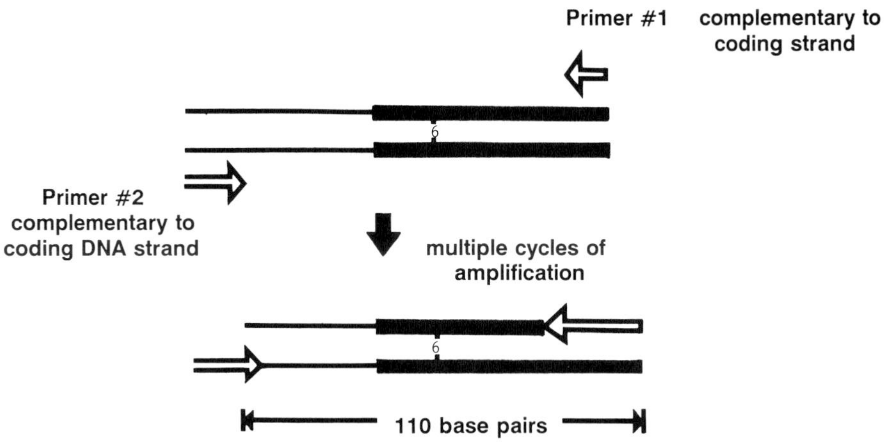

Figure 8.3 (Top) The two strands of the β-globin gene (dark lines) and its relation to polymerase chain reaction primers 1 and 2, which flank the sixth codon. The arrows show the direction in which the primers are extended by the polymerase reactions. Primers are provided as 20-base oligonucleotide DNA segments complementary to sequences that directly flank the target codon. Each oligonucleotide primer was extended using the Klenow fragment of *Escherichia coli* DNA polymerase I, so that each extension product serves as the template for a subsequent round of replication primed by the other primer. (Bottom) The major amplification products after many cycles of amplification. These products are discrete DNA segments (110 base pairs) that represent the genomic DNA between the outside or 5′ end of the primers and the sequences including the β codon of interest.

Sickle Cell Anemia

The method of DNA polymerase I amplification of β-globin DNA sequences has been particularly useful in the case of sickle cell anemia. This disease, which is associated with homozygous sickle cell gene expression, is relatively common in the United States, since it is estimated that 3000 pregnancies per year involve risk of producing a homozygous *ss* child. Several methods have been introduced for prenatal testing for sickle cell anemia, including recombinant DNA analysis.[22,23] Direct identification of DNA point mutations permits greater specificity and clinical certainty than the use of indirect methods involving linkage or haplotype analysis, which often require studies of parents and other family members.[14,24] The method of in vitro enzymatic amplification of genomic β-globin sequences was used for diagnosis of sickle cell anemia.[18] Previous methods that required DNA from fetal trophoblastic tissue were technically challenging procedures that usually took several days and could be done in only a few major centers. Application of the DNA polymerase I PCR amplification technique to the diagnosis of sickle cell anemia again involved hybridization of the amplified target DNA sequences with short synthetic oligonucleotide primers labeled at the 5′ end. Hybridized DNA was then digested sequentially with two different restriction enzymes,[25] which produced digestion products of different lengths, depending on whether the β^A or β^S allele was present in the digested target DNA. The restriction DNA fragments were separated by electrophoresis and then identified by autoradiography.[26] Embury et al.[26] used a new method of DNA analysis to differentiate the normal GAG codon at β^6 from the GTG sickle mutation. The method involved amplification of target DNA β^6 nucleotide sequences with a 40-nucleotide hybridization probe complementary to the β^A noncoding strand, following by labeling at the 5′ end with ^{32}P. The annealed hybrids were digested with *Dde*I and then *Hinf*I restriction enzymes. The selectivity of the method was based on the fact that the single base mismatch in the enzyme target restriction site of the hybridization probe–target DNA segment inhibited cleavage by the restriction endonuclease. Thus, sequential digestions by the two restriction enzymes produced DNA fragments of a size distribution that was characteristic of either β^A or β^S sequences. This differential process is shown in Figure 8.4. Using such an enzymatic amplification step followed by the restriction fragment length analysis, it was possible to identify accurately DNAs from subjects with AA, AS, SS, sickle β-thalassemia, and heterozygous β-thalassemia. Examples of this sort of analysis are shown in Figure 8.5. The major advantages of this method are that it is rapid and relatively sensitive. The high degree of amplification obtained with the PCR method (now automated) allows genetic analysis of very small

Figure 8.4 Steps in oligomer restriction. The top two diagrams show DNA molecules containing the β^A and β^S alleles, respectively. The two diagrams next to the top show the noncoding strands of these molecules after denaturation and annealing to the β^A-specific probe labeled at the 5′ end (*). The β^A target strand is perfectly complementary to the 40-base probe, but the β^S target strand contains a single-base mismatch, shown as a bubble (designated by the unaligned A). The two diagrams that are third from the top demonstrate the results of digesting the two duplexes with *Dde*I. The duplex with β^A target sequence contains a *Dde*I site, and its cleavage generates a 5′-labeled 8-base product that dissociates from the duplex. The duplex with β^S target sequences is not cleaved because of the mismatch within its *Dde*I site. Diagrams at the bottom demonstrate that subsequent digestion with *Hinf*I did not affect the 8-base cleavage product of β^A but generated a labeled 3-base product from the duplexes that contain β^S target sequences. *Reproduced from Embury et al,*[26] *with permission.*

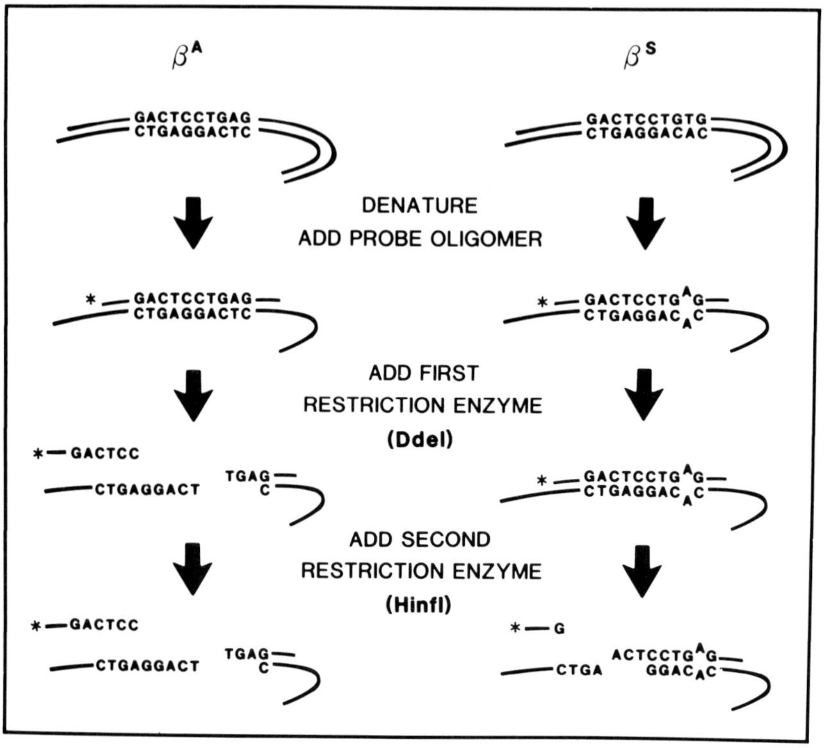

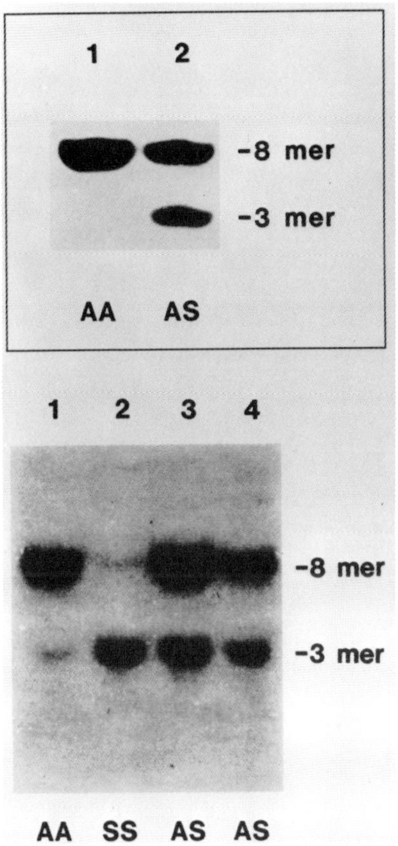

Figure 8.5 Autoradiograms showing results of polymerase chain reaction oligonucleotide restriction for diagnosis in pregnancy at risk for hemoglobin SS. (A) Results for couple 1. DNA from the cultured amniocytes of the fetus yielded only the 8-base cleavage product diagnostic of hemoglobin AA (lane 1). DNA from the leukocytes of the mother yielded both 8-base and 3-base products (lane 2), confirming the diagnosis of hemoglobin AS. The uncleaved 40-base probe is not shown in this or subsequent autoradiograms. (B) Results for couple 2 and for analysis of control DNA samples (hemoglobin AA and hemoglobin SS). Hemoglobin AA DNA produced to only an 8-base cleavage product (lane 1), and hemoglobin SS DNA only a 3-base product (lane 2). DNA from the leukocytes of the mother and amniocytes of the fetus resulted in both an 8-base and a 3-base product (lanes 3 and 4, respectively), demonstrating the hemoglobin AS genotype in each. *Reproduced from Embury et al,*[26] *with permission.*

amounts of fetal material, eliminating the previous need for cell culture. Moreover, detection of the hemoglobin SC genotype (which is indistinguishable from the AS genotype by direct restriction analysis of the β^6 codon affected by the single point sickle mutation) can also be done using the amplification γ double restriction digest procedure.

Another approach that can be used to detect the β^C allele is to prepare sequence-specific oligonucleotide probes to analyze amplified genomic DNA.[19,27] The general PCR DNA amplification procedure is a promising approach that will be applicable to a number of other genetic diseases. The only requirements are that the mutation being sought affects a known restriction enzyme site and that the adjacent nucleotide sequences are also known.

α-Thalassemia

In the normal state there are four copies of the α-globin gene on human chromosome 16. The most common cause of α-thalassemia is gene deletion; the α-thalassemia syndromes have been best defined in Asian populations, in which four clinical states of increasing severity have been described. These are shown in Table 8.1. A nonthalassemic individual has four α-globin structural genes per diploid cell; in α-thalassemia 1 and hemoglobin H disease, two and one α-globin genes, respectively, are still intact, and in homozygous α-thalassemia associated with hy-

Table 8.1 The α-Thalassemia (α-Thal) Syndromes (Asian Populations)

Syndrome	Genotype	Clinical manifestation	Number of α-globin genes[a]
Silent carrier state	α-Thal-2	None	3
Heterozygous α-thalassemia	α-Thal-1	Microcytosis	2
Hemoglobin-H disease	α-Thal-1 × α-Thal-2	Microcytosis, hemolysis	1
Homozygous α-thalassemia	α-Thal-1 × α-Thal-1	Hydrops fetalis (fatal)	0

[a] Number of globin genes in α-thalessemia-2 is inferred. All others have been shown by hybridization.

drops fetalis at birth, no α-globin genes remain intact.[28,29] Initially, homozygous α-thalassemia was detected using DNA hybridization from cultured amniotic fluid fibroblasts.[30] Subsequently this approach was improved to differentiate between homozygous α-thalassemia and hemoglobin H disease or α-thalassemia trait by using ^{35}S-labeled probes and slot blot analysis to measure actual α-globin gene dosage directly.[31] This assay is particularly useful because it does not require Southern blotting, restriction enzyme digestion, and gel electrophoresis. Moreover,the slot blot technique uses ^{35}S-labeled probes, which have a much longer half-life than those labeled with ^{32}P. In Southeast Asian countries such as Thailand, where approximately 15,000 hydropic pregnancies occur each year, application of this method should make prenatal diagnosis of α-thalassemia much more feasible than in the past.

Hemoglobin E

In 1954 Itano and co-workers[32] described a hemoglobin with previously unrecognized electrophoretic behavior in a child of Hindu and Mediterranean parentage. About the same time, Chernoff et al.[33] reported similar findings in patients studied in Thailand. This new abnormal hemoglobin was called hemoglobin E. It was found to result from the substitution of glutamic acid by lysine at position 26 in the β chain. At present, hemoglobin E may be the cause of the most prevalent hemoglobinopathy in the world, affecting an estimated 20 million individuals.[34] Because of the influx of Southeast Asian refugees into the United States and Canada since the Vietnam conflict, hemoglobin E is being recognized with increasing frequency.[35]

The single point mutation at codon position 26 of the β-globin gene involves substitution of an adenine for a guanine (GAG → AAG); the amino acid substitution occurs at the sensitive $\alpha_1\beta_1$ contact site, a region involved with other unstable hemoglobins.[36] Hemoglobin E is unusual because a single DNA base substitution produces both an unstable hemoglobin and erythrocytes with an apparent β^+-thalassemia phenotype. In homozygous individuals, hemoglobin E is associated with normal or mildly decreased hemoglobin levels in association with microcytosis; however, in doubly heterozygous individuals, the combination of hemoglobin E with β-thalassemia can produce an illness as severe as that seen with thalassemia major. The β^E mRNA appears to be unstable during erythrocyte maturation and abnormal qualitative and quantitative processing of β^E mRNA occurs, including

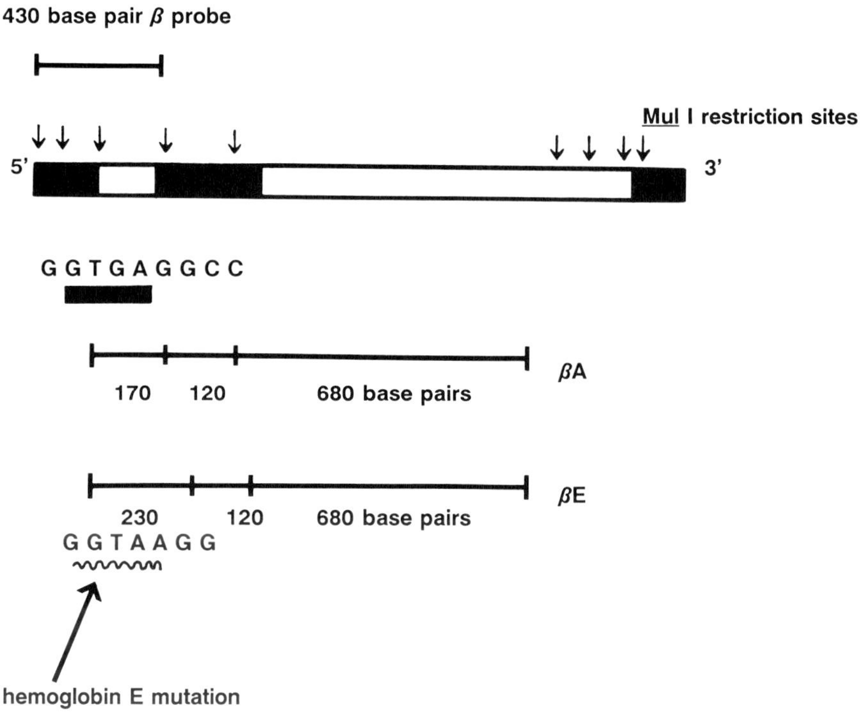

Figure 8.6 Schematic map of the β-globin gene showing the restriction cleavage sites (arrows) for *Mnl*I. After digestion and hybridization with the 430-bp β probe shown above, because the β^E chain shows the AAGG mutation at the *Mnl*I restriction site, a 230-bp fragment is produced instead of the 170-bp fragment seen in the normal β^A-globin digest.

abnormalities in mRNA splicing and production of an unstable β^E mRNA product.[37] Homozygous hemoglobin E is a clinically benign condition that does not appear to be associated with a survival disadvantage. Oxidant medications should be used with some caution because there have been a few reports of hemolytic anemia after use of drugs such as dapsone or para-aminosalicylic acid.

Direct detection of hemoglobin E using restriction endonuclease mapping has been reported[38] after digestion with the restriction enzyme *Mnl*I. This method takes advantage of the fact that the hemoglobin E single point mutation destroys the enzyme cleavage site at nucleotide sequences 3′-GGAG-5′. Figure 8.6 presents a schematic map of the β-globin gene along with the recognition sites for *Mnl*I restriction endonuclease. When normal $\beta^A\beta^A$ DNA is digested with *Mnl*I and hybridized with the 430-bp probe, fragments of 60, 170, 120, and 680 bp are observed. With the mutation of hemoglobin E, the cleavage site for *Mnl*I is removed and the 170-pb fragment increases in size to 230 bp. This method provides a convenient way to identify the abnormality.

Prospects for the Future

Because of the severe consequences of β-thalassemia and sickle cell anemia, considerable investigation and effort have gone into ways to treat the disease directly. Approaches that have been considered include reactivation of γ-globin gene expression and introduction of normal β genes into marrow of patients with severe homozygous forms of β-thalassemia or sickle cell anemia.

An approach that has long seemed reasonable is to influence the fetal switch process whereby during normal development γ gene expression is modulated down and β gene expression is increased late in gestation. Since relative hypomethylation of the 5′-flanking region of the γ genes in fetal erythroid cells has been noted,[39,40] an attempt was made in baboons to test the effects of 5-azacytidine, a compound known to produce extensive demethylation.[41] Treatment with this drug produced a considerable increment in hemoglobin F levels in the baboons, suggesting that drug treatments such as this might be used to modulate the fetal switch experimentally.[42] Small numbers of β-thalassemia and sickle cell anemia patients were treated with 5-azacytidine and dramatic increments in HbF expression were noted.[43–45] At the same time that HbF expression increased, hypomethylation of the γ-globin gene 5′-flanking regions was recorded. A strange feature of treatment with this drug was that the response appeared to be so rapid; after about 48 hours reticulocytes with the capacity to manufacture HbF were noted. It was therefore thought that the recorded modulation could have reflected cytotoxic effects of the drug involving preferential killing of cycling erythroid cells, permitting emergence of other precursors more committed to HbF synthesis.

Other cytotoxic drugs such as cytosine arabinoside and hydroxyurea, which do not specifically inhibit DNA methylation, have also been used experimentally, first in primates[46] and then in a few patients with sickle cell anemia.[47] After more extended use of 5-azacytidine in larger numbers of patients, enthusiasm for its protracted use was chilled by reports that the drug was probably a carcinogen in laboratory animals.[48] Hydroxyurea was employed because one of the theories of 5-azacytidine action suggested that it led to recruitment of early erythroid precursors capable of increased HbF production during recovery of early marrow suppression. Moreover, hydroxyurea was easily administered and reported to be noncarcinogenic in limited studies.[49] Since its initial trials, considerable long-term experience with this drug has been obtained, particularly in patients with sickle cell anemia.[50] In one study positive responses in several patients treated for longer than 2 years were encouraging and associated with signs of clinical improvement. It was noted that the HbF response was seen within 2 to 3 weeks but establishment of a large proportion of circulating F cells required treatment for a month or longer. Examples of positive responses are shown in Figure 8.7. During serial studies, several patients showed an increased number of chromosomal structural rearrangements or deletions, causing concern that patients placed on such regimes should avoid conception and pregnancy. One of the practical factors determining response to hydroxyurea was the patient's baseline level of reticulocyte production, which is an inherited trait and ranges from 4% to 50%. Patients with the highest initial baseline levels of recitulocyte production achieved the highest mean F reticulocyte counts during treatment.[50] The risk of carcinogenesis from long-term hydroxyurea treatment in humans is unknown. Studies to date have unproved safety but probably involve a relatively small risk.

Gene Therapy—Thalassemia Models

With the advent of recombinant methods and gene cloning, reintroduction of genes into various cell types and animals by transfection techniques has now become possible. One of the first successful attempts resulted in specific expression of a

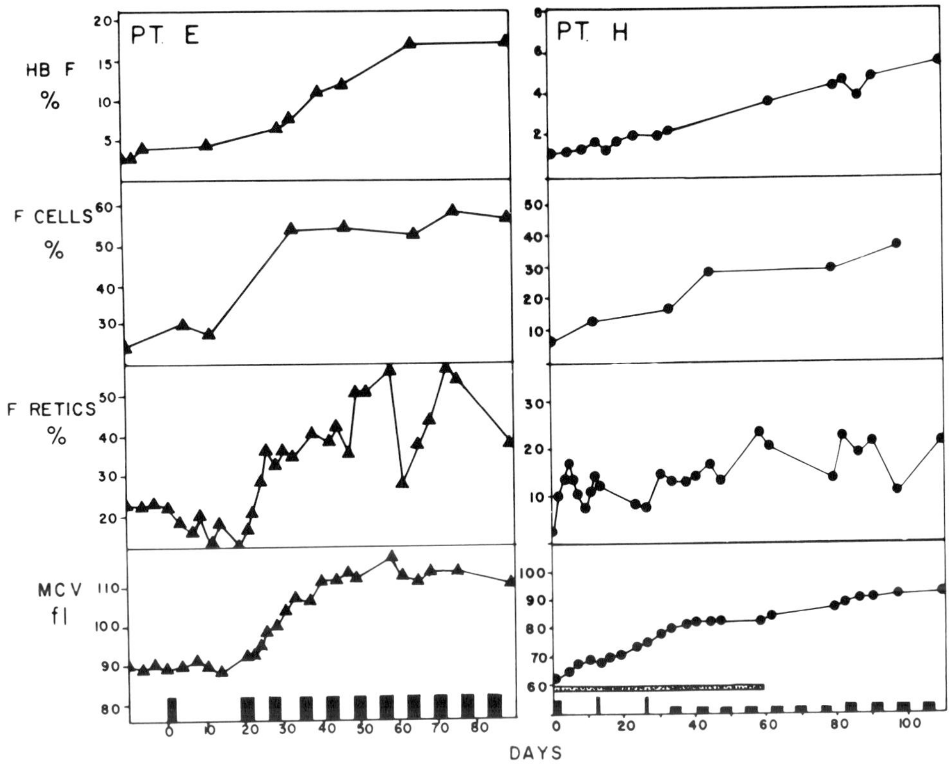

Figure 8.7 Response of patients E and H to hydroxyurea. Patient E had been transfused, and H was iron deficient before therapy. Black bars represent HU and the stippled bar $FeSO_4$ treatment. *Reproduced from Charache et al,*[50] *with permission.*

foreign β-globin gene in erythroid cells of transgenic mice.[51] Transgenic mice carrying a hybrid mouse–human adult β-globin gene showed gene expression predominantly within erythroid tissues. Most remarkable in these early studies was the fact that regulatory sequences closely linked to the transfected β-globin genes were able to specify correct patterns of tissue-specific expression in the experimental mice. More recently it was reported that a mouse model of β-thalassemia was corrected by transfer of cloned β-globin genes into the mouse germ line. The cloned mouse β^{maj}-globin gene or the cloned human β-globin gene was introduced into mice that were deficient in β-globin synthesis because of deletion of the β^{maj}-globin gene. Both artificially introduced genes produced functional β-globin chains leading to relative correction of the anemia and associated red blood cell abnormalities.[52] These interesting experiments were made possible by the availability of a murine model of thalassemia, which arose spontaneously in a mouse of the β-globin haplotype Hbb^{d}, which includes the two adult β-globin genes β^{maj} and β^{min}. In the mutation, $Hbb^{th\text{-}1}$, the β^{maj} gene is entirely deleted but the β^{min} gene remains intact. Animals homozygous for this deletion produced a reduced amount of β-globin (as in human thalassemia) and consequently also had a hemolytic anemia similar to that in human β-thalassemia of moderate severity. Although correction of genetic deficiencies by transfer of genes into the germ line itself instead of directly into developed somatic cells would not be a feasible strategy for human gene therapy, it was attempted as a trial in this mouse model. Two approaches were used, first with the cloned normal mouse β^{maj}-globin gene and second with the cloned normal human β-globin gene. In the first experiment, a 7.0-kb *Eco*RI fragment containing the mouse β^{maj}-globin gene was microinjected in the presence of calcium ions into homozygous $Hbb^{th\text{-}1}$ mouse zygotes and the eggs were transferred to foster mothers. Of 129 transferred eggs, 10 mice were born. Of these, two mice carried multiple copies of the original microinjected gene as determined by Southern blot analysis. Moreover, when peripheral blood samples from the transgenic mice were examined, one of the two mice was found to be producing a significant amount of the β^{maj}-type hemoglobin. These results are shown in Figure 8.8. When the transgenic male mouse was subsequently mated to homozygous $Hbb^{th\text{-}1}$ females, several of the progeny were found to produce β^{maj} hemoglobin, indicating that they had inherited and were capable of expressing the original microinjected gene (Figure 8.8). Although the ratio of β^{maj} to β^{min} hemoglobin is 4:1 in normal homozygous Hbb^b mice and 1:1 in heterozygotes, in the transgenic mice the ratio was found to be 1:2. Therefore, the transfected genes were producing fewer globin chains than a natural endogenous β^{maj}-globin gene.

Similar successful gene transfer experiments were performed in which a 7.7-kb

Figure 8.8 Expression of the cloned β^{maj}-globin gene in transgenic mice. (A) Cellulose acetate electrophoresis of cystamine-modified hemolysates from two normal mice (Hbb^d) and six mice (MB46–MB51) developed from homozygous Hbb^{th-1} zygotes that were microinjected with β^{maj} gene. Mice MB47 and MB51 are transgenic. (B) A similar analysis of eight offspring (1–8) of mouse MB47 mated to a homozygous Hbb^{th-1} female mouse. *Reproduced from Costantini et al,*[52] *with permission.*

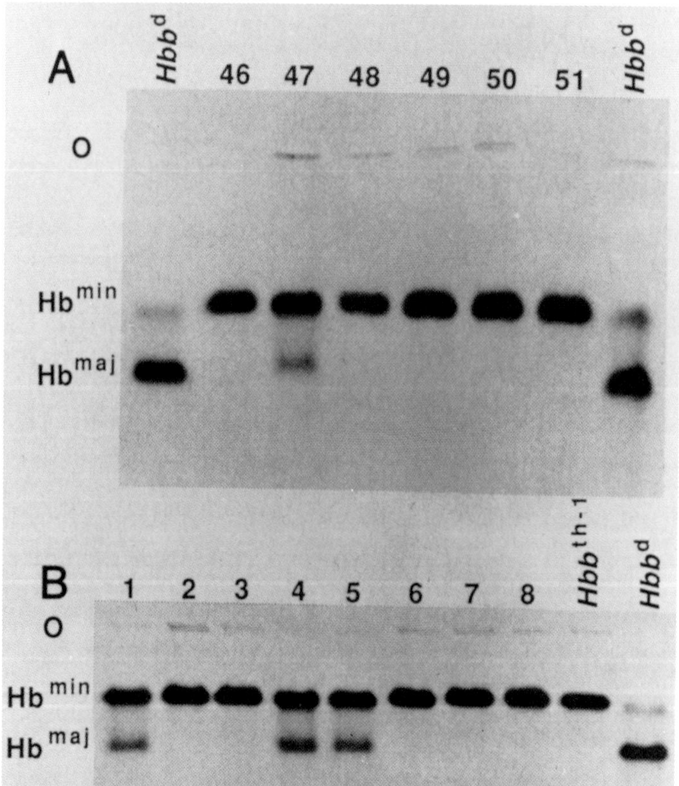

human β-globin gene fragment was transfected into the thalassemic mice. These experiments demonstrate the feasibility of transfection into germ-line DNA. Moreover, the thalassemia was essentially eliminated in the transgenic line carrying the human β-globin gene and partially corrected in the transgenic line carrying the cloned β^{maj}-globin gene. These differences were attributed to relative differences in level of expression of the two gene products. The actual levels of expression have been observed to vary considerably, apparently as a function of the chromosomal locus at which specific genes are integrated.[51–53] Cloned β-globin genes transferred into the erythroid cells of a thalassemic patient by any route would likely be expressed at variable levels from cell to cell. This potential problem will probably not be completely overcome until there is a much improved general understanding of regulatory mechanisms involving globin genes.

The idea that transgenic transfer therapy may be successful in human disorders of hemoglobin production was given impetus by the report by Dzierzak et al.[54] of successful expression of human β-globin genes in murine bone marrow transplant recipients reconstituted with retrovirus-transduced stem cells. A number of previous studies used retroviral vectors to introduce exogenous DNA sequences into target cells such as hematopoietic stem cells. In many of the systems previously studied there was no convincing evidence for significant levels of expression of the introduced genes. Until recently, all of the retroviral vectors employed with bone marrow cells were designed to promote eventual expression of inserted complementary DNA sequences. In most cases the vectors contained either viral transcriptional signals in the long terminal repeat (LTR) of the provirus or specific nonretroviral transcriptional control sequences to produce active chimeric RNAs. One reason for previous difficulties could be that the transcriptional sequences used lack elements specifically needed for hematopoietic stem cells to ensure gene expression at adequate levels during stem cell development. Another possibility is that the chimeric mRNAs generated by retroviral vectors produced structures that were abnormally processed or degraded prematurely.

In the studies by Dzierzak et al.,[54] three recombinant genomes were used. These are shown in detail in Figure 8.9. In two instances a 3-kb *Hpa*I-*Pst*I segment of human DNA encoding the human β-globin gene was inserted into the *Bam*HI site of the vectors in such a way that the direction of the β-globin transcription was *opposite* that of the proviral transcription. The third recombinant genome was constructed by insertion of a 4.1-kb *Hpa*I-*Xba*I DNA segment of the human β-globin gene into pSV(Z)neo. An additional 1.1 kg of 3′-flanking nucleotide sequence containing an enhancer element was thought to be responsible for increased levels of human β-globin expression in transgenic mice. Bone marrow cells from C3H/HeJ female mice were prepared and infected in vitro. Then 5×10^4 to 4×10^6 cells were injected intravenously into lethally irradiated male mice, and 1 to 2 months after transplantation peripheral blood samples were obtained from each recipient. DNA from blood samples was analyzed for the presence of proviral sequences using Southern blots. These studies confirmed successful engraftment in a substantial proportion of animals, indicating that the engrafted recipients expressed mRNA as well as human β-globin product in their erythroid cells. Tissue specificity of expression was also examined in the engrafted mice. It was found that human β-globin RNA expression was restricted to cell populations of erythroid lineage. These exciting preliminary results were encouraging about the prospect for transduced somatic cell engraftment as potential therapy for human disorders such as the hemoglobinopathies. The cDNA sequences of chromosomal human β-globin were

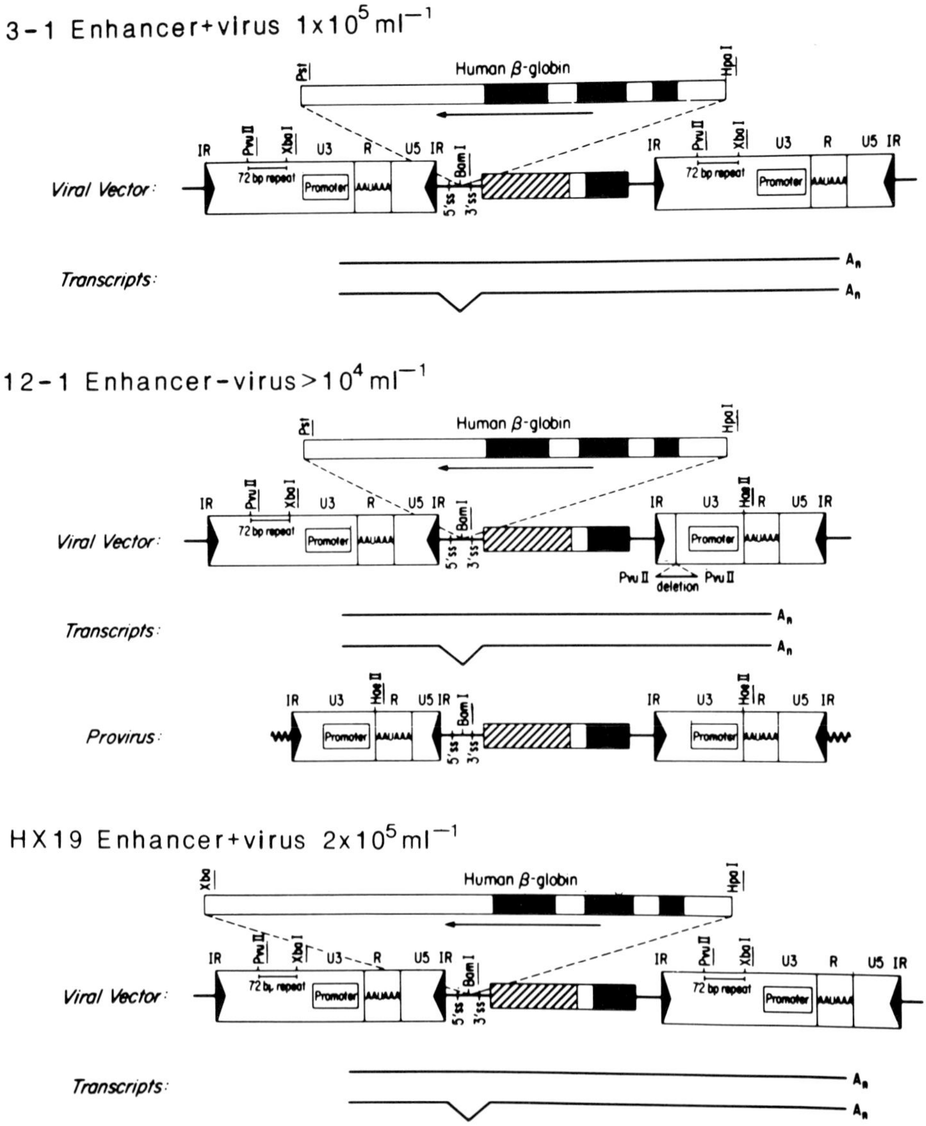

Figure 8.9 Structure of the human β-globin gene recombinant retroviruses and detection of recombinant proviral genomes in bone marrow–engrafted mice. HX19 was constructed by introducing a 4.1-kb *Hpa*I-*Xba*I fragment of human DNA containing the human β-globin gene into pSV(X)neo. Analysis of the three cell lines used in this study for helper virus in the XC assay yielded no plaques after multiple passages. *Reproduced from Dzierzak et al,*[54] *with permission.*

encoded as an intact gene that was expressed exclusively in erythroid cells of all transplant recipients, whether or not the viral transcriptional control signals in the vector were still present. Although expression of the transferred human β-globin gene was extremely tissue specific, the absolute level of β-globin transcription was relatively low. This could be improved by using additional sequences 50 kb upstream and 20 kb downstream of the β-globin structural gene, which dramatically boost expression in transgenic mice containing a single copy of the sequences.[55] Clearly, more work is now necessary to understand how the retroviral sequences directly influence transcription of β-globin sequences. It may be possible that *cis*-acting sequences of nucleotides within β-globin gene segments may actually activate proviral transcription. The results of the experiments of Dzierzak et al.[54] have a number of important implications for studies of gene expression in hematopoietic cells and the potential of human gene transfer therapy. Much previous effort had been directed at hematopoietic stem cell infection and the use of cDNA expression vectors. However, the work described above relied on the use of transcriptional signals normally used in hematopoietic cell differentiation. Perhaps gene therapy in the disorders of hemoglobin may not be as difficult or impossible as was thought a few years ago.

Sickle Cell Transgenic Models

An animal model of sickle cell disease has become available through the efforts of two different groups of investigators.[56,57] Greaves and co-workers[56] used the dominant control region sequences from the human β-globin locus together with human α- and β^S-globin genes in transgenic mice. The animals fell into two categories: one group that expressed the constructs at low levels because of deletions or various degrees of mosaicism and a second group that were fully transgenic and expressed the transgenes at high levels.[58,59] Among the fully transgenic mice, several showed high levels of human α-globin and β^S-globin mRNA in peripheral blood and showed a near-balanced synthesis of human α- and β^S-globin RNA. Analysis of peripheral blood hemolysates showed a large hemoglobin component that focused at the same isoelectric point as HbS from a sickle cell patient. These transgenic mice were not anemic, but in vitro sickling of their erythrocytes could be produced using the standard sodium metabisulfite assay. Some of the transgenic progeny showed sickling in vivo; others did not. One mouse with 83% Hb^S had normal somatic development, no obvious manifestations of disease, and no evidence for hemolytic anemia. No clear explanation for these findings was apparent; however, the authors suggested that the residual 17% of normal mouse hemoglobin may have been enough to protect the animal from disease manifestations. The same situation sometimes arises in the human disease state, where heterozygotes for β^S and HPFH also show HbS levels ranging from 65% to 90% but no anemia, normal red cell indices, and mostly normal reticulocyte counts.

In the second transgenic mouse model,[57] human α- and β^S-globin genes were inserted downstream of erythroid-specific, deoxyribonuclease I superhypersensitive sites coinjected into fertilized mouse eggs. Transgenic lines showed synthesis of human hemoglobin S. The same animals were then bred to β-thalassemic mice to reduce endogenous mouse globin levels. When red blood cells from these mice were subjected to deoxygenation, more than 90% showed the sickling phenomenon. Of great interest, these mice had decreased hematocrit values, elevated reticulocyte counts, and splenomegaly—all similar to human patients with sickle cell disease.

As more information is generated with these new models of sickle cell disorder it should be possible to test new drugs or therapeutic strategies with respect to affecting or ameliorating the underlying disease.

Stimulation of Fetal Hemoglobin Using Recombinant Erythropoietin

As noted earlier, stimulation of fetal hemoglobin production may help patients with sickle cell anemia by inhibiting sickling. Fetal hemoglobin appears to be limited in adults to a subpopulation of erythrocytes termed F cells.[60–62] Erythroid progenitor cells may have the ability to form F cells, but most of them fail to do so when the kinetics of erythroid differentiation and maturation are normal. However, when progenitor cells are pushed into their terminal differentiation prematurely, as in sudden bone marrow expansion, increased numbers of F cells may be formed. Erythropoietin produced by recombinant techniques has been used to stimulate F cell production in baboons.[63] The recombinant erythropoietin was administered to anemic baboons with the expectation that high doses would produce rapid erythroid regeneration. The animals were kept anemic by daily phlebotomy. These studies indicated that doses of 1500 to 3000 IU of erythropoietin rapidly produced substantial increments in F reticulocytes in both anemic and nonanemic animals. Levels of F reticulocytes increased from 1 to 2% to 40 to 50% after such treatment. Such stimulation of F cell production might be beneficial in the treatment of patients with sickle cell disease. If erythropoietin stimulates F cells in sickle cell disease, treatment spaced at 1- or 2-week intervals could result in accumulation of F cells in high enough numbers to inhibit sickling in vivo. Currently, it is estimated that about 20% hemoglobin F per red blood cell is required to protect against sickling.[64] The baboons treated with recombinant erythropoietin had levels of 12% to 26% hemoglobin F per red blood cell. Long-term human trials of this mode of therapy are in progress.

Coagulation Factors

Hemophilia A and B

Classical hemophilia or hemophilia A is a sex-linked disorder that occurs in approximately 20 per 100,000 males. In the blood, factor VIII:C is associated with another protein called von Willebrand factor VIII–related antigen. However, factor VIII:C is principally responsible for the functional clotting activity of the intrinsic complex, generically termed factor VIII. Injection of factor VIII preparations obtained from normal donor blood is currently the principal mode of therapy for hemophilia patients. These factor VIII preparations may carry a certain degree of risk for non-A, non-B hepatitis, and human immunodeficiency virus (HIV), or AIDS virus, has been shown to contaminate various plasma concentrates or preparations containing factor VIII used in the treatment of hemophilia. Most cases of hemophilia A are due to novel mutations of the factor VIII gene.[65] The relatively high frequency of hemophilia A in comparison to other autosomal clotting disorders is the result of localization of this gene on the X chromosome, so that only one affected chromosome is sufficient to produce the disease in males. Several decades ago the mean age of death of hemophiliacs was estimated to be less than 20 years; however, between 1950 and 1970 research on and improved understanding of the disorder led to successful treatment, first with whole plasma and later with fresh plasma concentrates. Human plasma from normal volunteer donors is an inefficient and relatively labile source of this protein, whose concentration is usually 100 to 200 ng/ml. Currently used commercial factor VIII–enriched plasma concentrates have usually been produced from pooled plasmas representing material derived in some instances from thousands of donors. Thus such products are associated with potential undesirable effects of transmitting hepatitis, particularly non-A, non-B, as well as HIV infection.

Recombinant Factor VIII

For these reasons it seemed urgent to provide factor VIII derived directly by recombinant techniques. This was first successfully accomplished in 1984 using factor VIII clones recovered from a bacteriophage λ genomic library.[66–69] The cloning strategy started with a λ phage cDNA library prepared from an individual with four X chromosomes (karotype 49,XXXXY). Initially the cDNA library was screened with a 36-base oligonucleotide probe synthesized to represent appropriate complementary nucleotides based on a sequenced peptide derived from highly purified factor VIII. When the initial hybridization procedure was successful, the clones containing clearly overlapping segments were expanded to cover 28 kb of the human chromosome. Later the clones were further amplified to include 200 kb of the human genome and the entire factor VIII gene. The structure of the human factor VIII gene is shown in Figure 8.10. The DNA sequence of the complete mRNA coding portion of the factor VIII gene, which spanned 9 kb, was then determined. Remarkably, in the initial analyses only two nucleotide differences between the genomic and the cDNA sequences were found within the entire 8860 nucleotides compared. These data suggested a much lower level of sequence polymorphism than had been documented in other genes and were consistent with similar findings related to other X chromosome loci.[70] The complete structural and sequence definition of the human factor VIII gene was a remarkable technical and scientific achievement, since at the time, in 1984, it was the largest gene yet characterized, including nearly 186 kb of the entire human X chromosome.

After the cloning of the gene for factor VIII, cDNA clones containing the complete 2351-amino-acid sequence for human factor VIII were isolated and used to produce biologically active functional factor VIII in cultured mammalian cells. When the recombinant work on factor VIII was begun, the exact site of synthesis was not known, although the liver was considered the most likely primary source. Also, because the concentration of factor VIII was known to be extremely low (200 ng/ml), or about 1/2,000,000 the concentration of serum albumin), it was not likely that cDNA libraries produced from mRNA of a tissue such as the liver would yield any clones producing factor VIII. For this reason the strategy noted briefly above was adopted. First, a genomic clone that expressed a portion of the sequenced part of the factor VIII protein was obtained. From this an exon-containing restriction fragment could be used to identify suitable tissue or cellular sources of factor VIII mRNA by Northern blotting or other hybridization procedures. Then mRNA from the latter tissue sources could be used to make cDNA clones, which could in turn be screened with previously cloned genomic fragments. With this approach, comparative nucleotide sequence analysis could be used to obtain overlapping genomic clones that would eventually include the entire factor VIII gene. Even if a direct tissue source of mRNA could not be found, exons from the genomic clones could be identified and spliced together using simian virus 40 (SV40) recombinant exon expression plasmids. In fact, the successful cloning and expression of factor VIII utilized information obtained from all of these approaches.

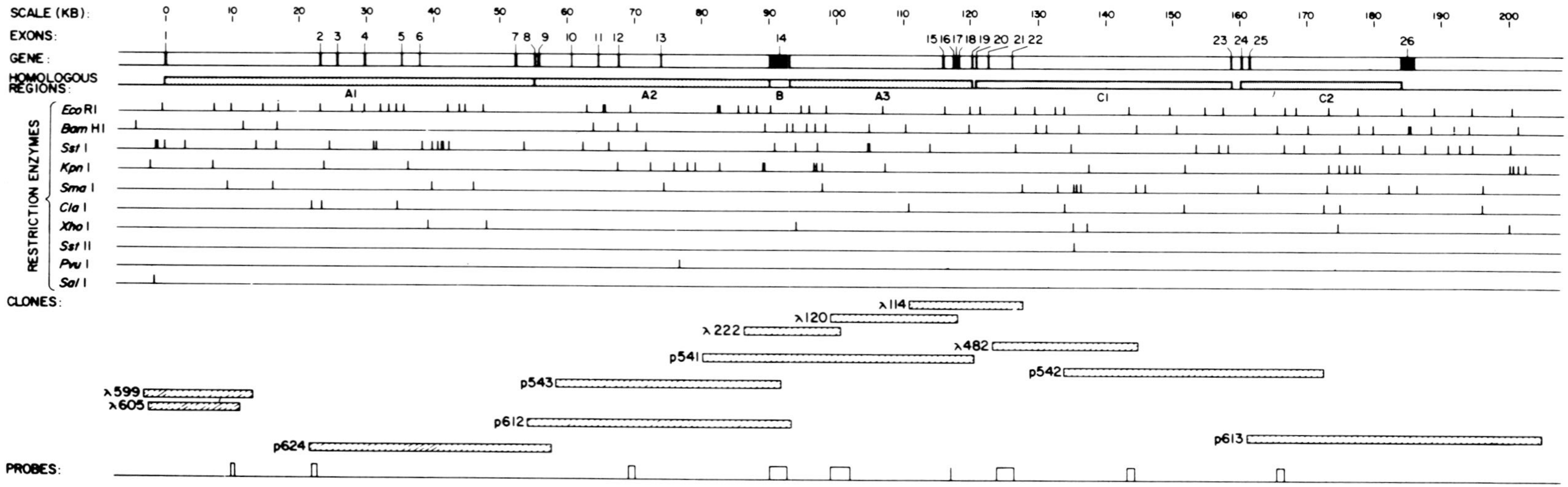

Figure 8.10 Map of human factor VIII gene. The structure of the gene is schematically represented by an open bar; the 26 exons are filled-in areas drawn to scale. The direction of transcription is from left to right. The size scale in kilobases is drawn above the gene; the initiating ATG is positioned at nucleotide 1. Immediately below the gene are the extent of the triplicated A domain, the unique B domain, and the duplicated C domain. The locations of the recognition sites for the 10 restriction enzymes used to map the factor VIII gene are given in the next series of lines. Crosshatched boxes represent the extent of human genomic DNA contained in each λ phage and cosmid clone. Other clones that encompass the same regions are not shown. No genomic clones span the 8.5 kb of intron located between clones λ599 and p624. Other *Eco*RI, *Bam*HI, or *Sst*I sites may lie in this region. The extent of this gap was determined by probing genomic Southern blots with single copy fragments isolated from the 3′ end of λ599 and the 5′ end of p624. The bottom line shows the locations of synthetic oligonucleotide or restriction fragment probes used in the genomic cloning; from left to right, these are the two probes used to map the uncloned region, three 5′ walk probes, the original synthetic 36-mer probe 8.3, and three 3′ walk probes. *Reproduced from Gitschier et al,*[66] *with permission.*

cDNA Sequence

When completely assembled and sequenced, the nucleotide arrangement of factor VIII predicted a polypeptide of 2351 amino acids that included a 19-amino-acid signal region, similar in some respects to the glycoprotein peptide hormones discussed in Chapter 7. Once the primary sequence was established, computer-aided search for internal homologies showed that the factor VIII sequence included three types of domains: a triplicated region of 330 to 370 amino acids referred to as the A domain, a unique B domain of 983 amino acids, and a duplicated C region of 150 amino acids. The arrangement of regions within the single long polypeptide chain was A1-A2-B-A3-C1-C2. Moreover, most of the potential asparagine-linked sites for glycosylation predicted by the primary sequence occurred within the B region. Of interest was the fact that the factor VIII protein sequence showed remarkable homology with the sequence of ceruloplasmin, a copper-containing protein that is known to behave as an acute phase reactant.

Expression of the Factor VIII Gene

The factor VIII gene was assembled and recombinant tailoring was performed to prepare it for expression[69] by placing it between a tandem SV40 early promoter/adenovirus-2 major late promoter and the hepatitis B virus surface antigen gene polyadenylation sequences. In addition, the plasmid contained a murine dihydrofolate reductase cDNA transcribed by an SV40 early promoter to provide a detectable selected marker when expressed in mammalian cells. The final assembled plasmid is shown in Figure 8.11. This plasmid was transfected into a hamster kidney cell line using the calcium phosphate coprecipitation method. To determine whether the transfected factor VIII gene was expressed in the kidney cell line, RNA was extracted from the cells and analyzed by Northern blot hybridization. These experiments clearly showed positive results; moreover, radioimmune assays of the hamster kidney cell line supernatants were positive for factor VIII. These results provided evidence for successful expression of the factor VIII gene. Coagulation assays of the factor VIII product itself indicated that it was active because it improved the clotting activity of plasma from patients with hemophilia A from 120 to 65 seconds and the activity was neutralized by monoclonal antibodies to factor VIII. These studies represented a monumental recombinant DNA achievement in the successful cloning and production of a trace coagulation protein essential for normal body function. Although these initial definitive experiments were carried out a number of years ago, we are still anticipating full-scale production of recombinant factor VIII. The initial work with the genetically engineered material showed a very low level of expression. Technical improvements have now been implemented and full-scale production of recombinant factor VIII is awaited with much enthusiasm. Effective large-scale tissue culture and purification methods need to be developed to ensure production of a high-purity product.

Genetic Mapping and Diagnosis of Hemophilia

Because hemophilia is the most common inherited bleeding disorder in humans, affecting one male in 10,000, it is important to be able to ascertain carriers of the disease and to assay fetal samples for diagnosis and genetic counseling. Initial studies directed at this problem built on the knowledge of the complete nucleotide

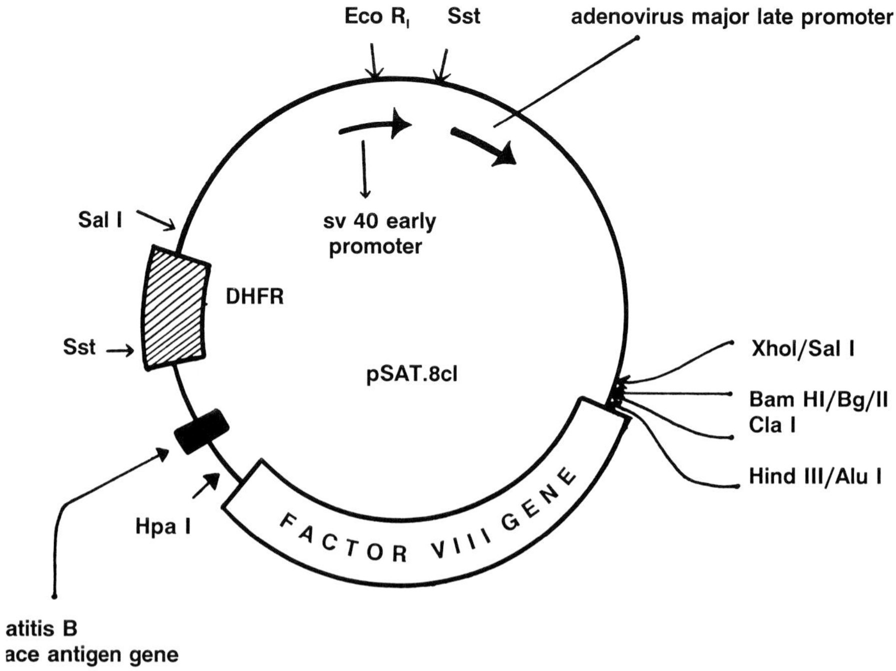

Figure 8.11 Simplified schematic diagram of full-length factor VIII expression plasmid derived from modification of pBR322. The factor VIII and dihydrofolate reductase genes are shown, as well as the tandem SV40 (early) and adenovirus major (late) promoters. The region between the *Sal*I and *Eco*RI restriction sites contains the ampicillin resistance gene. This plasmid was introduced or transfected into a hamster kidney cell line by the calcium phosphate coprecipitation method.

sequence and organization of the factor VIII gene. Early work showed an identifiable polymorphism revealed by DNA restriction digests using the enzyme *Bcl*I,[71] which could be used diagnostically in about 42% of families. Examples of this sort of analysis are shown in Figure 8.12. Interestingly, this study showed a relative paucity of restriction fragment length polymorphisms (RFLPs) with the factor VIII probe. As noted earlier, there are suggestions that DNA polymorphism may in general be more scarce in the human X chromosome than in autosomes. Within the entire 9000 base pairs making up the mRNA coding region of factor VIII genes sequenced from two individuals, only two nucleotide differences were found.[68,69]

Additional studies of individual hemophiliac patients began to turn up a pattern reminiscent of the β-thalassemia story, as it was shown that the disease stems from a heterogeneous collection of genetic lesions at different vulnerable sites within the nucleotide sequence of the gene for factor VIII. Moreover, because the disease can be fatal if untreated, new mutations must be constantly occurring to account for the observed frequency in the general population. Also, there appears to be a spectrum of disease severity, suggesting that a number of different individual mutations are involved in producing the disorder in individual patients.[71,72] As a corollary to this, only about 15% of hemophiliacs generate high levels of antibody to factor VIII. Studies of individual sequence mutations within patients' factor VIII genes have now confirmed that a number of different point mutations, including

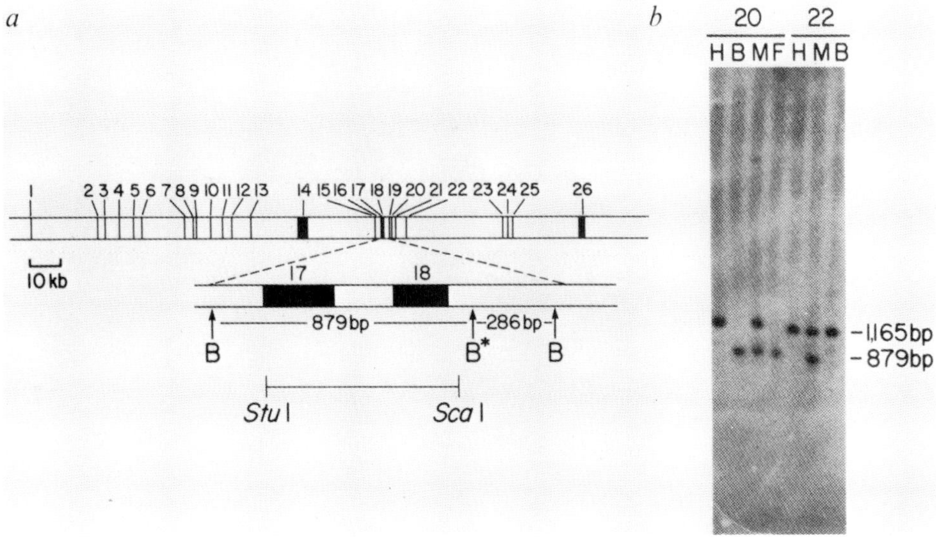

Figure 8.12 Position of the *BclI* polymorphism within the factor VIII gene and an example of its usefulness in the diagnosis of hemophilia. (a) The factor VIII gene, containing 26 exons, is shown schematically. Below it is an expanded portion of the gene covering exons 17 and 18; the positions of the *BclI* sites (B) in the gene are indicated. The presence or absence of the middle site (in the intron just 3′ of exon 18 and marked by an asterisk) is responsible for the two alleles of 879 bp and 1165 bp. The location of the 647-bp *StuI/ScaI* genomic probe fragment is also shown. (b) A Southern blot of *BclI*-digested DNA (5 mg per lane) from members of two different families affected by hemophilia was hybridized to the 647-bp *StuI/ScaI* genomic probe. Family 20 consists of a hemophiliac son (H), whose brother (B) is not a hemophiliac, his heterozygous mother (M), a known obligate carrier of hemophilia, and father (F). Family 22 consists of a hemophiliac (H), his heterozygous mother (M), and his brother (B), who is also a hemophiliac. Blood samples from both patients and families were obtained from the Royal Free Hospital, London. *Reproduced from Gitschier et al,*[71] *with permission.*

in some instances partial deletions, are associated with the disease.[72] In these studies the restriction enzyme *TaqI* proved particularly useful for detecting differences in the DNA of individual hemophiliacs. The recognition site for *TaqI* contains the CpG dimer, which is the major methylation site for human DNA, and these CpG dimers are probably mutation "hot spots" secondary to subsequent deamination of methylcytosine to thymidine, resulting in C → T transitions.[73] Examination of the sequence and general structure of the factor VIII gene indicated that five of the seven *TaqI* sites in coding exons could be converted to nonsense mutations by such a C → T transition. Therefore, when 92 hemophiliac DNA samples were digested with *TaqI*, case-specific *TaqI* site mutations were discovered in four severe hemophiliacs.[72] Of great interest was the finding that each mutation site was different. In one instance loss of a specific *TaqI* site was found, and in another a *TaqI* site was introduced by mutation in an intron. Moreover, two cases of partial deletions of the factor VIII gene were detected in this study. Again, these deletions occurred in case-specific sites. Thus, unlike the single point mutation in sickle cell anemia, the genetic causes of hemophilia A appear to encompass a wide spectrum of mutations and deletions in vulnerable regions of the factor VIII gene.

Sites for Factor VIII Synthesis

Cellular locations for factor VIII synthesis have been sought for some time. A number of studies suggested the liver as the primary site of synthesis, but other sources such as spleen and lung were also implicated.[74–78] Localization of factor VIII antigen by immunological techniques did not distinguish storage sites from actual sites of production. When the gene for factor VIII was cloned, hybridization probes for mRNA in tissues readily became available and were applied to a spectrum of human tissues.[78] Factor VIII mRNA was found in isolated human hepatocytes, in spleen, and in numerous other tissues including lymph nodes and kidney, but not in leukocytes or cultured endothelial cells. During this study spleens and hepatic tissues from several patients with severe hemophilia were studied for factor VIII mRNA. Surprisingly, normal levels were noted, suggesting that some cases of functionally severe hemophilia may be caused by other molecular defects affecting RNA or protein production and that only in some cases is the disease due to complete absence of factor VIII mRNA.

Examining the Spectrum of Hemophilia A Mutations

As work with the cloned factor VIII gene has progressed, various groups have pinpointed the cause of hemophilia in an increasing number of cases. It was long suspected that the disorder resulted from a diverse spectrum of mutations in the large factor VIII gene, and this has been borne out. Examination of several hundred hemophilia A patients by DNA blot hybridization has documented at least seven distinct mutations as shown in Figure 8.13.

As methodology and more experience are gained with genetic screening techniques after the gene corresponding to a disease has been isolated, RFLPs present within or very close to the gene can be employed for genetic screening. A human DNA fragment (St.14) that recognizes a very polymorphic locus in the q28 region of the X chromosome containing the hemophilia locus has been identified.[79] This probe appeared to be a useful tool for early prenatal diagnosis using chorionic biopsy material. The probe had previously been mapped to the distal end of the X chromosome, a short genetic distance from the locus for hemophilia A. Analysis of 13 families showed striking linkage of St.14 and hemophilia in 12, indicating high coefficients of linkage. Use of this probe was estimated to give an accuracy of 93%. Moreover, use of the probe combined with biological assays could provide an accuracy of at least 96% in carrier analysis. Although statistically useful, this approach is still not a foolproof method for carrier or prenatal identification, and additional refinements are clearly needed.

The polymerase amplification technique has also added a new dimension of accuracy and technical improvement. This technique has been utilized by Kogan et al.[80] with intragenic factor VIII polymorphisms recognized by *Xba*I and *Bcl*I and the *Taq* polymerase. It was found that an additional dividend could be achieved by raising the temperature of the polymerase reaction from 37°C to 63°C, which produced a vast majority of fragments of the correct size and sequence. This occurred because at the higher temperature priming occurred only when template and primer were exactly matched, whereas at 37°C the oligonucleotide primers might hybridize to nontarget sequences. This additional specificity made it possible to amplify segments of the human genome containing important disease-related allelic variations that alter defined enzyme restriction sites and simplify determi-

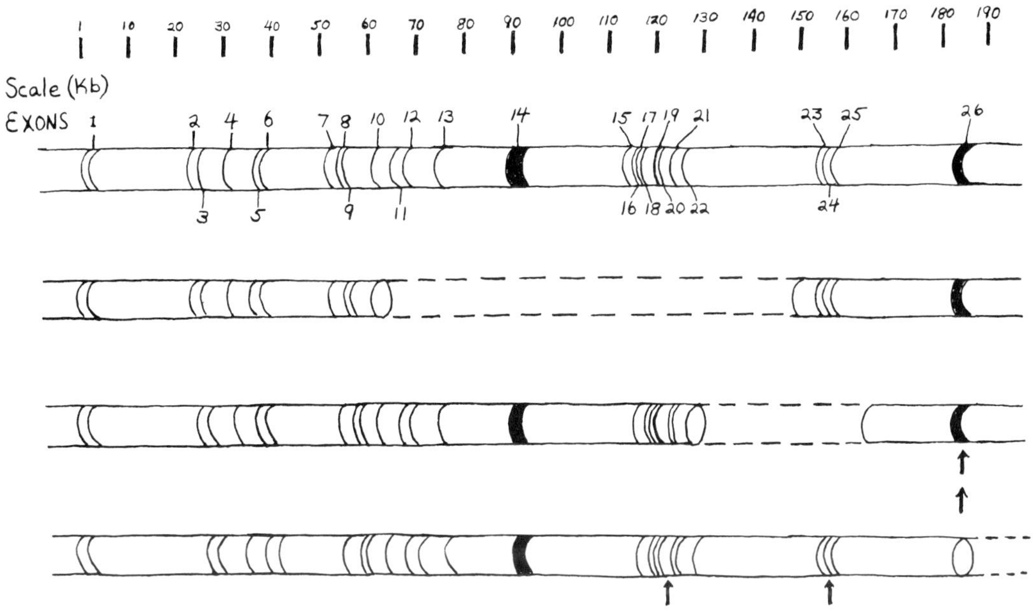

Figure 8.13 The factor VIII gene, one of the largest genes cloned to date (186 kb), includes almost 0.1% of the entire X chromosome. The unusually large exon 14 (3.1 kb) encodes the entire B domain of factor VIII, which is removed from the protein during proteolytic activation by thrombin. In addition to the intact structure (top), diagrams show locations of partial deletions (dashed lines) and single base mutations (arrows) found in individual patients. Two single base mutations at exon 26 occurred in the same codon. One led to production of a truncated protein in a patient with severe hemophilia, the other to a single amino acid substitution associated with mild disease. *Reproduced from Vehar GA, Lawn RM: The cloning of factor VIII and the genetics of hemophilia A. Hosp Pract, May 15, 1986;21:5, with permission.*

nation of alleles by restriction enzymatic digestion. The higher-temperature *Taq* polymerization procedure with its additional specificity could be applied to a number of other restriction site polymorphisms, such as those in cystic fibrosis and Duchenne's muscular dystrophy. The speed, ease, and relatively low cost of this approach appear promising.

Hemophilia B

Hemophilia B (Christmas disease) is an inherited, recessive sex-linked clotting disorder that is caused by a defect in clotting factor IX. It occurs in males at a frequency of about 1 in 30,000. As in hemophilia A, individual patients vary in the severity of clinical symptoms and in their clotting activity. Some heterogeneity in individual molecular factor IX defects is undoubtedly present. The initial characterization of the normal factor IX gene provided the means of studying hemophilia B.[81] It was observed that approximately 40% of hemophilia B patients have a normal plasma concentration of factor IX[82] and probably have point mutations in the coding sequence producing a single amino acid changes.[83] A few patients produce antibodies to the factor IX injected during replacement therapy, and in these instances the defect appears to be secondary to large deletions in the factor IX gene.[84] The remaining 60% of individuals with hemophilia B have either markedly reduced or undetectable factor IX levels. Using molecular cloning techniques similar to those described for hemophilia A, a number of point mutations have been described in hemophilia B. In one instance, this was done by using an oligonucleotide probe that detected a single $G \rightarrow T$ point mutation within the obligatory GT of a donor splice junction.[85] The same technique was capable of detecting the mutant allele expression in an asymptomatic heterozygote.

Using the cloned factor IX gene transfected into mammalian cells as an expression vector, factor IX has been produced by several groups.[86,87] The cell lines transfected were shown to produce small amounts of active factor IX.

Carrier Detection

Three RFLPs have been defined within the factor IX gene that allow inheritance of specific alleles to be determined.[88,89] This allows carrier detection and prenatal diagnosis with a very high degree of certainty. As new studies have appeared, additional probes and restriction fragments have been identified that will also prove useful in kindred and carrier analysis.[90,91]

Factor IX Locus in Hemophilia B Inhibitors

As structural and clinical knowledge concerning the genes and gene products in hemophilia A and hemophilia B increased, it appeared that large gene deletions were often associated with eventual production of antibodies against the normal factors given periodically during therapy. This has been directly assessed in the case of the factor IX locus in nine hemophilia B patients who produced anti–factor IX inhibitors.[92] Two of the inhibitor patients had deletions of the entire factor IX gene. Two other inhibitor patients who were brothers had presumably identical complex rearrangements of the factor IX gene involving two separate deletions. Five other hemophilia B inhibitor patients appeared to have a structurally intact factor IX gene. Originally, Giannelli et al.[84] had proposed that hemophilia B patients produced antibodies to factor IX because of their lack of immune tolerance to deletional epitopes in the autologous factor IX mutants. This could be due either to a lack of expression of any or a significant part of the factor IX antigens or to synthesis of an altered factor IX lacking certain immunodominant antigenic epitopes. In the report by Matthews et al.,[92] several noninhibitor patients did not produce any measurable plasma factor IX. It is possible that such patients synthesize minute amounts of factor IX that are undectable in conventional assays. Alternatively, these patients may have lost the ability to synthesize factor IX after the development of actual tolerance. In summary, of 14 hemophilia B inhibitor patients reported, 8 had a detectable deletion of the factor IX gene; in contrast, only 2 of 19 hemophilia A inhibitor patients studied had a detected gene deletion.[84,92–94]

Von Willebrand Disease

The general pattern of absence of gene product and correlation with development of reactive immune responses has also been examined in patients with von Willebrand disease who develop alloantibodies.[95–97] This disorder is associated with functional deficiency of a glycoprotein composed of 250-kd subunits essential for platelet adhesion to sites of vascular damage. As in the case of factors VIII and IX, the sequence of von Willebrand factor has been determined and its cDNA cloned.[98,99] In the study reported by Shelton-Inloes et al.[95] two unrelated patients with severe recessive von Willebrand disease and alloantibodies to von Willebrand factor were found to have large deletions in the von Willebrand gene. By contrast, 19 patients with autosomal dominant type I von Willebrand disease and 17 additional patients with recessive von Willebrand disease did not have detectable deletions, and none of these 36 patients showed antibodies to von Willebrand factor. These findings tend to support the idea that large gene deletions are associated with an increased tendency to produce antibody to the deficient clotting factor. Similar observations have been reported by Ngo et al.,[97] who found alloantibodies only in kindreds with homozygous rather than heterozygous deletions. This particular point is of special interest because the general interplay between specific alterations in gene structure and exposure to the exogenous gene product by transfusion and the host immune response is still not fully understood. Despite the occurrence of a gene deletion, some individuals might be unable to recognize the corresponding protein antigens as foreign—even though these same antigens are not produced endogenously—because of possible inherited human leukocyte antigen (HLA)-linked features of their immune response. Eventually, precise definition of the epitope specificities of these interesting alloantibodies may help to unravel the puzzle.

Other Clotting Factors

A number of other coagulation factors have been cloned and sequenced. Among them, human coagulation factor V is a high molecular weight plasma glycoprotein that is required for rapid thrombin formation and normal clotting.[100] This protein circulates in the blood as a large single polypeptide chain (330 kd) with little or no coagulant activity. During blood clotting, factor V is converted to factor Va by thrombin through a process of limited proteolysis; this exposes available binding sites for factor Xa and prothrombin. Factor V itself is composed of a 110-kd heavy chain and 76-kd light chain. The two chains are held together by calcium ions. The remainder of the factor V molecule is released as a large connecting fragment rich in carbohydrate. Factor V binds to cell surfaces and negatively charged phospholipid surfaces through the light chain and increases by about 10,000-fold the rate of prothrombin activation. Factor V is readily inactivated by activated protein C.

Human factor V was cloned from a phage λgt11 expression library using an affinity-purified antibody to factor V.[101] Clones were also identified using hybridization to an oligonucleotide complementary to the DNA coding for the amino acid sequence Met-Tyr-Glu-Gln-Glu-Trp-Val present in the light chain of human factor V. The isolated clones that contained the largest cDNA insert had 2970 nucleotides and coded for 938 amino acids. This region included 651 amino acids from the carboxyl terminus of the factor V light chain and 287 amino acids that were part of the connecting region of the protein. The amino acid sequence of the light chain of factor V is 40% homologous with the carboxyl-terminal fragment of human factor VIII. These fragments have a similar domain structure, which includes a single ceruloplasmin-related domain followed by two C domains. These similarities suggest that factor V, factor VIII, and ceruloplasmin are a family of related proteins that arose during evolution through gene duplication of the A domain.[102,103]

Factor XII, or Hageman factor, has also been successfully cloned and sequenced using a human cDNA library and mixtures of synthetic oligonucleotides.[104] Structural comparison showed extensive sequence identity with tissue-type plasminogen activator as well as with fibronectin. Factor XIII, the clotting factor essential for covalent stabilization of the fibrin clot, has been cloned and sequenced[105] from a human placental λgt11 cDNA library. The gene for factor XIII α chain mapped uniquely to chromosome 6. Northern blot analyses clearly demonstrated mRNA for factor XIII in human placental tissues as well as several monocyte-like cell lines. The availability of cDNA probes for clotting factors such as XIII should provide an invaluable tool for examination of their role in body homeostasis.

Gene Rearrangements in Hematological Disorders

Malignant conditions arising in cells of the hematopoietic system initially were classified on the basis of a histological and cytological system.[106] Later, with the rapid development of expression of various B or T cell lineage markers, an immunological classification emerged.[107–110] Subsequent analysis of immunoglobulin and T cell receptor genes has provided an additional dimension for classification and insight into the cellular origins of specific types of leukemias and lymphomas. Much of our current understanding of abnormal hematopoietic differentiation and expression naturally arose from careful studies of normal immunoglobulin[111–116] and later T cell receptor gene expression. New information in this area continues to emerge, particularly in the case of the various types or families of T cell receptors.

B Cell Gene Rearrangements

Studies of normal pre-B cells maturing to immunoglobin (Ig)-bearing B cells have shown recombination of the light chain variable (V_L) and joining (J_L) segments prior to or concomitant with eventual integration with light chain constant regions (C_L).[117,118] Since both κ and λ light chain immunoglobulins are used in approximately 60% and 40% of human B cells, respectively, an ordered rearrangement of human Ig light-chain genes occurs in which κ chain rearrangements precede those of λ chains.[117] This was demonstrated by study of highly purified λ-bearing B cells from normal subjects. It was found that over 95% of the collective κ genes in the λ B cells were no longer in their germ-line form; the majority were deleted and the remainder were present but in the rearranged state.

With the knowledge of normal Ig gene rearrangement sequences, this approach was applied to an interesting group of patients with B cell leukemia, which is generally regarded as a stage in early lymphoid development arrested before active Ig synthesis is initiated.[118,119] It was found that although most of the pre–B cell leukemias had not acquired the capacity to synthesize detectable cytoplasmic immunoglobulin, they had undergone Ig gene rearrangements that appeared to commit them to eventual B cell development. Moreover, the patterns of Ig gene rearrangement in the pre-B leukemic cells indicated the schedule for Ig gene rearrangements.[120,121] This showed that heavy-chain (μ) genes were first, followed by κ and then λ light chains.

In the original work eight patients were studied. Leukemic cells showed no sheep cell rosetting behavior nor any evidence of surface Ig. All patients showed acute lymphocytic leukemia (ALL)-associated antigen[118] and six patients were positive for Ia antigen. In this study genomic DNA was obtained from leukemic cells and compared to normal germ-line DNA harvested from the patients' fibroblast or normal-remission leukocytes. C_μ, C_κ, and C_λ DNA probes were used to detect Ig gene rearrangements. Representative results for these informative cases are shown in Figure 8.14. The leukemia patients could thus be placed in a scheme of development from primitive stem cell through pre–B cell leukemia to B cell leukemia, depending on the succession of Ig gene rearrangements and deletions. It was thus possible to devise a tentative scheme of cell development from stem cell to B cell leukemia. The basis for such a model is shown in Figure 8.15. In this scheme a stem cell would be expected to show all of its immunoglobulin genes in the germ-line configuration and to initiate immunoglobulin gene rearrangements by joining V_H, D_H, and J_H segments to provide the complete nucleotide units toward forming a μ chain. Because κ light chain genes precede λ light chain rearrangements, the two C_κ alleles could undergo a functional, aberrant, or deletional rearrangement. Cells containing rearranged μ and κ alleles should then be capable of maturation to B cells bearing surface Ig. In like manner, λ genes could rearrange to produce either complete surface Ig or aberrant patterns, thus preventing further progression toward full development.

On the basis of immunoglobulin gene rearrangements, patients with non-T, non-B leukemias showed malignant transformation of cells that were originally committed to B cell differentiation. This was a major step toward insight into cell differentiation pathways in this group of patients, because previously only 15% to 20% of them had been classified as pre-B on the basis of cytoplasmic μ chain staining.[122,123] Five of the eight leukemic patients studied by Korsmeyer et al.[118] had clear-cut Ig gene rearrangements but no detectable immunoglobulin protein product.

Acute Lymphocytic Leukemia

This general approach was next applied to a much larger group of 37 cases of acute lymphocytic leukemia.[121] All 12 cases of the T cell type showed germ-line κ and λ Ig gene patterns, and 11 of 12 showed germ-line heavy chain genes. By contrast, all 25 cases in the non-T, non-B classification, lacking definitive T cell markers or surface Ig, showed rearranged Ig genes. This indicated that they represented precursor cells already committed to the B cell lineage. When the patterns of immunoglobin gene rearrangements in this large group of patients were correlated with the results of cell surface marker studies, it appeared that expression of HLA-DR preceded reactivity with BA-1, a monoclonal antibody defining B cell populations.[122] Furthermore, BA-1 reactivity occurred before positive antibody reactions were detected with the common acute lymphoblastic leukemia antigen (CALLA).[123] The parallel analysis of cell surface markers and Ig rearrangements is shown diagrammatically in Figure 8.16. It was postulated that lack of Ig production by the majority of leukemic B cell precursors might be due to a variety of factors involved in B cell differentiation, including mistaken rearrangements or deletion of all D_H gene families.

Malignant Lymphomas

Gene analytic techniques quickly proved extremely useful in the analysis of malignant lymphomas. Since it had been recognized from cell marker studies that most non-Hodgkin's lymphomas were probably of B cell origin,[124] it was useful to explore this clonal B cell expression in terms of Ig gene rearrangement and expression patterns. In an analysis of patients with various lymphoproliferative diseases, the abnormalities found by cell surface staining and flow cytometry

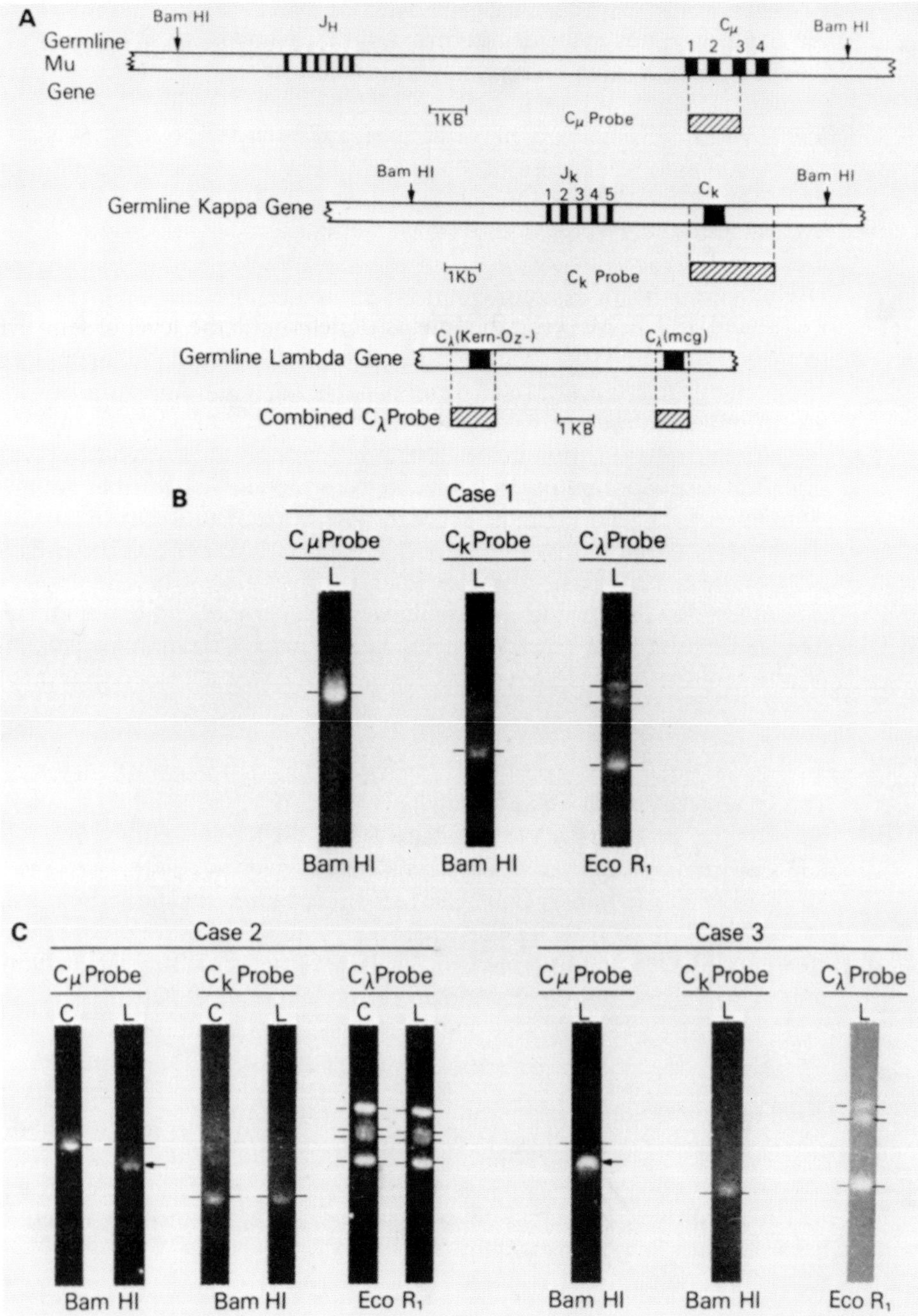

Figure 8.14 (A) Human C_μ, C_κ, and C_λ DNA probes utilized to detect germ-line and rearranged immunoglobulin genes. In order to demonstrate μ chain gene rearrangements, the *Bam*HI digests of the leukemic lymphocytes were hybridized with a 1.5-kb embryonic *Eco*RI probe. The *Bam*HI digests were also hybridized to a probe consisting of a 2.5-kb embryonic *Eco*RI C_μ-containing fragment. The combined C_λ probe used to detect λ light chain gene rearrangements in *Eco*RI digests consisted of a 1.2-kb embryonic *Bam*HI-*Eco*RI fragment containing the Mcg C_λ gene. This combined C_λ probe is capable of cross-hybridizing to all of the nonallelic C_λ genes. (B) Case 1: a germ-line pattern for μ, κ, and γ genes. The C_λ-containing gene fragment in the leukemic lymphocytes (L) of case 1 was a single band

showed a high degree of correlation with monoclonal Ig gene rearrangements as detected by Southern blot analysis.[125] These patients included individuals with non-Hodgkin's lymphoma and a few with hairy-cell leukemia, chronic lymphocytic leukemia, lymphosarcoma, and Hodgkin's disease. Earlier work had established the usefulness of this technique in studying human lymphoid neoplasms and in differentiating a malignant proliferative process from benign lymphoid hyperplasia.[126] In the past, monoclonality in lymphoid neoplasms had been identified by use of immunofluorescence or glucose-6-phosphate dehydrogenase markers.[127] Southern blot analysis of DNA from lymphoid tumors has proved to be of great usefulness in analyzing lymphomas or lymphoid neoplasms without a predominant cell immunotype. Mixing experiments to determine the level of sensitivity of this approach indicated that when genomic DNA from a T cell with germ-line Ig genes or from a polyclonal source of B cells such as normal human tonsils was admixed in various ratios with DNA from a monoclonal B cell line showing clear-cut Ig gene rearrangements, the clonal rearrangement was detectable when only 10% of the DNA hybridized contained rearranged Ig genes.[126] The great precision of this approach is illustrated in Figure 8.17 from the study by Arnold et al.[126] These studies also showed the value of Southern blotting in establishing the diagnosis of lymphoma in cancers of uncertain types, such as case 1 in Figure 8.17.

Immunoglobulin gene rearrangements in leukemias and lymphomas of the B cell series at all stages of maturation are useful because they serve as markers for individual tumors. Their identification after many courses of chemotherapy or radiation may indicate tumor persistence or of early recurrence. With other commonly used methods it is often very difficult or impossible to detect malignant cells present in small proportions in various body fluids or tissue specimens. The use of in situ or extracted DNA hybridization procedures capable of detecting 5% to 10% Ig gene rearrangements adds another dimension of sensitivity to diagnosis.

Application of Southern blotting to DNA from cells of patients with diffuse histiocytic lymphomas using Ig gene probes showed heavy and light chain gene rearrangements in a majority, thus placing these tumors in the B cell lineage.[128] The absence of Ig gene rearrangement in one patient indicated that the tumor might represent a true histiocyte or uncommitted stem cell.

identical to the 19.3-kb *Bam*HI germ-line fragment displayed by the non–B cell sources of DNA from 30 individuals (indicated by lines extending from the gel). Likewise, the C_κ probe revealed the κ genes to be in their germ-line configuration in a 12.0-kb *Bam*HI fragment similar to other non–B cell sources of DNA (indicated by lines). Case 1 displayed a germ-line configurations of λ chain genes referred to as a "type I" pattern, consisting of an upper band of 16.0 kb (actually a close doublet of 16.2 and 15.8 kb), a 14.0-kb fragment, and a lower 8.0-kb fragment. (C) Cases 2 and 3: μ genes rearranged, germ-line light chain genes. Peripheral leukocytes obtained during a clinical remission served as a control (C) for the germ-line pattern of immunoglobulin genes in case 2. The leukemic lymphocytes (L) of cases 2 and 3 showed a rearranged μ gene (arrow) with apparent loss of the other C_μ allele (the expected germ-line positions of the gene-containing fragments are indicated by lines extending from the gels). The κ and λ genes are in germ-line configurations. The type II germ-line configuration for C_λ genes is displayed by case 2 and consists of an upper 18.0-kb fragment, the close doublet at 16.0 kb (16.2- and 15.8-kb fragments), and the 14.0-kb fragment. These two germ-line configurations of C_λ genes in humans result from inherited differences in the location of *Eco*RI sites surrounding the C_λ genes. *Reproduced from Korsmeyer et al,*[118] *with permission.*

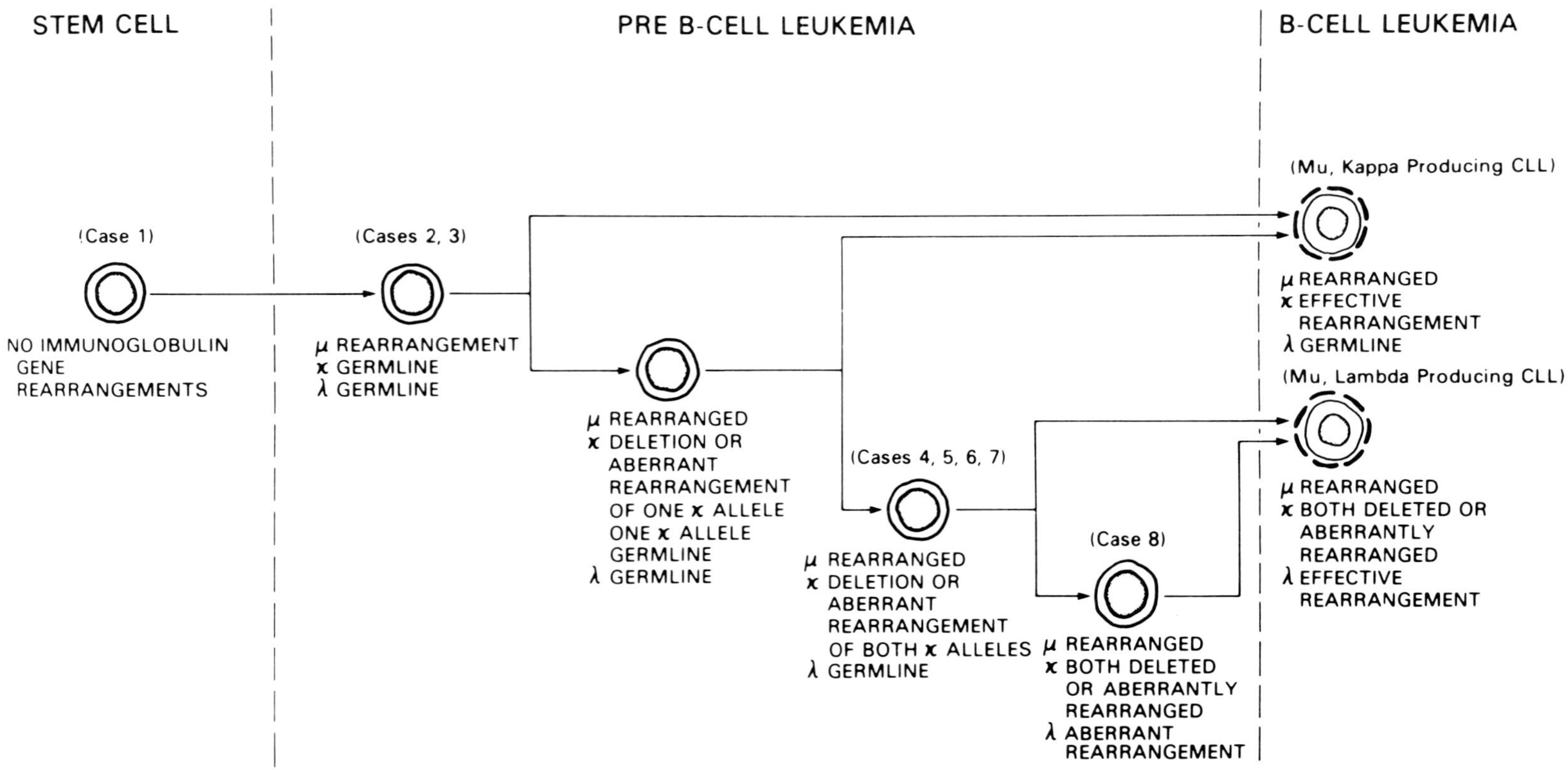

Figure 8.15 Patterns of immunoglobulin gene rearrangement in pre–B cell leukemia. In this hypothetical scheme of differentiation, a stem cell or B cell lymphoid precursor with germ-line immunoglobulin genes would initiate immunoglobulin gene activity by rearranging the μ gene (a pattern seen in two cases). In general, κ gene rearrangements would follow, and, if effective, the cell could mature into a B cell bearing surface μ and κ chains while retaining germ-line λ genes. If all the κ gene recombinations were aberrant or deletional, attempts at effective λ gene rearrangement could follow. *Reproduced from Korsmeyer et al,*[118] *with permission.*

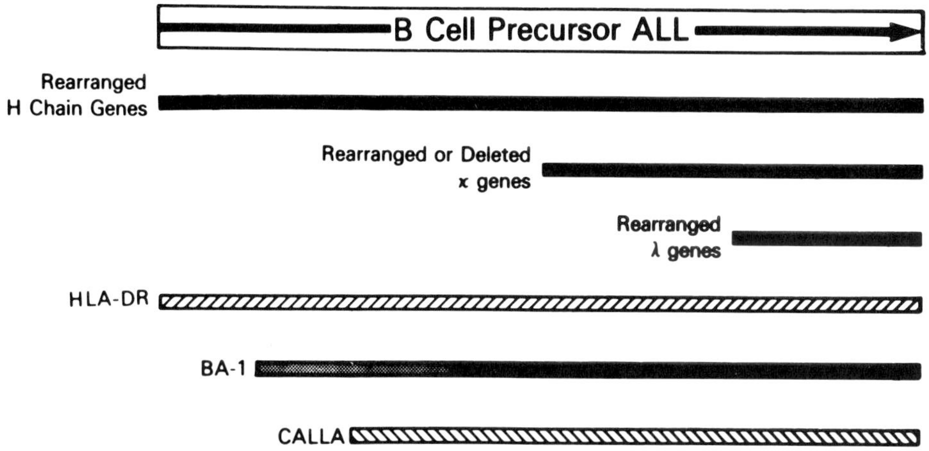

Figure 8.16 A coordinated sequence of Ig gene rearrangements and B cell surface antigen expression. Comparisons here suggest that rearranged H chain genes and Ia-DR molecules precede the appearance of BA-1 reactivity and CALLA. Cases with recombined light chain genes always had rearranged heavy chain genes and usually displayed all three of the surface antigens. *Reproduced from Korsmeyer et al,*[121] *with permission.*

Linkage Infidelity and Gene Rearrangements

There is evidence that Ig gene rearrangement may not always be restricted to leukemias or lymphomas of B lineage but may also occur in phenotypically defined acute myeloid leukemias[129] and acute T cell lymphocytic leukemia.[130] In some instances even biphenotypic expression of both myeloid and immunological markers could indicate that leukemic transformation occurred in an immature stem cell capable of differentiating into two distinct lineages.[131,132] A simple theory of irreversible lineage commitment to one differentiation line cannot explain apparent instances of anomalous myeloid or lymphoid markers occurring outside their expected lineages.[132,133] These phenomena, which include the identification of nuclear terminal deoxyribonucleotidyltransferase (TdT) in leukemic myeloblasts[134] and the inappropriate rearrangement of Ig or T cell receptor genes in lymphoid or myeloid lineage cells,[135–137] are now referred to under the collective term lineage infidelity.[138] It could be that this anomalous situation reflects normal stages of cell differentiation that are frozen or somehow selected by the leukemic transformation. Therefore lineage infidelity has also been called lineage promiscuity to indicate a physiological rather than a dyssynchronous process.[133]

This subject has been investigated by Seremetis et al.[139] with the idea that gene rearrangement and TdT expression might be associated and represent abortive activation of lymphoid line initiation before irreversible commitment. Results of this study indicated a high frequency (>60%) of Ig and/or Tβ gene rearrangements in a subset of acute myelogenous leukemia patients selected on the basis of expression of the lymphoid lineage TdT marker. These findings appear to indicate that TdT must be involved in both Ig and T cell receptor rearrangements and that true linkage infidelity is not present. Alternatively, leukemias with TdT expression and both Ig and T cell receptor rearrangements may represent a clonal expansion of a normal hematopoietic precursor that is scarce in normal hematopoietic differentiation. Furthermore, these findings raise a note of caution concerning the use of Southern

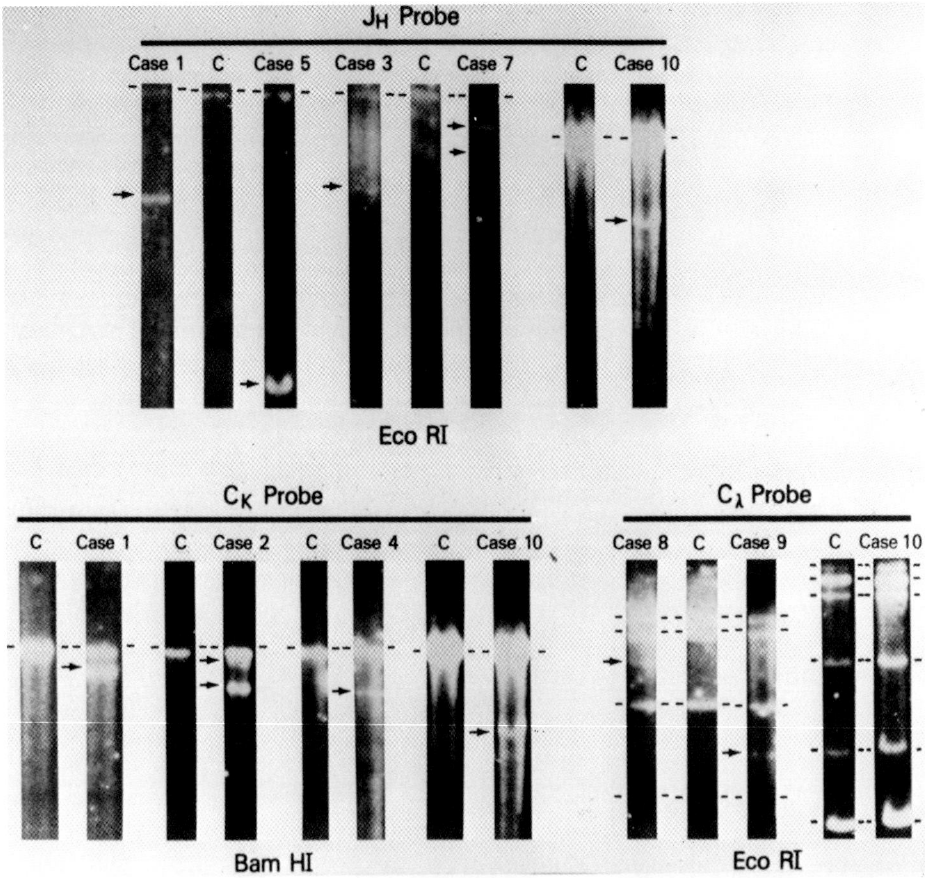

Figure 8.17 Representative Southern blot analyses of immunoglobulin genes from lymphoproliferative tissues and nonlymphoid control DNA(C) with germ-line immunoglobulin gene configurations. The dash marks indicate the germ-line position of the probed gene within the particular endonuclease digests and the arrows show the position of the rearranged alleles. Case 1, undifferentiated malignant tumor; case 5, follicular and diffuse mixed-cell lymphoma; case 3, diffuse large cell lymphoma; case 7, follicular small-cleaved cell lymphoma; case 10, atypical follicular hyperplasia; case 2, follicular large cell lymphoma; case 4, diffuse mixed-cell lymphoma; case 8, follicular small-cleaved cell lymphoma; case 9, diffuse large cell lymphoma. Note that C_g in case 10 is not rearranged, since the extra bands seen represent a restriction fragment length polymorphism shared by the lymph node DNA and the same patient's nonlymphoid control DNA. No such polymorphisms exist for the human J_H or C_k loci, so placental DNA is used as a control in the other cases. *Reproduced from Arnold et al,*[126] *with permission.*

blot analysis and Ig as well as Tβ gene rearrangements for absolute identification of clonal lymphoid populations; further comparative analyses are clearly in order.

Premalignant Conditions

Immunoglobulin gene rearrangement analysis has been applied to several human disorders thought to be premalignant or associated with eventual neoplastic transformation. Angioimmunoblastic lymphadenopathy was studied by DNA Southern blot analysis.[140] Lymph nodes and peripheral blood from these patients showed clones of lymphoid cells with either Ig or T cell receptor gene rearrangements, which in some instances regressed during the course of disease. These findings were of particular interest because the proliferating lymphoid clones did not always appear to be fixed in a rearranged gene phenotype but actually disappeared during the course of disease in several patients. A lymph node from one patient was involved by immunoblastic lymphoma and showed an additional gene rearrangement pattern not observed in premalignant samples from the same individual. These studies indicate that presence of one or more rearranged non–germ line bands in Southern blots of lymphoid tissues from patients with angioimmunoblastic lymphadenopathy do not always indicate malignancy. They are also pertinent to the problem of B cell clonal proliferation in patients with immune deficiency or profound immunosuppression, as in the case of marrow or cardiac transplantation.[141,142]

Sjögren's Syndrome and Immunoglobulin Gene Rearrangement

Immunoglobulin gene rearrangements have been uniformly detected in benign lymphoepithelial lesions found in association with Sjögren's syndrome.[143] Southern blot hybridization techniques were used to examine DNA from salivary gland tissues from eight patients with benign lymphoepithelial lesions. Ten samples from the eight patients all revealed rearrangement of both heavy-chain and light-chain Ig genes. One patient had two benign lymphoepithelial lesions removed 2 years apart. The heavy-chain rearrangements and the rearrangements detected with κ light chain genes in the two separate lesions were entirely different. Because detection of a gene rearrangement has been regarded as strong evidence for neoplastic lymphoid proliferation, these findings were of considerable interest. Patients with Sjögren's syndrome and lymphoepithelial tumors are generally regarded as having increased susceptibility to the development of non-Hodgkin's lymphoma. Although the clonal expansions in this study may reflect premalignant changes, this appears unlikely in view of their clinical regression. Another possible explanation is that the clonal expansions reflect a generalized defect in immune regulation that is intrinsic to this autoimmune disorder. Parallel observations on normal human tissues involved in polyclonal B cell activation or a normal immune response would be useful as a baseline for evaluating the significance of Ig gene rearrangements in disorders like Sjögren's syndrome, in which a small but significant proportion of patients eventually develop a lymphoid malignant process.

T Cell Receptor Genes and Their Rearrangements

In parallel with the work on Ig gene rearrangements and interpretations of clonality and potential malignancy, the same approach was applied to the analysis of T cell receptor genes in human leukemias and other lymphoproliferative disorders. One of the earliest aspects examined was T cell receptor β chain rearrangement.[144–147] Initially, virtually all T cell leukemias and most T cell lymphomas examined showed β chain gene rearrangement for the T cell receptor. As in the cases of Ig genes and Sjögren's syndrome, some anomalous results were recorded, including T cell receptor rearrangements in some chronic lymphocytic leukemias and immunoblastic lymphomas. A schematic diagram of the T cell receptor β chain gene is shown in Figure 8.18. Clear-cut rearrangements are seen in Figure 8.19. Beta chain rearrangements were readily detected in a patient with human lymphotropic virus type

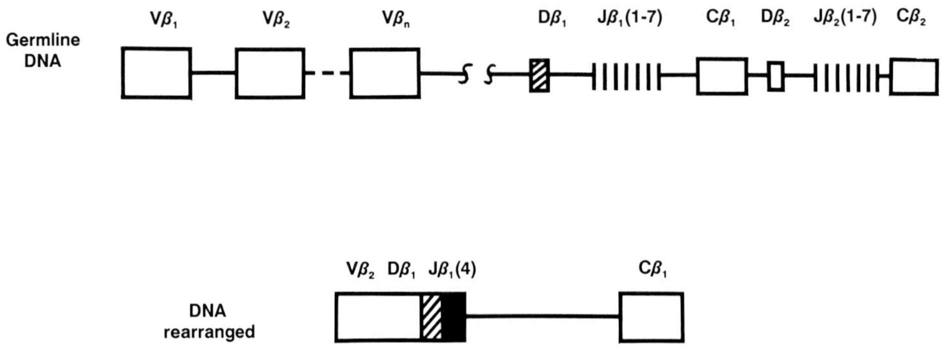

Figure 8.18 T cell receptor β chain gene in germ line (above) and reassembled (below). Variable (V_β), diversity (D_β), joining (J_β), and constant (C_β) segments are rearranged, followed by primary mRNA transcription from the rearranged segments below.

I (HTLV-1), as well as in a patient with cutaneous T cell lymphoma of the Sézary type.[147] A patient with T gamma lymphocytosis, a disorder that has not yet been defined in terms of a clonal T cell neoplasm, was studied by Bertness et al.,[146] who found no T cell receptor rearrangements. Moreover, the method proved useful in patients who were thought to develop a separate or new malignant process during long-term treatment and follow-up. In such instances, changes in T cell gene rearrangement patterns clearly identified emergence of a second or new malignant process.[146] Waldmann et al.[147] studied a patient with T_8 lymphocytosis and granulocytopenia and anemia (a syndrome characterized as a unique clinical entity by several groups[148–150]) and found definite T cell β chain rearrangement, making it likely that a malignant clonal expansion of cells was present.

T Cell Receptor Rearrangement Order

As in the case of Ig genes, T cell receptor genes undergo rearrangement in an ordered progression. The first to rearrange is the T cell γ chain gene, followed by T cell β chain gene and finally α chain gene rearrangement.[151] Studies of a number of

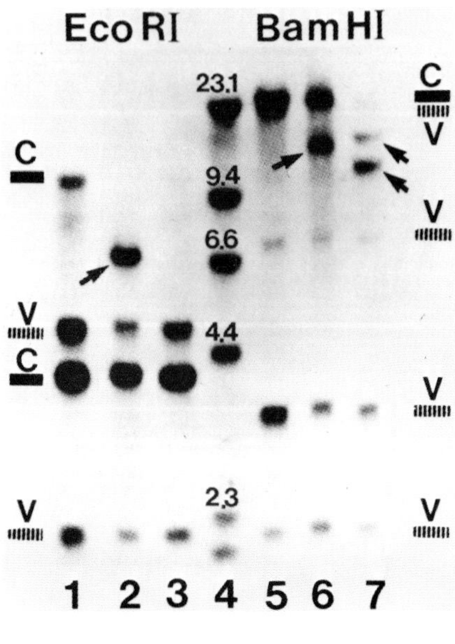

Figure 8.19 Rearranged DNA fragments bearing the gene (arrows) for the constant region of the β chain of the T cell receptor in a patient with T4 chronic lymphocytic leukemia (lanes 2 and 6) and in a patient with adult T cell (T4) lymphoma (lanes 3 and 7). The germ-line positions of the constant (C) and variable (V) β chain bands after *Eco*RI digestion (lanes 1 to 3) are indicated at the left and illustrated in DNA prepared from a thymoma (lane 1); the positions after *Bam*HI digestion (lanes 5 to 7) are indicated at the right. *Reproduced from Aisenberg et al,*[145] *with permission.*

leukemic cell lines have shown T_8 chain genes rearrangement in leukemic B cells as well as T cells.[152] Use of the T cell receptor α chain genes as evidence of rearrangement has not proved practical, primarily because of the dispersed organization of the J_α regions in the α chain locus.[153]

In a further assessment of the specificity of T cell receptor rearrangement for lineage of tumor cells in certain groups, T cell β and γ chain rearrangements were studied in 100 consecutive B cell lymphomas and B cell chronic lymphocytic leukemias by Aisenberg et al.[154] In only two cases were T cell receptor rearrangements detected, and γ chain rearrangement was not recorded in any of the patients. This is in sharp contrast to the findings in non-T, non-B acute lymphocytic leukemias, where almost half the patients examined showed rearrangement of one allele of the T γ chain and a significant proportion showed rearrangement of one T β allele.[155]

The question of T cell receptor expression and rearrangement in acute pre–B cell lymphoblastic leukemias was also addressed by examination of α, β, and γ T cell receptor gene activation in a relatively broad spectrum of patients with this disorder.[156] A developmental hierarchy of T cell receptor gene activation was suggested, in parallel with studies of Ig gene activation and cell surface marker studies. A coordinated sequence of gene expression based on these observations is shown in Figure 8.20. As in previous work, there was some evidence of what was termed lineage spillover, in that 10% of patients with acute T cell lymphocytic leukemia showed immunoglobulin heavy-chain gene rearrangements. Likewise, patients characterized as being pre-B showed a surprising amount (45%) of T γ chain receptor rearrangement as well as T β chain rearrangement (20%). Whether these findings are related to so-called lineage fidelity remains to be determined. As pointed out by Korsmeyer,[157] the final gold standards of lineage determination remain to be established, since lineage spillover or lineage promiscuity appears to

Figure 8.20 Proposed coordinated sequence of T cell receptor (TCR) rearrangement and expression plus cell surface antigen expression in T cell acute lymphoblastic leukemia (T ALL). 3A-1 appears to precede TCR gene rearrangement. The λ and β TCRs rearrange early in development, but in no fixed order. Most T ALLs are mature enough to have rearranged both λ and β TCRs and approximately one-half express the α TCR locus. The T11 molecule follows 3A-1 and T3 appears around the time of α TCR expression. *Reproduced from Felix et al,[156] with permission.*

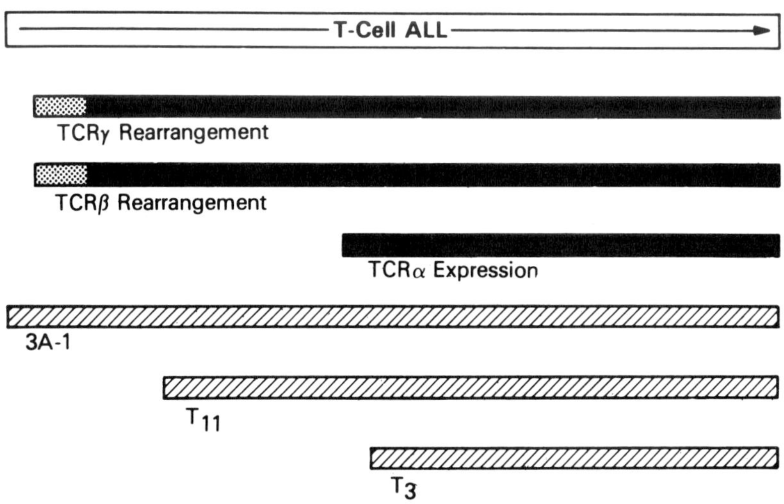

occur in several neoplasms, including pre-B acute lymphocytic leukemia and some cases of acute myelocytic leukemia. Basic genomic recombination signals (heptamer CACAGTG-spacer-nonamer [ACAAAAACC] that flank these gene segments are very similar in both Ig and T cell receptor genes and may be the recognition region for a recombinase enzyme capable of rearranging either Ig or T cell receptor genes.[157,158] Such a recombinase might help to explain the crossover of both Ig and T cell receptor gene rearrangements into the opposite lineage, perhaps representing vestigial elements of an early decision-making process that occurs before final commitment to B or T cell lineage differentiation.

Clinical Applications

One question about Ig or T cell receptor gene rearrangement analyses is whether these techniques can be helpful to the clinician who treats patients. In some patients clinical utility has already been demonstrated, since acute lymphocytic leukemias or lymphomas have been shown to display different profiles of rearrangement when followed sequentially over a period of time. In some instances, data from serial gene rearrangement studies appear to indicate that some tumors are biclonal or have evolved an entirely new neoplasm.[159,160] Whether such transformations actually reflect emergence of new tumors or evolution of subclones of a single malignant progenitor is still problematic. In many of these instances parallel analysis of multiple clonal markers can be interpreted merely to represent clonal or original tumor heterogeneity.[160] It is conceivable that mature B cell neoplasms could reactivate some sort of recombinase mechanism to result in these additional rearrangement modifications. In a parallel way, idiotypic variation has been observed to result from straightforward somatic mutations.[161]

Chromosomal Translocation

The relationship of specific chromosomal translocations to certain well-defined neoplasms has been discussed in Chapter 5 with respect to Burkitt's lymphoma and to the Philadelphia chromosome and chronic myelocytic leukemia. It is of considerable interest that in lymphoid neoplasms, the same gene segments that normally rearrange to produce antigen receptor molecules such as surface Ig in B cells or the T cell receptor in T cells also appear to be involved in these interchromosomal translocations. In Burkitt's lymphoma the chromosomal break point occurs at the Ig gene loci (8q 24.13 and 14q 32.33). Chromosomal break points at 14q32 occur repeatedly in other mature B cell tumors besides Burkitt's lymphoma.[162] Other studies now indicate that the most common lymphoma translocation is typical of follicular lymphoma and puts a new B cell–associated gene from chromosome location 18q21 into 14q32. Although follicular lymphomas are phenotypically mature B cells, translocations appear to occur at the D-J rearrangement step very early in pre–B cell differentiation. When such break points in chromosomes are carefully analyzed, it appears that the major site is on chromosome 18. Thus, over 70% of t(14;18) translocations occur within a 2- to 3-kb major break point region; moreover, most of these break points occur in a cluster within a region of 150 base pairs.[161,162] Such break point clustering must have a functional role. It is of interest that these occur in a 3′ exon of a B cell–associated gene, making a fusion transcript with the Ig J_H region. In summary, it now appears that malignancy-associated translocations in T cells are found classically at T cell re-

TABLE 8.2 Uses of Gene Rearrangement in Hematopathology

1. Distinguish clonal from polyclonal lymphoproliferations.
2. Determine the B or T cell identity of malignancies with admixed normal cells.
3. Determine the genetic lineage of neoplasms lacking definitive surface antigens.
4. Determine the developmental stage of early B or T cell precursors.
5. Identify unanticipated stages of development not appreciated within normal cell populations.
6. Tumor-specific markers to follow the natural history of neoplasms.
7. Search for minimal residual disease during clinical remission.
8. Molecularly mark chromosomal translocations and identify new transforming genes.

ceptor loci, whereas many of the B cell translocations, as in Burkitt's lymphoma, occur close to B cell receptor or Ig regions. A general summary of the use of gene rearrangement analysis in hematopathology is shown in Table 8.2.

Gene Rearrangements and the Sometimes Fuzzy Interface with Benign Conditions

Gene rearrangement analysis to establish clonality of various conditions probably has yet to reach its maximal use in clinical medicine. One of the most useful applications thus far has been in the diagnosis of mycosis fungoides or other malignant conditions of the skin. In many cases, clinical evaluation of patients suspected of having mycosis fungoides has been extremely difficult, even with expert histological review of tissue sections. This problem has now been addressed using Southern blot analysis for clonal arrangements of the T cell β receptor.[163] In this study evidence for T cell β chain rearrangements was closely correlated with histologically unambiguous positive tissue identification. Moreover, clonal rearrangements were found in seven of nine lymph nodes from mycosis fungoides patients that had been considered to show only benign lymphadenopathy.[163]

This approach has been rapidly adopted by oncologists and dermatopathologists, and it now represents a fundamental example of application of recombinant methods to initial precise clinical evaluation of problem patients. The same approach has been applied to lymphomatoid papulosis, which appears to be a cutaneous neoplastic process with a relatively benign course.[164] Studies of biopsy tissues in these patients showed multiple patterns of T cell receptor β gene rearrangement even in a single patient. These findings support the concept that the disease represents a multiclonal malignant T cell neoplasm. Similarly, T cell receptor β gene rearrangement has now been reported in pagetoid reticulosis, another rare skin disorder of unknown cause.[165] Such findings again present strong evidence that this condition is a T cell neoplastic process.

Monoclonal Cryoglobulinemia

Monoclonal cryoglobulinemia is a disease of unknown etiology with the clinical presentation of peripheral small vessel occlusions and arteritis, exacerbation by cold exposure, and the presence of distinct cryoglobulins in the serum. Several types of cryoglobulinemia have been described. Some patients produce monoclonal IgG, IgA, or IgM paraproteins that precipitate on cold exposure, whereas others have mixed cryoglobulins comprising a monoclonal IgM rheumatoid factor and autologous polyclonal IgG. Monoclonal cryoglobulinemia is generally not regarded as a malignant process. Four patients with monoclonal IgG cryoglobulins were studied in detail by Perl et al.[166] Southern Blot analysis of DNA extracted from peripheral blood of four patients showed Ig gene rearrangements in three of the four. In addition, the c-*myc* gene locus was altered in genomic DNA from peripheral blood leukocytes in two of the patients. These findings illustrate the high sensitivity of this approach. In patients with cryoglobulinemia it is estimated that 0.5% to 2% of the total peripheral blood B lymphocytes may be engaged in production of the monoclonal protein. The cryoglobulin is detected because of the profound biological effects it has in the circulation and particularly in vulnerable peripheral vessels. This disorder is not regarded as a cancer, yet it is assumed that it must have a monoclonal origin because of the homogeneous monoclonal immunoglobulin product. Similar Ig *myc* heavy chain rearrangements have been reported in patients with mixed IgM-IgG cryoglobulinemias.[167] It remains to be seen whether Ig gene rearrangement alone can be used in such patients as a criterion of malignancy.

Gene Rearrangement and Plasma Cell Myeloma

Because multiple myeloma clearly represents a malignant process, with production of osteolytic metastatic bone lesions and eventual overwhelming of the host, Ig gene rearrangement in the affected B cells is expected and has now been documented. No preferential use of any particular VH(V) region genes was found in the multiple myeloma patients studied.[168] A patient with both chronic lymphocytic leukemia (CLL) and multiple myeloma was studied to examine parallel B cell Ig gene rearrangements.[169] The CLL clonal character was IgM κ and the multiple myeloma IgA λ. Despite the different isotypes of the tumors, pretreatment studies of the heavy chain genes (J_H) showed a germ-line fragment and two identical rearrangements in blood and marrow. Both κ and λ light-chain genes showed rearrangement in the blood, suggesting involvement of peripheral blood lymphocytes by the myeloma. Analysis of the Ig genes after chemotherapy demonstrated no change in J_H or C_κ rearrangements, but the λ genes showed a germ-line configuration. These fascinating results suggested that both CLL and the plasma cell myeloma originated from the same B cell progenitors.

That myeloma may arise from a pluripotent stem cell is indicated by the finding of erythroid and megakaryocytic antigens on some myeloma cells and the demonstration of rearrangement of the T cell receptor γ genes in myeloma tumor cells with parallel rearrangement of Ig heavy-chain genes.[170]

Conventional search for Ig gene rearrangement in multiple myeloma, using comparative analysis of B cells from peripheral blood and bone marrow by Ig affinity and magnetized beads,[171] has clearly indicated uniform gene rearrangement in most patients studied. Of great interest would be analysis of patients with so-called benign monoclonal gammopathy, who display a monoclonal IgM serum component for years without developing clinical myeloma. Because such patients must harbor plasma cells with Ig genes rearranged to produce the monoclonal IgM component, studies of their marrow or peripheral blood B cells would be of great interest and might be predictive of subsequent development of malignancy. In this regard, some element of control or even autocrine regulation of plasma cell differentiation has been suggested for the lymphokine interleukin 6 (IL-6).[172,173] DNA rearrangement and constitutive expression of the IL-6 gene in a mouse plasmacytoma has been reported by Blankenstein et al.[174] The findings provided evidence that the IL-6 gene was expressed within the plasmacytoma and that the cell line actually produced IL-6 in the cell supernatants. Because IL-6 is a known growth factor for many types of immune cells, it may be related to the difference between benign monoclonal gammopathy and frank plasma cell myeloma.

Molecular Gene Rearrangements in Hematopathology

As noted in the foregoing sections, development of molecular biological techniques to supplement available information on patients with various malignancies of the immune or lymphoid system has progressed rapidly. Gene rearrangement analysis can be performed on virtually any tissue, including lymph node biopsy specimens, bone marrow aspirates, peripheral blood, body fluids, and cell suspensions. Frozen tissue in cryostats is also suitable for gene arrangement analysis. The limitations of gene rearrangement study are that it does not directly identify which cells are clonal because a mixture of DNA from a tissue is analyzed; the sensitivity of the method is 1% to 5%, and the testing in the current stage is labor intensive. The method enhances rather than supplants careful histologic and immunohistologic as well as cell-sorting analyses.[175]

Recombinant Products and Clinical Hematology

Erythropoietin

Successful cloning of the gene for human erythropoietin[176,177] was a major positive achievement for clinical medicine. Methods used included initial purification, amino acid sequence analysis, and subsequent construction of oligonucleotide probes based on the determined sequence. A human cDNA library prepared in bacteriophage λ was screened and clones showing positive hybridization were isolated. The structure of the erythropoietin gene was then deduced using overlapping sequences obtained by this methodology. Once the primary sequence and exact protein structure of this important hematopoietic hormone were established, precise studies of its biological activity and its tissue sources and metabolism became possible. The story of erythropoietin is a remarkable tale of the application of a recombinant technology, with an extremely short interval until the product was directly applied to important human clinical needs and problems. For some time it was known that erythropoietin is produced principally by the kidney in adults and by the liver during fetal life. Circulating levels of the hormone are normally approximately 20 mU/ml,[178,179] and they increase substantially during conditions of tissue hypoxia. Damage to the kidney such as occurs in chronic renal failure results in anemia due primarily to a deficiency in erythropoietin production.[180,181]

Cloning the Erythropoietin Gene

The isolation and cloning of the human erythropoietin gene proved to be a problem because initially no ready source of mRNA was known. Furthermore, no cell lines producing large amounts of erythropoietin were available, and although the kidney had been identified as the most likely source, it was not clear initially whether induction of erythropoietin resulted from de novo synthesis or activation of an inactive precursor molecule. Thus the whole cloning strategy was based on obtaining some initial accurate sequence data in order to construct oligonucleotide

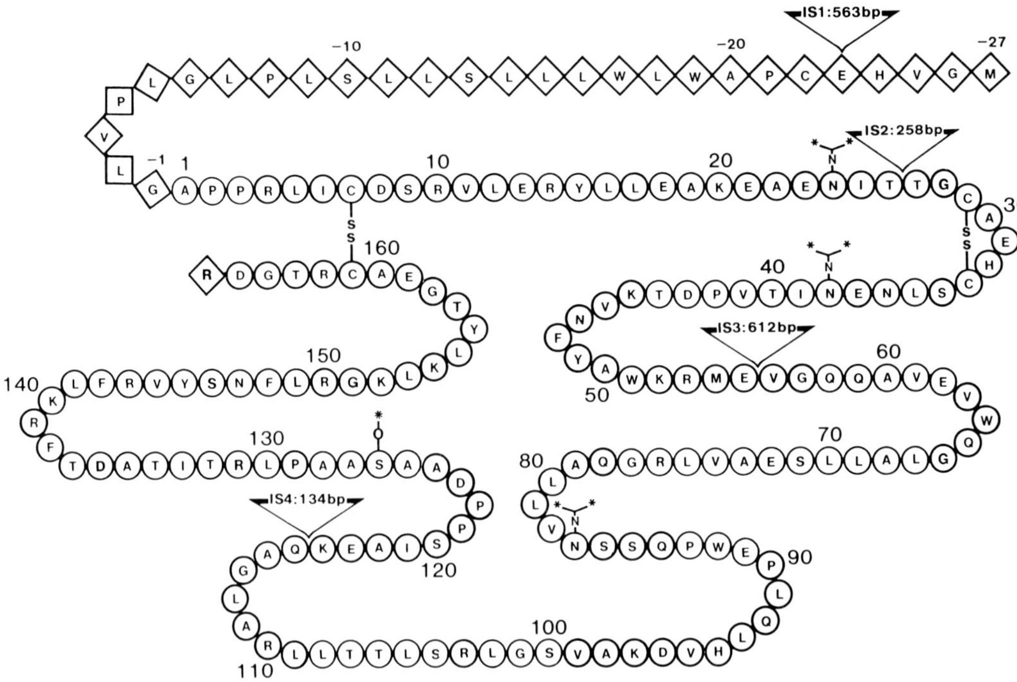

Figure 8.21 Amino acid sequence of the 193-amino-acid primary translation product of the human erythropoietin (EPO) gene is presented in one-letter code. The 27 amino-terminal amino acids of the putative signal peptide are boxed, and the residues in the 166-amino-acid mature hormone are circled. Positions of the two disulfide bonds (S—S) and the three N-linked glycosidation sites (Y) are noted. Positions at which the four intervening sequences (ISs) interrupt the coding position of the gene are given, along with the length of each IS.The C terminal arginine is removed post-translational yielding a mature product of 165 amino acids. *Reproduced from Browne JT et al. Erythropoietin: Gene cloning, protein structure, and biological properties. Cold Spring Harbor Symp Quant Biol 1986;51:693–702, with permission.*

probes to screen a human cDNA library. The erythropoietin that was finally produced proved to be a posttranslationally modified product of a single-copy gene that was highly conserved in mammalian species. These studies showed that the erythropoietin molecule is a 193-amino-acid primary translation product. The structure, including the 27-amino-acid putative signal peptide, is shown in Figure 8.21. N-linked glycosylation sites appear to be conserved in human, monkey, and mouse erythropoietin and are important to biological activity. When the gene is successfully expressed in somatic cell vectors, carbohydrate modifications apparently occur that enhance biological activity.

Clinical Applications

Patients with end-stage chronic renal failure were logical candidates for the first use of this recombinant hematopoietic hormone. Treatment of 10 patients on thrice-weekly hemodialysis with recombinant erythropoietin was reported by Winearls et al.[182] All patients showed encouraging rises in both reticulocyte numbers and hemoglobin concentration. Use in anephric end-stage patients with renal disease appeared to be particularly appropriate, because during normal homeostasis the hormone is produced by the kidney, and anephric patients have inappropriately low levels of circulating erythropoietin.[183] In such patients a possible complicating factor might be introduced, since increments in hematocrit might also increase blood viscosity and create additional clotting or coagulation problems in shunts commonly used for arterial access. Initial therapeutic trials appear to be encouraging, but a much longer period of trial and observation is needed before final evaluation of this form of treatment. Additional trials of recombinant erythropoietin given to 25 patients with end-stage renal disease have been reported.[184] Doses ranging from 15 to 500 U/kg of body weight were used, and dose-dependent increments of effective erythropoiesis were recorded. With 500 U/kg, changes in hematocrit of as much as 10% in 3 weeks were noted. Of 18 patients in whom erythropoietin doses proved effective, 12 who had previously required transfusions no longer needed them to maintain hemoglobin and hematocrit values. Four patients showed concomitant increments in serum creatinine and potassium values. No antibodies to the administered erythropoietin were detected. Some improvement in endogenous serum erythropoietin has been noted in a patient switched from hemodialysis to continuous ambulatory peritoneal dialysis.[185] Further long-term comparative assays will be necessary to determine whether the method of dialysis itself affects recombinant erythropoietin metabolism and availability.

The initial trials of recombinant erythropoietin in patients with end-stage renal disease are extremely promising, although other factors, such as shortened red cell survival and blood loss or mechanical trauma during dialysis, may aggravate the anemia in uremic patients. Long-term effects of increased hemoglobin and hematocrit values in patients with chronic renal failure also need to be evaluated. In both initial trials, there were difficulties with blood pressure control in some individual patients. Increasing hematocrit values in dialysis patients may be associated with an increase in peripheral resistance and blood pressure.[186] The possibility of increased problems with arterial and venous access because of increments in blood viscosity must also be considered. Long-term trials with monitoring for parallel changes in blood pressure and hypercoagulability and possible adverse effects of increments in total blood volume should provide a reasonable assessment of the overall benefits of recombinant erythropoietin therapy in patients with end-stage renal disease.

Insight into the effect of recombinant erythropoietin on the anemia of progressive renal failure was provided by a study of 17 patients with anemia and progressive renal failure who did not yet require dialysis.[187] In this trial, the dose of erythropoietin (50 to 150 U/kg body weight) was adjusted to the hematocrit response. All 17 patients showed a positive response to erythropoietin. The rate of beneficial response depended on the initial erythropoietin dose and was similar to what had been observed with patients on dialysis. The rate of decline in renal function did not change as the hematocrit rose, and all patients reported improvements in appetite, activity levels, and sense of well-being. This study showed that the anemia of renal failure was responsive to erythropoietin but that the renal failure itself did not improve as the anemia was corrected.

Erythropoietin Levels after Renal Transplantation

Further analysis of the interaction of decreased renal function, erythropoietin response, and restoration of renal function was reported by Sun and co-workers[188] in a group of 31 renal transplant recipients. These patients were treated with cyclosporine and human serum erythropoietin levels were measured by a radioimmunoassay procedure. The mean serum erythropoietin concentration in the patients before receiving the transplants was similar to that in normal subjects who were not anemic. After the transplant, a transient ninefold increase in serum erythropoietin was followed by a smaller (threefold) but sustained second elevation. The second increase in erythropoietin level was accompanied by graft recovery and complete resolution of previously existing anemia. Serum erythropoietin levels returned to normal as the patients' hematocrits continued to rise toward normal. Thus, once erythropoiesis is initiated in renal transplant patients, the process is sustained by normal serum erythropoietin levels.

Erythropoietin Treatment of Myeloma

Because anemia is a common complication of multiple myeloma and persists as the disease progresses, erythropoietin was administered to 13 patients with myeloma-associated anemia.[189] The patients received erythropoietin three times a week for 6 months. Eleven of the patients showed steady increases in hemoglobin levels and subsequent correction of the anemia. No adverse side effects were observed and no episodes of hypertension recorded. No changes were recorded during this study in levels of serum M component as a measure of overall tumor burden. Considerable improvement in sense of well-being was recorded by myeloma patients whose hemoglobin levels increased with erythropoietin therapy.

Erythropoietin and Megakaryocytes

The ready availability of recombinant (r) erythropoietin has afforded opportunities to examine its effect on other hematopoietic functions. Two groups have reported an apparent tropic or stimulatory effect of recombinant erythropoietin on human megakaryocyte colony formation[190] or differentiation of murine megakaryocytes in vitro.[191] These effects on other hematopoietic precursors besides erythroid cells raise some interesting questions concerning common signal transductions by many of the hematopoietic stimulatory factors or hormones. As work in this area progresses, it may become apparent that erythropoietin stimulates cells not necessarily restricted to the erythroid lineage.

Hematopoietic Colony-Stimulating Factors

Regulation of blood cell production and differentiation is one of the most complex examples of control of a system involving mutlilineage development. A common family of pluripotential stem cells located mainly in the bone marrow differentiates into large numbers of erythrocytes, neutrophils, basophils, eosinophils, monocytes, platelets, and lymphocytes, which circulate in the blood or perform selected functions in specific tissues. Many of these blood elements are short-lived and may be called on to expand their numbers rapidly in situations of physiological stress. Work on the specific factors that are necessary for differentiation of normal progenitor cells along multiple parallel but eventually distinct pathways has spanned at least two decades. Development of cell culture systems for the study of clonal growth of selected hematopoietic progenitor cells was essential in understanding the normal differentiation process. This sort of developmental scheme is illustrated in Figure 8.22.[192] Since the various tropic factors were identified through cell culture systems in which uniform colonies were identified, the factors were called hematopoietic colony-stimulating factors (CSFs). Work related to several of these factors including isolation of cDNAs and the genes of four of the major human

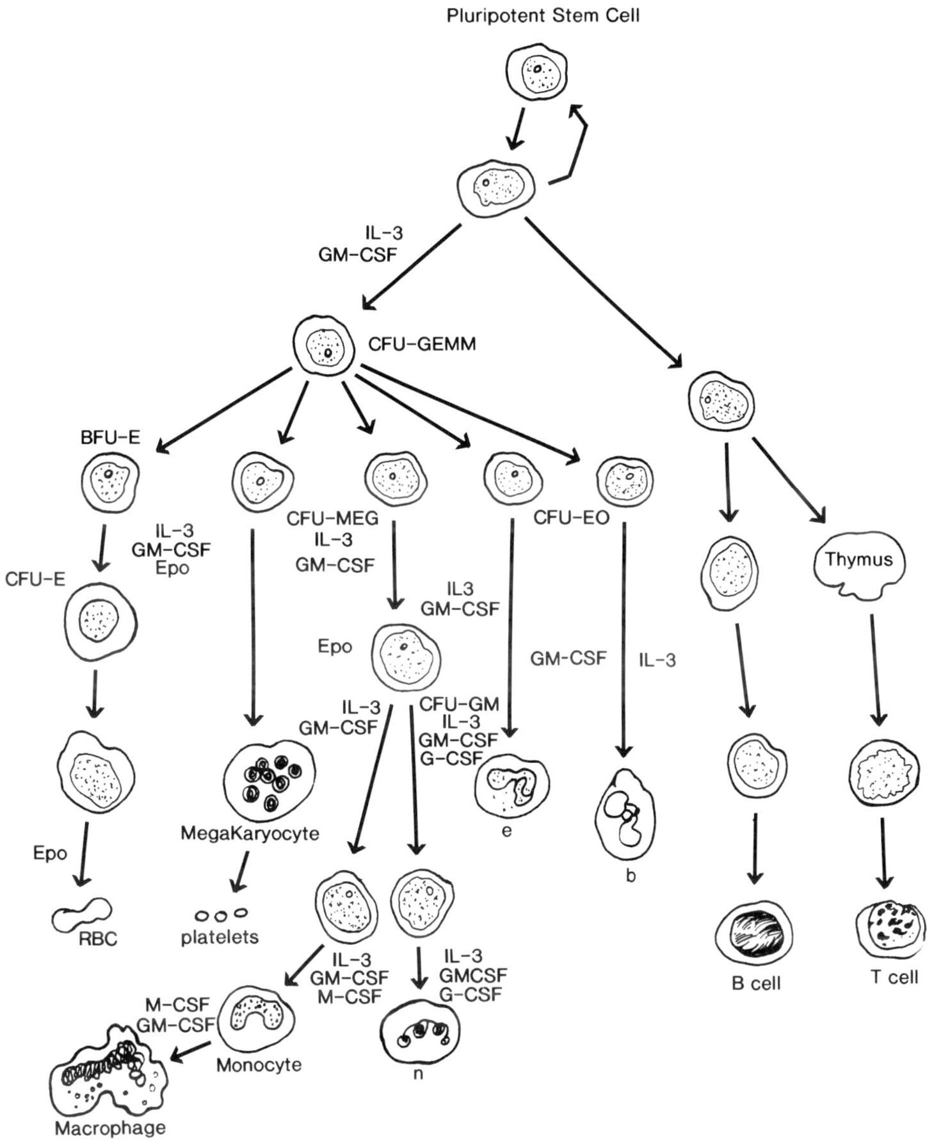

Figure 8.22 Interactions of the colony-stimulating factors with hematopoietic cells. The progenitor cells that are identified in the culture systems are CFU-GEMM, colony-forming unit granulocyte-erythrocyte-monocyte-megakaryocyte; CFR-Meg, CFU megakaryocyte; CFU-Eo, CFU eosinophil; CFU-GM, CFU granulocyte-monocyte); CFU-E, CFU erythroid, and BFU-E, burst-forming unit erythroid. The abbreviations for the hematopoietic lineages are n, neutrophil; e, eosinophil; b, basophil; m, monocyte/macrophage; E, erythrocyte; and M, megakaryocyte. The interactions of the different CSFs with the various lineages are as indicated. These interactions are based on analysis of mature cells from colonies grown in the presence of the CSFs. The sites of action indicate that at least some but not necessarily all of the progenitors of that lineage are responsive to the indicated CSF. Erythropoietin (Epo) is essential for development of erythroid cells and can promote the differentiation of megakaryocyte progenitors in vitro. *Reproduced from Clark and Kamen,*[192] *with permission.*

myeloid growth factors has rewarded the long and painstaking study of their actions. Moreover, application of several of these growth factors to pressing problems of clinical medicine is already beginning.

Colony-Stimulating Factors

Initially, several different CSFs were distinguished by analyses of individual activities in semisolid growth media. In the mouse system four major types were recorded; two showed relative lineage specificity for granulocytes (G-CSF) or macrophages (M-CSF). By contrast, colonies of cells grown with multi-CSF or interleukin 3 (IL-3) were sometimes found to contain many different cell lineages.[193] Stimulation of stem cells with both GM-CSFs (granulocyte-macrophage colony-stimulating factors) induced neutrophils, macrophages, eosinophils, and other cell types.[194] The theoretical model for the operation of the various growth factors presumes that G-CSF and M-CSF support growth and division of relatively late progenitors already committed to their own lineage differentiation pathways. The GM-CSF is thought to act on earlier progenitor cells that are still capable of differentiating into neutrophils, eosinophils, or monocytes. Within the human system, a family of four similar factors has been recognized as G-, B-, GM-, and multi-CSF.[194,195] A number of other growth factors that affect important differentiation pathways have also been recognized. Thus, as noted earlier, erythropoietin functions in early erythroid cell development and may have some effect on megakaryocytes. Interleukin 2 (IL-2) is an important growth factor for both T and B lymphocytes.[196] Another tropic factor, hematopoietin 1,[197] is thought to act on stem cells to allow them to become responsive to other CSFs; a similar function has also been attributed to interleukin I (IL-1).

Molecular Cloning of Human Colony-Stimulating Factor

Successful cloning of the four major CSFs has been a major practical achievement in the application of molecular biological techniques. Each separate factor was identified and cloned through a large collaborative effort involving cell biologists, protein chemists, and molecular biologists. Each CSF was known to exist in only trace amounts, making its biochemical characterization very difficult. The trace levels of the CSF molecules were naturally reflected in correspondingly low levels of mRNA in the tissues or cell lines examined, which made their cloning exceedingly difficult.

Complementary DNAs for G-CSF[198,199] and M-CSF[200] were obtained by a general strategy in which purification to homogeneity was followed by amino acid sequencing. The derived sequences were utilized to predict nucleotide sequence in several different mRNA regions, and synthetic oligonucleotides were prepared to identify cDNAs encoding the entire CSF molecules. The sequences derived were then checked by expression in mammalian cells to recover the respective biological activities. In the case of human M-CSF, much manipulation of the initial cDNA clones was necessary before a final functional product was obtained.

The cDNAs for human GM-CSF[201] and murine IL-3[202] were obtained by a different approach. Because of their original presence in trace or only picomolar amounts, purification to homogeneity proved very difficult. On the other hand, both of these factors showed uniquely potent biological activities. The high specific biological activity could be exploited by identifying cDNA clones that would direct expression of the desired CSF product after introduction into mammalian cells with appropriate expression vectors. cDNAs for both human GM-CSF and murine IL-3 were obtained in this manner. This was done by testing pools of plasmid cDNAs for their ability to express the biological activities when introduced into target COS-1 cells.

The human and murine IL-3 genes had very similar structures,[202] containing five relatively small coding exons separated by one large and three small introns. Nucleotide sequence homology was distributed relatively evenly over the 3000 base pairs making up the genes. The human GM-CSF and IL-3 genes have been mapped to the same band on the long arm of chromosome 5.[203] The same area has been found to contain the M-CSF gene and its receptor, the proto-oncogene, c-*fms*, as well as the genes for several other growth factors and receptors. The significance of this clustering of genes for growth factors on chromosome 5 is not clear, but it must be important in terms of basic mechanisms of cellular differentiation.

Regulation of Colony-Stimulating Factor Gene Expression

Much remains to be learned about the ways in which the CSF genes are regulated or modulated during normal events in vivo. It is well known that absolute numbers of various blood cells, such as polymorphonuclear leukocytes or monocytes, may change rapidly in response to extrinsic signals such as stress or infection. Many normal cells do not appear to have detectable levels of mRNA for G-CSF, GM-CSF, or IL-3, but several different cell types have been shown to express CSF genes after activation. Thus, monocytes activated by endotoxin or γ-interferon express high levels of mRNA for G-CSF and M-CSF but not for GM-CSF.[204] In parallel, IL-1 and tumor necrosis factor, which are potent lymphokine products of monocytes or macrophages, can activate G-CSF and GM-CSF in various body cells, including fibroblasts and endothelial cells.[205,206] Thus, signals released by stimuli such as infection, trauma, or other changes in the internal body milieu must be capable of up-regulating gene expression and mRNA coding for these potent CSF materials. The cellular source of these factors that maintain normal steady-state hematopoiesis has not been completely identified. By contrast, the mRNA for M-CSF is readily observed in a variety of normal cells and tissues. Such extensive representation in many tissues may indicate its importance in survival and the need to provide mechanisms for instant production of monocytes and macrophages at many tissue sites.

Recombinant Colony-Stimulating Factors—Structure and Biological Action

A number of studies have indicated that recombinant CSF products contain essential structural elements that are very similar to natural molecules produced in vivo. In several cases, such as GM-CSF and IL-3, the molecular size of the products studied indicates that up to 50% of the mass of the protein may be carbohydrate.[198–201] The functions of the extensive carbohydrate modifications of GM-CSF and IL-3 are still not understood. The addition of carbohydrate residues may influence body compartment distribution or catabolism. More information related to the latter is needed.

Availability of four human CSF proteins has made it possible to examine their biological activities in some detail. Recombinant G-CSF is a lineage-specific factor that supports proliferation of neutrophils. It also has a marked stimulatory effect on biological activities of neutrophils by increasing superoxide anion production, ability to kill tumor cells in antibody-dependent cytotoxicity, or ability to phagocytize particles.[207,208] Recombinant GM-CSF supports proliferation of macrophages and eosinophil colonies in vitro as well as colonies containing both neutrophils and macrophages.[207] It has also been shown to serve as a growth factor for the leukemia cells of many patients with chronic and acute myeloid leukemia.[208,209] There is no direct evidence that exaggerated expression of the GM-CSF gene is directly related to acute myelocytic leukemia, but patients have been reported whose early leukemic cells are autocrine for GM-CSF, indicating a built-in stimulus for continued growth.[208]

Studies with recombinant gibbon IL-3 indicate proliferation of myeloid progenitor cells as well as those of erythrocytes, neutrophils, eosinophils, basophils, macrophages, and megakaryocytes. It appears that this factor acts on earlier progenitor cells than GM-CSF or G-CSF. The activity of recombinant human M-CSF is still under study. Although the material strongly stimulates murine macrophage colony formation, it has been less active with human target cell systems. The recombinant form is a powerful activator of human macrophage cytotoxicity.

In Vivo Therapeutic and Clinical Trials

The potential usefulness of CSFs in a number of clinical situations has excited a great deal of hope and positive expectations. Such recombinant proteins might be useful in restoration of hematopoietic dysfunction after immunosuppression or irradiation and in augmenting host response to overwhelming infection.[210] The same factors might be very effective in increasing levels of primed effector cells that could help rid the host of malignant cells.[211] Suppression of myelopoiesis after chemotherapy is a common feature of many otherwise effective antitumor programs and is one of the major adverse effects of such treatment. Limited phase I trials of G-CSF have begun in such patients, and the results appear extremely promising.

Matoyoshi et al.[211] in Japan reported positive effects of partially purified human urinary colony-stimulating factor preparations on granulocytopenia after antitumor chemotherapy. This initial trial produced distinct increments in peripheral blood granulocytes in treated subjects but did not affect levels of erythrocytes, platelets, monocytes, or lymphocytes. Human granulocyte CSF was administered to normal and cyclophosphamide-treated primates.[212] The G-CSF was administered by two daily subcutaneous injections for a period of 2 to 4 weeks. A dramatic increase in neutrophils was noted 24 hours after initiation of treatment; increases in white blood cell counts were due primarily to increments in the absolute numbers of neutrophils. These changes are illustrated in Figure 8.23. Similar dramatic increases in circulating neutrophils were demonstrated in animals pretreated with cyclophosphamide. In these animals, treatment with G-CSF shortened the time period of white blood cell recovery to 1 week, compared to 4 weeks in control myelosuppressed monkeys. Recombinant GM-CSF was also given to 16 patients with acquired immunodeficiency syndrome (AIDS) and leukopenia by Groopman and co-workers.[213] Treatment consisted of an initial intravenous dose of 1.3×10^3 to 2.0×10^4 U/kg. Forty-eight hours after the initial dose, a continuous intravenous infusion was given. The GM-CSF produced a dose-dependent increase in counts of circulating white blood cells and of neutrophils, eosinophils, and monocytes, with peak leukocyte counts ranging from 4575 ± 2397 to 48,700 in patients on the lowest to highest doses. Mild side effects of flushing, myalgia, and phlebitis were noted in some patients but no severe adverse reactions were recorded. These

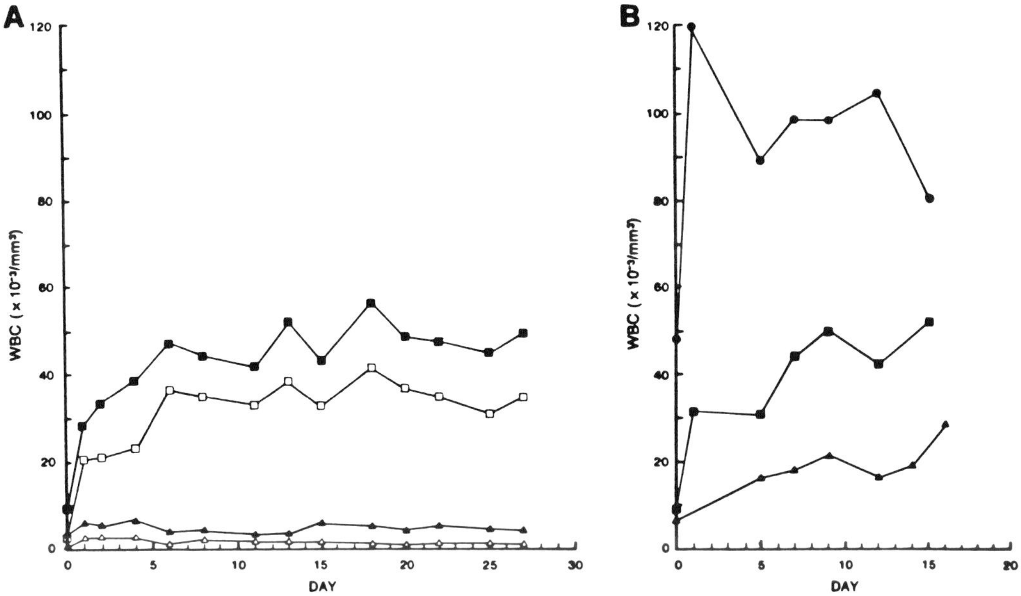

Figure 8.23 Time course of peripheral white blood cell (WBC) count and absolute neutrophil count (ANC) of healthy cynomolgus monkeys treated with recombinant human G-CSF and a control monkey receivng only buffer. (A) WBC level (■) and ANC (□) of a healthy monkey treated with rhG-CSF at 10 μg/kg/d subcutaneously in two daily doses. The WBC level (▲) and ANC (△) of the control monkey are also shown. (B) Effects of different concentrations of rhG-CSF: (▲) 1 μg/kg/d, (■) 10 μ/kg/d, and (●) 100 μg/kg/d. The monkey receiving 10 μg/kg/d was subsequently treated with 100 μg/kg/d, and therefore day 0 of 100 μg/kg/d corresponds to day 10 of 10 μg/kg/d. *Reproduced from Welte et al,*[212] *with permission.*

initial results were encouraging and suggested that GM-CSF was relatively well tolerated and could induce rapid increments in leukocyte counts in such immunosuppressed patients. Furthermore, it indicated that patients with serious infections could respond to the agent even though relative myelosuppression was a feature of the underlying AIDS disorder.

A second therapeutic trial of recombinant human GM-CSF has been reported in patients with myelodysplastic syndromes.[214] As part of a phase I trial, GM-CSF was administered to eight patients by continuous intravenous infusion for 2 weeks and again after a 2-week rest. Treatment was associated with 5- to 70-fold increases in peripheral blood leukocytes, including elevations of granulocytes in all eight patients. Absolute numbers of monocytes, eosinophils, and lymphocytes also increased in all patients. Three of the patients showed 2- to 10-fold increments in platelet counts and improved erythropoiesis to the extent that two patients who had required red cell and platelet transfusions were able to discontinue such replacement. Examination of bone marrow populations showed a relative decrease in percentages of blasts and an increase in ratio of differentiated myeloid cells to immune cells of the same lineage. Bone pain was experienced by some patients and was dose limiting when associated with high white blood cell counts. Of great interest in this study was the apparent multilineage stimulation in the patients treated. Although experimental in vitro and in vivo work had indicated that GM-CSF was potentially a multilineage hematopoietic stimulator, it is possible that such an effect was also mediated by other lymphokines such as IL-1 released in vivo as a result of the GM-CSF. As the authors point out, because acute leukemia develops in 15% to 40% of patients with myelodysplastic syndrome[215] and GM-CSF is capable of stimulating leukemic cells in vitro, there was some concern that GM-CSF administration might augment or exacerbate any underlying leukemic aspect of the disease. On the contrary, no changes in proportions of immature blast forms and an actual improvement in the spectrum of differentiation were recorded during this initial study.

The method of continuous intravenous administration was chosen in this study because previous studies in nonhuman primates had indicated a relatively short half-life of the active material. The initial results suggest that these recombinant products may have great therapeutic potential as their use is cautiously expanded. CSFs could potentially be utilized to supply large numbers of primed effector cells to sites of infection or tumor tissues.

Recombinant GM-CSF used after high-dose chemotherapy and autologous bone marrow transplantation[216] in 19 patients with breast cancer or melanoma accelerated total leukocyte and granulocyte recovery. Toxic effects in this trial group appeared to be relatively mild and included some edema, weight gain, or myalgias. Generalized edema, pleural effusions, and hypotension were observed in two patients. The results again appeared to be encouraging and showed that recombinant GM-CSF was capable of accelerating myeloid recovery in such patients. A similar study was reported by Sheridan and co-workers,[217] who administered G-CSF by continuous subcutaneous infusion to 15 patients with nonmyeloid malignancies treated with high-dose chemotherapy and autologous bone marrow infusion. Sustained levels of serum G-CSF were found. Recovery of neutrophils was accelerated in the G-CSF–treated patients compared with historical controls. This resulted in significantly fewer days of parenteral antibiotic therapy.

Recombinant G-CSF was used in parallel with chemotherapy for 27 patients with transitional T-cell carcinoma of the urothelium.[218] The protocol resulted in a dose-dependent increase in absolute neutrophil counts and reduced the number of days during which patients' absolute neutrophil counts remained below 1000 per microliter. It appeared to be capable of reducing both the hematopoietic and the oral toxicity of chemotherapy.

Another aspect of possible therapeutic benefit has been explored in experiments by Silberstein et al.[219] They found that recombinant GM-CSF enhanced eosinophil cytotoxicity against schistosomula sixfold. Moreover, GM-CSF increased the generation of leukotriene C4 by a mean of 135%. These findings indicated that GM-CSF could act to mediate T lymphocyte regulation of eosinophil functions. Because eosinophil-mediated cytotoxicity has been considered a possible protective mechanism whereby the host reacts against parasitic forms such as schistosomes, it is conceivable that administration of GM-CSF might find some use in the therapy of such parasitic diseases.

Cyclic Neutropenia

Six patients with cyclic neutropenia were treated with recombinant human granulocyte CSF for periods of 3 to 15 months.[220] All the patients showed aphthous stomatitis, pharyngitis, lymphadenopathy, febrile episodes, and frequent infections associated with periods of neutropenia. Daily intravenous or subcutaneous administration of G-CSF at 3 to 10 μg/kg body weight resulted in an increase in mean neutrophil counts from 717 ± 171 to 9814 ± 2198 per microliter ($P = 0.009$). Although in five of the six patients treated, cycling of blood cell counts continued, the length of the relative leukopenic period decreased from 21 to 14 days. The neutrophil turnover increased almost fourfold as measured by kinetics of labeled autologous blood neutrophils. Overall, the G-CSF therapy appeared to be effective in decreasing the frequency of oropharyngeal inflammation, fever, and infections in the cyclic neutropenia group. A few patients experienced mild bone pain after intravenous administration of G-CSF, and a consistent mild elevation of leukocyte alkaline phosphatase to levels above 200 was observed in all patients treated (normal 60 to 140 units). A moderate increase in splenic size was noted in all six patients.

Treatment success in a marrow stem cell disorder such as cyclic neutropenia is a substantial positive result in terms of therapeutic application. However, there are some lingering questions that a longer period of sequential observations may answer. The initial euphoria that prevailed shortly after the clinical introduction of GM-CSF and G-CSF has waned somewhat, but their widespread use clearly has led to beneficial results in a substantial proportion of patients. The early studies with GM-CSF indicated that this recombinant product could accelerate the leukemic process.[221] Also, in vitro studies have indicated that G-CSF may stimulate cell lines for colon cancer and small cell lung carcinoma.[222] Continued experience will help to assign their overall value in patients with various stem cell disorders.[222,223]

As more experience is gained with these agents a more seasoned evaluation of their usefulness and possible untoward effects will eventually become possible.

Tissue Plasminogen Activator

During the last several years there has been interest in dissolving clots in the therapy of acute vascular disorders such as myocardial infarction or deep vein thrombophlebitis. Therapeutic trials of streptococcus-derived products such as streptokinase are ongoing.[224] Plasminogen activators, streptokinase, and urokinase, which were the first agents available for clinical use, were effective in dissolving thrombi and emboli but were not specific for fibrin. These agents could activate plasminogen in the circulation as well as the plasminogen bound to fibrin in the clot. Activation of plasminogen in the circulation increased the proteolytic activity of the blood and produced degradation of fibrinogen and clotting factors V and VII. This led to an increased risk of bleeding complications in patients treated with these agents.

For these reasons, efforts were directed at manufacturing the naturally occurring body protein that activated blood clot dissolution, tissue plasminogen activator (t-PA). When small amounts became available, a number of trials were undertaken.[225] Experience thus far shows that this agent binds strongly to fibrin but avoids some of the hemostatic defects. Moreover, it has a very short half-life in the circulation. Despite the milder hemostatic defect observed in t-PA compared to streptokinase, the incidences of bleeding complications in parallel trials have been approximately equal.[226] Because t-PA represents a native recombinant protein, the potential problem with antigenicity and subsequent reactions appears to be relatively excluded. The rapid development and widespread use of this important recombinant product are yet another example of the remarkable spread of recombinant DNA methodology and its application to significant common medical problems. Whether recombinant t-PA will prove to be a more effective therapeutic approach to clot dissolution in acute myocardial infarction or pulmonary embolism than streptokinase will be known after a much longer period of evaluation and monitoring. The important aspect of t-PA in the context of the subjects addressed in this volume is that it is an outstanding example of the power of recombinant DNA technology to focus on an important clinical need.

References

1. Ingram VM: A specific chemical difference between the globins of normal human and sickle cell anaemia hemoglobins. *Nature* 1956;178:792–794.
2. Conley CL: Sickle cell anemia—the first molecular disease in Wintrobe M (ed): *Blood Pure and Eloquent*. New York, McGraw-Hill, 1980, pp 319–371.
3. Cooley TB, Lee P: A series of cases of splenomegaly in children with anemia and peculiar bone changes. *Trans Am Pediatr Soc*, 1925;37:29–40.
4. Maniatis T, Fritsch EF, Lauer J, Lawn RM: The molecular genetics of human hemoglobin. *Annu Rev Genet* 1980;14:145–178.
5. Fritsch EF, Lawn RM, Maniatis T: Molecular cloning and characterization of the human β-like globin gene cluster. *Cell* 1980;19:959–972.
6. Stamatoyannopoulos G, Nienhuis A: *Organization and Expression of Globin Genes*. New York, Alan R. Liss, 1981.
7. Fearon ER, Kazazian HH Jr, Weber PG, et al: The entire β-globin gene cluster is deleted in one form of αδβ-thalassemia. *Blood* 1983;61:1269–1273.
8. Kan YW, Forget BG, Nathan DG: Thalassemia: A cause of hemolytic disease of newborns. *N Engl J Med* 1972;286:129–134.
9. Pirastu M, Kan YW, Lin CC, et al: Hemolytic disease of the newborn caused by a new deletion of the entire β-globin cluster. *J Clin Invest* 1983;72:602–609.
10. Kioussis D, Vanin E, deLange T, et al: β-globin gene inactivation by DNA translocation in γβ thalassemia. *Nature* 1983;306:662–666.
11. Thein SL, Old JM, Wainscoat JS, et al: Population and genetic studies suggest a single origin for the Indian deletion form of β^0 thalassaemia. *Br J Haematol* 1984;57:271–278.
12. Kazazian HH Jr, Orkin SH, Boehm CD, et al: β-Thalassemia due to a deletion of the nucleotide which is substituted in the β^S-globin-gene. *Am J Hum Genet* 1983;35:1028–1033.
13. Chang JC, Alberti A, Kan YW: A β-thalassemia lesion abolishes the same *Mst*II site as the sickle mutation. *Nucleic Acids Res* 1983;11:7789–7794.
14. Orkin SH, Kazazian HH Jr: The mutation and polymorphisms of the human β-globin gene and its surrounding DNA. *Annu Rev Genet* 1984;18:131–171.
15. Thein SL, Old JM, Fiorelli G, et al: Feasibility of prenatal diagnosis of β-thalassaemia with synthetic DNA probes in two Mediterranean populations. *Lancet* 1985;2:345–347.
16. Rosatelli C, Falchi AM, Tuveri T, et al: Prenatal diagnosis of beta-thalassaemia with the synthetic oligomer technique. *Lancet* 1985;1;241–243.
17. Kazazian HH Jr, Orkin SH, Markham AF, et al: Quantification of the close association between DNA haplotypes and specific β-thalassemia mutations in Mediterraneans. *Nature* 1984;310:152–154.
18. Saiki RK, Scharf S, Faloona F, et al: Enzymatic amplification of β-globin genomic sequences and restriction site analysis for diagnosis of sickle cell anemia. *Science* 1985; 230:1350–1354.
19. Saiki RK, Bugawan TL, Horn GT, et al: Analysis of enzymatically amplified β-globin and HLA-DQa DNA with allele-specific oligonucleotide probes. *Nature* 1986;324:163–166.
20. Scharf SJ, Horn GT, Erlich HA: Direct cloning and sequence analysis of enzymatically amplified genomic sequences. *Science* 1986;233:1076–1078.
21. Wong C, Dowling CE, Saiki RK, et al: Characterization of β-thalassaemia mutations using direct genomic sequencing of amplified single copy DNA. *Nature* 1987;330:384–386.

22. Alter BP: Advances in the prenatal diagnosis of hematologic diseases. *Blood* 1984; 64:329–340.
23. Orkin SH: Prenatal diagnosis of hemoglobin disorders by DNA analysis. *Blood* 1984; 63:249–253.
24. Kan YW, Dozy AM: Polymorphism of DNA sequence adjacent to human β-globin structural gene: Relationship to sickle mutation. *Proc Natl Acad Sci USA* 1978;75:5631–5635.
25. Saiki R, Arnheim H, Erlich H: A novel method for the detection of polymorphic restriction sites by cleavage of oligonucleotide probes: Application to sickle cell anemia. *Biotechnology* 1985;3:1008–1012.
26. Embury SH, Scharf SJ, Saiki RK, et al: Rapid prenatal diagnosis of sickle cell anemia by a new method of DNA analysis. *N Engl J Med* 1987;316:656–661.
27. Studencki AB, Conner BJ, Impraim LC, et al: Discrimination among the human β^A, β^S, and β^C-globin genes using allele-specific oligonucleotide hybridization probes. *Am J Hum Genet* 1985;37:42–51.
28. Kan YW, Dozy AM, Varmus HE, et al: Deletion of α-globin genes in hemoglobin-H disease demonstrates multiple α-globin structural loci. *Nature* 1975;255:255–256.
29. Kan YW, Dozy AM, Varmus HE, et al: The molecular basis of the α-thalassemia syndromes. *Clin Res* 1975;23:398A.
30. Kan YW, Golbus MS, Dozy AM: Prenatal diagnosis of α-thalassemia: Clinical application of molecular hybridization. *N Engl J Med* 1976;295:1165–1167.
31. Rubin EM, Kan YW: A simple sensitive prenatal test for hydrops fetalis caused by α-thalassaemia. *Lancet* 1985;1:75–77.
32. Itano HA, Bergren WR, Sturgeon P: Identification of fourth abnormal human hemoglobin. *J Am Chem Soc* 1954;76:2278.
33. Chernoff A, Minnich V, Chongchareonsuk S: Hemoglobin E, a hereditary abnormality of human hemoglobin. *Science* 1954;120:605–606.
34. Flatz G: Hemoglobin E. Distribution and population dynamics. *Humangenetik* 1967; 3:189–234.
35. Lachant NA: Hemoglobin E: An emerging hemoglobinopathy in the United States. *Am J Hematol* 1987;25:449–462.
36. Frischer H, Bowman J: Hemoglobin E, an oxidatively unstable mutation. *J Lab Clin Med* 1975;85:531–539.
37. Orkin SH, Kazazian HH, Antonarakis SE, et al: Abnormal RNA processing due to the exon mutation of β^E-globin gene. *Nature* 1982;300:768–769.
38. Thein SL, Lynch JR, Old JM, Weatherall, DJ: Direct detection of haemoglobin E with *Mnl*I. *J Med Genet* 1987;24:110–112.
39. Busslinger M, Hurst J, Flavell RA: DNA methylation and the regulation of globin gene expression. *Cell* 1983;34:197–206.
40. van der Ploeg LHT, Flavell RA: DNA methylation in the γ-globin locus in erythroid and nonerythroid tissues. *Cell* 1980;19:947–958.
41. Jones PA, Taylor SM: Cellular differentiation, cytidine analogs, and DNA methylation. *Cell* 1980;20:85–93.
42. DeSimone J, Heller P, Hall L, Zwiers D: 5-Azacytidine stimulated fetal hemoglobin (HbF) synthesis in anemic baboons. *Proc Natl Acad Sci USA* 1982;79:4428–4431.
43. Ley TJ, DeSimone J, Anagnou N, et al: 5-Azacytidine selectively increased γ-globin synthesis in a patient with β^+-thalassemia. *N Engl J Med* 1982;307:1469–1475.
44. Ley TJ, DeSimone J, Noguchi CT, et al: 5-Azacytidine increases γ-globin synthesis and reduces the proportion of dense cells in patients with sickle cell anemia. *Blood* 1983; 62:370–380.
45. Charache S, Dover G, Smith K, et al: Treatment of sickle cell anemia with 5-azacytidine results in increased fetal hemoglobin production and is associated with nonrandom hypomethylation of DNA around the β-globin gene complex. *Proc Natl Acad Sci USA* 1983;80:4842–4846.
46. Veith R, Torrealba de Kon AT, Papayannopoulou T, et al: On the mechanism of 5-azacytidine induced HbF in baboons: Comparisons of data obtained from cytidine-arabinocide and 5′-azacytidine treatments. *Blood* 1983;62:76a.
47. Platt O, Orkin SH, Dover G, et al: Hydroxyurea enhances fetal hemoglobin production in sickle cell anemia. *J Clin Invest* 1984;74:652–656.

48. Carr BI, Reilly JG, Smith SS, et al: The tumorigenicity of 5-azacytidine in the male Fischer rat. *Carcinogenesis* 1984;5:1583–1590.
49. Muranyi-Kovacs I, Rudali G: Comparative study of carcinogenic activity in hydroxyurea and urethane in SVII/G mice. *Rev Eur Etudes Clin Biol* 1972;17:93–95.
50. Charache S, Dover GJ, Moyer MA, Moore JW: Hydroxyurea-induced augmentation of fetal hemoglobin production in patients with sickle cell anemia. *Blood* 1987;69:109–116.
51. Chada K, Magram J, Raphael K, et al: Specific expression of a foreign β-globin gene in erythroid cells of transgenic mice. *Nature* 1985;314:377–380.
52. Costantini F, Chada K, Magram J: Correction of murine β-thalassemia by gene transfer into the germ line. *Science* 1986;233:1192–1194.
53. Townes TM, Lingrel JB, Chen HY, et al: Erythroid-specific expression of human β-globin genes in transgenic mice. *EMBO J* 1985;4:1715–1723.
54. Dzierzak EA, Papayannopoulou T, Mulligan RC: Lineage-specific expression of a human β-globin gene in murine bone marrow transplant recipients reconstituted with retrovirus-transduced stem cells. *Nature* 1988;331:35–41.
55. Grosveld F, van Assendelft G, Greaves D, Kollias G: Position-independent high-level expression of the human β-globin gene in transgenic mice. *Cell* 1987;51:975–985.
56. Greaves DR, Fraser P, Vidal MA, et al: A transgenic mouse model of sickle cell disorder. *Nature* 1990;343:183–185.
57. Ryan TM, Townes TM, Reilly MP, et al: Human sickle hemoglobin in transgenic mice. *Science* 1990;247:566–568.
58. Blom van Asserdelft G, Hanscombe O, Grosveld F, Greaves DR: The β-globin dominant control region activates homologous and heterologous promoters in a tissue-specific manner. *Cell* 1989;56:969–977.
59. Talbot D, Collis P, Antoniou M, et al: A dominant control region from the human beta-globin locus conferring integration site–independent gene expression. *Nature* 1989;338:352–355.
60. Boyer SH, Belding TK, Margolet L, Noyes AN: Fetal hemoglobin restriction to a few erythrocytes (F cells) in normal human adults. *Science* 1975;188:361–363.
61. Galjaard H: *Genetic Metabolic Disease*. New York, Elsevier, 1980.
62. Wood WG, Stamatoyannopoulos G, Lim G, Nute PE: F-cells in the adult: Normal values and levels in individuals with hereditary and acquired elevations of HbF. *Blood* 1975;46:671–682.
63. Al-Khatti A, Veith RW, Papayannopoulou T, et al: Stimulation of fetal hemoglobin synthesis by erythropoietin in baboons. *N Engl J Med* 1987;317:415–420.
64. Schechter AW, Noguchi CT, Rodgers GP: Sickle cell disease, in Stamatoyannopoulos G, Nienhuis AW, Majerus P, Leder P (eds): *Molecular Basis of Blood Diseases*. Philadelphia, Saunders, 1987, pp 179–218.
65. Haldane JBS: Rate of spontaneous mutation of human gene. *J Genet* 1935;31:317–326.
66. Gitschier J, Wood WI, Goralka TM, et al: Characterization of the human factor VIII gene. *Nature* 1984;312:326–330.
67. Wood WI, Capon DJ, Simonsen CC, et al: Expression of active human factor VIII from recombinant DNA clones. *Nature* 1984;312:330–336.
68. Vehar GA, Keyt B, Eaton D, et al: Structure of human factor VIII. *Nature* 1984;312:337–342.
69. Toole JJ, Knopf JL, Wozney JM, et al: Molecular cloning of a cDNA encoding human antihaemophilic factor. *Nature* 1984;312:342–347.
70. Cooper DN, Schmidke J: DNA restriction fragment length polymorphisms and heterozygosity in the human genome. *Hum Genet* 1984;66:1–66.
71. Gitschier J, Drayna D, Tuddenham EGD, et al: Genetic mapping and diagnosis of haemophilia A achieved through a *Bcl*I polymorphism in the factor VIII gene. *Nature* 1985;314:738–740.
72. Gitschier J, Wood WI, Tuddenham EGD, et al: Detection and sequence of mutations in the factor VIII gene of haemophiliacs. *Nature* 1985;315:427–430.
73. Barker D, Shafer M, White R: Restriction sites containing CpG show a higher frequency of polymorphism in human DNA. *Cell* 1984;36:131–138.
74. Dodds WJ, Hoyer LW: Coagulation activities in perfused organs: Regulation by addition of animal plasmas. *Br J Haematol* 1974;26:497–509.

75. Owen CA, Bowie EJW: Generation of plasmatic coagulation factors by the isolated rat liver perfused with completely synthetic blood substitute. *Thromb Res* 1981;22:259–266.
76. Marchioro TL, Hougie C, Ragde H, et al: Hemophilia: Role of organ homografts. *Science* 1969;163:188–190.
77. Webster WP, Zukoski CF, Hutchin P, et al: Plasma factor VIII synthesis and control as revealed by canine organ transplantation. *Am J Physiol* 1971;220:1147–1154.
78. Wion KL, Kelly D, Summerfield JA, et al: Distribution of factor VIII mRNA and antigen in human liver and other tissues. *Nature* 1985;317:726–729.
79. Oberle I, Camerino G, Heilig R, et al: Genetic screening for hemophilia A (classic hemophilia) with a polymorphic DNA probe. *N Engl J Med* 1985;312:682–686.
80. Kogan SC, Doherty M, Gitshier J: An improved method for prenatal diagnosis of genetic diseases by analysis of amplified DNA sequences. Application to hemophilia A. *N Engl J Med* 1987;317:985–990.
81. Anson DS, Choo KH, Rees DJG, et al: The gene structure of human anti-hemophiilic factor IX. *EMBO J* 1984;3:1053–1060.
82. Bertina RM, Van der Linden: Detection and characterization of molecular variants of factor 1X. *Methods Hematol* 1982;5:137–155.
83. Noyes CM, Griffith MJ, Roberts HR, Lundblad RL: Identification of the molecular defect in factor IX Chapel Hill: Substitution of histidine for arginine at position 145. *Proc Natl Acad Sci USA* 1983;80:4200–4202.
84. Giannelli F, Choo KH, Rees DJG, et al: Gene deletions in patients with hemophilia B and antifactor IX antibodies. *Nature* 1983;303:181–182.
85. Rees DJG, Rizza CR, Brownlee GG: Haemophilia B caused by a point mutation in a donor splice junction of the human factor IX gene. *Nature* 1985;316:643–645.
86. Anson DS, Austen DEG, Brownlee GG: Expression of active human clotting factor IX from recombinant DNA clones in mammalian cells. *Nature* 1985;315:683–685.
87. Busby S, Kumar A, Joseph M, et al: Expression of active human factor IX in transfected cells. *Nature* 1985;316:271–273.
88. Camerino G, Grzeschik KH, Jaye M, et al: Regional localization on the human X chromosome and polymorphism of the coagulation factor IX gene (hemophilia B locus). *Proc Natl Acad Sci USA* 1984;81:498–502.
89. Winship PR, Anson DS, Rizza CR, Brownlee GG: Carrier detection in hemophilia B using two further intragenic restriction fragment length polymorphisms. *Nucleic Acids Res* 1984;12:8861–8872.
90. Hay CW, Robertson KA, Yong S-L, et al: Use of a *Bam*HI polymorphism in the factor IX gene for the determination of hemophilia B carrier status. *Blood* 1986;67:1508–1511.
91. Poon M-C, Chui DHK, Patterson M, et al: Hemophilia B (Christmas disease) variants and carrier detection analyzed by DNA probes. *J Clin Invest* 1987;79:1204–1209.
92. Matthews RJ, Anson DS, Peake IR, Bloom AL: Heterogeneity of the factor IX locus in nine hemophilia B inhibitor patients. *J Clin Invest* 1987;79:746–753.
93. Antonarakis SE, Waber PG, Kittur SO, et al: Haemophilia A: Detection of molecular defects and of carriers by DNA analysis. *N Engl J Med* 1985;313:842–848.
94. Phillips DG, Aronis S, Tsittis G, Kazazian HH: Detection of six new mutations in haemophilia A. *Clin Res* 1986;34:654A.
95. Shelton-Inloes BB, Chehab FF, Mannucci PM, et al: Gene deletions correlate with the development of alloantibodies in von Willebrand disease. *J Clin Invest* 1987;79:1459–1465.
96. Verwelj CL, Quadt R, Briet E, et al: Genetic linkage of two intragenic restriction fragment length polymorphisms with von Willebrand's disease type IIA: Evidence for a defect in the von Willebrand factor gene. *J Clin Invest* 1988;81:1116–1121.
97. Ngo KY, Glotz VT, Koziol JA, et al: Homozygous and heterozygous deletions of the von Willebrand factor gene in patients and carriers of severe von Willebrand disease. *Proc Natl Acad Sci USA* 1988;85:2753–2757.
98. Sadler JE, Shelton-Inloes BB, Sorace JM, et al: Cloning and characterization of two cDNA's coding for human von Willebrand factor. *Proc Natl Acad Sci USA* 1985;82:6394–6398.
99. Ginsburg DR, Hardin RI, Bonthron DT, et al: Human von Willebrand factor (vWF):

Isolation of complementary DNA (cDNA) clones and chromosomal location. *Science* 1985;228:1401–1406.
100. Owren PA: The coagulation of blood investigations on a new clotting factor. *Acta Med Scand Suppl* 1947;194:1–316.
101. Kane WH, Davie EW: Cloning of a cDNA coding for human factor V, a blood coagulation factor homologous to factor VIII and ceruloplasmin. *Proc Natl Acad Sci USA* 1986;83:6800–6804.
102. Fass DN, Hewick RM, Knutson GJ, et al: Internal duplication and sequence homology in factors V and VIII. *Proc Natl Acad Sci USA* 1985;82:1688–1691.
103. Church WR, Jernigan RL, Toole J, et al: Coagulation factors V and VIII and ceruloplasmin constitute a family of structurally related proteins. *Proc Natl Acad Sci USA* 1984; 81:6934–6937.
104. Cool DE, Edgell C-JS, Louie GV, et al: Characterization of human blood coagluation factor XII cDNA. *J Biol Chem* 1985;260:13666–13676.
105. Weisberg LJ, Shiu DT, Greenberg CS, et al: Localization of the gene for coagulation factor XIII a-chain to chromosome 6 and identification of sites of synthesis. *J Clin Invest* 1987;79:649–652.
106. Rappaport H: Tumors of the hematopoietic system, in *Atlas of Tumor Pathology*, Sect 3, Fascicle 8. Washington, D.C., Armed Forces Institute of Pathology, 1966.
107. Foon KA, Todd RF: Immunologic classification of leukemias and lymphomas. *Blood* 1986;68:1–31.
108. Greaves M, Janossy G, Francis G, Minowada J: Membrane phenotypes of human leukaemic cells and leukemic cell lines: Clinical correlates and biological implications, in Clarkson B, Till J, Marks P (eds): *Differentiation of Normal and Neoplastic Hematopoietic Cells*. Cold Spring Harbor, N.Y., Cold Spring Harbor Press, 1978, p 823.
109. Seligmann M, Preud'homme J-L, Brouet J-C: Membrane markers in human lymphoid malignancies: Clinicopathological correlations and insight into the differentiation of normal and neoplastic cells, in *Differentiation of Normal and Neoplastic Hematopoietic Cells*. Cold Spring Harbor, N.Y., Cold Spring Harbor Press, 1978, pp 859–876.
110. Newman RA, Sutherland R, Greaves MF: The biochemical characterization of a cell surface antigen associated with acute lymphoblastic leukemia and lymphocyte precursors. *J Immunol* 1981;126:2024–2030.
111. Hozumi N, Tonegawa S: Evidence for somatic rearrangement of immunoglobulin genes coding for variable and constant regions. *Proc Natl Acad Sci USA* 1976;73:3628–3632.
112. Seidman JG, Leder P: The arrangement and rearrangement of antibdy genes. *Nature* 1978;276:790–795.
113. Siden E, Alt FW, Shinefeld L, et al: Synthesis of immunoglobulin μ chain gene products precedes synthesis of light chains during B-lymphocyte development. *Proc Natl Acad Sci USA* 1981;78:1823–1827.
114. Leder P, Max EE, Seidman JG, et al: Recombination events that activate, diversify and delete immunoglobulin genes. *Cold Spring Harbor Symp Quant Biol* 1981;45:859–865.
115. Hieter PA, Korsmeyer SJ, Waldmann TA, Leder P: Human immunoglobulin κ light chain genes are deleted or rearranged in λ-producing B cells. *Nature* 1981;290:368–372.
116. Hieter PA, Max EE, Seidman JG, et al: Human mouse kappa immunoglobulin constant and J region genes conserve homology in functional segments. *Cell* 1980;22:197–207.
117. Korsmeyer SJ, Hieter PA, Sharrow SO, et al: Human B cells display ordered light chain gene rearrangements and deletions. *J Exp Med* 1982;156:975–985.
118. Korsmeyer SJ, Hieter PA, Ravetch JV, et al: Develomental hirarchy of immunoglobulin gene rearrangements in human leukemic pre–B-cells. *Proc Natl Acad Sci USA* 1981; 78:7096–7100.
119. Vogler LG, Crist WM, Bockman DE, et al: Pre–B-cell leukemia. *N Engl J Med* 1978; 298:872–878.
120. Brouet JC, Pred'homme JL, Penit C, et al: Acute lymphoblastic leukemia with pre–B-cell characteristics. *Blood* 1979;54:269–273.
121. Korsmeyer SJ, Arnold A, Bakhshi A, et al: Immunoglobulin gene rearrangement and cell surface antigen expression in acute lymphocytic leukemias of T cell and B cell precursor origins. *J Clin Invest* 1983;71:301–313.

122. Abramson CS, Kersey JH, Le Bien TW: A monoclonal antibody (BA-1) reactive with cells of human B lymphocyte lineage. *J Immunol* 1981;126:83–88.
123. Ritz J, Pesando JM, Notis-McConarty J, et al: A monoclonal antibody to human acute lymphoblastic leukemia antigen. *Nature* 1980;283:583–585.
124. Lukes RJ, Taylor CR, Parker JW, et al: A morphologic and immunologic surface marker study of 299 cases of non-Hodgkin's lymphomas and related disorders. *Am J Pathol* 1978;90:461–486.
125. Berliner N, Ault KA, Martin P, Weinberg DS: Detection of clonal excess in lymphoproliferative disease by κ/λ analysis: Correlation with immunoglobulin gene DNA rearrangements. *Blood* 1986;67:80–85.
126. Arnold A, Cossman J, Bakhshi A, et al: Immunoglobulin gene rearrangements as unique clonal markers in human lymphoid neoplasms. *N Engl J Med* 1983;309:1593–1599.
127. Fialkow PJ, Klein E, Klein G, et al: Immunoglobulin and glucose-6-phosphate dehydrogenase as markers of cellular origin in Burkitt lymphoma. *J Exp Med* 1973; 138:89–102.
128. Siminovitch KA, Jensen JP, Epstein AL, Korsmeyer SJ: Immunoglobulin gene rearrangements and expression in diffuse histiocytic lymphomas reveal cellular lineage, molecular defects, and sites of chromosomal translocation. *Blood* 1986;67:391–397.
129. Rovigatti U, Mirro J, Kitchingman G, et al: Heavy chain immunoglobulin heavy chain genes: Rearrangement in acute nonlymphocytic leukemia. *Blood* 1984;63:1023–1027.
130. Kitchingman GR, Rovigatt U, Mauer AM, et al: Rearrangement of immunoglobulin heavy chain genes in T cell acute lymphoblastic leukemia. *Blood* 1985;65:725–729.
131. Schkolnik T, Schlossman SF, Griffin JD: Acute undifferentiated leukemia: Induction of partial differentiation by phorbol ester. *Leuk Res* 1985;9:11–17.
132. Raghavachar A, Bartram CR, Ganser A, et al: Acute undifferentiated leukemia: Implications for cellular origin and clonality suggested by analysis of surface markers and immunoglobulin gene rearrangement. *Blood* 1986;68:658–662.
133. Greaves MF, Chan IC, Furley AJW, et al: Lineage promiscuity in hematopoietic differentiation and leukemia. *Blood* 1986;67:1–11.
134. Jani P, Verbe P, Greaves MF, et al: Terminal deoxynucleotidyltransferase in acute myeloid leukemia. *Leuk Res* 1983;7:17–29.
135. Ha K, Minden M, Hozumi N, Gelfand EW: Immunoglobulin chain gene rearrangement in a patient with T-cell acute lymphoblastic leukemia. *J Clin Invest* 1984;73:1232–1236.
136. Pelicci P-G, Knowles DM, Dalla-Favera R: Lymphoid tumors displaying rearrangements of both immunoglobulin and T-cell receptor genes. *J Exp Med* 1985;162:1015–1024.
137. Cheng GM, Minden M, Toyonaga B, et al: T-cell receptor and immunoglobulin gene rearrangement in acute myeloblastic leukemia. *J Exp Med* 1986;163:414–424.
138. Smith LJ, Curtis JE, Messner HA, et al: Lineage infidelity in acute leukemia. *Blood* 1983;61:1138–1145.
139. Seremetis SV, Pelicci P-G, Tabilio A, et al: High frequency of clonal immunoglobulin or T-cell receptor gene rearrangements in acute myelogenous leukemia expressing terminal deoxyribuonucleotidyltransferase. *J Exp Med* 1987;165:1703–1712.
140. Lipford EH, Smith HR, Pittaluga S, et al: Clonality of angioimmunoblastic lymphadenopathy and implications for its evolution to malignant lymphoma. *J Clin Invest* 1987; 79:637–642.
141. Shearer WT, Ritz J, Finegold M, et al: Epstein-Barr virus–associated B-cell proliferations of diverse clonal origins after bone marrow transplantation in a 12 year old patient with severe combined immunodeficiency. *N Engl J Med* 1985;312:1151–1159.
142. Cleary ML, Sklar J: Lymphoproliferative disorders in cardiac transplant recipients are multiclonal lymphomas. *Lancet* 1984;2:489–493.
143. Fishleder A, Tubbs R, Hesse B, Levine H: Uniform detection of immunoglobulin-gene rearrangement in benign lymphoepithelial lesions. *N Engl J Med* 1987;316:1118–1121.
144. O'Connor NTJ, Weatherall DJ, Feller AC, et al: Rearrangement of the T-cell receptor β-chain gene in the diagnosis of lymphoproliferative disorders. *Lancet* 1985;1:1295–1297.
145. Aisenberg AC, Krontiris TG, Mak TW, Wilkes BM: Rearrangement of the gene for the

beta chain of the T-cell receptor in T-cell chronic lymphocytic leukemia and relative disorders. *N Engl J Med* 1985;313:529–533.
146. Bertness V, Kirsch I, Hollis G, et al: T-cell receptor gene rearrangements as clinical markers of human T-cell lymphomas. *N Engl J Med* 1985;313:534–538.
147. Waldmann TA, Davis MM, Bongiovanni K, Korsmeyer SJ: Rearrangements of genes for the antigen receptor on T cells as markers of lineage and clonality in human lymphoid neoplasms. *N Engl J Med* 1985;313:776–783.
148. Aisenberg AC, Wilkes BM, Harris NL, et al: Chronic T-cell lymphocytosis with neutropenia: Report of a case studied with monoclonal antibody. *Blood* 1984;58:818–822.
149. Chan WC, Check I, Schick C, et al: A morphologic and immunologic study of the large granular lymphocyte in neutropenia with T lymphocytosis. *Blood* 1984;63:1133–1140.
150. Reynolds CW, Foon KA: Tγ-lymphoproliferative disease and related disorders in humans and experimental animals: A review of the clinical, cellular, and functional characteristics. *Blood* 1984;64:1146–1158.
151. Minden MD, Mak TW: The structure of the T cell antigen receptor genes in normal and malignat T cells. *Blood* 1986;68:327–336.
152. Greenberg JM, Quertermous T, Seidman JG, Kersey JH: Human T cell γ-chain gene rearrangements in acute lymphoid and nonlymphoid leukemia: Comparison with the T cell receptor β-chain gene. *J Immunol* 1986;137:2043–2049.
153. Baer R, Lefranc M-P, Minowada J, et al: Organization of the T-cell receptor a chain gene and rearrangement in human T-cell leukemias. *Mol Biol Med* 1986;3:265–277.
154. Aisenberg AC, Wilkes BM, Jacobson JO: Rearrangement of the genes for the beta and gamma chains of the T cell receptor is rarely observed in adult B cell lymphoma and chronic lymphocytic leukemia. *J Clin Invest* 1987;80:1209–1214.
155. Tawa A, Hozumi N, Minden M, et al: Rearrangement of the T cell receptor beta chain gene in non-T, non-B acute lymphocytic leukemia of childhood. *N Engl J Med* 1985; 313:1033–1037.
156. Felix CA, Wright JJ, Poplack DG, et al: T cell receptor α-, β-, and γ-genes in T cell and pre–B cell acute lymphoblastic leukemia. *J Clin Invest* 1987;80:545–556.
157. Korsmeyer SJ: Antigen receptor genes as molecular markers of lymphoid neoplasms. *J Clin Invest* 1987;79:1291–1295.
158. Yancopoulos GD, Blackwell TK, Suh H, et al: Introduced T cell receptor variable region gene segments recombine in pre-B cells: Evidence that B and T cells use a common recombinase. *Cell* 1986;44:251–259.
159. Sklar J, Cleary ML, Thielemans K, et al: Biclonal B-cell lymphoma. *N Engl J Med* 1984; 311:20–27.
160. Siegelman MH, Cleary ML, Warnke R, Sklar J: Frequency of biclonality and Ig gene alterations among B cell lymphomas that show multiple histologic forms. *J Exp Med* 1985;161:850–863.
161. Cleary ML, Meeker TC, Levy S, et al: Clustering of extensive somatic mutations in the variable region of an immunoglobulin heavy chain gene from a human B cell lymphoma. *Cell* 1986;44:97–106.
162. Yunis JJ: The chromosomal basis of human neoplasia. *Science* 1983;221:227–236.
163. Weiss LM, Hu E, Wood GS, et al: Clonal rearrangements of T-cell receptor genes in mycosis fungoides and dermatopathic lymphadenopathy. *N Engl J Med* 1985;313:539–544.
164. Weiss LM, Wood GS, Trela M, et al: Clonal T-cell populations in lymphomatoid papulosis. Evidence of a lymphoproliferative origin for a clinically benign disease. *N Engl J Med* 1986;315:475–479.
165. Wood GS, Weiss LM, Hu C-H, et al: T-cell antigen deficiencies and clonal rearrangements of T-cell receptor genes in pagetoid reticulosis (Woringer-Kolopp disease). *N Engl J Med* 1988;318:165–167.
166. Perl A, Wang N, Williams JM, et al: Aberrant immunoglobulin and c-*myc* gene rearrangement in patients with nonmalignant monoclonal cryoglobulinemia. *J Immunol* 1987;139:3512–3520.
167. Perl A, Hunt MJ, Abraham GN: Translocation of the c-*myc* proto-oncogene into the IgH locus in patients with nonmalignant mixed cryoglobulinemia. *Fed Proc* 1987; 46:916.

168. Clofent G, Brockly F, Commes T, et al: No preferential use of the VH(V) family in human multiple myeloma. *Br J Hematol* 1989;73:466–490.
169. Saltman DL, Ross JA, Banks RE, et al: Molecular evidence for a single clonal origin in biphenotypic concomitant chronic lymphocytic leukemia and multiple myeloma. *Blood* 1989;74:2062–2065.
170. Barlogie B, Epstein J, Selvanayagam P, Alexanian R: Plasma cell myeloma—new insights and advances in therapy. *Blood* 1989;73:865.
171. Van Riet I, Heirman C, Lacor P, et al: Detection of monoclonal B lymphocytes in bone marrow and peripheral blood of multiple myeloma patients by immunoglobulin gene rearrangement studies. *Br J Haematol* 1989;73:289–295.
172. Nordan RP, Potter M: A macrophage-derived factor required by plasmacytomas for survival and proliferation in vitro. *Science* 1986;233:566.
173. Van Snick J, Vink A, Cayphas S, Uyttenhove C: Interleukin-HP1, a T cell–derived hybridoma growth factor that supports the in vitro growth of murine plasmacytomas. *J Exp Med* 1987;165:641–649.
174. Blankenstein T, Qin Z, Li W, Diamantstein T: DNA rearrangement and constitutive expression of the interleukin 6 gene in a mouse plasmacytoma. *J Exp Med* 1990; 171:965–970.
175. Harrington DS: Molecular gene rearrangement analysis in hematopathology. *Am J Clin Pathol* 1990;93(Suppl 1):S38–S43.
176. Jacobs K, Shoemaker C, Rudersdorf R, et al: Isolation and characterization of genomic and cDNA clones of human erythropoietin. *Nature* 1985;313:806–810.
177. Lin F-K, Suggs C-H, Lin JK, et al: Cloning and expression of the human erythropoietin gene. *Proc Natl Acad Sci USA* 1985;82:7580–7584.
178. Koeffler HP, Goldwasser E: Erythropoietin radioimmunoassay in evaluating patients with polycythemia. *Ann Intern Med* 1981;94:44–47.
179. Garcia JF, Ebbe SN, Hollander L, et al: Radioimmunoassay of erythropoietin: Circulating levels in normal and polycythemic human beings. *J Lab Clin Med* 1982; 99:624–635.
180. Naets JP: Hematological disorders in renal failure. *Nephron* 1975;14:181–194.
181. Erslev AJ, Caro J, Kansu E, Silver R: Renal and extrarenal erythropoietin production in anemic rats. *Br J Haematol* 1980;45:65–72.
182. Winearls CG, Pippard MJ, Downing MR, et al: Effect of human erythropoietin derived from recombinant DNA on the anaemia of patients maintained by chronic hemodialysis. *Lancet* 1986;2:1175–1178.
183. Caro J, Brown S, Miller O, et al: Erythropoietin levels in uremic nephric and anephric patients. *J Lab Clin Med* 1979;93:449–458.
184. Eschbach JW, Egrie JC, Downing MR, et al: Correction of the anemia of end-stage renal disease with recombinant human erythropoietin. Results of a combined phase I and II clinical trial. *N Engl J Med* 1987;316:73–78.
185. Chandra M, McVicar M, Clemons G, et al: Role of erythropoietin in the reversal of anemia of renal failure with continuous ambulatory peritoneal dialysis. *Nephron* 1987; 46:312–315.
186. Neff MS, Kim KE, Persoff M, et al: Hemodynamics of uremic anemia. *Circulation* 1971; 43:876–883.
187. Eschbach JW, Kelly MR, Haley NR, et al: Treatment of the anemia of progressive renal failure with recombinant human erythropoietin. *N Engl J Med* 1989;321:158–163.
188. Sun CH, Ward HJ, Wellington LP, et al: Serum erythropoietin levels after renal transplantation. *N Engl J Med* 1989;321:151–157.
189. Ludwig H, Fritz E, Kotzmann H, et al: Erythropoietin treatment of anemia associated with multiple myeloma. *N Engl J Med* 1990;322:1693–1699.
190. Dessypris EN, Gleaton JH, Armstrong OL: Effect of human recombinant erythropoietin on human marrow megakaryocyte colony formation in vitro. *Br J Haematol* 1987; 65:265–269.
191. Ishibashi T, Koziol JA, Burstein SA: Human recombinant erythropoietin promotes differentiation of murine megakaryocytes in vitro. *J Clin Invest* 1987;79:286–289.
192. Clark SC, Kamen R: The human hematopoietic colony-stimulating factors. *Science* 1987;236:1229–1237.
193. Scharader JW: The panspecific hemopoietin of activated lymphocytes (interleukin-3). *Annu Rev Immunol* 1986;4:205–230.

194. Metcalf D: The molecular biology and functions of the granulocyte-macrophage colony-stimulating factors. *Blood* 1986;67:257–267.
195. Yang YC, Ciarletta AB, Temple PA, et al: Human IL-3 (multi-CSF): Identification by expression cloning of a novel hematopoietic growth factor related to murine IL-3. *Cell* 1986;47:3–10.
196. Smith KA: T-cell growth factor. *Immunol Rev* 1980;51:337–357.
197. Stanley ER, Bartocci A, Patinkin D, et al: Regulation of very primitive, multipotent, hematopoietic cells by hemopoietin-1. *Cell* 1986;45:667–674.
198. Nagata S, Tsuchiya M, Asano S, et al: Molecular cloning and expression of cDNA for human granulocyte colony-stimulating factor. *Nature* 1986;319:415–418.
199. Souza LM, Bonne TC, Gabrilove J, et al: Recombinant human granulocyte colony-stimulating factor: Effects on normal and leukemic myeloid cells. *Science* 1986;232:61–65.
200. Kawasaki ES, Ladner MB, Wang AM, et al: Molecular cloning of a complementary DNA encoding human macrophage-specific colony-stimulating factor (CSF-1). *Science* 1985;230:291–296.
201. Wong GG, Witek JS, Temple PA, et al: Human GM-CSF: Molecular cloning of the complementary DNA and purification of the natural and recombinant proteins. *Science* 1985;228:810–815.
202. Miyatake S, Yokota J, Lee F, Arai K-I: Structure of the chromosomal gene for murine interleukin 3. *Proc Natl Acad Sci USA* 1985;82:316–320.
203. Pettenati MJ, Le Beau MM, Lemons RS, et al: Assignment of CSF-1 to 5q33-1: Evidence for clustering of genes regulating hematopoiesis and for their involvement in the deletion of the long arm of chromosome 5 in myeloid disorders. *Proc Natl Acad Sci USA* 1987;84:2970–2974.
204. Zucali JR, Dinarello CA, Oblon DJ, et al: Interleukin 1 stimulates fibroblasts to produce granulocyte-macrophage colony-stimulating activity and prostaglandin E_2. *J Clin Invest* 1986;77:1837–1863.
205. Munker R, Gasson J, Ogawa M, Koeffler HP: Recombinant human TNF induces production of granulocyte-monocyte colony-stimulating factor. *Nature* 1986;323:79–82.
206. Platzer E, Welte K, Gabrilove JL, et al: Biological activities of a human pluripotent hemopoietic colony stimulating factor on normal and leukemic cells. *J Exp Med* 1985;162:1788–1801.
207. Lopez AF, Williamson DJ, Gamble JR, et al: Recombinant human granulocyte-macrophage colony-stimulating factor stimulates in vitro human neutrophil and eosinophil function, surface receptor expression, and survival. *J Clin Invest* 1986;78:1220–1228.
208. Griffin JD, Young D, Herrmann F, et al: Effects of recombinant human GM-CSF on proliferation of clonogenic cells in acute myeloblastic leukemia. *Blood* 1986;67:1448–1453.
209. Young DC, Wagner K, Griffin JD: Constitutive expression of the granulocyte-macrophage colony-stimulating factor gene in acute myeloblastic leukemia. *J Clin Invest* 1987;79:100–106.
210. Body GP, Buckley M, Sathe YS, Freireich E: Quantitative relationships between circulating leukocytes and infection in patients with acute leukemia. *Ann Intern Med* 1966;64:328–340.
211. Matoyoshi K, Takaku F, Maekawa T, et al: Protective effect of partially purified human urinary colony-stimulating factor on granulocytopenia after antitumor chemotherapy. *Exp Hematol* 1986;14:1069–1075.
212. Welte K, Bonilla MA, Gillio AP, et al: Recombinant human granulocyte colony-stimulating factor. Effects on hematopoiesis in normal and cyclophosphamide-treated primates. *J Exp Med* 1987;165:941–948.
213. Groopman JE, Mitsuyasu RT, DeLeo MJ, et al: Effect of recombinant human granulocyte-macrophage colony-stimulating factor on myelopoiesis in the acquired immunodeficiency syndrome. *N Engl J Med* 1987;317:593–598.
214. Vadhan-Raj S, Keating M, LeMaistre A, et al: Effects of recombinant human granulocyte-macrophage colony-stimulating factor in patients with myelodysplastic syndromes. *N Engl J Med* 1987;317:1545–1552.
215. Koeffler HP, Golde DW: Cellular maturation in human preleukemia. *Blood* 1978;52:355–361.
216. Brandt SJ, Peters WP, Atwater SK, et al: Effect of recombinant human granulocyte-

macrophage colony-stimulating factor on hematopoietic reconstitution after high-dose chemotherapy and autologous bone marrow transplanation. *N Engl J Med* 1988; 318:869–876.

217. Sheridan WP, Wolf M, Lusk J, et al: Granulocyte colony stimulating factor and neutrophil recovery after high-dose chemotherapy and autologous bone marrow transplantation. *Lancet* 1989;2:891–895.
218. Gabrilove JL, Jakubowski A, Scher H, et al: Effect of granulocyte colony-stimulating factor on neutropenia and associated morbidity due to chemotherapy for transitional-cell carcinoma of the urothelium. *N Engl J Med* 1988;318:1414–1422.
219. Silberstein DS, Owen WF, Gasson JC, et al: Enhancement of human eosinophil cytoxicity and leukotriene synthesis by biosynthetic (recombinant) granulocyte-macrophage colony-stimulating factor. *J Immunol* 1986;137:3290–3294.
220. Hammond WP, Price TH, Souza LM, Dale DC: Treatment of cyclic neutropenia with granulocyte colony-stimulating factor. *N Engl J Med* 1989;320:1306–1311.
221. Ganser A, Volkers B, Greher J, et al: Recombinant human granulocyte-macrophage colony-stimulating factor in patients with myelodysplastic syndromes: A phase I/II trial. *Blood* 1989;73:31–37.
222. Quesenberry PJ: Treatment of a marrow stem-cell disorder with granulocyte colony-stimulating factor. *N Engl J Med* 1989;320:1343–1345.
223. Metcalf D: Haematopoietic growth factors 2: Clinical applications. *Lancet* 1989;1:885–887.
224. TIMI Study Group: The thrombolysis in myocardial infarction (TIMI) trial: Phase I findings. *N Engl J Med* 1985;312:932–936.
225. Verstraete M, Bernard R, Bory M, et al: Randomized trial of intravenous recombinant tissue-type plasminogen activator versus intravenous streptokinase in acute myocardial infarction: Report from the European Cooperative Study Group for recombinant tissue-type plasminogen activator. *Lancet* 1985;1:842–847.
226. Sherry S: Tissue plasminogen activator (t-PA): Will it fulfill its promise? *N Engl J Med* 1985;313:1014–1017.

Genetic Disorders and Recombinant DNA

9

Ralph C. Williams, Jr., MD

Genetic Disorders and Molecular Biology

One of the most obvious applications of many of the molecular biological methods has been in the area of hereditary or genetically acquired disorders. The ready availability of Southern blotting techniques using selected probes has opened up a whole new area of prenatal diagnosis, informed genetic counseling, and genetic epidemiology. As more and more human diseases have been shown to have genetic factors that influence their expression, recombinant techniques are being applied to disorders that only a few decades ago were not considered to be genetic or even to have a hereditary component. Good examples of such leaps in our understanding are human leukocyte antigens HLA-B27 in the spondyloarthropathies[1–3] and HLA-DR4 in seropositive rheumatoid arthritis.[4] This chapter will provide an introduction to some of the precise mechanisms of inheritance in certain disease categories. Rather than presenting an encyclopedic compendium of everything known concerning all hereditary diseases, it seems more useful to review a limited spectrum of genetic disorders in some depth to illustrate how such problems lend themselves to investigation with recombinant methodology. Some important genetic diseases, such as the hemoglobinopathies and coagulation disorders, have already been covered in Chapter 8. The diseases addressed here represent hereditary conditions in which significant advances in understanding have been made through a molecular biological approach. As knowledge of many of these heritable disorders has increased, it has become apparent that some other common human disease states that are not considered to be genetic may actually have clear-cut underlying hereditary mechanisms.

The Human Genome

Increasing public, official, and corporate interest has been focused on the importance and technical feasibility of completely sequencing and defining the entire human genome.[5–7] A series of discussions and convocations were organized during which the technical problems and the potential leaps in scientific knowledge and understanding were discussed. It was initially estimated that obtaining a complete sequence of the 3 billion nucleotides in the entire human genome would require 30,000 person-years and $2 billion. Comparative figures for other important government-funded projects are $3 billion to build a superconducting supercollider and $8 billion to build a space station. It was estimated that with current methods a single person could sequence 100 kilobases of DNA per year at an approximate cost of about $1 a base. Introduction of new technology including automatic sequencing procedures would markedly accelerate and reduce the ultimate costs of such a herculean effort. One of the strongest motivations for such a project would be the ultimate ability to determine with great accuracy the exact loci for the 1000 or more known human genetic disorders. A related technical problem would be the handling and storage of the derived data. At present the major DNA data storage bases, located at Los Alamos National Laboratory in the United States and at the European Molecular Biology Laboratory in Heidelberg, have increased some 25-fold in size and considerably more expansion will undoubtedly be necessary.

During a number of public discussions an interesting legal or philosophical point emerged: when genomic sequences have been identified, can anyone or any group "own" such information in terms of a patent or a copyright?[8] That is, can anyone actually buy or own the human genome? This question was of interest because the situation appeared to be slightly different from that of allowing patents for recombinant microbes or animals. There seemed to be something intrinsically different about any group or individual claiming ownership of any part of the human genome. Discussions of various aspects of this question are ongoing and the issue is still unresolved.

During the next several years, we will probably hear considerably more about efforts to sequence the human genome, both in scientific journals and in the lay press. It promises to be a rapidly expanding and exciting story. One aspect of the proposal to sequence the entire genome that initially created some bureaucratic controversy was whether the task would be delegated to the National Institutes of Health or assigned to some other large federal office such as the Department of Energy (DOE).[9] There have been intense high-level discussions and debates about how to get the job done. Future developments and similar efforts in other countries such as Japan and West Germany undoubtedly will be aided by financial and technical momentum from private industry as well.

DNA Fingerprints

Application of routine recombinant analytic techniques showed that DNA fingerprints can be obtained which are as unique to individuals as are the whorls and patterns in the conventional hand fingerprints. The DNA fingerprint technique is based on the occurrence of simple tandem-repetitive regions of DNA, termed minisatellites, widely dispersed in the entire human genome. When individual DNA samples are obtained from a number of normal subjects, fragments of DNA produced after restriction enzyme digestion show substantial length and size polymorphism when hybridized with radiolabeled probes obtained from common core 10- to 15-base-pair sequences of these short tandem repeats.[10] This hybridization pattern appears to be unique for each individual. Identical twins have identical DNA fingerprints, but the patterns for siblings and other family members are individually specific. The hypervariable minisatellite nucleotide fragments involved in generating these DNA fingerprints are apparently inherited in a mendelian fashion with very low mutation rates to a new length allele.[10] There is considerable interest in the use of this technique for forensic purposes, such as in parental identification or in criminal cases.[11] For some time forensic scientists have sought a method for identifying the origin of blood and other body fluids with the same precision that has been possible with conventional hand fingerprinting. It appears that DNA fingerprinting could revolutionize forensic biological identification techniques.[12] The minisatellite nucleotide sequences in individual genomic DNA samples are scattered throughout the human genome. The satellites themselves do not appear to be derived from each other in the same way as a group of transposable sequences might be. On the other hand, they do appear to be related because they share similar "core" sequences of nucleotides that are approximately 15 bases long and form the structure of the repetitive unit. Two core sequences have been recognized, and probes hybridizing with these have been used in most of the human DNA fingerprinting work done thus far. Studies of large numbers of samples from kindreds indicate that a very low mutation frequency of 0.004 affects these patterns. Examples of the sorts of patterns obtained with this technique are shown in Figure 9.1. The potential usefulness of this method can be illustrated by details of an early case.[12] Dr. Jeffreys of the University of Leicester and colleagues who devised this DNA fingerprinting technique were called upon to examine evidence by the immigration authorities in Britain. A Ghanaian boy born in the United Kingdom had returned to Ghana to live with his father. Later he tried to return to England to be with his mother, brother, and two sisters. The immigration officials suspected that a switch had been made and that the boy applying for entry into Britain was not the actual son of the claimed mother. Legal counsel approached Dr. Jeffreys for help. The case was made more difficult by the fact that neither the boy's father nor any of the mother's sisters were available for direct testing. By using two probes and one restriction enzyme, *HinfI*, approximately 80 bands were identified in DNA from the boy, his putative mother, brother, two sisters, and an unrelated control. Since some of the DNA markers in such an analysis may reach almost 100% heterozygous expression, the boy's paternity could be identified by matching up the siblings' fingerprint bands that were absent in the mother. Of 39 bands then identified as paternal, about half were found in the boy's pattern. Because DNA fingerprints are seldom or virtually never shared by unrelated individuals, these findings strongly indicated that the boy had the same father as the two sisters. Moreover, the analysis showed that the boy and the mother shared 25 of 40 specific bands. The probability of two unrelated people sharing only one discrete fragment was calculated to be 2×10^{-15}. Clearly, the boy and the woman were closely related. This example illustrates the remarkable precision of this method.

As yet no inherited disease has been studied directly using the fingerprinting method during the several years that the technique has been available. The approach is similar to that used with restriction fragment polymorphism analyses with the exception that a single minisatellite probe can effectively scan as many as 20 different genetic loci at the same time. If one of the hypervariable segments is demonstrated to be fairly close to the disease locus or a gene involved in the disorder in question, it could be cloned in the usual way with an expression vector and then employed as a marker for the disease in studies of multiple large kindreds.

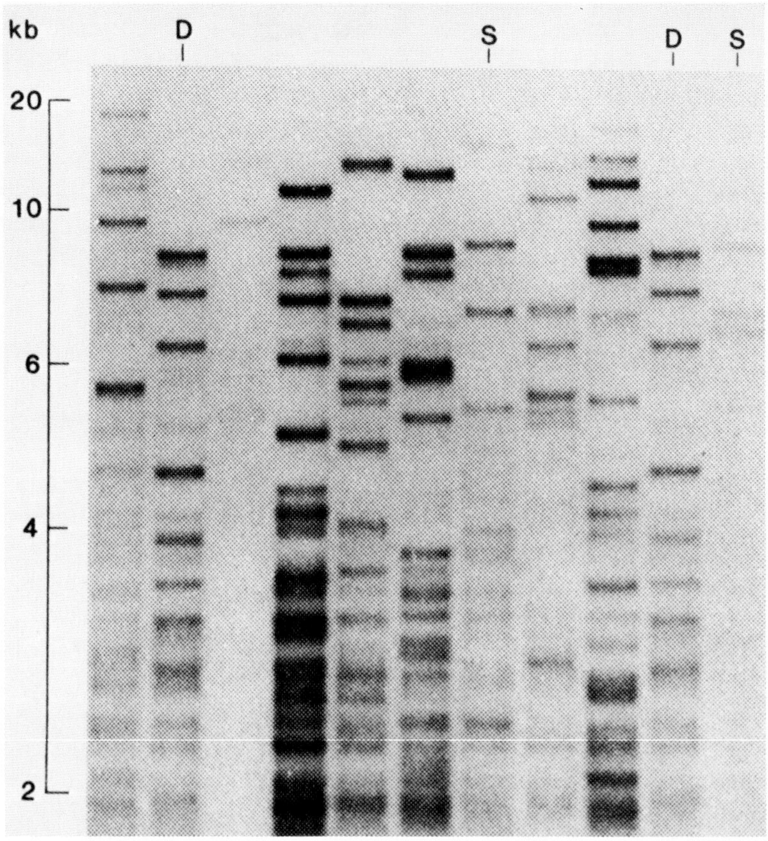

Figure 9.1 Individual identification using DNA fingerprints from small samples of blood. DNA was prepared by a rapid procedure using one or two drops of blood from a panel of 11 individuals and was digested with *Hinf* I; DNA fingerprints were prepared, using 33.15 as a probe. The panel consisted of nine unrelated individuals, two of whom had been previously DNA fingerprinted, plus two sisters. The autoradiograph was inspected by a colleague who was unaware of the order of samples on the autoradiograph. He correctly identified the two previously fingerprinted individuals on the basis of pattern identity and identified the two sisters, who have several bands in common (tracks S). He also correctly deduced the fact that duplicate samples had been taken from another individual (tracks D). *Reproduced from Jeffreys et al,*[11] *with permission.*

Restriction Fragment Length Polymorphisms

Recombinant DNA techniques have very rapidly revolutionized the study of inherited human disorders. Initially the technology was applied to diseases such as sickle cell anemia or thalassemia, in which there was already considerable chemical insight into the cause of the disease. However, it soon became apparent that molecular biological techniques could be used to detect DNA sequence variations in populations or disease states in which the exact biochemical lesion was not yet known.[13,14]

Previously, detection of DNA sequence differences in biologically important human loci had to rely on demonstration of electrophoretic or antigenic variation. However, using restriction enzymes and gel analyses of the DNA fragments produced, sequence differences in DNA can be analyzed and quantitated. The most direct approach to this type of analysis involves cutting genomic DNA using a restriction endonuclease, transferring the DNA to gels, and separating fragments by gel electrophoresis. Separated DNA fragments are then transferred to a membrane and probed with an appropriate cloned DNA segment for the region of interest to investigate the distribution of the sizes of DNA fragments in the locus to be studied. The restriction endonucleases used cleave DNA at highly specific recognition sequences and any change in these nucleotide sequences, particularly at the restriction enzyme site, will affect the sizes of DNA fragments produced. Inherited differences in the patterns of enzyme digestion fragments have been called restriction fragment length polymorphisms (RFLPs). Changes in such patterns can result from single base changes, simple deletions, or insertion or deletion of larger segments of nucleotide sequence.

RFLP analysis in genetic disease allows the observer to examine the two homologous chromosomes from an individual who is heterozygous at the particular polymorphic site. Thus, in family studies one can infer from the alleles of children which chromosome was transmitted from each or a single parent. The application of RFLPs in disorders in which loci are well marked out, such as thalassemia, sickle cell anemia, or phenylketonuria, has been discussed in Chapter 8.

RFLPs have also been extensively applied to the study of inherited disease in which the exact functional or biochemical defect is still unknown. Such analyses have in some instances allowed determination of chromosomal locations for disease loci.[13–15] The concept behind such analysis is that two genes situated close to one another on the same chromosome are likely to be transmitted together with much greater frequency than would be expected by chance. Actually, alleles of such closely spaced sites will be transmitted separately only if a recombination event involving the two loci occurs. The chance for such crossovers increases in proportion to the distance between the individual genes. Thus, the appearance of both genes together can often be utilized to estimate their spacing on the chromosome. In the same way, if a genetic marker and a disease are strongly correlated in kindred studies of inheritance, it is an indication that the disease-related gene is in the same chromosomal region as the marker.

When RFLPs were first used, it was not clear whether DNA polymorphisms were frequent enough to be readily identified in the human genome or whether the degree of polymorphism and individual allelic frequencies were high enough to make RFLPs useful as genetic markers. Subsequent work has firmly established the usefulness and practical application of this technique to a wide range of genetic disorders. The level of polymorphism in the human genome reflects a relatively high degree of sequence heterozygosity. Direct examination has indicated that an average value of 1 base in 250 to 500 differs between any two chromosomes chosen randomly.[16–19] This means that for any selected probe, a hybridization screen using multiple restriction enzymes will have a very high probability of picking up at least one RFLP. Also, probes containing larger sequences of nucleotides are more likely to identify polymorphisms than probes containing shorter sequences.

Most RFLPs have been identified by comparing the patterns of DNA fragments hybridized by a cloned probe in DNA digested with a single restriction enzyme from a number of unrelated individual subjects.[20] In many instances differences in patterns among individuals can often be related directly to the gain or loss of a particular restriction enzyme recognition site. If a similar pattern of differences is noted with many different restriction enzymes, an insertion or deletion can usually be associated with the particular polymorphism.[21] In order to identify a distinct

RFLP, an investigator must demonstrate a mendelian pattern of inheritance. The RFLP patterns must be established for a number of normal unrelated individuals in the populations being evaluated in order to identify the general level of allelic frequencies in the group studied.

Most individual RFLPs that have been described depend on the presence or absence of a restriction enzyme site and for this reason have two alleles.[22] When a small group of unrelated individuals is studied during the original screen for RFLPs, such an approach favors detection of polymorphisms with a high degree of heterozygosity. In general, the frequency of the less common allele for most RFLPs is usually between 0.15 and 0.5, corresponding to heterozygosity levels of 25% to 50%. Systems with two alleles are limited to an extent, because at least half of the individuals will be homozygous at the locus in question and thus it would be impossible to detect transmission of the two homologous alleles. A major advantage of identification of DNA markers is related to the fact that patterns of individual alleles from RFLPs within closely spaced DNA regions can be taken as characteristic of certain haplotypes, and in this way multiallelic systems can be analyzed in parallel. In cases such as this, it is often necessary to use multiple restriction enzymes in order to define several different two-allele systems.

The most readily applied DNA markers are those that detect multiallele systems that are recognized by variations in the lengths of DNA within certain defined fragments. These markers are less frequent than single-site variability. Analyses have indicated that the reason for length variations may be a high frequency of unequal crossovers within a segment of short direct nucleotide repetitive sequences.[10] RFLPs depending on insertions or deletions have been described adjacent to the insulin, Ha-*ras*, α-globin, and myoglobin genes in addition to several anonymous probes.[22]

RFLPs have now been identified on every human chromosome, and their presence does not seem to be restricted to any small region of the human genome. For some chromosomes, enough markers have now been defined to allow definition of preliminary genetic linkage maps by comparing the inheritance patterns of all known loci in large highly studied kindreds.[23,24] Within the next few years a complete linkage map may become available for the entire human genome, which will allow linkage analysis within all regions. This sort of map would allow localization of any disease gene that manifests a clear pattern of inheritance merely by applying a standard battery of selected linkage markers. Application of computer programs to the task of simultaneous analysis of multiple genetic markers with known linkage relationships has begun.[25] This approach may eventually provide a method for using the detailed human genome linkage map for analysis of diseases that may involve multiple genes, including some of the currently puzzling affective behavior disorders such as schizophrenia.

Mapping Diseases by Linkage to RFLPs

During the short period of their intensive application, RFLPs used as linkage markers have already helped in determining many disease loci. One of the areas in which this has been most effective involves the X chromosome and X-linked Duchenne muscular dystrophy.[26] A large number of linked genetic markers have been identified in Duchenne muscular dystrophy,[27–29] and progress continued in this area until final identification of the gene defect itself. This will be discussed later in more detail. An additional advantage of studying X-linked disorders is related to the presence of a single allele at each locus in males. This has led to detection of linked markers for Alport's syndrome, Becker's muscular dystrophy, X-linked Charcot-Marie-Tooth disease, choroideremia, fragile X mental retardation syndrome, Menkes' disease, ocular albinism, X-linked retinitis pigmentosa, and retinoschisis.[30] Because 68 polymorphic DNA markers have already been identified for the X chromosome, it seems certain that a specific linkage can be detected for virtually any X-linked disorder when these markers are applied to analyses of families.

In the case of autosomes, the first major breakthrough in the application of RFLPs was in 1983 with the definition of a marker closely linked to Huntington's disease.[31,32] This discovery subsequently allowed mapping of the defect to the short arm of chromosome 4. A number of other genetic disorders are being investigated with the hope that similar genetic linkages or markers can be identified. Genetic linkage analysis is one of the most direct and rapid methods of determining whether a particular cloned gene is the site of a primary defect.

Strategy Involved in the Use of RFLPs

How do RFLPs help to isolate defective genes? What is the precise logic behind their use? Once linkage of a single RFLP with a defective or disease-specific gene has been identified, other RFLPs that straddle the affected gene can be identified. Moreover, these flanking RFLPs (R_L and R_R to the left and right, respectively, of the gene in question) can be ordered by observing the recombination between R_L and R_R, R_L and the gene in question, or the gene and R_R. Thus, R_L and R_R define a segment of DNA that contains the gene of interest. The distance in nucleotides between R_L and R_R should be small if these two areas are to be useful in steps to isolate the defective gene. The proximity with which R_R and R_L can be mapped to the gene being sought is determined by the number of meiotic events available for the analysis. DNA probes that specifically define an RFLP provide indirect insight into the physical distance between R_R and R_L, since orthogonal or field-inversion gel electrophoresis can be used to separate DNA segments as large as 2×10^6 base pairs by cutting with relatively rare 6-, 7- or 8-base recognition restriction enzymes in combination with Southern blotting. The RFLPs define a segment of DNA that is about 2×10^6 base pairs long and located within or near a certain chromosomal band. In general, three approaches can be utilized to isolate a disease-specific or defective gene: (1) the candidate gene approach, (2) chromosome rearrangement, and (3) cloning strategies.

The candidate gene approach involves either initial isolation of mRNA from genes thought to be involved in the defect or use of random mRNAs from normal tissues known to be affected by the disease. This is followed by screening of somatic cell hybrid lines or by in situ chromosomal hybridization to demonstrate that the mRNA is encoded in a specific chromosomal band. Both approaches are now technically feasible but require a great amount of work, especially if the normal transcript of the gene is expressed at a relatively low level.

The chromosome rearrangement strategy makes use of chromosome deletions, insertions, and translocations to provide markers that must by definition be physically close to the affected gene.[33] Where these are present, the markers can be used to determine the order of genes or DNA probes within or near the chromosomal region of interest. When they are applicable, chromosome rearrangements are a powerful tool for analysis of specific disease-related loci. As we will show later in this section, Duchenne muscular dystrophy, Wilms' tumor, and retinoblastoma are diseases in which this strategy has been effectively and successfully applied.

The cloning strategy is based on isolating all the DNA interposed between R_L and R_R; this would include the disease-related gene as well. The next steps involve mapping, walking and jumping. Mapping was first employed by Steinmetz et al.,[34,35] who isolated cosmids containing mouse major histocompatibility complex (MHC) genes by cross-hybridizing with MHC class I and class II complementary DNA probes. The cosmids were mapped for endonuclease restriction sites, and the sequence for a large portion of the mouse MHC was defined by lining up cosmids that contained overlapping sequences. Similar strategies could be applied to a human chromosomal segment isolated from a cell hybrid containing the DNA of interest.

Walking is the technique of systematically isolating the DNA sequence located at the extreme left or right of a starting cloned DNA sequence and then using this fragment to screen a library of cDNA fragments for DNA adjacent to and slightly overlapping the starting sequence. The walk then proceeds as a linear sequence of steps of this kind. This is illustrated in Figure 9.2. One problem in this technique is that it is generally not possible to orient R_L and R_R with respect to each other, and therefore it is necessary to walk from R_R and R_L in both directions. This technique becomes much more practical when several different starting points for the walk are available in the DNA segment to be studied.

Figure 9.2 Strategy involved in walking using successive probes in sequence A, B, C, and D.

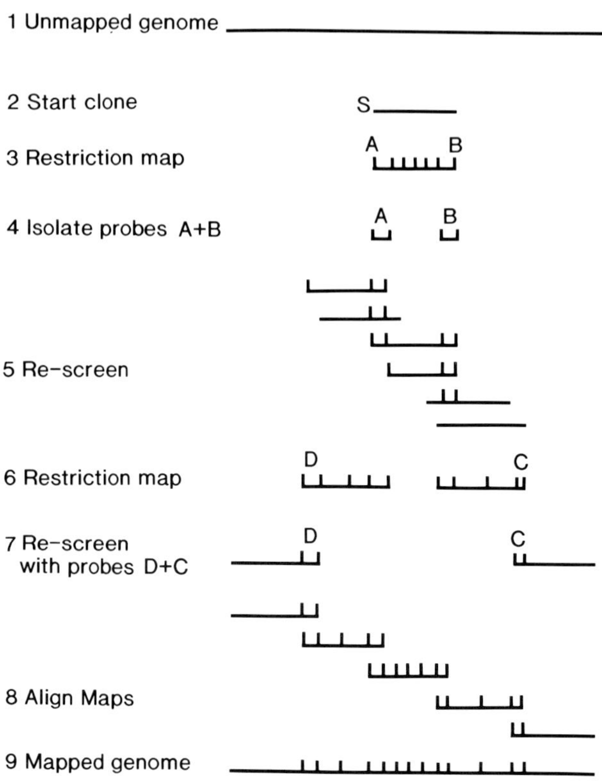

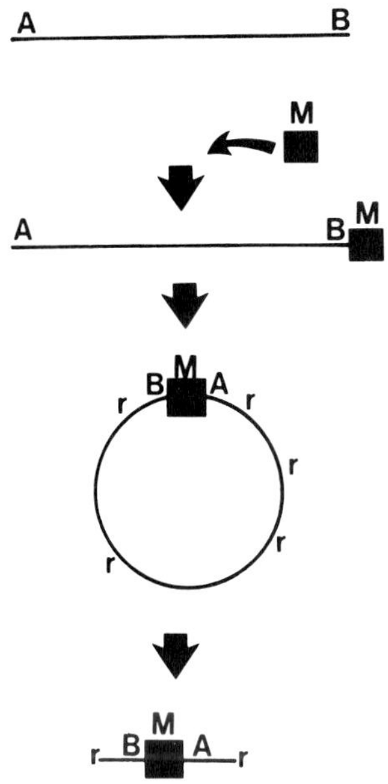

Figure 9.3 A and B represent sequences 200,000 base pairs apart. M is the marker DNA, which can be selected in *E. coli*, and r indicates restriction endonuclease cutting sites. The jumping library contains many fragments with r sites at each end containing M and different A's and B's. *Reproduced from Little,*[36] *with permission.*

Jumping is still another method that reduces the number of steps needed to walk large distances over the DNA. It is done by isolating a large, 200,000-bp piece of linear DNA, marking one end of this DNA with a marker (Figure 9.3), and joining the entire molecule in a circle. Thus, the pieces of DNA shown as A and B in Figure 9.3, which were ordinarily 200,000 bp apart, would end up on either side of the marker DNA. If many such random 200,000-bp circles can be put together and if the marker DNA plus its flanking sequences can be conveniently isolated, a "jumping" library representing DNA sequences 200,000 bp apart will be the product. This in turn can be used with R_L from the library, and the search involves jumping down the chromosome until R_R is identified.[36]

Why Should We Map Disease Genes?

Mapping of human disease genes in particular chromosomal settings is not merely an exercise in applied molecular biology; it has a great deal of practical importance as well. Being able to identify a particular genetic marker linked to a disease with great certainty allows the clinician to trace inheritance of the disease in families at risk even when no other phenotype directly related to the disease can be determined. Thus, the first major advantage of mapping disease genes would be improved diagnostic accuracy. Moreover, as noted in Chapter 8, improved prenatal diagnosis would become available. In late-onset disorders such as Huntington's disease, predisease diagnosis should also be possible. Another advantage of being able to identify a disease marker is that it enables the clinician to learn more about the genetic basis for the disease expressed in the patients. In many diseases a basic question is whether the disease is a result of a spectrum of *different* mutations or whether the mutation occurs at only *one* locus. It is important to be able to determine whether nonallelic heterogeneity is present, that is, whether more than one kind of mutation at the same broad locus may result in what appears to be the same phenotypic expression of the basic disease. As we saw in Chapter 8, different mutations provide a spectrum of disease severity in the β-thalassemias. Most important, the identification of defective proteins associated with hereditary disorders will improve fundamental understanding of the disease processes and, as the genes are identified, pave the way for effective gene or gene product replacement therapy.

Human Genetic Diseases—Why Do They Persist?

Persistence of infectious diseases such as acquired immunodeficiency syndrome (AIDS), for which there is still no effective treatment or cure, is easy to fathom. The persistence of genetically determined diseases in spite of evolutionary forces is more difficult to understand. It is a puzzle why the process of natural selection does not eliminate inherited diseases by which we are destroyed.[37] In an increasing number of examples, gene frequencies of such disorders appear to be maintained by possible compensating advantages. Some genetic disorders appear to persist by mutation rates alone; for example, Duchenne muscular dystrophy and achrondroplasia are so rare that their occurrence is thought to be adequately explained by mutation rates alone.[38] The highest mutation rate for a human genetic disease (Duchenne muscular dystrophy) is at the level of 10^{-4}; most other rates are considerably lower at 10^{-7}. In this way, mutation can maintain potentially lethal genes of high relative penetrance only if the disease persists at low frequencies.

It is more difficult to explain the widespread distribution of deleterious genes apparently present in high frequencies in large old populations, as appears to be the case in cystic fibrosis, peptic ulcer disease, and insulin-dependent diabetes mellitus. In these disorders some sort of compensating advantage has been postulated. The classical example of such a positive compensating advantage is sickle cell anemia. The homozygous carriers of this gene either die early or fail to reproduce because of their severely affected phenotype. However, the much more numerous heterozygotes have a much higher innate resistance to malaria than normal individuals. Thus, in endemic malarial regions the gene may be perpetuated by heterozygous individuals, who gain a relative biological advantage, while the unfortunate homozygous individual die without reproducing. A similar argument has been proposed for the persistance of genes coding for Tay-Sachs disease, Niemann-Pick disease, and adult-type Gaucher's disease—three lysosomal enzyme defect storage disorders. All have been reported to be associated with a lower incidence of death from tuberculosis in heterozygous affected individuals.[37]

A different positive compensatory advantage has been attributed to the gene associated with idiopathic hemochromatosis, in which men suffer and women benefit.[39] Idiopathic hemochromatosis is an autosomal recessive disease with a remarkable heterozygote frequency of 10% in the United States and Europe. The basic disorder involves abnormally increased rates of iron absorption through the gastrointestinal tract and the direct toxic tissue effects of iron overload. Serious symptoms are 5 to 10 times more common in men than in women because women routinely experience high iron losses through menstruation and pregnancy. Thus heterozygous women may benefit, whereas affected homozygous males pay the price.

Similarly, selective advantages have been suggested for genes predisposing to peptic ulcer disease, such as possible resistance to tuberculosis and cholera.[40] Moreover, improved prenatal survival seems to be a selective benefit of genes coding for insulin-dependent diabetes mellitus. In affected kindreds, normal fetuses are more likely to miscarry than those with diabetes-linked HLA antigens (DR3/DR4). Because only 20% of newborns with the insulin-dependent diabetes genotype actually become diabetic clinically, the other 80% of the progeny reap the prenatal benefits without subsequently paying the price by developing the disease. It seems possible that evolution favors genes that penalize homozygotes but benefit many of the heterozygotes by conferring some selective survival benefit.

DNA Restriction Fragment Length Polymorphisms and Application to Specific Human Disorders

Application of molecular biological approaches to human genetic diseases has rapidly advanced since the introduction of RFLP analysis. In the vast majority of instances this has involved comparative surveys of DNA polymorphisms after restriction enzyme digestion and Southern blotting. A polymorphism has generally been defined as a mendelian trait that exists in the population in at least two phenotypes, neither of which occurs at a frequency less than 1%.[41,42] An allele present at less than 1% frequency is termed a variant. A representative but by no means all-inclusive list of human diseases that have been studied using recombinant DNA technology is given in Table 9.1. Table 9.1A shows diseases studied by using direct gene probes to detect defects. Table 9.1B shows diseases in which gene probes have been used to detect closely linked polymorphisms.

Table 9.1 Human Diseases Studied by Recombinant DNA Techniques and RFLP

Disease	Gene probe
A. Direct analysis using gene probes to detect intragenic defects	
Antithrobin III deficiency	Antithrombin III
α-Antitrypsin deficiency	Synthetic oligonucleotide
Atherosclerosis	Apolipoprotein A-I
Diabetes	Insulin
Ehlers-Danlos syndrome	α(1) collagen
Growth hormone deficiency	Growth hormone
Hemophilia B	Factor IX
Hereditary persistence of fetal hemoglobin	β-Globin
Hypoxanthine-guanine phosphoribosyltransferase (HPRT) deficiency	HPRT
Lesch-Nyhan syndrome	HPRT
Osteogenesis imperfecta	Pro–α1 (1) collagen
Retinoblastoma	Chromosome 13 DNA
Sickle cell anemia	β-Globin synthetic neucleotide
Thalassemia	α-Globin and β-globin
B. Indirect analysis using gene probes to detect closely linked polymorphism	
Atherosclerosis	Apolipoprotein A-I
Diabetes type II	Insulin
Growth hormone deficiency type I	Growth hormone
Hypertriglyceridemia	Apolipoprotein A-I
Sickle cell anemia	β-Globin
Lesch-Nyhan syndrome	HPRT
Osterogenesis imperfecta	Pro–α2 (1) collagen
Phenylketonuria	Phenylalanine hydroxylase
Thalassemia	β-Globin

More specialized analytic techniques have been developed using RFLPs that can detect human recessive traits.[43] An example of this approach involved the analysis of RFLPs from affected children of consanguineous marriages, looking for patterns that might identify human recessive traits. The method was directed at identifying the disease locus by the fact that the adjacent DNA regions will preferentially be homozygous by descent in individual children with the recessive trait. It was postulated that a complete RFLP linkage map for the entire human genome will eventually be available, and such a map will make it possible to detect homozygosity by descent (and therefore a recessive trait) by looking for regions in which a number of adjacent RFLPs are all homozygous. In this way it might be feasible to map recessive traits without family studies. Rigorous testing of this approach deserves a trial in the future.

Cystic Fibrosis

Cystic fibrosis (CF) is one of the most common serious genetic diseases in white populations, with 1 in 2000 live births being affected. If it is assumed that homozygosity for a mutant allele at a single locus produces an affected individual, the predicted heterozygosity frequency is approximately 1 in 22. The major abnormalities in this disease are related to the exocrine secretory system, and a test for raised chloride and sodium concentrations in sweat is one of the best clinical diagnostic tests for the disease. Clinical problems in these patients are chronic obstructive lung disease; repeated respiratory infections, particularly with *Pseudomonas* species; and exocrine pancreatic insufficiency leading to malabsorption and neonatal intestinal obstruction. A number of attempts have been made over the years to identify a gene or gene product that can be used for identification of the primary defect in this disease. A method for detecting symptomless obligate heterozygotes has also long been needed.

Some initial success in identifying a possible CF marker was reported in 1985 by van Heyningen et al.[44] An immunologically indistinguishable protein was identified at high concentrations in granulocytes from normal individuals and individuals with CF, as well as in myeloid leukemia cells. Previous studies had shown that a serum protein producing a doublet at pI 8.4 on isoelectric focusing in the presence of urea was present in serum from CF homozygotes and heterozygotes but not in normal serum.[45] When used in a so-called rocket immunoelectrophoresis, the antiserum to the CF protein was capable of identifying homozygotes, heterozygotes, and normal subjects. In studies with somatic cell hybrids between mouse cells and human myeloid leukemic cells, it was found that the CF antigen was expressed only when human chromosome 1 was present.[44] This discovery was one of the first to suggest chromosomal and genetic localization of the CF gene.

Several observations reported in 1985[46–48] provided further evidence for detectable markers that showed promise as genetic linkage indicators. One study[46] utilized a genetic marker at the *met* locus after cloning a 1.6-kb *Sal*I/*Eco*RI fragment into plasmid pBR322. Studies of DNA from CF families using this probe digested with *Taq*I enzyme showed a high degree of segregation with the presumed CF gene and very tight linkage between CF and the *met* oncogene in the families studied. Moreover, the fact that no recombinants were found between the *met* gene locus and the CF locus suggested that the two loci were indeed very close. In this study it was even suggested that the product of the *met* gene itself could represent the CF gene. The argument was that the *met* product was a membrane-spanning receptor protein showing tyrosine kinase homology and that such characteristics might be expected of a gene causing a fundamental chloride ion-exchange defect.

The CF locus was definitely localized to human chromosome 7 in two other parallel studies reported simultaneously.[47,48] After analyzing the results of previous studies reporting extensive linkage exclusions (which excluded 40% of the human genome), one group studied possible linkage on chromosome 7 by multipoint linkage analysis.[47] The idea was to study human chromosomes for which few definite exclusions for CF had been obtained. Two different markers defined by an *Msp*I polymorphism and the T cell β chain receptor were used, with positive linkage results. The localization of these two probes (pJ3.11 and TCR_β) is shown in Figure 9.4. An entirely different randomly isolated cDNA probe was used by the other group of investigators,[48] who also localized the CF gene to chromosome 7. These findings provided valuable information and an excellent foundation for further RFLP investigation in this area.

Subsequently, the method of constructing a general human chromosome jumping library was applied to the CF problem.[49] The library allowed cloning of DNA sequences approximately 100 kb away from any starting point in human genomic DNA. The library was then searched for a jumping clone starting at the *met* oncogene (a marker tightly linked to the CF gene). Pulsed field gel electrophoresis showed that the CF gene was situated on chromosome 7 within 240 kb downstream of the *met* oncogene.

The exact method used in this study is of some interest because it illustrates the power of the chromosome jumping library approach to help narrow down the location of a disease-related gene. The library was produced from DNA from a human lymphoblastoid cell. Under the conditions selected, the distances of the chromosomal jumps in the library were chosen to be approximately 100 kb. Large genomic circular DNAs were used that had been size selected and contained a marker allowing for selection after *Eco*RI digestion. Using this approach, a library

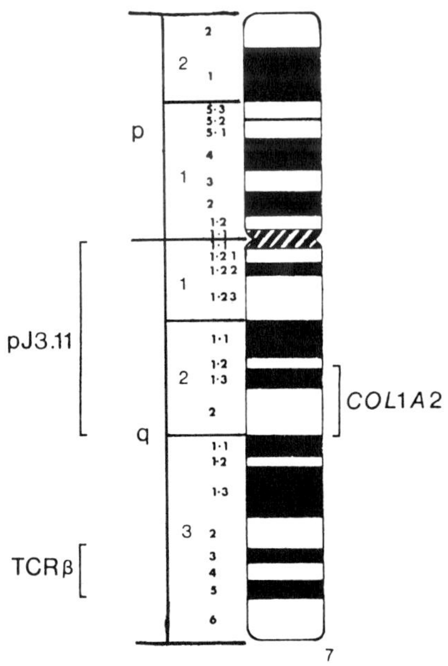

Figure 9.4 Map of human chromosome 7 showing localization of probes TCR_β and COL1A2 as well as pJ3.11 used as linking markers in family studies for localization of the CF gene.

of approximately 2 million jumping clones was produced. The library clones were then screened with a *met*G probe that represented a 2.2-kb *Pst*I genomic fragment near the 3′ end of the *met* oncogene, and positive g phage were purified. The general scheme for this superjumping approach is shown in Figure 9.5. Subsequent work with one clone isolated by this procedure indicated that it was located 50 to 250 kb to the 3′ side of *met* gene. Parallel evidence seemed to indicate a gene order of *met*-CF-pJ3.11. It was predicted that the general chromosome jumping method might, if successful in other applications, considerably speed up final identification of the CF gene.

Subsequently, a genomic sequence close to the CF locus was selectively cloned and characterized[50] by selection for methylation-free islands. A previous collaborative study had indicated that the recombination fraction between J3.11 and the CF gene was 0.003 and between *met* and CF was 0.004, indicating that J3.11 and *met* might actually flank the CF gene.[51] The technique of chromosome-mediated

Figure 9.5 Scheme for chromosome jumping. Circularization of genomic DNA is used to generate a jumping library. Each phage clone includes two genomic fragments that were originally located at opposite ends of an approximately 100-kb genomic fragment. One of the four clones obtained by screening 1 million phage from the primary library with *met*G is shown at the bottom. The heavy bar represents the *met*G sequence, the sawtooth mark represents the two sup^F genes with their internal *Ava*I sites that facilitate restriction mapping and subcloning, and the open bars represent the phage arms. The 0.7-kb genomic fragment denoted CF63 should be approximately 100 kb downstream from the 3′ end of *met*G. *Reproduced from Collins et al,*[49] *with permission.*

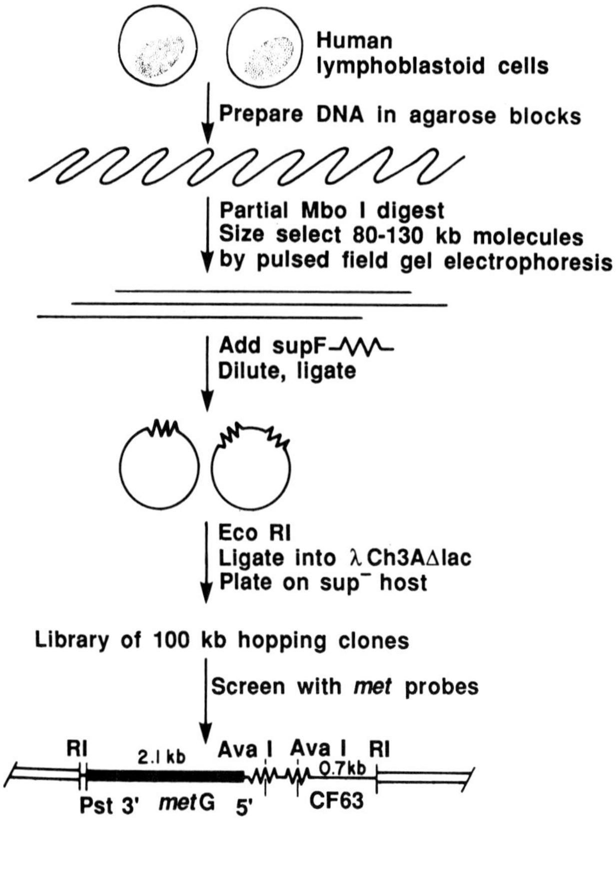

gene transfer was used to produce mouse-human hybrid cell lines that contained transgenomes including markers most closely related to the CF genes. These cell lines could be used to exclude possible candidate genes representing the CF mutation and to help localize the new markers for this region. Cosmid libraries were constructed to help analyze the human components present in these hybrids,[52] and a number of the cosmid clones were mapped into overlapping contiguous segments termed contigs. The contigs were further studied using pulsed-field gel electrophoresis and rare-cutter restriction enzyme mapping.[52] Using this approach, two cosmid clones that identified polymorphic loci in strong linkage disequilibrium with the CF locus were isolated.[50] Many vertebrate genes are preceded by a region of nucleotides rich in the nonmethylated dinucleotide CpG and containing clusters of restriction sites for CpG methylation–sensitive enzymes. These particular stretches of sequence are termed HTF (*Hpa*II tiny fragment) islands and generally include sequences of 500 to 2000 base pairs. They often contain the first exons as well as upstream segments 5′ and occasionally 3′ from the associated coding gene of interest. Thus HTF islands can be used as a set of markers for coding nucleotide sequences. This general approach was employed to identify genomic regions with the characteristics of an HTF island in high levels of linkage disequilibrium with the CF gene.[50] In the initial application, this approach appeared to show strong linkage to the CF gene.

Finally, in 1989, precise identification of the CF gene was accomplished using all of the previous data in conjunction with the chromosome walking and jumping and complementary DNA hybridization.[53–56] Analysis of 280 kb of contiguous DNA permitted cloning of the locus responsible for cystic fibrosis. The large size of the CF gene was a surprise; moreover, the lack of apparent genomic rearrangements in CF chromosomes and evidence indicating a limited number of CF mutations could have led to the expectation that a small mutation was responsible. It was found that the most common CF abnormality produces loss of a single amino acid (phenylalanine) in a key functional protein domain. Approximately 70% of the mutations in CF patients were found to correspond to a deletion of 3 base pairs that resulted in loss of a phenylalanine residue at position 508 in the gene product. Extensive haplotype analyses indicated that the remaining CF gene pool included multiple different mutations much like the profile of β-globin gene mutations in sickle cell disease or the thalassemias. Establishment of the precise locus for CF mutations is a major step toward understanding the molecular basis of many other genetic disorders.

Despite the initial lack of a direct gene probe that would identify the CF gene, prenatal analyses and genetic counseling were done using several of the cDNA probes that previously appeared to be linked to cystic fibrosis.[51–57] Tissue DNA was obtained from parents' blood cells as well as chorionic villus samples. In the initial report, which surveyed 30 pregnancies, nine children were born and shown by sweat tests and serum immune reactive trypsin assays to be free of CF. Eight terminations of pregnancies were performed on the basis of the Southern blot analysis. This early survey was done to test the reliability of prenatal diagnoses and to allow prospective parents to plan pregnancies with full knowledge of their respective genotypes based on Southern blot procedures using several closely linked probes such as pH3.11, *met*H, and *met*D. At that point it was felt that parents at risk should ideally plan their pregnancies after their genotypes have been confirmed.

After cloning of the CF gene, the polymerase chain reaction was employed using hybridization with allele-specific oligonucleotides in a large number of CF and normal chromosomes.[58] The mutation involving the 3-base deletion removing phenylalanine at position 508 was found in 75.8% of the CF chromosomes. Using two specific DNA markers, these workers found that 96% of CF chromosomes with the mutation showed a single DNA haplotype. However, the same haplotype was also found on 54% of the CF chromosomes without the 3-base deletion. At this juncture it was hoped that additional mutations that would account for the remaining CF chromosomes would be identified in the near future.

These and similar studies began to indicate that the CF mutations might be considerably more complex than they had appeared at first. The initial optimism that the 25% of mutations different from the major one would be quickly identified faded when, instead of a few other common mutations, more than 20 rare ones surfaced.[59] It appeared that most of these mutations were "private" ones that occurred in just one individual. The biggest threat to application of this sort of information to genetic screening is that once 30, 40, or hundreds of mutations have been duly found and identified, a final residual 10% to 15% may still be unaccounted for. These findings posed difficult questions for those who have been championing an accurate map of the entire human genome as the be-all and end-all of disease identification. If practically all unrelated people among the approximately 25% who do not show the common mutation have individually different mutations, accurate screening of the entire population for CF will not be definitive using current methodology. Essentially, the problem is that screening with a 75% test misses the opportunity to reduce the disease in 50% of those affected. These difficult issues are currently being debated by ethicists, geneticists, and many pilot programs.[60]

Muscular Dystrophy

Duchenne muscular dystrophy (DMD) is the most common of the muscular dystrophies, affecting 1 in 3000 live male births. Both DMD and the milder form, Becker muscular dystrophy (BMD), are X linked.

For some time, carrier detection in DMD was based on a serum creatine phosphokinase test that was not considered very reliable for this purpose.[61] The introduction of RFLP analysis has provided a tremendous impetus toward precise identification of the DMD gene locus, because the disease is X linked and therefore the chromosomal basis is automatically markedly restricted. Initial cloned DNA sequences, such as RC8 isolated by Murray and co-workers from a flow-sorted X library[26] and L1.28 isolated by Bakker and co-workers from a genomic plasmid library,[62] were derived from well-defined X chromosomal locations. Using linkage analysis, it was found that both RFLPs based on a *Taq*I polymorphism mapped to a distance of 15 centimorgans on either side of the DMD locus.[63] These and other probes were utilized in early attempts at prenatal diagnosis with some degree of success.[28] In the latter study, by Monaco and co-workers, a double crossover was detected in a DMD carrier and an affected male fetus was diagnosed at 12 weeks of gestation with what was calculated as probable accuracy of 99%. This survey utilized 11 RFLP markers available at the time on the short arm of the X chromosome (Figure 9.6). The results of this study demonstrated how crossovers in a region between two defined DNA markers could be useful in identifying the sequence of probes relative to each other in a small region containing the disease-related gene.

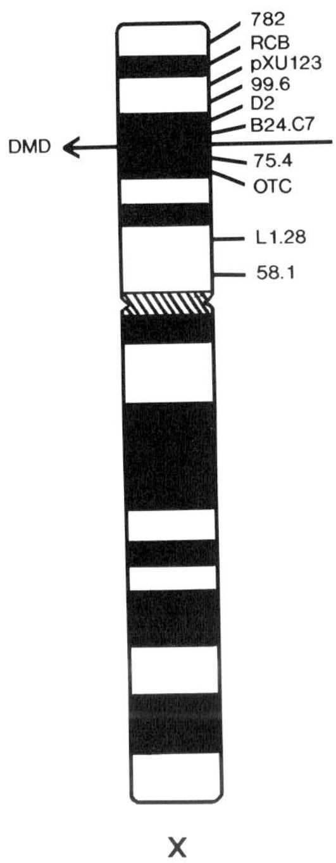

Figure 9.6 Schematic distribution of RFLP probes along the short arm of the X chromosome. The RFLP probes are as follows:

Clone	Xp region localization	Restriction enzyme	Heterozygote frequency
782	Xp222–Xp223	*Eco*RI	0'50
RC8	Xp221–Xp223	*Taq*I	0'23
pXUT23	Xp21–Xp223	*Bgl*II	0'42
99-6	Xp21–Tpter	*Pst* I	0'49
pD2	Xp21–Xpter	*Pvu*II	0'40
B24	XP213	*Msp*I	0'15
C7	Xp213	*Eco*RV	0'25
754	Xp211–Xp21'3	*Pst* I	0'46
OTC	Xp211	*Msp*I	0'67
1.1'28	Xp113	*Taq*	0'45
58'1	Xpcent-Xp11	*Msp*I	0'47

As work on muscular dystrophy progressed, an ingenious method that allowed specific cloning of DNA fragments absent from patients homozygous or hemizygous for chromosomal deletions was described.[28,64] The method utilized phenol-accelerated competitive DNA reassociation and subsequent molecular cloning of reassociated complementary molecules. The deletion DNA sample employed in the development of this method was obtained from a male child who was diagnosed as having DMD, chronic granulomatous disease, and retinitis pigmentosa. DNA containing the deletion from this male child was combined using a 200-fold excess with *Mbo*I-cleaved DNA obtained from a 49,XXXXY human lymphoblastoid cell line, and the mixture of DNAs was treated with a phenol enhancement technique. When the reassociated unique DNA segments were analyzed, four human DNA pieces were identified that were absent from the male patients' DNA. Subsequently, these clones were localized by hybridization of restricted fragments to three regions encompassing band 21 on the short arm of the normal human X chromosome. This general approach applied to deleted DNA from a patient with several X-linked disorders allowed direct identification of normal DNA fragments possibly involved in production of or protection from the disease in question.

With these and similar probes, this group then screened for other possible deletions in DNA isolated from 57 unrelated males with DMD.[28] Of great interest was the finding that five of the DMD male patients screened had deletions for one of the cloned DNA segments utilized. In addition, other subclones from the same region identified RFLPs that appeared to be close markers for the presence of DMD. All five patients with clear-cut deletions lacked a minimum of 38 kb of genomic sequences. The elements of this remarkable discovery are illustrated in Figure 9.7. The deletions detected by the pERT87 subclones subsequently provided a unique tool for defining the extent of the DMD locus itself.

Work with the cloned DNA segments that appeared to identify deletions in the

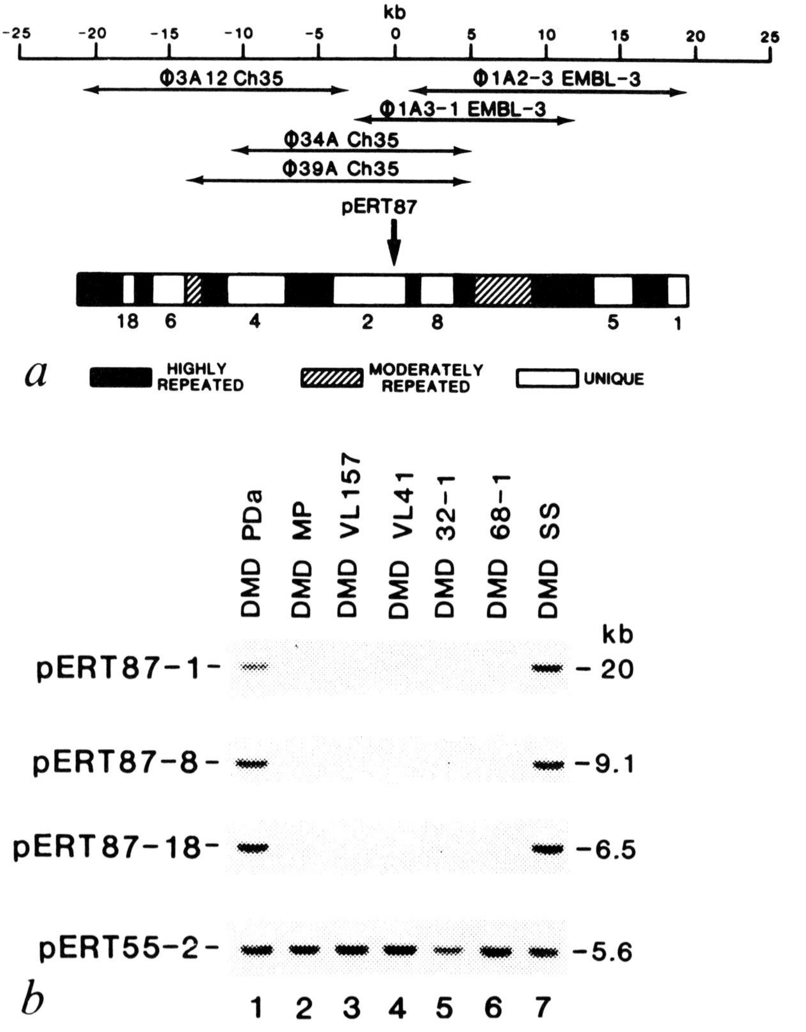

Figure 9.7 (a) Schematic map of 41 kb of genomic DNA around the original 200 base pairs, pERT87 (0 on kilobase scale). The subcloned segments that were free of human repeated sequence elements are indicated numerically below the block diagram, and the size of each segment is represented by an open block. Segments containing highly repeated sequences are indicated as darkened areas, and those containing moderately repeated segments as crosshatched areas. (b) Southern blot analysis of DNA from seven males with DMD. Lanes 1 to 7 contain *Pst*I-digested DNA of the patients. Two identical nitrocellulose filters were hybridized with pERT87-18 and pERT87-1. The same filters were then rehybridized with pERT87-8 and pERT55-2, respectively. The autoradiographs are presented as separate segments of each original autoradiograph. The fragment hybridized is indicated on the left; the sizes of hybridizing fragments are indicated on the right and were calculated from ^{32}P–end labeled *Hin*dIII-cleaved λ DNA. The numerical designation of the subclone used as probe and the probe's relative position in "walked" DNA are indicated in (a). The designation above each lane is the nomenclature for each DMD male. *Reproduced from Monaco et al,*[28] *with permission.*

DNA of unrelated DMD male patients was extended to provide a more precise definition of the disease locus, which included eight small regions spanning 130 kb.[65] This narrowed the range in which to search for the DMD gene. During this study, two segments in the gene region in question were identified that appeared to be relatively conserved in mammals and were therefore presumed to represent the coding regions. These two genes were then used after sequencing to look for RNA transcripts in fetal skeletal muscle cells. Remarkably, the probes picked out a 16-kb RNA, from which the investigators prepared cDNA. When the cDNA was mapped back to the chromosome, the cDNA probe hybridized to exons that extended across 130 kb of DNA! This finding was interpreted to indicate that the muscular dystrophy gene could be as long as 1 million to 2 million base pairs. Although the DMD gene itself was not yet directly identified, increasing insight was being gained into the genetic mechanisms behind the disease. The hybridization to fetal skeletal muscle RNA indicated that the key mutation linked to the muscular dystrophy process would be found in some as yet unidentified muscle protein. Precisely what that protein might be would become clear when the DMD gene was actually isolated and sequenced.

Since DMD is X-linked, females affected by the disease have an X-autosome translocation with the exchange point in the X chromosome always occurring within chromosome band Xp21.[66] This finding, along with the deletion data referred to above, localized the DMD gene to band Xp21.[67] In addition to the clones identifying the Xp21 deletion,[65] clones from a patient with a t(X;21) translocation have been identified.[29] These two sets of clones have identified the DXS164 locus and the DXS206 locus, respectively, and subclones from both regions have provided evidence for genomic X chromosomal deletions in approximately 11% of DMD patients.[28,29,68,69] As mentioned above, the fetal cDNA clone corresponding to exons in the DXS164 locus has been isolated and detects a 16-kb transcript.[65] Subsequently, an adult muscle cDNA clone was isolated from the DXS206 locus that hybridized with a 16-kb mRNA in adult human muscle.[70] The latter clone contained exons that mapped in the DXS206 locus, in the original DXS164 locus, and toward the centromeric side of these cloned genomic elements. It was deduced that the t(X;21) translocation exchange point occurred within a large 105-kb intron, indicating that the translocation had disrupted the DMD/BMD gene to produce the disease in the patient with the translocation. Again, these data were consistent with a gene of megabase proportions and extended the results reported by Monaco et al.[65]

In parallel with these observations and the presumption that the Duchenne and milder Becker muscular dystrophy genes might extend over 1000 to 2000 kb, a mouse model of human dystrophy (*mdx*) was examined using two human nucleotide sequences from the human Duchenne dystrophy work that cross-hybridized to murine X-linked sequences.[71] These studies were done on genes homologous to DMD in the mouse and showed closely localized patterns similar to those seen in the human disease. In the case of the mouse *mdx* model, a close genetic linkage to glucose-6-phosphate dehydrogenase (G6PD) and the hypoxanthine-guanine phosphoribosyltransferase (HPRT) locus on the mouse X chromosome was documented.[71,72] The latter findings indicated that the *mdx* mouse mutation might instead correspond to Emery-Dreifuss muscular dystrophy, which is linked to DNA markers in the region coding for G6PD on the human X chromosome. The mouse disorder may be helpful in elucidating genetic mechanisms in this group of diseases.

As more precise knowledge of the Duchenne and Becker types of muscular dystrophy has accumulated, it has been possible to attempt to analyze various mechanisms of hereditary transmission of these disorders. Since the gene itself extends over more than 1800 kb, a relatively high mutation rate of approximately 10^{-4} per generation is calculated. Small deletions detectable with one or more of the approximately 12 cloned DNA probes probably account for a significant segment of the observed mutations. Family studies using the group of cDNA clones have revealed partial deletions related to the muscular dystrophy gene that can be transmitted by an unaffected male.[73]

The pattern of exon deletions and its relationship to both Duchenne and Becker muscular dystrophies have been amplified by studies reported by Forrest et al.[74] In addition to the previous documentation of 10% to 11% of cases with detectable large deletions, using an additional DNA subclone of the DMD gene, 15 of 66 DMD and BMD patients (22%) showed deletions not detectable with previous screening procedures. Exon deletions were documented in 5 severely affected DMD patients and in 10 with BMD. Most important, most of these deletions occurred in the same region of the cDNA, possibly indicating a common mechanism for their production. It was estimated from the results of these studies that detection of significant deletions was possible in up to 40% of DMD and BMD patients. Another group,[75] using field inversion gel electrophoresis, provided evidence for large deletions in more than 50% of DMD patients. Thus, it appears clear that deletions involving a large genomic segment (1800 base pairs) in the Xp21 area of the X chromosome are associated with both DMD and BMD.

Still to be determined was when and how the gene related to both sorts of dystrophy was expressed.[76–78] Also, can the different phenotypes—Duchenne and Becker—be explained by differential deletions at different sites or by disruption of specific exons? Furthermore, what protein is encoded by the very long DMD transcript? A muscle protein of some type appeared to be a most likely candidate. Nebulin, a protein of the proper size (500 kb) and involved in structural maintenance of skeletal muscle, was first suggested,[79] but no proof of its central role was presented.

Because some 30% of DMD patients were known to suffer from various degrees of mental retardation and the giant DMD gene spanned about 2000 kb coding for a 14-kb mRNA and a 427-kd protein,[80] differential gene expression was examined in muscle as opposed to brain tissue.[81] It was found that the transcript of the DMD gene and the amino terminal of the encoded protein differed in brain and muscle and that the 5′ ends of these mRNA species were derived from different exons. These findings suggested that the two distinct kinds of mRNAs were transcribed from different promoters.

Dystrophin, the actual protein product of the DMD locus, was described in 1987 by Hoffman and co-workers.[82] These findings provided at least a distinct end point marker gene product associated with the disease. As it became clear that DMD resulted from a deficiency of dystrophin in the muscle fibers,[83] possible therapeutic strategies were explored by using grafts of normal muscle precursor cells to see if they were able to induce synthesis of dystrophin in the natural mouse model of dystrophin-deficient *mdx* muscle fibers.[84] This resulted in fusion of injected normal muscle precursor cells with preexisting *mdx* muscle fibers, rendering many of the latter dystrophin positive and thus partially or completely rescuing them from their endogenous dystrophic biochemical defect.

After it was established that deficiency of the protein dystrophin was related to the cause of DMD, this information was rapidly applied to strategies for carrier screening. Muscle biopsy samples from three carriers were examined with antisera against a synthetic peptide fragment of dystrophin[85]; in all three carriers most individual muscle fibers reacted either very positively or not at all to the antidystrophin antiserum. Such a mosiac staining pattern was present in all muscle specimens. Control studies with normal subjects showed strong staining with all muscle fibers. It was concluded that symptomatic carriers of DMD could be identified by this distinct mosiac pattern in the immunohistochemical staining of surface membranes of skeletal muscle preparations.

Alternative splicing of human dystrophin mRNA was shown by polymerase chain reaction to produce different isoforms at the carboxyl terminus,[86] a mechanism that might be related to the need for different dystrophin molecules specific for a broad spectrum of tissues. As structural knowledge concerning the mammoth dystrophin gene increased, pulsed field electrophoresis analysis suggested that it was 2.3 megabases in size—12 times the size of the coagulation factor VIII gene and 1500 times the size of the β-globin gene! The number of exons in the dystrophin gene apparently exceeds 75.

When cDNA probes were applied, deletions in the dystrophin gene were found in about two-thirds of Duchenne patients. Moreover, the size of the deletions did not always correlate with the severity of the dystrophy. In most instances, the numerous deletions that have been recognized with respect to the reading frame appear to lead to less clinically severe muscular dystrophy, whereas mutations that lead to frameshifts in the mRNA sequence downstream cause the severe form of Duchenne dystrophy.[87] Exceptions to the frame rule were first reported by Malhotra and coworkers[88] who demonstrated that out-of-frame deletions of exons 3–7 which remove most of the a-actinin-like amino-terminal domain, produce phenotypes that vary from a mild or severe Becker to a Duchenne type. Subsequently, a number of exceptions to the frame rule have been reported. In a large multicenter study, 258 independent deletions were analyzed and compared with subsequent clinical evolution and disease severity. Basic agreement with the reading-frame model was observed in 92% of cases; 4% were out-of-frame deletions leading to mild disease or in-frame deletions of more than 30 exons producing severe disease. Only 4% of the exceptions could not be accounted for. Additional diagnostic strategies are being explored throughout the world to provide a sound basis for genetic counseling.

Glycoprotein Component Deficiency Linked to Dystrophin Complex

With the molecular and subsequent protein structural characterization of dystrophin, it was discovered that four glycoproteins that are integral parts of the dystrophin complex of the muscle cell surface are affected in the disease.[89] Thus, loss of dystrophin appears to lead to loss of an important dystrophin-associated glycoprotein, and this may be one of the first lesions in the pathogenesis of the disease. From the results reported by a number of research groups, it appears that the DMD or BMD phenotype may arise from a variety of deletions or mutations over a relatively broad segment of the gene.

Unusual Hereditary Patterns

Several examples have been reported of a mother with two apparently normal X chromosomes who produced more than one offspring with the same DMD deletion. Several interpretations of this are possible. It cannot be assumed that a DMD boy born to parents of apparently normal genotype must be a new mutant, with the chance of an affected sibling no higher than in a normal family. The most likely explanation for a DMD offspring born to an apparently normal mother seems to be germ-line mosiacism, that is, a mutation occurring in a mitosis early in the differentiation of the mother's germ line. A second possibility is a "premutation" in the mother in which expression is silent for several generations. A third possibility is that in some rare cases the disease is autosomally transmitted. Considerably more work with families containing unusual expressions of disease is required before all of these questions can be answered. The experiences of Darras et al.[77] using DNA sequence polymorphisms with 12 different probes indicated that uncertainties still often remain in the prenatal diagnosis of DMD.

A portion of the DMD gene transcript from human fetal skeletal muscle and mouse adult heart was sequenced; the DNA segment sequenced represented about 25% of the total 14-kb transcript.[78] Of interest was the finding that the nucleic acid and predicted amino acid sequences for the two species were nearly 90% homologous. When the amino acid sequence predicted from the gene sequence was analyzed, it was apparent that the gene product itself might play a structural role in muscle; however, the relative abundance and tissue distribution of the mRNA indicated that the DMD protein was not nebulin, as had been previously proposed.[79] Searching for nebulin or other candidate gene products has now largely ceased since the definitive work on dystrophin.

The exciting discoveries concerning the molecular and genetic basis for disorders such as Duchenne or Becker muscular dystrophy are striking examples of how modern molecular biological techniques can provide rapid insight into structural and perhaps eventually the complete biochemical basis for a disease process.

Huntington's Disease

Huntington's disease (HD), also known as Huntington's chorea, is an autosomal dominant inherited disorder characterized by a progressive neurodegenerative defect including neurological impairment, behavioral changes, and finally dementia.[90,91] The first symptoms of HD usually occur in the third to fifth decades of life and the gene is completely penetrant. Intellectual deterioration is commonly accompanied by psychiatric symptoms including severe depression. The disease results from premature neuronal death, most marked within the basal ganglia and projection neurons in the striatum.[92] Although the prevalence of HD is only 5 to 10 per 100,000, because of its late onset of clinical expression, many individuals have children before they realize they have inherited HD, which greatly magnifies the practical problems of genetic counseling.[93] The basic or primary biochemical or structural defect in HD is still not known, and until recently there was no reliable method of presymptomatic or prenatal detection.

With the availability of recombinant DNA technology, attempts were begun to develop disease-related RFLP markers that would permit monitoring of the heritable process in individuals at risk for developing the disease. Initial progress in this direction was reported in 1983 by Gusella et al.[31] Family studies using an anonymous DNA probe from human chromosome 4 appeared to detect two different loci in a *Hin*dIII digest of human genomic DNA which showed very close genetic linkage to the HD gene. Representative pedigrees from families studied are illustrated in Figure 9.8.

The cDNA sequence found to be useful as a disease-linked marker was then

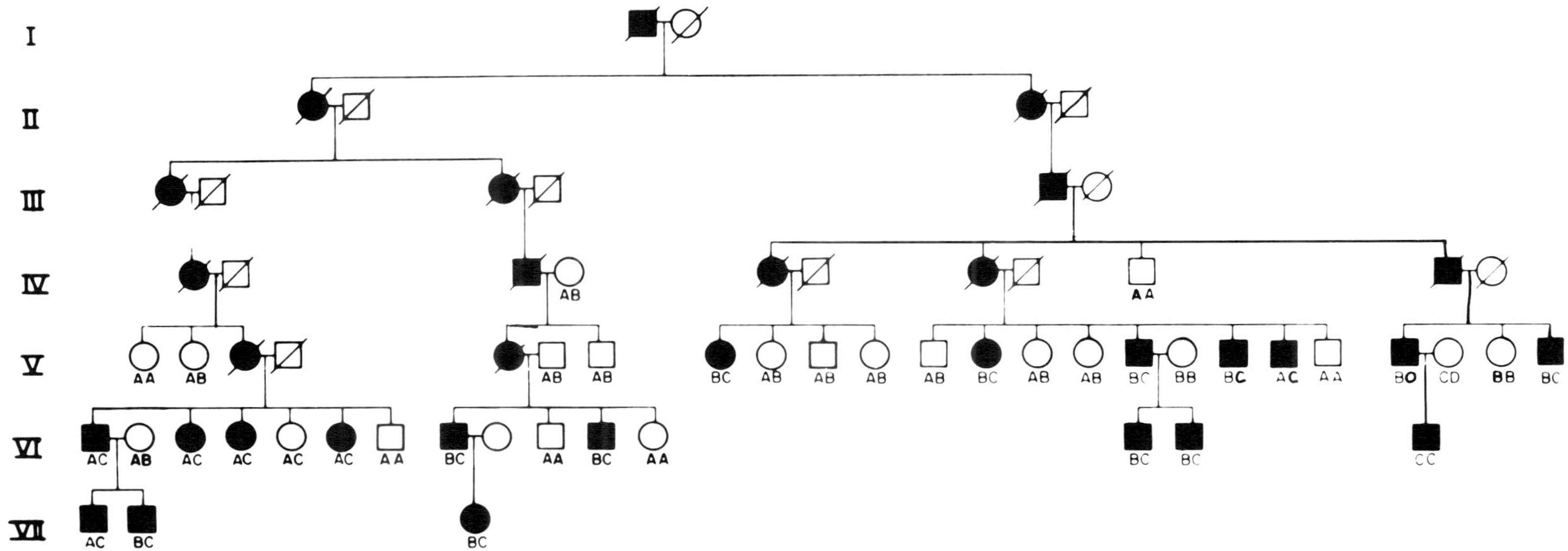

Figure 9.8 Pedigree of the Venezuelan Huntington's disease family. Permanent Epstein-Barr virus (EBV)-transformed lymphoblastoid cell lines were established from blood samples from these individuals. DNA prepared from the lymphoblastoid lines was used to determine the phenotype of each individual at G8 locus 3. The data were analyzed for linkage to the HD gene using the program LIPED with a correction for the late age of onset. Because of the high frequency of the HD gene in this population, some of the spouses of affected individuals have also descended from identified HD gene carrier. In none of these cases, however, was the unaffected individual at significantly greater risk for HD than a member of the general population. *Reproduced from Gusella et al,*[31] *with permission.*

mapped to human chromosome 4 using Southern blot analysis of human-mouse somatic cell hybrids. Positive hybridization signals were always recorded when chromosome 4 was present in the hybrid cell line. In the initial family studies it was thought that the disease was unlikely to be caused by a variety of mutations at separate loci, in view of its apparent exceedingly low mutation rate and epidemiologic data that traced the migrations of the HD gene from Europe to Venezuela and North America. The discovery of the cDNA that defined the G8 DNA probe was based on the fact that there is approximately a 1% chance that any base in a chromosome of an individual will differ from that of another individual or will differ between the same individual's two homologous chromosomes. Such differences, called polymorphisms, are inherited and lead to the development of differences in the patterns and sizes of DNA fragments produced by cleavage with various restriction enzymes. Thus, tracing the inheritance of numerous polymorphisms identified by hybridizing randomly selected cloned DNA segments will eventually lead to a pattern of polymorphism that should identify the gene of particular interest. In the original linkage analysis and family studies reported by Gusella et al.,[31] the researchers were remarkably lucky because the linked polymorphism was identified with the first 12 cloned sequences examined. Initial calculations indicated that the cDNA coding for G8 identified nucleotide sequences 3 million to 5 million base pairs from the actual HD gene.[94] These were initial steps using recombinant methodology aimed at eventual precise localization of the disease-related gene.

One of the next steps in defining the HD locus involved an ingenious application of information from patients with the rare Wolf-Hirschhorn syndrome, a birth defect resulting from partial heterozygous deletion of the short arm of chromosome 4.[32] Because chromosome 4 contains approximately 6.55% of the total human genomic DNA, the search for the HD gene had already been narrowed considerably. Cells from patients with Wolf-Hirschhorn syndrome were used for preparation of genomic DNA as well as production of a fibroblast cell. All the patients showed an apparent terminal deletion of 4p (the terminal band on the short arm of chromosome 4). These deletions are diagrammed in Figure 9.9. Because the cDNA identifying the original G8 marker had clearly localized this segment to chromosome 4, the following approach was adopted. It was postulated that if the G8 locus was outside the deleted region, this could be directly demonstrated by identifying heterozygosity for any one of several restriction enzyme sites close to G8. Alternatively, if the G8 region was the site of a specific deletion in a given individual, heterozygosity would never be observed. With this strategy in mind, Southern blots were performed with genomic DNA from each of seven patients with the deletional Wolf-Hirschhorn syndrome as well as their parents and normal siblings. None of the seven patients studied was heterozygous at any of the RFLPs in the G8 region, although each of the normal parents was heterozygous for at least one of the several restriction enzyme sites. In one patient the G8 locus had been deleted, and two others had a deletion in the same general region. These experiments focused attention on the 4p16 band of chromosome 4 as being relatively close to the HD locus. This localization in effect directed attention to approximately 0.5% of the total human genome. What was clearly needed at this point was another cDNA probe to bracket the HD gene on the other side of the chromosome.

Subsequent careful analysis of large affected kindreds has revealed another aspect of HD: there appears to be no clinically detectable difference between the disease phenotypes of heterozygous and homozygous affected individuals.[95] This was interesting because in virtually all other human disorders in which homo-

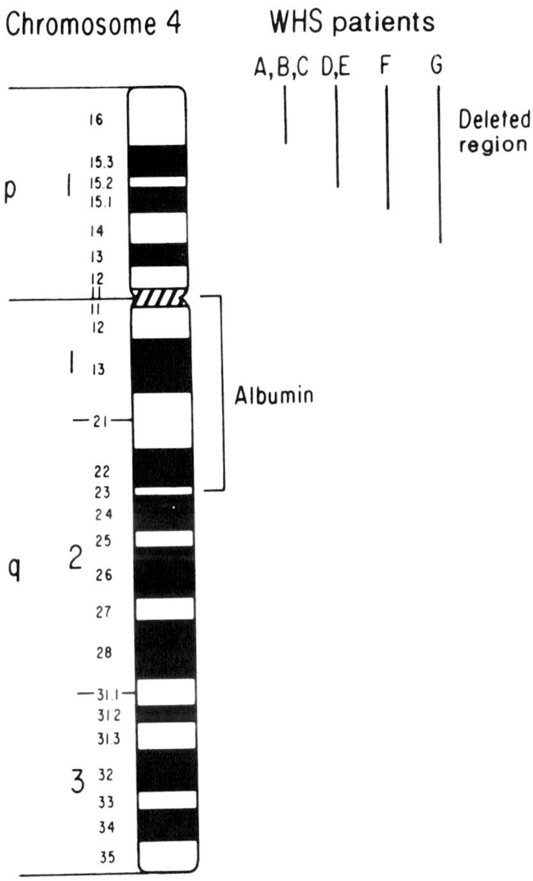

Figure 9.9 Extent of chromosome 4 deletion of Wolf-Hirschhorn syndrome (WHS) patients. The region of chromosome 4 deleted in each patient, as determined by standard trypsin-Giemsa banding techniques, is shown. Note that the application of more sophisticated prophase banding methods could potentially detect differences between the deletions in patients A, B, and C. *Reproduced from Gusella et al,[32] with permission.*

zygotes have been described, the symptoms and disease manifestations have been significantly more severe in the homozygotes than in heterozygotes. In this study homozygosity for HD was determined using the G8 cDNA probe. Therefore, HD appeared to be the first human disease of documented homozygosity that displayed complete phenotypic dominance. This finding has practical implications in terms of possible future attempts to correct the HD defect by introducing a normal gene counterpart. Since the normal allele at the HD locus in a heterozygous individual does not protect against eventual development of the disease, introduction of a normal allele into such an individual would probably not alter disease gene expression. A more profitable approach might be to introduce other genes that would directly block expression of the mutant allele.

An additional linkage marker, D4S43, was identified on the other side of the original G8 or D4S10 locus, and both showed extremely tight linkage to HD expression in family studies.[96] The flanking probe D4S43 was discovered by physically mapping more than 200 randomly chosen chromosome 4 cDNA probes. It was hoped that this new probe would provide much greater accuracy in predictive testing, since the position of D4S43 made it the closest available marker to the HD gene. Moreover, absence of crossovers with HD and the localization within the terminal 3% of the short arm of chromosome 4 made D4S43 a good starting point for eventual direct identification of the HD gene by chromosome walking or long-range jumping techniques. One RFLP was subsequently discovered by expansion of the D4S43 locus, and three additional ones at D4S111 and D4S115 were also recognized.[97] The final marker, D4S125, is a recently discovered variable number of tandem repeats (VNTR) locus. It was found that all of these markers map even closer to the HD gene and to the telomere than D4S10, the original linked HD marker. These new markers should dramatically improve the speed and accuracy of predictive testing for HD.

At present, preclinical testing for HD does not provide as high a probability for definite-at-risk identification as would be ideal or clinically useful. This is because the test is based on linkage analysis and so can never be 100% predictive. A positive or negative result can be interpreted only in the context of an altered probability that the individual tested may be carrying the HD gene, rather than direct identification of the disease gene itself. Moreover, many ethical issues arise in a situation such as testing for HD and making predictions based on the test, particularly because the disease is uniformly fatal and no successful treatment is available.[98]

Alzheimer's Disease

Alzheimer's disease, which is characterized by an accelerated premature process of aging, senility, and mental and physical debilitation, has received considerable attention in the lay press. Death usually occurs within 5 to 10 years after onset of symptoms. Alzheimer's disease is a major public health problem in many Western societies, where estimated age-dependent prevalence rates of up to 5.8 cases per 100 people per year are not uncommon. The disorder, along with its associated medical complications, is the fourth leading cause of death in Western societies. In the United States, an estimated $25 billion per year is spent on the institutional care of Alzheimer's patients. Moreover, as the relative size of the elderly population increases, Alzheimer's disease may become even more important as a public health problem. The disorder is characterized by a widespread functional disturbance in the human brain.[99] Pathological changes include fibrillar amyloid protein deposition inside neuronal cells in the form of neurofibrillary tangles and extracellular deposits of amyloid plaque cores and amyloid in blood vessels. Considerable attention has been focused on the major protein subunit, termed A4, comprising amyloid fibrils of tangles, plaques, and blood vessel deposits. The A4 protein is an insoluble, highly aggregating small polypeptide of relative molecular weight 4500.[100,101] The same material is also deposited in the brains of aged individuals with trisomy 21 (Down's syndrome).

Initially, the origin of protein A4 was unclear and it was thought that the polypeptide might be produced by neuronal cells as the cleavage product of a larger precursor protein. In order to characterize the protein more fully, a full-length cDNA clone coding for the A4 polypeptide was obtained and sequenced.[102] The predicted precursor molecule had 695 residues and features that were thought to be characteristic of glycosylated cell surface receptors. The cDNA was obtained from a library of mRNA derived from brain tissue of a 5-month-old fetus without trisomy 21.

The A4 precursor protein of 695 residues included a signal sequence for transport through the endoplasmic reticulum membrane. The region from residue 290 to 597, representing the amino terminus of A4, showed two potential glycosylation sites. Sequence data indicated that the amyloid subunit was 42 or 43 residues long and was located just before the postulated transmembrane region of residues 625 to 648. It appeared that carboxyl terminal cleavage of mature amyloid precursor would have to occur within the membrane. Southern blot analysis of DNA from human-mouse cell hybrids effectively localized human A4 precursor sequences to chromosome 21. It was also of great interest that chromosome 21 is the chromosome duplicated in Down's syndrome.

A subgroup of Alzheimer's disease—familial Alzheimer's disease (FAD)—was also localized to chromosome 21. This was done independently by another group of investigators using comparative analyses of multiple extremely complete kindreds along with genetic linkage markers for chromosome 21.[103]

The interrelationships between the deposits of brain amyloid in Alzheimer's disease and the known structure of brain amyloid peptides was used to select clones from an adult brain cDNA library using an oligonucleotide probe constructed from the first 20 amino acids of the β peptide of brain amyloid from Alzheimer's disease.[104] Computer analysis of the 28-amino-acid sequence of the amyloid-related polypeptide showed that it contained unique regions not found in the library of sequences recorded in computer banks of all known protein sequences. For this reason, this sequence was selected to fashion the oligonucleotide probe and used to screen the λgt11 cDNA library derived from human brain. Again, Southern blot screening of DNA from a series of human-mouse somatic hybrids provided definite localization to chromosome 21!

Similarly, another group[105] utilized a cDNA probe constructed from the first 24 amino acids of the amyloid protein, which was isolated from amyloid plaque cores and from the walls of intracortical blood vessels of a patient with Alzheimer's disease. The 4.2-kd amyloid polypeptide apparently was derived from proteolytic processing of a larger, as yet unidentified protein. Definition of the entire precursor protein would await cloning of larger cDNAs than those available. Observations confirming increased mRNA expression in Down's syndrome brain tissues indicated that amyloid deposition in Down's syndrome was probably the result of a dosage-related concentration increment of the specific gene product; however, it was not clear whether amyloid plaques characteristic of Alzheimer's disease were also a product of gene overexpression or were due to a change in the brain microenvironment that facilitated local precipitation of β protein. Of great interest at that point was the localization of the β protein gene also to chromosome 21. Whether the colocalization of the FAD and β-amyloid protein genes on similar areas of chromosome 21 meant that they were directly related remained to be determined. Amyloid deposition in the brain can be considered a normal age-linked phenomenon, but in Alzheimer's and aged Down's patients an additional factor must be related to large β protein deposits. In summary, the results of these studies by several large groups show that one or more genes causing FAD are localized on the same area of chromosome 21. One of the genes in this site codes for the β amyloid protein that makes up the cores of neuritic degenerative plaques in brains of both Alzheimer's and aged Down's patients. Moreover, mRNA for β protein can be identified in many human tissues besides brain, and abnormal deposits of amyloid occur in brains of a number of aged mammalian species. No specific cause for Alzheimer's disease has been identified, but an abnormality in the gene coding for amyloid β protein or its precursor appeared to be a strong positive lead.

As the amyloid β protein story continued to unfold, Whitson and co-workers[106] studied possible metabolic effects of a peptide ligand homologous to the first 28 residues of the β-amyloid protein (β1–28). This small peptide was tested in cultures of hippocampal pyramidal neurons for possible neurotropic or neurotoxic effects.

It appeared that the β1–28 peptide had neurotropic activity, since it enhanced neuronal survival in the cultures. The concentrations of β1–28 that produced neurotropic effects in this study were somewhat higher than those of other well-characterized tropic factors such as nerve growth factor, which is active at concentrations of only nanograms per milliliter. However, it was thought that the β1–28 peptide might be only a fragment of the actual physiological ligand and that it might lack primary sequences or structures that would further enhance binding and biological activity.

A subsequent report by Yankner and co-workers[107] examined the opposite hypothesis—namely, that amyloid itself contributes to the underlying neurodegenerative process in Alzheimer's disease. Pheochromocytoma and fibroblast cell lines were transfected with portions of the gene for the human amyloid precursor protein. Stable cell transfectants were obtained that expressed a specific amyloid-containing fragment of the precursor protein. These transfectants gradually degenerated when induced to differentiate into neuronal cells by in vitro addition of nerve growth factor. Moreover, the conditioned media or cell supernatants from these transfected cells were toxic to neurons in primary hippocampal cultures, and the toxic effects could be removed by immunoadsorption with an antibody against the amyloid polypeptide. These results strongly supported the idea that peptides derived from the amyloid precursor could be neurotoxic.

Additional studies of the mechanisms involved in A4 amyloid production showed that a novel mRNA of the A4 amyloid precursor gene appeared to code for a possibly secreted protein.[108] The gene that encoded the A4 peptide found in the amyloid core of senile plaques in Alzheimer's disease produced at least three precursors that resembled cell surface receptors. In the study by de Sauvage and Octave,[108] a clone isolated from a human brain DNA library included the structural sequence for an A4 amyloid peptide precursor with a serine protease inhibitor domain. In the latter region, 208 amino acids at the carboxyl-terminal end were replaced by 20 amino acids derived from nucleotide sequences showing homology to the *Alu* repeat family. It was therefore postulated that this protein devoid of the transmembrane domain could be a secreted form of the A4 amyloid peptide precursor.

Amyloid β Protein Precursor and Hereditary Cerebral Hemorrhage

Molecular approaches to β-amyloid and the mechanisms for deposition of amyloid fibrils in cerebral vessels in Alzheimer's disease and Down's syndrome have continued to produce exciting findings. Van Broeckhoven and co-workers[109] used RFLPs detected by the amyloid precursor protein (APP) gene to examine whether this gene is a candidate for the genetic defect in human hereditary cerebral hemorrhage with amyloidosis of the Dutch type. Data from this study indicated that the APP gene was tightly linked to the Dutch type of hereditary cerebral hemorrhage.

Genetic Linkage Studies of Alzheimer's Disease

Genetic linkage studies have been extended further by analyzing the patterns of segregation of the amyloid protein gene in several familial Alzheimer's disease kindreds using other RFLPs.[110,111] Failure of segregation together in these kindreds proved that the FAD gene was not tightly linked to the amyloid β-protein gene.

Also, the possibility was raised that amyloid β-protein gene duplication is an underlying abnormality in Alzheimer's disease. However, several reports from different laboratories did not support increased β-protein gene dosage or duplication when this was measured quantitatively by densitometric means in patients with Alzheimer's disease or Down's syndrome.[112,113]

Following a number of family studies in attempts to relate the occurrence of Alzheimer's disease to various genetic loci, a massive coordinated survey was reported by St. George-Hyslop and co-workers.[114] Some of the problems in interpreting the data from previous attempts at family linkage identification were due to differences in methods of pedigree analysis, effects of confounding variables in the various analyses, or true nonreplication. To resolve a number of these questions, St. George-Hyslop and co-workers examined the inheritance of five polymorphic DNA markers from the proximal long arm of chromosome 21 in a large unselected series of pedigrees with FAD. Their results suggested that Alzheimer's disease is *not* a single entity but rather results from genetic defects on chromosome 21 and from other genetic or nongenetic factors.

Olfactory Neurons and Alzheimer's Disease

A report by Talamo et al.[115] has provided a possible window for the diagnosis of Alzheimer's disease before death. These workers found pathological changes in the sensory epithelium of the nose in patients who died of Alzheimer's disease. Because such tissue is easy to access for biopsy, such changes, if confirmed by other workers, could be used to establish the diagnosis of Alzheimer's disease while the patient is still alive. In this study, antibodies were used that were specific for proteins present in the neurofibrillary tangles and neuritic plaques in the brains of Alzheimer's patients. The workers found abnormal masses of neurites in the olfactory sensory epithelium of eight of nine patients who also showed the typical tangles and plaques in the brain at autopsy. By contrast, neuritic masses were present in only 2 of 14 age-matched controls who died with other neurologic disorders. Longitudinal studies are clearly needed, as well as an exhaustive study of other patients with various types of dementia, before such an approach can be applied to practical clinical evaluations.

Lesch-Nyhan Syndrome

Lesch-Nyhan syndrome is a severe, X-linked recessive neurological disorder caused by a defect in the hypoxanthine phosphoribosyltransferase (HPRT) gene that results in HPRT deficiency.[116,117] When cDNA probes complementary to HPRT are used to analyze patients with this disease, in 85% of cases the gene for HPRT appears normal by both Southern and Northern blotting techniques.[118] This puzzling finding indicated that most cases are due to point mutations or small DNA deletions or rearrangements. A technique involving ribonuclease A (RNase A) cleavage has been used to detect discrete mutations in genomic DNA, and this approach has been successfully applied to the Lesch-Nyhan problem.[119] The cleavage assay is based on the fact that some single base mismatch sites in RNA hybrids with RNA or DNA can be cleaved by RNase A. A single RNA probe can be employed to identify a single base substitution, or a pair of overlapping probes can be used together to map mutation sites. Gibbs and Caskey[119] used the RNase A cleavage method to study HPRT mutations with a single HPRT antisense RNA probe produced from HPRT cDNA cloned into appropriate transcription vectors. Because HPRT mRNA

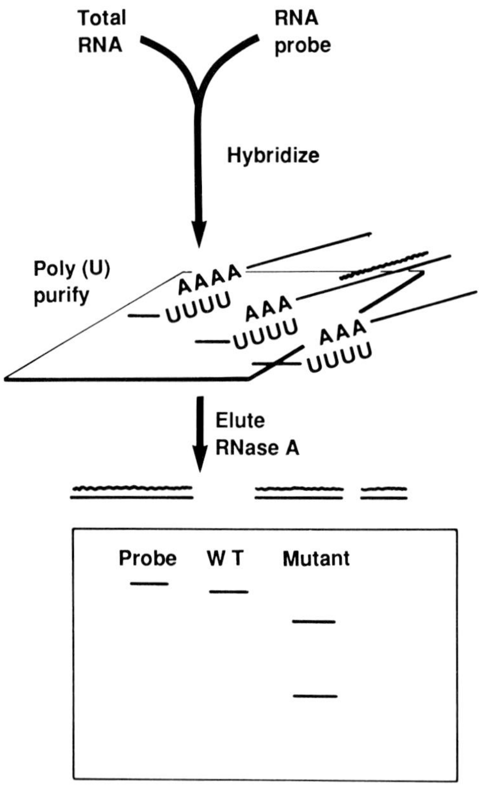

Figure 9.10 Flow diagram for RNase A cleavage analysis with messenger affinity paper purification. Total cellular RNA extracts were hybridized with radiolabeled RNA probe and the HPRT mRNA-probe hybrids were partially purified by passage through polyuridylic acid [poly(U)]–bound affinity paper. The hybrids were eluted and treated with RNase A to digest single-stranded regions and internal mismatch sites, and the fragments were analyzed by denaturing polyacrylamine gel electrophoresis and autoradiography. A polyacrylamide gel is shown in schematic form to illustrate the appearance of the probe alone (Probe), probe hybridized to wild-type HPRT mRNA (WT), and probe hybridized to a mutant HPRT mRNA that yields two fragments (Mutant). *Reproduced from Gibbs and Caskey,*[119] *with permission.*

was only 0.01% of the total mRNA population, the sensitivity of the procedure was increased by using polyuridylic acid affinity paper as a "messenger affinity paper" support for the recovery of RNA probes, which were then hybridized to polyadenylated HPRT mRNA. The nonpolyadenylated HPRT including the unbound probe can be washed away. This procedure is diagrammed in Figure 9.10.

When RNase A cleavage was applied to Lesch-Nyhan material, distinct RNase cleavage patterns were identified in 5 of 14 patients who had shown no detectable Northern or Southern blot changes. This method therefore holds great promise for identifying other mutational changes in situations of very low abundance mRNAs.

Manic-Depressive Illness or Bipolar Disorder

For a number of years, the influence of genetic factors in mental illness has been a subject of intense study and debate. The separate effects of environmental and innate biological features in individual subjects have made this a difficult area for investigation. The major affective disorders are illnesses characterized by disturbances in mood and various physiological, cognitive, and endocrine abnormalities. The primary divisions include unipolar major depression and bipolar disorder, the latter characterized by manic episodes including an elevated, expansive, euphoric, or irritable mood. Depressive episodes are characterized by a mood of utter hopelessness, decreased appetite, sleep disturbances, loss of interest, and often feelings of worthlessness or guilt. Bipolar disorder (manic-depressive illness) is estimated to have a prevalence of approximately 1% of the population of North America and Western Europe.[120]

The onset of illness is generally between 15 and 35 years of age. Major depression alone has a much higher estimated prevalence, with a wide range of age of onset as well. Twin surveys, adoption studies, and studies of affected families have demonstrated that genetic factors may play an important role in the etiology of these disorders.[121–123] Nevertheless, establishing the mode of inheritance and identifying a well-defined genetic locus have posed extremely difficult problems.

A study of the role of genetic mechanisms in bipolar affective disorder took advantage of the relatively genetically isolated Old Order Amish (OOA) population in southeastern Pennsylvania.[123] The Amish population afforded a unique opportunity to examine hypotheses of genetic transmission. Detailed family records on all 12,000 members of the OOA population indicated that all were descended from 30 original progenitors who emigrated from Europe in the early eighteenth century. Prevalence of affective disorder was found to be similar to that in the general population of North America.[123] The study concentrated on families of bipolar patients; this work clearly indicated that an autosomal dominant mode of transmission was most consistent with the patterns of illness expression among the Amish kindreds. Blood samples were collected from a multigenerational pedigree highly affected with bipolar illness, and permanent lymphoid cell lines were established for all members studied. DNA from 51 of these cell lines was used for initial recombinant DNA studies with RFLP analyses. The initial work suggested a link between bipolar illness and two markers on chromosome 11, insulin (INS) and the cellular oncogene Ha-*ras*-1.[124–125] Using DNA from 30 additional family members' cell lines, a strong linkage was discovered with specific alleles of the INS and Ha-*ras*-1 (HRAS1) loci. An example of this striking degree of segregation is shown in Figure 9.11. Statistical significance of the cosegregation of these markers

Figure 9.11 INS and HRAS1 typings on selected members of pedigree 110. Both the INS and HRAS1 loci have multiple alleles as a result of an insertion-deletion sequence in the immediate vicinity of the coding region. Four different alleles at the INS locus were observed when genomic DNA of these individuals was digested with the enzyme *Pvu*II. The alleles are: A, 760 base pairs (bp); B, 800 bp; C, 830 bp; D, 2,300 bp. Alleles of HRAS1 are as follows: 1, 5.2 kb; 2, 5.7 kb; 3, 6.2 kb; 4, 6.8 kb—on *Sac*I-digested DNA. Filled symbols, individuals with affective disorder; open symbols, individuals without affective disorder to date. Squares, males; circles, females; superimposed square and circle, other siblings (number given in center). Note that alterations in sex designation and reductions in sibship size have been made to preserve anonymity of individuals at risk of affective disorder. *Reproduced from Egeland et al,*[124] *with permission.*

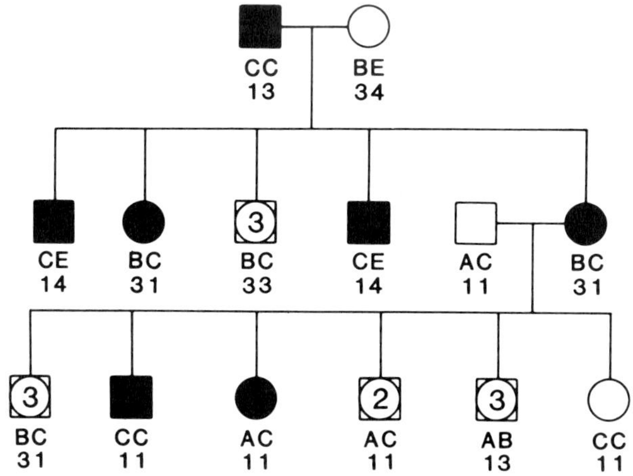

was determined by linkage analysis using a well-defined computer program (LIPED) modified for age-dependent penetrance. This analysis provided a lod score of 3.340 and thus fairly strong evidence for linkage to the gene for bipolar illness.

This study provided what appeared to be convincing evidence that RFLPs and careful applications of recombinant methods could localize a gene implicated in the etiology of a common clinical disorder. It was felt that these findings had broad implications for research in human genetics and psychiatry. By using linkage analysis, it was shown that a simple genetic mechanism could account for the transmission of bipolar affective disorders.

More information is needed to determine whether the linkage uncovered with the INS and HRAS1 markers in the Amish families can be applied to other kindreds in which multiple members have shown affective bipolar disorders. It was of interest that in these initial analyses, particularly of pedigree 110, a number of clinical variant forms of affective disorder, including bipolar illness, major depressive, and schizoaffective disorder, shared a common chromosome 11 haplotype. It may well be that in the future DNA-based diagnostic procedures might be able to identify patients or premorbid individuals at risk. Because not all individuals who inherit the gene will manifest symptoms or full-blown expression of the affective illness phenotype, prospective evaluation studies of this sort raise important social and ethical questions. When the gene coding directly for the disorder has been identified, more insight will be possible into what biochemical step or enzyme is compromised or defective, causing the disease.

A study from Iceland that appeared at the same time as that of the Amish kindreds dealt with the possible relevance of mutations affecting the tyrosine hydroxylase gene,[126,127] which codes for the rate-limiting enzyme for the three neurotransmitters: epinephrine, norepinephrine, and dopamine. These workers analyzed linkages between the clinical expression of bipolar affective disorder and genes coding for tyrosine hydroxylase, Ha-*ras*-1 (HRAS1), and the variable region of the insulin gene, INS.[128] All three markers were found to be closely linked on chromosome 11. Disappointingly, within three Icelandic kindreds no evidence was obtained for linkage to these three markers. It therefore appeared that mutations at different loci were responsible for manic-depression in kindreds from the Amish community and those in Iceland. Moreover, absence of close linkage was also found by a third group who examined bipolar American families outside the Amish group.[129] From these collective data it appears that, similar to the situation in β-thalassemia (Chapter 8), multiple mutations perhaps at the same general locus may account for the bipolar illness phenotype in the general population. Studies of the familial distribution of the disease in general suggest that only 60% to 70% of those who inherit the manic-depression gene manifest the disease, and there has always been speculation about the role of inciting environmental factors in disease expression.[130] Despite their discrepant results, the three studies examining marker linkage to the bipolar illness gene are quite encouraging in that they represent the first concerted applications of molecular biological approaches to study important aspects of a common mental disorder.

The complexity of the problems is further emphasized by subsequent studies from Israel, in which close linkage of bipolar affective illness was demonstrated in a kindred with color blindness and glucose-6-phosphate dehydrogenase deficiency—both X-linked mutations.[131,132] Maximum lod scores in this study ranged from 7.52 to 9.17. The population surveyed in Israel was perhaps more similar to those in the Amish studies because it was in general characterized by a relatively high lifetime prevalence of affective disorder, large multigenerational pedigrees, and a high frequency of G6PD deficiency in the non-Ashkenaszi Jews (60% in Kurdish males). From the results of these studies, it was estimated that a proportion of the bipolar population at risk might carry the X-linked gene.

Manic-Depression Gene Still Elusive

An additional chapter in the study of the genetics of bipolar illness has been added by the work of Kelsoe et al.,[133] who sought to confirm the original linkage studies of Older Order Amish by Egeland and co-workers.[123,124] On reexamination of the same material it was found that 40 new subjects had been added to the 81 in the first study and that 2 of the original group had become newly ill. With these changes the lod scores plunged by about 1.5. In addition, the reexamination indicated that the original markers might not yet have identified a significant linkage. The lessons to be derived from this are many—among which that it is extremely important for results in such studies to be checked and rechecked and for the study group to be followed long enough for ascertainment of any new diagnoses.[134,135] Work on this complex and important problem continues.

α_1-Antitrypsin Deficiency

α_1-Antitrypsin deficiency is a genetic disorder that leads to early pulmonary damage and severe manifestations of emphysema.[136] In addition, several lines of evidence indicate that the P:Z phenotype may be associated with cirrhosis of the liver and liver cancer.[137] The α_1-protease inhibitor itself (AAT) is a major plasma protein that is an efficient inhibitor of leukocyte elastase and bacterial proteases. This protein is encoded at the Pi locus on chromosome 14 (14q24.3–q32.1).[138] A deficiency for the Pi Z allele occurs in 1 of 2000 to 8000 whites[139] and confers an increased risk of early-onset adult emphysema as well as liver disease in childhood. DNA polymorphisms associated with the AAT gene have now been applied to define the origin of the Pi Z allele. This was done by Cox et al.[140] using two probes encompassing the 5′ and 3′ flanking regions. These studies indicated that a single unique DNA haplotype could be identified for most Pi Z alleles. This finding of a single disease-associated haplotype was considered to be exceptional. For example, multiple restriction sites associated with the sickle cell hemoglobin gene indicate three independent origins in Africa.[141] Moreover, a specific restriction fragment haplotype for thalassemia is usually associated with a specific mutant gene, and the ranges of haplotypes tend to be characteristic of or defined for various ethnic populations.[142] By contrast, in AAT the mutant gene is almost always associated with a single haplotype, probably indicating a single, possibly recent origin for the Pi Z allele. Many different racial groups have now been surveyed, and it appears that the Pi Z allele is limited to whites and is not present among black or Oriental populations. Furthermore, because the unique *Ava*II haplotype for the Pi Z was clearly defined, it provided a reasonable approach to prenatal diagnosis. A schematic representation of the α_1-antitrypsin gene and flanking regions, along with the gel appearance of the unique *Ava*II restriction pattern identifying the Pi Z haplotype, is shown in Figure 9.12.

The molecular basis for α_1-antitrypsin deficiency can be defined for mutant alleles Z and S, both of which result from point mutations in the coding region for the gene. In both cases, the protein product translated by the mRNA differs from

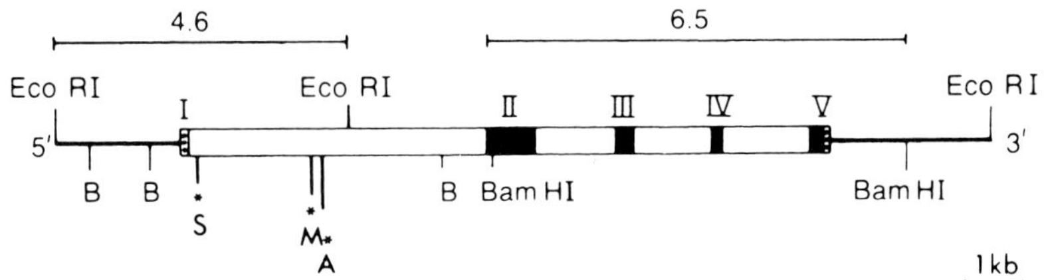

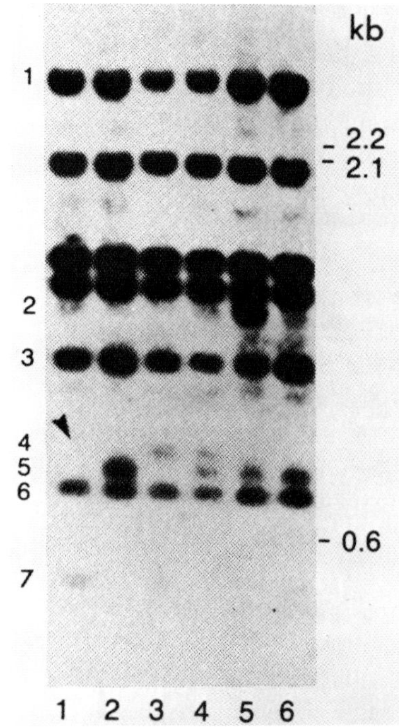

Figure 9.12 (A) The α_1-antitrypsin gene and adjacent flanking regions. Solid areas are coding regions, dots indicate untranslated regions, and open areas are introns. Restriction sites for the enzymes *Eco*RI and *Bam*HI, B, are indicated. (*) Polymorphic restriction sites for *Sst*I (S), *Msp*I (M), and *Ava*II (A). The two genomic probes used in this study are shown above the gene. (B) DNA fragments observed using restriction enzyme *Ava*II and the 6.5-kb probe. Size of fragments are in kilobases; anode is at the bottom. The seven polymorphic fragments are numbered at the left. Three strong bands are present between polymorphic bands 1 and 2 and appear to be constant in all samples observed to date. The *Ava*II DNA phenotypes observed in the lanes are as follows: (1) 11336677; (2) 11336655; (3) 10334677; (4) 10334657; (5) 11236657; (6) 11336657. The arrow in lane 1 shows the unique region of the pattern associated with Pi type ZZ. *Reproduced from Cox et al,[140] with permission.*

the normal protein by a single amino acid substitution.[143–145] It is thought that cells producing α_1-antitrypsin are unable to process and secrete the mutant proteins normally, and therefore the amounts released by cells making the protein are markedly reduced in persons of the ZZ and ZS phenotypes, resulting in α_1-antitrypsin deficiency. In a rare example of α_1-antitrypsin deficiency, described as the "null-null" state, no α_1-antitrypsin is detectable at all.[146] The "null" α_1-antitrypsin gene represents a different class of mutants because it is unable to direct synthesis of any detectable mRNA transcript.[147]

Hybridization studies of the family and of the propositus who expressed the null-null allele indicated that the disease in this instance was not due to a major deletion, addition, or even rearrangement. In addition, no distinct abnormality such as hypermethylation of the promoter regions for the α_1-antitrypsin gene was detected. Garver et al.[147] reported that grossly intact α_1-antitrypsin genes were apparently present in the rare patients with the null-null phenotype and that there was no definable reason for nonexpression of the protein.

With probes available for M- and Z-specific oligonucleotides, 16 pregnancies at risk have been screened for α_1-antitrypsin deficiency by Hejtmancik et al.[148] It was found that the strong linkage disequilibrium between the *Ava*II RFLP and the Z allele was extremely useful. Three affected pregnancies were terminated, and analysis of fetal tissues in two was confirmed by repeating recombinant studies on fetal tissues. RFLP analyses under routine conditions took 7 days in this study and studies using oligonucleotides took 5 days. However, with the oligonucleotides a number of assays had to be repeated to ensure clarity, and differentiation between M and Z bands was more difficult. Progress in refining several probes for certainty in prenatal diagnosis continues.

Additional studies by Cox and Mansfield[149] have indicated that the unique haplotype identified with *Ava*II digests, now termed 371, has extended the apparent usefulness of this method for detecting Pi ZZ patients. Thus this approach should prove useful in detecting Pi ZZ α_1-antitrypsin deficiency simply by analysis of fetal DNA without recourse to patients or siblings. Other RFLPs have also been identified, including one that occurs after restriction enzyme digestion with *Mae*III,[150] which appears to be promising for use with chorionic villus biopsy samples.

There is a potential problem of interpreting the results of prenatal diagnosis in that only a small proportion of deficient subjects have serious health problems in early life and the pulmonary disease, including premature development of emphysema, may be modified by attention to environmental factors such as avoidance of smoking. Also, a large prospective population survey in Sweden, where the Pi ZZ genotype has an incidence of 1 in 500, indicated that although 14 of 120 Pi ZZ infants showed prolonged neonatal jaundice, by the end of four years only 4 of these children had clinical evidence of liver disease.[151,152]

Brantly and co-workers[153] in 1988 made a unique observation of considerable interest. It was known that subjects homozygous for the Z-type mutant form of the α_1-antitrypsin gene often show the most common type of α_1-antitrypsin deficiency. The α_1-antitrypsin–synthesizing cells of patients with the Z gene have normal α_1-antitrypsin mRNA levels, but their secretion of the protein is markedly reduced secondary to accumulated newly produced α_1-antitrypsin in the cellular rough endoplasmic reticulum. Crystallographic studies of α_1-antitrypsin had predicted that in the normal protein a negatively charged Glu^{342} was adjacent to a positively charged Lys^{290}. The $Glu^{342} \rightarrow Lys^{342}$ Z mutation causes loss of a normal salt bridge, which results in the intracellular aggregation of the Z molecule. It was therefore predicted that a second mutation in the α_1-antitrypsin gene that changed the positively charged Lys^{290} to a negatively charged Glu^{290} might correct the secretion defect associated with the disease. When the second mutation was produced in the Z-type DNA, the derived gene directed synthesis and secretion of α_1-antitrypsin that was similar to that produced by the normal α_1-antitrypsin DNA—all in an in vitro eukaryotic expression system. These findings indicated that in this case a human hereditary disease could be corrected by inserting an additional mutation in close proximity on the same gene.

Aerosolized Recombinant α_1-Antitrypsin

Because the major defect in α_1-antitrypsin deficiency is an insufficient supply of the protein in the lower respiratory tract, where theoretically it protects the normal structures against neutrophil elastase, aerosols of recombinant α_1-antitrypsin have been studied, first in sheep and later in α_1-antitrypsin–deficient subjects.[154,155] When the aerosols were administered, no adverse side effects were apparent and reasonable levels of the recombinant material were subsequently detected in serum and lung epithelial lining fluid. Practical implementation of this approach in patients may be limited by the cost of the recombinant protein, of which approximately 1400 mg/week would be needed for aerosol therapy.[155]

α_1-*Antitrypsin and Liver Damage*

Circulating α_1-antitrypsin is produced primarily in the liver and then secreted into the bloodstream, where it is the major protease inhibitor. In the Pi Z variant of α_1-antitrypsin, serum levels of the protein are decreased because it is retained in hepatocytes. Patients who are homozygous for the variant allele are predisposed to the development of pulmonary emphysema and liver disease. The Pi Z variant of the human α_1-antitrypsin gene was introduced into the germ line of mice to determine whether the mutant protein would accumulate in mouse hepatocytes and whether this would produce liver damage in the transgenic animal model.[156] It was found that the mutant protein was synthesized in large amounts in the liver of transgenic mice and accumulated in the rough endoplasmic reticulum of hepatocytes, much as in human patients. The degree of liver damage correlated with the amount of Pi Z α_1-antitrypsin accumulated in the livers of the animals. These findings appear to relate hepatic accumulation of the mutant protein directly to liver damage.

Tay-Sachs and Gaucher's Diseases

Several hereditary disorders are associated with storage or deposition of normal metabolic products secondary to mutations of key enzymes in normal metabolic pathways. Tay-Sachs disease is an inherited disorder caused by mutation in the α chain of β-hexosaminidase A, a lysosomal enzyme that contains two polypeptides called α and β chains.[157] The defect in the enzyme results in abnormal piling up and storage of its major substrate, ganglioside GM_2. Progressive nonphysiological accumulation of this substrate produces the characteristic neurodegenerative lesions in patients with Tay-Sachs disease. The clinical phenotypic expression of the disorder shows a rather broad spectrum of both severity and age at onset. For the early-onset and fatal form, or classical Tay-Sachs disease, the gene frequency is 10 times higher in Ashkenazi Jews than in the general population. A carrier frequency equivalent to that in Ashkenazi Jews is found in a non-Jewish community of French Canadians living in eastern Quebec.[158] Identity of mutations has been proposed as an explanation for the clinical similarities between these groups at risk for Tay-Sachs disease.

Molecular biologic analysis of these two populations was performed by Myerowitz and Hogikyan.[159] DNA was isolated from cultured fibroblasts of two non-Jewish French Canadian patients and two Ashkenazi patients in parallel. After restriction digestion with *Eco*RI, each sample was hybridized by Southern blotting with labeled cDNA probes for the entire coding sequence of the α chain of human β-hexosaminidase. Identical patterns were seen in normal nonaffected controls and Ashkenazi Tay-Sachs patients; however, a 4-kb DNA fragment present in the former two groups was absent from the patterns for the French Canadian Tay-Sachs patients. These results indicated that a large deletion of 5 to 8 kb in the 5′ end of the α chain gene had occurred only in the French Canadian group. This is reminiscent of findings in Duchenne muscular dystrophy subsets who showed similar clinical phenotypes but differences in localization of deletions where mutations were identified.

Gaucher's disease, the most common of the lipid storage disorders, is caused by a deficiency of lysosomal glucocerbrosidase.[160,161] Three major clinical phenotypes of the disease have been identified. Type 1 is characterized by hepatosplenomegaly, anemia, thrombocytopenia, and bone involvement; the nervous system is not affected. Patients with type 2 Gaucher's disease have cranial nerve and brain stem abnormalities and hepatosplenomegaly, usually with onset of symptoms before 6 months of age. Patients with type 3 have systemic symptoms similar to those of type 1 but neurological signs that appear early in childhood or adolescence. Direct biochemical studies have shown that the Gaucher's mutations result in synthesis of catalytically defective glucocerebrosidases that cannot be processed normally or metabolized. The differences in clinical manifestations and ethnic predilection only in type 1 Gaucher's disease support the idea that the types are caused by different allelic mutations.

Cloning of the cDNA for the glucocerebrosidase gene permitted analysis of the genomic basis for the three different phenotypes. The glucocerebrosidase gene has been cloned from a patient with type 2 Gaucher's disease and the distinct mutation has been identified.[162] Genomic DNA from cultured fibroblasts from the patient was restricted by *Eco*RI, hybridized to glucocerebrosidase cDNA, and ligated into phage. Recombinants carrying the glucocerebrosidase gene were isolated by hybridization with glucocerebrosidase cDNA. Oligonucleotide probes constructed on the basis of the normal gene sequence and part of the type 2 mutant genomic sequence were used as templates to synthesize labeled oligodeoxynucleotide probes by primer extension techniques. Clones of genomic DNA from a normal control subject and the patient with type 2 Gaucher's disease showed 15-kb inserts and identical restriction patterns with *Bam*HI; however, sequence analysis showed that the two genomic clones were identical except for a T → C single base substitution in exon X of the type 2 gene. This is illustrated in Figure 9.13. This isolated base change in which leucine was replaced by proline at residue 444 and could be predicted to destroy the alpha-helical configuration of the protein in this area. Advantage was taken of the fact that this single base substitution produced a new restriction enzyme cleavage site for *Nci*I, and three successive generations of the family of the type 2 patient were examined by Southern blot hybridization. The presence of the mutation and thus the new *Nci*I site generated two new DNA fragments (1.6 kb and 1.0 kb) that clearly identified the mutation in other members of the kindred. The remarkable insight this yielded into the biochemical makeup of the family is illustrated in Figure 9.14. These observations were extended to patients with type 1 and type 3 Gaucher's disease, and it was found that 4 of 5 patients with type 2 Gaucher's disease and all 11 patients with type 3 had at least one allele for the same mutation. By contrast, only 4 of 20 patients with type 1 disease had the same mutant allele and none were homozygous for this mutation.

These results again illustrate the profound biological effects which may accompany single base mutations in biologically potent or potentially vulnerable spots. It is possible that the proline-for-leucine substitution in this example is responsible for the loss of glucocerebrosidase activity. The substitution at position 444 occurs very close to the proposed catalytic site for the glucocerebrosidase enzyme. These findings are reminiscent of the single leucine → proline mutations in various hemoglobinopathies and of a case of α-thalassemia in which a leucine → proline mutation produced a profound difference in stability of the hemoglobin tetramer.[163]

The mechanism of nervous system involvement in neuronopathic Gaucher's is still unknown; however, the RFLP revealed by *Nci*I endonuclease may prove

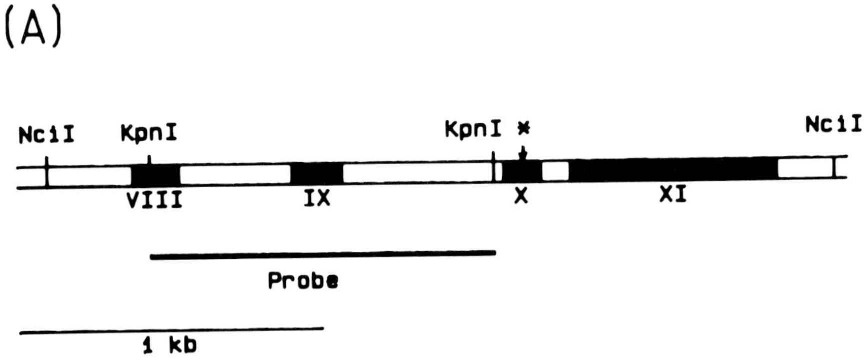

(B)

```
                 lys phe ile pro glu gly ser gln arg val gly leu
···ccacattcag C AAG TTC ATT CCT GAG GGC TCC CAG AGA CTG GGG CTG

val ala [ser gln lys asn asp leu asp ala val ala leu met his] pro
GTT GCC  AGT CAG AAG AAC GAC CTG GAC GCA GTG GCA CTG ATG CAT  CCC
                             ↓
                             CCG
                             Pro

asp gly ser ala val val val val leu asn
GAT GGC TCT GCT GTT GTG GTC GTG CTA AAC CG gtgagggcaa··········
```

Figure 9.13 Schematic diagram of exons VIII through XI, including the flanking introns (A), and nucleotide sequence of exon X (uppercase letters) and its flanking intervening sequences (lowercase letters) (B). In (A) exons and introns are indicated by filled and open areas, respectively. The location of the single base substitution (T→C) in exon X of type 2 genomic DNA is indicated by an asterisk. This single base substitution (CCTGG → CCCGG) generates a new *Nci*I site. The two adjacent *Nci*I sites are in introns 5′ and 3′ to exons VIII and XI, respectively. A 1.1-kb *Kpn*I genomic DNA fragment of the normal gene in λPCN9 was used as the probe in the Southern blot hybridization analyses. Panel (B) shows that a comparison of the nucleotide sequences of the entire coding sequence, splice junctions, and 5′-flanking region revealed only a single base substitution (T→ C). This mutation results in the replacement of leucine at residue 444 by proline. The region in which alpha-helical structure is predicted for the normal protein sequence is enclosed by a box. *Reproduced from Tsuji et al,*[162] *with permission.*

useful in the future as a diagnostic predictor for patients who may eventually show neurological signs. Also, since the mutation identified in this study was present in all three phenotypes, though with different degrees of expression, it seems likely that each phenotype of Gaucher's results from more than one mutation. If both type 2 disease and type 3 disease can be caused by homozygosity for this allele, it will eventually be necessary to look at other possible mechanisms, such as modifier genes or nongenetic factors, to explain the differences between these two Gaucher's disease phenotypes.

Experimental approaches to the treatment of Gaucher's disease have included

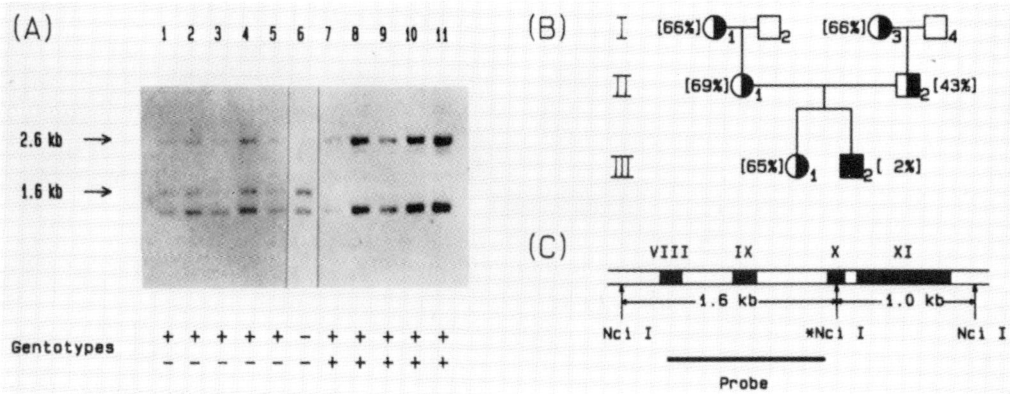

Figure 9.14 Southern blot hybridization analysis of the patient with type 2 Gaucher's disease (III-2), his family members, and five normal controls (A); pedigree chart of the patient's family (B); and schematic diagram showing the location of the single base substitution that generates a new *Nci*I site, as well as the two adjacent *Nci* sites in the intervening sequences (C). In (A), *Nci*I-digested fibroblast or white-cell genomic DNA was probed with the 1.1-kb *Kpn*I genomic DNA fragment from the normal gene in λPCN9. Lane 1, paternal grandmother [I-3]; lane 2, maternal grandmother [I-1]; lane 3, sister [III-1]; lane 4, mother [III]; lane 5, father [II-2], lane 6, proband [III-2]; and lanes 7 to 11, normal controls. The genotype of each sample is shown below the autoradiogram; a plus sign denotes the presence of the 2.6-kb band (upper row) and a minus sign the presence of the 1.6-kb band (lower row). In (B), the glucocerebrosidase activities (in brackets) of cultured skin fibroblasts are shown as percentages of normal control values. *Reproduced from Tsuji et al,*[162] *with permission.*

bone marrow transplantation and enzyme infusion.[164] Central nervous system involvement is unlikely to be prevented by bone marrow transplantation. Another approach—somatic gene therapy—needs additional consideration. Functional cDNA for the human enzyme has been successfully introduced into a retroviral vector and expressed in human fibroblasts.[165] It is possible that such retroviral infection of autologous marrow stem cells might provide an effective treatment for Gaucher's disease with less risk than transplantation of heterologous marrow. Much remains to be explored in this genetic disorder.

Adult Polycystic Kidney Disease

Adult polycystic kidney disease (APKD) is a relatively common and often lethal multiple organ disease that is inherited in an autosomal dominant pattern. Roughly 1 in every 1000 people carry the mutant gene for this disorder.[166] The pathology of the disease consists of development and progressive enlargement of cysts in a number of organs including the liver, pancreas, spleen, and kidneys. The essential defect that leads to cyst formation is still unknown. The progressive development of cysts leads to irreversible renal failure and death at a mean age of 51 unless chronic dialysis or transplantation is introduced. Individuals with APKD account for about 9% of chronic dialysis patients.[167] The disease resembles Huntington's chorea in that the first symptoms tend to occur in the fourth decade after most patients have already had children. Early identification of subjects who will eventually develop the disease has not been ideal. Ultrasonographic detection is not foolproof because 34% of at-risk patients in the second decade and 14% in the third will go on to develop cysts after a prior negative diagnostic workup.[168] Thus, the low sensitivity of predictive diagnostic techniques produces great problems for genetic counseling. For this reason, the initial report of an identified linkage marker for APKD to the α-globin locus on the short arm of chromosome 16 was hailed as a major breakthrough in this field.[169]

The strategy in the initial study was to utilize highly polymorphic DNA probes to maximize searching for the APKD locus.[169,170] A highly polymorphic region encompassing 8 kb beyond the 3′ end of the α-globin gene cluster was chosen. It consisted of a tandem repeat sequence, and polymorphism occurred because of variability in the copy number of the elements in the repeat. The probe utilized a *Hin*I fragment hybridized to *Pvu*II-digested genomic DNA. Studies of large kindreds with this simple technique showed disease segregation with one of the 10 electrophoretically distinguishable alleles. No crossovers were observed between the probed area and the polymorphic loci in the α-globin cluster used as a probe in additional families. Lod scores of 7.31 to 16.41 demonstrated striking apparent linkage for this probe. The gene for APKD was mapped to within 11 centimorgans of the α-globin cluster on chromosome 16.[171] The finding of close linkage in all of the nine families studied suggested that a single locus might be involved; however, since all of the families represented a common phenotype, additional surveys are needed.

Genes and Heredity of Specific Tumors

Retinoblastomas

As noted briefly in Chapters 5 and 8, it has been recognized for some time that retinoblastomas are a group of childhood cancers that appear to be inherited as an autosomal dominant trait.[172] The tumor appears to develop after inheritance of a predisposing germ-line gene in about 10% of cases and as a de novo germ-line mutation in about 30%. In either case, multifocal presentation of tumors appears to be characteristic of heritable disease. Predisposition to tumor mutations is completely penetrant, since nearly 50% of progeny of bilaterally affected persons develop the tumors.[172] Also, approximately 10% to 15% of patients with unilateral retinoblastomas have multifocal disease and are at risk of passing the trait to their children.

The accepted clinical approach to this disorder in possibly affected or at-risk children involves repeated ophthalmologic examinations under anesthesia until at least $2\frac{1}{2}$ years of age. Another important feature of this disease is that there is a threefold increased incidence of osteosarcoma in survivors of hereditable retinoblastoma.[172]

Previous studies had placed the locus for hereditary retinoblastoma on human chromosome 13 at band q14. Studies of karyotypes of retinoblastoma patients had indicated occasional deletions of parts of chromosome 13 often involving band q14, and molecular analyses of retinoblastoma tumors had shown loss of constitutional heterozygosity for chromosome 14.[173–175]

A method for predicting whether retinoblastoma will develop in a child at risk would be of great value in both genetic counseling and clinical treatment of the disorder. Cavenee et al.,[176] using cDNA probes for loci on chromosome 13 including those for esterase D (a convenient enzymatic marker), studied families at risk for retinoblastoma. They were successful in identifying parentally transmitted mutations in children at risk. It appeared that separate mutations at the locus in question were identified for almost every tumor.

Insight into the retinoblastoma-osteosarcoma association was provided by studies showing complete or partial deletions in a DNA segment spanning at least 70 kb in human chromosome band 13q14 in many retinoblastomas and osteosarcomas.[177,178] In another survey, many of 40 retinoblastomas examined with a cDNA probe representing what was presumed to be the retinoblastoma (Rb) gene showed clearly identifiable Rb gene structural changes including homozygous internal deletions. An osteosarcoma also showed a homozygous internal deletion with a truncated transcript. In tumors in which no obvious structural changes could be identified, there was either absence or abnormal expression of the Rb gene transcript.[179] Some 20 years ago Knudson[180] had proposed that the development of retinoblastomas requires two mutational events. In hereditary cases in which the germ-line mutation was already present in all cells, it was postulated that a second random mutation occurring in a target retinoblast was required for development of tumor. Following this, studies with RFLPs showed that reduction to hemizygosity or homozygosity for all or part of an allele on chromosome 13 was a common event in the development of retinoblastoma.[172–174] All available data therefore suggested that loss of activity in both normal alleles at the Rb locus was one of the key events in tumor development. This hypothesis was confirmed in the report by Fung et al.,[179] who used a probe representing the putative Rb locus. In several instances one mutation was detected in fibroblast cell lines from the patient and a second mutation in DNA from the retinoblastoma tumor. This work appeared to confirm the hypothesis suggested by Knudson's analysis two decades earlier!

As an extension of these concepts, it was proposed that retinoblastoma tumor formation required complete inactivation of the Rb gene, and therefore the Rb gene functioned in an opposite manner to an oncogene and might be a *cancer suppressor* gene or anti-oncogene.[181] Careful analyses of the type of protein encoded by the Rb gene showed features suggestive of a DNA-binding protein.[182] Antisera prepared against a fusion protein derived by recombinant techniques immunoprecipitated a protein doublet of 110 to 114 kd. The protein could be detected in all cell lines expressing normal Rb in RNA but not in retinoblastoma cell lines. The protein appeared to contain phosphorus and was localized to the nucleus. Affinity for DNA cellulose indicated that the Rb gene product was a DNA-binding protein.

Using molecular cloning techniques and the Rb gene, Wiggs et al.[183] made an extended analysis of 20 families at risk. Correct predictions were possible in 19 of the 20 kindreds, and in 18 families a consistent association was made using RFLP markers showing the retinoblastoma mutation. A schematic diagram of the genomic sites for the retinoblastoma gene is shown in Figure 9.15. One of the most useful probes in this study was derived from the DNA sequence cloned in p68RS2.0, which contained a 50- to 53-base-pair sequence that is repeated approximately 30 times. This is illustrated in Figure 9.16. A portion of this repeated sequence is homologous to core sequences in the variable number of tandem repeats (VNTR) loci observed in many human genes.[184] Because such tandemly repeated sequences tend to be genetically unstable, the actual number in a given DNA segment is highly variable. In the case of the Rb gene, eight distinct alleles were detected. Because of the high number of common alleles, it was estimated that 80% of unrelated individuals would be heterozygous for this polymorphism, which would therefore be useful as a marker. Of the 20 families examined, 14 showed positive linkage and tumor associations with this polymorphism. A series of polymorphism analyses for all 20 families made it possible to calculate linkage with the retinoblastoma

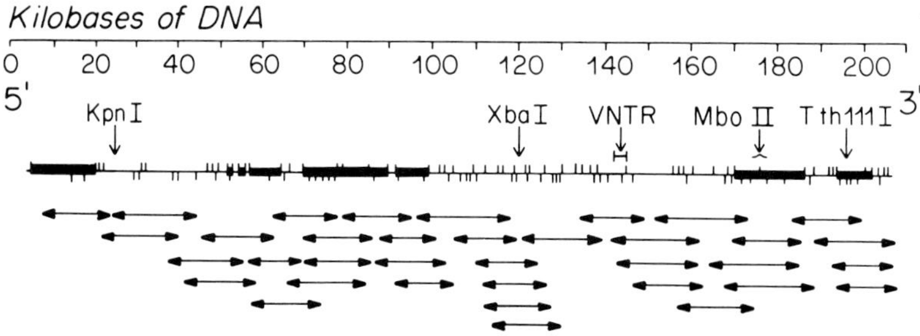

Figure 9.15 Restriction enzyme map of the genomic locus of the retinoblastoma gene. Vertical marks above the map represent the locations of *Hin*dIII sites; those below the map represent the locations of *Eco*RI sites. The boxed areas represent *Hin*dIII fragments that contain sequences found in the cDNA (exons). Each double-headed arrow beneath the map represents a distinct recombinant bacteriophage clone. The positions of the polymorphic DNA sites are indicated with arrows. The location of the polymorphic *Mbo*II site or sites has not been determined precisely but is shown at approximately 175 kb on this map. *Reproduced from Wiggs et al,*[183] *with permission.*

trait. Such calculations assumed that retinoblastoma would develop in 90% of persons with a mutant Rb gene allele and lod scores varied between 8.9 and 12.2 when all families were considered together. These findings clearly indicated that the DNA polymorphisms observed were very tightly linked to the retinoblastoma trait. Thus, in this instance RFLPs helped to identify which children of a carrier parent had inherited the mutant allele. The presence of such an allele carried a high risk (≈90%) of retinoblastoma as well as other cancers such as osteosarcoma. The studies were also useful in identifying asymptomatic carriers who might pass the mutant allele to their children and in reassuring unaffected family members that they did not carry the mutant retinoblastoma allele.

Dryja and co-workers[185] have examined the parental origin of mutations of the Rb gene. By using the cloned Rb gene along with other linked polymorphic loci, it was possible to follow sequential loss of the two homologous copies of the gene, which immediately precedes the development of the tumors. In the case of nonhereditary tumors the loss of each of the two homologous genes occurs somatically, whereas in the hereditary cases the initial mutation occurs in the germ line. A report by Toguchida et al.[186] had indicated that the copy of the gene derived from the father preferentially showed the first mutation during genesis of nonhereditary osteosarcomas. A similar analysis by Dryja and co-workers showed no apparent predilection for initial somatic mutations. By contrast, when an initial mutation occurred as a new germ-line mutational event, it was in all instances

Figure 9.16 Sequence of the 50- to 53-bp repeat at the VNTR locus detected by p68RS2.0. The brackets denote regions of variability within the repeat unit. The 11-bp sequence above the repeat unit represents the core sequence reported for some VNTRs. *Reproduced from Wiggs et al,*[183] *with permission.*

```
                                      GGGNNGTGGGG
          TG      A                   |||||||  ||                AG
CGTGTTGAA{ A}ACAC{C}TCCCCAATGCTGGAGGTGAGGTTTGGT{AGAAG}ATGACT
           G
```

examined derived from the father. This pattern appeared to be consistent with new germ-line mutations arising primarily during the process of spermatogenesis. In Dryja's study eight informative patients showed the new germ-line mutation on the paternally derived chromosome. The hypothesis that paternal mutations arose predominantly during spermatogenesis would be supported by a paternal age effect; new cases of bilateral retinoblastoma would occur more frequently in the children of older fathers. A retrospective review of available data concerning this point appeared to support such an effect. Results of a similar study confirming that the germ-line mutation in retinoblastoma occurred in the paternal allele were published shortly thereafter by Zhu et al.[187] When the data from studies by Dryja et al. and Zhu et al. were combined, 13 of 14 bilateral retinoblastoma tumors retained the paternal chromosome ($P < 0.01$).

Wilms' Tumor

Wilms' tumor is the most common abdominal neoplasm in children and occurs in both hereditary and nonhereditary forms. In the hereditary form, the tumor is often associated with other specific malformations such as aniridia, genitourinary abnormalities, and mental retardation. Previous work indicated that mutations on loci within the p13 band on chromosome 11 may predispose to this tumor. When DNA samples from patients with Wilms' tumors were analyzed using chromosome 11 cDNA probes, heterozygosity was often noted to be preserved in tumor DNAs. However, one case of homozygosity was recorded by Orkin et al.,[188] implying that mechanisms for a recessive locus similar to that described for retinoblastoma might be involved in Wilms' tumor as well. Other observations, including the finding of loss of the Ha-*ras* allele marker in an individual with Wilms' tumor by Reeve et al.,[189] also provided evidence of possible deletion or rearrangements along a specific portion of chromosome 11. This was confirmed by observations by Fearon et al.[190] indicating somatic deletions of genes at 11p loci in four of six Wilms' tumors. In all four cases, deletions were associated with duplications leading to homozygosity of the nondeleted alleles in the tumor cells. These findings were reminiscent of similar mechanisms reported for retinoblastoma. Thus the Wilms' tumor studies provided an additional example of how recessive mutations might contribute to oncogenesis.

Further definition of markers helpful in studies of Wilms' tumors was afforded by identification of deletion of the gene for the β subunit of follicle-stimulating hormone (FSH) in patients with aniridia and Wilms' tumor.[191] These results indicated a close physical linkage between the gene encoding the β subunit of FSH and the Wilms' locus. The approach was subsequently extended by several groups using material from individual patients with Wilms'-aniridia-genitourinary-mental retardation (WAGR) syndrome.[192–194] In one study chromosome-mediated gene transformants were used to select genomic fragments from WAGR material expressing the closely linked Ha-*Ras*-1 (HRAS1) gene locus.[193] Analysis of the recombinants suggested that the break point associated with genitourinary dysplasia may be causal and that Wilms' tumor and genitourinary dysplasia may be alternative manifestations of mutation at the same locus. However, no successful cloning or isolation of the single Wilms' tumor gene has been accomplished.

Of parallel interest was the report in 1987 that introduction of a normal human chromosome 11 into a Wilms' tumor cell line by a microcell transfer technique markedly suppressed the ability of the cells to produce tumors in nude mice.[195]

Moreover, transfer of other chromosomes such as X or 13 had no effect on the tumor-producing capacity of the Wilms' tumor cells. This provided support for the idea that normal DNA on chromosome 11 can control or suppress malignant expression on Wilms' tumor cells. Because Wilms' tumors with overexpression of N-*myc* oncogene have been reported,[196] efforts were made to determine whether N-*myc* expression had been altered in the Wilms' tumor cell line by introduction of the normal human chromosome 11 material. However, no N-*myc* expression was detected before or after the introduction of chromosome 11. Therefore these experiments support the concept of normal suppressor or anti-oncogene DNA in some human genomic elements. When parts of such regions are deleted, as in the case of Wilms' tumors or retinoblastomas, increased potential for tumor development may result.

A homozygous deletion of a zinc finger gene has now been identified as a possible genetic marker in Wilms' tumors.[197] This study employed a chromosome jumping technique. Cloning of the break points in one tumor revealed a deletion which resembled a zinc finger gene, which might serve some regulatory function in the tumors.

Acoustic Neurinomas and Other Central Nervous System Tumors

A situation similar to that described in retinoblastomas and Wilms' tumors has also been found in the case of acoustic neuromas. One of the most common tumors in the human nervous system, the acoustic neuroma appears to be specifically associated with loss of genes on human chromosome 22.[198] Acoustic neuromas arise from Schwann cells covering the vestibular branch of the vestibulocochlear nerve. Most of them are solitary noninherited unilateral tumors. In contrast, bilateral tumors are characteristic of familial acoustic neuromas or bilateral acoustic neurofibromatosis. Polymorphic DNA markers were employed to search for loss or deletions of chromosomal DNA in patients with acoustic neuromas. DNA derived from both tumors and peripheral blood leukocytes was studied. Previous studies had indicated an increased frequency of meningiomas in patients with bilateral acoustic neurofibromatosis and loss of genes on chromosome 22 in patients with meningioma.[199–201] For this reason attention was first directed at chromosome 22. Three markers were used: the *sis* platelet-derived growth factor, an anonymous DNA locus on chromosome 22, and the immunoglobulin λ constant region at 22q11. In 7 of 16 cases (44%), loss or marked reduction of one of the two alleles in the tumor tissue was detected using these chromosomal markers. Examples of these striking differences are shown in Figure 9.17. Loss of genes on chromosome 22 was not restricted to unilateral sporadic cases of acoustic neuroma but was also observed in a patient with bilateral acoustic neuroma. Two other patients with bilateral tumors showed no detectable chromosome 22 deletions with these three probes. Again, this discovery was similar to the findings with retinoblastomas and Wilms' tumors, where recessive mutant alleles were unmasked when deletions occurred on their respective chromosomes.

Subsequent studies of bilateral acoustic neuromas by the same group have proved illuminating. Of three neurofibromata from the same patient, two showed loss of identical alleles, indicating that the same chromosome suffered deletion in both tumors.[202] The third tumor from the patient showed no detectable loss of heterozygosity, suggesting a more subtle mutational event affecting chromosome 22. The localization of deletions on chromosome 22 was narrowed to regions distal

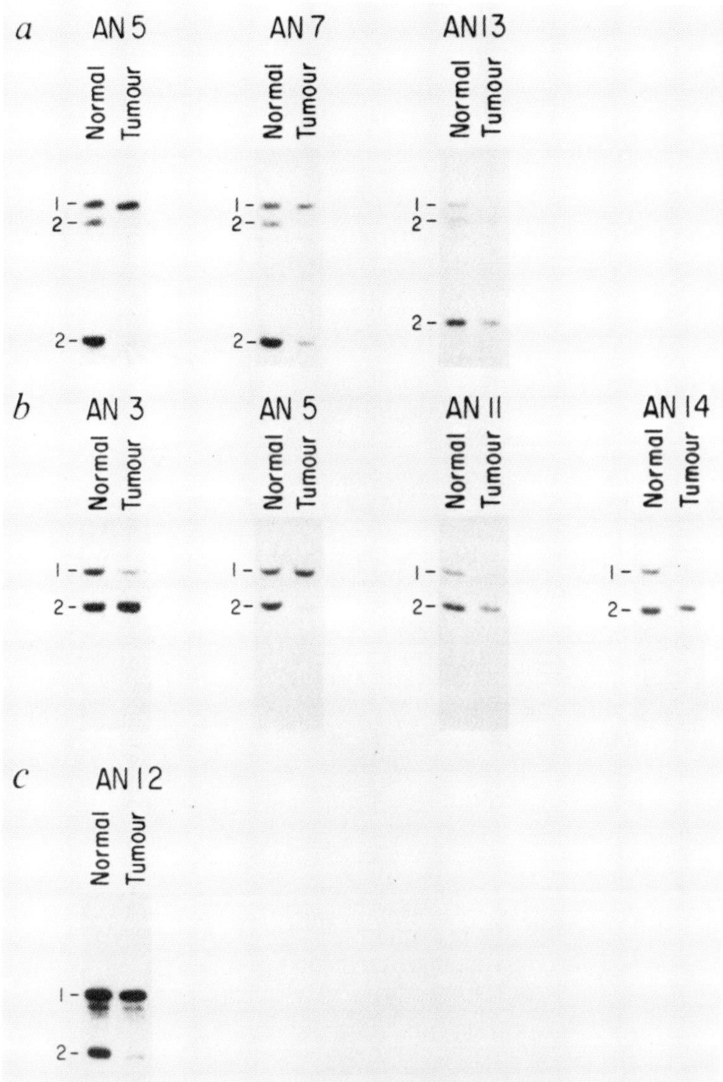

Figure 9.17 Loss of constitutional heterozygosity of loci on human chromosome 22 in acoustic neuromas. Designations of patients are shown above the autoradiograms. Numbers at the left indicate the observed alleles, with 1 and 2 referring to the larger and smaller allelic restriction fragments, respectively. Southern blots were hybridized to ^{32}P-labeled DNA probes for the following loci on chromosome 22: a, Psis.1 (Oncor) (*sis* oncogene); 6, pMS3-18 (D22S1); c, HugC2 (IGLC). Psis.1 reveals an RFLP with fragments of 21 kb (1 allele) and 14.5 and 6.5 kb (2 allele) in *Hin*dIII-digested human genomic DNA. The pMS3-18 probe detects an RFLP in *Bgl*II-digested human DNA with fragments of 9.5 kb (1 allele) and 6.5 kb (2 allele). The huλC2 probe reveals a multiallele RFLP represented by fragments of 8, 13, and 23 kb in *Eco*RI-digested human genomic DNA. To simplify the nomenclature, the allele of higher relative molecular mass is called the 1 allele and that of lower relative molecular mass is termed the 2 allele in each case. *Reproduced from Seizinger et al,*[198] *with permission.*

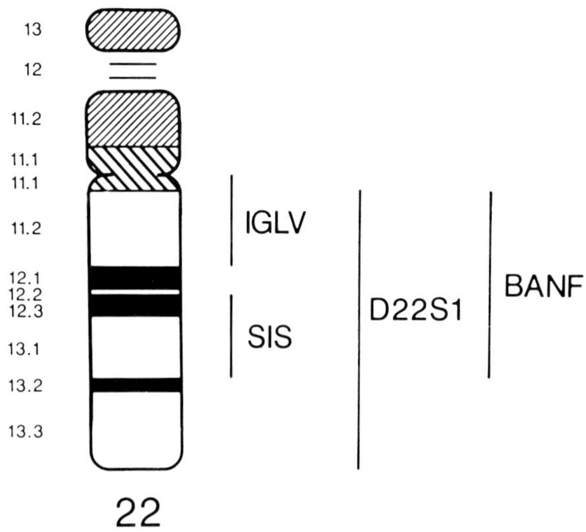

Figure 9.18 Location of DNA markers on chromosome 22. The physical locations of D22S1, IGLV (the immunoglobulin λ chain variable-region genes), and *sis*, the gene for platelet-derived growth factor, are shown. The approximate location of BANF was inferred from the previous genetic mapping of D22SI relative to the two other markers. *Reproduced from Rouleau et al,*[203] *with permission.*

to the D22S9 locus on band 22q11 by using selected probes. Loss of genetic material on chromosome 22 or a mutation in this location appears to be directly related to this interesting syndrome.

Studies of von Recklinghausen's disease or hereditary neurofibromatosis have localized the gene for this disorder to chromosome 17. Examination of families affected by bilateral acoustic neuromas in conjunction with the chromosome 22 markers have indicated that the disorder is clearly associated with the chromosomal abnormality.[203] The lod score for linkage in kindreds examined was +3.24, which exceeded the critical value (+3.0) accepted as proof of linkage. Thus, the gene for bilateral acoustic neuromas mapped to the center of the long arm of chromosome 22, as shown in Figure 9.18.

Cloning the NF1 Gene

Extensions of the original localization of the von Recklinghausen's gene (NF1) to chromosome 17 included studies of two patients with balanced translocations involving chromosome 17q11.2.[204] Identification of these break points suggested that the NF1 gene might be identified by cloning the region around them. Accordingly, a series of chromosome 17 Not-I-linking clones was mapped to proximal 17q and studied using pulsed field electrophoresis. One clone clearly identified the break point in a neurofibromatosis patient who showed a t(1;17) translocation. A pulsed field map of this region extending over 2.3 megabases was made, and analysis showed that the von Recklinghausen's break point was only 10 to 240 kb away from 17L1. Subsequently, the same group identified and cloned the neurofibromatosis gene by using chromosome jumping and yeast artificial chromosome technology.[205] This achievement represents a major advance in the application of recombinant technology to identification of the specific gene involved in a hereditable disorder. It was found that the disease gene encompassed a large ($\approx$13 kb) segment. Moreover, a new patient with NF1 was identified who showed a de novo 0.5-kb insertion in the NF1 gene.

The identification of the NF1 gene illustrates the value of a detailed DNA map of a candidate disease-associated region. At present, the clinical value of the cloning and identification of the NF1 gene is unclear. Because the gene locus has a high mutation rate, many kindreds contain only one NF1 individual. Prenatal identifications may be possible, but as in the case of cystic fibrosis or Duchenne muscular dystrophy, this will depend on whether most mutations are identified within a limited region of the large gene.

Meningiomas

Meningiomas have been studied with respect to their relationship to acoustic neuromas and also because cytogenetic studies have indicated loss of one copy of chromosome 22.[206] This study compared DNA markers in tumors with those in normal tissues in the same subjects. Of 51 patients, 40 were constitutionally heterozygous for at least one chromosome 22 DNA marker. Of the 40 heterozygous patients, 17 showed homozygosity for the corresponding marker in their meningioma tumor tissues. This pattern appeared very similar to that in the syndrome of bilateral acoustic neuromas and suggested a similar underlying mechanism. With this baseline information, it seems likely that refinements in chromosome 22 probes may be able to detect the underlying defect directly. Also, despite this similarity, it is not yet clear whether meningioma and acoustic neuroma arise from a mutation or deletion at the same locus on chromosome 22.

Familial Adenomatous Polyposis

Colorectal cancer is now the second most common neoplasm in the United States and other developed Western countries. Individuals with a condition termed familial adenomatous polyposis (FAP) have a markedly increased susceptibility to colon cancer. Patients with this familial disorder frequently require prophylactic colectomy to avoid subsequent development of widespread colonic adenocarcinoma. Based on a single case report of an interstitial deletion of chromosome 5 in a mentally retarded individual with multiple other developmental abnormalities and FAP, it has been demonstrated that the FAP gene is on chromosome 5, probably near bands 5q21–q22.[207] In parallel with this interesting report, it was found that chromosome 5 allelic loss occurred in human colorectal carcinomas.[208] Thus, sporadic colorectal cancers were examined for loss of alleles on chromosome 5 using a highly polymorphic minisatellite probe which was known to map to chromosome 5. Of interest in this study was the finding that at least 20% of these tumors had lost one of the alleles present in normal matched tissues. These findings suggested that the loss of an allele or an allele becoming recessive for this gene could be directly implicated in the genesis of colorectal cancer.

The final localization of the gene for familial polyposis was amplified by extensive linkage studies and analyses using a number of well-defined chromosome 5 markers.[209] A primary genetic map of chromosome 5 was constructed using 16 DNA markers, and linkage analysis was done with data from a number of affected families. Using a computer program in conjunction with the three chromosome 5 markers that appeared to be most tightly linked to the occurrence of polyposis, the region of the chromosome c11p11 was directly calculated. Localization of the polyposis gene to a specific chromosomal region should be a major step in its eventual identification and isolation. With such information in hand, it would be possible to clone segments from the regions that may identify nucleotide segments expressed as mRNA in normal mucosa, polyps, or carcinomas.

Lipid Disorders

Familial hypercholesterolemia (FH) is a relatively common disease with a heterozygote frequency of approximately 1 in 500 in the general population.[210] The clinical recognition of FH is founded on high levels of serum cholesterol and low-density lipoprotein (LDL) cholesterol, variable clinical findings such as xanthomas, and a positive family history.[211,212] The disease is inherited in an autosomal dominant fashion and is clearly associated with an increased risk of myocardial infarction after age 35. For these reasons, it is a causative factor in many individuals with coronary artery disease. Early definitive studies had shown that the principal defect in FH was a mutation in the gene that codes for the LDL receptor.[213,214] Patients showing a variety of defects in the receptor protein have been studied, and more information has been obtained concerning some of the alterations in gene expression that can cause the disease.[215,216]

It seems important, as advances in knowledge of genetic variations in LDL receptor structure are being explored, to attempt to identify individual patients with FH before they develop elevated lipid levels and premature onset of coronary artery disease. Techniques for determining the number of LDL receptors on cultured fibroblasts or lymphocytes have been highly refined,[217,218] but such techniques are difficult to establish and not yet suitable for identification of heterozygotes in large populations. A cDNA clone for the LDL receptor isolated several years ago by Yamamota et al.[219] has proved more suitable for kindred studies and early screening in order to establish a definitive early diagnosis of FH. Application of this probe has been reported by Humphries et al.,[220] using RFLP analysis.

The probe used in this study was a 1.9-kb fragment from the 3′ end of the LDL receptor cDNA clone.[219] Samples from normal unrelated individuals as well as kindreds with FH were analyzed. RFLPs were traced in two extended families in which FH was known to occur, and segregation of the V1 allele obtained with *Pvu*II digests with FH was observed. Schematic diagrams of the alleles and a simplified map of the LDL receptor gene are shown in Figure 9.19. The distribution of the alleles is shown in two FH families in Figure 9.20. In both families the gene causing the FH segregated with the V1 allele. Because the frequency of the V2 allele was about 20% and there appeared to be no strong linkage disequilibrium between either allele and the gene for FH, it was expected that in approximately one in five informative families the FH phenotype would segregate with the V2 allele.

It has been shown that several different mutations at the LDL receptor gene locus can cause HL,[216] and more extended surveys of these RFLPs are needed. This is particularly true since one of the mutations represents a deletion of the membrane-spanning and cytoplasmic domains of the LDL receptor protein.[221] These studies of polymorphism for LDL receptor expressions were an initial step in defining familial patterns, which will be supplemented as additional RFLPs and patterns are collected.

As work has proceeded on the LDL receptor gene, heterogeneous molecular

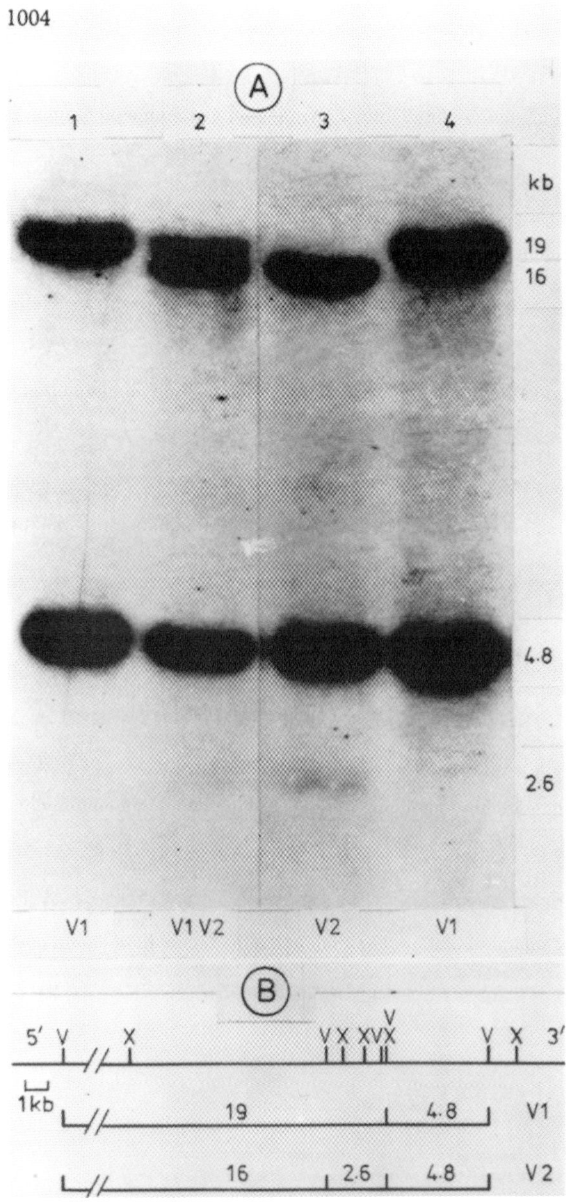

Figure 9.19 DNA polymorphism detected with LDL receptor probe pLDLR-2HH1. (A) Southern blot analysis of hybridization pattern obtained from a *Pvu*II digest of 5 μg of DNA from four individuals of types *V*1, *V*1*V*2, and *V*2. (B) Diagram of how DNA fragments detected with the probe can be produced by presence of an extra *Pvu*II site in the DNA sequence of gene. Map was derived from double digests of DNA from an individual heterozygous for the polymorphism. Sizes of relevant fragments: *Xba*I (X), 10 kb + 7 kb; *Pvu*II (V), 19 kb + 16 kb + 4.8 kb; *Xba*I plus *Pvu*II, 10 kb + 9.0 kb + 4.8 kb. *Reproduced from Humphries et al,*[220] *with permission.*

defects have been described, including insertion, deletion, and missense mutations.[222] The types of mutations described are summarized in Figure 9.21. Because the different types of mutations affect different domains of the LDL receptor protein, they have different effects on binding or receptor-mediated endocytosis of LDL. Four deletion mutations have now been studied by cloning and sequencing; three of the four involve homologous recombinations between two *Alu* sequences. The fourth is a recombination between one *Alu* sequence and a sequence in an exon that by chance is homologous to the *Alu* sequence. Recombination between *Alu* sequences in the same orientation caused other deletions—probably by homologous recombinations in which unequal crossing over followed mispairing at chro-

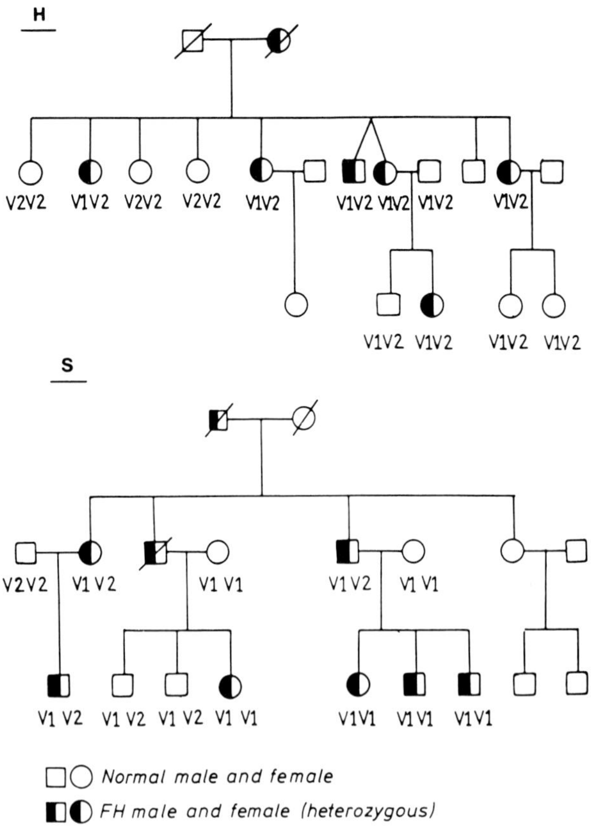

Figure 9.20 Segregation of FH and the LDL receptor gene polymorphism in two families. For clarity, deceased individuals who from their clinical histories were inferred to have had FH are shown as FH individuals. *Reproduced from Humphries et al,*[220] *with permission.*

matids. In all, seven *Alu* sequences have been shown to be involved in LDL receptor gene mutations; this is similar to *Alu* sequence involvement in mutations in globin genes.[223] Eight of the 10 deletion points mapped within a 36-bp sequence in the left arm of the *Alu* repeat. This region constitutes the bipartite RNA polymerase III promoter in the *Alu* element, and it may be more susceptible to recombination events when interacting with transcription factors.

The types of mutations in the LDL receptor gene have been grouped with respect to their impact on LDL function, as shown in Figure 9.22. Briefly the class 1 mutations are represented by null alleles, some of which contain detectable mRNA, while others show no detectable mRNA. Class 2 mutations produce defects in the LDL transport system with abnormal accumulation in the endoplasmic reticulum. Several of these mutations involve in-frame deletions of binding domains. Class 3 mutations produce defects in LDL binding. In several of these, relatively large deletions have been characterized that drastically affect LDL binding. Class 4 mutations produce defects in the internalization of LDL receptors. All of these mutations studied to date show alterations that affect the cytoplasmic domains of the LDL receptor.[222]

The extensive work on the LDL receptor mutations has provided some interesting information of general application. The four large deletions initially studied

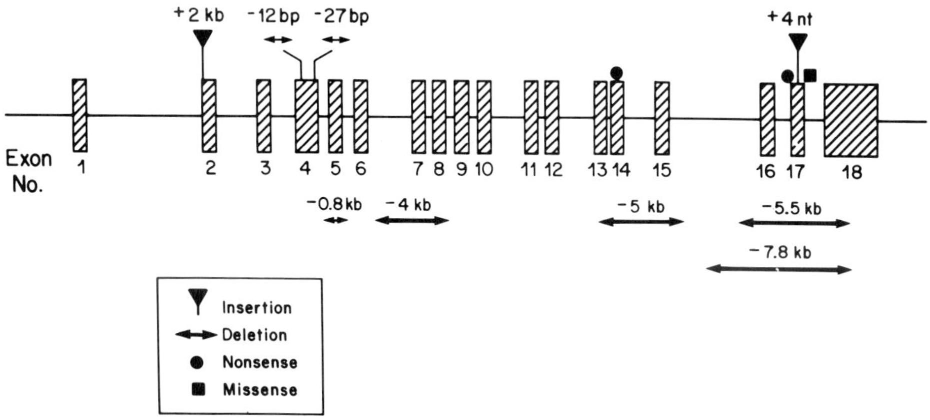

Figure 9.21 Mutations in the LDL receptor gene. Exons are shown as hatched boxes and introns as the lines connecting them. The map is not drawn to scale. Mutations that have been mapped or cloned are indicated above and below the gene. The mutations are a 2-kb insertion in intron 1 (FH 295), 12-bp deletion in exon 4 (WHHL rabbit), 27-bp deletion in exon 4 (FH 563), nonsense mutation in exon 14 (FH 264, 550, 786, 793), nonsense mutation in exon 17 (FH 683), 4-bp insertion in exon 17 (FH 763), missense mutation in exon 17 (FH 380), deletion of 0.8 kb (FH 626), deletion of 4 kb (FH 359 and FH 454), deletion of 5 kb (FH 381), deletion of 7.8 kb (FH 781), and deletion of 5.5 kb (FH 274). *Reproduced from Russell et al,*[222] *with permission.*

revealed an unexpected frequent relationship with *Alu* repeats, possibly indicating that repetitive DNA sequences can destabilize a gene through homologous recombination. The LDL receptor gene is a mosaic of exons shared with at least several other proteins (factor IX, factor X, protein C, and epidermal growth factor), and it is possible that early shuffling of exons involved recombinations between these repetitive elements.

Additional observations of LDL receptor gene mutations reemphasize the probable role of *Alu* repeats in many mutations. A defective LDL receptor gene associated with FH was shown to result from duplication of seven exons in the gene caused by *Alu-Alu* recombination.[224] A deletion also resulted from a recombination between two repetitive sequences of the *Alu* family, with deletion of splice acceptor sites.[225]

Correlation of deletion mutations in the LDL receptor with heterozygous LDL receptor status has been described in 63% of a population of French Canadians.[226] Since at least 12 different mutations of the LDL receptor gene have been recognized, the plethora of these potential defects makes it impractical to screen DNA of hypercholesterolemia patients for mutations. To date, all mutations in the LDL receptor gene are ones in which mRNA produced by the resulting receptor protein is somehow defective, with the exception of the null alleles mentioned above, where no detectable receptor protein is produced.

In the study of the French Canadian population by Hobbs et al.[226] a mutation was found in the LDL receptor gene that also produced a null allele and removed the 5′ end of the gene necessary for mRNA production. The mutation was originally detected in four homozygotes who inherited the mutant LDL receptor gene from both parents. The patients therefore had extremely high LDL levels and even showed evidence of myocardial infarctions in childhood. Later the deletion was

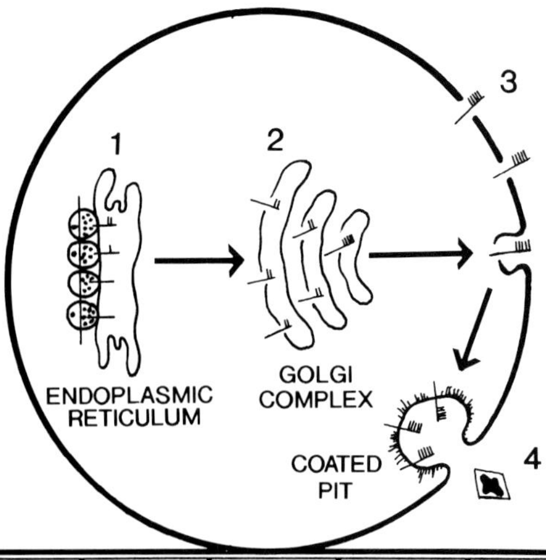

Class of Mutation	Synthesis	Transport from ER to Golgi	Binding of LDL	Clustering in Coated Pits
1	✕			
2	→	✕		
3	→	→	✕	
4	→	→	→	✕

Figure 9.22 Four classes of mutations at the LDL receptor locus. These mutations disrupt synthesis, transport, binding, and clustering of the LDL receptor.

also detected in almost two-thirds of heterozygotes of French Canadian descent. Representative Southern blot analyses from this study are shown in Figure 9.23.

Similar observations have been reported from South Africa, where FH at a prevalence of more than 1 in 100 is at least five times more common than in North American or European populations.[227] The RFLP analyses in this study were thought to be compatible with a "founder gene effect" based on the original small genetic pool of the Afrikaner settlers.

The biological potency of the LDL receptor was further explored by injecting the cDNA encoding the human receptor (under control of the mouse methallothionein-I promoter) into fertilized mouse eggs using a transgenic approach.[228] These experiments produced a strain of mice expressing high levels of LDL receptors. A total of 674 eggs were injected with the transgene and transplanted into pseudopregnant females; in 80 offspring 25% showed successful integration of the transgene. Partial hepatectomy was conducted on transgenic mice at 5 to 6 weeks of age and production of mRNA for LDL receptor was assayed in liver tissues. One mouse showed extremely high levels of mRNA for LDL receptor protein, and this animal was used as a founder to establish a line of mice expressing the human LDL receptor. Quantitative analysis of DNA from tail blot samples showed that the founder animal contained 11 copies of the transgenic DNA. Pretreatment with

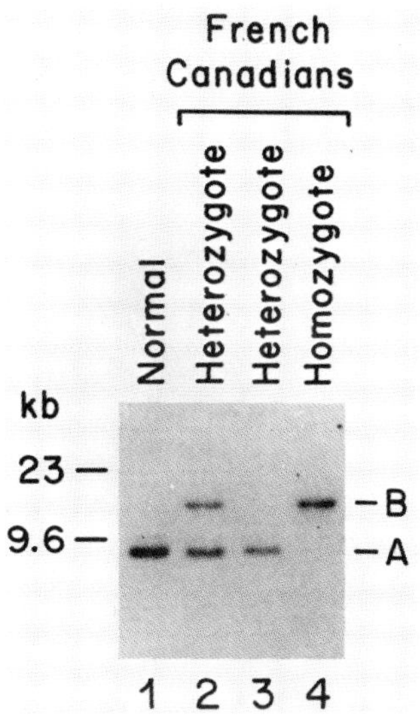

Figure 9.23 Southern blot hybridization of genomic DNA from four French Canadians with homozygous familial hypercholesterolemia. Total genomic DNA from each person was subjected to Southern blot hybridization. The DNA in lane 1 is from a normal person and shows the normal band (A). The DNAs in lanes 2 and 3 are from two unrelated heterozygotes of French Canadian descent. The DNA in lane 4 is from a patient who is homozygous for the French Canadian deletion and thus has two copies of the abnormal band (B). A restriction map of the relevant region of the LDL receptor gene is shown below with the *Kpn*I and *Xba*I sites noted. *Reproduced from Hobbs et al,*[226] *with permission.*

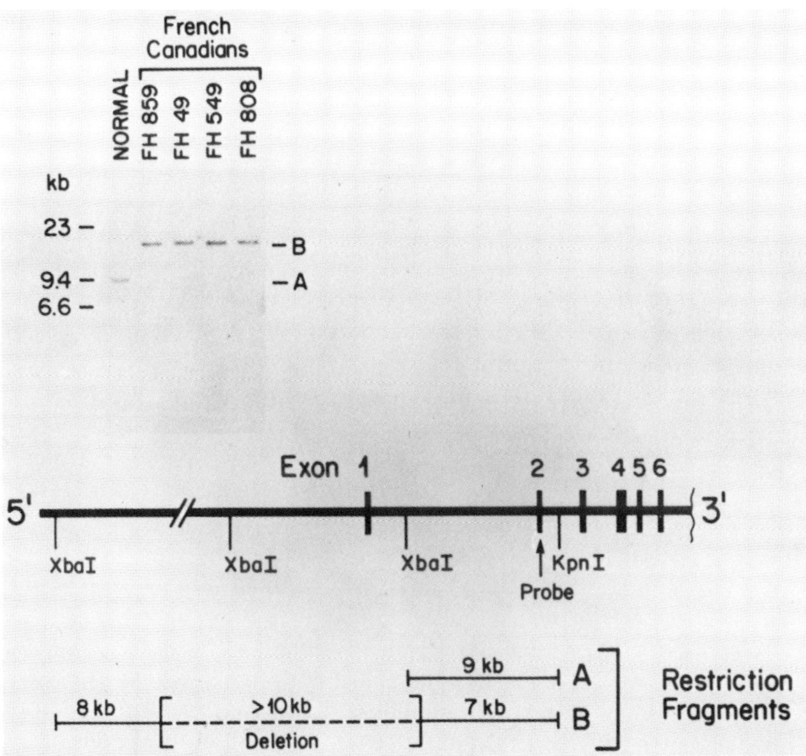

cadmium was used to stimulate transcription from the metallothionein promoter.[229] In the livers of the transgenic mice, the calculated amounts of LDL receptor mRNA increased from 630 molecules per cell to 3300. Moreover, functional studies showed that, when pretreated with $CdSO_4$ and injected intravenously with ^{125}I-labeled LDL, transgenic mice cleared more than 50% of labeled LDL within 10 minutes, indicating remarkable physiological activity of the transgenic protein

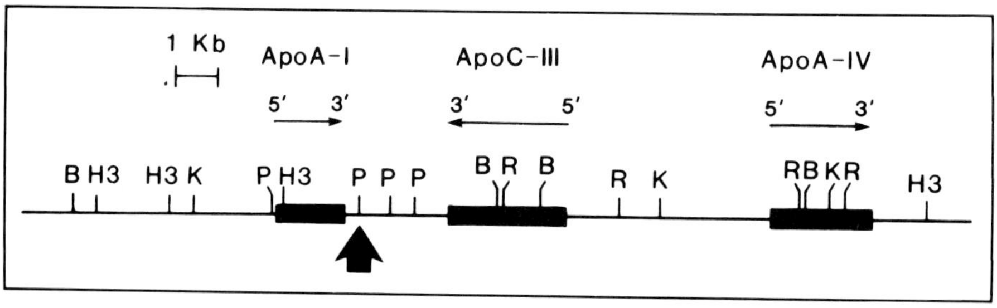

Figure 9.24 Diagram of the apolipoprotein A-I (*ApoA*-I), C-III (*ApoC*-III), and A-IV (*ApoA*-IV) gene complex on the long arm of chromosome 11, showing the locations of restriction enzyme cleavage sites for *Bam*HI (B), *Hin*dIII (H), *Kpn*I (K), *Pst*I (P), and *Eco*RI (R). The arrow indicates the site of the polymorphism. *Reproduced from Ordovas et al,*[231] *with permission.*

product. These data indicate that overexpression of an endocytosis-mediating receptor in transgenic mice can lower the content of ligand for that receptor. Additional studies will be illuminating when this approach is applied to other common receptor systems.

Other Lipid Disorders

Several other lines of study have been directed at examination of genes affecting lipids and therefore prevalence of heart disease. A survey of DNA polymorphisms that appeared to be adjacent to the apolipoprotein A-II gene in humans was reported in 1985 by Scott and co-workers.[230] This study found that the human apolipoprotein A-II gene was closely linked to a DNA polymorphism that appeared to affect high-density lipoprotein (HDL) composition. Because higher levels of HDL were thought to correlate with relative protection from coronary artery disease, the apolipoprotein A-II polymorphisms may prove to be clinically useful. Similar approaches were subsequently recorded by Ordovas et al.[231] Since decreased plasma HDL cholesterol and apolipoprotein A-I levels have in the past been associated with premature coronary artery disease, DNA close to the apolipoprotein A-I gene at its 3′ end was studied for possible linked polymorphisms using *Pst*I digests, and an apolipoprotein A-I gene probe producing 3.3-kb and 2.2-kb hybridization bands. Whereas the 3.3-kb band was identified in 4.1% of 123 random control subjects and in 3.3% of 30 patients with no angiographic evidence of coronary artery disease, the 3.3-kb band was identified in 32% of 88 subjects with severe coronary artery disease occurring before the age of 60. This 3.3-kb band was also found in 8 of 12 index cases of kindreds with familial hypoalphalipoproteinemia. When kindreds of patients with familial hypoalphalipoproteinemia were studied along with relatives of patients with coronary artery disease, the 3.3-kb band was associated with decreased HDL levels. These findings again indicated that polymorphism in the region between the apolipoprotein A-I and apolipoprotein C-III genes might be a useful marker for risk of premature coronary artery disease. The apolipoprotein A-I gene has been cloned and found to be directly adjacent to the genes for apolipoproteins C-III and A-IV. Apparently these genes form a multigene family or complex situator on the long arm of chromosome 11.[232–234] This clone arrangement is illustrated in Figure 9.24. The polymorphisms detected in this study were due to absence of a *Pst*I restriction site at the intergenic region of apolipoproteins A-I and C-III. The absence of this site apparently was not due to any major DNA insertion, deletion, or rearrangement. Nevertheless, there appeared to be a clear-cut association with the 3.3-kb *Pst*I band and coronary disease as well as with an HDL cholesterol value below the 10th percentile of normal in 58% of the patients with coronary artery disease studied.

Several other common genetic disorders have been definitely associated with premature coronary artery disease. These include familial combined hyperlipidemia, familial hypertriglyceridemia, and familial hypercholesterolemia. The data in the HDL apolipoprotein linkage study suggested that HDL deficiency might be the commonest genetic lipoprotein abnormality in patients with coronary artery disease.[235] How the polymorphisms related to the absence of the *Pst*I restriction site are related to differences in HDL levels is not yet clear, but the findings in this study provide an interesting basis for further work in this area.

Another enzyme that has long been the subject of study in relation to obesity and arteriosclerosis is lipoprotein lipase. This enzyme plays a major role in the regulation of lipid metabolism, as its principal function is to hydrolyze the core triglycerides of circulating chylomicrons and very low density lipoprotein (VLDL). Viewed broadly, lipoprotein lipase can influence the maturation and development of several major classes of lipids, including lipoprotein particles, and the transfer of cholesterol esters into endothelial cells. To a certain extent the ultimate disposition of plasma cholesterol is dependent on proper activity of lipoprotein lipase. The enzymatic activity itself is synthesized by parenchymal cells of many tissues but acts mainly at the luminal source of vascular endothelium, where it hydrolyzes dietary or endogenous triglycerides to monoglycerides and free fatty acids.

Lipoprotein lipase has now been cloned from a mouse macrophage cDNA library.[236] This is a major step forward, since previously the human enzyme had been difficult to purify and no protein sequence was available. Comparative analysis of the human lipoprotein lipase indicated a protein of 448 amino acids; moreover, human lipoprotein lipase, hepatic lipase, and pancreatic lipase appeared to be members of a distinct gene family. The cloning of this key enzyme is an important step in eventual understanding of lipid disorders.

Similarly, the human cholesterol ester transfer protein has also now been cloned.[237] Cholesterol synthesized in tissues in excess of that needed for production of cell membranes or steroid hormones must be returned to the liver for catabolism. Reactions that are involved in this process have been termed reverse cholesterol transport. A major link in this pathway is cholesterol ester transfer protein. The cloning of this 74-kd protein was accomplished using a partial amino acid sequence and DNA derived from human liver that was hybridized and completely sequenced. This cholesterol transfer protein may play an essential part in the exchange of cholesterol esters between cells and plasma, as in the unloading of macrophages containing free and esterified cholesterol into the so-called foam cells of atherosclerotic plaques.

Will the New Genetics Pay Off?

As already illustrated in this chapter, the number of genetic disorders currently detectable antenatally by using DNA probes has risen quite rapidly. A review of present-day applications of recombinant techniques in a number of common disorders has summarized the present approach.[238] The popular demand for diagnosis and, if indicated, termination of affected pregnancies is likely to increase as these techniques are refined and improved. An attempt at cost-benefit analysis has begun, and clearly in some diseases in which the probes are extremely closely linked to or able to identify directly the disease-related gene, considerable human benefit or saving in long-term health care resources may be possible. In less well-defined situations in which disease probes are close but not infallible, more refinement and improvement will be needed. Some initial projections have already been made,[239] and it is hoped that the financial aspects of savings in health care costs will be thoroughly and judiciously modulated by a humane approach and recognition of the real limitations in prognostication in the case of some disorders.

References

1. Brewerton DA, Hart FD, Nicholls A, et al Ankylosing spondylitis and HL-A27. *Lancet* 1973;1:904–907.
2. Schlosstein L, Terasaki PI, Bluestone R, et al: High association of an HL-A antigen, W27, with ankylosing spondylitis. *N Engl J Med* 1973;288:704–706.
3. Arnett FC: Seronegative spondylarthropathies. *Bull Rheum Dis* 1987;37:1–12.
4. Stasny P: Association of the B-cell alloantigen DRw4 with rheumatoid arthritis. *N Engl J Med* 1978;298:869–871.
5. Lewin R: Proposal to sequence the human genome stirs debate. *Science* 1986; 232:1598–1600.
6. Lewin R: Molecular biology of *Homo sapiens*. *Science* 1986;233:157–158.
7. Lewin R: National Academy looks at human genome project, sees progress. *Science* 1987;235:747–748.
8. Roberts L: Who owns the human genome? *Science* 1987;237:358–361.
9. Roberts L: Agencies vie over human genome project. *Science* 1987;237:486–488.
10. Jeffreys AJ, Wilson V, Thein SL: Hypervariable "minisatellite" regions in human DNA. *Nature* 1985;314:67–73.
11. Jeffreys AJ, Wilson V, Thein SL: Individual-specific 'fingerprints' of human DNA. *Nature* 1985;316:76–79.
12. Lewin R: DNA fingerprints in health and disease. *Science* 1986;233:521–522.
13. Housman D, Gusella JF: Molecular genetic approaches to neural degenerative disorders, in Schmitt FO, Bird S, Bloom FE (eds): *Molecular Genetic Neuroscience: A New Hybrid*. New York, Raven Press, 1982, pp 415–424.
14. Botstein D, White RL, Skolnick M, Davis R: Construction of a genetic linkage map in man using restriction fragment length polymorphisms. *Am J Hum Genet* 1980;32:314–331.
15. Gusella JF: Recombinant DNA techniques in the diagnosis of inherited disorders. *J Clin Invest* 1986;77:1723–1726.
16. Murray JC, Mills KA, Demopolous CM, et al: Linkage disequilibrium and evolutionary relationships of DNA variants (restriction enzyme fragment length polymorphisms) at the serum albumin locus. *Proc Natl Acad Sci USA* 1984;81:3486–3490.
17. Cooper DN, Schmidtke J: DNA restriction fragment length polymorphisms and heterozygosity in the human genome. *Hum Genet* 1984;66:1–16.
18. Cooper DN, Smith BA, Cooke HJ, et al: An estimate of unique DNA sequence heterozygosity in the human genome. *Hum Genet* 1985;69:201–205.
19. Watkins PC, Tanz RE, Gibbons KT, et al: Isolation of polymorphic DNA segments from human chromosome 21. *Nucleic Acids Res* 1985;13:6075–6088.
20. Aldridge J, Kunkel L, Bruns G, et al: A strategy to reveal high-frequency RFLPs along the human X chromosome. *Am J Hum Genet* 1984;36:546–564.
21. Wyman A, White R: A highly polymorphic locus in human DNA. *Proc Natl Acad Sci USA* 1980;77:6754–6758.
22. Willard HF, Skolnick MH, Pearson PL, Mandel J-L: Human gene mapping by recombinant DNA techniques. *Cytogenet Cell Genet* 1985;40:360–498.
23. White R, Leppert M, Bishop DT, et al: Construction of linkage maps with DNA markers for human chromosomes. *Nature* 1985;313:101–105.
24. de la Chapelle A (ed): Human gene mapping 8: Eighth International workshop in human gene mapping. *Cytogenet Cell Genet* 1985;40:1–823.

25. Lathrop GM, Lalouel JM, Julier C, Ott J: Multilocus linkage analysis in humans: Detection of linkage and estimation of recombination. *Am J Hum Genet* 1985;37:482–499.
26. Murray JM, Davies KE, Harper PS, et al: Linkage relationship of a cloned DNA sequence on the short arm of the X chromosome to Duchenne muscular dystrophy. *Nature* 1982;300:69–71.
27. Bakker E, Goor N, Wrogemann K, et al: Prenatal diagnosis and carrier detection of Duchenne muscular dystrophy with closely linked RFLPs. *Lancet* 1985;1:655–658.
28. Monaco AP, Bertelson CJ, Middlesworth W, et al: Detection of deletions spanning the Duchenne muscular dystrophy locus using a tightly linked DNA segment. *Nature* 1985; 316:842–845.
29. Ray P, Belfall B, Duff C, et al: Cloning of the breakpoint of an X;21 translocation associated with Duchenne muscular dystrophy. *Nature* 1985;318:672–675.
30. Goodfellow P, Davies K, Ropers HH: Report of the committee on the genetic constitution of the X and Y chromosomes. *Cytogenet Cell Genet* 1985;40:296–352.
31. Gusella JF, Wexler NS, Conneally PM, et al: A polymorphic DNA marker genetically linked to Huntington's disease. *Nature* 1983;306:234–238.
32. Gusella JF, Tanzi RE, Bader PI, et al: Deletion of Huntington's disease–linked G8 (D4510) locus in Wolf-Hirschhorn syndrome. *Nature* 1985;318:75–78.
33. Kunkel LM, Monaco AP, Middlesworth W, et al: Specific cloning of DNA fragments absent from the DNA of a male patient with an X chromosome deletion. *Proc Natl Acad Sci USA* 1985;82:4778–4782.
34. Steinmetz M, Stephan D, Lindahl KF: Gene organization and recombinational hotspots in the murine major histocompatability complex. *Cell* 1986;44:895–904.
35. Steinmetz M, Winoto A, Minard K, Hood L: Clusters of genes encoding mouse transplantation antigens. *Cell* 1982;28:489–498.
36. Little D: Finding the defective gene. *Nature* 1986;321:558–559.
37. Rotter JI, Diamond JM: What maintains the frequencies of human genetic diseases? *Nature* 1987;329:289–290.
38. Vogel V, Motulsky AG: *Human Genetics*. Berlin, Springer-Verlag, 1986.
39. Crosby WW: Hemochromatosis: Current concepts and management. *Hosp Pract* 1987; 22(2):173–192.
40. Petersen GM, Rotter JI: Genetic and evolutionary implications in peptic ulcer disease. *Am J Phys Anthrop* 1983;62:71–79.
41. Harris H, Hopkinson DA: Average heterozygosity per locus in man: An estimate based on the incidence of enzyme polymorphisms. *Ann Hum Genet* 1972;36:9–20.
42. Vogel F, Motulsky AG: *Human Genetics*, 2nd ed. Berlin, Springer-Verlag, 1982.
43. Lander ES, Botstein D: Homozygosity mapping: A way to map human recessive traits with the DNA of inbred children. *Science* 1987;236:1567–1570.
44. van Heyningen V, Hayward C, Fletcher J, McAuley C: Tissue localization and chromosomal assignment of a serum protein that tracks the cystic fibrosis gene. *Nature* 1985;315:513–515.
45. Wilson GB, Fudenberg HH, John TL: Studies on cystic fibrosis using isoelectric focussing. I. An assay for detection of cystic fibrosis homozygotes and heterozygote carriers from serum. *Pediatr Res* 1975;9:635–640.
46. White R, Woodward S, Leppert M, et al: A closely linked genetic marker for cystic fibrosis. *Nature* 1985;318:382–384.
47. Wainwright BJ, Scambler PJ, Schmidtke J, et al: Localization of cystic fibrosis locus to human chromosome 7cen-q22. *Nature* 1985;318:384–385.
48. Knowlton RG, Cohen-Haguenauer O, Van Cong N, et al: A polymorphic DNA marker linked to cystic fibrosis is located on chromosome 7. *Nature* 1985;318:380–382.
49. Collins FS, Drumm ML, Cole JL, et al: Construction of a general human chromosome jumping library, with application to cystic fibrosis. *Science* 1987;235:1046–1049.
50. Estivill X, Farrall M, Scambler PJ, et al: A candidate for the cystic fibrosis locus isolated by selection for methylation-free islands. *Nature* 1987;326:840–845.
51. Beaudet A, Bowcock A, Buchwald M, et al: Linkage of cystic fibrosis to two tightly linked DNA markers: Joint report from a collaborative study. *Am J Hum Genet* 1986; 39:681–693.
52. Estivill X, Williamson R: A rapid method to identify cosmids containing rare restriction sites. *Nucleic Acids Res* 1987;15:1415–1425.

53. Rommens JM, Iannuzzi MC, Kerem B-S, et al: Identification of the cystic fibrosis gene: Chromosome walking and jumping. *Science* 1989;245:1059–1065.
54. Riordan JR, Rommens JM, Kerem B-S, et al: Identification of the cystic fibrosis gene: Cloning and characterization of complementary DNA. *Science* 1989;245:1066–1073.
55. Kerem B-S, Rommens JM, Buchanan JA, et al: Identification of the cystic fibrosis gene: Genetic analysis. *Science* 1989;245:1073–1080.
56. Marx JL: The cystic fibrosis gene is found. *Science* 1989;245:923–925.
57. Super M, Schwarz M, Elles RG, et al: Clinic experience of prenatal diagnosis of cystic fibrosis by use of linked DNA probes. *Lancet* 1987;2:782–784.
58. Lemna WK, Feldman GL, Kerem B-S, et al: Mutation analysis for heterozygote detection and the prenatal diagnosis of cystic fibrosis. *N Engl J Med* 1990;322:291–296.
59. Roberts L: CF screening delayed for awhile, perhaps forever. *Science* 1990;247:1296–1297.
60. Workshop on population screening for the cystic fibrosis gene. *N Engl J Med* 1990; 323:70–71.
61. Thomson WHS: The biochemical identification of carrier state in X-linked (Duchenne) muscular dystrophhy. *Clin Chim Acta* 1969;26:207–221.
62. Bakker E, Wieacker P, Beverstock GC: Recombinant DNA techniques for mapping the human X chromosome. *Clin Genet* 1983;23:225. (abstract).
63. Davies KE, Pearson PL, Harper PS, et al: Linkage analysis of two cloned DNA sequences flanking the Duchenne muscular dystrophy locus on the short arm of human X chromosome. *Nucleic Acids Res* 1983;11:2303–2312.
64. Kunkel LM, Hejtmancik JF, Caskey CT, et al: Analysis of deletions in DNA of patients with Becker and Duchenne muscular dystrophy. *Nature* 1986;322:73–77.
65. Monaco AP, Neve RL, Colletti-Feener C, et al: Isolation of candidate cDNAs for portions of the Duchenne muscular dystrophy gene. *Nature* 1986;323:646–650.
66. Boyd Y, Buckle V, Holt S, et al: Muscular dystrophy in girls with X; autosome. *J Med Genet* 1986;23:484–490.
67. Francke U, Ochs H, de Martinville B, et al: Minor Xp21 chromosome deletion in a male associated with expression of Duchenne muscular dystrophy, chronic granulomatous disease, retinitis pigmentosa, and MacLeod syndrome. *Am J Hum Genet* 1985; 37:250–267.
68. Kunkel LM, Hejtmancik JF, Caskey CT, et al: Analysis of deletions in DNA from patients with Becker and Duchenne muscular dystrophy. *Nature* 1986;322:73–77.
69. Thomas NST, Ray PN, Worton RG, Harper PS: Molecular deletion analysis in Duchenne muscular dystrophy. *J Med Genet* 1986;23:509–515.
70. Burghes AHM, Logan C, Hu X, et al: A cDNA clone from Duchenne/Becker muscular dystrophy gene. *Nature* 1987;328:434–437.
71. Heilig R, Lemaire C, Mandel J-L, et al: Localization of the region homologous to the Duchenne muscular dystrophy locus on the mouse X chromosome. *Nature* 1987; 328:168–170.
72. Brockdorff N, Cross GS, Cavanna JS, et al: The mapping of a cDNA from the human X-linked Duchenne muscular dystrophy gene to the mouse X chromosome. *Nature* 1987;328:166–168.
73. Darras BT, Francke U: A partial deletion of the muscular dystrophy gene transmitted twice by an unaffected male. *Nature* 1987;329:556–558.
74. Forrest SM, Cross GS, Speer A, et al: Preferential deletion of exons in Duchenne and Becker muscular dystrophies. *Nature* 1987;329:638–640.
75. den Dunnen JT, Bakker E, Breteler EGK, et al: Direct detection of more than 50% of the Duchenne muscular dystrophy mutations by field inversion gels. *Nature* 1987; 329:640–642.
76. Robertson M: Muscular dystrophy. Mapping the disease phenotype. *Nature* 1987; 327:372–373.
77. Darras BT, Harper JF, Franke U: Prenatal diagnosis and detection of carriers with DNA probes in Duchenne's muscular dystrophy. *N Engl J Med* 1987;316:985–992.
78. Hoffman EP, Monaco AP, Feener CC, Kunkel LM: Conservation of the Duchenne muscular dystrophy gene in mice and humans. *Science* 1987;238:347–350.
79. Wood DS, Zeviani M, Prelle A, et al: Is nebulin the defective gene product in Duchenne muscular dystrophy? *N Engl J Med* 1987;316:107–108.

80. Koenig M, Monaco AR, Kunkel LM: The complete sequence of dystrophin predicts a rod-shaped cytoskeletal protein. *Cell* 1988;53:219–228.
81. Nudel U, Zuk D, Einat P, et al: Duchenne muscualr dystrophy gene product is not identical in muscle and brain. *Nature* 1989;337:76–78.
82. Hoffman EP, Brown RH, Kunkel LM: Dystrophin: The protein product of the Duchenne muscular dystrophy locus. *Cell* 1987;51:919–928.
83. Hoffman EP, Hudecki MS, Rosenberg PA, et al: Cell and fiber-type distribution of dystrophin. *Neuron* 1988;1:411–420.
84. Partridge TA, Morgan JE, Coulton GR, et al: Conversion of *mdx* myofibres from dystrophin-negative to -positive by injection of normal myoblasts. *Nature* 1989;337:176–179.
85. Arahata K, Ishihara T, Kamakura K, et al: Mosaic expression of dystrophin in symptomatic carriers of Duchenne's muscular dystrophy. *N Engl J Med* 1989;320:138–142.
86. Feener CA, Koenig M, Kunkel LM: Alternative splicing of human dystrophin mRNA generates isoforms at the carboxy terminus. *Nature* 1989;338:509–511.
87. Mandel JL: Dystrophin, the gene and its product. *Nature* 1989;339:584–586.
88. Malhotra SB, Hart KA, Klamut HJ, et al: Frame-shift deletions in patients with Duchenne and Becker muscular dystrophy. *Science* 1988;242:755–759.
89. Ervasti JM, Ohlendieck K, Kahl SD, et al: Deficiency of a glycoprotein component of the dystrophin complex in dystrophic muscle. *Nature* 1990;345:315–319.
90. Martin JB: Huntington's disease: New approaches to an old problem. *Neurology* 1984; 34:1059–1072.
91. Hayden MR: *Huntington's Chorea.* New York, Springer-Verlag, 1981.
92. Martin JB, Gusella JF: Huntington's disease. Pathogenesis and management. *N Engl J Med* 1986;315:1267–1276.
93. Wexler NS: in Kessler S (ed): *Genetic Counseling: Psychological Dimensions* New York, Academic Press, 1979, pp 190–220.
94. Cantor CR: Huntington's disease: Charting the path to the gene. *Nature* 1984;308:404–405.
95. Wexler NS, Young AB, Tanzi RE, et al: Homozygotes for Huntington's disease. *Nature* 1987;326:194–197.
96. Gilliam TC, Bucan M, MacDonald ME, et al: A DNA segment encoding two genes very tightly linked to Huntington's disease. *Science* 1987;238:950–952.
97. MacDonald ME, Cheng SV, Zimmer M, et al: Clustering of the multiallele DNA markers near the Huntington's disease gene. *J Clin Invest* 1989;84:1013–1016.
98. Wexler NS, Conneally PM, Houseman D, Gusella JF: A DNA polymorphism for Huntington's disease marks the future. *Arch Neurol* 1985;42:20–24.
99. Alzheimer A: Uber eine eigenartige Erkrankungder Hinrinde. *Allg Z Psychiat* 1907; 64:146–148.
100. Masters CL, Simms G, Weinman NA, et al: Amyloid plaque core protein in Alzheimer disease and Down syndrome. *Proc Natl Acad Sci USA* 1985;82:4245–4249.
101. Glenner GG, Wong CW: Alzheimer's disease: Initial report of the purification and characterization of a novel amyloid protein. *Biochem Biophys Res Commun* 1984; 120:885–890.
102. Kang J, Lemaire H-G, Unterbeck A, et al: The precursor of Alzheimer's disease amyloid A4 protein resembles a cell-surface receptor. *Nature* 1987;325:733–736.
103. St. George-Hyslop PH, Tanzi R, Polinsky RJ, et al: The genetic defect causing familial Alzheimer's disease maps on chromosome 21. *Science* 1987;235:885–890.
104. Goldgaber D, Lerman MI, McBride OW, et al: Characterization and chromosomal localization of a cDNA encoding brain amyloid of Alzheimer's disease. *Science* 1987; 235:877–880.
105. Tanzi RE, Gusella JF, Watkins PC, et al: Amyloid β protein gene: cDNA, mRNA distribution, and genetic linkage near the Alzheimer locus. *Science* 1987;235:800–884.
106. Whitson JS, Selkoe DJ, Lotman CW: Amyloid β protein enhances the survival of hippocampal neurons in vitro. *Science* 1989;243:1488–1490.
107. Yankner BA, Dawes LR, Fisher S, et al: Neurotoxicity of a fragment of the amyloid precursor associated with Alzheimer's disease. *Science* 1989;245:417–420.
108. de Sauvage F, Octave J-N: A novel mRNA of the A4 amyloid precursor gene coding for a possibly secreted protein. *Science* 1989;245:651–653.
109. van Broeckhoven C, Haan J, Bakker E, et al: Amyloid β protein precursor gene and

hereditary cerebral hemorrhage with amyloidosis (Dutch). *Science* 1990;248:1120–1122.
110. Tanzi RE, St. George-Hyslop PH, Haines JL, et al: The genetic defect in familial Alzheimer's disease is not tightly linked to the amyloid β-protein gene. *Nature* 1987; 329:156–157.
111. van Broeckhoven C, Genthe AM, Vanderberghe A, et al: Failure of familiar Alzheimer's disease to segregate with the A4-amyloid gene in several European families. *Nature* 1987;329:153–155.
112. Tanzi RE, Bird ED, Latt SA, Neve RL: The amyloid β protein gene is not duplicated in brains from patients with Alzheimer's disease. *Science* 1987;238:666–669.
113. Podlisny MB, Lee G, Selkoe DJ: Gene dosage of the amyloid β precursor protein in Alzheimer's disease. *Science* 1987;238:669–671.
114. St. George-Hyslop, PH, Haines JL, Farrer LA, et al: Genetic linkage studies suggest that Alzheimer's disease is not a single homogeneous disorder. *Nature* 1990;347:194–197.
115. Talamo BR, Rudel RA, Kosik KS, et al: Pathological changes in olfactory neurons in patients with Alzheimer's disease. *Nature* 1989;337:736–739.
116. Lesch M, Nyhan WL: A familial disorder of uric acid metabolism and central nervous system function. *Am J Med* 1964;36:561–670.
117. Kelley WN, Wyngaarden JB: in Wyngarden JB, Frederickson DS, Goldstein JL, et al. (eds): *The Metabolic Basis of Inherited Disease*, 5th ed. New York, McGraw-Hill, 1978, pp 1010–1036.
118. Yang TP, Patel PI, Chinault AC: Molecular evidence for new mutation at the *hprt* locus in Lesch-Nyhan patients. *Nature* 1984;310:412–414.
119. Gibbs RA, Caskey CT: Identification and localization of mutations at the Lesch-Nyhan locus by ribonuclease A cleavage. *Science* 1987;236:303–305.
120. Robbins LN, Helzer JE, Weissman MM, et al: Lifetime prevalence of specific psychiatric disorders in three sites. *Arch Gen Psychiatry* 1984;41:949–958.
121. Cadoret RJ: Evidence for genetic inheritance of primary affective disorder in adoptees. *Am J Psychiatry* 1978;135:463–466.
122. Mendlewicz J, Rainer JD: Adoption study supporting genetic transmission in manic-depressive illness. *Nature* 1977;268:327–329.
123. Egeland JA, Hostetter AM: Amish study. I. Affective disorders among the Amish 1976–1980. *Am J Psychiatry* 1983;140:56–61.
124. Egeland, JA, Gerhard DS, Pauls DL, et al: Bipolar affective disorders linked to DNA markers on chromosome 11. *Nature* 1987;325:783–787.
125. Harper PS, Youngman S, Anderson MA, et al: Genetic linkage between Huntington's disease and the DNA polymorphism G8 in South Wales families. *J Med Genet* 1985; 22:447–450.
126. Kidd KK, Gerhard DS, Kidd JR: Recombinant DNA methods in genetic studies of affective disorders. *Clin Neuropharmacol* 1984;7:198–199.
127. Lamouroux A, Biguet AN, Samolyk D, et al: Identification of cDNA clones coding for rat tyrosine hydroxylase antigen. *Proc Natl Acad Sci USA* 1982;79:3881–3885.
128. Mallet J, Berod A, Blanot F, et al: Molecular genetics of tyrosine hydroxylase. in Guiditta A, Kaplan BB, Zomzely-Neurath C, et al. (eds): *Role of RNA and DNA in Brain Function: A Molecular Biological Approach*. Norwell, Mass., Nijhoff, 1987, pp 57–70.
129. Hodgkinson S, Sherrington R, Gurling H: Molecular genetic evidence for heterogeneity in manic depression. *Nature* 1987;325:805–806.
130. Detera-Wadleigh SD, Berrettini WH, Goldin LR: Close linkage of c-Harvey-*ras* 1 and the insulin gene to affective disorder is ruled out in three North American pedigrees. *Nature* 1987;325:806–808.
131. Kolata G: Manic-depression gene tied to chromosome 11. *Science* 1987;235:1139–1140.
132. Baron M, Risch N, Hamburger R, et al: Genetic linkage between X-chromosome markers and bipolar affective illness. *Nature* 1987;326:289–291.
133. Kelsoe JR, Ginns EI, Egeland JA, et al: Re-evaluation of the linkage relationship between chromosome 11p loci and the gene for bipolar affective disorder in the old order Amish. *Nature* 1989;342:238–243.
134. Barinaga M: Manic depression gene put in limbo. *Science* 1989;246:886–887.
135. Robertson M: False start on manic depression. *Nature* 1989;342:222.

136. Eriksson S: Studies in alpha-1-antitrypsin deficiency. *Acta Med Scand* 1965;[*Suppl*]432:1–85.
137. Eriksson S, Carlson J, Velez R: Risk of cirrhosis and primary liver cancer in alpha 1-antitrypsin deficiency. *N Engl J Med* 1986;314:736–739.
138. Cox DW, Markovic VD, Teshima IE: Genes for immunoglobulin heavy chains and α_1-antitrypsin are localized to specific region of chromosome 14q. *Nature* 1982;297:428–430.
139. Fagerhol MK, Cox DW: The Pi polymorphism: Genetic, biochemical, and clinical aspects of human alpha 1-antitrypsin. *Adv Hum Genet* 1981;11:1–62.
140. Cox DW, Woo SLC, Mansfield T: DNA restriction fragment associated with α_1-antitrypsin indicate a single origin for deficiency allele Pi Z. *Nature* 1985;316:79–81.
141. Kan YW, Dozy AM: Antenatal diagnosis of sickle-cell anaemia by D.N.A. analysis of amniotic fluid cells. *Lancet* 1978;2:910–911.
142. Orkin SH, Kazazian HH Jr, Antonarakis SE, et al: Linkage of β-thalassemia mutations and β-globin gene polymorphisms with DNA polymorphisms in human β-globin gene cluster. *Nature* 1982;296:627–631.
143. Yoshida A, Lieberman J, Gaidulis L, Ewing C: Molecular abnormality of human alpha 1-antitrypsin variant (Pi ZZ) associated wtih plasma activity deficiency. *Proc Natl Acad Sci USA* 1976;73:1324–1328.
144. Yoshida A, Ewing C, Wessels M, et al: Molecular abnormality of Pi S variant of human alpha 1-antitrypsin. *Am J Hum Genet* 1977;29:233–239.
145. Long GL, Chandra T, Woo SLC, et al: Complete sequence of the cDNA for human α_1 antitrypsin and the gene for the S variant. *Biochemistry* 1984;23:4828–4837.
146. Talamo RC, Langley CE, Reed CE, Makino S: α_1-Antitrypsin deficiency: A variant with no detectable α_1-antitrypsin. *Science* 1973;181:70–71.
147. Garver RI, Mornex J-F, Nukiwa T, et al: Alpha$_1$-antitrypsin deficiency and emphysema caused by homozygous inheritance of non-expressing alpha$_1$-antitrypsin genes. *N Engl J Med* 1986;314:762–766.
148. Hejtmancik JF, Sifers RN, Ward PA, et al: Prenatal diagnosis of α_1-antitrypsin deficiency by restriction fragment length polymorphisms, and comparison with oligonucleotide probe analysis. *Lancet* 1986;2:767–770.
149. Cox DW, Mansfield T: Prenatal diagnosis of α_1 antitrypsin deficiency and estimates of fetal risk for disease. *J Med Genet* 1987;24:52–59.
150. Cox DW, Billingsley GD: Restriction enzyme *Mae*III for prenatal diagnosis of α_1-antitrypsin deficiency. *Lancet* 1987;2:741–742.
151. Sveger T: Liver disease in alpha$_1$-antitrypsin deficiency detected by screening of 200,000 infants. *N Engl J Med* 1987;294:1316–1321.
152. Sveger T, Thelin T: Four-year old children with α_1-antitrypsin deficiency. *Acta Pediatr Scand* 1981;70:171–177.
153. Brantly M, Courtney M, Crystal RG: Repair of the secretion defect in the Z form of α_1-antitrypsin by addition of a second mutation. *Science* 1988;242:1700–1702.
154. Hubbard RC, Casolaro MA, Mitchell M, et al: Rate of aerosolized recombinant DNA–produced α_1-antitrypsin: Use of the epithelial surface of the lower respiratory tract to administer proteins of therapeutic importance. *Proc Natl Acad Sci USA* 1989;86:680–684.
155. Hubbard RC, McElvaney NG, Sellers SE, et al: Recombinant DNA–produced α_1-antitrypsin administered by aerosol augments lower respiratory tract antineutrophil elastase defenses in individuals with α_1-antitrypsin deficiency. *J Clin Invest* 1989;84:1349–1354.
156. Carlson JA, Rogers BB, Sifers RM, et al: Accumulation of PiZ α_1-antitrypsin causes liver damage in transgenic mice. *J Clin Invest* 1989;83:1183–1190.
157. Sandhoff K, Conzelmann E: The biochemical basis of gangliosidoses. *Neuropediatrics* 1984;15(Suppl.):85–92.
158. Anderman E, Scriver CR, Wolfe LS: Genetic variants of Tay-Sachs disease—Tay-Sachs disease and Sandhoff's disease in French Canadians, juvenile Tay-Sachs disease in Lebanese Canadians, and a Tay-Sachs screening program in French Canadian population. *Prog Clin Biol Res* 1977;18:161–188.
159. Myerowitz R, Hogikyan ND: Different mutations in Ashkenazi Jewish and non-Jewish Canadians with Tay-Sachs disease. *Science* 1986;232:1646–1648.

160. Patrick AD: A deficiency of gluco-cerebrosidase in Gaucher's disease. *Biochem J* 1965; 97(Suppl):17c–18c.
161. Brady RO, Kanfer JR, Shapiro D: Metabolism of glucocerebrosides. II. Evidence of an enzymatic deficiency in Gaucher's disease. *Biochem Biophys Res Commun* 1965;18:221–225.
162. Tsuji S, Choudary PV, Martin BM, et al: A mutation in the human glucocerebrosidase gene in neuronopathic Gaucher's disease. *N Engl J Med* 1987;316:570–575.
163. Goossens M, Lee KY, Liebhaber SA, Kan YW: Globin structural mutant α^{125} Leu → Pro is a novel cause of α-thalassemia. *Nature* 1982;296:864–865.
164. Rappeport JM, Barranger JA, Ginns EI: Bone marrow transplantation in Gaucher disease. *Birth Defects* 1986;22:101–109.
165. Choudary PV, Barranger JA, Tsuji S, et al: Retrovirus-mediated transfer of the human glucocerebrosidase gene to Gaucher fibroblasts. *Mol Biol Med* 1986;3:293–299.
166. Danovitch GM: in Gardner KD (ed): *Cystic Disease of the Kidney*. New York, Wiley, 1976, pp 125–150.
167. Lowrie EG, Hampers CL: The success of Medicare's end-stage renal disease program. The case for profits and the private marketplace. *N Engl J Med* 1981;305:434–438.
168. Bear JC, McManamon P, Morgan J, et al: Age at clinical onset and ultrasonographic detection of adult polycystic kidney disease: Data for genetic counselling. *Am J Med Genet* 1984;18:45–53.
169. Reeders ST, Breuning MH, Davies KE, et al: A highly polymorphic DNA marker linked to adult polycystic kidney disease on chromosome 16. *Nature* 1985;317:542–544.
170. Botstein D, White RL, Skolnick M, Davis RW: Construction of a genetic linkage map in man using restriction fragment length polymorphisms. *Am J Hum Genet* 1980; 32:314–331.
171. Barton P, Malcolm S, Murphy C, Ferguson-Smith MA: Localization of the human alpha-globin gene cluster to the short arm of chromosome 16 (16p21–16pter) by hybridization in situ. *J Mol Biol* 1982;156:269–278.
172. Vogel F: Genetics or retinoblastoma. *Hum Genet* 1979;52:1–54.
173. Abramson DH, Ellsworth RM, Kitchin FD: Second monocular tumors in retinoblastoma survivors: Are they radiation-induced? *Ophthalmology* 1984;91:1351–1355.
174. Cavenee WK, Dryja TP, Phillips RA, et al: Expression of recessive alleles by chromosomal mechanisms in retinoblastoma. *Nature* 1983;305:779–784.
175. Dryja TP, Cavenee W, White R, et al: Homozygosity of chromosome 13 in retinoblastoma. *N Engl J Med* 1984;310:550–553.
176. Cavenee WK, Hansen MF, Nordenskjold M, et al: Genetic origin of mutations predisposing to retinoblastoma. *Science* 1985;228:501–503.
177. Cavenee WK, Murphree AL, Shull MM, et al: Prediction of familial predisposition to retinoblastoma. *N Engl J Med* 1986;314:1201–1207.
178. Friend SH, Bernards R, Rogelj S, et al: A human DNA segment with properties of the gene that predisposes to retinoblastoma and osteosarcoma. *Nature* 1986;323:643–646.
179. Fung Y-KT, Murphree AL, T'Ang A, et al: Structural evidence for the authenticity of the human retinoblastoma gene. *Science* 1987;236:1657–1661.
180. Knudson AG: Statistical study of retinoblastoma. *Proc Natl Acad Sci USA* 1971;68:820–823.
181. Lee WH, Bookstein R, Hong F, et al: Human retinoblastoma susceptibility gene: Cloning, identification, and sequence. *Science* 1987;235:1394–1399.
182. Lee WH, Shew J-Y, Hong FD, et al: The retinoblastoma susceptibility gene encodes a nuclear phosphoprotein associated with DNA binding activity. *Nature* 1985;329:642–645.
183. Wiggs J, Nordenskjold M, Yandell D, et al: Prediction of the risk of hereditary retinoblastoma, using DNA polymorphisms within the retinoblastoma gene. *N Engl J Med* 1988;318:151–157.
184. Nakamura T, Leppert M, O'Connell P, et al: Variable number of tandem repeat (VNTR) markers for human gene mapping. *Science* 1987;235:1616–1622.
185. Dryja TP, Mukai S, Petersen R, et al: Parental origin of mutations of the retinoblastoma gene. *Nature* 1989;339:556–558.
186. Toguchida J, Ishizaki K, Sasaki MS, et al: Preferential mutation of paternally derived RB gene as the initial event in sporadic osteosarcoma. *Nature* 1989;338:156–158.

187. Zhu X, Dunn JM, Phillips RA, et al: Preferential germline mutation of the preferential allele in retinoblastoma. *Nature* 1989;340:312–313.
188. Orkin SH, Goldman DS, Sallan SE: Development of homozygosity for chromosome 11p markers in Wilms' tumor. *Nature* 1984;309:172–174.
189. Reeve AE, Housiaux DJ, Gardner RJM, et al: Loss of a Harvey *ras* allele in sporadic Wilms' tumor. *Nature* 1984;309:174–176.
190. Fearon ER, Vogelstein B, Feinberg AD: Somatic deletion and duplication of genes on chromosome 11 in Wilms' tumors. *Nature* 1984;309:176–178.
191. Glaser T, Lewis WH, Bruns GAP, et al: The β-subunit of follicle-stimulating hormone is deleted in patients with aniridia and Wilms' tumour, allowing a further definition of the WAGR locus. *Nature* 1986;321:882–886.
192. Housman DE, Glaser T, Gerhard DS, et al: Mapping of human chromosome 11: Organization of genes within the Wilms' tumor region of the chromosome. *Cold Spring Harbor Symp Quant Biol* 1986;51(2):837–841.
193. Porteous DJ, Bickmore W, Christie S, et al: HRAS1-selected chromosome transfer generates marker that colocalize aniridia- and genitourinary dysphasia–associated translocation breakpoints and the Wilms tumor gene within band 11p13. *Proc Natl Acad Sci USA* 1987;84:5355–5359.
194. Mannens M, Slater RM, Heyting C, et al: Regional localization of DNA probes on the short arm of chromosome 11 using aniridia–Wilms tumor-associated deletions. *Hum Genet* 1987;75:180–187.
195. Weissman BE, Saxon PJ, Pasquale SR, et al: Introduction of a normal human chromosome 11 into a Wilms' tumor cell line controls its tumorigenic expression. *Science* 1987;236:175–180.
196. Nisen PD, Zimmerman KA, Cotter SV, et al: Enhanced expression of the N-*myc* gene in Wilms' tumor. *Cancer Res* 1986;46:6217–6222.
197. Gessler M, Poustka A, Cavenee W, et al: Homozygous deletion in Wilms tumors of a zinc-finger gene identified by chromosome jumping. *Nature* 1990;343:774–778.
198. Seizinger BR, Martuza RL, Gusella JF: Loss of genes on chromosome 22 in tumorigenesis of human acoustic neuroma. *Nature* 1986;322:644–647.
199. Zarg KD: Cytological and cytogenetical studies on human meningioma. *Cancer Genet Cytogenet* 1982;6:249–274.
200. Mark J: Chromosomal abnormalities and their specificity in human neoplasms: An assessment of recent observations by banding techniques. *Adv Cancer Res* 1977;24:165–222.
201. Zankl H, Zang KD: Correlations between clinical and cytogenetical data in 180 human meningiomas. *Cancer Genet Cytogenet* 1980;1:351–356.
202. Seizinger BR, Rouleau G, Ozelius LJ, et al: Common pathogenetic mechanism for three tumor types in bilateral acoustic neurofibromatosis. *Science* 1987;236:317–319.
203. Rouleau GA, Wertelecki W, Haines JL, et al: Genetic linkage of bilateral acoustic neurofibromatosis to a DNA marker on chromosome 22. *Nature* 1987;329:246–248.
204. Fountain JW, Wallace MR, Bruce MA, et al: Physical mapping of a translocation breakpoint in neurofibromatosis. *Science* 1989;244:1085–1087.
205. Wallace MR, Marchuk DA, Andersen LB, et al: Type 1 neurofibromatosis gene: Identification of a large transcript disrupted in the three NF1 patients. *Science* 1990; 249:181–186.
206. Seizinger BR, de la Monte S, Atkins L: Molecular genetic appproach to human meningioma: Loss of genes on chromosome 22. *Proc Natl Acad Sci USA* 1987;84:5419–5423.
207. Bodmer WF, Bailey CJ, Bodmer J, et al: Localization of the gene for familial adenomatous polyposis on chromosome 5. *Nature* 1987;328:614–616.
208. Solomon E, Voss R, Hall V, et al: Chromosome 5 allele loss in human colorectal cancer. *Nature* 1987;328:616–619.
209. Leppert M, Dobbs M, Scambler P, et al: The gene for familial polyposis coli maps to the long arm of chromosome 5. *Science* 1987;238:1411–1413.
210. Fredrickson DS, Goldstein JD, Brown MS: The familial hyperlipoproteinemias, in Stanbury JB, Wyngaarden JB, Fredrickson DS (eds): *The Metabolic Basis of Inherited Disease*. New York, McGraw-Hill, 1978. pp 604–655.
211. Harlan WR, Graham JB, Estes EH: Familial hypercholesterolemia, a genetic and metabolic study. *Medicine* 1966;45:77–110.

212. Goldstein JL, Schrott HG, Hazzard WR, et al: Hyperlipidemia in coronary heart disease, part 2 (genetic analysis of lipid levels in 176 families and delineation of a new inherited disorder, combined hyperlipidemia). *J Clin Invest* 1973;52:1544–1568.
213. Brown MS, Goldstein JL: Expression of the familial hypercholesterolemia gene in heterozygotes—mechanism for a dominant disorder in man. *Science* 1974;185:61–63.
214. Brown M, Goldstein JL: Familial hypercholesterolemia—defective binding of lipoproteins to cultural fibroblasts associated with impaired regulation of 3-hydroxy-3-methylglutaryl coenzyme A reductase activity. *Proc Natl Acad Sci USA* 1974;71:788–792.
215. Tolleshang H, Goldstein JL, Schneider WJ, Brown MS: Post translational processing of the LDL receptor and its genetic disruption in familial hypercholesterolemia. *Cell* 1982;30:715–724.
216. Tolleshang H, Hobgood KK, Brown MS, Goldstein JL: The LDL receptor locus in familial hypercholesterolemia—multiple mutations disrupt transport and processing of a membrane receptor. *Cell* 1983;32:941–951.
217. Goldstein JL, Basu SK, Brunschede GY, Brown MS: Release of low density lipoprotein from its cell surface receptor by sulfated glycosaminoglycons. *Cell* 1976;7:85–95.
218. Ho YK, Faust JR, Biheimer DW: Regulation of cholesterol synthesis by low density lipoprotein in isolated human lymphocytes. *J Exp Med* 1977;145:1531–1549.
219. Yamamoto T, Davis LG, Brown MS: The human LDL receptor: A cysteine-rich protein with multiple *Alu* sequences in its mRNA. *Cell* 1984;39:27–38.
220. Humphries SE, Horsthemke B, Seed M, et al: A common DNA polymorphism of the low-density lipoprotein (LDL) receptor gene and its use in diagnosis. *Lancet* 1985; 1:1003–1005.
221. Lehman MA, Schneider WJ, Sudhof TC, et al: Mutations in LDL receptor: *Alu-Alu* recombination deletes exons encoding transmembrane and cytoplasmic domains. *Science* 1985;227:140–147.
222. Russell DW, Lehrman MA, Sudhof TC, et al: The LDL receptor in familial hypercholesterolemia: Use of human mutations to dissect a membrane protein. *Cold Spring Harbor Symp Quant Biol* 1986;51:811–819.
223. Jagadeeswaran D, Tuan D, Forget BG, Weissman SM: A gene deletion ending at the midpoint of a repetitive DNA sequence in one form of hereditary persistance of fetal hemoglobin. *Nature* 1982;296:469–470.
224. Lehrman MA, Goldstein JL, Russell DW, Brown MS: Duplication of seven exons in LDL receptor gene caused by *Alu-Alu* recombination in a subject with familial hypercholesterolemia. *Cell* 1987;48:827–835.
225. Lehrman MA, Russell DW, Goldstein JL, Brown MS: *Alu-Alu* recombination deletes splice acceptor sites and produces secreted low density lipoprotein receptor in a subject with familial hypercholesterolemia. *J Biol Chem* 1987;262:3354–3361.
226. Hobbs HH, Brown MS, Russell DW, et al: Deletion in the gene for the low-density-lipoprotein receptor in a majority of French Canadians with familial hypercholesterolemia. *N Engl J Med* 1987;317:734–737.
227. Brink PA, Steyn LT, Coetzee GA, Van der Westhuyzen DR: Familial hypercholesterolemia in South African Afrikaners *Pvu* and *Stu*I DNA polymorphisms in the LDL-receptor gene consistent with a predominating founder gene defct. *Hum Genet* 1987; 77:32–35.
228. Hofmann SL, Russell DW, Brown MS, et al: Over-expression of low density lipoprotein (LDL) receptor eliminates LDL from plasma in transgenic mice. *Science* 1988; 239:1277–1281.
229. Palmiter RD, Norstedt G, Gelinas RE, et al: Matallothionein–human gH fusion genes stimulate growth of mice. *Science* 1983;222:809–814.
230. Scott J, Knott TJ, Priestley LM, et al: High-density lipoprotein composition is altered by a common DNA polymorphism adjacent to apoprotein AII gene in man. *Lancet* 1985;1:771–773.
231. Ordovas JM, Schaeffer EJ, Salem D, et al: Apolipoprotein A-I gene polymorphism associated with premature coronary artery disease and familial hypoalpha lipoproteinemia. *N Engl J Med* 1987;314:671–677.
232. Karathansis SK, Zannis VI, Broslow JL: Isolation and characterization of the human apolipoprotein A-I gene. *Proc Natl Acad Sci USA* 1983;80:6147–6151.

233. Karathanasis SK, McPherson J, Zannis VI, Breslow JL: Linkage of human apolipoproteins A-I and C-III genes. *Nature* 1983;304:371–373.
234. Karathanasis SK: Apolipoprotein multigene family: Tandem organization of human apolipoprotein A-I, C-III and A-IV genes. *Proc Natl Acad Sci USA* 1985;82:6374–6378.
235. Schaefer EJ, Levy RI: Pathogenesis and management of lipoprotein disorders. *N Engl J Med* 1985;312:1300–1310.
236. Wion KL, Kirchgessner TG, Lusis AJ, et al: Human lipoprotein lipase complementary DNA sequence. *Science* 1987;235:1638–1641.
237. Drayna D, Jarnagin AS, McLean J, et al: Cloning and sequencing of human cholesterol ester transfer protein cDNA. *Nature* 1987;327:632–634.
238. Antonarakis SE: Diagnosis of genetic disorders at the DNA level. *N Engl J Med* 1989; 320:153–163.
239. Chapple JC, Dale R, Evans BG: The new genetics: Will it pay its way? *Lancet* 1987; 1:1189–1192.

Recombinant DNA and the Immune System

10

Ralph C. Williams, Jr., MD

Recombinant Immunoglobulins

When monoclonal antibodies produced by fusion of plasma cell lines with actively secreting antigen-stimulated cells became available, the idea of using this technique or an analogous approach to produce further tailor-made antibody molecules began to emerge. The concept of applying principles of genetic engineering to the synthesis of precisely constructed antibody molecules was quite appealing. Perhaps one could use the mechanisms of normal immunoglobulin (Ig) heavy and light chain segment rearrangement to fashion antibody molecules with specific sorts of combining sites joined to constant regions that might offer less immunogenic stimulus than would be expected of a polyclonal heterologous antibody molecule.

This approach has now been taken in experiments designed to produce chimeric antibodies containing mouse variable region and human constant region sequences.[1] One of the problems encountered in applying mouse or other heterologous monoclonal antibodies to diagnosis or therapy in human problems has been rapid neutralization of the administered monoclonal antibodies by the appropriate human host immune response.[2–4] It was thought that such a neutralizing immune response might be avoided or perhaps modulated by constructing chimeric immunoglobulins consisting of mouse variable regions and human constant regions. In order to do this, a retrovirus DNA was constructed that contained genomic heavy-chain (H) variable-diversity-joining (V_H-D-J_H) and constant region (C_H) genes from the different species. This was accomplished, and the resulting mRNAs were spliced out precisely to produce chimeric Ig molecules. It was found that a slight unexpected variation occurred in the assembly of the chimeric Ig molecules in that a hidden splice donor site in the 5′-flanking region of a human V_H gene was used in place of the donor site of the leader exon sequence, producing a V region without the leader sequence.

In the report by Takeda and co-workers[1] the chimeric Ig molecules were prepared from Cγl mouse genes and a human V_H region reacting with the surface antigen of hepatitis B virus (HBV). These two genes were inserted into a retrovirus vector plasmid. Another pair of genes representing the human Cγl gene and the mouse V_H gene of the TEPC15 myeloma was inserted into another retrovirus-activated vector. In each case the retroviral vector sequences were derived from spleen necrosis virus. The viral long terminal repeat sequences and promoters were used to ensure proper expression of the continguous human or mouse Ig gene segment sequences. In order to recover infectious viruses carrying the Ig genes, the vectors containing plasmids were used to transfect chicken embryo fibroblasts with or without one-tenth as much DNA of a provirus clone from an additional non-defective helper virus (reticuloendotheliosis virus strain A). Before the transfection procedure, the plasmid containing the Ig genes was digested with the *Sal*I restriction enzyme to remove the plasmid pBR322 sequence, and virus production was measured using the reverse transcriptase activity assay.[5] The enzyme activity of the culture supernatants of cells transfected with recombinant and helper viral DNAs showed 10-fold increments 5 to 7 days after transfection. Virus from culture supernatants was then used to infect fresh chicken embryo fibroblasts, and 3 days later viral DNA was isolated and assayed for integration of the inserted Ig genomic segments by Southern blot procedures. No significant viral DNA was recovered from the transfectants without addition of the helper virus DNA. To analyze the structure of the progeny viruses, the inserts were cloned into chann 28 H vector and, after DNAs had been cut with appropriate restriction enzymes, studied in additional Southern blotting experiments with appropriate human or mouse Ig segment probes.

Subsequent nucleotide sequence analysis of the regions surrounding the junctions of the V and C gene regions in the progency clones showed exact splicing of the parental V and C gene exons, indicating precise slicing according to the GT-AG immunoglobulin gene pattern.[6] This was rather remarkable because it indicated that interspecies (human-mouse) pairing of exons followed precise removal of introns without any significant deletions or unexpected rearrangements. A scheme of this chimeric strategy is shown in Figure 10.1.

This remarkable achievement may eventually have practical therapeutic or diagnostic usefulness, since most clinically useful antibodies are more easily produced as hybridomas in the mouse system. Generally, when such purified murine antibodies are administered to humans—for instance, in attempts to image tumors or metastases or in the case of transplant patients to effect potent immunosuppression—the human host rapidly makes an immune response to the heterologous mouse antibody. However, because the constant region–encoded segments of the mouse antibody are the most antigenic (Fc and constant region–encoded H chain segments), in the chimeric antibody (in which the constant region is of human origin) a weaker immune response and consequently less host neutralization would occur. Refinement of various mouse monoclonal antibodies has allowed a wide spectrum of antibody specificities to be identified, and it seems possible that with more extended human use the technique of chimeric antibody administration may hold considerable promise.

When monoclonal antibodies were first discovered and produced,[7] there was great excitement about their potential diagnostic and therapeutic use. As therapeutic agents, however, monoclonal antibodies have not proved to be as helpful as had been hoped.[8–10] In studies of their immunosuppressive and antitumor effectiveness, it was quickly recognized that their ability to deplete T cell populations or interact with tumor target cells was hampered by two major problems. First,

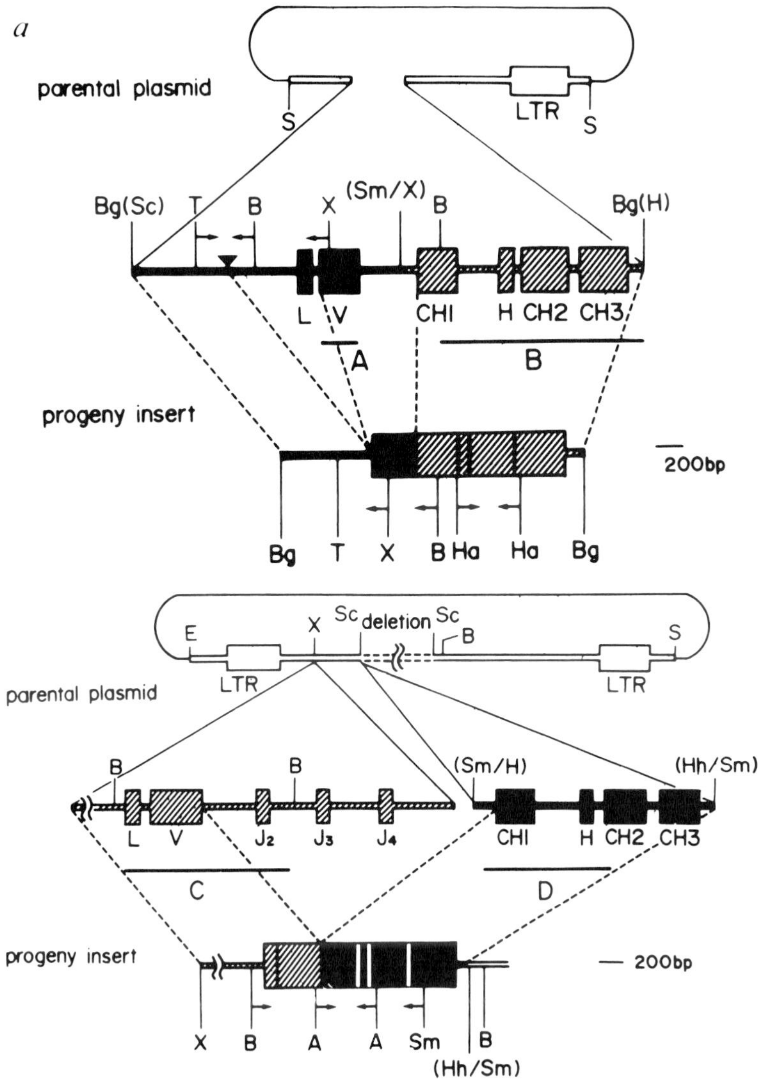

Figure 10.1 Structure of parental and progeny recombinant immunoglobulin genes. (a) Plasmid pSNV$_1$-hV$_{HBC}$-mCγ1 containing human V regions with specificity for hepatitis B and mouse constant regions (mCλ1). (b) Plasmid pSNV$_2$-mV$_{T15}$-hCγ1 with mouse variable regions and human constant genes (hCγ1). Retroviral, mouse, and human Ig sequences are shown as open, oblique, and solid bars, respectively. The long terminal repeat (LTR) and exons are indicated as wider rectangles and the original plasmid pBR322 by a solid line. Letters below the final gene indicate restriction enzyme sites. *Reproduced from Takeda et al.,*[1] *with permission.*

monoclonal antibody sensitization occurred in most patients who received the same antibody for several consecutive days.[10–13] Second, host sensitization to the administered antibody was often intense and rapid enough to nullify its beneficial or therapeutic effects and also possibly to produce an eventual serum sickness–like reaction. For instance, the human immune response to the OKT3 monoclonal antibody against pan–T cell determinants has been extensively studied in human recipients.[14] Although in some instances a relatively oligoclonal response was recorded, most results confirm that these products are highly antigenic. The development of tailor-made chimeric monoclonal or even polyclonal antibodies containing human constant regions represents a salutary and effective compromise.

Changing Monoclonal Antibodies—Other Strategies

Besides the interesting technique used in producing chimeric antibodies, several other recombinant approaches can be employed to tailor changes in monoclonal antibodies that would make them more suitable as therapeutic or diagnostic agents. One of the most practical techniques involves changing the structure of the genes encoding an already useful monoclonal antibody to produce an even more effective one. This can be done by cloning the gene for a potentially useful antibody and then reconstructing it or mutating it using recombinant DNA technology before introducing the modified gene into an animal cell or microorganism to produce large amounts of the tailor-made antibody molecule.

Figure 10.2 The μ heavy chain can switch class to any of the downstream 3′ C regions. At the DNA level, the switch is mediated by switch regions (S) located upstream 5′ to each of the C regions except δ. The class switches of μ to γ1 and of γ1 to $\gamma 2_a$ are shown. The other switch regions (Sγ3, $S\gamma 2_b$, Sϵ, and Sα) are represented by vertical lines. For example, a μ- to-λ1 switch results when a recombination occurs between Sμ and Sγ1 followed by deletion of the intervening DNA. This places the Cγ1 coding region close to the expressed VDJ. Similar class switching can take place successively to move downstream C regions. This succession of class switching occurs during the normal differentiation of B cells. Because C_μ is the most upstream constant region, it is expressed first at the pre–B cell stage. As the cell matures, it begins to form dimers of heavy and light chains, at which point the cell presents the resulting IgM on its surface. Once a class switch occurs, the cell is capable of producing surface antibody with the new class or subclass. Ultimately, the cell becomes fully differentiated into a plasma cell that dvotes most of its cellular mechanisms to secreting antibody. *Reproduced from DePinho, RA, Scharff M: Tailor-made monoclonal antibodies. Ann Intern Med 1986;104:225–233, with permission.*

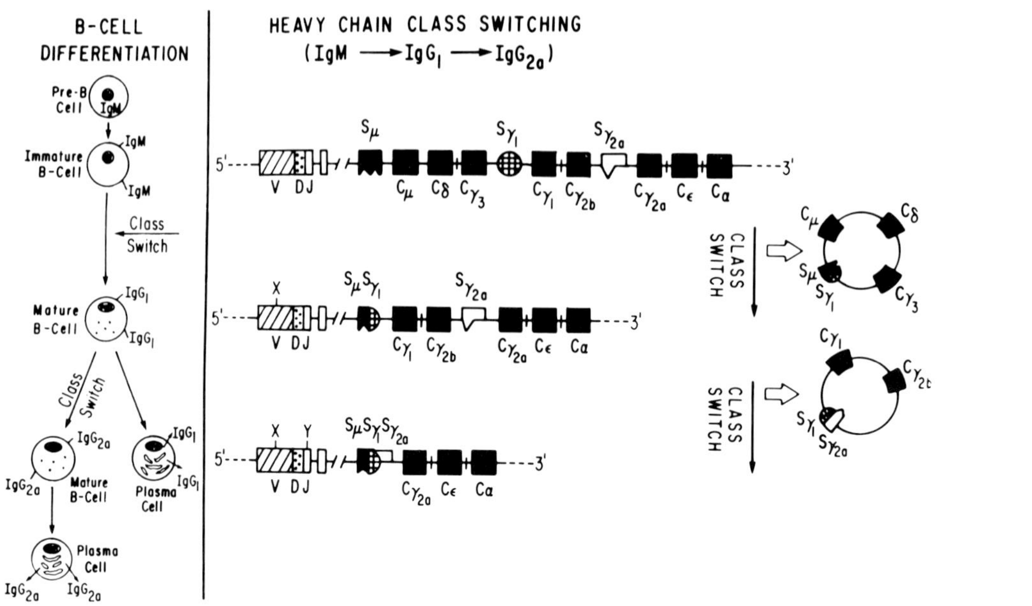

A number of screening techniques have been developed which are capable of identifying different subclones of cultured antibody-forming cells.[15,16] With many antigens of interest, the majority of the monoclonal antibodies produced contain only one or a restricted number of H-chain groups or subclasses. Thus, most murine antibodies made by BALB/c mice against pneumococcal C polysaccharides are of the IgM class.[17] To make a monoclonal antibody of the IgG2_a subclass, one would take a hybridoma making IgM anti–C polysaccharide and, through screening procedures, identifying a rare subclone that had rearranged its V_HDJ_H from the M to the $\gamma 2_a$ constant region so that an IgG2_a antibody was produced.[18] These switch variants would have the same anti–C polysaccharide specificity as the original antibody but could be used to produce different effector functions. Use of switch variants detected by subcloning is illustrated in Figure 10.2.

Another example of a recombinant technique used to obtain antibodies designed for a particular function is the subclass switching technique employed by Parham and co-workers.[18] These workers had previously generated a set of mouse monoclonal antibodies that reacted with a group of different human transplantation antigens. However, many of the mAbs were of the IgG1 subclass and did not fix complement, so target cells were not lysed in vitro. Since tissue typing is often done with cytotoxicity assays, the IgG1 monoclonal antibodies were switched to IgG2_b production by the subclass çloning techniques mentioned above. The resulting antibodies, which carried the original antibody specificity, could then be employed for direct tissue typing experiments.

Still another example of this use of recombinant approaches is related to the problem presented in using a particular monoclonal antibody produced by Steplewski and co-workers at Wistar Institute.[19] The original antibody reacted with an antigenic epitope present on human colon, pancreatic, and gastric cancer cells; however, since it was of the IgG1 subclass it did not bind well to Fc receptors on human monocytic cells and thus was not effective in facilitating antibody-dependent killing of human tumor cells by monocyte effector cells. This original hybridoma was subcloned and carefully selected out to provide cells that produced IgG2_a of the original antibody specificity. This IgG2_a subclass had considerably increased potential therapeutic usefulness.[19]

Other Novel Functions Produced in Recombinant Antibodies

Several other recombinant approaches have been utilized to construct antibodies with novel properties. It was felt that chimeric proteins in which an antigen-binding portion of an Ig molecule was fused to an enzymatic moiety might be of use in immunoassays.[20] Also, the chemical coupling of various toxins to cell-specific antibodies has been advocated from time to time as a possible way to improve results with chemotherapeutic drugs. It seemed possible that antibody fusion proteins containing cytotoxic carboxyl-terminal end segments might be employed in this area.

Neuberger and co-workers[20] demonstrated that it was possible to prepare $F(ab')_2$-like fragments of a specific antibody by using mouse Ig genes cloned into a suitable vector. This was done by transfecting μ immunoglobulin genes for the V_H region of the specific antibody into a plasma cell line producing $\lambda 1$ light chains but no heavy-chain products. Stably transfected cells were then demonstrated to produce homogeneous antigen-specific antibody. Using these cells as a base, a new vector was constructed that replaced the exons coding for the μ-chain region (C_μ)

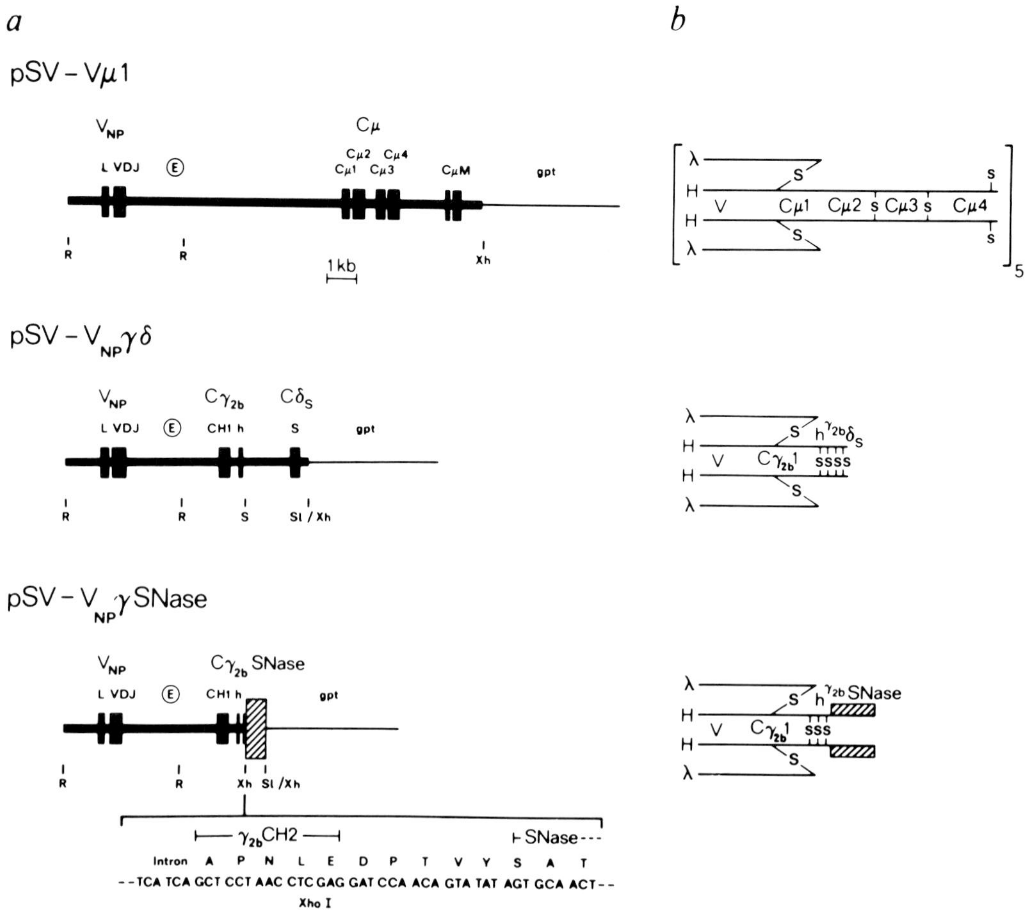

Figure 10.3 Structures of plasmids used in immunoglobulin gene constructs. The thin horizontal lines denote the pSV2gpt vector and thick lines indicate the Ig gene DNAs. The boxes show exons and the hatched box in the diagram at the bottom shows the *S. aureus* nuclease coding region. Heavy chain enhancer is labeled E, and gpt indicates the *gpt* gene. *Reproduced and adapted from Neuberger et al., with permission.*[20]

with genes for the C_Hl and hinge regions of the mouse $\gamma 2_b$ IgG gene. In order to provide proper translation termination and polyadenylation nucleotide sequences, a $C\delta_s$ exon from the gene coding for secreted mouse δ chains was inserted at the 3′ exon from the gene. This construct is illustrated in Figure 10.3. The plasmid pSV-$V_{NP}\gamma\delta$ shown in the middle diagram of Figure 10.3 was designed to direct the synthesis of F(ab′)$_2$-like molecules disulfide-linked via the $\gamma 2_b$ hinge region with a 21-amino-acid tail segment encoded by the $C\gamma_S$ exon. When transfected into a cell line, this plasmid showed that the stably transfected cells secreted high levels of a 110-kd protein that appeared to behave as an F(ab′)$_2$ on agarose gel analyses. These workers then fused the nuclease from *Staphylococcus aureus* to the secreted antibody F(ab′)$_2$ fragments. The *S. aureus* nuclease was chosen because it is an active enzyme present as a monomer containing no thiol residues. A DNA restriction fragment containing the gene for *S. aureus* nuclease was inserted into the C_H2 exon of the mouse $\gamma 2_b$ gene (lower portion of Figure 10.3).

Supernatants from cells transfected with the F(ab′)$_2$-nuclease construct were then shown to contain definite nuclease activity. Moreover, the antibody-nuclease conjugate produced by the fusion gene could be utilized as an effective reagent in enzyme-linked immunosorbent assays (ELISA). These studies demonstrated convincingly that hybrid molecules containing the combining sites of monoclonal antibodies linked to other nonantibody proteins could be constructed genetically. Further extension or amplification of this approach in other systems is clearly feasible.

Fusion Proteins as Magic Bullets

Extensive investigations during the past decade have been directed to engineering what can be looked on as magic bullets for use in cancer chemotherapy.[21] The object would be to combine the specificity of monoclonal antibodies with the selective killing power of potent cell toxins such as the ricins or diphtheria toxin. A major obstruction to making a practical immunotoxin conjugate of this kind is related to the fact that diphtheria toxin and ricin are extremely potent toxins composed of two disulfide-linked polypeptide chains.[22] The B chains bind the toxin to the cell surface and facilitate transport of the A chain into the cytosol. The A chains of these toxins are powerful catalytic inhibitors of protein synthesis, and a single molecule of either diphtheria toxin or ricin A chain within the cell cytosol is enough to kill the cell. Conjugates of monoclonal antibodies to toxins or toxin subunits have been called immunotoxins.[23,24] A number of studies have found that toxin A chains linked to monoclonal antibodies show a high degree of cell type selectivity in vitro but very little cytotoxic effect against solid tumors in vivo,[25] presumably because of slow transport of A chains into the cell cytosol.[26] Target cell toxicity of immunotoxins can be markedly increased by including the B chain in the antibody-toxin conjugate, however, for maximal in vitro target cell selectivity it is necessary to add lactose to the system to block nontarget cell (innocent bystander) binding and toxicity of the immunotoxin via the B chain. Such an approach has proven feasible in the setting of bone marrow transplantation,[27] where the target cell population can actually be incubated in vitro in the presence of lactose. However, without blockade of B chain binding regions, whole toxin conjugates have a very high proportion of nontarget cell killing, which reduces their usefulness to that equivalent to A chain plus antibody alone.

A number of strategies have been employed in attempts to separate the A chain toxicity from the problems involved in B chain binding to nontarget cells. Many chemical modifications that block binding activities of ricins also block their cell entry function. Studies of partial deletion mutants of diphtheria toxin[28] show a good deal of promise in this regard. Laird and Groman[29] intentionally mutagenized *Corynebacterium* with nitrosoguanidine and ultraviolet irradiation and from these preparations isolated several classes of diphtheria mutants. Further characterization of the mutants indicated that three showed full enzymatic and therefore cell-killing activity but had defective receptor binding. Amino acid analysis of the mutations in these three strains was of interest and is shown diagramatically in Figure 10.4. It can be seen that the location for the induced mutations for each of the CRM strains of altered diphtheria organisms was limited. In strain CRM103 a single point mutation was noted at position 508 (Ser → Phe). CRM102 contained a similar point mutation at position 508 but also showed a mutation at position 308 (Pro → Ser). Strain CRM107 had two mutations, one at position 390 (Leu → Phe) and one at position 525 (Ser → Phe). The binding affinity of CRM103 and CRM102

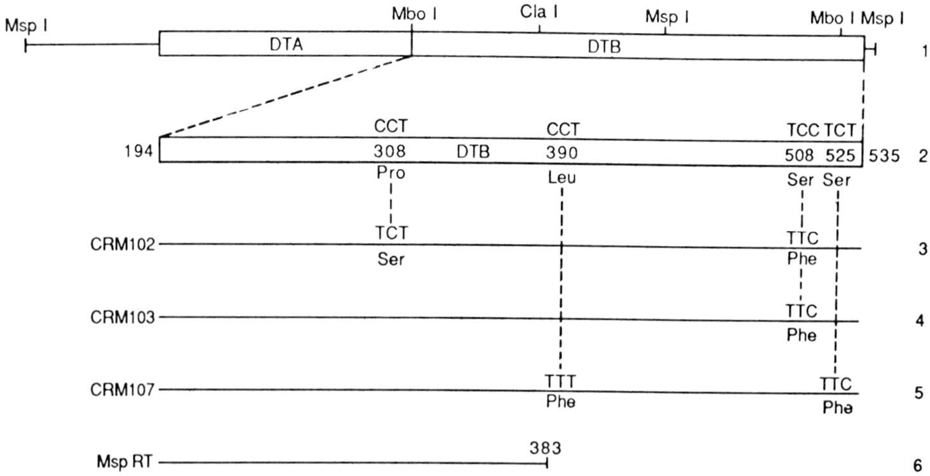

Figure 10.4 Location of the point mutations in diphtheria strains within the structural gene for diphtheria toxin (DT). The restriction enzyme map for DT gene 2 (above) shows the B chain structural region in more detail, indicating the original native amino acid sequence and the point mutations. Line 3 shows mutations found in the B chain of strain CRM102, line 4 shows those in the B chain of strain CRM103, and line 5 shows those in strain CRM107. *Adapted from Greenfield et al,*[28] *with permission.*

was 1/100 that of the original diphtheria toxin, showing that the serine at position 508 was important for toxin binding to cells. Moreover, the conservative change at position 390 in strain CRM107 indicates that the additional mutation at 525 caused an 8000-fold difference in toxin-binding activity.

Additional studies with these toxin mutants further elucidated the sites for binding and cell entry and showed that toxin cell binding and translocation into the cytosol required different segments of the toxin molecule. It was shown that the mutant toxin produced from CRM107 was 1/100 as toxic as native diphtheria toxin but that it also had a large decrease in bystander nontarget cell toxicity both in vitro and in vivo. Thus, the binding activity diphtheria toxin can be separated from the translocation activity and a greatly improved cell type–specific toxin made. This new type of immunotoxin had the potential for combining full killing efficiency of the native toxin with the high specificity of monoclonal antibody. This work was a good example of the impact of recombinant techniques in clinical medicine because it used cloning and DNA analytic techniques together with observations of properties of diphtheria mutants made 12 years before.

Applications of Chimeric Antibodies

The potential usefulness of chimeric antibodies in both tumor imaging techniques and directed chemotherapy was mentioned in the early work[1] on making chimeric antibody molecules from hybrid constructions of mouse and human Ig genes. Other reports have confirmed the practical applicability of this approach[30,31] A chimeric antibody was produced with human constant regions and mouse variable regions directed against carcinoma-associated antigen 17-1A expressed on human colorectal carcinoma cells.[32] The original mouse monoclonal antibody 17-1A had been raised against colorectal carcinoma cells and appeared to react with a cancer-associated surface antigen on the cells.[33] Transfected expression vectors encoding the V regions from the heavy and light chains of the original mouse hybridoma were joined to the DNA segments coding for human $\gamma 3$ and κ constant regions, resulting in production of chimeric IgG with the same binding specificity as the original mouse hybridoma antibody. Since the mouse monoclonal antibody 17-1A had already produced some encouraging results in immunotherapy trials in patients with gastrointestinal cancer,[34,35] a chimeric antibody possessing the same anti–tumor antigen specificity may provide an important additional therapeutic tool. In addition, it was expected that the presence of human C regions would result in an antibody with less immunogenic potential than the whole monoclonal mouse antibody.

Interleukin 2—*Pseudomonas* Exotoxin

A variation on the principle employed in linking ricin or diphtheria toxin to monoclonal antibodies has been developed using genes for human interleukin 2 (IL-2) and a gene for a modified *Pseudomonas* exotoxin.[36] The chimeric protein, called IL-2–PE_{40}, was produced in *E. coli* and was found to be extremely toxic to IL-2 receptor–positive cells but had no measurable effect on cells lacking IL-2 receptors. Previous studies of the *Pseudomonas* exotoxin (PE) had indicated, that as in the case of diphtheria toxin, the gene coding for the entire PE product contained domains encoding cell recognition, a sequence coding for translocation of toxin across the cell membrane, and a sequence for production of the protein responsible for ADP-ribosylation of elongation factor 2 and cell death.[37] Previous studies had shown that construction of a fusion gene product encoding domains II and III of the *Pseudomonas* exotoxin responsible for cell death with DNA coding for transforming growth factor α (TGA-α) produced a fusion protein product that was capable of killing cells expressing TGF-α receptors.[38] It was thought that chimeric product of the *Pseudomonas* exotoxin and human IL-2 would be capable of destroying cells possessing IL-2 receptors, including activated T cells. After construction of the plasmid vector containing the genes for both the *Pseudomonas* exotoxin and IL-2, the cytotoxic activity of the fusion protein product was tested on T cell lines actively expressing IL-2 receptors. Marked cytotoxicity of the fusion protein product was recorded and could be reversed by treatment of target cells with monoclonal antibodies to IL-2 receptors. In the design of the fusion protein it was shown that an intact amino terminus is required for full biological activity of the IL-2 molecule.[39] Also, the cell recognition site for the PE protein was located at the 5′ end of the gene and the amino terminus of the protein. For these reasons the IL-2 was placed at the amino terminus of the PE protein so that the IL-2 receptor–binding sequences were left freely exposed. When a different fusion protein was constructed with the IL-2 at the carboxyl terminus of PE_{40}, no cytotoxic activity for IL-2 receptor–bearing cells was demonstrable.

Activated IL-2 receptors have been found in activated lymphocytes in the inflammatory tissue infiltrates associated with a number of presumed autoimmune disorders, including autoimmune thyroid disease,[40] rheumatoid arthritis,[41] and inflammatory bowel disease,[42] and the IL-2–PE conjugates might eventually be considered in a therapeutic approach to some of these disorders. Moreover, successful treatment of allograft rejection is still the major problem in clinical organ transplantation. Because treatment with antibody to IL-2 receptor prolongs cardiac allograft survival in mice[43] while sparing suppressor T cells and prevents or reverses acute allograft rejection in rats,[44] the IL-2–PE fusion proteins could represent an effective approach to immunosuppressive therapy in transplantation. The preliminary observations indicate that IL-2–PE_{40} is a powerful cytotoxic agent for cells bearing an IL-2 receptor and has relatively low nonspecific toxicity, and this fusion protein might eventually be a useful reagent for treatment of disorders involving activation of large numbers of IL-2 receptor–positive lymphoid cells.

Interleukin 1 Antagonist Protein

Interleukin 1 (IL-1) appears to play a central role in host responses to infections, tissue destruction, autoimmune processes, and the basic inflammatory process.[45,46] Two distinct forms of IL-1, IL-1α and IL-1β, bind equally to IL-1 receptors to produce its biological effects. Because in some chronic inflammatory reactions high levels of IL-1 in conjunction with other potent cytokines, such as tumor necrosis factor (TNF), may lead to tissue destruction and inflammation, several natural down-modulating loops are also present which serve to maintain body homeostasis. One counterregulatory pathway involves stimulation of corticotropin (adrenocorticotrophic hormone, ACTH) and glucocorticoid hormones through IL-1 stimulation of the pituitary-adrenal axis with subsequent inhibition by the glucocorticoids of IL-1 production as well as production of IL-2. In addition, specific inhibitors of IL-1 activity are apparently released from immune cells or macrophages and may down-modulate effects of IL-1. An IL-1 receptor antagonist protein has been purified, cloned, and characterized by Carter and co-workers.[47] This material was originally derived from a human myelomonocytic cell line. After a cDNA coding for the IL-1 receptor antagonist was cloned and sequenced, the mature 22-kd translation product was expressed in *E. coli*. Material derived from this vector acted as a potent inhibitor of binding of IL-1 to IL-1 receptors. This inhibitor was also capable of inhibiting IL-1–induced corticosteroid production in vivo. Such a material, when present in sufficient quantity, might prove useful in many clinical situations characterized by overabundant activation of the immune system, as in generalized connective tissue diseases such as rheumatoid arthritis or systemic lupus erythematosus. The general physiological response to this material is under study.

Cell Surface Receptors for Activated Complement and Immune Complexes

An area at the interface between clinical medicine and the basic research laboratory concerns various cell surface receptors that interact with activated complement components and other portions of immune complexes such as Ig antibody bound to antigen. Basic understanding of such cell surface receptors has been markedly increased by the cloning of cDNA segments coding for several of these receptors, including the C3b/C4b receptor.[48] This receptor is present on the surface of erythrocytes, polymorphonuclear leuckoytes, monocytes, macrophages, B lymphocytes, some T lymphocytes, follicular dendritic cells, and glomerular podocytes.[49,50] These receptors, collectively termed CR1, are involved in the binding of particles or immune complexes to these cells, where C3b has been deposited by complement activation. Since approximately 90% of the CR1 receptors in blood are located on erythrocytes, they are thought to play a major role in the processing and clearance of circulating immune complexes.[51,52]

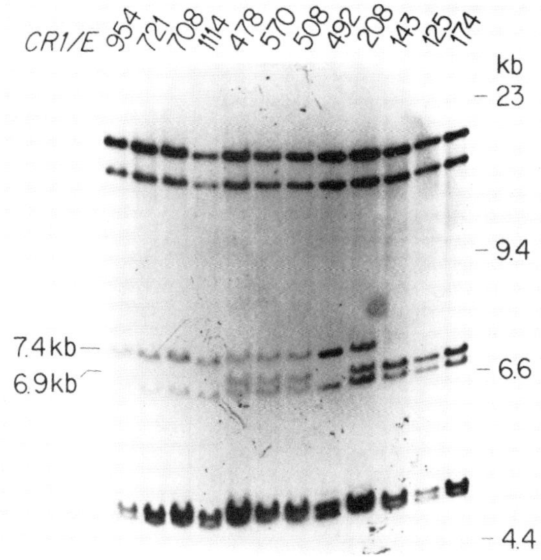

Figure 10.5 Comparison of *Hin*dIII RFLP with number of CR1 expressed on erythrocytes. Genomic DNA from normal individuals was digested with *Hin*dIII. Southern blots of the DNA were probed with ^{32}P-labeled CR1-1, and hybridizing fragments were detected by autoradiography. Numbers of CR1 per erythrocyte are shown above the lanes and sizes of DNA fragments (kilobases) are indicated to the right and left of the autoradiogram. *Reproduced from Wilson et al,[56] with permission.*

A quantitative variation in the number of CR1 receptors per cell has been the subject of intense study. The actual numbers of CR1 expressed by different cell types can vary from 500 sites per cell for erythrocytes to 40,000 per cell for neutrophils, monocytes, or B lymphocytes. Also, erythrocytes from normal individuals may show up to a 10-fold variation in the number of CR1 per cell.[53] A genetic influence on the differences in number of these receptors was noted in analyses of normal families[53] as well as lupus erythematosus kindreds.[53–55] The possibility that this genetically determined difference in the number of CR1 per erythrocyte could be better defined was examined by Wilson and co-workers[56] using restriction fragment length polymorphisms (RFLPs) and cDNA for CR1. A correlation was noted between the RFLP patterns and numbers of CR1 per erythrocyte in DNA samples from individual subjects. This is illustrated in Figure 10.5. Furthermore, family studies showed segregation of certain RFLP patterns with what were considered specific allotypic expressions of CR1. Additional studies by the same group supported a role for genetic factors in the lowered number of CR1 in systemic lupus erythematosus (SLE) patients and their relatives.[57] Direct evidence for inheritance of low CR1 erythrocyte expression was provided by the finding that first-degree relatives had fewer CR1 per erythrocyte than did normal individuals. Using the CR1 probe and *Hin*dIII digestion of genomic DNA, it was found that the 6.9-kb band correlated with low expression of CR1 on erythrocytes but not on polymorphonuclear cells or B lymphocytes, and the 7.4-kb band correlated with high CR1 expression.

Although the reduction in CR1 expression in SLE may clearly be influenced by genetic factors, it may also be influenced by acquired factors.[58] There is a correlation between CR1 levels and disease activity[59] as well as reduced C4 levels.[60] Loss of CR1 from erythrocytes has been directly demonstrated on cells transfused into SLE patients or patients with hemolytic anemias.[61] There is still controversy about the relative influence of genetic factors and acquired factors on expression of these receptors, but the general area again provides an example of how recombinant technologies yield useful new information pertinent to a practical problem.

Fc-receptors

In addition to activated complement receptors, macrophages, granulocytes, and many lymphocytes express or even secrete receptors for the Fc domains of immunoglobulins.[62] A great deal of interest has been focused on Fc receptors in general as possible central switch points or regulating signals for many processes in the immune system.[63] It has been postulated, for instance, that B cell activation may be regulated by several distinct mechanisms, including feedback of secreted antibody products that bind to antigen receptors and to Fc receptors on B cells. This early B cell repressive action could be blocked by rheumatoid factors which, by binding the IgG portion destined to react with Fc receptors on cells and by blocking negative or down-regulating Fc signals, could permit burgeoning immune responses and potential autoimmune responses. This possible central role for rheumatoid factors is illustrated in Figure 10.6. Because of their potential influence on a number of facets of immunoregulation, much interest has been focused on Fc receptors. Complementary DNAs have been cloned from representative genomic libraries for Fc receptor proteins of macrophages and lymphocytes.[64] The Fc receptor gene products were transmembrane glycoproteins of 47 to 60 kd containing four N-linked oligosaccharide chains. Sequence analyses indicated a number of sequence homologies with Ig molecules, indicating that certain Fc receptor components appeared to be members of an immunoglobulin superfamily. In many ways these findings resembled those of the investigations comparing various polypeptide hormones with their specific cell surface or nuclear receptors. These fundamental observations should contribute to understanding as the whole story of Fc receptors unfolds.

Thus, a rough analogy between bacterial surface Fc receptors and the binding sites of human or animal rheumatoid factors has begun to emerge. In brief, for some time human and animal rheumatoid factors have been recognized as probable autoantibodies with reactivity against Fc portions of autologous immunoglobulins.[65,66] Protein A, a bacterial surface component that reacts with the Fc portions

Figure 10.6 (a) B cell negative feedback secondary to cross-linking of antigen receptors and Fc receptors. This shows an initial positive signal (signal 0), which is followed by a turn-off (negative) signal (signal 1 −). (b) Derepression by T cells. This is produced by T cells or their lymphokines arising from cancellation of the negative signal 1. (c) Blockage of negative-feedback Fc signals through interference by Fc-reactive IgM-rheumatoid factors. *Reproduced from Sinclair et al,[63] with permission.*

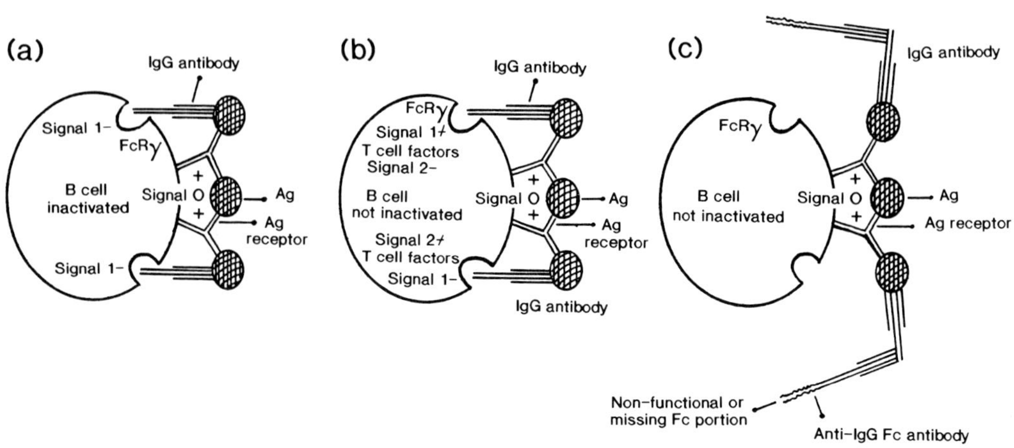

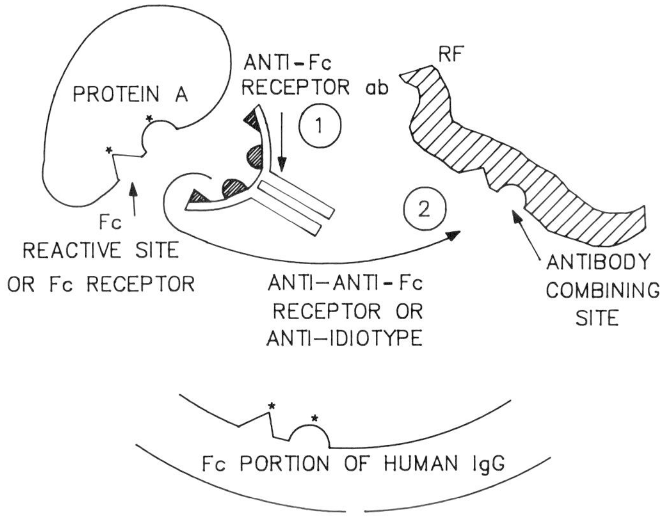

Figure 10.7 Protein A and rheumatoid factor (RF) show virtually identical combining sites for the same conformations on Fc of IgG. In the case of infective endocarditis, an initial antibody response to protein A Fc receptor 1 is followed by an anti-idiotypic response to anti–Fc receptor 2, which could provide for induction of rheumatoid factor by direct idiotypic stimulation of B cells or alternatively could be rheumatoid factor itself.

of human and animal IgGs, has now been found to react with virtually the same epitopes on IgG as does human rheumatoid factor.[67] In fact, human rheumatoid factors have been shown in a series of elegant immunochemical studies to bear the same internal image as the Fc binding region of staphylococcal protein A.[68] There is a large literature on the immunochemical and structural aspects of bacterial Fc receptors such as staphylococcal protein A.[69,70] It also appears that Fc receptors induced on cell membranes by virus infections are similar to Fc receptors already identified on bacteria such as streptococci or staphylococci. A convergence of several diverse lines of thought has therefore suggested that rheumatoid factors might be anti-idiotypic antibodies against virus-induced anti–Fc receptor antibodies.[71,72] This hypothesis is illustrated schematically in Figure 10.7.

Genetic Deficiency of Complement Component C2

Among Western European populations, genetic deficiency of the second component of complement (C2) is one of the most frequently encountered complement deficiencies.[73] The C2 deficiency occurs in 1 in 10,000 whites, and approximately half of the patients described have had associated autoimmune diseases including Henoch-Schönlein purpura, systemic lupus erythematosus, and polymyositis.[74]

Genes coding for the complement proteins C2, C4, and factor B are located in the major histocompatibility complex (MHC) on the short arm of chromosome 6. Family studies suggested that C2 deficiency might be due to a silent allele at the C2 structural locus.[75] Subsequent studies of C2-deficient subjects using a factor B complement cDNA probe (within 1000 base pairs of the C2 gene) provided no evidence for genomic deletions in these patients.[76] Instead, the patients had a specific pretranslational regulatory defect in C2 gene expression that led to lack of detectable C2 mRNA expression and no C2 protein synthesis. The defect was interpreted to result from a defect in production or processing of C2 RNA transcripts. These studies were some of the first to demonstrate a genetic defect based on defective RNA processing rather than a mutation affecting structural genes or a deletion.

Lymphokines—Recombinant Effector Molecules

Perhaps no other area in immunobiology has been strengthened and enriched more in basic recombinant DNA knowledge with strong implications for practical application than that related to lymphokines. Lymphokines are polypeptide products of activated lymphocytes or other immune system cells, such as monocytes and macrophages, that participate in a wide variety of cellular responses, including those that basically regulate the entire immune system. Many lymphokine molecules are released by cells in direct response to antigen. They differ from antibodies in that their makeup does not vary depending on the structure of the inducing antigen but is uniformly the same. Lymphokines may be thought of as intercommunicating hormonelike molecules used by cells of the immune system. Since lymphokines, strictly speaking, are not produced only by lymphoid cells, the general term cytokines has also been used for this class of polypeptides.

Lymphokines are produced throughout the body. Within local areas such as lymph nodes, an inflamed joint, or sections of the body involved in a localized immune reaction such as a tuberculin delayed-type hypersensitivity reaction, lymphokines may act on other cells in the immediate area, producing a paracrine effect. They may also act on the type of cell that is actually producing them in what is called an autocrine effect. Other lymphokines are produced in large quantities and may gain access to the circulation and act in the same way as endocrine hormones with powerful systemic effects. At present, no disease is known to be directly caused by any isolated lymphokine; however, their lack or ineffective action has been implicated in several specific disease states. During the past 10 years the precise cellular origin and cellular effects of a number of lymphokines have been elucidated. With the increase in knowledge of their chemical composition, structure, and primary amino acid sequence, it has become possible to clone cDNAs that will produce recombinant products in appropriate vector systems. Several lymphokines, such as the interferons, have been tried extensively in human phase I trials for efficacy and safety, and others, such as IL-2, have undergone extensive preliminary trials in certain clinical situations. A number of other cytokines and growth factors are being considered as possible therapeutic agents. This section will provide a brief overview of the lymphokine/cytokine field with particular emphasis on the advances generated by recombinant technology.

A number of lymphokines have been named on the basis of their intrinsic biological effect. Thus, B cell–differentiating factor causes activated B cells to differentiate into antibody-producing cells, and what was originally called T cell growth factor (IL-2) causes T cells to proliferate. Naming lymphokines on the basis of a single biological property has produced problems, since most have more than one activity. Therefore, it was agreed that any new lymphokine described would first be named according to the demonstrable biological activity, and once the amino acid sequence of the human form was established, it would be assigned an interleukin number.[77] Thus, the cDNA for human B cell–stimulating factor has now been reported and it is now called interleukin 4 (IL-4). Table 10.1 shows a partial list of lymphokines and their biological properties. The great promise and beginning

Table 10.1 Prominent Biological Properties of Human Lymphokines

Lymphokine	Biologic properties
Interleukin 1 (α and β)	Activates resting T cells; is cofactor for hematopoietic growth factors; induces fever, sleep, ACTH release, neutrophilia, and other systemic acute phase responses; stimulates synthesis of lymphokines, collagen, and collagenases; activates endothelial and macrophagic cells; mediates inflammation, catabolic processes, and nonspecific resistance to infection
Interleukin 2	Is growth factor for activated T cells; induces the synthesis of other lymphokines; activates cytotoxic lymphocytes
Interleukin 3	Supports the growth of pluripotent (multilineage) bone marrow stem cells; is growth factor for mast cells
Colony-stimulating factor (CSF)	
Granulocyte-macrophage CSF	Promotes neutrophile, eosinophile, and macrophage bone marrow colonies; activates mature granulocytes
Granulocyte CSF	Promotes neutrophilic colonies
Macrophage CSF	Promotes macrophagic colonies
Interleukin 4 (B cell–stimulating factor 1)	Is growth factor for activated B cells; induces DR expression on B cells; is growth factor for resting T cells; enchances cytolytic activity of cytotoxic T cells; is mast cell growth factor
Interleukin 6 (B cell–stimulating factor 2) (B cell–differentiating factor)	Induces the differentiation of activated B cells into immunoglobulin-secreting plasma cells; is identical to interferon β, plasmacytoma growth factor, and hepatocyte-stimulating factor
Interferon γ	Induces class I, class II (DR), and other surface antigens on a variety of cells; activates macrophages and endothelial cells; augments natural killer cell activity; has fever-inducing and antiproliferative properties
Tumor necrosis factor (α and β)	Is directly toxic for some tumor cells; induces fever, sleep, and other systemic acute phase responses; stimulates the synthesis of lymphokines, collagen, and collagenases; activates endothelial and macrophagic cells; mediates inflammation, catabolic processes, and septic shock

(Reprinted from Dinarello and Mier,[77] with permission.)

clinical applications of the colony-stimulating factors in terms of marrow rescue after intensive chemotherapy have already been discussed in Chapter 8.

Lymphokines and the Immune Response

Lymphokines provide many interconnecting links during the immune response. As mentioned briefly in Chapter 4, T cells interact with antigens after the antigens have been presented on the surface of macrophages or other antigen-presenting cells in physical association with class II histocompatibility (DR) molecules. When antigens are properly presented to the T cell antigen receptor, activation of the T cell occurs and as a secondary event the activated T cell produces a number of proteins, some of which are secreted and others become components of the cell membrane. Interleukins 1, 2, 3, and 4 (B cell–differentiating factor) and interferon γ are released. In addition, the receptors for IL-2 and other important lymphokines are produced during an antigenic stimulus. Cell activation and lymphokine release are shown in Figure 10.8.

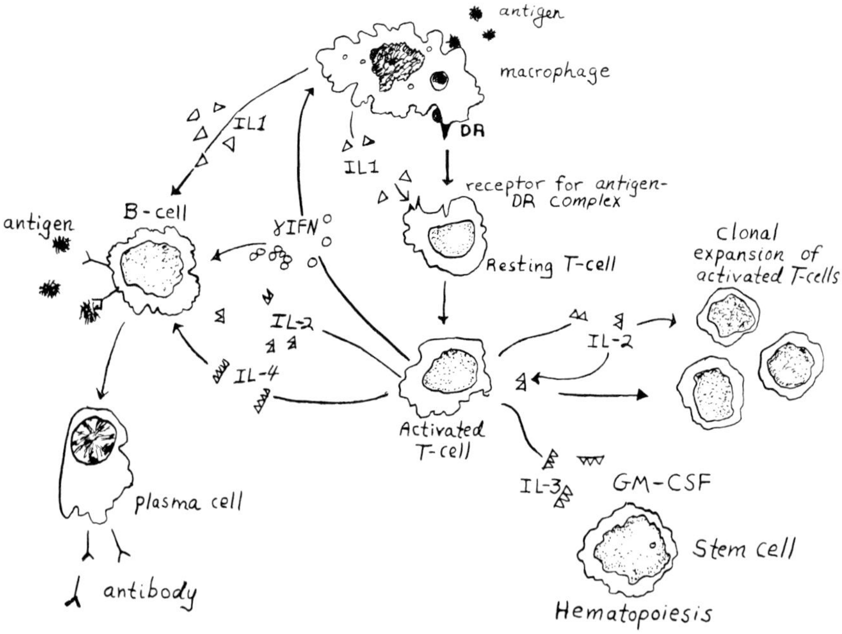

Figure 10.8 Interactions of various immune cells during antigenic activation with functions of various lymphokines. After antigen is processed by macrophages and presented to the resting T cell, in conjunction with the proper DR molecule and with IL-1, the T cell is activated and produces interferon γ (IFN-γ), IL-2, IL-3, and IL-4 plus B cell–differentiating factors. These factors provide additional stimuli for B cells to mature into plasma cells making antibody. Activated T cells also expand under the stimulus of IL-2.

Interferons

Among the first lymphokines to be explored with respect to therapeutic potential were the interferons. These materials were originally identified by their apparent ability to interfere with viral replication in infected cells. Interferon α (originally called leukocyte interferon) and interferon β (initially termed fibroblast interferon) were shown to be structurally related polypeptides stimulated by viral infection and having potent antiviral, antiproliferative, and other immune features. The γ interferons were shown to be structurally different from the α and β interferons and had diverse effects on various immune functions. They showed powerful ability to induce the expression of both class I and class II histocompatibility molecules and acted as potent activators of macrophage function.

Most of the clinical applications of the use of interferons have been in the area of cancer. The rationale for the use of recombinant interferons in tumor patients was based on their ability to activate natural killer (NK) cells as well as their apparent antiproliferative activity. Positive results have been reported in the treatment of lymphomas and some leukemias, including hairy-cell leukemia and chronic myelogenous leukemia.[78–80] Interferon α has also been used in the treatment of metastatic breast cancer, malignant lymphomas, and multiple myeloma and resulted in remissions or partial improvement in approximately one-third of patients.[81,82] The side effects of interferon treatment have been fatigue, asthenia, weight loss, and leukopenia in some patients, but toxicity was reversible on discontinuation of treatment.

Recombinant human interferon α has also been used as an adjuvant to increase the amounts of tumor antigen expressed on cancer cells and augment the localization of radiolabeled monoclonal antibody to collections of tumor cells.[83] This approach took advantage of the ability of interferon to augment expression of various cell surface antigens, thereby rendering tumor cell populations more homogenous and reactive with monoclonal antitumor antibodies. Since as important aspect of the use of monoclonal antitumor antibodies is related to the intrinsic tumor heterogeneity, increasing tumor antigen expression and therefore tumor homogeneity should theoretically make such tumors more vulnerable to therapy with monoclonal antibody linked to chemotherapeutic or toxin molecules. At present, this type of approach appears to offer more promise than the use of interferon preparations as a adjunct to conventional radiation and chemotherapy in cancer.

Some success has been observed in the use of human interferon α in the treatment or modulation of cytomegalovirus reactivation syndrome, observed among renal transplant patients.[84] Clinical signs of cytomegalovirus infection were greatly reduced in interferon recipients, but survival of patients and grafts was approximately equivalent in both treatment and control groups. It appeared that in seropositive renal transplant recipients, interferon α provided effective prophylaxis against complicating cytomegalovirus infection.

As might be expected, a defect in interferon production has been demonstrated in patients with acquired immunodeficiency syndrome (AIDS).[85] Mononuclear cells from most patients studied did not generate effective production of interferon γ in response to mitogenic stimuli; however, after addition of exogenous recombinant interferon γ, the AIDS patients' macrophages showed enhanced intracellular antimicrobial activity. The authors of this study suggested that interferon γ might be considered an ancillary immunotherapy in these patients.

Interferon in Chronic Hepatitis

The fluctuating balance between chronic viral infection and the host that must exist in chronic viral hepatitis made this a reasonable area for trials of recombinant interferon preparations. In one pilot study, eight patients with chronic hepatitis B virus (HBV) infections were treated with interferon γ and interferon α in combination.[86] It appeared that interferon γ by itself had minimal inhibitory effects on serum levels of HBV as determined by analysis of HBV DNA and DNA polymerase activity. There were a number of unpleasant side effects when interferon γ was given, including pain and erythema at the site of injections, aching and malaise, and mild decreases in red blood cell, white blood cell, and platelet counts. By contrast, interferon α had more pronounced inhibitory effects on serum HBV levels and far fewer side effects. Results of a multicenter, randomized, controlled trial of interferon α in the treatment of hepatitis C were published by Davis and collaborators in 1989.[87,88] In this study 166 patients with chronic hepatitis C were randomly assigned to receive treatment with either 3 million or 1 million units of recombinant interferon α three times weekly for 24 weeks or to receive no treatment.

Normalization or near normalization of serum alanine aminotransferase levels after 6 months of treatment was noted in 46% of patients treated with 3 million units of interferon α ($P < 0.001$) and in 28% of the patients who received 1 million units, compared with only 8% of untreated control patients. Patients who received

3 million units showed histological improvement with regression of lobular and periportal inflammation. However, relapse within 6 months after completion of treatment was noted in 51% of patients treated with 3 million units and in 44% of those receiving 1 million units. It was concluded that a 24-week course of interferon therapy was effective in controlling disease activity in many patients with hepatitis C, though relapse after interruption or discontinuance of interferon α injections was common.

Side effects in patients receiving interferon α included nausea, irritability, and depression but were equally as common in the untreated patient group. However, flu-like symptoms, including myalgia, headache, and fever, occurred more frequently in the patients treated with interferon ($P < 0.001$). Typically these symptoms resolved or lessened after the first few doses of the drug. Moreover, diarrhea and mild alopecia were seen in 34% and 24%, respectively, of patients receiving 3 million and 1 million units of interferon α. This study indicated that interferon α might hold out considerable hope for patients afflicted with chronic hepatitis C. However, knowledge of the long-term effects of interferon α therapy on viral replication, infectivity, and the evolution of histological changes in this disease awaits longer periods of study and observation.

A second study published at the same time by the group at National Institute of Health[88] confirmed the apparent efficacy of interferon α in the treatment of chronic hepatitis C infection. This study was conducted with a double-blinded, placebo-controlled protocol. Mean serum alanine aminotransferase levels and the histological picture in the liver improved significantly in the patients receiving interferon but not in the patients injected with placebos. Of the patients treated with interferon α, 48% showed what was classified as a complete response, namely a decline in mean serum alanine aminotransferase levels to the normal range during treatment. Again, however, after treatment was discontinued, serum alanine aminotransferase usually returned to pretreatment levels.

Interferon in Autoimmune Disease

Very little evidence has emerged that interferons might be a useful modality in the treatment of autoimmune disorders. Early reports from several groups provided clear-cut evidence of accelerated progression of autoimmune disease when leukocyte interferon was administered to NZB mice.[89,90] Marked increases in autoimmune hemolytic anemia and renal immune complex deposition were noted after interferon treatment in these mice.

A number of presumed autoimmune disorders have now been treated with interferon α, including multiple sclerosis,[91] rheumatoid arthritis,[92] and immune thrombocytopenia.[93,94] Eight patients suffering from either discoid lupus erythematosus or subacute cutaneous lupus erythematosus were treated by Thivolet and co-workers[95] with low or intermediate doses of interferon α ($18–45 \times 10^6$ U/week) for a short time (4–8 weeks). A marked improvement in the skin lesions was observed in six, an exacerbation in one, and no change in one. Two patients with subacute cutaneous lupus erythematosus were given high doses of interferon α ($100–120 \times 10^6$ U/week) for 12 weeks, with complete clearing of skin lesions in one and marked improvement in the other. Again, positive responses were short-lived, and within a few weeks after treatment was discontinued all patients who had improved or cleared showed relapse.

Interleukin 2

Interleukin 2 is a polypeptide lymphokine of relatively low molecular mass (15 kd) that serves diverse immunological functions. The most widely studied biological activity is the capacity to amplify proliferative responses in activated T cells.[96,97] High-affinity receptors for IL-2 are absent on resting T cells but can be demonstrated within several hours of T cell activation. Interleukin 2 has been used for a number of years as an ancillary agent to help in long-term cultures of T cell lines or to study activated T cell helper, cytotoxic, and suppressor functions. More recently, interest has focused on the ability of IL-2 to stimulate a population of natural killer cells to kill tumor cells through mechanisms that do not require specific interaction of HLA-DR–like histocompatibility antigens. Phase I therapeutic trials have been initiated in which either recombinant IL-2 is administered intravenously to selected cancer patients or their own autologous lymphocyte populations are incubated in vitro with IL-2 before being returned to the patients.[98,99] When autologous lymphokine-activated killer (LAK) cells are used in the treatment of cancer patients, it is thought that the recombinant IL-2 activates endogenous NK or cytotoxic cells, which in turn destroy tumor cells.[99] Initially the most favorable results were observed in patients with melanoma and renal cell cancer. Phase I trials have now been completed in many centers and a number of patients have received this form of treatment—some with striking benefit. To date, the treatment has been associated with side effects that in some patients include fever, hypotension, hepatic and renal failure, myocardial infarctions, and a strange capillary leak syndrome associated with massive edema. More experience concerning the potential benefits of this mode of therapy in parallel with possible side effects is now being accumulated.

The IL-2 receptor was also cloned,[100,101] providing cDNA clones and the sequence for the primary structure of the IL-2 receptor precursor. The availability of this gene has been helpful in developing additional innovative approaches to medical treatment, such as the IL-2–*Pseudomonas* toxin gene fusion product described earlier.[36]

Interleukin 1 and Tumor Necrosis Factor

Since the beginnings of research on the lymphokines, there has been considerable interest in IL-1. As noted previously, two structurally similar forms of IL-1, α and β are produced by virtually all cell types (T and B lymphocytes, natural killer cells, skin keratinocytes, brain astrocytes, microglia, epithelial cells, mesangial cells, and cells of vascular origin.[102–105] Interleukin-1 production can be initiated by antigenic stimuli, toxins, acute injury, and chronic inflammation. This lymphokine acts as an immune modulator and amplifier because it produces a stimulus for the synthesis of other lymphokines such as IL-2. It is also the principal mediator for a wide spectrum of acute phase reactions, including fever through effects on the central nervous system, production in the liver of a number of acute phase reactant proteins, and the release of neutrophils. It also causes release of ACTH, cortisol, and insulin. It may also play an important mediating role in the pathogenesis of the synovitis in such chronic diseases as rheumatoid arthritis by activating proteases and basic mechanisms of bone resorption.

Another lymphokine that has been extensively characterized is tumor necrosis factor (TNF), or cachexin. It has a number of potent systemic effects but, in contrast to IL-1, appears to have no direct influence on lymphocyte activation. Although IL-1 and TNF show the same breadth of effects in terms of acute phase changes, they have different gene and amino acid sequences. A number of studies have indicated that IL-1 and TNF may interact during various systemic reactions such as the Shwartzman reaction and other inflammatory states. Lymphotoxin (sometimes called TNF-β) is a lymphocyte product that has considerable sequence homology to TNF-α and has similar biological effects in many assay systems.

During the past 5 years there has been a virtual explosion of knowledge and insight related to TNF. From a historical standpoint, the potential importance of this biological activity has been recognized for some time. TNF activity was first described in 1975 by Carswell and co-workers,[106] who described a protein in the serum of bacille Calmette-Guérin (BCG)–primed animals, treated with endotoxin, that produced a hemorrhagic necrosis of tumors in vivo or cytostatic/cytotoxic activity against tumor cells in vitro. Several lines of evidence subsequently indicated that monocytes and macrophages are a principal source of TNF.[107,108]

Human TNF is a nonglycosylated protein of 17 kd. It is made up of 157 amino acids and contains two cysteine residues probably involved in a single intramolecular disulfide bond. TNF amino acid sequences are highly conserved in different animal species, which may explain why the biological activity is not species specific. Thus, mouse TNF polypeptide shows approximately 80% homology with the human TNF product.[109] The molecular cloning of human TNF has now been successfully accomplished.[110,111]

Tumor Necrosis Factor Kills Tumor Cells

The first important biological activity ascribed to TNF was hemorrhagic necrosis of tumors. A number of experimental observations indicated that this lymphokine had remarkable effectiveness in producing necrosis in a series of transplanted solid tumors including sarcomas, leukemias, and mastocytomas.[106] With the availability of recombinant TNF, considerable evidence has accumulated that this material may be cytotoxic against a number of human tumor cell lines, although some human tumors were strangely resistant. Direct evidence has also been presented that TNF may be a potent mediator for monocyte lysis of tumor cells. Work with murine or human monocytes/macrophages indicates that sensitivity of tumor cells to monocyte-mediated cytotoxicity correlates closely with their sensitivity to the killing action of purified recombinant TNF. Also, antibodies that neutralize TNF inhibit monocyte/macrophage cytotoxicity against tumor cells. Finally, incubation of target tumor cells with recombinant interferon increased in parallel their susceptitibility to monocyte killing and the cytotoxic action of TNF. The therapeutic potential of recombinant TNF in human tumors is being extensively explored. The diversity of its profound biological effects on a number of systems is truly remarkable. These are listed briefly in Table 10.2.

Now that good polyclonal and monoclonal anti-TNF antisera are available, a number of clinical and laboratory studies have related changes in TNF to various human disease states. An ELISA assay showed that 50% of fresh serum samples from cancer patients had detectable levels of TNF.[112] Elevations of serum TNF were most likely to be detected in samples from patients with ovarian and oat cell carcinoma (69% and 63%) or in lymphoma patients (26%). RNA preparations from peripheral blood mononuclear cells as well as solid tumors frequently expressed TNF complementary sequences, which were virtually absent in parallel controls.

Table 10.2 Biological Effects of Tumor Necrosis Factor (TNF)

Biological effects	Clinical relevance
Inhibition of proliferation or differentiation of hematopoietic precursor cells	May affect results of bone marrow transplants, aid colony-stimulating factors
Cachectin activity	Possibly related to cachexia observed in malignant conditions, and parasitic infections
Endogenous pyrogen	Similar to IL-2 in ability to induce rapid-onset fever (? by central action)
Neutrophil activation	Causes neutrophils to discharge secondary granules and a marked respiratory burst
Effects on endothelial cells	Produces tissue factor–like procoagulant activity
Osteoclast activation and cartilage resorption; antiviral activity	Induces cartilage resorption in vivo; induces interferon β activity
Collagenase and prostaglandin synthesis	Stimulates collagenase and prostaglandin E production
Cross-induction between IL-1 and TNF	Each lymphokine under appropriate circumstances can induce synthesis of the other

In this study it was not clear whether elevated serum levels of TNF reflected host response to the endogenous tumor or inability to metabolize and clear secreted amounts of TNF.

Tumor Necrosis Factor and Gram-Negative Sepsis

The intricate biological mechanisms operative in septic shock appeared to involve TNF when use of monoclonal anti-TNF antibodies was demonstrated to prevent septic shock.[113] These observations were of general interest because previous studies had indicated that bacterial lipopolysaccharides released from gram-negative bacteria were often involved in the genesis of septic shock. Hypotension and irreversible shock were blocked in primates by prior passive administration of monoclonal antibody to TNF. Complete protection against shock, critical organ dysfunction, and death was provided to experimental animals by $F(ab')_2$ fragments of monoclonal anti-TNF antibody given 2 hours before septic bacterial challenge. Furthermore, these interesting studies suggested that antibodies against cachectin/TNF might be potent therapeutic agents for some patients experiencing septic shock.

Another group of patients with marked elevations of serum TNF levels have been those with parasitic (malarial or visceral leischmaniasis) infections.[114] Since 60% to 70% of these patients showed marked elevations of serum TNF, it was thought that the cytokine might play some role in host defense against parasitic infection. In later experimental studies of cerebral malaria in mice, TNF appeared to play a key role in the pathogenesis of the disorder,[115] since antibody to TNF fully protected mice from cerebral malaria without modifying the degree of parasitemia. Furthermore, accumulation and activation of TNF-producing macrophages in the central nervous system lesions was responsible for the virulent course of cerebral malaria.

Further correlations between TNF and clinical infections were provided in a study of patients with meningococcal disease.[116] Serum samples from 79 patients with meningococcal meningitis, septicemia, or both showed TNF in 10 of 11 patients who died but in only 8 of 68 survivors. Again, these findings indicated that changes in serum TNF levels might be helpful as a prognosticator in this disease.[116]

Thus far, discovery and cloning of TNF have led to a number of potentially important insights into disease pathogenesis. Rather than being a vital therapeutic agent, this potent lymphokine may help to clarify some poorly understood areas of disease pathogenesis. The relationship of abnormalities in the handling of TNF-α and TNF-β (lymphotoxin) to various other diseases associated with immune response dysfunction is now being studied. Apparently, the genes coding for tumor necrosis factor and lymphotoxin map are close to the H-2D region of chromosome 17 in the mouse MHC.[117] The TNF genes were closely linked to the HLA-B locus (HLA-B27 allele and ankylosing spondylitis), and it is of great interest that the genes for two potent inflammatory lymphokines are closely linked to this region. In particular, as noted in Table 10-2, TNF-α stimulates collagenase production by human synovial cells[106] and resorption of proteogylcans from cartilage.[118,119]

Other Potentially Important Cytokines or Factors

A number of other lymphokines have been identified, cloned, and sequenced. These include human B cell growth factor (12 kd), which supports the clonal proliferation of B lymphocytes.[120] Further studies with this material will be of interest because of its possible potential use in some forms of immune deficiency.

Recombinant proteins similar in some ways to the lymphokines already mentioned include epidermal growth factor, fibroblast growth factor, and the hormone/growth factors—insulin-like growth factor and nerve growth factor. These products may have some eventual therapeutic usefulness in wound healing, bone repair, and nerve repair. Availability of these materials in homogeneous, well-characterized recombinant form should provide experimental biologists and eventually practicing clinicians with a whole new therapeutic armamentarium.

Currently, there seems to be a gap between the precise knowledge available through recombinant techniques, such as primary gene sequences and therefore amino acid sequences, and knowledge of the way in which many of the recently characterized messenger molecules are metabolized or exert their biological effects in the intact organism. Good examples of the hiatus between highly refined knowledge of lymphokine sequence and structure and understanding of the intrinsic actions of these materials in specific disease processes are related to the studies of IL-2 production in several connective tissue diseases, such as systemic lupus erythematosus (SLE) and rheumatoid arthritis (RA). In SLE, it was found that cultured lymphocytes from patients with the disease had a defect or decreased production and response to IL-2[121,122] Since IL-2 is required for clonal expansion of both helper and suppressor T cells, it was not clear whether decreased IL-2 production in SLE was due to suppressive influences in the disease or to a synthetic defect in diseased cells. Subsequently it was found that removal of OKT8$^+$ lymphocytes (presumably of suppressor phenotype) increased IL-2 production in SLE to normal or above-normal levels. Thus, in a disease of presumed fundamental immune dysregulation, decreases in IL-2 production appeared to be secondary to spontaneous or intrinsic suppressor cell activation. Although initially defective IL-2 production was documented in SLE, most clinicians would have been reluctant to administer recombinant IL-2 to SLE patients for fear of activating harmful self-directed immune responses, such as precursor B lymphocytes capable of producing increasing amounts of anti-DNA antibody.

Similarly, specific abnormalities have been documented with respect to IL-2 in RA. Surprisingly little direct evidence of IL-2 was obtained when synovial tissues from patients with RA were examined for IL-2 directly by immunofluorescence.[124] Studies by Emery and others[125] indicated that depressed cell-mediated immune responsiveness in patients with RA could be reversed by in vitro addition of IL-2. Later it was demonstrated that peripheral blood and synovial fluid suppressor cells appeared to be capable of reducing IL-2 responsiveness in RA patients.[126] These apparent abnormalities in IL-2 have not as yet been fitted together to help specific therapy in the disease, and it is clear that a great deal more needs to be learned about the networks by which lymphocytes communicate before these observations can provide a new therapeutic approach.

Oncogenes and Autoimmunity

Several of the oncogenes, such as c-*myc* and c-*myb*, appeared to show marked increased in expression during the course of a number of autoimmune diseases in mice. For some time it had been recognized that normal events in cell activation, including B cell stimulation of T cell activation, were associated with increased c-*myc* expression.[127] Moreover, because several strains of mice developing murine lupus also appeared to be more susceptible to various malignancies, including leukemias and malignant lymphomas, it was of interest to study the behavior and expression of *myc* and other oncogenes in these animal models. The first indication that an increase in oncogene expression might occur in autoimmune disease was reported by Mountz et al.,[128] who studied the lymphoid organs of *lpr*/*lpr* mice, which ordinarily develop an autoimmune-lymphoproliferative syndrome. Lymph node tissues of these animals contained increased amounts of *myb* RNA, but in parallel T cells activated by mitogens did not express large amounts of *myb* RNA. Spleens of NZB and BXSB mice (other SLE-prone murine strains) also showed two to three times more c-*myc* RNA than did those of nonautoimmune mice. This is illustrated by the patterns in Figure 10.9. These findings appeared to reemphasize the presence of abnormal lymphoid proliferation in autoimmune diseases. Additional studies of oncogene expression in autoimmune mouse strains revealed that cyclophosphamide (Cytoxan) therapy markedly reduced the abnormal expression of the *myb* gene in the *lpr*/*lpr* mouse in parallel with a decrease in both lymphadenopathy and disease activity.[129] Despite the large increment in lymph node *myb* RNA, these *lpr*/*lpr* mice showed diminished thymic *myb* RNA. These findings were considered compatible with a basic thymic abnormality in such mice that might be related to their subsequent deranged immunological development. Thus, low *myb* expression in the primitive thymus was later associated with an increase in peripheral T cells that showed disordered maturation and high *myb* RNA expression. No evidence was obtained for any intrinsic mutation in the original *myb* RNA, and no corresponding increases in *myc*, *ras*, *fos*, *abl*, or *erb* B oncogene RNA were found in parallel.

Human Systemic Lupus Erythematosus

When human SLE was also studied for oncogene expression,[130] it was found that peripheral blood mononuclear cells had marked elevations of c-*myc* and that SLE patients with active disease had significantly elevated N-*ras* mRNA expression. In parallel, diminished levels of c-*myb* and c-*fos* were found in the SLE group. Patients

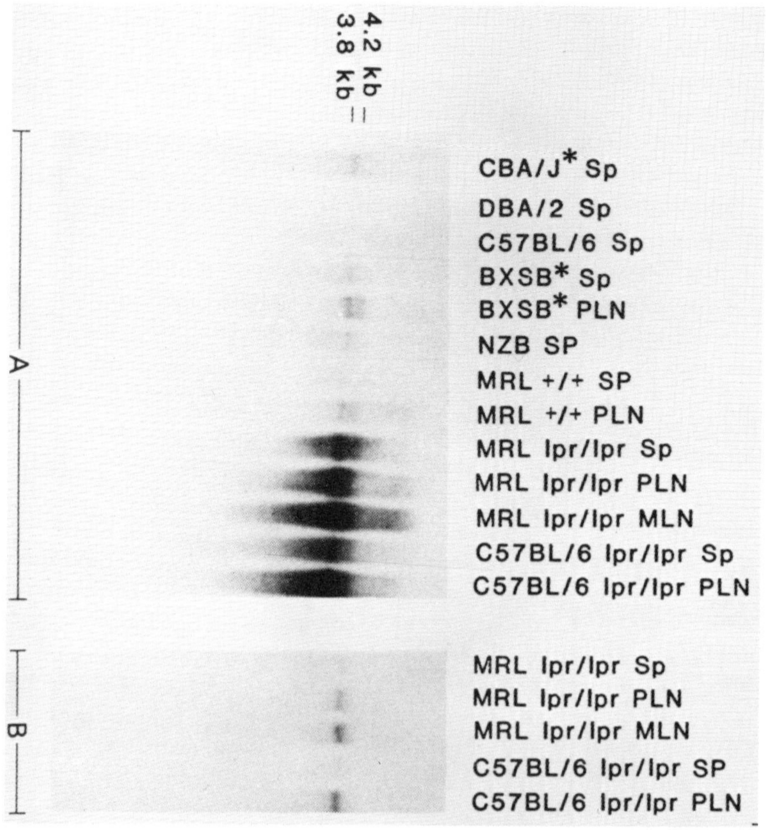

Figure 10.9 Hybridization of v-*myb* probe to poly(A) RNA from spleens (Sp), peripheral lymph nodes (PLN), and mesenteric lymph nodes (MLN) of 6-month-old mice of strains indicated. The v-*myb* probe was a 1.3-kb fragment obtained from cloned avian myeloblastosis virus. All mice were female except those designated with an asterisk. *Reproduced from Mountz et al,*[128] *with permission.*

with Sjögren's syndrome, rheumatoid arthritis, and angioimmunoblastic lymphadenopathy also had elevations of c-*myc* RNA in peripheral blood lymphocyte populations. These findings were thought to reflect the intense B cell activation phenomena previously documented in patients with SLE. Whether the abnormalities in c-*myc* and N-*ras* mRNA expression in SLE represented intrinsic defects in gene regulation, increased lymphocyte activation, or amplification of particular lymphoid subsets has not been determined. However, these early studies emphasized the involvement of oncogene activation in the immune system stimulation.

C-*fos* Gene Expression in Human Monocytes

As with the c-*myc* activation during certain phases of normal lymphocyte activation, it has been shown that normal monocytes, when activated by a chemoattractant stimulus, may show a marked up-regulation of c-*fos* gene expression.[131] For some time it has been known that the cellular responsiveness of polymorphonuclear leukocytes or monocytes is mediated by a guanine nucleotide regulatory G protein, which in turn activates a membrane-associated phospholipase C to cleave phosphatidylinositide 4,5-bisphosphate to inositol 1,4,5-triphosphate and 1,2-diacylglycerol. The last compound activates protein kinase C, which in turn triggers O_2 production and release of lysosomal enzymes.[132,133] When normal human monocytes were treated with chemoattractants such as the peptide FMLP or platelet-activating factor (PAF), there were marked increments in gene expression for the c-*fos* proto-oncogene.[131] Also, the activation of c-*fos* expression was mediated through a pertussis toxin–sensitive G protein. Additional studies showed that virtually all agonists that could activate protein kinase C in human monocytes also elevated c-*fos* mRNA levels, which indicated that protein kinase C had an effect on gene expression in these cells. Interestingly, parallel measurement of another proto-oncogene, c-*myc* showed a decrease in gene expression during chemoattractant monocyte activation. Again, these studies showed that normal nonmalignant cellular activation of various components of the immune system can cause major increments in expression of proto-oncogenes. They also show the great sensitivity of recombinant techniques for defining key steps in such areas.

T Cell Receptors and Autoimmune Disease

As discussed in Chapter 8, rearrangements of T cell receptors have proved valuable in many cases of leukemia and lymphoma for making a diagnosis in terms of lineage assignment and in other perplexing clinical disorders, particularly those involving cutaneous lesions. For instance, in Sézary's syndrome they have helped in determining whether a malignant process is present.

Thus far, little is known about possible abnormalities of T cell receptor genes in other nonmalignant conditions. In 1985 Kotzin and co-workers[134] described a large deletion in the T cell receptor β chain gene complex in New Zealand white (NZW) mice. The deletion was identified as an 8.8-kb segment of DNA containing the Cb, Db2, and Jb2 gene cluster. In spite of the fact that all NZW T cell receptors must originally be derived from a single set of β chain gene segments, this strain has functional T cells and appears phenotypically normal. It was of great interest that a deletion of T cell receptor β chain genes occurred in a strain of mice known to contribute genetic features associated with lupus-like autoimmune disease in mice of the New Zealand black (NZB) × NZW F_1 hybrids.

The murine mouse model of systemic lupus erythematosus is best exemplified in the NZB × NZW F_1 hybrid. However, the NZB strain alone shows mostly an autoimmune hemolytic anemia as a manifestation of autoimmune diathesis. The NZW mice appear normal and have normal life spans without autoimmune hemolytic anemia, immune complex glomerulonephritis, antinuclear or anti-DNA antibodies, or other features of the murine lupus syndrome. Precisely how the NZW strain contributes to the induction of SLE in the NZB × NZW hybrid strains has never been clearly defined.[135]

T cell antigen receptors located in the MHC are disulfide-linked heterodimer molecules made up of α and β chains.[136] Their structure is very similar to those of the Ig genes, with variable (V), joining (J), and constant (C) region gene segments (Chapter 4). Beta chain cDNAs contain a set of diversity (D) gene segments as well. The β chain complex contains two highly similar C region genes, $C\beta_1$ and $C\beta_2$, each joined to a cluster of Jβ gene segments and a single Dβ segment. Duplication of Dβ, Jβ, and Dβ segments is distributed throughout all mouse strains and in humans as well. However, the significance of this duplication is not yet clear. The deletion discovered in the β chain locus of the NZW mice[134] was uncovered on routine Southern blot screening of DNA samples using a labeled β chain cDNA

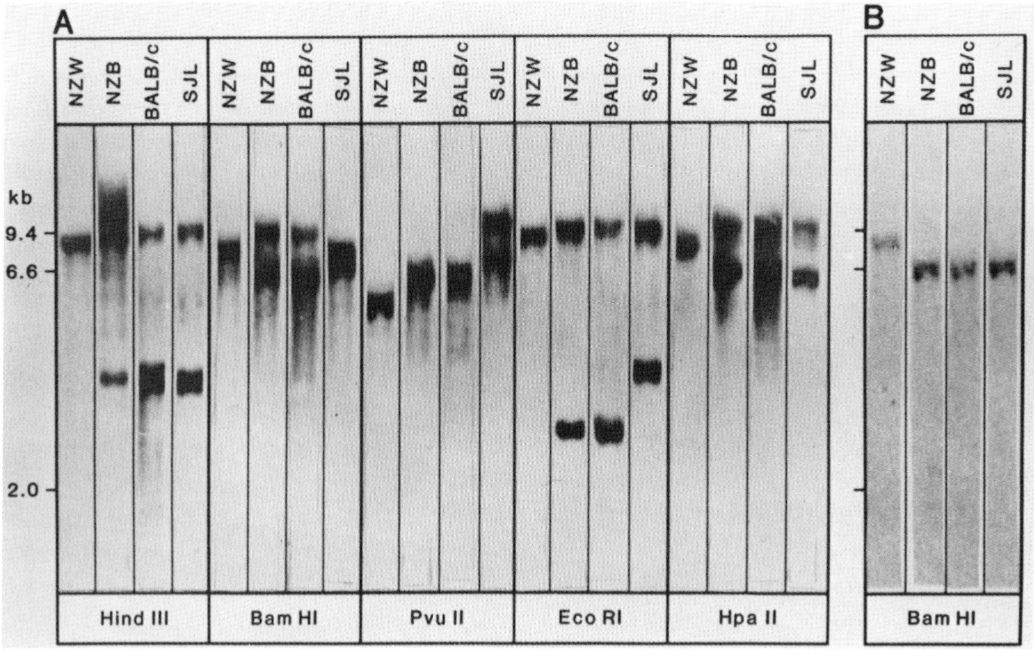

Figure 10.10 Southern blot analysis of liver DNA with a Cβ probe using various strains of mice and digestion with different restriction enzymes. The Cβ probe used was derived from a $C\beta_2$ cDNA that hybridizes to restriction fragments containing $C\beta_1$ and $C\beta_2$. The probe produces only six different restriction enzymes. *Reproduced from Kotzin et al.*[134] *with permission.*

probe, which hybridized with *both* $C\beta_1$ and $C\beta_2$ germ-line genes. When this probe was tested with restriction fragment enzyme digests of DNA from a number of mouse strains, in all cases the NZW DNA produced only one band. This is illustrated in Figure 10.10. Since it was extremely unlikely that five different restriction enzymes would cut the different strains' DNAs so as to produce two Cβ fragments of similar length, it became apparent that the NZW mice carried a single Cβ gene. This was confirmed using probes that distinguished $C\beta_2$ from $C\beta_1$. Additional β region probes were utilized to define the size of the β chain deletion in the NZW strain. When this deletion was discovered, T cell phenotypes, in vitro responsiveness to mitogens, and primary in vitro antibody responses to various antigens were compared in NZW mice and BALB/c strains and no differences were detected. No specific T cell defect that could be ascribed directly to the β chain T cell receptor deletion could be defined.

Subsequent studies of the T cell β gene defect in NZW mice included screening of 18 strains of mice for other gross alterations in T cell receptor genes.[137] Within the limitations of RFLP analysis, the C region loci of the T cell receptor—α and β chains—appeared to be highly conserved among autoimmune, *lpr* congenic, and normal mouse strains. The NZW T cell β chain deletion was further characterized and found to originate in the first 167 base pairs of exon 1 of the $C\beta_1$ gene (residues 18 through 149) and extend to the complementary region of the first exon of the $C\beta_2$ gene. The deletion was thought to be a limiting factor in the ability of the NZW strain to generate T cell diversity in this aspect of their immune response.

Again, it was not really clear whether the β gene defect in the NZW mice was related to their autoimmune predisposition.

Other quantitative studies of T cell receptor gene expression in various autoimmune and *lpr/lpr* mice by several groups have provided more background information on c-*myb* expression in *lpr/lpr* animals[138] and the clonal diversity of T cell receptor β chain variable gene expression[139] in other autoimmune mouse strains. Thus far, no other T cell receptor abnormalities have been identified in these or other forms of autoimmune disease.

T Cell Receptor Restriction in Human Autoimmune Disorders

To date, there seems to be only limited evidence for any oligoclonality or specificity in the T cell receptor (TCR) repertoire in the human autoimmune diseases that have been examined. Oksenberg and co-workers[140] have used the polymerase chain reaction (PCR) to search for restrictions or unique characteristics of TCR V_α transcripts in the brains of patients with multiple sclerosis. In each of three multiple sclerosis brains examined, only two to four rearranged TCR V_α transcripts were detected. Moreover, no V_α transcripts could be detected in control brains studied in parallel. Sequence analysis of the transcripts encoded by the V_α 12.1 region showed rearrangements limited to a small repertoire of J_α region segments. These results indicated that TCR V_α gene expression was to a certain extent restricted in lesions in brains of multiple sclerosis patients. These findings are important because in the mouse experimental model for multiple sclerosis—experimental allergic encephalomyelitis (EAE)—the animals express a predominant or restricted set of TCR α and β gene products in their response to injected myelin basic protein.[141–143] Moreover, treatment of experimental mice who have EAE in vivo with monoclonal antibody specific for the major or predominant TCR V region product reversed the experimental EAE. Elucidation of TCR restriction or specific elements in TCR patterns of human patients might eventually result in a similar strategy for elimination of the disease.

γδ T Cell Receptors and Human Disease

As indicated in Chapter 4, the discovery a few years ago of a second major molecular form of TCR consisting of γδ (gamma delta) subunits was not immediately applied to analysis of specific human disease states, although cells with γδ receptors appeared earlier in ontogeny and fetal development than did cells with conventional αβ TCR structure. Early studies indicated, for instance, that T cells expressing the γδ TCR were capable of cytolytic function, particularly against antigens related to the so-called heat shock proteins.[144,145] A series of observations by Rajagopalan and co-workers[146] appears to point to T helper cells with γδ TCRs as being important in production of pathogenic anti-DNA autoantibody in patients with active lupus nephritis. These workers established T helper cell lines from T cells activated in vivo from five patients with lupus nephritis. Only 15% of the 396 IL-2–dependent T cell lines could augment production of pathogenic anti-DNA autoantibodies. The majority of the T helper cell lines studied showed the αβ T cell receptor; however, about one-sixth of the T helper cell lines were $CD4^-CD8^-$, and of these 7 of 10 showed the γδ TCR. It appeared that heat shock proteins or epitopes simulating the latter might be involved in stimulating autoreactive proliferation of the γδ T cells. More information relating specific oligoclonality or restriction of TCRs in various autoimmune disorders in clearly needed.

Recombinant Approaches to Autoantigens

Antinuclear antibodies are a prominent feature of the human lupus syndrome, and over the years there has been intensive study directed at characterizing their immunochemical makeup and their sites of origin in several of the connective tissue diseases. The spectrum of potential lupus autoantigens has been reviewed by several authors[147,148] and appears to focus mainly on nuclear components. Progress has been made in characterizing nuclear antigens that react with antibodies in the sera of patients with lupus or mixed connective tissue disease, particularly in studies of the small ribonucleoproteins such as RNP and Sm.[149–153.]

A molecular cloning approach antigen identification has been applied to only a few of these putative lupus autoantigens. Isolation and analysis of cDNA clones expressing human lupus La antigen were reported by Chambers and Keen.[154] Several clones producing human La antigen and reacting with lupus autoantibodies were isolated from λgtll expression libraries made from human liver. Protein translated in vitro reacted with anti-La sera from patients and appeared to be identical to the original La protein by peptide mapping. Using overlapping cDNA clones, an antigenic site of the La protein at the terminal 12% of the carboxyl end of the molecule was identified. Shortly thereafter, SLE sera reacting with the Sm antigen were used to isolate a cDNA clone for the mRNA of one of these nuclear antigens.[155] In this instance a HeLa cell cDNA library was screened for autoantibody reaction with cell-free translation products, and a cDNA clone containing sequences reacting with presumed Sm autoantibody was identified. The cloned Sm antigen reacted with several lupus anti-Sm antisera; however, precise sequence data or other structural analyses are still forthcoming.

One nuclear antigen previously implicated as an autoantigen has now been more precisely identified; the so-called proliferating cell nuclear antigen (PCNA) has been identified as a DNA polymerase S auxiliary protein.[156] In these studies a 36-kd cellular protein was identified that was previously shown to be a component necessary for origin of elongation in SV40. The antigen, PCNA, closely resembled a protein that regulated the activity of calf thymus DNA polymerase. Because both proteins, PCNA and the polymerase S auxiliary protein, had similar electrophoretic behavior and reacted with human anti-PCNA autoantibodies, it appeared that PCNA was the same as the DNA polymerase S protein. This work illustrates how parallel studies of the molecular biological processes involved in a particular reaction contributed to final identification of the nuclear autoantigen. Subsequent molecular work on native antigenic epitopes of the PCNA has revealed an amazing degree of unique specificities using individual sera from SLE patients.[157]

Cloning of Putative Thyroid Autoantigens

In the area of thyroid disease, studies using recombinant techniques have focused on pig and human thyroid peroxidase (TPO) genes as a probable source of the thyroid microsomal antigen. As work progressed in this area, it became evident that a major part of the microsomal antigen did appear to be TPO.[158] A pig thyroid cDNA library was constructed in λgtll because an antibody against pig TPO was readily available, and oligonucleotides were made to screen the same library. The probes were synthesized from sequences derived from four peptides isolated after tryptic digestion of highly purified pig TPO. Final sequencing of the antigen has now been completed. Three antigens of great potential interest in relation to thyroid autoimmune disease are the thyroid microsomal antigen, thyroglobulin, and the thyroid-stimulating hormone (TSH) receptor. Screening of expression libraries with human autoantibodies as well as highly characterized monoclonal antibodies may speed the search and help in the identification of all human thyroid autoantigens of interest.

Recombinant Studies of Autoantibody–Rheumatoid Factor

Recombinant DNA approaches to understanding the genetic basis for the production of several autoantibodies have provided insight into immunoglobulin gene selection in this process. For some time it had been known that certain human anti-γ-globulin antibodies or rheumatoid factors shared cross-idiotypic structural antigenic features.[159] This shared cross-idiotypic specificity was similar to that shown originally for the IgM cold agglutinins.[160] In the cross-idiotypic studies of human IgM rheumatoid factors two separate antigenic groups (the major Wa and minor Po designations) had been defined with anti-idiotypic antisera.[159] Subsequently it was shown by Ledford et al.[161] that a number of human monoclonal autoantibodies including rheumatoid factors utilized the relatively uncommon V region subgroup κ IIIb light chains in making up the antibody structure. Amino acid sequences on the amino ends of 10 κ chains isolated from such monoclonal IgM autoantibodies were virtually identical. These findings were an early indication that monoclonal IgM human rheumatoid factors appeared to have a relatively marked restriction in their selection of V region light-chain κ genes.

Based on the previously defined antigenic cross-reactivity between isolated human rheumatoid factors, Chen and co-workers[162] utilized sequence data defined by Andrews and Capra[163] on the variable regions of two Wa^+ IgM rheumatoid factors. A monoclonal antibody was produced against a cross-reactive antigenic epitope shared by both Wa^+ rheumatoid factors. In addition, a synthetic peptide corresponding to the known sequence of the third complementarity-determining region (CDR) of the IgM Wa heavy chain was prepared and used to generate antisera. This peptide, which constituted a 13-mer, was useful in further studies of idiotypic antigens associated with a group of monoclonal rheumatoid factors. The antibody to the peptide reacted with IgM rheumatoid factor proteins known to share a cross-reactive (Wa) idiotype; it also reacted with the isolated κ chains from these proteins. These studies, therefore, sharpened the impression of marked antigenic and structural restriction among monoclonal human rheumatoid factors.

The use of synthetic peptides corresponding to the third hypervariable regions of human monoclonal IgM rheumatoid factors was subsequently extended.[164] Only antisera against the third CDR consistently bound to the intact IgM rheumatoid factors studied, suggesting that this third CDR of human μ chains is an immunodominant antigenic site and must be important in generation of idiotypic diversity. Antisera against synthetic peptides were further used in an analysis of a panel of human IgM rheumatoid factors by Chen et al.[165] A peptide corresponding to the second CDR of one rheumatoid factor light chain induced an antibody against a cross-reacting idiotype present on 10 of 12 human monoclonal rheumatoid factors.[165] Five other anti-idiotypic antibodies were generated by immunization with synthetic peptides identical to the third L chain CDR of the same rheumatoid factor and the second and third H chain CDRs of a second and third rheumatoid factor. It was concluded that only one or very few V_L genes were involved in production of IgM rheumatoid factors but a larger number of V_H genes might be involved.

The complete amino acid sequences of five light-chain variable regions from human monoclonal IgM-κ rheumatoid factors were analyzed in parallel with assays of their reactivity with the antipeptide antisera against the second and third CDR sequence.[166] Similar to previous findings, it was shown that all seven rheumatoid factors belonged to the V_κIIIb subgroup. These studies indicated that the majority of human IgM rheumatoid factor light chains were derived from a single germ-line V_κ gene or gene family and expressed two structure-dependent cross-reactive idiotypes that were largely dependent on the amino acid sequences at the second and third light chain CDRs.

Genomic studies were next carried out on the molecular origins of rheumatoid factors. In order to identify human germ-line genes related to the putative rheumatoid factor V_κ sequence, genomic DNA from a normal human placenta was examined using an 865-bp probe of a human V_κIII germ-line gene.[167] This probe shared 92 of 96 amino acids with the prototype V_κ rheumatoid factor sequence. Clones obtained by hybridization were subcloned and sequenced to establish the sequence of the entire putative V_κ rheumatoid factor gene, which is shown in Figure 10.11. This sequence identification from genomic human DNA proved that the light-chain variable region of the IgM human rheumatoid factor was encoded by a germ-line gene without any somatic mutation. Thus, the V_κ rheumatoid factor gene is not confined to patients with monoclonal IgM rheumatoid factor cryoglobulins but was also identified in DNA from normal individuals.

By contrast, studies of the makeup of the V_H genes encoding human IgM rheumatoid factor autoantibodies are less indicative of a major restriction. Amino acid sequences of two rheumatoid factor heavy-chain variable regions belonging to the Wa idiotypic group showed differences.[163] The findings suggested that rheumatoid factor V_H regions may be encoded by several V_H genes in the germ-line repertoire. Experiments on the gene precursors for an entirely different autoantibody—anti-DNA antibody—have indicated that both germ-line and somatically mutated Ig variable region genes may function to encode these anti-DNA antibody molecules.[168,169] More extensive analysis is needed to determine what proportions of many other autoantibodies are derived from germ-line genes and/or somatic mutations.

When complete amino acid sequences of variable regions of two human IgM rheumatoid factors of the Wa idiotypic family were carefully analyzed,[170] some relative restriction in heavy-chain variable and joining segments was apparent. It appeared that the heavy-chain variable regions were all derived from the V_H1 subgroup of human V region genes. The IgM rheumatoid factors studied showed relative conservation in the size of the D region, and all used the J_H4 gene segment. It then began to appear that besides the previously noted V_κIIIb light-chain restriction, there might be some relative restriction with H chain variable region genes as well.

Murine Models of Rheumatoid Factors

Parallel studies of the immunoglobulin genes responsible for production of rheumatoid factors in mice have provided a slightly different interpretation.[171] Studies in mice had indicated that during the secondary immune response to protein an-

```
                                                                                                       ******** pentadecamer
   1 T C A C C T G G G C A T G G G C T G C T G A G A G C A G A A A G G G G A G G C A G A T T G T C T C T G C A G C T G C A A  60
                       .                   .       T           G                   .       G           .   A           G   .

     *******                                     ***** decamer *****
  61 G C C C A G C A C C C G . . . C C C C A G C T G C T T T G C A T G T C C C T C C C A G C C G C C C T G C A G T C C A G A 120
                       .     C C C         .                   .                   .                   .                   .

             ** TATA box ***
 121 G C C C A T A T C A A T G C C T G G G T C A G A G C T C T G G A G A A G A G C T G C T C A G T T A G G A C C C A G A G G 180
                       .                   .                   .   G     G     A   .                   .                   .

               Met   Glu   Thr   Pro   Ala   Gln   Leu   Leu   Phe   Leu   Leu   Leu   Leu   Trp   Leu   Pro
 181 G A A C C A T G G A A A C C C C A G C G C A G C T T C T C T T C C T C C T G C T A C T C T G G C T C C C A G G T G A G G 240
                       .                   .                   .                   .                   .                   .

 241 G G A A C A T G G G A T G G T T T T G C A T G T C A G T G A A A A C C C T C T C A A G T C C T G T T A C C T G G C A A C 300
                       .                   .                   .                   .                   .                   .

 301 T C T G C T C A G T C A A T A C A A T A A T T A A A G C T C A A T A T A A A G C A A T A A T T C T G G C T C T T C T G G 360
                 G   A .                   .                   .                   .                   .                   .

 361 G A A G A C A A T G G G T T T G A T T T A G A T T A C A T G G G T G A C T T T T C T G T T T T A T T T C C A A T C T C A 420
                   G   .                   .                   .                   .     A             .                   .

  -4 Asp   Thr   Thr   Gly   Glu   Ile   Val   Leu   Thr   Gln   Ser   Pro   Gly   Thr   Leu   Ser   Leu   Ser   Pro   Gly    16
 421 G A T A C C A C C G G A G A A A T T G T G T T G A C G C A G T C T C C A G G C A C C C T G T C T T T G T C T C C A G G G 480
                       .                   .                   .               C   .                   .                   .
                                                                              Ala

                                               24------------------------------CDR1-------30A-------------------34
  17 Glu   Arg   Ala   Thr   Leu   Ser   Cys   Arg   Ala   Ser   Gln   Ser   Val   Ser   Ser   Ser   Tyr   Leu   Ala   Trp    35
 481 G A A A G A G C C A C C C T C T C C T G C A G G G C C A G T C A G A G T G T T A G C A G C A G C T A C T T A G C C T G G 540
                       .                   .   G               .                   .                   .                   .
                                              Gly

                                                                                         50-------------------------CDR2
  36 Tyr   Gln   Gln   Lys   Pro   Gly   Gln   Ala   Pro   Arg   Leu   Leu   Ile   Tyr   Gly   Ala   Ser   Ser   Arg   Ala    55
 541 T A C C A G C A G A A A C C T G G C C A G G C T C C C A G G C T C C T C A T C T A T G G T G C A T C C A G C A G G G C C 600
                       .                   T       G           .                   .       A           .                   .
                                          Leu                                             Asp

     --56
  56 Thr   Gly   Ile   Pro   Asp   Arg   Phe   Ser   Gly   Ser   Gly   Ser   Gly   Thr   Asp   Phe   Thr   Leu   Thr   Ile    75
 601 A C T G G C A T C C C A G A C A G G T T C A G T G G C A G T G G G T C T G G G A C A G A C T T C A C T C T C A C C A T C 660
                       .                   .                   .                   .                   .                   .

                                                                                   89-----------------CDR3------------95
  76 Ser   Arg   Leu   Glu   Pro   Glu   Asp   Phe   Ala   Val   Tyr   Tyr   Cys   Gln   Gln   Tyr   Gly   Ser   Ser   Pro    95
 661 A G C A G A C T G G A G C C T G A A G A T T T T G C A G T G T A T T A C T G T C A G C A G T A T G G T A G C T C A C C T 720
                       .                   .                   .                   .                   .                   .

         **heptamer***                         **** nonamer ****
 721 C C C A C A G T G A T T C A G C T T G A A A C A A A A A C C T C T G C A A G A C C T T C A T T G T T T A C T A G A T T A 780
                       .                   .                   .                   .                   .                   .

 781 T A C C A G C T G C T T C C T T T A C A G A T A G C T G C T G C A A T G A C A A C T C A A T T T T A G C A T C T C T T C 840
                       .                   .                   .                   .                   .                   .

 841 T C T G C T T G G G C A T T T T G G G G A T C   863
```

Figure 10.11 Sequence of V_κ rheumatoid factor (RF) compared with probe for V_κIII, Humkv305/1. Nucleotides and the deduced amino acids of the V_κ RF gene appear above, and differences from Humkv305/1 are shown below. Dots at positions 73 to 75 were introduced to maximize homology with Humkv305/1. The pentadecamer, decamer, TATA box, leader sequence, heptamer, and nonamer are indicated. Complementarity-determining regions (CDRs) are indicated, and the extra codon for CDR1 is placed at position 30A. *Reproduced from Radoux et al,*[167] *with permission.*

tigens, immune complexes could stimulate normal B cells to produce IgM anti-IgG and, the amount of IgM anti-IgG could equal or even exceed the amount of antigen-specific antibody.[172] Moreover, the fraction of B cells in normal mice that can produce IgM rheumatoid factor is estimated to be relatively high. Three to ten percent of hybridomas produced from mice stimulated by the B cell mitogen lipopolysaccharide produce IgM that binds IgG.[173] Thus, production of rheumatoid factors by normal mice stimulated in these ways resulted in IgM rheumatoid factor production in a relatively large proportion of the B cells available. Studies of nine such murine IgM rheumatoid factors based on amino acid sequence data showed that a large number of different V_H genes were used in association with a variety of D_H, J_H, and J_κ genes in combination with what appeared to be a limited set of V_κ genes.[174] Further studies were performed on 15 IgM anti-IgG1 mouse hybridomas producing rheumatoid factors. The monoclonal antibodies were sequenced to obtain comparative information on their complement of V region, J region, and D region genes. It was found that multiple germ-line genes were used. Of 25 rheumatoid V_κ sequences, four were $V_\kappa 8$, four were $V_\kappa 19$, and two were $V_k 1$; in addition, two were $V_\kappa 24$ and two were $V_\kappa 4$. In parallel, looking at the repertoire of mouse V_H genes used, 12 different V_H J558 family sequences were detected among 15 rheumatoid factor samples. In mice the V_H germ-line genes fall into 10 sequence homology families. The V_H J558 family includes about 60% of the genes; the V_H 7183, V_H J606, and V_H Q52 families each contain at most 10% of the genes. The 100 to 200 V_κ genes appear to be organized into approximately 30 groups, each of which includes 1010 genes. When these baseline data were compared to the V region representation in the 15 mouse IgM rheumatoid factors studied, there appeared to be some restriction to the $V_\kappa 1$, $V_\kappa 8$, and $V_\kappa 19$ light-chain genes; by contrast, the representation of each of the four major V_H families was consistent with random selection and did not favor any degree of prior restriction. Several laboratories are now examining the precise arrangement and construction of additional mouse and human rheumatoid factors to determine whether any additional pattern of selective Ig gene construction may be involved in their production.

Thyroid-Stimulating Antibodies and IgG Restriction

Another autoantibody of interest with respect to restricted gene participation during manufacture is found in the group of IgG thyroid-stimulating antibodies (TsAb) in patients with Graves' disease. Patients with high levels of long-acting thyroid stimulator (LATS) showed this particular biological activity associated with λ IgG molecules, as established by selective immunoabsorption.[175] We have confirmed this interesting λ IgG TsAb predominance in some Graves' patients,[176] although in others κ and a mixture of κ and λ IgG molecules showed TsAb activity. Studies of the C_λ gene allelic frequencies, moreover, showed moderate linkage disequilibrium and overrepresentation of the 8/8 C_λ allele in Graves' patients, particularly those with clinically obvious eye involvement. In the case of Graves' patients, the precise relationship between shifts in Ig gene alleles and the production of relatively restricted TsAb of IgG λ type in some patients has not yet been clarified. Moreover, apparent oligoclonal restriction of thyroid-stimulating antibodies has been shown to segregate not only in IgG λ molecules but also in IgG of the IgG1 H chain subgroup.[177] The genetic or molecular basis for this has not yet been defined. Studies of Ig gene representations in other autoimmune disorders are clearly indicated, since a similar shift in C_λ gene representation has also been reported in association with rheumatoid arthritis.[178]

Recombinant Methods as an Approach to Identifying Putative Autoantigens

Recombinant strategies have been employed to characterize and further identify what are presumed to be potential autoantigens responsible for disorders such as polymyositis, ankylosing spondylitis, and Reiter's syndrome. In the case of polymyositis, a disease of unknown etiology, several lines of evidence suggest that viruses may activate molecular mimicry mechanisms, eventually resulting in an autoimmune inflammatory reaction against autologous host muscle and connective tissue structures.[179] The anti–Jo-1 antibody response in some patients with polymyositis may be an example of a virus-induced change in an autologous antigen.[180] It has been suggested that viral RNA complexed with histidyl-tRNA synthetase might be capable of overcoming immune tolerance and inducing a potentially self-directed immune response.

Another possible explanation for the association between polymyositis and anti–histidyl-tRNA synthetase is molecular mimicry. This is based on the presumption that many viruses probably share antigenic epitopes with normal host tissue secondary to direct homologies in primary amino acid sequence. The probability of identical pentapeptide sequences occurring in completely unrelated proteins is high, and even when the sequence homology requirement is extended to hexapeptides, such similarities are not uncommon.[181]

Polymyositis is a rare autoimmune disease whose primary site of injury is skeletal muscle. When antibodies to the synthetases of histidyl-tRNA,[179] alanyl-tRNA, and threonyl-tRNA[180,181] were identified in association with this disorder, it did not help to explain why the disease affected primarily muscle, since these enzymes appear within cells in many tissue locations. Studies of primary sequence data in *E. coli* histidyl-tRNA synthetase were performed in an attempt to test the limits of molecular mimicry.[182] Utilizing a computer program, a number of possible homologies were found between sequences in histidyl-tRNA and portions of both myosin and Epstein-Barr virus protein.[183] A linear sequence of 20 amino acids of the Epstein-Barr virus protein showed 13 amino acids with homology to two linear stretches of histidyl-tRNA synthetase separated by an 8-amino-acid gap and 12 amino acids homologous with two linear portions of myosin light chain separated by a 57-amino-acid gap. It is difficult to draw conclusions based on such comparisons because conformation of cross-reacting epitopes may not depend only on the primary sequence of comparable peptides. The use of primary sequence data in this instance provided only indirect support for the idea that molecular mimicry might be responsible at least in part for initiation of the disease. It is therefore of interest that antibodies to human skeletal muscle myosin have now been documented in serum from a large number of patients with polymyositis.[184]

Ankylosing Spondylitis, Reiter's Syndrome, and Molecular Mimicry

Another possible molecular mimicry detected by computer analysis of primary protein sequence data has been reported for ankylosing spondylitis and Reiter's syndrome and a protein from *Klebsiella*.[185] In this instance, since the complete amino acid sequence for the HLA-B27 gene product was known, a computer search using the Dayhoff data bank showed a 6-amino-acid exact homology with *Klebsiella* nitrogenase (QTDRED). When synthesized with slight appropriate extensions at either end, this peptide showed positive ELISA reactions with sera from patients with ankylosing spondylitis as well as Reiter's syndrome.[185] Moreover, similar peptides containing the QTDRED sequence prepared from the sequence of HLA-B27.1 itself reacted in ELISA with serum samples from the same patients. These findings show the new types of insights that molecular approaches may bring to rather old established problems.

Examination of the molecular mimicry hypothesis in ankylosing spondylitis

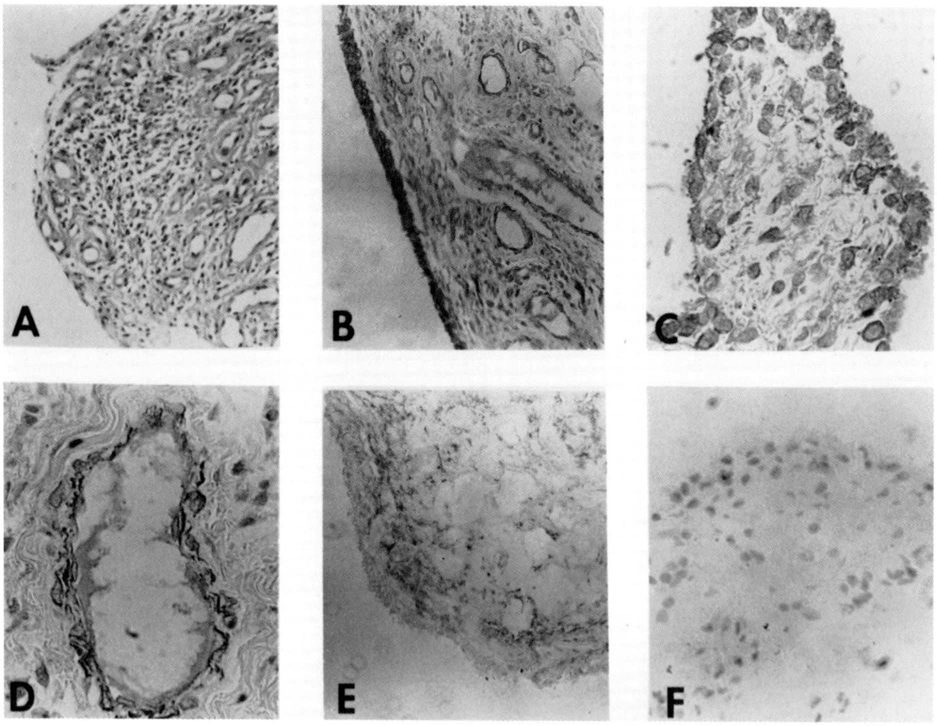

Figure 10.12 Photomicrographs of synovial tissue sections from patient 4N who had ankylosing spondylitis (AS). (A) Formalin-fixed and paraffin-embedded section showing mild inflammation and hypervascularity, but no marked synovial lining cell proliferation (hematoxylin and eosin stained, original magnification × 160). (B) Parallel section of the same synovial specimen, stained with rat anti-HLA-B27.1 peptide (antiserum A), showing marked staining of synovial lining cells but less staining of endothelial cells. Nuclei are counterstained with 1% methyl green. The same staining pattern was obtained with anti-HLA-B27.1 peptide 46 (antiserum B) (original magnification × 160). (C) Higher power view of synovial villus from the same section shown in (B) stained with antipeptide 43 (antiserum A) and showing distinct expression of HLA-B27.1 peptide by synovial lining cells (original magnification × 400). (D) Small vessel in a deeper portion of the same AS synovium demonstrating endothelial cell membrane localization of HLA-B27.1 peptide (antiserum A stained, original magnification × 400). (E) The same synovial tissue after absorption of anti-B27.1 peptide antiserum with B27.1 peptide showing abolition of positive staining. (F) The same synovial tissue stained with normal rat serum as a control. Only cell nuclei counterstained with methyl green are shown (original magnification × 400). *Reproduced from Husby et al,* [187] *with permission.*

and rheumatoid arthritis by members of our group has confirmed that HLA-B27–positive patients with ankylosing spondylitis have that react with peptides sharing the QTDRED sequence in the hypervariable region of the HLA-B27 molecule.[186] Moreover, direct immunoperoxidase and immunofluorescence studies of synovial tissues from patients with both ankylosing spondylitis and rheumatoid arthritis clearly showed cross-reactive staining for antibodies to both HLA-B27 peptides or *Klebsiella* peptides.[187] An example of these reactions is shown in Figure 10.12. Much is still to be explored with respect to molecular mimicry mechanisms and many of the so-called autoimmune disorders. Application of oligonucleotide probes to detect possible shared homologies of gene sequences in affected or diseased tissues may eventually provide another dimension in which to examine this hypothesis.

Cloning of Common Allergens

Since they were first identified, common allergens used in clinical medicine for skin testing and eventual systematic desensitization procedures have presented a problem in isolation and purification. Allergy to mites of the genus *Dermatophagoides* is associated with allergic disorders such as asthma, severe rhinitis, and atopic dermatitis.[188,189] A number of studies have focused on the allergens present in material produced by these mites, and one of the major mite allergens, designated *Der* p 1, has been purified and shown to react specifically with anti–mite IgE antibodies in up to 80% of sera from allergic individuals.[190,191] The mite allergen, which appears to be a glycoprotein of 27 kd, is an excretory product associated with fecal particles. Until recently, the diagnosis and immunotherapy of house dust mite allergy had been based on the use of crude mite preparations. In order to provide purified materials that could be more easily standardized, cDNA clones coding for the *Der* p 1 allergen were digested with restriction enzymes and ligated to M13-derived sequencing vectors.[192,193] The gene was then sequenced by the dideoxynucleotide chain termination method.[194] The complete sequence was shown to be 857 bases in length with a coding region for the *Der* p 1 protein of 222 amino acids. Comparative sequence analyses showed that the *Der* p 1 protein exhibited considerable homology with a group of cysteine proteases. These comparative analyses are shown in Figure 10.13. Homology analysis showed a 29% correlation with actinidin, 24% with papain, 28% with cathepsin H, and 21% with

Figure 10.13 Composite alignment of the amino acid sequences of *Der* p 1, papain, actinidin, cathepsin H, and cathepsin B (pre and pro regions omitted). Gaps were added to maximize similarity, and the arrows indicate residues making up the active site of papain as deduced by x-ray crystallography. *Reproduced from Chua et al,*[193] *with permission.*

```
                                       10        20             30        40        50
                      Der p 1: TNACSINGNAPAEIDLRQMR..TVT..PIRMQGGCGSCWAFSGVAATESAYLAHRNQS
            Rat cathepsin H:          YPSSMDWRKKG.NVVS..PVKNQGACGSCWTFSTTGALESAVAIAS.GK
Chinese gooseberry actinidin:          LPSYVDWRSAG..AVV..DIKSQGECGGCWAFSAIATVEGINKITS.GS
              Papaya papain:          IPEYVDWRQKG..AVT..PVKNQGSCGSCWAFSAVVTIEGIIKIRT.GN
            Human cathepsin B:          LPASFDAREQWPQCPTIKEIRDQGSCGSCWAFGAVEAISDRICIHTN..
                                                             ▲   ▲▲▲

                                   60          70        80        90
                      Der p 1: LD..LAEQELVDC...ASQHGCHGDTIPRGIEYIQHNGVVQ................
            Rat cathepsin H: MM.TLAEQQLVDCAQNFNNHGCQGGLPSQAFEYILYNK....GI.............
Chinese gooseberry actinidin: LI.SLSEQELIDCGRTQNTRGCDGGYITDGFQFIINDG....GI.............
              Papaya papain: LN.QYSEQELLDC..DRRSYGCNGGYPWSALQL.VAQY....GI.............
            Human cathepsin B: VSVEVSAEDLLTCCGSMCGDGCNGGYPAEAWNFWTRKGLVSGGLYESHVGCRPYSIPP

                                           100       110       120           130
                      Der p 1: ......ESY..YRYVAREQSCRRPNAQRFGISNYCQIYPPNANK...IRE.ALAQPQR
            Rat cathepsin H: ...MGEDSY.PYI..GKNGQCKFNPEKA...VAF.VKNVVNITL...NDEAAMVEAVA
Chinese gooseberry actinidin: ...NTEENY.PYT..AQDGDCDVALQDQKY.VTI.DT.YENVPY...NNEWALQTAV.
              Papaya papain: ...H.YRNT.PYY.EGVQRYCRSR.EKGPY.AAKTDGVRQVQPY...NQG.ALLYSIA
            Human cathepsin B: CEHHVNGSRPPCTGEGDTPKCSKICEPGYSPTYKQDKH.GYDSYSVSNSEKDIMAEI.

                               140          150       160           170       180
                      Der p 1: YCRHYWT....IKDLDAFRHYDGRTIIQRDNGYQ.PN..Y.HAVNIVGYSNAQGVDYW
            Rat cathepsin H: LY..NPVSFAFEV.TEDFMMYKS.GVYSSNSCHKTPD.KVNHAVLAVGYGEQNGLLYW
Chinese gooseberry actinidin: TY..QPVSVALDAAGDAFKQYAS.GIFTGP.C...GT.AVDHAIVIVGYGTEGGVDYW
              Papaya papain: N...QPVSVVLQAAGKDFQLYRG.GIFVGP.C...GN.KVDHAVAAVGY..NPG..YI
            Human cathepsin B: .YKNGPVEGAFSV.YSDFLLYKS.GVYQ....HVTGEMMGGHAIRILGWGVENGTPYW
                                                                      ▲

                               190       200       210       220
                      Der p 1: IVRNSWDTNWGDNGYGYFAANIDLMMIEEYPYVVIL
            Rat cathepsin H: IVKNSWGSNWGNNGYFLIERGK....NMCGLAA..CASYPIPQV
Chinese gooseberry actinidin: IVKNSWDTTWGEEGYMRILRNVG.GAGTCGIAT..MPSYPVKYNN
              Papaya papain: LIKNSWGTGWGENGYIRIKRGTGNSYGVCGLYT..SSFYPVKN
            Human cathepsin B: LVANSWNTDWGDNGFFKILRGQ....DHCGIESEVVAG..IPRTD
                                  ▲▲▲
```

cathepsin B. Sequence homologies with these cysteine proteases were particularly high at both the amino-terminal and carboxyl-terminal ends. It appeared that the residues making up the active site of *Der* p 1 based on those derived from papain were very highly conserved. This homology and sharing of sequences between the house dust mite antigen and cysteine proteases of both plant and animal origin was quite a surprise, although it was reported that cysteine protease activity had been discovered in the growth medium of the mites as well as in purified *Der* p 1 preparations.[193] Now that considerable sequence data on the mite *Der* p 1 allergen are available, it may be possible to further define the properties of the molecule that make it so allergenic. Also, its proposed proteolytic activity may have some relationship to the production of allergic reactions. In this regard, it has also been reported that clinical use of chymopapain to treat discogenic root pain or domestic exposure to papain may be associated with a high level of hypersensitivity reactions.[195]

Genetic Probes for Immune Disorders

During the past 10 years there has been a great deal of effort to link specific HLA or MHC markers to diseases thought to occur on the basis of immunological mechanisms. Although this subject could be a topic for a complete book itself, a few comments are in order on the importance and recent directions of this area of investigation. Thus, a disease such as insulin-dependent diabetes mellitus (IDDM) has evolved from being perceived as an endocrine hormone deficiency disorder to one clearly involving disordered immunological mechanisms—principally by the growing appreciation of a well-defined HLA association[196] as well as the involvement of activated T cells and anti–islet cell antibodies early in the course of the disease. In IDDM, the high predisposition of HLA-DR4$^+$ subjects, alluded to in Chapter 7, can be enhanced in terms of disease expression by the presence of HLA-B8 or DR3. The same HLA-B8-DR3 combination has also been associated with myasthenia occurring in young adulthood, Graves' disease, and celiac disease.

Since T lymphocytes recognize foreign antigens that are presented by antigen-presenting cells, which function by physically associating antigenic peptides or components with clefts formed in HLA class II antigens, these associations between HLA-DR types and diseases may be crucial for T helper cell activation and a fundamental part of the immune response. In studies of mice, amino acid substitutions in the class IIA-A_β chain molecules led to loss of immune responsiveness to acetylcholine receptor, bovine insulin, and the H-Y antigen by unaltered helper T cells.[197] Several immunologically mediated diseases appear to have strong association with class II HLA genes; it might be postulated that in some situations particular variations of class II molecules generate the ability of helper T cells to induce self-directed immune responses.

In many disorders, despite an apparent genetic predisposition to the disease, only a few potentially susceptible individuals actually go on to develop the disease. This has led to numerous attempts to subdivide various HLA-DR associations. In the case of HLA-B27, this has not yet provided any additional useful insight, but in other disorders such as myasthenia, HLA-DQ subtypes within HLA-DR2 have shown extremely strong associations with manifest disease.[198] Attempts to subclassify HLA-DR associations with other diseases have shown some promise. For some time it has been apparent that there is an association between seropositive rheumatoid arthritis and HLA-DR4.[199–201] A number of approaches have been tried

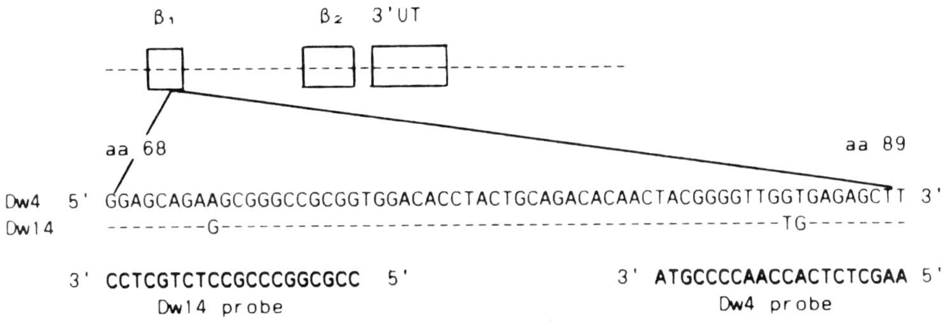

Figure 10.14 Construction of allele-specific oligonucleotide DNA probes to distinguish between the Dw4 and Dw14 DR β genes. The nucleotide sequence for the DR4 β gene corresponding to amino acids 68 to 69 in the 111.A β chain is shown. The three nucleotide differences between Dw4 and Dw14 genes within the exon and the 20-nucleotide probes used for detection of Dw4- and Dw14-specific sequences are indicated. *Reproduced from Nepom et al,[206] with permission.*

in attempts to understand this association. Structural analyses of the HLA-DR4 antigenic specificity have provided evidence that HLA-DR4 represents a serological specificity that can be identified on multiple allelic products of a single gene within the β locus of HLA-DR.[202,203] Subset specificities in DR4 have therefore been sought in an attempt to provide a stronger link with rheumatoid arthritis. DNA sequence analyses have indicated that certain DR4 alleles may differ by 3 to 12 nucleotides, which in turn produce differences in 2 to 7 amino acids in the HLA-DR β polypeptide sequence.[204]

Although rheumatoid arthritis is a Dw4-associated disease, seropositive juvenile rheumatoid arthritis (JRA) is associated with both Dw4 and Dw14.[205] To study the possible contribution of Dw14 in adult rheumatoid arthritis, Nepom et al.[206] constructed allele-specific oligonucleotide probes to detect nucleotide differences in genomic DNA that could be used to discriminate between Dw4 and Dw14 DR β genes. The construction of the oligonucleotide probes in this study was based on sequences shown in Figure 10.14. The Dw4 cDNA probe corresponded to a 20-nucleotide sequence coding for amino acids 83 and 84 in the Dw4 gene but was mismatched by 2 nucleotides from Dw14. On the other hand, the Dw14 DNA probe corresponded to a 20-nucleotide area encoding amino acids 68 to 74 and showed only a 1-nucleotide mismatch with the Dw4 sequence. The hybridization conditions employed with these probes allowed a positive result only when the probe was 100% homologous with the tested sample. Studies with these highly specific probes in patients with rheumatoid arthritis indicated that six of seven individuals homozygous for DR4 showed a Dw4 DR β gene; one of the DR4/5 heterozygotes was positive for Dw4. Of the seven DR4-homozygous patients, six had a Dw14 DR β gene and five of the individuals were heterozygous for Dw4 and Dw14. The unexpectedly high frequency of Dw14 in the DR4 rheumatoid arthritis patients suggested that the Dw14 DR β gene in addition to the Dw4 gene might be linked with susceptibility to the disease. Work along these lines continues in efforts to subdivide HLA associations with particular disease states.

Another interesting investigation involved the possible influence of HLA-DR–related human genes on specific autoantibody formation in patients with Sjögren's syndrome.[207] It has been demonstrated that Sjögren's syndrome is associated with

B8, DR3[208]; antibodies to Ro/SSA and La/SSB are found in virtually all Sjögren's syndrome patients, and many of the extrasalivary gland aspects of the disease are often associated with elevations of these two apparent autoantibodies.[209] Analysis of the occurrence of antibodies to Ro/SSA and La/SSB indicated that particular combinations of the DQ1 and DQ2 (HLA-DR) haplotypes were associated with production of high titers of these autoantibodies. Only patients who were heterozygous for the DQ1 and DQ2 haplotypes showed a positive correlation. A difference was also noted in IgG levels in the DQ1/DQ2-positive patients and was reflected in the incidence of rheumatoid factor and antinuclear antibody. From these observations, a number of possible explanations emerged. First, it seemed possible that gene interaction at the HLA-DQ or a contiguous locus could lead to marked enhancement of immunological responsiveness. Second, there could be an important interaction with DR3, also known to be associated with autoimmune disorders. When the Sjögren's patients were examined in parallel for the expression of DR3, a remarkable concurrence with DQ2 was shown. These findings suggested a model in which DR3 or a subset of DQ2 molecules closely associated with DR3 increases the risk of developing primary Sjögren's syndrome. On the other hand, the specific interaction between DQ1 and DR3 or a particular DQ2 markedly increases the relative amounts of autoantibodies such as Ro/SSA and La/SSB related to Sjögren's syndrome.

Genetic Marker for Amyloidosis

Systemic amyloidosis is a rare but extremely serious chronic inflammatory disease. With the deposition of amyloid, blood vessels become surrounded by dense fibrous deposits, and amyloid accumulation in the kidney frequently leads to chronic irreversible renal failure. Generalized amyloid deposition is an often fatal complication of systemic-onset juvenile rheumatoid arthritis. Previous long-term studies have indicated that amyloidosis may develop in up to 10% of these children within 10 years after onset of the disease; moreover, mortality remains extremely high despite steroid or cytotoxic therapy. The systemic manifestations of the arthritis often include high fevers, marked elevations in sedimentation rate, and high levels of acute phase proteins including C-reactive protein and the serum amyloid A (SAA) protein, which is thought to be the precursor protein of amyloid fibrillar deposits.

An attempt has been made to predict which individuals afflicted with systemic-onset JRA might be predisposed to develop amyloidosis.[210] Genomic DNA samples prepared from peripheral blood were digested with a large panel of restriction endonucleases, but only *Msp*I yielded a polymorphic pattern. Restrictions digests of DNA were examined with labeled SAA or SAP (serum amyloid P) cDNA probes. The amyloid patients showed only one 8.8-kb band, whereas other groups, including normal controls and JRA patients without amyloidosis, had both 8.8- and 5.6-kb bands or homozygosity for the 5.6-kb allele. Moreover, 71% of the amyloid group were homozygous for the 8.8-kb band. Thus the polymorphic *Msp*I site appeared to be related to the eventual development of amyloidosis in systemic-onset JRA. When the SAA and SAP probes were used, correlation of RFLP polymorphisms with amyloidosis was observed only with the SAP probe. Since SAP is always associated with amyloid fibers, these results suggested that SAP is directly involved in the development of amyloid deposition in these patients. Whether this correlation is actually related to the amyloidosis remains to be determined; however, these studies represent an interesting application of recombinant methodology to an important clinical problem.

Allergic Diathesis

Any clinician who has dealt with patients with allergic disorders, such as hay fever, asthma, or eczema, is aware that in many instances a strong genetic or family tendency must be involved in contributing the disease. Cookson et al.[211] analyzed a group of family responses to common inhaled antigens in parallel with molecular genetic analysis. In seven kindreds the transmission of atopy was linked with a maximum load score of 5.58 to an RFLP DNA polymorphism that was defined by p lambda MS.51. This appeared to confirm dominant inheritance and the gene locus on chromosome 11 in these kindreds. Considerable extensions of this sort of family analysis will be necessary before the relevance of these observations becomes apparent. One is reminded of the burst of enthusiasm when the cystic fibrosis gene was cloned, followed by the apparent bursting of the bright bubble when so many private mutations began to surface.

Gene Defects in Diseases of the Immune System

Although X-linked agammaglobulinemia was one of the earliest of the major immunodeficiency syndromes described,[212] the precise gene defect associated with this condition is not yet fully delineated. X-linked agammaglobulinemia is usually associated with onset of repeated severe bacterial infections during the first few years of life, marked reduction in the concentrations of serum immunoglobulins, and a proportion of B cells in the peripheral circulation that is less than 1% of normal.[213,214] Despite this, approximately normal numbers of pre-B cells are detected in the bone marrows of affected individuals.[215] The inability of the pre-B cells to differentiate into mature B cells may be secondary to an intrinsic defect in the B stem cell, or it might result from an abnormality in accessory cells necessary for normal B cell differentiation.

Studies of two sisters who were heterozygous for both X-linked agammaglobulinemia and another X-linked trait, the alleles of glucose-6-phosphate dehydrogenase,[216] were conducted to examine possible genetic leads in this disorder. During normal female differentiation, random inactivation of part of one X chromosome occurs in each somatic cell and is then transmitted to all the descendants of that cell.[214] Thus, in a woman who is heterozygous for an X-linked gene, an average of half the cells in the each tissue will express each allele. Exceptions to this sort of segregation or mosaic expression would indicate origin from a common precursor such as a clonal tumor or a possible selective advantage of cells expressing the genes on one X chromosome over cells expressing the genes on the other.[217,218] If the fundamental gene defect in X-linked agammaglobulinemia is contained within the B cell line and the gene product is not transportable between cells, one would expect B lineage cells, from women heterozygous for a normal and an X-linked agammaglobulinemia gene, to express only the X chromosome carrying the normal allele, whereas the other lineage would show normal mosaicism. Studies of the two heterozygous sisters mentioned above indicated that T lymphocytes and neutrophils showed approximately equal amounts of the A and B alleles of glucose-6-phosphate dehydrogenase enzymatic activity, but B cells showed activity of only the A form of glucose-6-phosphate dehydrogenase.[216] A

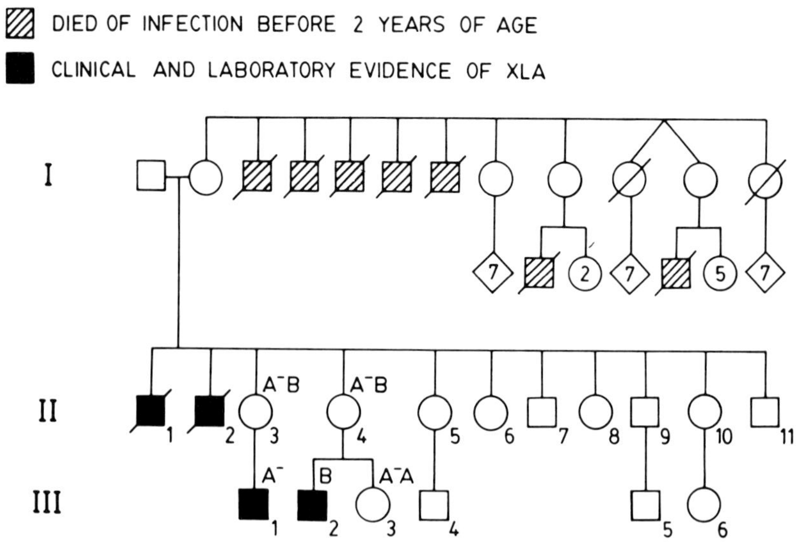

Figure 10.15 Pedigree demonstrating X-linked inheritance of agammaglobulinemia (XLA). The squares denote males and the circles females. Diagonal lines indicate subjects who have died. The glucose-6-phosphate dehydrogenase phenotype is shown above and to the right of the symbol for the subjects studied. *Reproduced from Conley et al,*[216] *with permission.*

diagram of the kindred studied including the two afflicted sisters is shown in Figure 10.15. The unbalanced expression of glucose-6-phosphate dehydrogenase alleles in the B cell lines of the subjects studied but not in the T cells suggested that the gene defect in X-linked agammaglobulinemia was intrinsic to the B cell lineage and did not influence survival of T cells. If the X-linked agammaglobulinemia gene product were derived from cells other than B cells (such as T cells or other cells in the bone marrow microenvironment), the cells with the normal X-linked agammaglobulinemia gene on the active X chromosome could provide the message necessary for normal differentiation of B cell precursors. The finding that a single glucose-6-phosphate dehydrogenase allele was expressed in both uncultured and transformed B cells strongly supported the concept that normal function of the X-linked agammaglobulinemia gene was necessary in each developing B cell for B cell differentiation to occur.

Based on these observations, another group of investigators used the technique of X chromosome inactivation analysis to attempt to detect female members of families with X-linked agammaglobulinemia who might be carriers of the disorder.[219] To do this, they attempted to determine whether one of the two X chromosomes was preferentially active in the peripheral B cells of female carriers. The analysis was based on three assumptions. First, it was assumed that a single X chromosome would be active in each female cell and that the pattern of X chromosome inactivation in each stem cell would be transmitted in a stable fashion to progeny cells. Second, it was assumed that certain genes on the X chromosome had different DNA methylation patterns depending on whether they were located on the active or inactive X chromosome.[220,221] Differences in methylation could be monitored by using restriction enzymes with the capacity to recognize methylated cytosine residues. Third it was assumed that maternal and paternal copies of the X chromosome in female cells would be recognizable at the DNA level in RFLP analyses.

Analysis of agammoglobulinemic families involved a cloned polymorphic X chromosome gene and two restriction enzymes. It was not necessary that the X-chromosome gene be directly related to the mutation causing the disease; rather it served as a marker gene for identification of the patterns of X chromosome inactivation. Analyses were performed on DNA from females who were heterozygous for an RFLP within the polymorphic gene. The first restriction enzyme distinguished maternal from paternal copies of the gene through the RFLP pattern, and the second enzyme distinguished active from inactive gene copies on the basis of differences in methylation. In normal female somatic tissues, random patterns of X chromosome inactivation are present. The ratio of active to inactive maternal X chromosomes varies but is normally distributed around a mean of 50% active and 50% inactive maternal X chromosomes. In the analysis of these parameters, random X inactivation patterns were recorded in the isolated peripheral blood granulocytes, T lymphocytes, and B lymphocytes of three women not considered to be carriers of X-linked agammaglobulinemia. When two obligate carriers were examined, the X inactivation patterns in the granulocytes and T cells appeared to be random; however, preferential X inactivation was observed in the B cells. In addition, the mother of a child afflicted with X-linked agammaglobulinemia showed preferential X inactivation in her circulating B cells. These studies provided strong support for the idea that X-linked agammaglobulinemia results from an intrinsic B cell lineage defect. In the future, linkage analysis in addition to methylation and X inactivation may prove extremely useful in identifying potential carriers of the trait in affected families.

Chronic Granulomatous Disease

Chronic granulomatous disease (CGD) is an interesting X chromosome–linked immunodeficiency syndrome that has been the subject of intense study since its initial description.[222] Affected individuals show marked impairment of host defenses against infection with a wide variety of microorganisms, and phagocytic cells from such patients fail to generate superoxide and activated oxygen derivatives after ingestion of microorganisms. Family studies have defined both X-linked and autosomal recessive patterns of inheritance. A consistent biochemical abnormality identified in the X-linked form of the disease has been absence of the heme spectrum derived from cytochrome *b*.[223] This important lead was used in studies of the basic genetic defect in this disease.[224] This approach initially involved studies of the CGD gene by both deletion and formal linkage analysis. Using DNA from two patients found to have deletion of Xp21 and two parallel Xp21 DNA markers, the CGD locus was assigned to Xp21.1.[225,226] This had delimited the X-CGD locus to approximately 0.1% of the genome, or an area of perhaps 3000 kb of DNA. Using chromosome walking techniques in the defined DNA region and hybridization with an enriched cDNA probe, the genetic locus of CGD was defined without reference to a specific protein product.[224] The key to the puzzle was identification of a relatively limited RNA transcript that had a high level of expression in the phagocytic cell lineage but was deleted in several patients with CGD.

Subsequent studies were aimed at identifying the predicted CGD protein by raising antibodies to a synthetic peptide derived from the cDNA sequence and to a fusion protein produced in *E. coli*.[227] Studies with Western blots showed a 90-

kd protein that was absent in CGD patients. The antisera also were shown to react with the larger component of cytochrome *b*, which had been purified from neutrophil plasma membranes as a complex of glycosylated and nonglycosylated polypeptides. It was proposed that the X-CGD protein must interact with the cytochrome *b* polypeptides to form an active functional complex. These remarkable studies illustrate the power of recombinant approaches in elucidating complex human disorders. The findings provided strong immunological and biochemical proof that the previously defined cloned mRNA X-CGD sequence was the authentic missing product of the disease-specific gene. They also provided support for the functional role of the neutrophil cytochrome *b* in the respiratory burst of phagocytes, as initially suggested by Segal and co-workers.[223]

Hereditary Angioneurotic Edema

Hereditary agioneurotic edema (HANE) is a rare disorder due to deficiency of the inhibitor of the first component of human complement (C1-INH) and is inherited as an autosomal dominant trait.[228] The disease is manifested in afflicted individuals by episodic localized swelling of the subcutaneous tissue or of the gastrointestinal or laryngeal mucosa.[229,230] Attacks can be reversed by administration of C1-INH plasma concentrate. Some heterogeneity of the basic parameters of the disease has been noted with respect to both levels of C1-INH protein and mRNA levels.[231–233] Most patients with HANE have decreased serum concentrations of both antigenic and functional C1-INH, and mononuclear phagocytes from these patients have C1-INH mRNA levels that are approximately half of normal levels (type 1 HANE). Type 2 HANE (which affects 15% of afflicted kindreds) is characterized by normal or increased serum concentrations of a mutant C1-INH. Analysis of the C1-INH gene in normal subjects using the restriction enzyme *Hgi*A[234] has revealed an RFLP. Moreover, using a 170-bp probe it was demonstrated that three different mutations were responsible for the disease and were present in the same region of the C1-INH gene. The findings indicated that most mutations leading to C1-INH deficiency were not the result of major chromosomal rearrangements or deletions. Studies by other groups were interpreted as indicating that gene alterations responsible for HANE might arise more frequently from DNA rearrangements than from nucleotide substitutions.[235,236] A final answer on what constitute the definitive mutations will await detailed sequence analyses in much larger numbers of patients and their kindreds.

Transgenic Approaches and the Immune System

Experimental transgenic approaches, particularly in mice, have been employed to help decipher how immunoglobulin and in some instances T cell receptor genes are regulated and interact. Antibody-forming B cells display a special form of internal regulation whereby each cell makes immunoglobulins from only one of its two sets of antibody genes. This phenomenon, which has been called allelic exclusion, is thought to be controlled by the product of one heavy-chain allele that restricts the expression of the other. Immunoglobulin heavy chains are produced in two distinct molecular forms—secreted and membrane bound. Experiments were devised to ascertain whether the specific membrane-bound form of the immunoglobulin M (IgM) heavy chain (μ) mediated this sort of regulation or allelic exclusion.[237] Transgenic mice were produced which carried a human μ chain gene that had been altered so that it could direct the synthesis of only the membrane-bound IgM protein. The membrane-bound form of the human μ chain was produced by most of the B cells in these transgenic animals, as detected by assays of mRNA and expression of surface immunoglobulins. The many B cells that expressed the human Ig gene did not in parallel express the native endogenous mouse IgM. Moreover, the small proportion of B cells that did express endogenous mouse μ chain did not have the transgene. These experiments, therefore, indicated that the membrane form of the μ chain was sufficient to control allelic exclusion. They also indicated that an immunoglobulin molecule could signal the genomic machinery for its maturation by way of two pathways. One pathway could signal B cell maturation and depended on light-chain expression and initial antigen binding. The other pathway must be antigen independent because it was activated before light-chain expression and therefore before an interaction between the B cell and antigen. The latter pathway involved the membrane form of IgM heavy chain and established μ heavy-chain allelic exclusion.

Studies of the transfer of genomic immunoglobulin genes into mammalian cells by retrovirus vectors have also been quite instructive. In experiments reported by Cone et al.[238] recombinant retroviral genomes were constructed to encode a functionally rearranged murine λ1 immunoglobulin gene. Four constructs were prepared in which the λ1 sequences were introduced in both orientations into each of two sites in a plasmid vector. In each constructed gene, the insert constituted a 4.4-kb rearranged λ1 gene with a leader, joined V-J gene segments, and the C gene segment with the two normal λ1 introns and approximately 250 base pairs of 5′ untranslated nucleotides and 2000 base pairs of 3′ untranslated nucleotides. Several of these genomes could be transmitted intact to recipient cells by viral infections. However, successful placement depended on both correct orientation of the λ1 sequences and correct placement within the retroviral vector nucleotide sequences. Of interest was the finding that the λ1 gene transmitted by retroviral infection appeared to be expressed in a cell lineage–specific manner. However, expression was at a lower level than normal endogenous λ1 gene expression in cells from the B lymphocyte background. These studies provided additional useful information on the behavior of immunoglobulin genes transfected by viral vectors, and they may eventually be helpful in strategies aimed at gene therapy for immune deficiencies.

Immune Deficiency After Transgenic μ Heavy Chain Introduction

A strain of transgenic mice (M54) was generated by introducing a functionally rearranged immunoglobulin μ heavy-chain gene into the germ line of a C57B1/6 inbred mouse.[239,240] Studies of hybridomas from these transgenic mice indicated that the presence of the μ transgene grossly changed the immunoglobulin capacity of unimmunized animals and suggested that such animals suffered from a serious immune defect.[241] Lymphoid tissues from the transgenic mice lacked virtually all conventional bone marrow–derived B cells. Also, marrow from the transgenic mice was depleted of pre-B lymphocytes, indicating a basic problem in early B cell development. By contrast, $CD5^+$ B cells, representing a second B cell population derived from different precursor cells in the peritoneum, were found in normal frequencies in transgenic animals. Thus, the μ chain transgene introduced in this experimental model was capable of inducing a selective developmental defect more serious than that seen in human X-linked agammaglobulinemia. The results with this model might be used to study underlying genetic defects in several of the poorly understood human immunodeficiency disorders.

Immunological Tolerance—Some Beginning Insight from Transgenic Mice

An ingenious set of transgenic experiments has begun to shed some light on a paradox that has long vexed students of the immune system, namely that of immunological tolerance.[242] Autoimmunity is a rare event indeed, despite the possible genetic potential of every individual to mount immune reactions to many autologous antigens. Exactly how self-tolerance develops has never been precisely understood, although it is thought that it may depend on the time of exposure to antigen during ontogeny, the tissue site of antigen expression, and possibly the actual structure of the putative antigen. An early hypothesis was that self-reactive lymphocytes may undergo a clonal deletion from the preimmune repertoire or remain silent from a functional standpoint as a result of repeated exposure to self-antigen (clonal anergy) or of activation of T suppressor cells or the anti-idiotypic network. Several of these mechanisms could be operating together. Goodnow and co-workers[242] employed double transgenic mice that expressed the genes for a neo-self-antigen—hen egg lysozyme—*and* a high-affinity antilysozyme antibody. The majority of antilysozyme B cells did *not* undergo clonal deletion but they could no longer secrete antilysozyme antibody. Moreover, the antilysozyme B cells in these animals showed markedly reduced levels of surface or cell membrane IgM while continuing to express high levels of IgD. These findings indicated that self-tolerance may result from mechanisms different from clonal deletion and also that IgD could have a unique place in maintaining B cell tolerance.

The approach of transfection may be helpful in understanding the role of many components of the immune system. As this method becomes more widely used, it should provide data with which to interpret many experiments of nature that have long perplexed both the clinician and the basic investigator.

References

1. Takeda S, Naito T, Hama K, et al: Construction of chimaeric processed immunoglobulin genes containing mouse variable and human contant region sequences. *Nature* 1985;314:452–454.
2. Ritz J, Schlossman S: Utilization of monoclonal antibodies in the treatment of leukemia and lymphoma. *Blood* 1982;59:1–11.
3. Ritz J, Pesando JM, Sallan SE, et al: Serotherapy of acute lymphoblastic leukemia with monoclonal antibody. *Blood* 1981;58:141–152.
4. Miller RA, Maloney DG, Warnke R, Levy R: Treatment of B-cell lymphoma with monoclonal anti-idiotype antibody. *N Engl J Med* 1982;306:517–522.
5. Hagino K, Kano K, Ikegami S, et al: Effects of aphidicolin on retrovirus induction by halogenated pyrimidines. *Virology* 1980;107:174–179.
6. Breathnach R, Benoist C, O'Hare K, et al: Ovalbumin gene: Evidence for a leader sequence in mRNA and DNA sequences at the exon-intron boundaries. *Proc Natl Acad Sci USA* 1978;75:4853–4857.
7. Kohler G, Milstein C: Continuous cultures of fused cells secreting antibody of predefined specificity. *Nature* 1975;256:495–497.
8. Borst J, Coligan JE, Oettgen H, et al: The S- and E-chains of the human T3/T-cell receptor complex are distinct polypeptides. *Nature* 1984;312:455–458.
9. Dillman R, Shawler DL, Sobol RE, et al: Murine monoclonal antibody therapy in two patients with chronic lymphocytic leukemia. *Blood* 1982;59:1036–1045.
10. Cosimi AB, Colvin RB, Burton RC, et al: Use of monoclonal antibodies to T-cell subsets

for immunologic monitoring and treatment in recipients of renal allografts. *N Engl J Med* 1981;305:308–314.

11. Chatenoud L, Boudrihaye MF, Kreis H, et al: Human *in vivo* antigenic modulation induced by the anti-T cell OKT3 monoclonal antibody. *Eur J Immunol* 1982;12:979–982.
12. Jaffers GF, Colvin RB, Cosimi AB, et al: The human immune response to murine OKT3 monoclonal antibody. *Transpl Proc* 1983;15:646–648.
13. Jonker M, Goldstein G, Balner H: Effects of in vivo administration of monoclonal antibodies specific for human T cell subpopulations on the immune system in a rhesus monkey model. *Transplantation* 1983;35:521–526.
14. Chatenoud L, Jonker M, Villemain F, et al: The human immune response to the OKT3 monoclonal antibody is oligoclonal. *Science* 1986;232:1406–1408.
15. Spira G, Bargellesi A, Teilland JL, Scharff MD: The identification of monoclonal class switch variants by sib selection and an ELISA assay. *J Immunol Methods* 1985;74:307–315.
16. Zack DS, Scharff MD: Identification of somatic mutations in immunoglobulin expression and structure, in Ansari AA, De Serres FS (eds): *Single-Cell Mutation Monitoring Systems. Methodologies and Applications.* New York, Plenum Press, 1984, pp 223–263.
17. Desaymard C, Giusti AM, Scharff MD: Rat anti-T15 monoclonal antibodies with specificity for V_H and V_H-V_L epitopes. *Mol Immunol* 1984;21:961–967.
18. Parham P, Kipps TJ, Ward FE, Herzenberg LA: Isolation of heavy chain class switch variants of a monoclonal anti-DC1 hybridoma cell line: Effective conversion of noncytotoxic IgG_1 antibodies to cytotoxic IgG_2 antibodies. *Hum Immunol* 1983;8:141–151.
19. Steplewski Z, Spira G, Blaszcyk M, et al: Isolation and characterization of antimonosialoganglioside monoclonal antibody 19-4 class-switch variants. *Proc Natl Acad Sci USA* 1985;82:8653–8657.
20. Neuberger MS, Williams GT, Fox RO: Recombinant antibodies possessing novel effector functions. *Nature* 1984;312:604–608.
21. Thorpe PE, Edwards DC, Davies AJS, Ross WCJ: Monoclonal antibody-toxin conjugates: Aiming the magic bullet. in McMichael AJ, Fabre JW (eds): *Monoclonal Antibodies in Clinical Medicine.* London, Academic Press, 1982, pp 167–201.
22. Collier RJ: Diphtheria toxin: Mode of action and structure. *Bacteriol Rev* 1975;39:54–85.
23. Olsnes S, Pihl A: Chimeric toxins. *Pharmacol Ther* 1982;15:355–381.
24. Moolten FL, Schreiber BM, Zujdel SH: Antibodies conjugated to potent cytotoxins as specific antitumor agents. *Immunol Rev* 1982;62:47–73.
25. Youle RJ, Colombatti M: in Roth JA (ed): *Monoclonal Antibodies in Cancer.* Mt. Kisco, N.Y., Futura, 1986, pp 173–213.
26. Casellas P, Bourrie BJP, Gros P, Jansen FK: Kinetics of cytotoxicity induced by immunotoxins. *J Biol Chem* 1984;259:9359–9364.
27. Vallera DA, Ash RC, Zanjani ED, et al: Anti–T-cell reagents for human bone marrow transplantation: Ricin linked to three monoclonal antibodies. *Science* 1983;222:512–515.
28. Greenfield L, Johnson VG, Youle RJ: Mutations in diphtheria toxin separate binding from entry and amplify immunotoxin selectivity. *Science* 1987;238:536–539.
29. Laird W, Groman N: Isolation and characterization of *tox* mutants of corynebacteriophage beta. *J Virol* 1976;19:220–227.
30. Morrison SL, Johnson JJ, Herzenberg LA, Oi VT: Chimeric human antibody molecules: Mouse antigen-binding domains with human constant region domains. *Proc Natl Acad Sci USA* 1984;81:6851–6855.
31. Morrison SL: Transfectomas provide novel chimeric antibodies. *Science* 1985; 229:1202–1207.
32. Sun LK, Curtis P, Rakowicz-Szulczynska E, et al: Chimeric antibody with human constant regions and mouse variable regions directed against carcinoma-associated antigen 17-1A. *Proc Natl Acad Sci USA* 1987;84:214–218.
33. Herlyn M, Steplewski Z, Herlyn D, Koprowski H: Colorectal carcinoma–specific antigen: Detection by means of monoclonal antibodies. *Proc Natl Acad Sci USA* 1979;76:1438–1442.

34. Sears HF, Mattis J, Herlyn D, et al: Phase-I clinical trial of monoclonal antibody in treatment of gastrointestinal tumours. *Lancet* 1982;1:762–765.
35. Sears HF, Herlyn D, Steplewski Z, Koprowski H: Phase II clinical trial of a murine monoclonal antibody cytotoxic for gastrointestinal adenocarcinoma. *Cancer Res* 1985;45:5910–5913.
36. Lorderboum-Galski H, Fitzgerald D, Chaudhary V, et al: Cytotoxic activity of an interleukin 1–*Pseudomonas* exotoxin chimeric protein produced in *Escherichia coli. Proc Natl Acad Sci USA* 1988;85:1922–1926.
37. Hwang J, FitzGerald DJP, Adhya S, Pastan I: Functional domains of *Pseudomonas* exotoxin identified by deletion analysis of the gene expressed in *E. coli. Cell* 1987;48:129–136.
38. Chaudhary VK, FitzGerald DJP, Adhya S, Pastan I: Activity of a recombinant fusion protein between transforming growth factor type α and *Pseudomonas* toxin. *Proc Natl Acad Sci USA* 1987;84:4538–4542.
39. Ju G, Collins L, Kaffka KL, et al: Structure-function analysis of human interleukin-2. *J Biol Chem* 1987;262:5723–5731.
40. Wall JR, Baur R, Schleusener H, Bandy-Dafoe P: Peripheral blood and intrathyroidal mononucelar cell populations in patients with autoimmune thyroid disorders enumerated using monoclonal antibodies. *J Clin Endocrinol Metab* 1983;56:164–168.
41. Lemm G, Warnatz J: Evidence for enhanced interleukin 2 (IL-2) secretion and IL-2 receptor presentation by synovial fluid lymphocytes in rheumatoid arthritis. *Clin Exp Immunol* 1986;64:71–79.
42. Selby WS, Janossy G, Bosil M, Jewell DP: Intestinal lymphocyte subpopulations in inflammatory bowel disease: An analysis by immunohistological and cell isolation techniques. *Gut* 1984;25:32–40.
43. Kirkman RL, Barrett LV, Gaulton GN, et al: Administration of an anti-interleukin 2 receptor monoclonal antibody prolongs cardiac allograft survival in mice. *J Exp Med* 1985;162:358–362.
44. Kupiec-Weglinski JW, Diamantstein T, Tilney NL, Strom T: Therapy with monoclonal antibody to interleukin 2 receptor spares suppressor T cells and prevents or reverses acute allograft rejection in rats. *Proc Natl Acad Sci USA* 1986;83:2624–2627.
45. Dinarello CA: Biology of interleukin 1. *FASEB J* 1988;2:108–115.
46. Oppenheim JJ, Kovacs EJ, Matsushima K, Durum SK: There is more than one interleukin 1. *Immunol Today* 1986;7:45–56.
47. Carter DB, Deibel MR, Dunn CJ, et al: Purification, cloning, expression and biological characterization of an interleukin-1 receptor antagonist protein. *Nature* 1990;344:633–638.
48. Wong WW, Klickstein LB, Smith JA, et al: Identification of a partial cDNA clone for the human receptor for complement fragments C36/C46. *Proc Natl Acad Sci USA* 1985;82:7711–7715.
49. Fearon DT, Wong WW: Complement ligand-receptor interactions that mediate biological responses. *Annu Rev Immunol* 1983;1:243–271.
50. Reyes M, Aubert JP, Cohen JHM, et al: Human follicular dendritic cells express CR1:CR2/CR3 complement receptor antigens. *J Immunol* 1985;135:2687–2694.
51. Cornacott JB, Hebert LA, Smead WL, et al: Primate erythrocyte–immune complex–clearing mechanisms. *J Clin Invest* 1983;71:236–247.
52. Medof ME, Iida K, Mold C, Nussenzweig V: Unique role of complement receptor CR1 in the degradation of C3b associated with immune complexes. *J Exp Med* 1982;156:1739–1754.
53. Wilson JG, Wong WW, Schur PH, Fearon DT: Mode of inheritance of decreased C3b receptors on erythrocytes of patients with systemic lupus erythematosus. *N Engl J Med* 1982;307:981–986.
54. Miyakawa Y, Yamada A, Kosaka K, et al: Defective immune adherence (C3b) receptor on erythrocytes for patients with systemic lupus erythematosus. *Lancet* 1981;2:493–497.
55. Minota S, Terai C, Nojima Y, et al: Low C3b receptor reactivity on erythrocytes from patients with systemic lupus erythematosus detected by immune adherence hemagglutination and radioimmunoassays with monoclonal antibody. *Arthritis Rheum* 1984;27:1329–1335.
56. Wilson JG, Murphy EE, Wong WW, et al: Identification of a restriction fragment length

polymorphism by a CR1 cDNA that correlates with the number of CR1 on erythrocytes. *J Exp Med* 1986;164:50–59.

57. Wilson JG, Wong WW, Murphy EE III, et al: Deficiency of the C3b/C4b receptor (CR1) of erythrocytes in systemic lupus erythematosus: Analysis of the stability of the defect and of a restriction fragment length polymorphism of the CR1 gene. *J Immunol* 1987;138:2706–2710.
58. Walport MJ, Lachmann PT: Erythrocyte complement receptor type 1, immune complexes, and the rheumatic diseases. *Arthritis Rheum* 1988;31:153–158.
59. Ross GD, Yount WJ, Walport MJ, et al: Disease-associated loss of erythrocyte complement receptors (CR_1 C3b receptors) in patients with systemic lupus erythematosus and other diseases involving autoantibodies and/or complement activation. *J Immunol* 1985;135:2005–2014.
60. Iida K, Mornaghi R, Nussenzweig V: Complement receptor (CR_1) deficiency in erythrocytes from patients with systemic lupus erythematosus. *J Exp Med* 1982;155:1427–1438.
61. Walport MJ, Ng YC, Lachmann PJ: Erythrocytes transfused into patients with SLE and haemolytic anaemia lose complement receptor type 1 from their surface. *Clin Exp Immunol* 1987;69:501–507.
62. Unkeless JC, Fleit H, Mellman IS: Structural aspects and heterogeneity of immunoglobulin Fc receptors. *Adv Immunol* 1981;31:247–270.
63. St C Sinclair NR, Panoskaltsis A: Immunoregulation by Fc signals. *Immunol Today* 1987;8:76–79.
64. Lewis VA, Koch T, Plutner H, Mellman I: A complementary DNA clone for a macrophage-lymphocyte Fc receptor. *Nature* 1986;325:372–375.
65. Franklin EC, Holman HR, Muller-Eberhard HJ, Kunkel HG: An unusual protein component of high molecular weight in the serum of certain patients with rheumatoid arthritis. *J Exp Med* 1957;105:425–438.
66. Pope RM, Teller DC, Mannik M: The molecular basis of self-association of antibodies to IgG (rheumatoid factors) in rheumatoid arthritis. *Proc Natl Acad Sci USA* 1974;71:517–521.
67. Nardella FA, Teller DC, Barber CV, Mannik M: IgG rheumatoid factors and staphylococcal protein A bind to a common molecular site on IgG. *J Exp Med.* 1985;162:1811–1824.
68. Oppliger IR, Nardella FA, Stone GC, Mannik M: Human rheumatoid factors bear the internal image of the Fc binding region of staphylococcal protein A. *J Exp Med* 1987;166:702–710.
69. Longone JJ: Protein A of *Staphylococcus aureus* and related immunoglobulin receptors produced by streptococci and pneumocci. *Adv Immunol* 1982;32:157–252.
70. Boyle MDP, Reis KJ: Bacterial Fc receptors. *Biotechnology* 1987;5:697–703.
71. Mouritsen S: Rheumatoid factors are anti-idiotypic antibodies against virus-induced anti-Fc receptor antibodies. A hypothesis for the induction of some rheumatoid factors. *Scand J Immunol* 1986;24:485–490.
72. Williams RC Jr: Hypothesis: Rheumatoid factors are anti-idiotypes versus anti-bacterial or anti-viral receptors. *Arthritis Rheum* 1988;31:1204–1207.
73. Glas D, Raum D, Gibson D, et al: Inherited deficiency of the second component of complement: Rheumatic disease associations. *J Clin Invest* 1976;58:853–861.
74. Agnello V: Complement deficiency states. *Medicine* 1978;57:1–23.
75. Pariser KM, Raum D, Berkman EM, et al: Evidence of a silent or null gene in hereditary C2 deficiency. *J Immunol* 1978;121:2580–2581.
76. Cole FS, Whitehead AS, Auerbach HS, et al: The molecular basis for genetic deficiency of the second component of human complement. *N Engl J Med* 1985;313:11–16.
77. Dinarello CA, Mier JW: Lymphokines. *N Engl J Med* 1987;317:940–945.
78. Foon KA, Roth MS, Bunn PA Jr: Alpha interferon treatment of low-grade B-cell non-Hodgkin's lymphomas, cutaneous T-cell lymphomas, and chronic lymphocytic leukemia. *Semin Oncol* 1986;13(Suppl. 2):35–42.
79. Talpaz M, McCredie K, Kantajian H, et al: Chronic myelogenous leukemia: Haematological remissions with alpha interferon. *Br J Haematol* 1986;64:87–95.
80. Quesada JR, Reuben J, Manning JT, et al: Alpha interferon for induction of remission in hairy-cell leukemia. *N Engl J Med* 1984;310:15–18.

81. Borden EC, Holland JF, Dao TL, et al: Leukocyte-derived interferon (alpha) in human breast carcinoma. *Ann Intern Med* 1982;97:1–6.
82. Quesada JR, Hawkins M, Horning S, et al: Collaborative phase I-II study of recombinant DNA–produced leukocyte interferon (clone A) in metastatic breast cancer, malignant lymphoma, and multiple myeloma. *Am J Med* 1984;77:427–432.
83. Greiner JW, Guadagni F, Noguchi P, et al: Recombinant interferon enhances monoclonal antibody-targeting on carcinoma lesions in vivo. *Science* 1987;235:895–898.
84. Hirsch MS, Schooley RT, Cosimi AB, et al: Effects of interferon-alpha on cytomegalovirus reactivation syndromes in renal transplant recipients. *N Engl J Med* 1983;308:489–493.
85. Murray HW, Rubin BY, Masur H, Roberts RB: Impaired production of lymphokines and immune (gamma) interferon in the acquired immunodeficiency syndrome. *N Engl J Med* 1984;310:883–889.
86. DiBisceglie AM, Rustgi VK, Kassianides C, et al: Therapy of chronic hepatitis B with recombinant human alpha and gamma interferon. *Hepatology* 1990;11:266–270.
87. Davis GL, Balart LA, Schiff ER, et al: Treatment of chronic hepatitis C with recombinant interferon alfa. A multicenter randomized, controlled trial. *N Engl J Med* 1989;321:1501–1506.
88. DiBisceglie AM, Martin P, Kassianides C, et al: Recombinant interferon alfa therapy for chronic hepatitis C. A randomized, double blind, placebo-controlled trial. *N Engl J Med* 1989;321:1506–1510.
89. Heremans H, Billiau A, Colombatti A, et al: Interferon treatment of NZB mice: Accelerated progression of autoimmune disease. *Infect Immun* 1978;21:925–930.
90. Engleman EG, Sonnenfeld G, Dauphinee M, et al: Treatment of NZB/NZW F_1 hybrid mice with *Mycobacterium bovis* strain BCG or type II interferon preparations accelerates autoimmune disease. *Arthritis Rheum* 1981;24:1396–1402.
91. Knobler RL: Systemic interferon therapy of multiple sclerosis: The pros. *Neurology* 1988;38:58–61.
92. Peddinani MV, Savery F, and Hang LM: Human leukocyte interferon in the treatment of rheumatoid arthritis. *Clin Ther* 1986;9:39–43.
93. Lever AML, Brook MG, Yap I, Thomas HC: Treatment of thrombocytopenia with alpha interferon. *Br Med J* 1987;295:1519–1520.
94. Proctor SJ, Jackson G, Carey P, Stark A: Short-course alpha interferon in severe unresponsive immune thrombocytopenic purpura. *Lancet* 1988;1:947.
95. Thivolet J, Nicholas JF, Kanitakis J, et al: Recombinant interferon α2a is effective in the treatment of discoid and subacute cutaneous lupus erythematosus. *Br J Dermatol* 1990;122:405–409.
96. Morgan DA, Ruscetti FW, Gallo RC: Selective in vitro growth of T lymphocytes from normal human bone marrows. *Science* 1976;193:1007–1008.
97. Mier JW, Gallo RC: Purification and some characteristics of human T-cell growth factor from phytohemagglutinin-stimulated lymphocyte-conditioned media. *Proc Natl Acad Sci USA* 1982;77:1823–1841.
98. Grimm EA, Mazumder A, Zhang HZ, Rosenberg SA: Lymphokine-activated killer cell phenomenon: Lysis of natural killer–resistant fresh solid tumor cells by interleukin 1–activated autologous human peripheral blood lymphocytes. *J Exp Med* 1982; 155:1823–1841.
99. Rosenberg SA, Lotze MT, Muul LM, et al: Special report. Observations on the systemic administration of autologous lymphokine-activated killer cells and recombinant interleukin-2 to patients with metastatic cancer. *N Engl J Med* 1985;313:1485–1492.
100. Nikaido T, Schimizu A, Ishida N, et al: Molecular cloning of cDNA encoding human interleukin-2 receptor. *Nature* 1984;311:631–635.
101. Leonard WJ, Depper JM, Crabtree GR, et al: Molecular cloning and expression of cDNAs for the human interleukin-2 receptor. *Nature* 1984;311:626–631.
102. Dinarello CA: Interleukin-1: Amino acid sequences, multiple biological activities, and comparison with tumor necrosis factor (cachectin). *Year Book Immunol* 1986;2:68–89.
103. Scala G, Kuang YD, Hall RE, et al: Accessory cell function of human B cells. I. Production of both interleukin 1–like activity and an interleukin 1 inhibitory factor by an EBV-transformed human B cell line. *J Exp Med* 1984;159:1637–1652.
104. Libby P, Ordovas JM, Birinyi LK, et al: Inducible interleukin-1 gene expression in human vascular smooth muscle cells. *J Clin Invest* 1986;78:1432–1438.

105. Bell TV, Harley CB, Stetsko D, Sauder DN: Expression of mRNA homologous to interleukin 1 in human epidermal cells. *J Invest Dermatol* 1987;88:375–379.
106. Carswell EA, Old LJ, Kassel RL, et al: An endotoxin-induced serum factor that causes necrosis of tumors. *Proc Natl Acad Sci USA* 1975;72:3666–3670.
107. Kelker HC, Oppenheim JD, Stone-Wolff D, et al: Characterization of human tumor necrosis factor produced by peripheral blood monocytes and its separation from lymphotoxin. *Int J Cancer* 1985;36:69–73.
108. Mannel DN, Moore RN, Morgenhayen SE: Macrophages as a source of tumoricidal activity (tumor necrosis factor). *Infect Immun* 1980;30:523–530.
109. Pennica D, Hayflick JS, Bringman TS, et al: Cloning and expression in *Escherichia coli* of the cDNA for murine tumor necrosis factor. *Proc Natl Acad Sci USA* 1985;82:6060–6064.
110. Shirai Y, Yamaguchi H, Ito H, et al: Cloning and expression in *Escherichia coli* of the gene for human tumor necrosis factor. *Nature* 1985;313:803–806.
111. Marmenout A, Fransen L, Tavernier J, et al: Molecular cloning and expression of human tumor necrosis factor and comparison with mouse tumor necrosis factor. *Eur J Biochem* 1985;152:515–522.
112. Balkwill F, Burke F, Talbot D, et al: Evidence for tumor necrosis factor/cachectin production in cancer. *Lancet* 1987;2:1229–1232.
113. Tracey KJ, Fong Y, Hesse DG, et al: Anti-cachectin/TNF monoclonal antibodies prevent septic shock during lethal bacteremia. *Nature* 1987;330:662–664.
114. Scuderi P, Lam KS, Ryan KJ, et al: Raised serum levels of tumor necrosis factor in parasitic infections. *Lancet* 1986;2:1364–1365.
115. Grau GE, Fajardo LF, Piquet PF, et al: Tumor necrosis factor (cachectin) as an essential mediator in murine cerebral malaria. *Science* 1987;237:1210–1212.
116. Waage A, Halstensen A, Espevik T: Association between tumour necrosis factor in serum and fatal outcome in patients with meningococcal disease. *Lancet* 1987;1:355–357.
117. Muller U, Jongeneel CV, Nedospasov SA, et al: Tumor necrosis factor and lymphotoxin genes map close to H-2D in the mouse major histocompatibility complex. *Nature* 1987;325:265–267.
118. Dayer JM, Beutler B, Cerami A: Cachectin/tumor necrosis factor stimulates collagenase and prostaglandin E_2 production by human synovial cells and dermal fibroblasts. *J Exp Med* 1985;162:2163–2168.
119. Saklatvala J: Tumour necrosis factor α stimulates resorption and inhibits synthesis of proteoglycan in cartilage. *Nature* 1986;322:547–549.
120. Sharma S, Mehta S, Morgan J, Maizel A. Molecular cloning and expression of a human B-cell growth factor gene in *Escherichia coli*. *Science* 1987;235:1489–1492.
121. Varela JA, Alarcon-Segovia D: Decreased production of and response to interleukin 2 by cultured lymphocytes from patients with systemic lupus erythematosus. *J Clin Invest* 1982;69:1388–1395.
122. Linker-Israeli M, Bakke AC, Kijtridou RC, et al: Defective production of interleukin 1 and interleukin 2 in patients with systemic lupus erythematosus (SLE). *J Immunol* 1983;130:2651–2655.
123. Linker-Israeli M, Bakke AC, Quismorio FP Jr, Horwitz DA: Correction of interleukin-2 production in patients with systemic lupus erythematosus by removal of spontaneously activated suppressor cells. *J Clin Invest* 1985;75:762–768.
124. Husby G, Williams RC Jr: Immunohistochemical studies of interleukin-2 and gamma interferon in rheumatoid arthritis. *Arthritis Rheum* 1984;28:174–181.
125. Emery P, Panayi GS, Nouri AME: Interleukin-2 reverses deficient cell-mediated immune responses in rheumatoid arthritis. *Clin Exp Immunol* 1984;57:123–129.
126. Combe B, Andary M, Klein B, et al: Regulation of interleukin-2 production in rheumatoid arthritis. *J Rheumatol* 1980;14:226–229.
127. Kelly K, Cochran BH, Stiles CD, Leder P: Cell-specific regulation of the c-*myc* gene by lymphocyte mitogens and platelet-derived growth factor. *Cell* 1983;35:603–610.
128. Mountz JD, Steinberg AD, Klinman DM, et al: Autoimmunity and increased c-*myb* transcription. *Science* 1984;226:1087–1089.
129. Mountz JD, Mushinski JF, Smith HR, et al: Modulation of c-*myb* transcription in autoimmune disease by cyclophosphamide. *J Immunol* 1985;135:2417–2422.

130. Klinman DM, Mushinski JF, Honda M, et al: Oncogene expression in autoimmune and normal peripheral blood mononuclear cells. *J Exp Med* 1986;163:1292–1307.
131. Ho Y-S, Lee WMF, Snyderman R: Chemoattractant-induced activation of c-*fos* gene expression in human monocytes. *J Exp Med* 1987;165:1524–1538.
132. Berridge MJ, Irvine RF: Inositol triphosphate, a novel second messenger in cellular signal transduction. *Nature* 1984;312:315–321.
133. Nishizuka Y: The role of protein kinase C in cell surface signal transduction and tumor promotion. *Nature* 1984;308:693–698.
134. Kotzin BL, Barr VL, Palmer E: A large deletion within the T-cell receptor beta-chain gene complex in New Zealand white mice. *Science* 1985;229:167–170.
135. Steinberg AD, Huston DP, Taurog JD, et al: The cellular and genetic basis of murine lupus. *Immunol Rev* 1981;55:121–154.
136. Marrack P, Hannum C, Harris M, et al: Antigen-specific major histocompatibility complex–restricted T cell receptors. *Immunol Rev* 1983;76:131–145.
137. Noonan DJ, Kofler R, Singer PA, et al: Delineation of a defect in T cell receptor β genes of NZW mice and predisposed to autoimmunity. *J Exp Med* 1986;163:644–653.
138. Mountz JD, Huppi K, Seldin MF, et al: T cell receptor gene expression in autoimmune mice. *J Immunol* 1986;137:1029–1036.
139. Singer PA, McEvilly RJ, Noonan DJ, et al: Clonal diversity and T-cell receptor β-chain variable gene expression in enlarged lymph nodes of MRL-*lpr*/*lpr* lupus mice. *Proc Natl Acad Sci USA* 1986;83:7018–7022.
140. Oksenberg JR, Stuart S, Begovich AB, et al: Limited heterogeneity of rearranged T cell receptor Vα transcripts in brains of multiple sclerosis patients. *Nature* 1990;345:344–346.
141. Acha-Orbea H, Mitchell DJ, Timmermann L, et al: Limited heterogeneity of T cell receptors from lymphocytes mediating autoimmune encephalomyelitis allows specific immune intervention. *Cell* 1988;54:2163–2173.
142. Urban JL, Kumar V, Kono DH, et al: Restricted use of T cell receptor V genes in murine autoimmune encephalomyelitis raises possibilities for antibody therapy. *Cell* 1988; 54:577–582.
143. Sakai K, Sinha A, Mitchell DJ, et al: Involvement of distinct murine T cell receptors in the autoimmune encephalitogenic response to nested epitopes of myelin basic protein. *Proc Natl Acad Sci USA* 1988;85:8608–8612.
144. Procelli S, Brenner MB, Greenstein JL, et al: Recognition of cluster differentiation 1 antigen by human CD4-CD8-cytolytic lymphocytes-T. *Nature* 1989;341:447–450.
145. Rajasekar R, Sim G-K, Augustin A: Self heat shock and γδ T-cell reactivity. *Proc Natl Acad Sci USA* 1990;87:1767–1771.
146. Rajagopalan S, Zordan T, Tsokos GC, Datta SK: Pathogenic anti-DNA autoantibody–inducing T helper cell lines from patients with active lupus nephritis: Isolation of CD4(−) CD8(−) T helper cell lines that express the γδ T cell antigen receptor. *Proc Natl Acad Sci USA* 1990;87:7020–7024.
147. Tan EM: Autoantibodies to nuclear antigens (ANA): Their immunobiology and medicine. *Adv Immunol* 1982;33:167–240.
148. Hardin JA: The lupus autoantigens and the pathogenesis of systemic lupus erythematosus. *Arthritis Rheum* 1986;29:457–460.
149. Lerner MR, Boyle JA, Hardin JA, Steitz JA: Two novel classes of small ribonucleoproteins detected by antibodies associated with lupus erythematosus. *Science* 1981;211:400–402.
150. Hardin JA, Rahn DR, Shen C, et al: Antibodies from patients with connective tissue disease bind specific subsets of cellular RNA-protein particles. *J Clin Invest* 1982;70:141–147.
151. Lerner EA, Lerner MR,Janeway CA, Steitz JA: Monoclonal antibodies to nucleic acid–containing constituents: Probes for molecular biology and autoimmune diseases. *Proc Natl Acad Sci USA* 1981;78:2737–2742.
152. Wolin S, Steitz JA: The Ro small cytoplasmic ribonucleoproteins: Identification of the antigenic protein and its binding site on the Ro RNAs. *Proc Natl Acad Sci USA* 1984;81:1996–2000.
153. Chan EKL, Francoeur AM, Tan EM: Epitopes, structural domains, and asymmetry of amino acid residues in SS-B/La nuclear protein. *J Immunol* 1986;136:3744–3749.

154. Chambers JC, Keene JD: Isolation and analysis of cDNA clones expressing human lupus La antigen. *Proc Natl Acad Sci USA* 1985;82:2115–2119.
155. Wieben ED, Rohleder AM, Nenninger JM, Pederson T: cDNA cloning of a human autoimmune nuclear ribonucleoprotein antigen. *Proc Natl Acad Sci USA* 1985;82:7914–7918.
156. Prelich G, Tan C-K, Kostura M, et al: Functional identity of proliferating cell nuclear antigen and a DNA polymerase-2 auxiliary protein. *Nature* 1987;326:517–520.
157. Hutt JP, Roos G, Peebles CL, et al: Insights into native epitopes of proliferating cell nuclear antigen using recombinant DNA protein products. *J Exp Med* 1990;172:419–429.
158. Rapoport B: Recombinant DNA technology in the study of autoimmune thyroid disease. Endocrinol Metab Clin North Am 1987;16:445–468.
159. Kunkel HG, Winchester RJ, Joslin FG, Capra, JD. Similarities in the light chains of anti-γ-globulins showing cross-idiotypic specificities. *J Exp Med* 1974;139:128–136.
160. Williams RC Jr, Kunkel HG, Capra JD: Antigenic specificities related to the cold agglutinin activity of gamma M globulins. *Science* 1968;161:379–381.
161. Ledford DK, Goni F, Pizzolato M, et al: Preferential association of κ IIIb light chains with monoclonal human IgM autoantibodies. *J Immunol* 1983;131:1322–1325.
162. Chen PP, Fong S, Normansell D, et al: Delineation of a cross-reactive idiotype on human autoantibodies with antibody against a synthetic peptide. *J Exp. Med* 1984;159:1502–1511.
163. Andrews DW, Capra JD: Complete amino acid sequence of variable domains from two monoclonal human anti–gamma globulins of the Wa cross-idiotypic group: Suggestion that the J segments are involved in the structural correlate of the idiotype. *Proc Natl Acad Sci USA* 1981;78:3799–3803.
164. Goldfien RD, Chen PP, Fong S, Carson DA: Synthetic peptides corresponding to third hypervariable region of human monoclonal IgM rheumatoid factor heavy chains define an immunodominant idiotype. *J Exp Med* 1985;162:756–761.
165. Chen PP, Goni F, Houghten RA, et al: Characterization of human rheumatoid factors with seven anti-idiotypes induced by synthetic hypervariable region peptides. *J Exp Med* 1985;162:487–500.
166. Goni F, Chen PP, Pons-Estel B, et al: Sequence similarities and cross-idiotypic specificity of L chains among human monoclonal IgM κ with anti–γ-globulin activity. *J Immunol* 1985;135:4073–4079.
167. Radoux V, Chen PP, Sorge JA, Carson DA: A conserved human germline V_κ gene directly encodes rheumatoid factor light chains. *J Exp Med* 1986;164:2119–2124.
168. Diamond B, Scharff MD: Somatic mutation of the T15 heavy chain gives rise to an antibody with autoantibody specificity. *Proc Natl Acad Sci USA* 1984;81:5841–5844.
169. Naparstek Y, Andre-Schwartz J, Manser T, et al: A single germline V_H gene segment of normal A/J mice encodes autoantibodies characteristic of systemic lupus erythematosus. *J Exp Med* 1986;164:614–626.
170. Newkirk MM, Mageed RA, Jefferis R, et al: Complete amino acid sequences of variable regions of two human IgM rheumatoid factors POR and Kas of the Wa idiotypic family reveal restricted use of heavy and light chain variable and joining region gene segments. *J Exp Med* 1987;166:550–564.
171. Shlomchik M, Nemazee D, von Snick J, et al: Varible region monoclonal autoantibodies (rheumatoid factor). II. Comparison of hybridomas derived by lipopolysaccharide stimulation and secondary protein immunization. *J Exp Med* 1987;165:970–987.
172. Nemazee D, Sato V: Induction of rheumatoid antibodies in the mouse. Regulated production of autoantibody in the secondary humoral response. *J Exp Med* 1983;158:529–545.
173. Van Snick J, Coulie P: Monoclonal anti-IgG autoantibodies from lipopolysaccharide-activated spleen cells of 129/sv mice. *J Exp Med* 1982;155:219–230.
174. Shlomchik M, Nemazee D, Sato V, et al: Variable region sequences of murine IgM anti-IgG monoclonal autoantibodies (rheumatoid factors): A structural exploration for the high frequency of IgM anti-IgG B cells. *J Exp Med* 1986;164:407–427.
175. Knight J, Laing P, Knight A, et al: Thyroid-stimulating autoantibodies usually contain only λ-light chains: Evidence for the "forbidden clone" theory. *J Clin Endocrinol Metab* 1986;62:342–347.

176. Williams RC Jr, Marshall NJ, Kilpatrick K, et al: Kappa/lambda Ig distribution of Graves' thyroid stimulating antibodies simultaneous analysis of Cλ gene polymorphisms. *J Clin Invest* 1990;82:1306–1312.
177. Weetman AP, Yateman ME, Ealey PA, et al: Thyroid-stimulating antibody between different immunoglobulin subclasses. *J Clin Invest* 1990;86:723–727.
178. Sidebottom D, Grennan DM, Sanders PA, Read A: Immunoglobulin lambda light chain genes in rheumatoid arthritis. *Ann Rheum Dis* 1987;46:587–589.
179. Mathews MB, Bernstein RM: Myositis autoantibody inhibits histidyl-tRNA synthetase: A model for autoimmunity. *Nature* 1983;304:177–179.
180. Bernstein RM, Bunn CC, Hughes GRV, et al: Cellular proteins and RNA antigens in autoimmune disease. *Mol Biol Med* 1984;2:105–120.
181. Mathews MB, Bernstein RM, Bunn CC: Jo-1 and other autoantibody systems in myositis. XVth International Congress of Rheumatology R89, 1985.
182. Walker EJ, Jeffrey PD: Polymyositis and molecular mimicry, a mechanism of autoimmunity. *Lancet* 1986;2:605–607.
183. Kabash W, Sander C: On the use of sequence homologies to predict protein structure: Identical pentapeptides can have completely different conformations. *Proc Natl Acad Sci USA* 1984;81:1075–1078.
184. Wada K, Ueno S, Hazawa T, et al: Radioimmunoassay for antibodies to human skeletal muscle myosin in serum from patients with polymyositis. *Clin Exp Immunol* 1983;52:297–304.
185. Schwimmbeck PL, Yu DTY, Oldstone MBA: Autoantibodies to HLA B27 in the sera of HLA B27 patients with ankylosing spondylitis and Reiter's syndrome. Molecular mimicry with *Klebsiella pneumoniae* as potential mechanism of autoimmune disease. *J Exp Med* 1987;166:173–181.
186. Tsuchiya N, Husby G, Williams RC Jr: Studies of humoral and cell-mediated immunity to peptides shared by HLA-B27.1 and *Klebsiella pneumoniae* nitrogenase in ankylosing spondylitis. *Clin Exp Immunol* 1989;76:354–360.
187. Husby G, Tsuchiya N, Schwimmbeck PL, et al: Cross-reactive epitope with *Klebsiella pneumoniae* nitrogenase in articular tissue of HLA-B27(+) patients with ankylosing spondylitis. *Arthritis Rheum* 1989;32:437–445.
188. Voorhorst R, Spieksma-Boezeman MIA, Spieksma FTM: Is a mite (*Dermatophagoides* sp.) the producer of the house dust allergen? *Allerg Asthma* 1964;10:329–334.
189. Voorhorst R, Spieksma FTM, Varekamp H, et al: The house dust mite (*Dermatophagoides pteronyssinus*) and the allergens it produces. Identity with the house dust allergen. *J Allergy* 1967;39:325–339.
190. Krillis S, Baldo BA, Basten A: Antigens and allergens from the common house dust mite *Dermatophagoides pteronyssinus*. Part II. Identification of the major IgE binding antigen by crossed radioimmunoelectrophoresis. *J Allergy Clin Immunol* 1984;74:142–146.
191. Chapman MD, Platts-Mills TAE: Measurement of IgG, IgA, and IgE antibodies to *Dermatophagoides pteronyssinus* by antigen binding assay using a partially purified fraction of mite extract (F4P1). *Clin Exp Immunol* 1978;34:126–136.
192. Stewart GA, Thomas WR: In vitro translating of messenger RNA from the house dust mite *Dermatophagoides pteronyssinus*. *Int Arch Allergy Appl Immunol* 1987;83:384–389.
193. Chua KY, Stewart GA, Thomas WR, et al: Sequence analysis of cDNA coding for a major house dust mite allergen *Der p* 1, homology with cysteine proteases. *J Exp Med* 1988;167:175–182.
194. Sanger F, Nicklen S, Coulson AR: DNA sequencing with chain-terminating inhibitors. *Proc Natl Acad Sci USA* 1977;74:5463–5467.
195. Baur X, Konig G, Bencze K, Fruhmann G: Clinical symptoms and results of skin test, RAST and bronchial provocation test in thirty-three papain workers. Evidence for strong immunogenic potency and clinically relevant proteolytic effects of airborne papain. *Clin Allergy* 1982;12:9–17.
196. Bodmer WFL (ed): The HLA system. *Br Med Bull* 1978;34:213–316.
197. Mengele-Gaw L, Conner S, McDevitt HO, Fathman CG: Gene conversion between murine class II major histocompatibility complex loci: Functional and molecular evidence from the bm12 mutant. *J Exp Med* 1984;160:1184–1194.
198. Bell J, Rassenti L, Smoot S, et al: HLA-DQ beta-chain polymorphism linked to myasthenia gravis. *Lancet* 1986;1:1058–1060.

199. Tiwari J, Terasaki P: HLA and disease associations. New York, Springer-Verlag, 1985.
200. Stastny P: Association of the B-cell alloantigen DRw4 with rheumatoid arthritis. *N Engl J Med* 1978;298:869–871.
201. Thomsen M, Morling M, Snorrason E, et al: HLA-Dw4 and rheumatoid arthritis. *Tissue Antigens* 1979;13:56–60.
202. Nepom BS, Nepom GT, Mickelson E, et al: Electrophoretic analysis of human HLA-DR antigen from HLA-DR homozygous cell lines: Correlation between beta-chain diversity and HLA-D. *Proc Natl Acad Sci USA* 1983;80:6932–6966.
203. Nepom GT, Nepom BS, Antonelli P, et al: The HLA-DR family of haplotypes consists of a series of distinct DR and DS molecules. *J Exp Med* 1983;159:394–404.
204. Gregersen PK, Shen M, Song QL, et al: Molecular diversity of HLA-DR4 haplotypes. *Proc Natl Acad Sci USA* 1986;83:2642–2646.
205. Nepom BS, Nepom GT, Michelson E, et al: Specific HLA-DR4–associated histocompatibility molecules characterize patients with seropositive juvenile rheumatoid arthritis. *J Clin Invest* 1984;74:287–291.
206. Nepom GT, Seyfried C, Holbeck SL, et al: Identification of HLA-Dw14 gene in DR4+ rheumatoid arthritis. *Lancet* 1986;2:1002–1004.
207. Harley JB, Reichlin M, Arnett FC, et al: Gene interaction at HLA-DQ enhances autoantibody production in primary Sjögren's syndrome. *Science* 1986;232:1145–1147.
208. Wilson RW, Provost TT, Bias WB, et al: Sjögren's syndrome. Influence of multiple HLA-D region alloantigens on clinical and serologic expression. *Arthritis Rheum* 1984;27:1245–1253.
209. Harley JB, Alexander EL, Bias WB, et al: Anti-Ro (SS-A) and anti-La (SS-B) in patients with Sjögren's syndrome. *Arthritis Rheum* 1986;29:196–206.
210. Woo P, O'Brien J, Robson M, Ansell B: A genetic marker for systemic amyloidosis in juvenile rheumatoid arthritis. *Lancet* 1987;2:767–769.
211. Cookson WOCM, Faux JA, Sharp PA, Hopkin JM: Linkage between immunoglobulin E responses underlying asthma and rhinitis and chromosome 11q. *Lancet* 1989;1:1292–1294.
212. Bruton OC: Agammaglobulinemia. *Pediatrics* 1952;9:722–728.
213. Cooper MD, Lawton AR: Agammaglobulinaemia with B lymphocytes: Specific defect of plasma-cell differentiation. *Lancet* 1971;2:791–795.
214. Conley ME: B cells in patients with X-linked agammaglobulinemia. *J Immunol* 1985;134:3070–3074.
215. Pearl ER, Vogler LB, Okos AJ, et al: B lymphocyte precursors in human bone marrow: An analysis of normal individuals and patients with antibody-deficiency states. *J Immunol* 1978;120:1169–1175.
216. Conley ME, Brown P, Pickard AR, et al: Expression of the gene defect in X-linked agammaglobulinemia. *N Engl J Med* 1986;315:564–567.
217. Nyhan WL, Bakay B, Connor JD, et al: Hemizygous expression of glucose-6-phosphate dehydrogenase in erythrocytes of heterozygotes for the Lesch-Nyhan syndrome. *Proc Natl Acad Sci USA* 1970;65:214–218.
218. Migeon BR, Moser HM, Moser AB, et al: Adrenoleukodystrophy: Evidence for X-linkage, inactivation, and selection favoring the mutant allele in heterozygous cells. *Proc Natl Acad Sci USA* 1981;78:5066–5070.
219. Fearon ER, Winkelstein JA, Civin CI, et al: Carrier detection in X-linked agammaglobulinemia by analysis of X-chromosome inactivation. *N Engl J Med* 1987;316:427–431.
220. Yen PH, Patel P, Chinault AC, et al: Differential methylation of hypoxanthine phosphoribosyltransferase genes on active and inactive human X chromosomes. *Proc Natl Acad Sci USA* 1984;81:1739–1763.
221. Wolf SF, Jolly DJ, Lunnen KD, et al: Methylation of the hypoxanthine phosphoribosyltransferase locus on the human X-chromosome: Implications for X-chromosome inactivation. *Proc Natl Acad Sci USA* 1984;81:2806–2810.
222. Tauber AF, Borregaard N, Simons E, Wright J: Chronic granulomatous disease: A syndrome of phagocyte oxidase deficiencies. *Medicine* 1983;62:286–309.
223. Segal AW, Cross AR, Garcia RC, et al: Absence of cytochrome b_{-245} in chronic granulomatous disease. A multicenter European evaluation of its incidence and relevance. *N Engl J Med* 1983;308:245–251.
224. Royer-Pokora B, Kunkel LM, Monaco AP, et al: Cloning the gene for an inherited

human disorder—chronic granulomatous disease—on the basis of its chromosomal location. *Nature* 1986;322:32–38.
225. Baehner RL, Kunkel LM, Monaco AP et al: DNA linkage analysis of X chromosome–linked chronic granulomatous disease. *Proc Natl Acad Sci USA* 1986;83:3398–3401.
226. Kunkel LM, Monaco AP, Middlesworth W, et al: Specific cloning of DNA fragments absent from the DNA of a male patient with an X chromosome deletion. *Proc Natl Acad Sci USA* 1985;82:4778–4782.
227. Dinauer MC, Orkin SH, Brown R, et al: The glycoprotein encoded by the X-linked chronic granulomatous disease locus is a component of the neutrophil cytochrome *b* complex. *Nature* 1987;327:717–720.
228. Donaldson VH, Evans RR: A biochemical abnormality in hereditary angioneurotic edema. *Am J Med Sci* 1963;35:37–44.
229. Cicardi M, Bergamaschini L, Marasini B, et al: Hereditary angioedema: An appraisal of 104 cases. *Am J Med Sci* 1982;284:2–9.
230. Brickman CM, Tsokos GC, Balow JE, et al: Immunoregulatory disorders associated with hereditary angioedema. I. Clinical manifestations of autoimmune disease. *J Allergy Clin Immunol* 1986;77:749–757.
231. Rosen FS, Alper CA, Pensky J, et al: Genetically determined heterogeneity of the C1 esterase inhibitor in patients with hereditary angioneurotic edema. *J Clin Invest* 1971;50:2143–2149.
232. Gadek JE, Hosea SW, Gelfand JA, Frank MM: Response of variant angioedema phenotypes to danazol therapy. *J Clin Invest* 1979;64:280–286.
233. Cicardi M, Igarashi T, Rosen FS, Davis AE III: Molecular basis for the deficiency of complement 1 inhibitor in type 1 hereditary angioneurotic edema. *J Clin Invest* 1987;79:698–702.
234. Bock SS, Skriver K, Neilsen E, et al: Human C1 inhibitor: Primary structure, cDNA cloning, and chromosomal localization. *Biochemistry* 1986;25:4292–4301.
235. Cicardi M, Igarashi T, Kim MS, et al: Restriction fragment length polymorphism of the C1 inhibitor gene in hereditary angioneurotic edema. *J Clin Invest* 1987;80:1640–1643.
236. Stoppa-Lyonnet D, Tosi M, Laurent J, et al: Altered C1 inhibitor genes in type I hereditary angioedema. *N Engl J Med* 1987;317:1–6.
237. Nussenzweig MC, Shaw AC, Sinn E, et al: Allelic exclusion in transgenic mice that express the membrane form of immunoglobulin μ. *Science* 1987;236:816–819.
238. Cone RD, Reilly EB, Eisen HN, Mulligan RC: Tissue-specific expression of functionally rearranged λ1 Ig gene through a retrovirus vector. *Science* 1987;236:954–957.
239. Grosschedl R, Weaver D, Baltimore D, Costantini F: Introduction of a μ immunoglobulin gene into the mouse germ line: Specific expression in lymphoid cells and synthesis of functional antibody. *Cell* 1984;38:647–658.
240. Herzenberg LA, Stall AM, Braun J, et al: Depletion of the predominant B-cell population in immunoglobulin μ heavy chain transgenic mice. *Nature* 1987;329:71–73.
241. Weaver D, Costantini F, Imanshi-Kari T, Baltimore D: A transgenic immunoglobulin mu gene prevents rearrangements of endogenous genes. *Cell* 1985;42:117–127.
242. Goodnow CC, Crosbie J, Adelstein S, et al: Altered immunoglobulin expression and functional silencing of self-reactive B lymphocytes in transgenic mice. *Nature* 1988;334:676–682.

Implications for Extension of Molecular Biology: Gene Transfer, Gene Therapy, and Other New Directions

11

Ralph C. Williams, Jr., MD

A Broad Base of Interconnections

A great diversity of theoretical implications and branchings has now emerged from a virtual avalanche of scientific thought, speculation, and experimentation—all based on the original Watson-Crick DNA discoveries. The impact is difficult to evaluate, but it extends from new theories concerning the origins and divisions of the human species to direct insight into the basic mechanism of action of structures and fundamental physiological mechanisms such as the cellular potassium channel. In this chapter selected applications of molecular biology and recombinant methods will be discussed with the intent of illustrating the numerous potentially important vistas that have begun to open up and of stimulating imagination of what might be possible when current approaches are further refined or when we learn more about the elements that control gene expression.

Human Origins

Chapter 1 as an introduction indicated that human DNA is for the most part concentrated in chromosomes and the nuclear material. However, a small fraction of DNA localized within cell mitochondria has always been of interest to investigators and those involved with questions of evolution. A number of molecular approaches have already been implemented to document humans' divergence from apes,[1–4] as well as the way humans are genetically related to each other.[5,6] Most of the early work in this field was based on comparisons of nuclear genes. These genes, of course, are inherited from both parents and are mixed in every generation. Because of this mixing, recombinations occur, and these recombination events often make it difficult to trace the history or lineage of particular genomic segments unless very tightly linked adjacent markers are available.

For these reasons, a number of workers have focused attention on the DNA contained within mitochondria (mtDNA). First, mtDNA provides a magnified scheme of human gene pool diversity because mutations occur in this particular kind of DNA several times faster than in chromosomal DNA. Most important, because mtDNA is inherited maternally and does *not* recombine,[7] it could potentially be used for relating individuals to each other. Finally, there are approximately 10^{16} mtDNA molecules in a typical human and they are all usually identical to one another.[8] Thus, mammalian females behave as haploids because of a bottleneck or limitation in the genetically effective size of the number of mtDNA molecules in each oocyte.[9] A pair of breeding individuals can transmit only one type of mtDNA but carry four haploid sets of nuclear genes, all of which are transmissible to offspring. The rapid evolution and peculiar modes of inheritance of mtDNA have provided new insights into where and when the human gene pool may have arisen and spread.

Using this approach, mtDNA was obtained from 145 human placentas derived from widely dispersed sources: black Americans, Australia, New Guinea, Asia, the Philippines, Indonesia, Europe, North Africa, and the Middle East.[10] Purified mtDNAs were studied with 12 restriction enzymes and restriction sites were mapped. A total of 467 independent restriction sites were identified, 195 of which were found to be polymorphic. Using restriction mapping analyses, 147 mtDNAs were divisible into 133 distinct types. Seven of the types were found in more than one individual. None of the seven shared types occurred in more than one of the five major geographic regions.

Using the restriction site analysis, a hypothetical tree was constructed relating the 133 types of human mtDNA. Two assumptions were made: that mtDNA is strictly transmitted on the maternal side and that each individual is homogenous for its multiple mtDNA genomic copies. The tree could therefore be viewed as a genealogy linking maternal lineages in modern human populations to a common ancestral female bearing mtDNA type a. On the basis of this analysis, it was inferred that Africa was the likely primary source of the human mitochondrial gene pool. Moreover, it appeared that multiple lineages were present in each race. It was possible to calculate the minimum numbers of female lineages that colonized Australia, Asia, and Europe. Time scales could also be estimated by assuming that mtDNA sequence divergence accumulated at a constant rate in humans. An example of the types of calculations made and interconnections deduced using these methods is shown in Figure 11.1.

Sequence data and juxtaposition of sequences of nucleotides within mtDNA have been utilized to infer the sites of origin of current human populations[10] based on the phylogeny of mtDNA types. Rates of nucleotide substitution were assumed to be 2% to 4% per single nucleotide site per million years;[10] however, other workers on the same problem have estimated that 0.71% per site per million years is probably a more reasonable figure.[11] From such estimates, a tentative phylogenetic scale of the human species has been developed. Several interpretations have been proposed that support the origin of human populations in Africa or elsewhere. Considerable controversy still exists about the interpretation of mtDNAs within various human populations and its relevance to evolutionary steps.[12,13] However, the mere fact that mitochondrial cellular DNA has been seized upon as a clue to the details of human evolution illustrates the power of magnification and theoretical application of this unique recombinant approach.

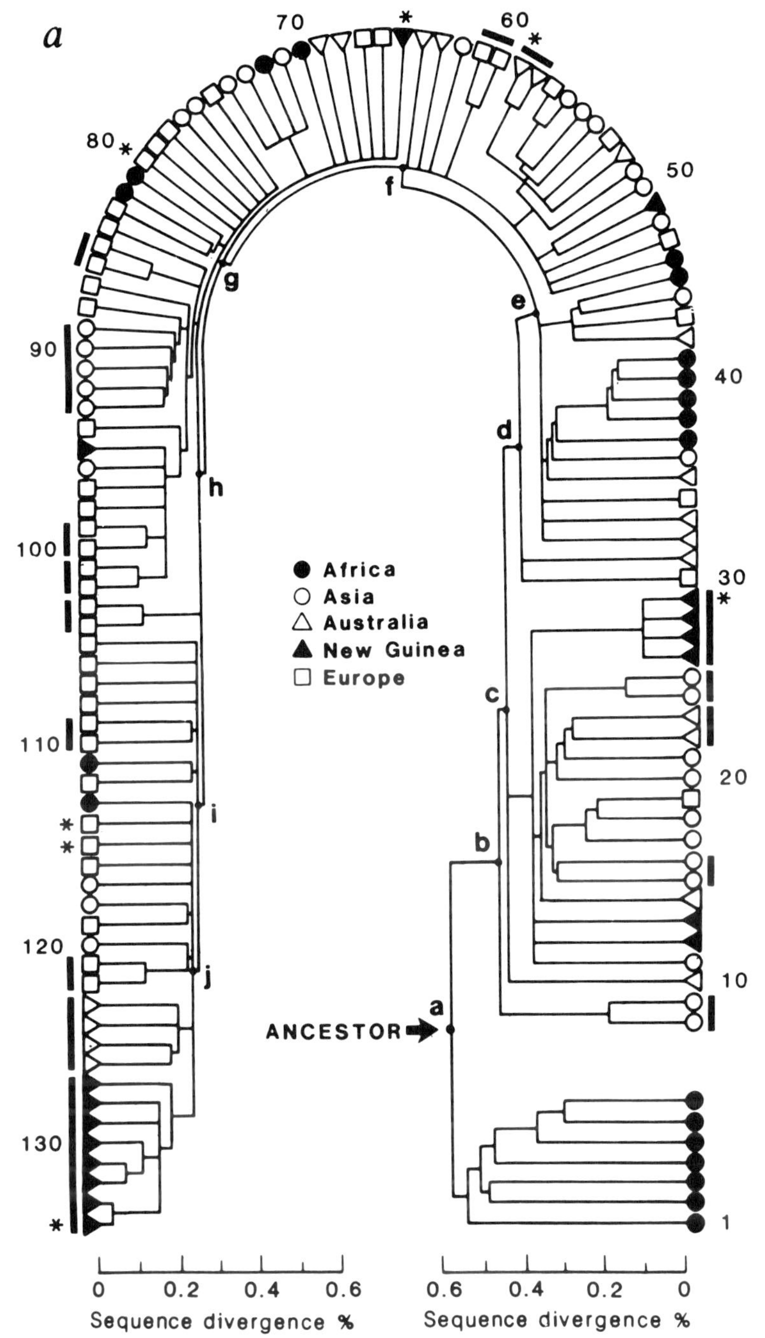

Figure 11.1 Genealogical tree for 134 types of human mtDNA (133 restriction maps plus reference sequence). The tree accounts for the site differences observed between restriction maps of these mtDNAs with 398 mutations. No other order of branching tested is more parsimonious than this one. This order of branching was obtained (using the computer program PAUP, designed by Dr. David Swofford) by ignoring every site present in only one type of mtDNA or absent in only one type and confining attention to the remaining 93 polymorphic sites. The computer program produces an unrooted network, which was converted to a tree by placing the root (arrow) at the midpoint of the longest path connecting the two lineages. The numbers refer to mtDNA types, 1 being from the aboriginal South African (!Kung) cell line (GM 3043), 45 being from the HeLa cell line, and 110 being the published human sequence. Black bars represent clusters of mtDNA types specific to a given geographic region; asterisks, mtDNA types found in more than one individual. Type 134 was found in six individuals; types 29, 65, and 80 each occurred thrice; and other types flagged with asterisks occurred twice. To place the nodes in the tree relative to the percent divergence scale, account was taken of the differences observed at all 195 polymorphic sites. *Reproduced from Cann et al,[10] with permission.*

Mitochondrial DNA and Disease

Although mtDNA was first identified four decades ago, it has only recently been appreciated that mitochondrial mutations may be associated with human disease.[14–16] Human mtDNA has been completely sequenced[17] and is known to encode 13 proteins, which include subunits of the electron transport chain as well as ATP synthetase.[18] Each of the thousands of mitochondria that are found in every human cell contains 2 to 10 identical copies of the 16.5-kb circular genome. Mitochondrial myopathies are a category of skeletal muscle disorders in which skeletal muscle mitochondria are abnormal. Affected patients may show only mild muscle weakness or severe multisystemic disease. Most cases appear to be sporadic, and there appears to be a high ratio of maternal to paternal transmission in the familial cases studied,[19,20] which indicates that some cases could be due to mutations in mtDNA. The early inference that mitochondrial mutations were responsible for the mitochondrial myopathic process was based on the finding of deleted mtDNA from skeletal muscle samples of some patients. Because the abnormal mtDNAs were limited to muscle and apparently normal mtDNAs were present in all other patient's tissues, it appeared that a state of heteroplasmy—the existence of more than one defined type of mtDNA in a single organism—was present. Poulton and coworkers[21] reported two patients with mitochondrial myopathy and involvement of multiple systems in whom true heteroplasmy for an abnormal mtDNA was found in white blood cells and other tissues. Other workers identified an autosomal dominant disorder characterized by myopathy and multiple deletions of mtDNA.[22] In this disorder a mendelian trait was clearly associated with lesions in the human mitochondrial genome. Because knowledge of the protein products specified by mammalian mitochondrial genes is far from complete, the precise biochemical or physiological lesions in this disorder remain to be determined.

The Other Genome—mtDNA

Because mtDNA did not follow the usual rules of mendelian inheritance, it attracted little attention for a time. More recently, it has become apparent that more knowledge of this subject will help in understanding a number of as yet unrecognized human diseases.[23,24]

Hemoglobin Genes in Plants

Hemoglobin has been described in plants in the nitrogen-fixing nodules involved in symbiotic associations between *Rhizobium* or *Frankia* and legume or nonlegume hosts.[25–28] Structural studies of these plant hemoglobins and various animal hemoglobins at the primary amino acid and gene sequence levels have suggested a common evolutionary origin.[29,30] Thus, hemoglobin genes conceivably inherited from an ancestor common to both animals and plants might be identifiable in all plants. A report by Boguez et al.[31] presents convincing evidence that yet another plant (in this instance *Trema tomentosa,* a nonnodulating relative of *Parasponia*) has hemoglobin genes. The genes have three introns situated at the same positions to those previously described in nodulating plant species. This appeared to strengthen the argument for a common original source of all plant hemoglobin genes. Moreover, the data appear to argue against what could be thought of as horizontal hemoglobin gene transfer from animals to plants. The *Trema* gene showed a tissue-specific pattern of both transcription and translation producing a monomeric hemoglobin in roots. The findings suggested that this plant hemoglobin might function as a respiratory protein within such root structures, similar to functions later served by animal hemoglobins as they evolved.

The plant hemoglobin genes in the newly described locations in *Trema tomentosa* were identified using a soybean hemoglobin cDNA or an oligonucleotide probe designed on the basis of known sequence data. Southern blot hybridization showed hemoglobin-related gene sequences in several of the new plant species studied.[30,31] After isolating the complete DNA segments, it was shown that the plants actually contained a complete functional hemoglobin sequence. When various plant tissue sites were hybridized in order to localize plant hemoglobin mRNAs, detectable levels were identified in plant nodules and roots but not in leaves, indicating tissue-specific gene expression. The function of hemoglobin gene products in plant roots was presumed to be associated with oxygen transport. These fascinating observations again illustrate the amplification and focus provided by recombinant methods for the study of such complex evolutionary problems.

Antibody Production in Transgenic Plants

In 1989 Hiatt and co-workers[32] reported production of mouse antibody in plants. Complementary DNAs from a mouse hybridoma mRNA were used to transform tobacco leaf segments, followed by regeneration and regrowth of mature plants. Plants that expressed single γ or κ immunoglobulin chains were then crossed, resulting in progeny in which both chains were expressed at the same time. Functional antibody was produced to 1.3% of total leaf protein in plants that expressed full-length cDNAs containing leader sequences. Plant-derived antibody binding specificity was similar to that of the original hybridoma from which the mRNA had been derived. These remarkable results demonstrated that production of immunoglobulins could occur very efficiently in tobacco. Practical applications of these transgenic approaches await further study.

Potassium Channel Genes Uncovered by Study of *Drosophila*

Again, remarkable insight into fundamental physiologic mechanisms was illustrated when highly focused studies of the so-called *Shaker* locus in *Drosophila* uncovered the basic gene structure and organization of a potassium channel.[33]

For some time it had been apparent that potassium currents are crucial for the repolarization of electrically excitable membranes. This function makes potassium channels important relay stations that are capable of modifying synaptic efficiency as well as other vital physiological functions. Because a number of important structural and functional loci had been studied for many years in the fruit fly, *Drosophila melanogaster*, advantage was taken of the previous extensive knowledge of various loci in this model. A genomic DNA and several cDNA clones from the *Shaker* region were selected.[33] Both clones selected showed at least seven exons distributed over 65 kb of DNA. Since there was little prior biochemical information about any K^+ channel, proof that the cDNA clones represented structural genes for such a component awaited functional testing; however, the presence of multiple hydrophobic sequences characteristic of integral membrane proteins suggested that they were involved in K^+ channel functions.

A computer search of the Dayhoff protein sequence database revealed signif-

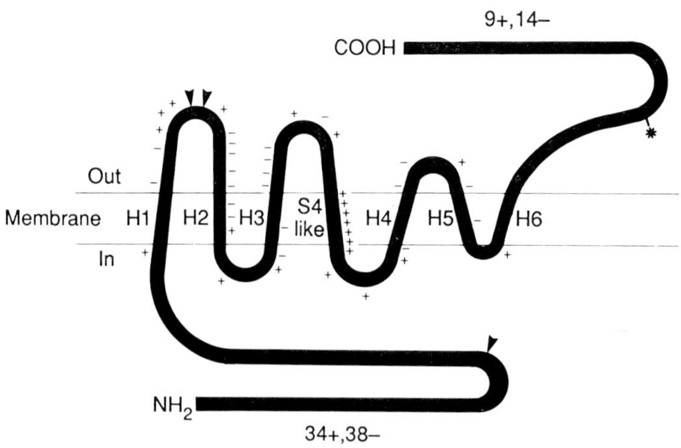

Figure 11.2 Proposed model for the transmembrane topography of ShA1 protein. Potential sites for N-glycosylation (arrowheads) and cyclic AMP–mediated phosphorylation (*) are shown. Only the basic (Arg and Lys are +) and acidic (Asp and Glu are −) residues found in or near the hydrophobic regions are shown individually. The total numbers of charged residues found in the hydrophilic amino and carboxyl domains are indicated. *Reproduced from Temple et al,*[33] *with permission.*

icant homology between the predicted *Shaker* protein and the already established Na^+ channel from the electric eel,[34] with 27% identical homologies including an identical stretch of 121 amino acids. The center of the homologous sequence had an arginine-rich region that had been previously linked to voltage-dependent gating of the Na^+ channel.[35,36] Further structural comparisons showed that this arginine-rich region of the *Shaker* gene product had features very similar to those of the dominant S4 elements in the Na^+ channel, which appears to act as a voltage sensor.[35,37] In models of the operation of the Na^+ channel it had been proposed that the S4 elements make up a transmembrane helix in which basic residues constitute a layer of positive charges across the membrane. Positive charges are then stabilized inside the hydrophobic interior of the protein by ion pairing with negatively charged residues provided by other transmembrane structures. When membrane potentials change, the four highly charged S4 regions would move, producing conformational alterations that would open the channel. Using projections from the somewhat analogous Na^+ channel structure, a tentative model for the *Shaker* protein K^+ channel was proposed and is shown in Figure 11.2. Of the three potential sites on the *Shaker* protein for N-linked glycosylation, two appeared to be extracellular. Negatively charged residues predominated in the extracellular loops near membrane-spanning regions; by contrast, positively charged residues predominated on cytoplasmic loops between membrane-crossing regions. As in the previous models of the Na^+ channel, the S4-like region of the *Shaker* protein represents the fourth membrane-spanning region, which is flanked by hydrophobic regions and oriented so that it appears to cross the membrane from the extracellular to the cytoplasmic side.

Whether the *Shaker* protein is part of a multimeric cell membrane unit is not known. Muscle acetylcholine receptors are composed of five subunits, only two of which are identical. Many such channels appear to be composed of repetitive or homologous subunits. There has been considerable interest in this area since Hille[38] advanced the hypothesis that all ion channels are evolutionarily related. A good deal of current clinical medical work has involved studies of other ionic channels such as calcium channels; however, it is not yet known whether S4-like sequences function similarly in calcium channels.

Results of Alternative Splicing

Subsequent work on the potassium channel components, based largely on the successful cloning of the *Shaker* locus of *Drosophilia,* revealed that this locus is a complex transcriptional unit that encodes several different proteins using separate mechanisms of alternative splicing[39]—a mechanism that occurs as a repetitive theme. In this system, a family of K^+ channel proteins were formed by splicing of different combinations of exons as mature mRNA was processed. The production of multiple channel components by alternative splicing was quite new, although the same mechanism has frequently been encountered in other genes (calcitonin and calcitonin gene–related peptide; Chapter 7). Complex transcription units such as those involved in alternative splicing often require multiple initiation units for transcription and the use of different promoters in different tissues, which are then differentiated by further alternative splicing. It is not known why there are multiple *Shaker* proteins, but their variety suggests that a spectrum of K^+ channels exists. Basic physiological studies established that K^+ channels are activated by membrane depolarization and influence both the rate of firing and duration of action potentials. They also influence the amounts of transmitter released and have even been implicated in several model systems of learning.[40] Much still remains to be learned about how the genes that regulate these channel proteins are expressed.

Expression in Oocytes

The *Shaker* gene encoding a potassium-selective ion channel was introduced into *Xenopus* oocytes to test whether the isolated gene could direct full expression.[41] This was done by subcloning both *Shaker* cDNA clones into a plasmid expression vector, transcribing *Shaker* mRNA in vitro, and injecting each type of mRNA into groups of *Xenopus* oocytes. The positive physiological result was that injected oocytes showed transient outward currents that were in every way similar to those described for the model A current–associated K^+ channels. It therefore appears that if the *Shaker* channels turn out to be multimeric, homo-oligomers of at least two and possibly four peptides form functional channels in the cell membrane. The *Shaker* locus apparently provides maximum flexibility in potassium channel expression. A final commitment to form a particular channel subtype is delayed at least until the RNA splicing stage and possibly until translation or even subunit assembly. The splicing mechanism that selects the combination of exons to be used probably depends on developmental, tissue-specific, and neuromodulatory regulation. Several possible outcomes of alternative mRNA splicing are shown in Figure 11.3.[42] These findings related to basic neurophysiological phenomena are important because they resulted from studies of a genetic locus in *Drosophila* and may point the way to a better understanding of many aspects of human neurophysiology.

Pulmonary Surfactant Apoprotein

Another recent development has linked recombinant techniques with studies of pulmonary surfactant, generally regarded as a complex phospholipid-protein complex that is essential to the body in lowering alveolar surface tension at the air-

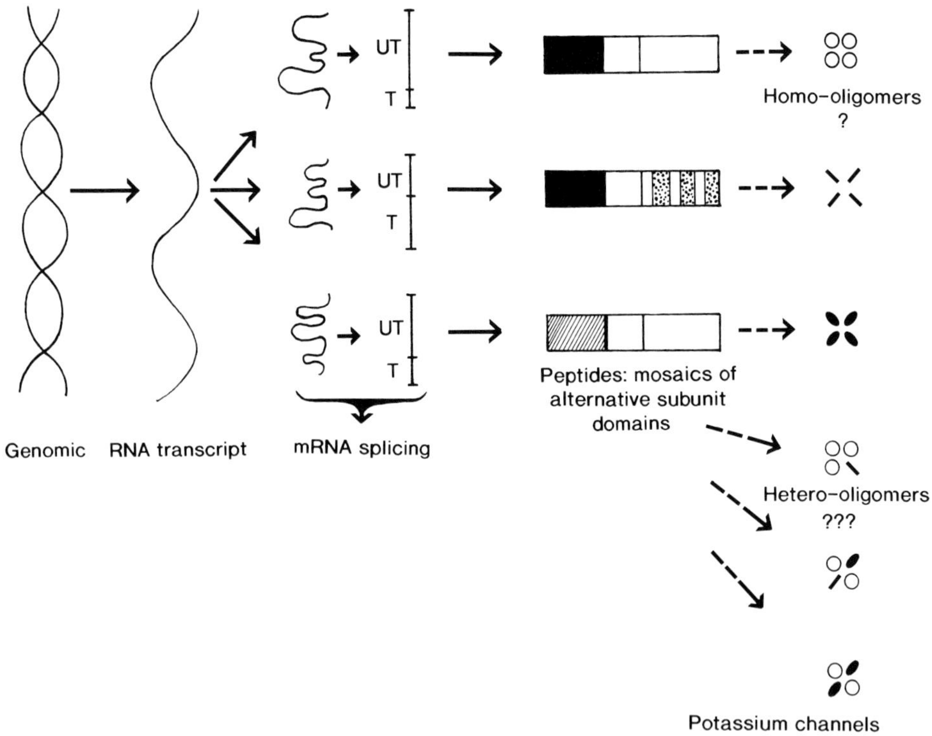

Figure 11.3 Possible biogenesis of potassium A channels. Dashed lines emphasize that multisubunit assemblies are hypothetical. Regulation of potassium channel expression may occur at transcription, at splicing of both untranslated (UT) and translated (T) domains, and at subunit assembly. *Reproduced from Agnew,[42] with permission.*

liquid interface in mammalian lung. This material appears to be essential for normal respiration; inadequate levels of surfactant at birth, which frequently accompanies prematurity, may result in respiratory failure and rapid demise. Chemical studies of surfactant composition have indicated that the material consists of dipalmitoylphosphatidylcholine and two major protein components of 32 and 10 kd.[43–45] It has been possible to reconstitute a surfactant-like material with surface tension–lowering activity by mixture in vitro of purified 32-kd surfactant apoprotein with synthetic lipids to form a lipoprotein complex.[46]

A full-length cDNA encoding the canine pulmonary surfactant apoprotein was isolated and used to screen a human genomic library in bacteriophage charon 28.[47] One clone showed intense hybridization and was chosen for further study. A partial sequence analysis of this genomic isolate appeared to show extensive homology with the canine clone. The positive clone appeared to encode the carboxyl-terminal half of the pulmonary surfactant apoprotein and a long 3′ untranslated region of mRNA. Next a cDNA library prepared with poly(A)$^+$ RNA isolated from adult human lung was screened. This procedure produced a marked relative increment in positively hybridizing clones, suggesting that the adult human lung was a more abundant source. With two independently isolated cDNA clones encompassing overlapping sequences, the continuous and nearly full-length sequence of the pulmonary surfactant apoprotein was derived. Approximately 100 base pairs upstream from the TATAAA sequence preceding exon I of the surfactant protein gene, a potential binding site for the glucocorticoid receptor sequence was identified. This finding was of interest because previous evidence had indicated that glucocorticoids could regulate surfactant synthesis.[48]

This study also showed that the 22-amino-acid sequence involving the amino terminus of a 32-kd protein isolated from lung lavage of a patient with alveolar proteinosis corresponded exactly to a segment of the protein product encoded by the pulmonary surfactant gene. Thus, the alveolar protein present in great excess in patients with pulmonary alveolar proteinosis is identical to that present in normal lung surfactant. Additional studies of relative degrees of glycosylation of this surfactant protein indicated some size heterogeneity based mainly on differences in degrees of glycosylation. With the successful cloning of the surfactant protein, abundant supplies of this material should eventually be available for the treatment of human disorders such as prematurity and possibly certain forms of chronic lung disease.

Human Prion Protein cDNA

Diseases attributed to slow virus or prion infections include scrapie in sheep and goats and Creutzfeldt-Jakob disease and kuru in humans. These diseases have been proposed to exemplify a group of degenerative encephalopathies caused by a small proteinaceous infectious particle called the prion. Prions are resistant to inactivation by procedures that are known to modify nucleic acids.[49,50] A major protein material of the prion cosediments with scrapie infectivity for hamsters. This 27- to 30-kd sialoglycoprotein, which has been designated the prion protein, appears to aggregate extracellularly, forming rodlike structures that resemble amyloid fibrils or rods.[51–56] It has been suggested that this protein, termed PrP27–30, is altered pathological protein or disease product and is not an integral part of the underlying infectious agent.[54,55] In general, it is an understatement to say that there has been heated controversy about what slow infections such as scrapie or Creutzfeldt-Jakob disease actually are and what can be considered an infectious agent for the disease.

A PrP27–30 complementary DNA was cloned from scrapie-infected hamster brain,[56] and a scrapie prion protein–specific mRNA was found in both scrapie-infected and uninfected brain samples.[57] This was followed by isolation of homologous human cDNA clones, identification of the primary sequences, and deduction of the presumed primary structure of the protein by Liao et al.[58] Approximately 100 rat and human clones cross-hybridized with two different synthetic oligonucleotide probes; this was calculated to represent mRNA abundances of 0.02% and 0.018% in human retinal and rat brain tissues. Northern blot studies identified PrP mRNA in placenta, kidney, vipoma, and neurofibroma tissues. Overlapping g clones with the largest inserts were subcloned into M13 and the inserts were sequenced. The sequenced human PrP27–30 cDNA included 2430 nucleotides with eight poly(A) tail bases. The deduced final PrP product contained 245 amino acids and had a molecular weight of 26,818 without glycosylation. A stop codon TGA was found at nucleotide 818, and the noncoding region extended to nucleotide 2427, followed with a poly(A) tail. Using a *Bam*HI restriction fragment, chromosomal mapping indicated that the prion gene is on human chromosome 20.[58] Comparative Southern blot and sequence analyses showed 89% sequence homology between the primate and rodent (hamster and mouse) PrP27–30 proteins, indicating marked conservation of this gene through more than 50 million years of evolution. Additional analyses of PrP27–30 showed elements of a β-pleated sheet conformation thought to be common to many forms of amyloid.[59] Strangely, purified PrP27–30 has been reported to be noninfectious,[60] although the aggregated scrapie-associated fibrils are clearly associated with the ability to produce infections. A number of hypotheses have been proposed to explain these perplexing findings: (1) that the prion is an infectious agent devoid of any nucleic acid, (2) that the PrP27–30 protein is a component of the infectious prion, (3) that the actual infectious scrapie agent is a virus which exists at extremely low levels and is copurified with scrapie-associated fibrils, or (4) that the prion is an accumulated or slightly changed protein that is normally present in cell membranes. The localization of the prion protein gene to chromosome 20 suggests that the amyloid-like rod structures associated with scrapie are probably unrelated to the amyloid plaques of Alzheimer's or some Down's syndrome patients (see Chapter 9). These observations show how definition of one part of the difficult puzzle of slow infections such as scrapie and kuru by cloning the prion protein may eventually help in understanding this baffling class of human neurodegenerative diseases.

Sequencing the Human Genome

As noted at the beginning of Chapter 9, a major effort is under way to make sequencing the entire human genome—all 3 billion base pairs—technically feasible. The Department of Energy (DOE) and the National Institutes of Health (NIH) have now launched into a program to accomplish this task.[61] There is little doubt that the eventual benefits to medicine and basic knowledge will be substantial; however, one of the stumbling blocks has been the very scale of the project. Initially it was estimated that the work might take 15 to 20 years to complete and cost several billion dollars. In addition, from the outset of planning this herculean task, issues of ethics were a potential central concern. Questions still exist about the privacy of DNA or the patentability of parts of the human genome.

Since the original plans to sequence the human genome were announced, a number of technical innovations have been developed that should make the task easier and more rapid. In 1986 a group at California Institute of Technology developed the first automated DNA sequencer, which theoretically had the capacity to sequence 10,000 to 15,000 bases a day. Since 1981 the Japanese government has been investing $1 million a year in the development of automatic DNA sequencing techniques that would be able to sequence 1 million bases per day at 17 cents per base.[62] Currently, several groups in the United States are working hard on improving methodology and speed while keeping the chance of inadvertent errors very low. Several alternative DNA sequencing methods are under study, including application of flow cytometry to DNA sequence procedures by tagging gases with fluorescent labels and cleaving them off one at a time. Another method that is being explored might employ scanning transmission electron microscopy and labeling of bases with clusters of gold or tungsten atoms. Many of these methods are still highly experimental, and time will tell whether they speed the tasks at hand.

A number of parallel forces have been at work as plans were laid to map and sequence the human genome. These have been admirably summarized by McKusick.[63] It was essential to establish a map for efficient sequencing. Accurate localization of complementary DNA (including messenger RNA, exon, and transcript) will be extremely helpful in identifying all the genes as well as describing their functions in different developmental stages or in specialized tissues, ultimately providing candidate genes as regions of mutation in various hereditary disorders or neoplasms.

Moreover, as experts began to ponder the problem, it became obvious that other model organisms besides the human genome needed to be examined. The then-NIH director, James Wyngaarden, created an Office of Human Genome Research at NIH in March 1988 and in September of the same year appointed James D. Watson, the original codiscoverer of DNA structure, to head the office of Human Genome Research.

The ultimate objective now will be to put together the best encyclopedia of the human genome available. It has been pointed out that merely recording the sequence of the 3 billion base pairs of the haploid genome will fill up as many pages as at least 13 sets of the *Encyclopedia Britannica.*[63] One might ask, if the human genome is to be sequenced, whose will be chosen? Certainly, keeping track of the source of the DNA worked on will be important, but DNA can come from different individuals chosen for study of various parts of the whole.

Besides eventual leaps in understanding of the basis of a number of mendelian or heritable disorders, considerable insight into other fundamental differentiation problems such as cancer is bound to increase.[64]

Variable Number of Tandem Repeat Markers

Considering the size and complexity of the task, a number of specific approaches have emerged that may simplify or help to organize gene mapping and studies of human DNA. One of the most promising areas is related to the use of variable number of tandem repeat (VNTR) markers for gene localization.[65] All authorities agree that a large assortment of reliable genetic markers is essential for accurately mapping genes that cause various diseases. With the widespread use of restriction fragment length polymorphisms (RFLPs), as discussed extensively in Chapter 9, about 400 polymorphic DNA markers for human chromosomes have now been developed. Most of these contain only two alleles and therefore may not be very helpful for determination of linkage in family surveys. A few known DNA markers are dependent on restriction fragments that are of many different sizes. This sort of polymorphism, discussed in Chapter 9, depends on variation in the number of tandem repeats of a short DNA sequence. Since most individuals are heterozygous at such DNA segments, these markers may produce linkage information in most kindreds studied. Ten oligomeric sequences generated from the tandem repeat segments of the myoglobin gene, the ζ-globin pseudogene, the insulin gene, and the X-gene area of hepatitus B virus were used to produce a battery of single-copy probes.[65] These probes showed a number of new, extremely polymorphic loci whose allele size distribution reflected considerable variation in the number of tandem repeats.

Chapter 9 provided a number of examples of how linkage analysis in families has helped to focus on specific locations on certain chromosomes for hereditary disorders such as Huntington's disease, Duchenne muscular dystrophy, polycystic kidney disease, and cystic fibrosis. As noted in the examples, one has to determine whether there is a positive correlation of an identifiable DNA marker with the way the disease segregates in the families studied. Moreover, it is often essential to ascertain which of the two copies of a particular chromosomal marker region has been inherited by an individual from each parent. Thus, if the DNA marker is represented in an affected parent by two distinguishable alleles (i.e., the parent is

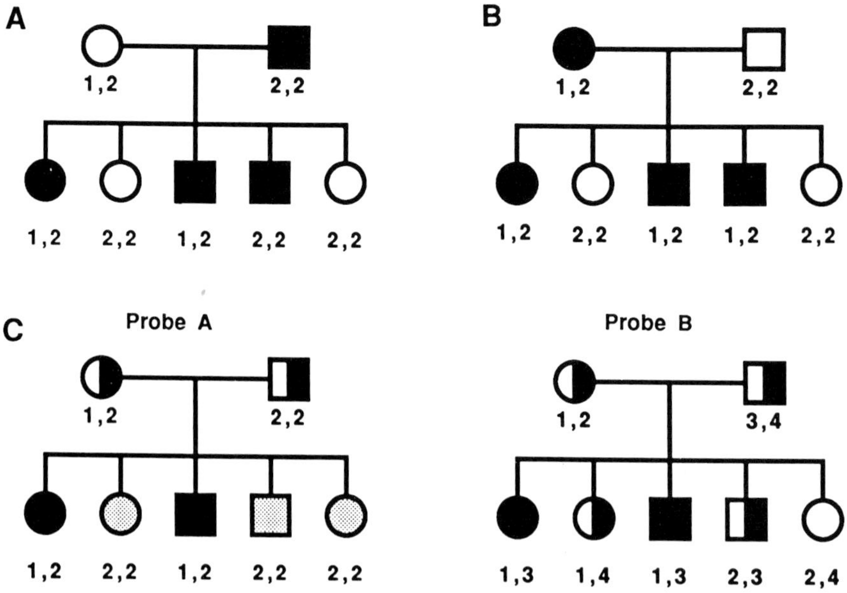

Figure 11.4 Informativeness of the probes. Pedigree charts illustrate the models of autosomal dominant (A and B) and recessive (C) genetic disease in a family. Probe A is a site-polymorphic DNA marker that has two alleles, and probe B is a highly polymorphic DNA marker that has many alleles. ●, ■: Affected individuals; ○, □: normal individual; ◑, ◨: carrier; ◍, ▩: carrier status uncertain. Round symbols designate females; square, males. *Reproduced from Nakamura et al,*[65] *with permission.*

heterozygous for the marker), it may be possible to determine quite easily that one particular allele is cosegregating with the disease process in children and grandchildren. The importance of having DNA markers that are heterozygous is also illustrated by a family in which there is an autosomal recessive disorder. Using probe 1, for which both parents are heterozygous, would be much more informative than using probe 2, for which one parent is homozygous, in associating the marker with the autosomal recessive trait. This is illustrated by reference to Figure 11.4.

For these reasons the frequency of heterozygosity with respect to a particular DNA marker in the population to be studied becomes extremely important in predicting the precision with which linkage information can be obtained from family studies. The expected frequency of heterozygosity at a particular locus or segment of DNA is directly proportional to the number of common alleles present in the population at large. The more alleles a locus has, the greater the probability that a single individual will be heterozygous.

Since most genetic diseases are relatively rare, it is clear that the development of a large number of highly polymorphic DNA markers for the human genetic map will help to speed up localization of disease genes of interest.

DNA Markers

Variations in sequence in human DNA are common. Most of the known DNA sequence variations are due to base pair changes that either produce or take away restriction enzyme cleavage sites, resulting in changes in the lengths of DNA fragments produced. If a DNA segment is based on a single such variation, only two alleles will be present because the enzyme cleavage site is either present or absent. In such an instance the chance that a parent will have two different alleles at this locus is less than 50%. In contrast, for a few well-studied DNA markers highly variable lengths of fragments are obtained after restriction enzyme digestion. Such systems have been observed with the insulin gene,[66] the Ha-*ras* oncogene,[67] the ζ-globin pseudogene,[68] and the myoglobin gene loci.[69] The nucleotide sequences of these hypervariable loci indicated that the DNA restriction fragments always contained a set of tandem repeats of a short 11- to 60-base-pair sequence, and therefore the relative lengths of the restriction fragments were a function of the number of copies of the tandem repeats in the fragment. A DNA marker system based on this sort of polymorphism instead of a change in a single base pair can be highly informative in family linkage analysis, especially when associated with multiple alleles. When a DNA sequence has been identified that contains tandem repeats but represents only a single gene locus, it has been termed a variable number of tandem repeats (VNTR) locus. As noted in Chapter 9, a method based on the distribution of these tandem repeats has been used to develop a system of DNA fingerprinting unique to each individual.

Using this general approach, Nakamura et al.[65] constructed a series of oligonucleotides corresponding to the consensus sequences of the tandem repeats present at the loci for the insulin gene, the pseudogene of ζ-globin, the myoglobin gene, and part of the core sequence of pYNZZZ, a plasmid identified by the ζ-globin oligonucleotide. In addition, 16- and 20-base oligonucleotides were synthesized that corresponded to a portion of the X-gene region of hepatitis B virus. The sequences of these synthetic oligonucleotides are shown in Table 11.1. These probes were screened for hybridization with human phage libraries and human cosmid libraries, as well as in individual families. Screening of the human genomic cosmid libraries yielded more than 1000 candidate clones. Fewer than 4% of these appeared to hybridize with more than one oligonucleotide probe. It seems possible that cloning a large number of these VNTR segments may provide additional precisely localized markers that would show heterozygosity in most individuals studied.

Table 11.1 DNA Sequences and Hybridization Conditions for Oligonucleotide Probes[a]

		Temperature C° of:	
Probe	Sequences	Hybridization	Washing
Zeta-globuin	TGGGGCACAGG$_{C}^{T}$GTGAG	42	48
Insulin (14mer)	ACAGGGGTGTGGGG	30	37
Myoglobin-1 (16mer)	GGAGGTGGGCAGGAAG	37	44
Myoglobin-2 (14mer)	GGAGGCTGGAGGAG	37	42
HBV-1 (16mer)	GGAGTTGGGGGAGGAG	37	44
HBV-2 (20mer)	GGCTGGGAGGAGTTGGGGG	50	60
HBV-3 (15mer)	GGTGAAGCA$_{C}^{G}$AGGTG	37	42
HBV-4 (15mer)	GAGAGGGGTGTAGAG	37	42
HBV-5 (15mer)	GGTGTAGAGAGGGGT	37	42
YNZ22 (15mer)	CTCTGGGTGTCGTGC	37	42

Hybridization was carried out in a solution of 5 × SSC (1 × SSC = 0.15M NaCl and 0.016M sodium citrate); 50mM tris-HCl 9 μH 7.4); 1 × Denhardt's solution (0.02% bovine serum albumin, 0.02% polyvinylpyrrolidone, 0.02% Ficoll); yeart tRNA (10 μg/ml); and 1 × 10^5 cpm/ml; ^{32}P 5′-end labeled probe (specific activity 2 × 10^8 cpm/pmol) for 16 hours. Washing was done three times in 5 × SSC-0.1% SDS for 5 minutes.

(Reprinted from Nakamura et al,[65] with permission.)

The origin of the VNTRs in DNA is, of course, unknown, but it has been suggested that the tandem repeats themselves may increase or decrease in number through an unequal frequency of crossing over. Another suggestion is that the VNTRs encode relative hot spots for recombinational activity—thus in part accounting for their origin and high mutability by way of unequal crossing over. One core sequence, GGGCAGGAXG, seems to be common to virtually all of the VNTR sequences described. When more is known about the precise distribution of the VNTR sequences, they may act as signposts or major reference areas on human genetic linkage maps. Moreover, they should prove useful in family studies because of the high frequency of heterozygosity found in all individuals studied thus far.

Pulsed Field Gel Electrophoresis

For some time studies of mammalian chromosomes and the relationship between cloning experiments and family linkage have revealed a gap in ability to focus on the physical basis for some genetic traits and their phenotypic expression. From time to time this gap has actually discouraged full-speed-ahead efforts to put together complete genetic maps of various mammalian chromosomes. However, several methods have now been developed for the molecular analysis of DNA sequences spread out on the genome over distances of up to 1000 kb.[70,71] One of the most widely employed approaches is pulsed field gel electrophoresis, which can separate fragments of DNA as large as yeast chromosomes and thus help to generate an expanded full-scale map of the entire *Escherichia coli* genome.[72]

In the latter work it was thought that the first step in complete characterization of *E. coli* DNA was to arrange an accurate map of the whole genetic structure. It was found that the restriction enzyme *Not*I cuts the *E. coli* genome into 22 DNA segments ranging from 20,000 to 1,000,000 base pairs and that these segments could be separated by pulsed field gel electrophoresis. The enzymes chosen to cut genomic DNA were those that recognized infrequent nucleotide sites such as 8-base-pair sequences and/or those that contained CpG in their recognition sites. The two enzymes *Not*I and *Sfi*I, with 8-base-pair recognition sites, seemed ideal for these studies. Finally, *Not*I was chosen because it produced 22 fragments that gave the best electrophoretic separation in the pulsed field technique. Since a detailed genetic map of the *E. coli* genome was already available, a large repertoire of cloned DNA probes for genes whose position was already established were chosen and studied with Southern blotting techniques.

A similar approach has been applied to initial analyses of the human major histocompatibility complex, also using pulsed field gel electrophoresis.[73] Again, enzymes with 8-base-pair cleavage sites or CpG in their recognition sequences were selected, because CpG occurs rarely in mammalian genomes. Successful application of these techniques should allow initial alignment of megabase human DNA fragments. This approach, together with the standard cloning techniques and judicious use of gene walking or jumping libraries (Chapter 9), should help in establishing molecular maps of the human genome.

Transgenic Approaches and Gene Therapy

In the initial days of recombinant DNA research, biologists throughout the world asked for a moratorium to appraise the immediate and long-term risks of the application of all recombinant techniques to biological problems in the laboratory.

Much more is now known about the sequence of events that leads to uncontrolled growth of cells and the safe application of recombinant DNA methodology. Moreover, a number of lines of investigation have revealed that genetic engineering has been occurring in the natural environment for millions of years without major catastrophes or upheavals. Thus investigators were initially intrigued by the opportunity to manipulate some of the features in the recombinant systems that were being uncovered. One of the earliest approaches to this was in the various models available in transgenic mice.

Transgenic Mice as Models for Gene Transfer

One of the first models for gene transfer was developed using an infectious retroviral vector to transfer a bacterial gene coding for resistance to the neomycin analogue of G418 into pluripotent hematopoietic murine stem cells contained in bone marrow.[74] Transplantation of these cells into lethally irradiated mice resulted in engraftment of the recipient animals with donor hematopoietic tissue that expressed the bacterial gene. It was soon recognized that this method afforded an efficient and relatively rapid way to reintroduce genetically modified tissue into intact organisms; moreover, the experimental system could be used to study actual expression and regulation of cloned genes in the whole animal or host.

The strategy involved in the use of retroviral vectors was relatively straightforward and took advantage of several growth-enhancing and intrinsic self-generating features of the viral genome. The Moloney murine sarcoma viral genome utilized contained substitutions of viral sequence with both a mouse cDNA sequence coding for dihydrofolate reductase and a DNA segment coding for neomycin resistance in bacteria and G418 resistance in mammalian cells. The built-in efficiency of the viral promoter elements and other features such as the reverse transcriptase enabling the virus to survive in infected cells were used to ensure effective transfer of resident genes (neomycin resistance) into the transfected cells. The MSV genome, along with its introduced dihydrofolate reductase and neomycin-resistance gene inserts, is shown in Figure 11.5. Successful transfer of the genes by this procedure was confirmed by Southern blotting. When these experiments were reported, it was rapidly recognized that similar methods might be used with a number of other systems for introducing new DNA into target cells.[75]

Figure 11.5 Structure of the Moloney sarcoma virus (MSV) genome containing the dihydrofolate reductase (dhfr) and neomycin-resistance (neo) gene inserts. Sequences of the Moloney sarcoma virus that are retained are shown as long terminal reports (LTRs) which are involved in the reverse transcription process and sequences necessary for encapsidation of the viral genome (ψ). The 5′ splice site used for the envelope message is shown as 5′ ss. Sequences have also been introduced at the 3′ ss areas necessary for generation of spliced *env* transcripts. Restriction sites are shown by arrows above.

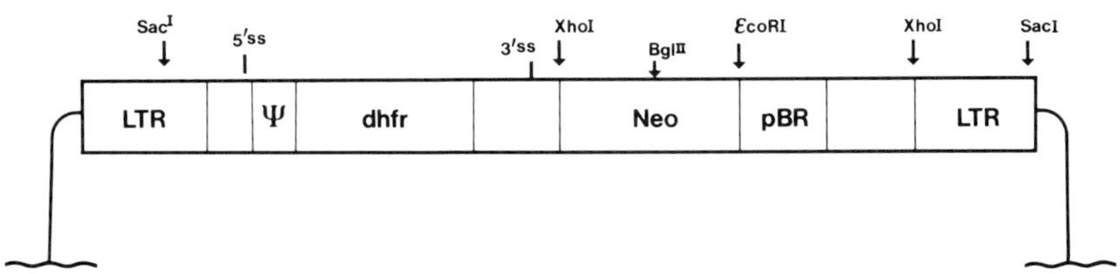

Other Methods of Gene Transfer

Another method of gene transfer, transfection, introduces genes into cells by co-precipitation with calcium phosphate. Studies by Austin and co-workers[76] showed that HLA class II antigens could be expressed in previously negative mouse fibroblasts after introduction of the extraneous genes directly into cultured fibroblast cell lines. Positive expression of the new class II (DP) antigens was demonstrated in these studies by reactivity with highly specific monoclonal antibodies and in functional experiments where fibroblasts expressing DP antigens activated T cells in the presence of appropriate antigen.

A third method for introducing new genes into somatic cells involves microinjection. Initial studies with mouse ova indicated that integration of a foreign gene into host chromosomes was much more efficient with nuclear rather than cytoplasmic injection.[77] The microinjection can be performed using interference contrast microscopy to visualize the cell nucleus directly at the time when linear fragments of the genes of interest are studied[78] (Figure 11.6). The injected ovum is then transferred to foster mothers and pregnancies are carried to term. The number of copies integrated can then be estimated by quantitative dot blot hybridization. When the animals mature, gene expression can be measured directly or by appropriate function. The studies by Hammer et al.[78] indicated that with these techniques foreign genes could be introduced into several large animals, including rabbits and pigs, in which eventual gene expression was documented. It seems amazing to the amateur that merely putting genes with other cells in the presence of calcium or injecting fragments of foreign DNA into nuclei of cells will result in successful integration and eventual expression of specific mRNA for the injected or introduced gene. However, the techniques have now been repeatedly shown to work and much of the methodology has been substantially improved over the past decade.

Viewed broadly, virtually all mice could be considered transgenic, since all show foreign viral DNA that has been stably integrated into their genomes throughout evolution.[79] However, with the microinjection technique a nearly infinite number of different DNA molecules can be introduced into all cells of a number of experimental animals.

The foreign DNA introduced by either microinjection or retroviral infection can integrate directly into host chromosomal DNA and is then carried in germ cells as well as somatic cells, so it can be transmitted to subsequent progeny.[80,81] The efficiency of producing transgenic mice averages approximately 25% when linear DNA molecules are microinjected, and comparable efficiency is possible with viral vectors as well when they are used to infect preimplantation embryos. The advantage of the microinjection techniques is that there appears to be no constraint on the size or sequence of DNA that is to be introduced. DNA that is microinjected at the one-cell stage is usually integrated at random sites prior to cell division and appears in all subsequent cells of the embryo and embryonic tissues. When more than one copy of the microinjected gene integrates (which is usually observed with microinjected DNA), the multiple copies are usually arranged in tandem head-to-tail lineups that may contain several hundred copies. This ordering is presumed to be the result of homologous recombinations between the injected molecules.[82] Most genes that have been microinjected are expressed, but the quantitative level of expression appears to be unpredictable and does not directly correspond to the number of copies of the gene in the host.

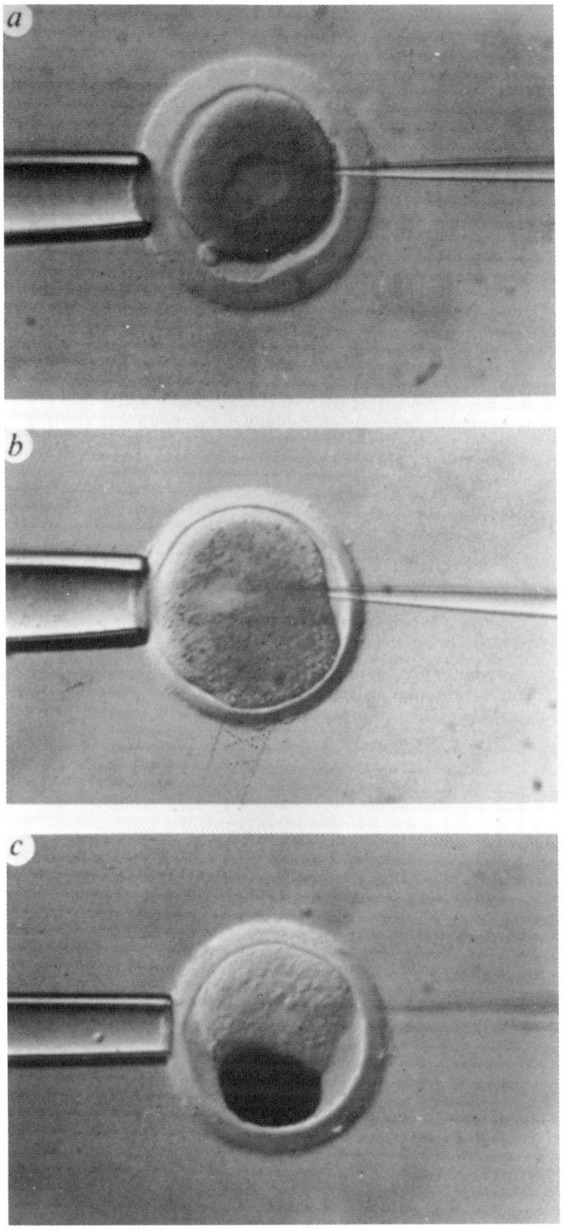

Figure 11.6 Interference-contrast photomicrographs of one-cell fertilized ova from rabbit (a), sheep (b), and pig (c) after microinjection. Ova are held by a blunt holding pipette (diameter 50 μm); injecting pipette (diameter 1.5 μm) has penetrated the zona pellucida, plasma membrane, and pronuclear envelope. The tip is seen within the nucleoplasm immediately following injection of buffer containing DNA. The porcine ova (c) has been centrifuged at 15,000*g* for 3 min to reveal the normally obscure pronuclei. Visualization of nuclear structures is aided by the use of interference-contrast optics, and microinjection is carried out under 250× magnification using a Leitz microinjecting apparatus. Injection is monitored by observing the diameter of the pronucleus or nucleus, which was expanded 50%. *Reproduced from Hammer et al,*[78] *with permission.*

By contrast, in the case of retroviral gene introduction, only single copies of the viral DNA integrate with the host genome and integration always occurs at a set locus in the viral DNA itself, so that genes cloned into retroviral vectors are usually flanked by the long terminal repeats (LTRs) after integration.[83] One problem with retroviral vectors is that infection usually occurs at later embryonic stages, resulting in mosaic founder animals. Also, more than one cell may be infected, which produces mice with several different areas of integration of the new gene. For this reason, it may be necessary to outbreed the founder animals to be sure that one has pure lines hemizygous for the single gene originally inserted.

Tissue-Specific Gene Expression

A number of studies of transgenic mice have indicated that introduced genes appear to be expressed in a tissue-specific pattern.[84–89] This allows assessment in many instances of whether cis-acting nucleotide sequences are involved in the programming of such gene expression in the recipient host. Among the genes shown to be expressed in a tissue-specific manner are those for α-fetoprotein in the yolk sac, β-globin in erthyroid cells, T cell receptor in the T cell, and α-actin in muscle. The exact tissue specificity assigned, of course, depends on the sensitivity of the biochemical or functional assays employed.

Consequences of Foreign Gene Expression

In many instances the expression of foreign genes can be used to study important biological systems of interest. Thus, introduction of metallothionein–growth hormone fusion genes into mice stimulates the production of growth hormone in tissues that normally synthesize metallothionein, and as a result some of the mice grow to double their normal size.[90–92] In such mice the growth hormone is produced mostly in the liver, under control of the metallothionein gene promoter, so the normal physiological control mechanisms regulating growth hormone production are bypassed. Secondary results of the disordered network of interrelated control mechanisms in such animals are that the somatotrophe cells in the pituitaries fail to develop properly and the fertility of females is impaired.[92,93]

Oncogene Expression in Transgenic Mice

Production of transgenic mice containing insertions of various potent oncogenes is an exciting new approach for analysis of the role of these elements in both normal cell development and tumor production. When DNA containing the simian virus 40 (SV40) early region genes was introduced into mice, choroid plexus papillomas developed within 6 months of birth.[94] These tumors expressed SV40 mRNA and T antigen, but the level of expression of these markers was quite low in other tissues. No tumor developed when the large T antigen gene was truncated, and deletion of the SV40 72-base-pair repeats (which included the gene enhancer) greatly reduced the incidence of choroid plexus tumors. Thus, T antigen expression and the enhancers were important for the induction of tumors.

Similarly, transgenic mice bearing the *myc* gene fused to the mouse mammary tumor virus enhancer elements developed mammary tumors during the second or third pregnancy.[95] Tumors were not seen at other times, although the *myc* genes were expressed at significant levels elsewhere.

Insertional Mutagenesis

Since integration of foreign DNA appears to occur randomly, insertional mutagenesis of endogenous genes may occur. Three lines of transgenic mice have now been described in which the introduction of foreign DNA disrupted genes essential for normal embryonic development.[96,97] The frequency of insertional mutations is difficult to estimate accurately, since limited numbers of lines of transgenic mice have been inbred and some mutations may not show new, clearly identifiable phenotypes. Currently it is estimated that 10% to 20% of transgenic mice may harbor recessive mutations of essential genes.

Gene Modification

When cloned DNA is introduced into cells by calcium phosphate precipitation or microinjection, it is incorporated into the recipient host genome at random sites in a process that apparently does not involve homologous recombination. Evidence for recombination between the introduced gene and its natural chromosomal counterpart has sometimes been difficult to obtain. A different approach involving gene modification was described by Smithies et al.,[97] who developed a strategy for introducing cloned human β-globin genes into chromosomal DNA by homologous recombination. In this case the donor or test plasmid contained DNA sequences that allowed it to be ascertained both before and after integration from the normal chromosomal copy of the endogenous β-globin gene. Also, a cell line was used in which the β-globin gene was active, since insertion of the plasmid into an inactive locus could shut off expression of the linked selectable marker gene for neomycin resistance. Finally, the donor plasmid was cleaved by *Bst*XI within the 5′ flanking sequences for the β-globin gene to maximize the frequency of recombination. A bacteriophage λ gene rescue protocol was employed to identify and isolate the recombinant DNA segments from a mixture of total genomic DNA. The strategy employed is illustrated in Figure 11.7, which shows that recombination between the donor plasmid and the chromosomal copy of the β-globin gene generates an *Xba*I fragment that contains the *supF* bacterial suppressor tRNA gene and the intact β-globin gene plus 5′ and 3′ flanking nucleotide sequences. This DNA segment

Figure 11.7 Strategy for introducing cloned human β-globin gene into chromosomal DNA by homologous recombination. Numbered boxes are exons; thick lines are introns (intervening sequences, or IVS). Restriction endonuclease cleavage sites for the enzymes *Bsi*XI and *Xba*I are indicated. Dashed lines indicate the site of recombination between chromosomal DNA and the donor plasmid. *Reproduced from Maniatis,[98] with permission.*

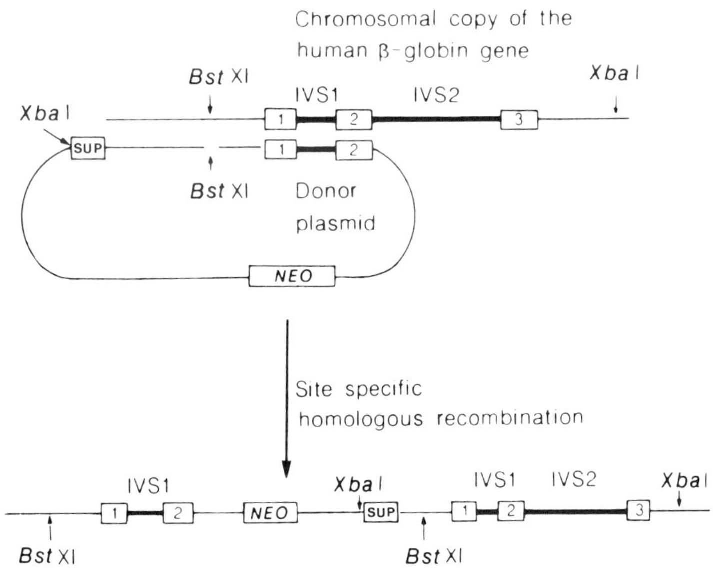

can then be rescued from total genomic DNA of the host cell by digestion with the *Xba*I restriction enzyme and ligation to the *Xba*I phage vector packaging the phage and plating on Su^- bacteria. The recombinant phage that grow on Su^- bacteria will have integrated the *supF* gene and contain either randomly integrated globin genes or the targeted gene. The phage that contain the whole targeted β-globin gene can be identified by the presence of the second intron (IVS 2) of the β-globin gene.

These observations were important because they showed that recombination could be observed and therefore methods of increasing the frequency of recombination could next be explored. A possible outgrowth of such approaches would be to develop similar or parallel methods for altering genes in their normal chromosomal environment. It is still difficult to decipher the rules whereby cloned genes are regulated after introduction into mammalian cells. As noted earlier, many studies have shown that there may be large qualitative as well as quantitative differences between the expression of transfected genes and that of their endogenous counterparts.[99] Thus, β-globin genes randomly introduced into mouse erythroleukemia cells are somehow regulated when these cells are induced to differentiate into erythroid cells; however, the amounts of β-globin mRNA are 10- to 100-fold smaller than those seen with normally placed β-globin genes. A number of other examples of inappropriate regulation of transfected genes have been documented.[100]

Precise gene replacement would be preferable to random integration if it could be accomplished. The initial results of Smithies et al.[97] indicated that this could be partially accomplished using the procedure outlined. Thus, as indicated in Figure 11.7, the integrated β-globin gene encompasses a partially duplicated β-globin gene, vector sequences, and the selectable gene. In this situation the chromosomal environment of the integrated β-globin gene is different from that in the normal situation. Since β-globin genes linked to viral vector sequences may be inappropriately regulated after introduction into transgenic mice, it would be ideal to have the vector sequences removed.

By contrast to the picture in mammalian cells, exact gene replacement has been possible for some time in the yeast system.[101,102] This is illustrated in Figure 11.8. It can be seen that the first step in this procedure is similar to that proposed by Smithies et al.,[97] since homologous recombination between the wild-type donor gene plasmid (containing Ura3 + as a selectable marker nucleotide sequence) and a chromosomal gene containing the mutation produces integration of the wild-type copy of the gene depending on the site of crossover. Since the first phase of possible approaches to gene replacement has been illustrated by Smithies et al.,[97] it might be feasible to achieve the second phase by a plan similar to that used with the yeast system. The major problem is the tremendous number of background random integration events occurring in mammalian cells in comparison to those in yeast. This makes it difficult to pick out mammalian cells in which recombination has occurred on the basis of experimental manipulation.

If technology improves, it might be possible to target genes directly, thereby, for instance, allowing modification of defective β-globin genes in bone marrow stem cells from patients with β-thalassemia or sickle cell anemia. At present, there appear to be formidable difficulties to overcome. The frequency of transformation and homologous recombination is still several orders of magnitude too low to be able to target genes in hematopoietic stem cells, which occur in relatively low abundance in bone marrow. Also, at present, it is not technically possible to select

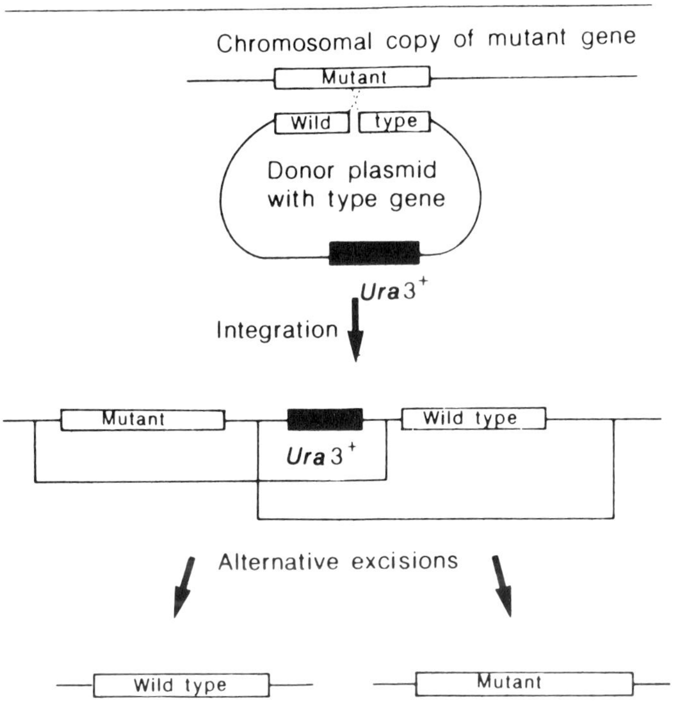

Figure 11.8 Strategy for performing precise gene replacements in the yeast system. *Reproduced from Maniatis,[98] with permission.*

for homologous recombination over the loud background noise of randomly occurring insertional events. Thus, almost all of the originally cloned β-globin genes introduced into primordial stem cells might not be targeted and successfully modified.

Electroporation

Several other methods besides calcium phosphate precipitation, microinjection, and retroviral gene transfer have been developed to transfer genes or nucleotide sequences into cells. One method, electroporation, involves exposure of biological membranes to a high-voltage discharge that reversibly destabilizes or permeabilizes the cell membrane. Neumann and co-workers[102] were the first to employ this method to introduce DNA into cultured mammalian cells. In the initial work, mouse L cells were suspended in an isotonic salt solution together with the DNA, and the mixture was subjected to a series of electrical impulses of high field strength. It appeared that the number of L cells expressing the marker DNA depended on the initial field strength and the ionic composition of the salt solution. Later, other workers[103,104] modified the method by using a single electric pulse of lower field strength. It was shown that electroporation could be employed to introduce DNA into a variety of cell types that had proved refractory to the calcium phosphate precipitation method of DNA cell insertion. Actual electric waveforms, optimal peak voltages, and ideal specific experimental conditions have been established by Knutson and Yee,[105] using a circuit monitoring procedure. It was found that different cell types varied considerably in survival and percentages of the cells taking up and expressing the transferred DNA. Using different conditions, it was determined that for 50% cell survival the percentage of surviving cells expressing the transferred gene product might vary from 5% to 20%.

Examples of Transgenic Transfers

The introduction of foreign genes into the germ-line DNA of transgenic mice provides a model for investigating how such transferred genes might be expressed or regulated in their new environment. In the past, one of the limitations of retroviral vectors was apparent lack of provirus expression once the genes were introduced into the target early embryo or embryonic stem cells.[106,107] Moreover, this apparent block in gene expression in certain circumstances was maintained throughout later stages of development and postnatal life by a cis-acting mechanism that was thought to be related to methylation of the proviral genome.[107,108] One way to circumvent this blockage of transfected gene expression might be to use a gene transcribed from an internal promoter that would be regulated independently of the factors that produced inhibition deriving from the proviral LTR sequences. It was postulated that transfer of such a gene might produce developmentally regulated gene activation. With this in mind, Soriano et al.[109] prepared retroviral vectors containing a genomic human β-globin gene that was under the control of its own intrinsic promoter. The transferred genes also contained a neomycin phosphotransferase (*neo*) gene under the control of the viral LTR. It was found that the human β-globin gene was expressed in a tissue-specific pattern in three transgenic mouse lines. Unexpectedly, the LTR-drived *neo* gene was strongly expressed in muscle tissues of one mouse line. Retroviral constructs used in this study to infect embryos are shown in Figure 11.9. Preimplantation mouse embryos were infected with the two viral constructs. Southern blot analyses showed that 3 transgenic mouse lines carrying the $en^{(+)}$ virus (Glob 1) or the $en^{(-)}$ virus (Glob 2 and Glob 3) were obtained from a total of 69 mice exposed to virus as embryos. This 5% efficiency of embryo infection was judged to be quite low in comparison to more than 50% successful infection generally obtained with murine Moloney leukemia virus transfers.

Transgenic Approaches to β-Thalassemia

In order to study lineage-specific globin expression, Glob-strain mice were treated with phenylhydrazine to induce anemia, resulting in a marked increment in reticulocyte numbers.[109] Human β-globin–specific transcripts were then detected by RNase mapping to define accurate transcription initiation from the β-globin promoter. In two Glob 1 animals, transcription of the globin gene was detected in the blood, bone marrow, and spleen. The third Glob 1 animal expressed lower amounts of human β-globin mRNA. All three animals showed variable levels of expression in nonhematopoietic tissues, including heart, skeletal muscle, and smooth muscle. In animals transfected for Glob 2 and Glob 3, human β-globin transcripts were detected mainly in the blood, suggesting that the pattern of foreign gene expression was more specific for hematopoietic tissues.

The levels of human β-globin expression in all mice were considerably lower than the level of expression of the mouse β-globin gene or of the β-globin present in human hematopoietic tissues. It was thought that the low levels of expression

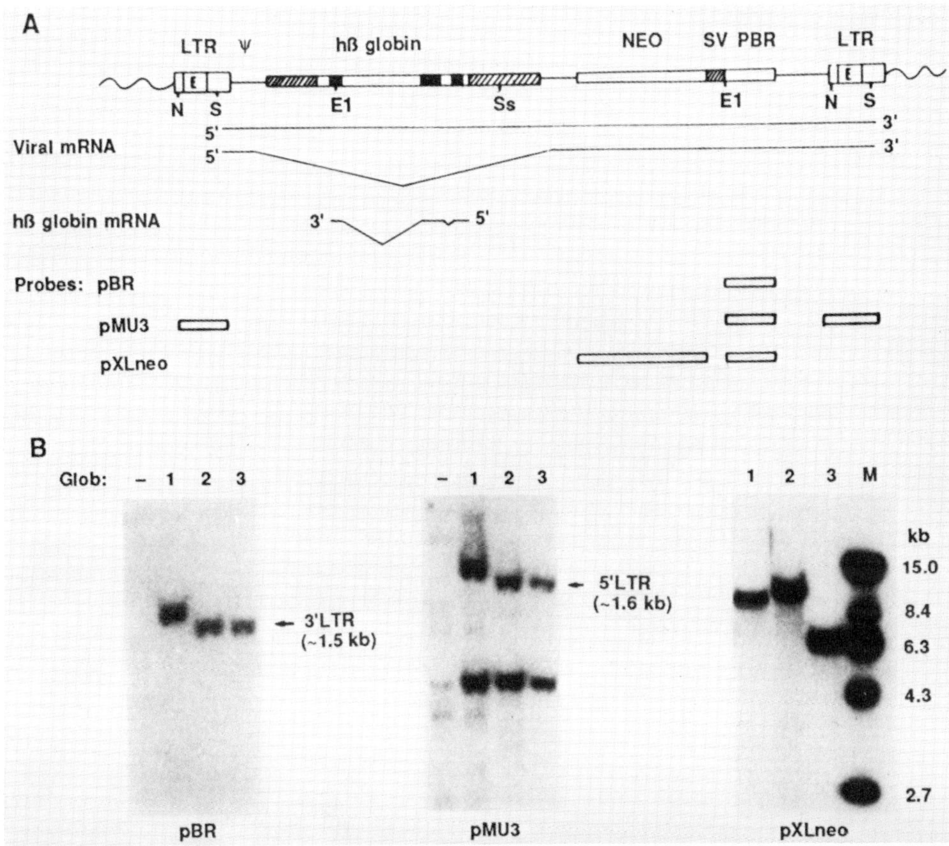

Figure 11.9 Structure of proviral genomes. (A) Structure of retroviral vectors. A 3-kb human (h) β-globin sequence, extending from an *Hpa*I site 815 nucleotides upstream of the mRNA cap site to a *Pst*I site 2.2 kb downstream from this site, was inserted in the reverse orientation into the retroviral vector pZip-neoSV(X) or pZip-neoSV(X) and introduced by transfection into ψ2 cells. A deletion of 7933 to an *Xba*I site at nucleotide 8111 and is transferred to the 5′ LTR upon integration of the provirus. ψ, the psi sequence necessary for encapsidation of retroviral genomes; hatched sequences, hβ-globin; black boxes, flanking cellular sequences; open boxes, noncoding 3′ and 5′ sequences and intervening sequences; black boxes, coding sequences. Inserts corresponding to the *neo* gene (NEO), the SV40 (SV), and pBR322 (pBR) origins of replication are indicated. N, S, Ss, and E1 under the boxes represent cleavage sites for *Nhe*I, *Sac*I, *Ssp*I, and *Eco*RI, respectively. Sequences hybridizing to pBR322 (pBR), pMU3, and pXLneo are indicated; pMU3 is a plasmid containing a head-to-tail dimer of an *Rsa*I fragment (nucleotides 7762–8297) from the LTR inserted in pUC9; pXLneo contains the origin of replication of pBR322 and the *neo* gene. (B) Provirus structure in the Glob 1, 2, and 3 strains. DNAs isolated from tails of control (−) or Glob animals (1 to 3) were digested with *Sac*I and *Eco*RI, with *Nhe*I and *Eco*RI, or with *Ssp*I and hybridized with pBR322, pMU3, or pXLneo, respectively. Hybridization with pBR322 reveals sequence specific for 3′ LTR. Hybridization with pMU3 reveals the 5′ LTR (top band) and internal segment (bottom band) and single 3′ cellular flanking sequences (not shown); weak bands in control animals are due to cross-hybridization of the probe with endogenous retroviral genomes. Hybridization with pLXneo demonstrates the presence of single proviral insertions in each strain. Addition *Ssp*I sites at the 5′ end of the provirus have been omitted from the map. *Reproduced from Soriano et al,*[109] *with permission.*

might be due to prokaryotic sequences in the retroviral vector, which are known to inhibit expression of the β-globin genes in transgenic mice. Another possible explanation is that the sites in the genome where the transfected genes ended up were simply not permissive for high levels of β-globin gene expression. These studies provided an interesting experimental database for further work in this area.

Similar approaches to the basic problem of β-thalassemia were reported by Costantini et al.[110] using a mouse model of β-thalassemia. The murine model for human β-thalassemia was based on a mouse strain with a spontaneous mutation in which the β^{maj} globin gene was entirely deleted but the β^{min} gene remained intact. Animals homozygous for this deletion produced a reduced amount of β-globin and had a hemolytic anemia similar to that in human β-thalassemia of intermediate degree. It is theoretically possible that β-thalassemia in humans might eventually be treated by insertion of the normal β-globin gene into hematopoietic stem cells with subsequent successful expression in erythrocytes of affected individuals. Therefore, the first-step mouse model was employed to see if cloned normal β-globin genes could be introduced into the germ line of β-thalassemic mice. At the time it was recognized that such transfer of genes into the germ line was not a feasible approach for human gene therapy, but information gained through such trials might prove useful in planning alternative approaches in the human sphere.

Two experimental strategies were employed—one using the cloned normal mouse β^{maj}-globin gene and the other using the cloned human β-globin gene. In the first instance a 7.0-kb *Eco*RI fragment containing the mouse β^{maj}-globin gene was microinjected into homozygous mutant mouse zygotes. Of 129 eggs transferred to foster mothers, 10 produced mice and 2 of these showed multiple copies of the microinjected gene as established by Southern blot analysis. Functional expression of β-globin production in two successfully transfected mice showed that one was indeed able to synthesize a significant amount of β^{maj}-type hemoglobin. This animal was then mated to homozygous β-thalassemic females, and several of the progeny synthesized β^{maj} hemoglobin. This indicated that they had inherited and could express the original microinjected normal β^{maj}-globin gene. The mouse β^{maj} gene thereafter continued to be transmitted as a mendelian trait and was expressed at similar levels in subsequent generations. However, the transgenic mice with the introduced β^{maj}-globin genes seemed to produce fewer globin chains than animals with a single endogenous normal β^{maj} gene. Nevertheless, expression of the introduced genes was associated with a significant reduction in the anemia in the transgenic mice, which had hematocrits and reticulocyte counts closer to normal.

In parallel experiments, human β-globin genes were used to replace homozygous β-thalassemic mice, producing a hybrid human-mouse hemoglobin composed of equal proportions of mouse α-globin chains and human β-globin chains. This also resulted in an apparent correction of the previous mouse thalassemic defect! The hematocrit and hemoglobin levels in these transgenic mice approached normal values, and the severe morphological abnormalities (anisocytosis and poikilocytosis characteristic of the thalassemic state) were entirely eliminated. A photograph of the two types of mice—homozygous (deficient) and successful human transgenic animals—is shown in Figure 11.10. Similar experiments directed at transgenic correction in mouse models of sickle cell disease have been described in Chapter 8.

Clearly, direct transfer of normal genes into the germ line would not be a practical approach for correction of human thalassemia or other genetic diseases. Even gene transfer into hemopoietic stem cells presents severe practical problems,

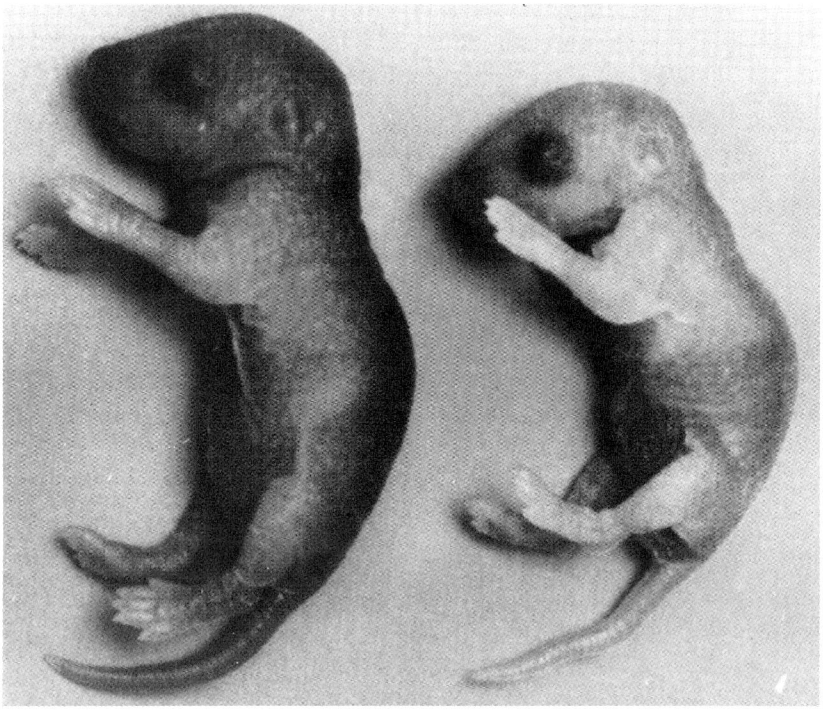

Figure 11.10 Newborn mice inheriting (left) or failing to inherit (right) the human β-globin transgene. Both mice are homozygous for the $Hbb^{th\text{-}1}$ deficiency at the mouse β-globin locus. *Reproduced from Costantini et al,*[110] *with permission.*

particularly in a disease such as thalassemia, where the α/β globin chain balance must be restored. Using any methodology, cloned normal β-globin genes transferred to erythroid precursor cells of a thalassemic patient are likely to be expressed at variable levels from cell to cell. This tissue-specific regulatory problem may not be addressed effectively until we have a better fundamental understanding of globin gene regulation in general. Nevertheless, the work with the murine model for β-thalassemia demonstrates the dramatic potential of a cloned normal β gene to substitute for the defective gene in thalassemic erythrocytes. It also points up the apparent relative lack of intrinsic limitations on effective functional expression of transferred normal β-globin genes.

Growth Hormone Fusion Genes in Transgenic Models

One of the earliest dramatic examples of the potential in vivo effects of the transgenic approach was reported in 1982 by Palmiter et al.,[90] concerning effects of microinjection of the mouse metallothionein I gene fused to the structural gene for rat growth hormone (GH). In this early study a fragment of the cloned rat GH gene with the 5′ regulatory region deleted was fused to the mouse metallothionein (MT) promoter. The specific construction of the fusion gene retained the initiation codon of the GH gene and the 5′ regulatory region of the MT-I gene. Thus, the fusion gene contained nucleotides sufficient to direct transcription of mRNA containing 68 bases contributed by the MT-I promoter followed by one base from the

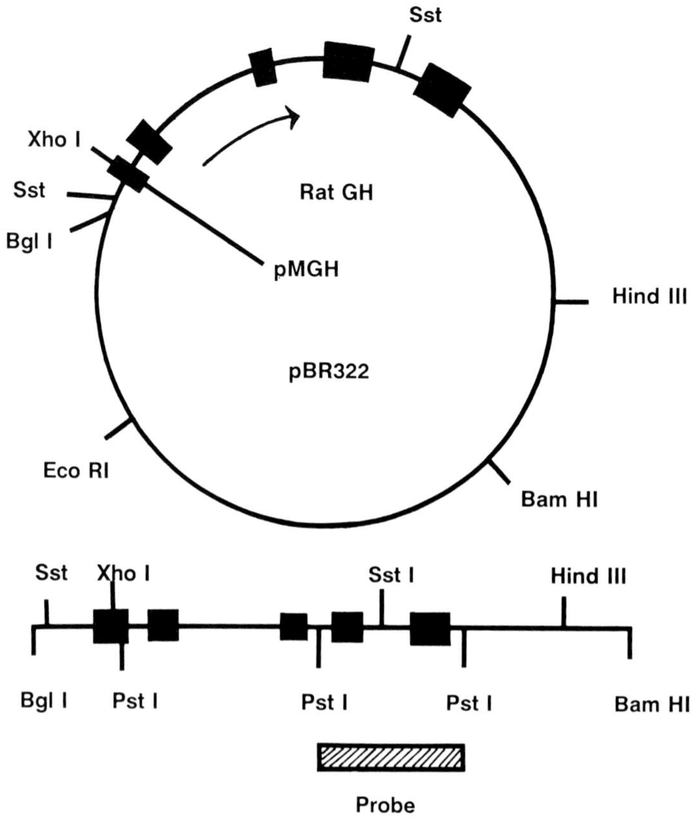

Figure 11.11 Construction of the fusion plasmid pMGH. The *Bgl*II restriction site of the MT-I genomic clone was converted to an *Xho*I site by digesting the *Bgl*II followed by filling in the sticky ends with Klenow fragment of DNA polymerase. The fragment used for microinjection was a 5.0-kb *Bgl*II-*Bam*HI fragment. For genomic Southern blots, a 1.0-kb *Pst*I fragment spanning exons 4 and 5 was isolated and used as a probe (labeled probe on bottom diagram).

*Xho*I linker and then the rat GH mRNA sequence beginning at nucleotide 7. This design preserved the AUG initiation codon for GH synthesis at positions 124 to 126, the four intervening sequences, and the poly(A) site of the rat GH gene. The design of the fusion gene is illustrated in Figure 11.11. The 5.0-kb fragment was injected into eggs, which were inserted in the reproductive tracts of foster mothers. Twenty-one animals developed from 170 eggs. DNAs from the tail tissues of these animals were examined by Southern blotting to determine which animals carried metallogrowth hormone (MGH) sequences. Seven animals were positive. One of the mice transmitted the genes to half its offspring, suggesting that the genes were stably integrated into one of its chromosomes.

The mice were then given zinc to enhance the zinc-sensitive MT-I promoter that had been fused to the original transfected GH gene. Marked increases in size in the transfected mice occurred in proportion to the number of copies of transfected GH genes. Moreover, analysis of liver tissue mRNA for MGH showed a parallel between growth of the transfected mice and quantitative amounts of detectable MGH mRNAs. Also, the marked growth increase in successfully transfected mice was paralleled by 100- to 800-fold increases in serum GH levels as measured by radioimmunoassay. These high levels were associated with growth of the animals to twice the weights of their nontransfected control littermates. It was felt that the high levels of GH in these animals probably also reflected loss of normal feedback regulatory control mechanisms. These experiments were some of the first to illustrate the power of the microinjection method as observed by demonstrable in vivo effects.

Two years later, as an extension of this early work on GH fusion genes, the same group demonstrated that partial correction of a murine hereditary growth disorder was possible using transfection of the same metallothionein–rat growth hormone fusion gene.[92] In these experiments 41 animals developed from microinjected eggs and 7 were shown by Southern blot analysis of genomic DNA to contain the fusion GH gene. By 5 weeks of age six of the seven transgenic mice were already twice the size of the dwarf mutant mice. In this instance there was an apparent absence of correlation between copy number or gene dosage and growth rate, but all of the transgenic mice showed serum GH elevations of 20 to 400 times those of the mutant dwarf controls. These two sets of experimental results were of great interest in illustrating the power of the transgenic method. It was clear from these early reports that the techniques were inefficient, since only about 1% of the injected eggs developed into mice that expressed the gene. Also, since integration of the foreign injected DNA was unpredictable, it might not always be possible to replace the mutant sequences with normal genes. It seemed possible that the normal gene and the mutant gene involved in a particular defect might segregate independently in the next generation and that the mutant gene would not be eliminated from the gene pool at large. There was also the possibility, mentioned earlier, that introduction of the new gene could in turn create a new mutation.

α_1-Antitrypsin Genes

The problem of α_1-antitrypsin (AAT) deficiency has been studied in a cell transfer system, again using a retroviral vector to insert human AAT cDNA into the genome of mouse fibroblasts.[111,112] The retroviral vectors were chosen in this instance because they utilized their two LTRs for efficient integration of the nonviral genes into the genomes of the target mammalian cells. Using a retroviral N2 vector and an incomplete Moloney murine leukemia virus, human cDNA for AAT produced molecules that functioned quite well as inhibitors of human neutrophil elastase, had a half-life in serum similar to that produced by native AAT mRNA, and secreted AAT. These characteristics were unchanged after the original clone was amplified and transplanted into nude mice and then recovered 4 weeks later. The clonal population recovered was identical to the original fibroblast cell line and showed no evidence of proviral rearrangement. In the nude mice that had harbored the transplanted cell line, AAT was readily detected in the serum as well as on lung epithelial surfaces.

Additional studies of tissue-specific expression of AAT have been reported by Carlson and co-workers.[113] Hepatocytes are considered to be the main source of AAT, which is the major antiprotease in human plasma. It is also well established (as discussed in Chapter 9) that patients with AAT deficiency appear to be at increased risk for development of cholestasis during infancy and chronic liver disease as adults. Transgenic mice that express the normal human AAT gene show considerable AAT in liver cells and also in other cell types of nonhepatic tissues. In the work by Carlson et al.,[113] the issue of local AAT synthesis versus endocytosis or uptake of AAT was addressed. It was found that AAT mRNA was present in a variety of tissue locations in transgenic mice, including renal tubular and small intestinal epithelial cells. These findings suggest that widespread local production of AAT in tissues outside the liver could help to maintain body homeostasis in other sites beside lung.

These interesting results indicated that future gene therapy of such diseases as AAT deficiency did not have to be targeted toward cells that normally produce the protein gene product in vivo, such as the hepatocytes and mononuclear phagocytes making AAT. Moreover, modifications of the original transfection procedure to allow introduction of multiple copies of the integrated gene might make the protocol outlined in the fibroblast transfection assay or other models feasible as a practical approach.

Transgenic Mice and Tumors

In 1984 Stewart and co-workers[94] produced 13 strains of transgenic mice that carried an otherwise normal mouse *myc* gene but with increasingly large proportions of the *myc* promoter region replaced with a hormonally inducible mouse mammary tumor virus (MTV) promoter. The female founders of these transgenic strains spontaneously developed mammary adenocarcinomas during one of their early pregnancies. Moreover, both the tumors and the surrounding breast tissue of these founder animals expressed RNA transcripts corresponding to the fusion gene. In one strain examined, all the F_1 female progeny who inherited the MTV/*myc* gene also developed mammary adenocarcinomas during their second or third pregnancies. It appeared that the fusion gene containing the *myc* sequence acted as a heritable predisposing factor that accelerated development of a tissue-specific adenocarcinoma later in adult life.

These studies were of interest because some of the earliest work with the *myc* gene had demonstrated that it appeared to be involved in various stages of normal cell growth. It was shown to be markedly up-regulated as lymphocytes progress from the resting (G0) to the growth phase of the cell cycle under the influence of growth factors or plant mitogens.[114] Since certain oncogenes are apparently derived from genes coding for normal growth factors or growth factor receptors, the ways in which *myc* gene regulation might affect neoplastic processes were of interest in attempts to understand how other oncogenes might be linked to neoplasia.

With this background, the *myc* oncogene was chosen with the objective of replacing the normal *myc* promoter gene region with a hormonally inducible mouse mammary virus promoter. It was hoped that expression of the *myc* gene could be induced by prolactogenic stimuli or glucocorticoids and its effect on normal development or oncogenesis assessed directly. Furthermore, if the *myc* gene participated in transformation of a wide variety of cell types and if the MTV promoter retained its usual degree of hormonal responsiveness, one might expect malignant changes in the breasts of animals bearing these fusion gene products. Also, if the *myc* gene were sufficient for malignant transformation of these cells, its effect should be to induce a uniform malignant neoplasm involving the entire mammary epithelium. If additional somatic mutations were necessary for neoplastic transformation, one might expect to observe solitary mammary adenocarcinomas.

In fact, apparent tissue-specific induction of mammary adenocarcinoma was found in both founder females and many of their F_1 female progeny who expressed the fusion *myc* gene. There appeared to be variation in the tissue-specific expression of the MTV/*myc* gene, since 11 of 12 lines tested showed gene expression in the salivary gland and intestine as well as the mammary gland. It was thought that these differences in expression might be explained by differences in either chromosomal location or the structure of the integrated genes.

That the constitutional MTV/*myc* fusion genes contributed directly to the susceptibility of carrier animals to malignancies was supported by the finding of malignant breast adenocarcinomas in 2 of the 10 founder transgenic animals. A direct relationship between the transfected MTV/*myc* fusion gene and the genesis of tumors was also indicated by the finding that each of the breast adenocarcinomas showed evidence of MTV/*myc* fusion gene expression in amounts far exceeding those of the other endogenous genes. The hormonal environment produced by pregnancy also appeared to be important in activating the mammary carcinoma process.

These studies were followed 2 years later[115] by additional observations relating the deregulation of the c-*myc* gene in transgenic mice and multiple additional neoplasms. One of the original mouse strains that had been transfected with the MTV/c-*myc* fusion gene anomalously expressed mRNA corresponding to the fusion gene. When more experience with this anomalous strain and its progeny became available, it was found that it also contributed to the development of malignant tumors in breast, testis, mast cells, pre-B cells, and T cells. It therefore appeared that the fusion gene was biologically active in a wide variety of organ and tissue substrates. Also, the c-*myc* gene had the capacity to produce transformation in a wide spectrum of tissues, but malignant transformation was not seen in all tissues in which the fusion gene was expressed. It was therefore postulated that besides the c-*myc* gene, additional factors were probably required to induce malignant transformation in the precursor cells of many of the tumors.

In the particular transgenic strain studied it was noted that MTV/c-*myc* mRNA was expressed not only in mature breast and salivary gland but also in spleen, testis, lung, brain, and preputial glands. This strain was treated with dexamethasone as an inducing agent, and increased amounts of mRNA fusion gene transcript were noted. As later generations of mice were bred and developed from this strain, animals with testicular neoplasms and B cell lymphomas were recognized in addition to the previously described breast adenocarcinomas. In all, 14 of 35 transgenic mice (40%) of this pedigree eventually developed a wide variety of malignant neoplasms at a mean age of 14 months. In contrast, there was a low spontaneous tumor incidence among 200 control nontransgenic mice of the same genetic background. Analyses of tissue mRNA distribution showed fusion gene transcripts within various tumors but also in other tissues that did not show evidence of malignant transformation. The fact that the tumors that appeared in this unusual transfected strain were solitary masses amidst normal cells expressing the deregulated c-*myc* gene indicated that additional events were probably necessary for maligant cellular transformation. Additional somatic or transgenic mutations could play some role in the development of the wide assortment of tumors observed. In 13 tumors studied in detail, most showed no evidence of gross rearrangement or amplification of the c-*myc* sequences.

The fact that tumors were not observed in every organ in which the c-*myc* transgene was aberrantly expressed has helped to bring the transforming spectrum of the c-*myc* gene into focus. Although very high levels of c-*myc* expression were recorded in both salivary glands and lung in some animals, no tumors were ever recorded in these locations. Such results suggest that other influences besides simple c-*myc* activation are essential for malignant transformation. The experimental model developed by Leder et al.[115] provides a unique opportunity to study additional, possibly dominantly acting oncogenes that might provide the second "hit" necessary for development of neoplasia.

Transgenic Neurofibromatosis

Although much recent attention has been directed at human T lymphotropic virus type I (HTLV-I) and its association with lymphoma and leukemia in humans, fewer than 0.1% of individuals with antibodies to the virus go on to develop lymphoid malignancies.[116] During the basic studies of HTLV-I it was found that the *tat* gene of this virus acts under control of its own LTR and can induce two phenotypes in transgenic mice in the absence of other viral genes.[117] Tumors were recorded in all three founder *tat* transgenic mice studied, and the pattern of tumor development suggested that the tumors themselves were unrelated to the anatomic site of integration of the transgene. All tumors had a spindle cell morphology and were thought to be of mesenchymal derivation. Remarkably, the offspring of the original *tat* transgenic mice showed faithful transmission of tumor development to 30 of 30 F_1 transgenic progeny. Transmission to the F_2 and F_3 generations was observed as well.[118] Primary tumors had high concentrations of the *tat* protein, several times above those in any normal tissues examined. Some animals had tumors in cranial nerves composed of spindle-shaped cells admixed with ganglion cells. Many of the tumors were closely associated with nerves or nerve trunks. All of these features closely resembled those of a neurofibroma-like lesion. Electron microscopic studies of early tumors showed origin in the perineural sheaths. Neurofibromatosis in humans occurs in two main clinical forms—von Recklinghausen neurofibromatosis, manifested by dermal café au lait spots, and dermal neurofibromas and central or bilateral acoustic neuromas (NF). In humans neurofibromatosis is thought to be inherited as an autosomal dominant single-gene disorder, and with the cloning of the NF gene this experimental model has become particularly interesting (Chapter 9). In the *tat* transgenic mice reported by Hinrichs and co-workers[118] the tumors were composed of fibroblastic cells that appeared to originate in the perineurium. Several other parallels with human neurofibromatosis were observed in the *tat* mice. Puberty and pregnancy in humans accelerate the growth of human neurofibromas, suggesting a tropic hormonal effect. In parallel, the tumors in *tat* mice rapidly enlarged during pregnancy, necessitating transfer of offspring to foster females.

Hinrichs et al.[118] found that the *tat* gene controlled by the HTLV-I LTR has neurotropic properties and is capable of generating multiple similar tumors in diverse sites. In view of the remarkable similarities to NF, it is conceivable that a proportion of cases of the latter disorder could be due to HTLV-I infection. An infectious etiology for all cases of NF appears less likely in that the disease is uniformly distributed throughout the world, although clustering of NF in HTLV-I endemic regions such as the Caribbean or parts of Japan has not been scrutinized. These studies illustrate the insight into some human disease states that the transgenic approach may provide.

The Hypogonadal Mouse

Studies using genes capable of producing various hormones may eventually provide a compendium of information on gene expression and possible uses for ultimate gene therapy. One instructive study in this regard concerned mice homozygous for the hypogonadal (*hpg*) mutation.[119] These animals are sexually immature and show arrested germ cell development.[120] A number of immunocytochemical and radioimmunoassays have indicated that these hypogonadal mice lack detectable levels of hypothalamic gonadotropin-releasing hormone (GnRH), which naturally leads to low levels of luteinizing hormone (LH) and follicle-stimulating hormone (FSH). It has now been established that this mutation involves a deletion of at least 33.5 kb of DNA, including the distal half of the gene responsible for the biosynthetic precursor of GnRH and GnRH-associated peptide (GAP).[121,122] This deletion does not seem to affect the number and location of the GnRH neurons in the *hpg* mouse hypothalamus, since mRNA transcripts from the deleted *hpg* gene can be detected by in situ hybridization in the proper locations in the *hpg* brain.[122]

Mason and co-workers[121] attempted to restore a functional hypothalamic-pituitary-gonadal axis by introduction of a DNA fragment containing the mouse GnRH gene with a 5-kb stretch of 5′ flanking and a 3.5-kb segment of 3′ flanking nucleotides into a wild-type mouse. By a series of planned matings with *hpg*/+ heterozygotes, the transgene was successfully introduced into an *hpg*/*hpg* homozygous background. Remarkably, *hpg*/*hpg* homozygotes showed tissue-specific expression of the GnRH transgene and reached full sexual maturity. Subsequently they were capable of mating and raising healthy litters of progeny.

This remarkable use of transgenic techniques to restore reproductive functions of the hypogonadal (*hpg*) mice first included, as noted, introduction of the GnRH-GAP gene into the homozygous background by a series of matings. This was necessary because the original *hpg*/*hpg* homozygotes were sterile. The strategy of this approach is illustrated in Figure 11.12. The genotype of the progeny of the first matings was checked by Southern blot DNA hybridization with probe 2 (Figure 11.12B and D). This *Bam*HI-*Eco*RI fragment (which includes the 3′ break point of the *hpg* deletion and therefore does not pick out the presence of the transgene) could differentiate between genomic DNA from wild-type, *hpg*/+, and *hpg*/*hpg* mice. Subsequent hybridization of the mouse DNAs with probe 1 was then used to define which of the mice contained the introduced transgene. Initially, 200 copies of a DNA segment carrying the GnRH gene plus flankers were injected into 250 fertilized, wild-type mouse eggs; of 27 pups born, 2 were demonstrated by Southern blotting analysis to be transgenic. These were then used to introduce the GnRH gene into a *hpg*/+ heterozygous background mouse. Finally, surgical examination of the progeny of cross-matings showed that the transgenic *hpg*/*hpg* homozygotes and their five nontransgenic *hpg*/*hpg* homozygous littermates had well-developed gonads and accessory sexual glands, whereas in the *hpg* littermates gonadal tissues were developmentally arrested.

When the successful transgenic animals were mated, they were able to produce normal healthy progeny, all heterozygous for the *hpg* allele. The finding that the transgenic *hpg* homozygotes had fully developed gonads and produced offspring indicated that an apparently normally functioning reproductive system had been restored in these animals. This was confirmed by establishing normal levels of FSH in these transgenic animals by radioimmunoassay. The LH levels in serum of transgenic *hpg* females and males were somewhat lower than normal, but were considerably higher than in the original *hpg*/*hpg* homozygotes.

Tissue localization of the expression of the introduced GnRH-GAP gene in a number of transgenic mice showed normal brain and hypothalamic distribution. Immunological identification of GnRH-containing neurons indicated that all the normal expected neuronal sites were involved in GnRH expression.

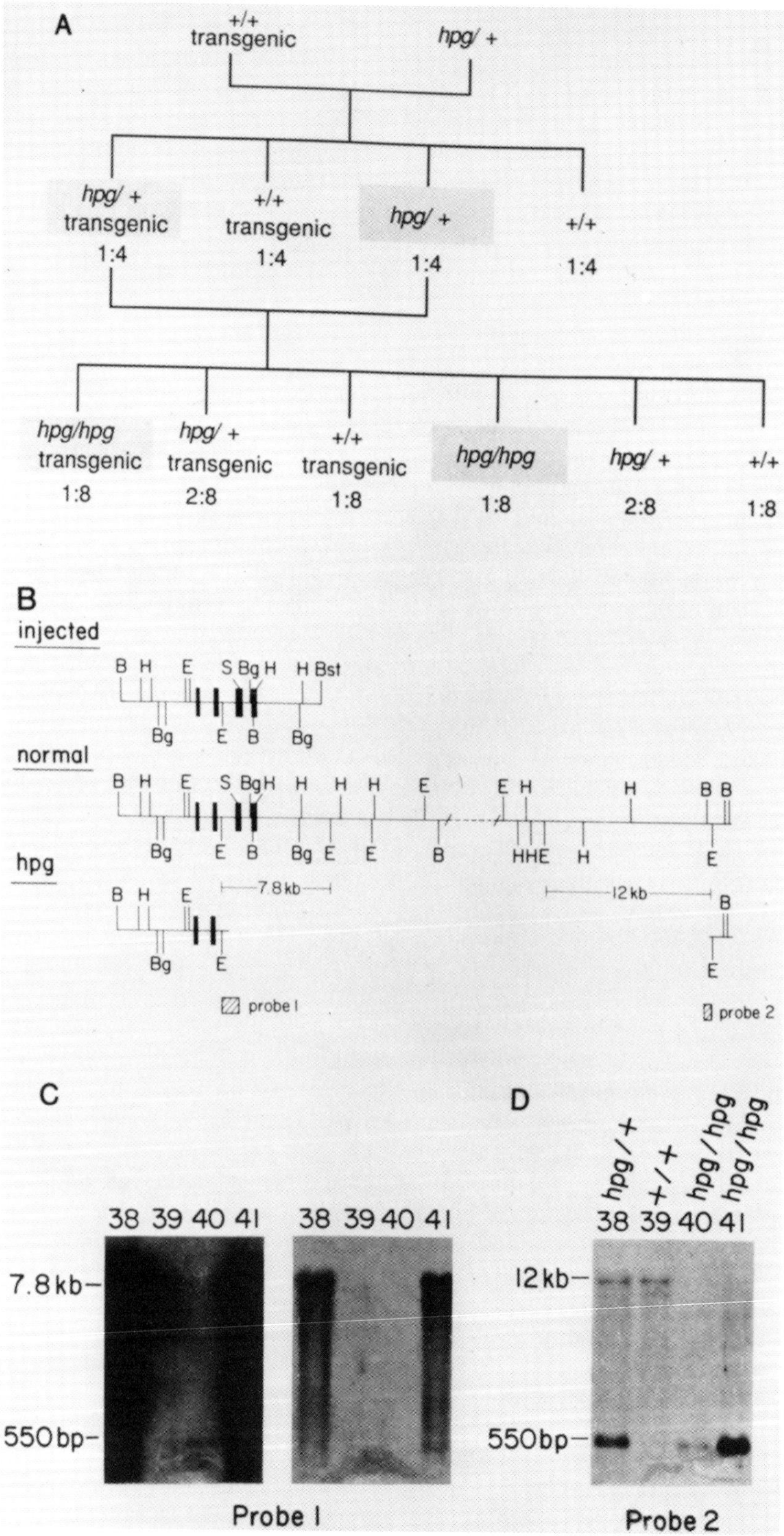
A
+/+ transgenic
hpg/ +
hpg/ + transgenic 1:4
+/+ transgenic 1:4
hpg/ + 1:4
+/+ 1:4
hpg/hpg transgenic 1:8
hpg/ + transgenic 2:8
+/+ transgenic 1:8
hpg/hpg 1:8
hpg/ + 2:8
+/+ 1:8
B
injected
normal
hpg
7.8 kb
12 kb
probe 1
probe 2
C
38 39 40 41
38 39 40 41
7.8 kb–
550 bp–
Probe 1
D
hpg/+
+/+
hpg/hpg
hpg/hpg
38 39 40 41
12 kb–
550 bp–
Probe 2

These experiments again established the ability of transgenic techniques to correct mutational defects in an apparently tissue-specific manner. Once the deleted GnRH gene had been properly restored, normal endocrinological and reproductive functions ensued through a complex set of neuroendocrine events initiated by the GnRH-secreting cells. These results are a striking example of neural-specific as well as testes- and ovarian-specific gene expression. Work on extending the transgenic approach into a multitude of other areas continues.[123]

Gene Therapy

Because of the rapid growth in basic knowledge of how genes function and what mutations or deletions might be responsible for certain widespread human diseases, the prospect for ultimate gene therapy in patients has become an exciting topic. Both ethical and legal issues are, of course, involved and in some circumstances have become popular areas for general discussion. From time to time physicians and basic scientists will be called upon to present the facts and problems to political, religious, or other broader audiences in such a way as to provide sound information that can be used for an honest appraisal of the benefits and possible pitfalls of such an approach. As shown in the foregoing section on transgenic transfers in mice, the virtual explosion of fundamental knowledge and understanding in molecular genetics clearly indicates the ultimate possibility of inserting human genes into

←

Figure 11.12 Generation and identification of transgenic *hpg/hpg* mice. (A) Schematic of protocol for generation of *hpg/hpg* homozygous mice carrying the GnRH-GAP transgene. The 13.5-kb *Sal*I-*BstE*II GnRH-GAP gene fragment was generated from a *Bam*HI-*Eco*RI subclone of the entire insert of the λ phage gmGP-1. The *Sal*I restriction site is adjacent to the *Bam*HI site of the plasmid polylinker sequence. The fragment was gel purified and injected into a pronucleus of one-cell fertilized mouse eggs. The eggs were derived by mating C57BL/6J females and SJL males. Transgenic mice were identified as described above and are indicated as being wild type at the hypogonadal locus (+/+) and heterozygous for the GnRH transgene (transgenic). The transgenic mice identified were backcrossed to nontransgenic mice heterozygous for the *hpg* allele (hpg/+). Mice resulting from this mating that were heterozygous *hpg*/+ and transgenic were backcrossed to *hpg*/+ heterozygous mates. Among the offspring of this mating were animals that were homozygous *hpg/hpg*; approximately half of these were transgenic and half were nontransgenic. The indicated genotypes are expected to occur at a frequency represented by the ratios shown below each. (B) Restriction enzyme map of the GnRH genes and probes used. The normal and mutant alleles of the GnRH gene are more fully described in ref. 106. The injected gene was a *Sal*I-*BstE*II fragment derived from a subclone of the normal GnRH gene. Probe 1 is a 767-bp *Eco*RI-*Sac*I fragment (hatched square) that contains sequences for a portion of intron B and exon III of the mouse GnRH gene. Probe 2 (hatched oblong) is a 360-bp *Bam*HI-*Eco*RI fragment containing the 3′ break point of the *hpg* deletion. (C) Four third-generation mice (mouse 38 to mouse 41) from the B family were analyzed for the presence of the GnRH-GAP transgene by Southern analysis with probe 1 hybridized to the filter used in (D). At short exposures (right panel, 2 hours) only the transgene is detected in mouse 38 and mouse 41. At longer exposures (left panel, 20 hours), the expected endogenous 7.8-kb band is seen in mouse 39. As expected from the results in (C), this 7.8-kb band is not seen in mouse 40. (D) The same four mice as in (C) were tested for genotype at the hypogonadal locus by Southern analysis with probe 2. The DNA was extracted from the tails, digested with *Eco*RI, purified by electrophoresis through1.2% agarose, transferred to a nylon membrane, and hybridized with probe 2. The mutant GnRh gene/*hpg* allele is revealed by a band at 550 bp and the normal gene by a band at 12.0 kb. Thus mouse 38 is an *hpg*/+ heterozygote, mouse 39 is wild type (+/+), and mouse 40 and mouse 41 are *hpg/hpg* homozygotes. *Reproduced from Mason et al,*[121] *with permission.*

human cells. It seems likely that this approach will afford lifesaving or even life-enhancing treatment to patients with severe disabling hereditary disorders. Gene therapy could conceivably be of real value to sick or impaired people, particularly if no other forms of effective treatment were available. However, the measures used in any approach to gene therapy should be safe and not cause new problems or mutations that might threaten the health of those receiving such treatment. At this juncture it is important to consider a few methodological aspects of possible human gene therapy and the practical problems as well as applications of what eventually may be an exciting new form of medical therapy.

General Methods of Gene Therapy

From the outset it must be recognized that two points are important: (1) the method through which a modification will be made in the individual patient's genomic makeup and (2) whether the intervention is to be directed at germ cells or embryonic cells or somatic cells of the subject.[124] Several specific approaches to gene therapy have been contemplated. First, a gene could be introduced at an undefined site in the genome to replace or complete a defect in which the functional gene is missing or deleted. This could be regarded as direct gene insertion. A second general approach would involve selective replacement of an abnormal gene with an autologous or foreign normal gene. This would involve gene substitution. A third approach would require selective reverse mutation of an abnormal gene back to its normal sequence, as in polynucleotide site-directed mutagenesis or modification of the abnormal gene. Fourth, gene therapy might be primarily directed at regulation of a specific gene. Finally, a second mutation might be intentionally inserted in the affected gene to correct the primary defect. Such an example has been suggested by Crystal and co-workers as a strategy for correcting α-antitrypsin deficiency (see Chapter 9). With current methodology it is possible to isolate a defined human gene and, after analysis, to insert it (usually randomly) into the genome of another human cell. However, methods have generally not yet been refined to the point where we are able to exchange or manipulate all single genes of interest in the intact organism.

Actual Techniques of Gene Insertion

Genes can be introduced into human cells, as noted in the section on transgenic mice, by injection into the cell with a very fine needle. Drawbacks are related to the fact that only a small number of cells can be treated. When genes are injected into the pronuclei of fertilized mouse eggs, only about 10% to 25% of the fertilized ova show a positive take. Also, the ova often die as a result of the microinjection. DNA viruses such as SV40 can be used to transfer genes; however, it is possible that the viruses will kill the treated cells.[125,126] RNA viruses can also be employed to transfer human genes. In many instances the RNA virus nucleotide sequences have built-in intrinsic promoters or enhancers which will ensure a reasonable level of expression for the transferred genes. Several older methods of DNA transfer have also been used, including calcium precipitation. In this method, the cells to which the gene is to be transferred are treated with DNA that has been precipitated by addition of $CaCl_2$. Unfortunately, yields in gene transfer are relatively small, especially in the case of relatively large DNA molecules. Electroporation involving a brief electrical impulse with a specific field strength has been used for gene transfer, as have membrane fusion procedures using liposomes containing DNA.[127] In the latter method DNA is introduced into the recipient cells by fusion of the cell membrane with the liposomes. Thus far, none of these methods allows transfer of a tightly defined number of genes into a given cell or special cell type. For the time being, this practical disadvantage may limit some applications of gene therapy.

Use of Human Germ and Embryonic Cells

Direct interventions whereby genes or human genomic segments are introduced directly into the genome of human gametes such as ova or spermatozoa or into zygotes or early human embryos are probably not feasible at present, although being able to do this with great precision in the future has not been ruled out. There are several major problems with introduction of human genes into human gametes and embryos. Even if one parent carries a high genetic risk, only a portion of his or her gametes can be expected to represent carriers of the abnormal disease-related gene. Treatment or gene therapy would be indicated only for gametes or embryos that were genetically abnormal, and it is not possible now to detect a single abnormal gene in an individual living cell or a small collection of cells. With most current methodologies, if one were to look at the DNA of these cells for a mutation or a defect, it would be necessary to destroy the cells for DNA extraction. If one were dealing with an embryo, it might be possible to separate a small number of cells from it. Then one could multiply the cells in culture and analyze their DNA before deciding whether an attempt at gene therapy might be feasible. Also, before the treated embryo was returned to the uterus it would be necessary to determine whether the transferred gene had been properly integrated into the embryonic genome so that it would be functional and would have no undesirable effects on adjacent genes.

These requirements for ideal conditions are hard to meet, since at present it would not be possible unequivocally to demonstrate a clinical cure for the treated gene-injected embryo. For instance, regulation for the genes coding for α or β chains of hemoglobins is not established during early embryogenesis but much later in fetal development, since these genes are physiologically inactive at the early embryo stage. Also, at present it is not possible to predict the consequences of inserting a gene on other genes in its vicinity. There seems to be some risk of disrupting the workings of other cellular genes by insertional mutagenesis. The data on transgenic mice indicate a 10% to 20% incidence of recessive mutations in essential genes.[79] Moreover, a fetus that had not been successfully treated or whose ultimate health status had been impaired by the gene therapy might still be identifiable by other prenatal diagnostic maneuvers. One pauses to wonder what would constitute informed consent, or where the concept of benign intent and malpractice jurisdiction might be applied in such instances. Right to life versus parental control over conception arguments begin to surface.

An important aspect of any gene therapy attempt involving human embryos would be embryo transfer to the mother. Although this technique, recently introduced into reproductive medicine, has enjoyed some limited success, it has also faced bioethical problems. At present after in vitro fertilization the successful pregnancy rate is approximately 10% to 15%. Of these pregnancies slightly more than half result in a fetus carried successfully to term and a viable birth. Many of the in vitro embryo transfers often achieve success by transferring several embryos together. This might per se limit the chances of human gene therapy under such circumstances.

Also to be considered is the possibility that the initiation of planned gene therapy in human embryos might result somehow in a sick or otherwise deformed child. With the current furor about birth defects, million-dollar malpractice awards, and the controversy concerning nonapplicability of informed consent, it is hard to imagine a system that could be devised to protect the rights of the gene-treated embryo, the prospective parents, and the well-meaning physician or practitioner of gene therapy.

A method of treatment that can achieve only rare success and that carries with it the risk of killing the treated embryo or seriously threatens the health of the growing child seems unacceptable. Another consideration is the possible long-term consequences of inserting genes into individuals now and not being able to predict the effects on future generations of individuals arising from the product of present well-meaning intentional gene therapy.

Gene Therapy of Somatic Cells

The concept of somatic cell therapy is much more limited and not quite as far-reaching as that of introducing DNA segments into fertilized human eggs, gametes, or embryos. The approach in somatic cell therapy would involve removal of cells that are not functioning properly from the patient and transferring a normal or corrective gene into these cells before returning them to the patient. After the cells have been removed from the patient, they are maintained in culture, and when it has been determined that the gene insertion worked, the cells can be multiplied before being returned to the patient.

Initial trials of this sort of approach would probably be directed at treating a disorder such as adenosine deaminase deficiency.[128] It has been demonstrated that transfer of a normal adenosine deaminase gene can normalize the enzyme defect in hematopoietic cells maintained in culture.[129] A number of other conditions, including nucleotide phosphorylase deficiency and the enzymatic defect associated with Lesch-Nyhan syndrome, are also possible prospects for somatic gene therapy. The same approach might also be tried for children with sickle cell disease or β-thalassemia. The major limitation of somatic cell gene therapy is that only a few cell types can be maintained and cultured in vitro; therefore, it is not presently possible to adapt somatic cell therapy to tissues such as brain, muscle, or liver, which are involved in a number of relatively common hereditary diseases. Also, it would not be practical to treat only one cell type in a genetic disorder that involves defects of cells in many different organs. Of more than 3000 diseases now probably linked to single gene defects, only a small number are amenable to somatic cell gene therapy in the immediate future. Since the transferred genes are not integrated into the germ line of the host, correction of the original mutation ceases at the death of the individual recipient and, of course, is not transmitted to progeny.

Fundamental Requirements for Gene Therapy of Somatic Cells

Several basic requirements must be addressed before somatic cell gene therapy can be implemented. First, the defect or mutation should have been precisely identified at the nucleotide sequence level and the normal replacement gene should be cloned and isolated. A listing of cloned human DNAs has been published[130] and is presently being expanded.

The second requisite for effective somatic gene therapy is that an adequate number of suitable target cells be collectable. In the case of human hematopoietic cells, a few stem cells may give rise to 1000 or 100,000 bone marrow cells. If the gene is inserted into cells already on their way to completed differentiation, the cell line will die out. Therefore, it is essential to introduce the functional genes into a sufficient number of true primordial stem cells in such a way that they will be expressed. Ideally, it would also be important to know what mechanisms regulate the activity of the introduced genes, particularly in the specific cell types into which the genes have been introduced. At present there is very little knowledge of the mechanisms of such gene regulation or inactivation in specific human cell systems.

Rather than emphasizing the hope that eventual successful applications of gene therapy may be able to bring, unfortunately the lay press and media have tended to present or overemphasize the possibilities of abuse of gene therapy. At times it has been suggested that many such avenues can be automatically equated with so-called Frankenstein experimentation. With luck and a positive presentation of the facts and new discoveries to the public, a more balanced attitude toward gene therapy and its potential role in medical treatment will certainly supervene.

Prospects and Background for Successful Somatic Gene Therapy

Not all hereditary diseases would be suitable potential candidates for gene therapy. Several sets of criteria which would help to focus the most effective and successful efforts at actual implementation. Ideally, the disease to be treated should be a single-gene disorder inherited in an autosomal recessive manner. Recessive disorders are more apt to be completely cured by replacement of a missing gene product than dominant disorders, in which removal or inactivation of a dominant gene product might be required.

Second, there should be preliminary in vitro and in vivo experimental evidence that the normal recombinant replacement gene can in fact be transferred to a mutant cell and thereafter be expressed at the tissue level adequately enough to play the desired restorative role. Safety of transfer of the replacement normal gene into the deficient host should be evaluated whenever possible in an animal model of the human disease. Unfortunately, at present there are only a few well-studied animal models of human genetic disease. Therefore, extensive assessment and appraisal of gene therapy results may not be feasible in many human disease states, in which animal models have yet to be established.

Different diseases present a wide spectrum of problems in the sites, local control, and degree of expression of various gene products. In some disorders, the genes must be present in very special cell types and the gene products available in a particular anatomic location or extracellular compartment. In some conditions, correction of a biological defect will necessitate not only replacement of a defective gene product but also the activity of other special metabolic components or factors working in the same system. In other diseases, precise regulation of gene expression is necessary along with extremely high levels of gene expression. Finally, the clinician who is contemplating the use of gene therapy must be convinced that the potential benefits clearly outweigh the risks for an individual patient and represent a distinct improvement over the results of currently available conventional therapy. Human diseases that might be potentially treatable by somatic gene therapy are listed in Table 11.2.[131]

Table 11.2 Potential Candidates for Human Somatic Gene Therapy

Immune deficiency syndromes resulting from deficiency in bone-marrow derived cells
Adenosine deaminase
Purine nucleoside phosphorylase
Chronic granulomatous disease
Carbonic anhydrase II deficiency
Complement component deficiencies
Hematopoietic diseases
Thalassemic syndromes
Sickle cell disease
Serum protein deficiencies
Hemophilia
α-1 anitrypsin deficiency
C1 esterase inhibitor deficiency
Errors of metabolism
Urea cycle deficiency (carbonyl phosphate synthetase, ornithine transcarbanylase, argininosuccinate synthestase, argininosuccinate lyase, arginase)
Phenylketonuria
Galactosemia
Homocystinuria
Maple syrup urine disease
Storage diseases
Gaucher's disease
Fabray disease
Glycogen storage disease
Mucopolysaccharidosis
Central Nervous System Diseases
Lesch-Nyhan syndrome
Tay-Sachs disease

Immunodeficiency Secondary to Adenosine Deaminase or Purine Nucleoside Phosphorylase Deficiency

As briefly noted earlier, adenosine deaminase deficiency and purine nucleosidase phosphorylase deficiency are rare errors of metabolism that may be associated with severe combined immunodeficiency. Adenosine deaminase activity is usually normally expressed at the maximal levels in T lymphocytes. When this enzyme is absent, thymic development is deficient and T cell insufficiency results in ineffective immune system maturation and severe combined cell-mediated and humoral immunodeficiency. This condition may be effectively treated using bone marrow transplantation, but it could conceivably be treated with gene transfer to autologous bone marrow and return of the treated stem cells to the patient with enough cells to express adenosine deaminase enzymatic activity and normal immunological development. Successful animal models involving retrovirus-mediated gene transfer have been developed,[111] and a trial of this approach in humans appears to be imminent. A mouse model with adenosine deaminase deficiency[131] may improve the effectiveness of a gene therapy approach in preparation for human trials.

Hemoglobin Disorders

Hemoglobinopathies such as sickle cell anemia and thalassemia[132] are curable by bone marrow transplantation; therefore, effective introduction of normal globin genes into marrow stem cells appears to be a possible therapeutic alternative. In this group of disorders, the globin genes are expressed at very high levels in erythroid precursor cells and the proportions of α- and β-globin genes may be critical for correct assembly of a functional hemoglobin molecule. Murine experiments discussed above and in Chapters 8 and 9 have shown effective expression of β-globin genes after retrovirus-mediated gene transfer. The availability of a mouse model for thalassemia should greatly enhance the knowledge of specific aspects of this potential area for gene transfer in humans.

Lesch-Nyhan Syndrome

The hypoxanthine phosphoribosyltransferase gene was one of the first human genes to be cloned, and there has been some exploration of the possibility of gene transfer to treat patients with this disorder. Retroviral vectors that effectively carry the normal gene are available and have been used to transfer the gene to bone marrow cells as well as other cell types.[133,134] The major problem with the gene therapy approach to treating Lesch-Nyhan patients is that the disease principally involves cells of the central nervous system, where the enzymatic deficiency produces the neurological disorder. Apparently, trials of bone marrow transplantation have not been successful in ameliorating the clinical course of the disorder.[135] It appears that methods will have to be developed for replacing the normal gene in cells of the central nervous system to be able to effect clinical improvement.

Storage Diseases

Storage diseases that affect bone marrow–derived cells such as monocytes, macrophages, and Kupffer cells often respond favorably to conventional bone marrow transplantation.[136] Thus one might expect somatic gene therapy to be a possible alternative. Gaucher's disease could be a good starting point because the gene for glucocerebrosidase has been cloned[136] and retroviral gene transfer demonstrated in cell cultures. Other storage diseases, including the gangliosidoses (Tay-Sachs, Sandhoff disease) or the mucopolysaccharidoses, may represent a much more difficult problem because the accumulation of the metabolite in many cells besides those in bone marrow produces eventual cell death or severe dysfunction. Gene replacement in such diseases would probably require reconstitution of virtually every cell in affected organs.

Current Strategy Projections

At present, the objective for gene therapy in a genetic disease such as adenosine deaminase deficiency is to introduce a functional new gene elsewhere in the genome rather than attempt to repair the defective gene site in vivo. When techniques for in situ gene repair have been improved, direct alteration of the mutant gene may also prove feasible. Of all the methodologies available at present for the actual gene transfer into tissues such as bone marrow cells, the use of retroviral vectors seems most practical.[137,138] Retroviruses are RNA viruses that produce exact DNA copies after initial infection of the target cell. The DNA copies then integrate apparently randomly within the host genome and produce RNA copies and proteins from constituent intrinsic viral genes. Features that favor retroviruses as vectors for transferring genes are their high rate of stable integration, low copy number of virus per cell, wide host range including many animals plus humans, and flexible genomic capacity for addition of up to 13,000 new base pairs.[139,140] Possible problems related to the use of retroviruses are death of the cell by random integration of the transferred gene into a critical endogenous gene region or production of infectious viruses that might ultimately spread to other organs and tissues. Another theoretical disadvantage of retroviral vectors is the remote possibility that malignant transformation could occur as a result of viral integration and inadvertent secondary oncogene activation. With this in mind, a number of special in vitro manipulations have been employed to render the retrovirus gene segments used incapable of self-replication. This is usually done by removing critical portions of the viral genome necessary for intrinsic replication. Even though great pains have been taken to make noninfectious viral vectors by excising key viral gene sequences, there remains the remote possibility that the virus could repair its infective capacity by hybridizing with adjacent normal host cell nucleotide sequences. In most strategies involving viral vectors, the gene to be transferred is a construct prepared of coding sequences from one gene and promoters from another. This strategy has been employed because endogenous promoters of many normal genes are relatively weak and produce relatively small amounts of RNA. Thus bone marrow cells may be initially infected with a viral vector carrying the gene of interest; however, the vector might prove incapable of replication. The marrow cells would then be returned to the host to allow the transfected stem cells to home to hematopoietic sites and to begin dividing. In a few initial trials in humans this was apparently done in a two-stage procedure involving lethal whole-body irradiation followed by intravenous injection of the genetically altered cells to restore hematopoietic activity.[140] This sort of approach appears rather heroic for gene therapy of a chronic genetic disease, but it is favored by some. Initial limited applications of this general approach in humans employed much more limited skeletal irradiation and apparently allowed engraftment of cells with new genes.[141]

Thus far, early work on somatic cell gene replacement in experimental animals including primates has provided encouraging data. In general, very little is known about the basic stability of the viral vectors. For this reason long-term integration of virus vector plus gene insert beyond several months has not been extensively studied. Also, expression of the inserted gene has been apparent only when the gene itself appears to carry a selective advantage. Nonselective genes, such as the adenosine deaminase nucleotide sequence, generally appear to be expressed only temporarily or at a very low level. Abundant expression of transferred genes is a necessity in many systems and needs to be improved. Whether irradiation of the host will be necessary to ensure adequate homing of gene-treated marrow stem cell populations has also not yet been settled.

Actual First Implementation of Human Gene Therapy

As discussions went back and forth concerning potential benefits measured against risks and possible inadvertent effects of using various retroviral constructs to facilitate gene expression, the emphasis in initial human gene therapy shifted toward patients with cancer. In a number of animal and human experimental trials, Rosenberg and co-workers[142–144] found that treatment of patients with far-advanced metastatic cancer using tumor-infiltrating lymphocytes primed by the potent lymphokine energizer interleukin 2 (IL-2) could mediate regressions of metastatic tumors such as malignant melanoma. Accordingly, the subject was extensively reviewed by a special committee termed the Recombinant DNA Advisory Committee of the U.S. National Institutes of Health, and an experiment was approved in which lymphocytes containing a bacterial gene were to be introduced into terminally ill cancer patients.[145] This historic first trial was reported on August 30, 1990.[146] In this trial five patients with metastatic melanoma received the gene-modified tumor-infiltrating lymphocytes (TIL). Presence and expression of the neomycin resistance gene were subsequently demonstrated in the TIL from all patients by Southern blot analysis plus enzymatic assay for the neomycin phosphotransferase encoded by the bacterial gene. Using polymerase chain reaction amplification, gene-modified cells were repeatedly found within the circulation and from tumor deposits as long as 64 days after initial cell administration. These studies demonstrated the feasibility and safety of retroviral gene transduction for human gene therapy.

As noted in an editorial that accompanied this initial report, during the preceding several years great progress had been made in constructing genes with various defective viral vectors.[147] Replication-defective retroviral vectors and various cell lines that are capable of putting the vectors into viral particles have now been developed.[148,149] Certainly, other clinical trials must follow with future analysis and careful evaluation of the clinical results.

Adenosine Deaminase Gene Therapy

No sooner had the tumor-infiltrating lymphocyte experiments been initiated, when plans were made to attempt reconstitution of children or infants with adenosine deaminase (ADA) deficiency.[150] Because introduction of the gene for ADA represented the first proposal to attempt gene therapy to correct an inherited enzyme deficiency, the protocol received long scrutiny. In parallel, a new drug called PEG-ADA was approved by the Food and Drug Administration. This material contained the deficient enzyme treated with polyethylene glycol to enhance its survival after administration. It has now been agreed to administer the treated enzyme in conjunction with the ADA gene in what promises to be an exciting era of gene replacement trial therapies.[150]

Gene Expression Within Endothelial Cells of the Arterial Wall

Besides the exciting new developments with respect to deficiencies of enzymes like ADA, a number of animal experiments have been reported that may eventually lead to the use of gene replacement or gene therapy for vascular disease. Two reports of the use of recombinant techniques appeared in June 1989.[151,152] In one study a recombinant β-galactosidase gene was expressed in a specific segment of an arterial vessel by direct infection with a murine retroviral vector or by DNA transfection using liposomes.[151] A number of cell types in the experimental vessel walls were transduced, including endothelial as well as vascular smooth muscle cells. After the retroviral introduction of the new gene, a recombinant reporter gene was expressed for at least 5 months and no helper virus was detected. The recombinant gene expression induced in this experimental model seemed to be limited to the vascular wall site of infection and was not found in liver, lung, kidney, or spleen of experimental animals. Such studies may hold great promise for site-specific gene expression in such human diseases as atherosclerosis.

The second study[152] also utilized vascular endothelial cells as the target for gene replacement therapy. Recombinant retroviruses were employed to transduce the *lacZ* gene into endothelial cells harvested from mongrel dogs. Then prosthetic vascular grafts seeded with the genetically altered cells were placed as carotid grafts into the dogs from which the original endothelial cells had been harvested. Examination of grafts 5 weeks after implantation showed that the genetically altered endothelial cells were still lining the surface of the grafts. Again, when modified for use in humans, this approach might be applicable to the problem of atherosclerosis and defective low-density lipoprotein receptors.

Cystic Fibrosis Now Corrected in Vitro

Probably as harbingers of future directions in human gene therapy, two separate teams of workers have successfully employed gene transfer to correct the cystic fibrosis defect in cells in culture.[153,154] Although the research teams used different recombinant strategies to place the normal gene within the defective cystic fibrosis cells, the result was the same in that the chloride transport channel was opened. Lack of this channel is thought to be related to one of the most prominent features of cystic fibrosis, which involves a buildup of thick tenacious mucus in the small air passages of the lungs. Exciting as the initial in vitro experiments are, a number of additional steps are now needed. It must be demonstrated that the normal gene can be expressed in vivo, and an animal model is clearly needed. The next problem will be how to put the vector bearing the new gene where it will do the most good, namely into the small airways of a living, breathing patient. This may be a temporary problem, since currently it is not clear precisely which airway cells are affected. Work in this rapidly developing area continues.

Regulation and Ethics of Gene Therapy

The manifest possibilities for practical application of recombinant DNA technology and gene therapy in humans raise a number of ethical and legal questions. Many popular books, scientific articles, and government reports have considered the social and ethical problems involved.[155–158] Suggestions about the specific committees appropriate for approval and review of all proposals for gene therapy have been outlined.[158] They include initial approval by a local Institutional Review Board with referral subsequently to a Recombinant Advisory Committee, composed ideally of three laboratory scientists, three clinicians, three ethicists, three lawyers, two specialists in public policy, and a representative of the public. Initially it was suggested that final approval of all human gene therapy protocols should be given by the NIH director. Such a chain of command would probably not be practical in many institutions; however, obviously it is necessary to address the potential serious questions and problems that will arise.

Naturally, any experimental protocols involving gene therapy will involve decisions involving uncertain benefits versus potential risks to treated individuals. Adequacy of informed consent in critically ill patients or in retarded, minor, or severely impaired subjects is clearly a major problem in many possible gene therapy situations. Results of gene therapy and of conventional treatment must be compared objectively and a number of difficult value judgments entertained.

When gene therapy is considered, the individuals and advisory committees involved must also consider risks of vertical (inheritable) transmission of recombinant genes to offspring through inadvertent mutations induced in germ-line cells. Also to be considered are the risks of infectious or horizontal transmission of recombinant viral agents to other individuals and of contaminating the human gene pool at large with recombinant genes that were not previously present.

It is clear that a fundamental difference exists between the use of recombinant products to treat human diseases and the direct introduction of new genes into autologous human cells to alleviate or cure a disease caused by a well-defined nucleotide mutation. Somatic gene therapy appears to offer a bright perspective for some human afflictions. As with all forms of medical treatment, the final judgment on the utility or practical application of such gene therapy will be based on the palpable results weighed against the possibility of inadvertent harm. As more is learned about the entire system of gene regulation and expression, more reasoned expectations concerning gene therapy will undoubtedly emerge.

References

1. Goodman M: Serological analysis of systematics of recent hominoids. *Hum Biol* 1963; 35:377–424.
2. Sarich VM, Wilson AC: Immunological time scale for hominid evolution. *Science* 1967; 158:1200–1203.
3. King M-C, Wilson AC: Evolution at two levels in humans and chimpanzees. *Science* 1975;188:107–116.
4. O'Brien SJ, Nash WG, Wildt DE, et al: A molecular solution to the riddle of the giant panda's phylogeny. *Nature* 1985;317:140–144.
5. Cavalli-Sforza LL: The genetics of human populations. *Sci Am* 1974;231(3):80–89.
6. Wainscoat JS, Hill AVS, Boyce AL, et al: Evolutionary relationships of human populations from an analysis of nuclear DNA polymorphisms. *Nature* 1986;319:491–493.
7. Olivo PD, Van de Walle MJ, Laipis PJ, Hauswirth WW: Nucleotide sequence evidence for rapid genotypic shifts in the bovine mitochondrial DNA D-loop. *Nature* 1983; 306:400–402.
8. Monnat RJ, Loeb LA: Nucleotide sequence preservation of human mitochondrial DNA. *Proc Natl Acad Sci USA* 1985;82:2895–2899.
9. Hauswirth WW, Laipis PJ: Transmission genetics of mammalian mitochondria: A model and experimental evidence. in Quagliariello E, Slater EC, Palmieri F, et al. (eds): *Achievements and Perspectives in Mitochondrial Research,* Vol. 2: *Biogenesis*. New York, Elsevier, 1985, pp 49–59.
10. Cann RL, Stoneking M, Wilson AC: Mitochondrial DNA and human evolution. *Nature* 1987;325:31–36.
11. Nei M: Implications for evolutionary theory. in *Molecular Evolutionary Genetics.* New York, Columbia Univ. Press, 1987.
12. Saitou M, Omoto K: Time and place of human origins from mt.DNA data. *Nature* 1987;327:288.
13. Takahata N, Nei M: Gene genealogy and variance of interpopulational nucleotide differences. *Genetics* 1985;110:325–344.
14. Holt IJ, Harding AE, Morgan-Hughes JA: Deletions of muscle mitochondrial DNA in patients with mitochondrial myopathies. *Nature* 1988;331:717–719.
15. Lestienne P, Ponsot G: Kearns-Sayre syndrome with muscle mitochondrial DNA deletion. *Lancet* 1988;1:885.
16. Zeviani M, Moraes CT, DiMauro S, et al: Deletions of mitochondrial DNA in Kearns-Sayre syndrome. *Neurology* 1988;38:1339–1346.
17. Anderson S, Bankier AT, Barrell BG, et al: Sequence and organization of the human mitochondrial genome. *Nature* 1981;290:457–465.
18. Attardi G: Animal mtDNA: An extreme example of genetic economy. *Int Rev Cytol* 1984;93:93–145.
19. Petty RKH, Harding AE, Morgan-Hughes JA: The clinical features of mitochondrial myopathy. *Brain* 1986;109:915–938.
20. Egger J, Wilson J: Mitochondrial inheritance in a mitochondrially mediated disease. *N Engl J Med* 1983;309:142–146.

21. Poulton J, Deadman ME, Gardner RM: Duplications of mitochondrial DNA in mitochondrial myopathy. *Lancet* 1989;1:236–240.
22. Zeviani M, Servidei S, Gellera C, et al: An autosomal dominant disorder with multiple deletions of mitochondrial DNA starting at the D-loop region. *Nature* 1989;339:309–311.
23. Palca J: The other human genome. *Science* 1990;249:114–115.
24. Grossman LI: Invited editorial: Mitochondrial DNA in sickness and in health. *Am J Hum Genet* 1990;46:415–417.
25. Appleby CAA: Leghemoglobin and rhizobium respiration. Review. *Annu Rev Plant Physiol* 1984;35:443–478.
26. Appleby CA, Tjepkema JD, Trinick MJ: Hemoglobin in a nonleguminous plant, *Parasponia:* Possible genetic origin and function in nitrogen fixation. *Science* 1983;220:951–953.
27. Tjepkema JD: Hemoglobins in the nitrogen fixing root nodules of actinorhizal plants. *Can J Bot* 1983;61:2924–2929.
28. Fleming AI, Wittenberg JB, Wittenberg BA, et al: The purification, characterization and ligand-binding kinetics of hemoglobins from root nodules of the non-leguminous *Casuarina glauca-Frankia* symbiosis. *Biochim Biophys Acta* 1987;911:209–220.
29. Landsmann J, Dennis ES, Higgins TJV, et al: Common evolutionary origin of legume and non-legume plant haemoglobins. *Nature* 1986;324:166–168.
30. Dikerson RE, Geis J: *Hemoglobin: Structure, Function, Evolution, and Pathology.* Menlo Park, Calif., Benjamin/Cummings, 1983.
31. Boguez D, Appleby CA, Landsmann J, et al: Functioning hemoglobin genes in nonnodulating plants. *Nature* 1988;331:178–180.
32. Hiatt A, Cafferkey R, Bowdish K: Production of antibodies in transgenic plants. *Nature* 1989;342:76–78.
33. Tempel BL, Papazian DM, Schwarz TL, et al: Sequence of a probable potassium channel compoent encoded at *Shaker* locus of *Drosophilia. Science* 1987;237:770–775.
34. Noda M, Shimizu S, Tanabe T, et al: Primary structure of *Electrophorus electricus* sodium channel deduced from cDNA sequence. *Nature* 1984;312:121–127.
35. Greenblatt RE, Blatt Y, Montal M: Hypothesis. The structure of the voltage-sensitive sodium channel. *FEBS Lett* 1985;193:125–134.
36. Noda M, Ikeda T, Kayano T, et al: Existence of distinct sodium channel messenger RNAs in rat brain. *Nature* 1986;320:188–192.
37. Armstrong CM: Sodium channels and gating currents. *Physiol Rev* 1981;61:644–683.
38. Hille B: *Ionic Channels of Excitable Membranes.* Sunderland, Mass., Sinauer, 1984.
39. Schwarz TL, Tempel BL, Papazian DM, et al: Multiple potassium-channel components are produced by alternative splicing at the *Shaker* locus in *Drosophila. Nature* 1988;331:137–142.
40. Alkon DL, Lederhendler I, Shoukimas JL: Primary changes of membrane currents during retention of associative learning. *Science* 1982;215:693–695.
41. Timpe LC, Schwarz TL, Tempel BL, et al: Expression of functional potassium channels from *Shaker* cDNA in *Xenopus* oocytes. *Nature* 1988;331:143–145.
42. Agnew WS: Molecular neurobiology. A Rosetta stone for K channels. *Nature* 1988;331:114–115.
43. King RJ, Klass DJ, Gikas EG, Clements JA: Isolation of apoproteins from canine surface active material. *Am J Physiol* 1973;224:788–795.
44. Katyal SL, Singh G: Analysis of pulmonary surfactant apoproteins by electrophoresis. *Biochim Biophys Acta* 1981;670:323–331.
45. Phelps DS, Taeusch HW, Benson B, Hawgood S: An electrophoretic and immunochemical characterization of human surfactant-associated proteins. *Biochim Biophys Acta* 1984;791:226–238.
46. Hawgood S, Benson BJ, Hamilton RL: Effects of a surfactant-associated protein and calcium ions on the structure and surface activity of lung surfactant lipids. *Biochemistry* 1985;24:184–190.
47. White IT, Damm D, Miller J, et al: Isolation and characterization of the human pulmonary surfactant gene. *Nature* 1985;317:361–363.
48. Smith BT, Sabry K: Glucocorticoid-thyroid synergism in lung maturation: A mechanism involving epithelial-mesenchymal interaction. *Proc Natl Acad Sci USA* 1983;80:1951–1954.

49. Prusiner SB, Cochran SP, Groth DF, et al: Measurement of the scrapie agent using an incubation time interval assay. *Ann Neurol* 1982;11:353–358.
50. Prusiner SB. Novel proteinaceous infectious particles cause scrapie. *Science* 1982; 216:136–144.
51. Diringer H, Gelderblom H, Hilmert H, et al: Scrapie infectivity, fibrils, and low molecular weight protein. *Nature* 1983;306:476–478.
52. McKinley MP, Bolton DC, Prusiner SB: A protease-resistant protein is a structural component of the scrapie prion. *Cell* 1983;35:57–62.
53. Merz PA, Rohwer RG, Kascsak R, et al: Infection-specific particle from the unconventional slow virus disease. *Science* 1984;225:437–439.
54. Manuelidis L, Valley S, Manuelidis EE: Specific proteins associated with Creutzfeldt-Jakob disease and scrapie share antigenic and carbohydrate determinants. *Proc Natl Acad Sci USA* 1985;82:4263–4267.
55. Multhaup G, Diringer H, Hilmert H, et al: The protein component of scrapie-associated fibrils is a glycosylated low molecular weight protein. *EMBO J* 1985;4:1495–1501.
56. Oesch B, Westaway D, Walchli M, et al: A cellular gene encodes scrapie PrP27–30 protein. *Cell* 1985;40:735–746.
57. Chesebro B, Race R, Wehrly K, et al: Identification of scrapie prion protein–specific mRNA in scrapie-infected and uninfected brain. *Nature* 1985;315:331–333.
58. Liao Y-C, Lebo RV, Clawson GA, Smucker EA: Human prion protein cDNA: Molecular cloning, chromosomal mapping, and biological implications. *Science* 1986;233:364–367.
59. Puchtler H, Sweat F, Levine M: On the binding of congo red by amyloid. *J Histochem Cytochem* 1962;10:355–364.
60. Bolton DC, McKinley MP, Prusiner SB: Molecular characteristics of the major scrapie prion protein. *Biochemistry* 1984;23:5898–5906.
61. Roberts L: Genome project underway, at last. *Science* 1989;243:167–168.
62. Roberts L: New sequencers to take on the genome. *Science* 1987;238:271–273.
63. McKusick V: Mapping and sequencing the human genome. *N Engl J Med* 1989; 320:910–915.
64. Friedmann T: Opinion: The human genome project—some implications of extensive "reverse genetic" medicine. *Am J Hum Genet* 1990;46:407–414.
65. Nakamura Y, Leppert M, O'Connell P, et al: Variable number of tandem repeat (VNTR) markers for human gene mapping. *Science* 1987;235:1616–1622.
66. Bell GI, Selby MJ, Rutter WJ: The highly polymorphic region near the human insulin gene is composed of simple tandemly repeating sequences. *Nature* 1982;295:31–35.
67. Capon DJ, Chen EY, Levinson AD, et al: Complete nucleotide sequences of the T24 human bladder carcinoma oncogene and its normal homologue. *Nature* 1983;302:33–37.
68. Proudfoot NJ, Gil A: The structure of the human zeta-globin gene and a closely linked, nearly identical pseudogene. *Cell* 1982;31:553–563.
69. Jeffreys A, Wilson V, Thein S: Hypervariable 'minisatellite' regions in human DNA. *Nature* 1985;314:67–73.
70. Lawrance SK, Srivastava R, Rigas B, et al: Molecular approaches to the characterization of megabase regions of DNA: Applications to the human major histocompatibility complex. *Cold Spring Harbor Symp Quant Biol* 1986;51:123–130.
71. Schwartz DC, Cantor CR: Separation of yeast chromosome-sized DNAs by pulsed field gradient electrophoresis. *Cell* 1984;37:67–75.
72. Smith CL, Econome JG, Schutt A, et al: A physical map of the *Escherichia coli* K12 genome. *Science* 1987;236:1448–1453.
73. Lawrance SK, Smith CL, Srivastava R, et al: Megabase-scale mapping of the HLA gene complex by pulsed field gel electrophoresis. *Science* 1987;235:1387–1390.
74. Williams DA, Lemischka IR, Nathan DG, Mulligan RC: Introduction of new genetic material into pluripotent haematopoietic stem cells of the mouse. *Nature* 1984; 310:476–480.
75. Weatherall D: A step nearer gene therapy. *Nature* 1984;310:451.
76. Austin P, Trowsdale J, Rudd C, et al: Functional expression of HLA-DP genes transfected into mouse fibroblasts. *Nature* 1985;313:61–64.
77. Brinster RL, Chen HY, Trumbauer ME, et al: Factors affecting the efficiency of intro-

ducing foreign DNA into mice by microinjecting eggs. *Proc Natl Acad Sci USA* 1985; 82:4438–4442.
78. Hammer RE, Pursel VG, Rexroad CE Jr, et al: Production of transgenic rabbits, sheep and pigs by microinjection. *Nature* 1985;315:680–683.
79. Palmiter RD, Brinster RL: Transgenic mice. *Cell* 1985;41:343–345.
80. Jaenisch R: Germ line integration and mendelian transmission of the exogenous Moloney leukemia virus. *Proc Natl Acad Sci USA* 1976;73:1260–1264.
81. Costantini F, Lacy E: Introduction of a rabbit β-globin gene into the mouse germ line. *Nature* 1981;294:92–94.
82. Brinster RL, Chen HY, Trumbauer M: Somatic expression of herpes thymidine kinase in mice following injection of a fusion gene into eggs. *Cell* 1981;27:223–231.
83. Harberts K, Jahner D, Jaenisch R: Microinjection of cloned retroviral genomes into mouse zygotes: Integration and expression in the animal. *Nature* 1981;293:540–542.
84. Swift GH, Hammer RE, MacDonald RJ, Brinster RL: Tissue-specific expression of the rat pancreatic elastase I gene in transgenic mice. *Cell* 1984;38:639–646.
85. Townes RM, Lingrel JB, Chen HY, et al: Erythroid-specific expression of human β-globin genes in transgenic mice. *EMBO J* 1985;4:1715–1723.
86. Chada K, Magram J, Raphael K, et al: Specific expression of a foreign β-globin gene in erythroid cells of transgenic mice. *Nature* 1985;314:377–380.
87. Shani M: Tissue-specific expression of rat myosin light-chain 2 gene in transgenic mice. *Nature* 1985;314:283–286.
88. Storb U, O'Brien RL, McMullen MD, et al: High expression of cloned immunoglobulin K gene in transgenic mice is restricted to B lymphocytes. *Nature* 1984;310:238–241.
89. Grosschedl R, Weaver D, Baltimore D, Costantini F: Introduction of a μ immunoglobulin gene into the mouse germ line: Specific expression in lymphoid cells and synthesis of functional antibody. *Cell* 1984;38:647–658.
90. Palmiter RD, Brinster RL, Hammer RE: Dramatic growth of mice that develop from eggs microinjected with metallothionen–growth hormone fusion genes. *Nature* 1982; 300:611–615.
91. Palmiter RD, Norstedt G, Gelinas RE, et al: Metallothionen–human GH fusion genes stimulate growth of mice. *Science* 1983;222:809–814.
92. Hammer RE, Palmiter RD, Brinster RL: Partial correction of murine hereditary growth disorder by germ-line incorporation of a new gene. *Nature* 1984;311:65–67.
93. Brinster RL, Chen HY, Messing A, et al: Transgenic mice harboring SV40 T-antigen genes develop characteristic brain tumors. *Cell* 1984;37:367–379.
94. Stewart TA, Pattengale PK, Leder P: Spontaneous mammary adenocarcinomas in transgenic mice that carry and express MTV/c-*myc* fusion genes. *Cell* 1984;38:627–637.
95. Jaenisch R, Harbers K, Schnieke A, et al: Germ line integration of Moloney murine leukemia virus at the Mov13 locus leads to recessive lethal mutation and early embryonic death. *Cell* 1983;32:209–216.
96. Wagner EF, Covarrubias L, Stewart TA, Mintz B: Prenatal lethalities in mice homozygous for human growth hormone gene sequences integrated in the gene line. *Cell* 1983;35:647–655.
97. Smithies O, Gregg RG, Boggs SS, et al: Insertion of DNA sequences into the human β-globin locus by homologous recombination. *Nature* 1985;317:230–234.
98. Maniatis T: Gene modification targeting in mammalian cells. *Nature* 1985;317:204–206.
99. Palmiter R: Transgenic mice. *Cell* 1985;41:343–345.
100. Hinnen A, Hicks JB, Fink GR: Transformation of yeast. *Proc Natl Acad Sci USA* 1978; 75:1929–1933.
101. Scherer S, Davis RW: Replacement of chromosome segments with altered DNA sequences constructed in vitro. *Proc Natl Acad Sci USA* 1979;76:4951–4955.
102. Neumann E, Schaefer-Ridder M, Wang Y, Hofschneider PH: Gene transfer into mouse myeloma cells by electroporation in high electric fields. *EMBO J* 1982;1:841–845.
103. Potter H, Weir L, Leder P: Enhancer-dependent expression of human κ immunoglobulin genes introduced into mouse pre-B lymphocytes by electroporation. *Proc Natl Acad Sci USA* 1984;81:7161–7165.
104. Graham FL, Van der Eb AJ: A new technique for the assay of infectivity of human adenovirus 5 DNA. *Virology* 1973;52:456–467.

105. Knutson JC, Yee D: Electroporation: Parameters affecting transfer of DNA into mammalian cells. *Anal Biochem* 1987;164:44–52.
106. Jaenisch R, Fan H, Crocker B: Infection of preimplantation mouse embryos and of newborn mice with leukemia virus: Tissue distribution of viral DNA and RNA and leukomogenesis in the adult animal. *Proc Natl Acad Sci USA* 1975;72:4008–4012.
107. Gautsch J, Wilson M: Restriction of Moloney murine leukemia virus growth in teratocarcinoma: Involvement of factors other than DNA methylation. *Cold Spring Harbor Conf Cell Prolif* 1983;10:363–378.
108. Jahner D, Jaenisch R: Chromosomal position and specific demethylation in enhancer sequences of germ line–transmitted retroviral genomes during mouse development. *Mol Cell Biol* 1985;5:2212–2220.
109. Soriano P, Cone RD, Mulligan RC, Jaenisch R: Tissue-specific and ectopic expression of genes introduced into transgenic mice by retroviruses. *Science* 1986;234:1409–1413.
110. Costantini F, Chada K, Magram J: Correction of murine β-thalassemia by gene transfer into the germ line. *Science* 1986;233:1192–1194.
111. Garver RI, Chytil A, Courtney M, Crystal RG: Clonal gene therapy: Transplanted mouse fibroblast clones express human α_1-antitrypsin gene in vivo. *Science* 1987; 1237:762–764.
112. Garver RI, Chytil A, Karlsson S, et al: Production of glycosylated physiologically "normal" human α_1-antitrypsin by mouse fibroblasts modified by insertion of a human α_1-antitrypsin cDNA using a retroviral vector. *Proc Natl Acad Sci USA* 1987;84:1050–1054.
113. Carlson JA, Rogers BB, Sifers RN, et al: Multiple tissues express $alpha_1$-antitrypsin in transgenic mice and man. *J Clin Invest* 1988;82:26–36.
114. Kelly K, Cochran BH, Stiles CD, Leder P: Cell-specific regulation of the c-*myc* gene by lymphocyte mitogens and platelet-derived growth factor. *Cell* 1983;35:603–610.
115. Leder A, Pattengale PK, Kuo A, et al: Consequences of widespread deregulation of the c-*myc* gene in transgenic mice: Multiple neoplasms and normal development. *Cell* 1986;45:485–495.
116. Gallo RC: The first human retrovirus. *Sci Am* 1986;255 (December):88–98.
117. Nerenberg M, Hinrichs SH, Reynolds RK, et al: The *tat* gene of human T-lymphotropic virus type 1 induces mesenchymal tumors in transgenic mice. *Science* 1987;237:1324–1329.
118. Hinrichs SH, Nerenberg M, Reynolds RK, et al: A transgenic mouse model for human neurofibromatosis. *Science* 1987;237:1340–1343.
119. Mason AJ, Pitts SL, Nikolics K, et al: The hypogonadal mouse: Reproductive functions restored by gene therapy. *Science* 1986;234:1372–1378.
120. Cattanach BM, Iddon CA, Charlton HM, et al: Gonadotrophin-releasing hormone deficiency in a mutant mouse with hypogonadism. *Nature* 1977;269:338–340.
121. Mason AJ, Hayflick JS, Zoeller RT, et al: A deletion truncating the gonadotropin-releasing hormone gene is responsible for hypogonadism in the *hpg* mouse. *Science* 1986;234:1366–1371.
122. Nikolics K, Mason AJ, Szonyi E, et al: A prolactin-inhibiting factor within the precursor for human gonadotropin-releasing hormone. *Nature* 1985;316:511–517.
123. Hanahan D: Transgenic mice as probes into complex systems. *Science* 1989;246:1265–1274.
124. Muller H: Human gene therapy: Possibilities and limitations. *Experientia* 1987;43:375–378.
125. Mulligan RC, Berg P: Expression of a bacterial gene in mammalian cells. *Science* 1980; 209:1422–1427.
126. Schaffner W: Direct transfer of cloned genes from bacteria to mammalian cells. *Proc Natl Acad Sci USA* 1980;77:2163–2167.
127. Anderson WF: Prospects of human gene therapy. *Science* 1984;226:401–409.
128. Williams DA, Orkin SH, Mulligan RC: Retrovirus-mediated transfer of human adenosine deaminase sequences into cells in culture into murine hematopoietic cells in vivo. *Proc Natl Acad Sci USA* 1986;83:2566–2570.
129. Baudet AI: Bibliography of cloned human DNAs. *Am J Hum Genet* 1985;37:386–406.
130. Ledley FD: Somatic gene therapy for human disease: Background and prospects. Part II. *J Pediatr* 1987;110:167–174.

131. Abbott CM, Skidmore CJ, Searle AG, Peters J: Deficiency of adenosine deaminase in the wasted mouse. *Proc Natl Acad Sci USA* 1986;83:693–695.
132. Bank A: Genetic defects in thalassemias. *Curr Top Hematol* 1985;5:1–23.
133. Miller AD, Jolly DJ, Friedman T, Verma IM: A transmissible retrovirus expression human hypoxanthine phosphoribosyl transferase (HPRT): Gene transfer into cells from humans deficient in HPRT. *Proc Natl Acad Sci USA* 1983;80:4709–4717.
134. Miller AD, Eckner D, Jolly DJ, et al: Expression of a retrovirus encoding human HPRT in mice. *Science* 1984;225:630–632.
135. Bayever E, Feig SA: Bone marrow transplantation for inherited metabolic disorders, in Gale PR (ed): *Recent Advances in Bone Marrow Transplantation.* New York, Liss, 1983, pp 185–198.
136. Sorge J, West C, Westwood B, Beutler E: Molecular cloning and nucleotide sequences of human glucocerebrosidase cDNA. *Proc Natl Acad Sci USA* 1985;82:7289–7293.
137. Cline MJ: Gene therapy: Current status. *Am J Med* 1987;83:291–297.
138. Varmus HE: Form and function of retroviral proviruses. *Science* 1982;216:812–820.
139. Cline MJ, Stang J, Mercola K, et al: Gene transfer in intact animals. *Nature* 1980; 284:422–425.
140. Cline MJ: Perspectives for gene therapy. Inserting new genetic information into mammalian cells by physical techniques and viral vectors. *Pharmacol Ther* 1985;29:69–92.
141. Friedmann T: *Gene Therapy: Fact and Fiction in Biology's New Approaches to Disease.* Cold Spring Harbor, N.Y., Cold Spring Harbor Laboratory, 1983.
142. Rosenberg SA, Spiess P, Lafreniere RA: New approach to the adoptive immunotherapy of cancer with tumor-infiltrating lymphocytes. *Science* 1986;233:1318–1321.
143. Rosenberg SA: The development of new immunotherapies for the treatment of cancer using interleukin-2: A review. *Ann Surg* 1988;208:121–135.
144. Rosenberg SA, Packard BS, Aebersold PM: Use of tumor-infiltrating lymphocytes and interleukin-2 in the immunotherapy of patients with malignant metastatic melanoma: A preliminary report. *N Engl J Med* 1988;319:1676–1680.
145. Palca J: Gene transfer to human approved in face of advice. *Nature* 1988;335:577.
146. Rosenberg SA, Aebersold P, Cornetta K, et al: Gene transfer into humans—immunotherapy of patients with advanced melanoma, using tumor-infiltrating lymphocytes modified by retroviral gene transduction. *N Engl J Med* 1990;323:570–578.
147. Cournoyer D, Caskey CT: Gene transfer into humans a first step. *N Engl J Med* 1990; 323:601–603.
148. Mann R, Mulligan RC, Baltimore D: Construction of a retrovirus packaging mutant and its use to produce helper-free defective retrovirus. *Cell* 1983;33:153–159.
149. Joyner A, Keller G, Phillips RA, Bernstein A: Retrovirus transfer of a bacterial gene into mouse hematopoietic progenitor cells. *Nature* 1983;305:556–558.
150. Culliton BJ: ADA gene therapy enters the competition. *Science* 1990;249:975–976.
151. Nabel EG, Plautz G, Boyce FM, et al: Recombinant gene expression *in vivo* within endothelial cells of the arterial wall. *Science* 1989;244:1342–1344.
152. Wilson JM, Birinyi LK, Salomon RM, et al: Implantation of vascular grafts lined with genetically modified endothelial cells. *Science* 1989;244:1433–1446.
153. Drumm ML, Pope HA, Cliff WH, et al: Correction of the cystic fibrosis defect in vitro by retrovirus-mediated gene transfer. *Cell* 1990;62:1227–1233.
154. Rich DP, Anderson MP, Gregory RJ, et al: Expression of cystic fibrosis transmembrane conductance regulator corrects defective chloride channel regulation in cystic fibrosis airway epithelial cells. *Nature* 1990;347:358–363.
155. Anderson WF, Fletcher JC: Gene therapy in humans: When is it ethical to begin? *N Engl J Med* 1980;303:1293–1296.
156. Mercola KE, Cline MJ: The potentials of inserting new genetic information. *N Engl J Med* 1980;303:1297–1300.
157. Anderson WF: Prospects for human gene therapy. *Science* 1984;226:401–403.
158. President's Commission for the Study of Ethical Problems in Medicine and Biomedical and Behavioral Research: *Splicing Life.* Washington, D.C., Government Printing Office, 1982.

Index

Page numbers followed by t *represent tables; those followed by* f *represent figures.*

H

N

O

U

V